WITHDRAWN

INDUSTRIAL ELECTRONICS

INDUSTRIAL ELECTRONICS

ALLAN LYTEL
Manager, U.S. Electronic Publications, Inc., Syracuse, New York

McGRAW-HILL BOOK COMPANY, INC.
New York Toronto London

INDUSTRIAL ELECTRONICS

 Library of Congress Catalog Card Number 62-12483.

Preface

Electronic systems are used by industry to sense, measure, warn, and control. The list of industrial applications is so long that it appears all industry has some use for the magic of electronics.

This book is intended as a text covering the important industrial applications of electronics. Because electronic devices, products, and systems are ubiquitous, perhaps the scope of this book can best be defined by exclusion: no broadcast or entertainment radio or television receivers or transmitters are included. Certain aspects of television, two-way radio, and data processing are included to show how they are related to and used by industry. A minimum of material has been included on motors and generators, as the great detail given them in many other texts is not necessary here.

This book has been prepared for use by students at the technical-institute level. The student should have had courses in basic electronics, vacuum tubes, and possibly AM radio receivers and transmitters to use this book most effectively. No prior knowledge of semiconductors is assumed. A minimum of mathematics is used in this text, and this is limited to algebra and some trigonometry, plus graphs and charts to show interrelationships between variables.

Borderlines for industrial electronics are not easy to draw. Emphasis has been given to broad techniques which have wide applicability to a variety of industrial equipment. Many of the devices and circuits described are those actually used in industrial electronics, but fundamental principles have been given first-order importance. This will permit transfer of knowledge from one circuit, device, or technique to other situations which are related. To provide for a broad understanding of industrial electronics, emphasis has been placed on basic devices and circuits.

Industrial electronics promises to be the fastest-growing segment of the electronics industry in coming years. Widening uses of computers, industrial controls, communications—particularly in microwaves—and testing and measuring equipment have been largely responsible for the past growth in the field. The total sales of equipment are still relatively small, but the impact of electronics on industrial and commercial markets has hardly begun.

As electronics technology progresses and costs of advanced products are reduced, more and more applications of electronics in industry and commerce will be seen. The basic reason for this expectation is the pressure on American industrial management to reduce expenses in the face of constantly rising material and labor costs and increasing foreign competition.

ALLAN LYTEL

Contents

CHAPTER

1

Industrial Electronics

1·1 Electronics in industry Electronics is an exciting field. The electron stream can warn, inspect, detect, measure, and also do a host of other things in modern industry. At one time, not so long ago, industrial electronics meant only motor control and small switching systems. Now, industrial electronics offers its techniques to almost every industry and process in many ways. The age of automation has brought a new meaning to electronics.

From simple switches and solid-state devices to X-ray beams, solar cells in orbit (Fig. 1·1), and heavy-duty gas tubes, electronics has many roles to play in industry and science.

An electronic counter, for example, as shown in Fig. 1·2, has a number of different applications, such as counting objects, measuring frequency, or measuring time. Counters are also used in modern computers.

In addition to their applications in computers there are many uses in industrial electronics for high-speed, direct-reading electronic counters. These can be used to control any operation or to activate an alarm after a preselected total count has been reached.

Any electrical, mechanical, or optical events which can be converted into electric impulses can be counted and controlled. Devices to effect this conversion may be photocells, magnetic coils, switches, and suitable transducers for pressure, temperature, velocity, acceleration, and displacement.

Many circuits used in electronic counters are similar to those found in radio, TV, and regular test equipment. Counting and frequency measurement are closely related. A number of electronic and electromechanical configurations are possible. Tubes, transistors, and relays are used in multivibrator, flip-flop, staircase, and blocking-oscillator circuits.

Computers, both digital and analog, are significant in their growing importance as tools; both types are shown in Fig. 1·3. They use certain logic circuits for specific functions.

Although logical design is employed for large-scale digital computers, it is not limited to this area. The same principles have a direct bearing on, and increasing importance in, control circuits for other types of electronic devices. The automatic control of manufacturing processes is an area where logic plays an important role. Wherever relays, transistors, or vacuum tubes are used for control, logic is an aid to design and understanding of the circuit operations.

Fig. 1·1 Solar cells used to generate electric current directly from the sun's rays on Tiros satellite. (*International Rectifier Corporation*)

1·2 Electronic control systems Many of the industrial electronic systems discussed in this book may be considered as having three related blocks in a diagram of the system as shown in Fig. 1·4. Something is manufactured, tested, or packed in the process. A transducer monitors this process and creates control signals, which are amplified and acted upon by the control circuits. One of the outputs from the control circuits can be an indicator. Another output is the controlling action itself, which acts upon the process.

One of several ways of classifying automatic control systems is defining their action in terms of proportional, floating, or two-position control.

Two-position control provides full-on or full-off operation of the controlled device. There are no intermediate positions. Many applications such as simple on-off motor control are best served by two-position arrangements.

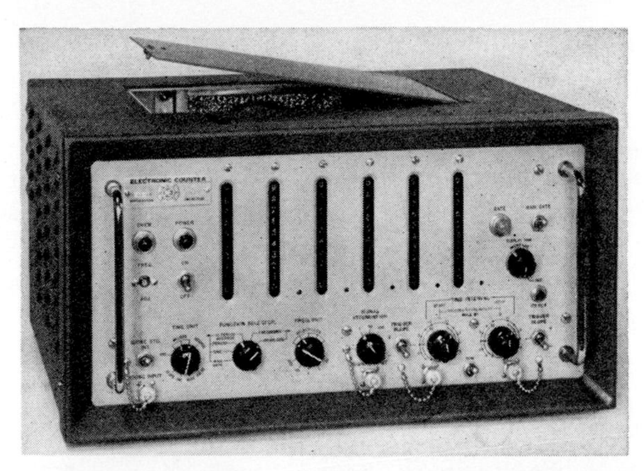

Fig. 1·2 Counter used to measure units of time, to count events or electric pulses, and to measure frequency. (*Northeastern Engineering*)

A

B

Fig. 1·3 (*A*) IBM 7090 Data Processing System, a large-scale digital computer used for industrial, business, and scientific calculations. (*B*) Philbrick analog computer as installed in the mechanical engineering department of Massachusetts Institute of Technology.

In floating control the position of the controlled device is varied as required to maintain conditions at the control set point. Floating control permits the system to stop at any position between "full on" and "full off," but applications for floating control are limited because of the tendency to overshoot. Overshooting is likely to set up a condition of cycling, also called "hunting," in the system. Both overshooting and hunting are the result of time lags in the overall system.

In proportional control the actuators assume a position proportional to the change in conditions. These conditions can be almost any other controlled variable, such as temperature, pressure, or light. For a given

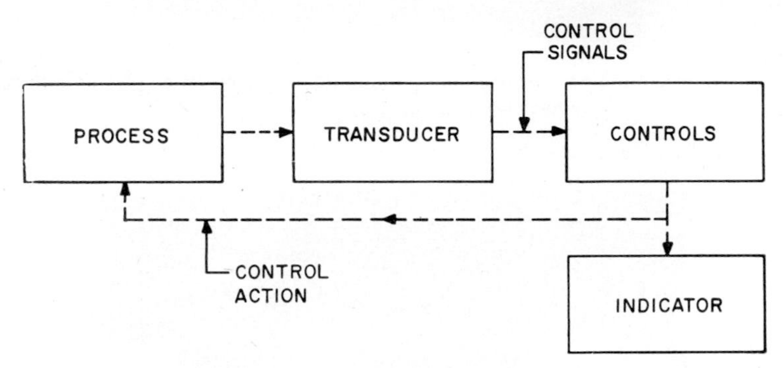

Fig. 1·4 Process-control loop showing the transducer, actuated by the process, feeding control signals which are fed back for control.

increment of change in the variable, the controlled device will move to a definite and corresponding position which is proportional to that increment. In this way the cycling of floating control is avoided since the position reached by the actuator is always the same for any given load on the system. Proportional action uses feedback, which is a link from the actuator to the controlled device. As the actuator moves in response to a change in load, a signal is returned or fed back. The effect of this feedback is opposite to that of the change caused by the load change. When these opposite effects balance, movement in the system is stopped, usually before it can reach the extreme "open" or "closed" position.

Figure 1·5 shows several simple systems. The control in *A* is two-position since closing the switch *S* turns on the load by providing power from the source and opening the switch cuts off power to the load. *B* shows a potentiometer used to vary the amount of power applied to the load; this is a type of floating control. Any amount of power from "full on" to "full off" may be applied. *C* illustrates the principle of feedback and proportional control. Load *L* is a light, part of which goes to a photocell. In darkness, when there is no outside light, the photocell calls for more light from *L*. When there is sunlight, the control turns *L* off.

1·3 Transducers Industrial electronic systems can measure, detect, count, control, warn, and inspect; but for any of these operations, a sensor is required to produce the electric signal. In a home or industrial heating system, for example, closing the switch in a simple open-loop system will cause current flow and the resulting heating of the resistive element, but as long as the switch is closed there will be heat. Adding a bimetallic thermostat makes this a closed-loop system in which the bimetal is a sensor which controls the heating and turns the circuit on and off;

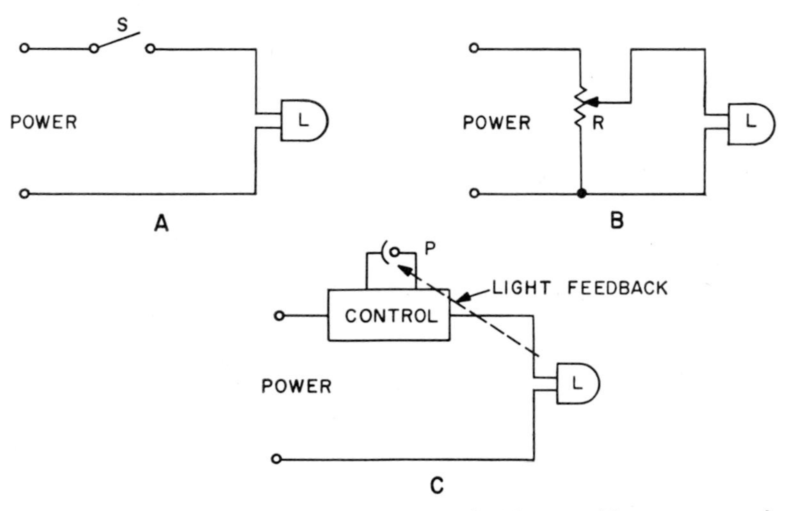

Fig. 1·5 Types of control: (A) a simple off-on switch; (B) a variable power control; and (C) the use of a light-path feedback.

in a manner of speaking it is a "decision maker" since it senses the temperature and either opens or closes the circuit.

This temperature device is a sensor or transducer which changes some physical property such as heat, light, or temperature into an electrical quantity such as current, voltage, or resistance. There is a wide range of possibilities. A transducer, which is somehow mechanically or electrically coupled into an electronic circuit, can vary capacitance, inductance, or resistance to produce a change in current, voltage, or frequency by means of various techniques including the use of piezoelectric, photoelectric, or magnetic devices. By this system it is possible to measure, among other things, pressure, temperature, or humidity.

Many industrial systems use transducers between the process and the electronic control. The control itself has a feedback action to the process and, often, an indicator as well.

The list of transducers is very long and may be divided according to the parameter being measured, the type of measurement, or the technique for producing the electric signal.

Not included specifically in this discussion are the more obvious types such as magnetic pickups, photoelectric sensors, and thermosensitive devices, which are all discussed in detail in other chapters.

Humidity can be measured easily with hygroscopic materials, such as calcium chloride, zinc chloride, or similar compounds in a gridlike structure or an insulating base. As the humidity increases, the material absorbs water from the air and changes its resistance.

The strain gauge is a transducer of special importance because of its wide use. Made of fine wires, about 0.001 in. in diameter, this gauge is usually mounted on a piece of paper for support. The paper is, in turn, cemented onto the structure to be measured. When, under physical stress, the object bends or twists, the wire in the gauge changes its resistance because it is stretched a small amount. As its length is increased the total resistance, of course, is also increased. Typical wires are made of constantan, Nichrome, and nickel. A strain gauge forms one leg of a Wheatstone bridge which is unbalanced to produce an a-c signal output to an amplifier. Strain gauges are in wide use because of their small, flexible, easy-to-use nature and their high stability.

In another transducer, the diaphragm (which can measure barometric pressure) can be used to force a coil core to move. Here alternating current is coupled to the two push-pull tubes. Motion of the core, either in or out, creates an unbalance which can be read, as voltage, on the meter between the anodes of the amplifiers.

1·4 Control signals A variable current or voltage is used to signal the state and action of control systems. Control signals are of three main types: changing d-c levels, a-c sine waves, and complex waveshapes.

Changing d-c levels are amplified by being chopped into a series of segments which are then sent through a-c amplifiers. D-c amplifiers are also used, but not often, because they tend to drift and because they cannot be used for more than two or three stages without serious voltage-supply problems.

Sine-wave signals appear frequently in industrial electronics. Either the sine-wave phase or its amplitude can be varied to cause a change in the signal.

In a simple RC circuit, with applied alternating current, current and voltage are not in phase. They differ by an angle θ whose tangent is X/R. If $X \gg R$, the current leads the voltage by almost 90°. If $R \gg X$, the current and voltage are almost in phase and θ is small.

For a capacitive reactance, current leads voltage; for an inductive reactance, voltage leads current.

To obtain a phase shift, usually R is variable while C is fixed. The amount of phase shift is a function of both X and R, as shown in Fig. 1·6. For example, if $R = 1{,}000$ ohms and $X = 100$ ohms, the phase angle is about 6°.

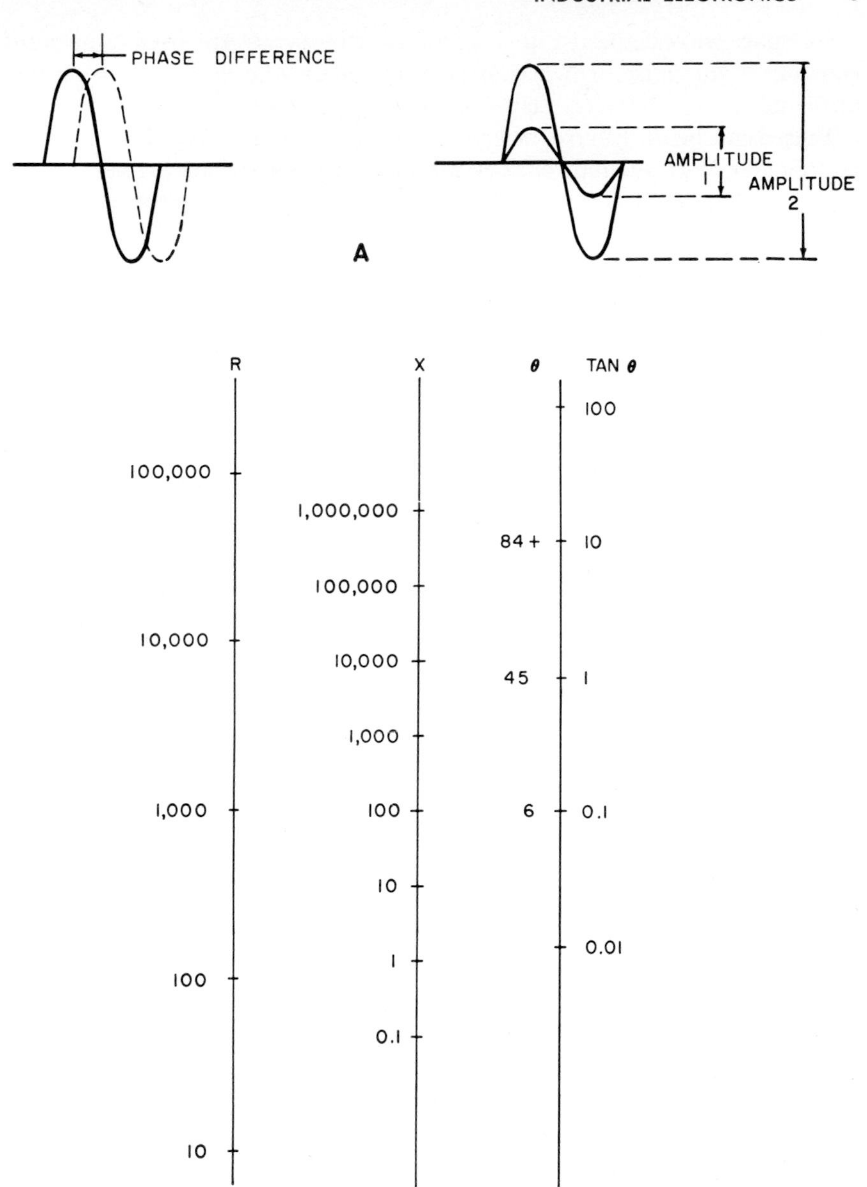

Fig. 1·6 (*A*) A-c variables: a phase difference and an amplitude difference. (*B*) Phase-angle nomograph showing the relationship between resistance, reactance, and phase angle.

Complex waveshapes (Fig. 1·7) of various types are used. The most common is the pulse, which is often obtained by the action of the transducer itself.

Pulses can be of different shapes, as shown in Fig. 1·7*A*. They may be rectangular or triangular, or they may appear as half-sine waves.

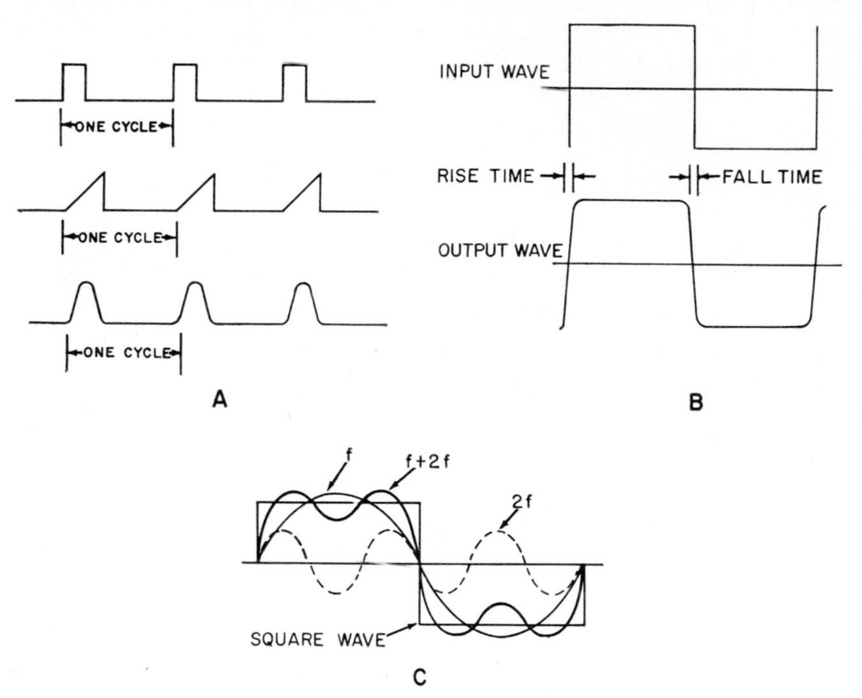

Fig. 1·7 (*A*) Three waveshapes which may be used for control signals; these are rectangular pulses, triangular waves, and half-sine waves. (*B*) A bipolar square wave which is between a negative and positive voltage suffers during passage through amplifiers so that the rise time and fall time are not zero but require a certain short time to change from negative to positive. (*C*) A square-wave control signal of frequency f is composed of frequencies f and $2f$, plus other harmonics of f. The wave shape of $f + 2f$ only approximates the perfect square wave.

The square wave (Fig. 1·7*B*) is common. The rise time and fall time are important in some switching applications and must be controlled to ensure sharp switching. Components of this square wave are the harmonics of the fundamental frequency, as in Fig. 1·7*C* where the fundamental, first harmonic, and second harmonic are shown adding to produce a square wave.

1·5 Circuit elements A circuit is made up of active and passive elements. The active elements are electron tubes or solid-state devices.

Electron Tubes. These tubes are in many different forms for various jobs.

One possible classification of vacuum and gas tubes is a primary division into four groups based upon the cathode types, as shown in Fig. 1·8. Four different cathode sources of electrons are shown. Heated cathodes, using direct or indirect heating, emit electrons when their temperature is sufficiently high. Cold cathodes use gas ionization, which frees electrons, as the source of current flow. Photocathodes emit electrons under the excitation of light. The final type has a pool of mercury as the electron source. Most of these tubes are discussed in later chapters.

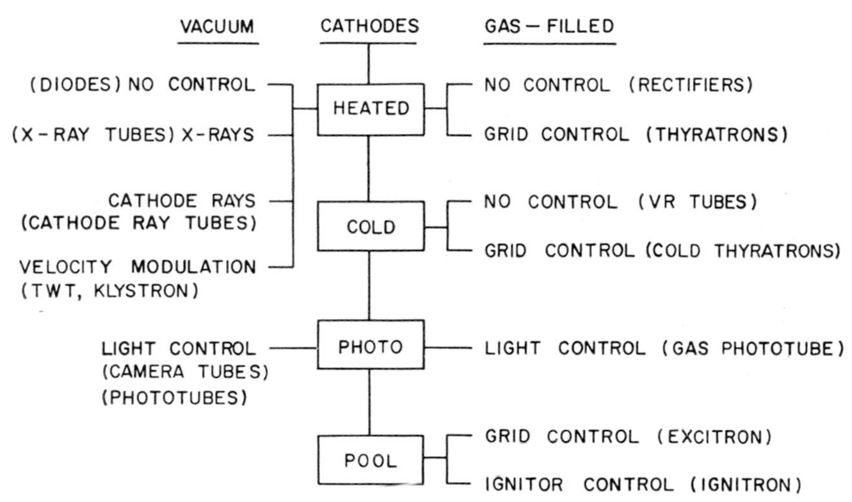

Fig. 1·8 The industrial tube family tree showing vacuum and gas-filled tubes in the four groups with different cathode types.

Solid-state Devices. These include semiconductor devices which cover many active components. There are three states of matter: solid, liquid, and gas. In the vacuum tube, electrons travel through an atmosphere of very thin gas; in a solid-state device such as a diode or a transistor, the electrons travel through a solid.

Transistors are current-operated devices, whereas tubes are voltage-operated devices. The tube plate, grid, and cathode are like the transistor collector, base, and emitter (respectively) in some of their actions. Transistor circuits use both NPN and PNP configurations. NPN transistors, in a grounded-base arrangement, have a negative voltage applied to the emitter and a positive voltage to the collector. In the PNP transistor these voltages (or, more accurately, current sources) are of opposite polarity. A grounded-grid tube corresponds to the grounded-base transistor, the grounded-cathode tube to the grounded-emitter transistor, and the grounded-plate tube to the grounded-collector transistor. These circuits are all used in one way or another in industrial electronics.

One of the greatest uses of transistors is in multivibrator switches, of which there are several variations. A multivibrator may be free-running, which means it produces an output without an input signal. If required, a synchronization signal may be introduced to lock in this self-excited oscillator to a desired frequency. Or the multivibrator may be a flip-flop, which has two stable states. Here each input switches the circuit to the other state. There are also other types.

But the family of solid-state devices grows ever bigger. Both controlled rectifiers and photodevices are used and are important. Diodes have appeared in various types; they can amplify, control, detect, and measure. Magnetics, in their several configurations, are also important and are covered in detail in Chap. 11.

A special form of solid-state device, the field-effect transistor may be considered as a stepping stone between a vacuum tube, which is a voltage-operated device, and the true transistor, which is a current-operated device. The unit consists of an N-type silicon bar with two ohmic contacts (cathode and anode) on either end of the bar. Two PN junctions are built into the middle of the bar and connected in parallel to serve as the grid of the device.

A negative bias applied to the grid increases the effective resistance between the anode and cathode of the unit and produces a triode-type output characteristic. As the anode voltage is increased, the grid junctions are reverse-biased by the voltage drop occurring because of the anode current. At some point, further increase in anode voltage will not result in any appreciable increase in anode current, causing the output characteristics of the unit to closely resemble those of the vacuum-tube pentode.

The anode potential, at which the saturation of anode current occurs, is known as the pinch-off voltage. The anode current flowing through the device after the pinch-off voltage has been reached is known as pinch-off current. With zero grid bias, the pinch-off current is the maximum specified anode current of the transistor. The device is in the triode region before the pinch-off occurs and in the pentode region after the pinch-off potential has been reached.

The anode and cathode terminals of the field-effect transistor are, in a manner quite unlike that of a tube, interchangeable, although a somewhat higher transconductance and lower noise figure are generally obtained if the unit is used in the normal way.

The grid normally requires a negative potential. Positive bias on the grid will increase the anode current, but this potential should remain below 0.6 volt. Substantial grid currents capable of destroying the device will be drawn if the grid is biased with a positive voltage in excess of this.

Circuitry for use of the field-effect transistor as an amplifier is identical with that of triodes and pentodes, except that lower anode voltages can

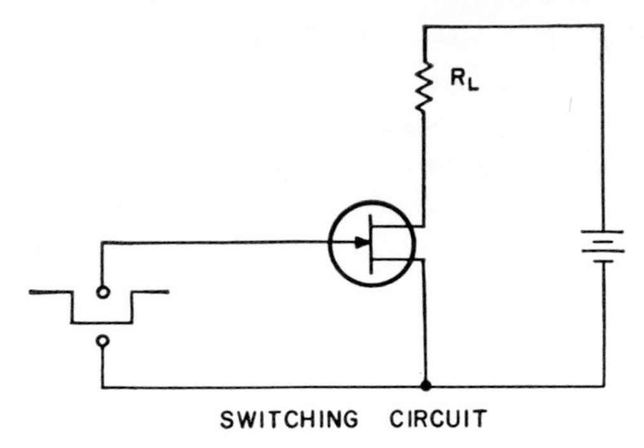

Fig. 1·9 Switching circuit using a field-effect transistor.

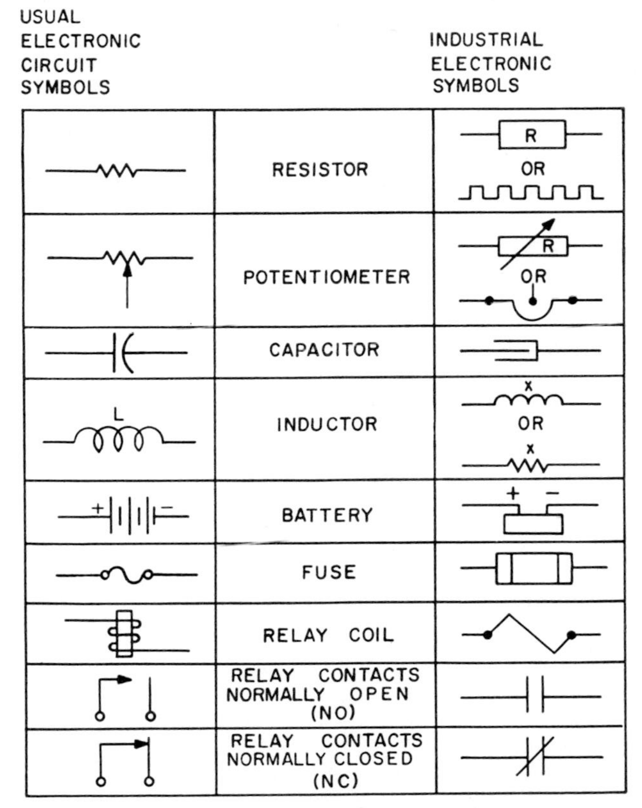

Fig. 1·10 Industrial electronic circuit symbols compared with the more usual electronic circuit symbols.

be used for the field-effect devices. Polarities of applied voltages are also the same.

The noise parameters of the field-effect transistor are extremely low even with very high source impedance, provided that the anode voltage is maintained at approximately 3 volts. The device then operates in the triode region, and it is extremely useful as the input stage in very low-noise, high-impedance transistor amplifiers. The output impedance of the device in the triode region is approximately 2.5 kilohms, which is the ideal source impedance for conventional low-noise transistors following the input stage.

Field-effect transistors can be used as switches in digital applications, as shown in Fig. 1·9. Unlike transistors, they require pulses of only one polarity for full on-off operation. The device is on when no voltage appears on the grid, and off when the grid is pulled negative. In the off condition, the resistance of the device is in the order of 100 megohms. In the on condition, resistance of the device is approximately 2 kilohms. The switching speeds are directly affected by the impedance of the generator driving the switch. For high switching speed, the driving generator is of low impedance during both the buildup and the collapse of the driving pulse.

Passive Circuit Elements. Circuit elements which are passive include resistors, coils, capacitors, and like elements. Figure 1·10 shows the symbolism used in industrial circuits as compared with that of other electronic circuits.

CHAPTER

2

Power Supply and Control

A basic need for every industrial electronic system is some type of power supply or source. The devices used for power supply and power control may be of several types. Only vacuum tubes and gas tubes are covered in this chapter while Chap. 3 covers solid-state devices, including those used for power control.

2·1 Gas tubes

Thyratrons. Thyratrons are gas-filled triodes, but their construction is quite different from that of vacuum-tube triodes. Figure 2·1 illustrates a

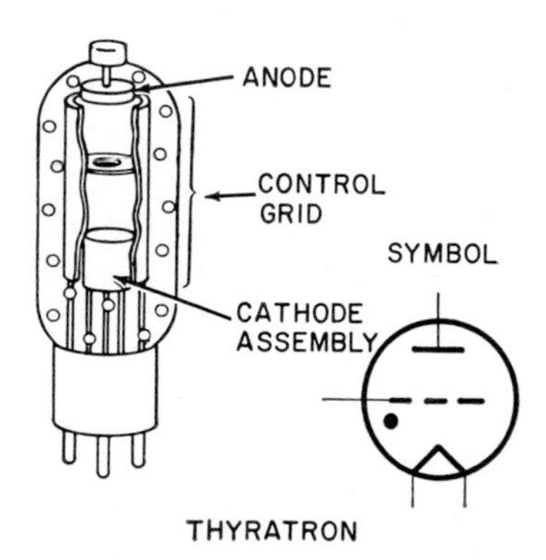

Fig. 2·1 Triode thyratron showing cathode, anode, and tubular grid. (*General Electric*)

thyratron with the tubular grid structure. In a gas-filled tube such as this the grid must prevent the flow of *all* electrons; in a vacuum tube a minute current flow is not damaging. But even a very small electron flow can ionize the gas and create the conducting condition. The grid is designed to serve also as a cathode shield. In this way it completely controls electron flow, helps contain the cathode heat, and also shields the cathode from external stray fields. Because of this type of structure there is a heavy grid current.

Some typical thyratrons are shown in Fig. 2·2. These tubes are used to control small amounts of power directly or to drive larger tubes which ultimately control the power.

The grid in a thyratron draws very little current, which makes it possible for the tube to control large amounts of power while, at the same time, the thyratron itself expends very little power.

Four important characteristics of thyratrons are discussed below.

A

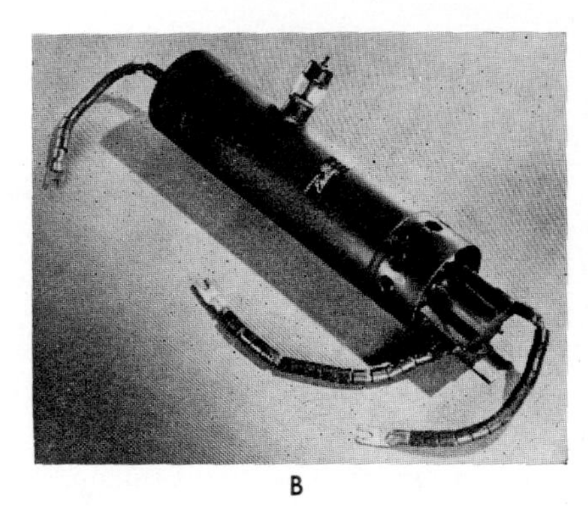

B

Fig. 2·2 (*A*) Glass-bulb thyratron; (*B*) metal thyratron. (*General Electric*)

1. *Maximum peak forward voltage.* This is the instantaneous value that would be observed on an oscilloscope. Exceeding the peak forward voltage may result in loss of grid control.

2. *Grid-control characteristic.* This characteristic is defined in the technical data for any individual thyratron type. A range of characteristics defines the grid characteristics within which tubes will fall. In use, the electronic equipment usually will provide sufficient grid-control voltage and a sufficiently steep wave. Both are needed to provide accurate control of the firing point.

3. *Deionization time.* This is usually known as recovery time and is defined as the minimum time after anode current has ceased before the grid can once again regain control. This time is required for the positive ions in the tube to reach the walls and electrodes. In so doing, they lose their charges.

4. *Maximum commutation factor.* This limitation applies particularly to inert gas-filled thyratrons and must be observed if gas cleanup is to be avoided. It is concerned with the amount of ionization present in the tube at the moment anode current ceases and inverse voltage is applied. High inverse voltages attract the remaining positive ions and generate great velocities which may result in ion bombardment of the ele-

ments and gradual gas cleanup. Commutation factor is defined as the rate of change of current (just before anode current ceases) in amperes per microsecond times the rate of application of inverse voltage in volts per microsecond.

Figure 2·3 shows a shield-grid thyratron. Note that this grid shields not only the cathode but also the new element, the control grid. Less grid current flows in this tube (cathode to control grid) and the tube is also more sensitive to changes in control-grid voltage. This shield grid, in a small tube, is usually tied to a tube-base pin. If this grid is at cathode potential, the shield-grid thyratron and the triode thyratron have about

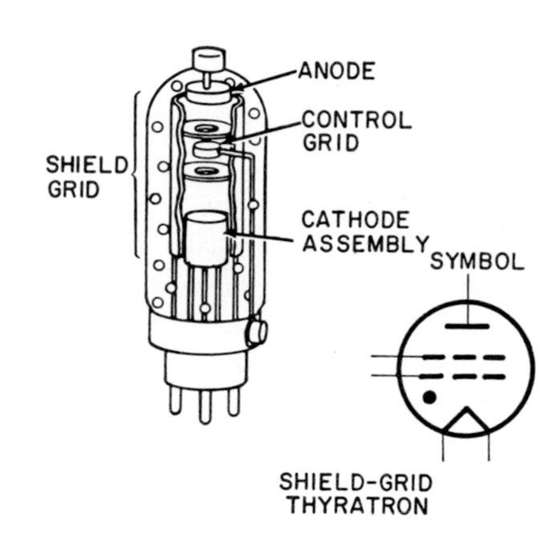

Fig. 2·3 Shield-grid thyratron showing tubular shield grid and disklike control grid. (*General Electric*)

the same grid-voltage and plate-voltage relationship. But if the shield grid is more negative than the cathode, the control grid must be slightly positive to cause conduction.

In a thyratron the grid controls a very large plate current, but a thyratron grid can only start conduction. It cannot stop conduction. There is a given value of negative voltage required to prevent conduction for every value of plate voltage, as illustrated in Fig. 2·4 where one half-cycle of plate and grid voltage is shown.

As the plate voltage increases from zero to 300, 500, and 600 volts, the grid voltage must increase from zero to 3, 6, and 9 volts negative, to prevent conduction. Since any grid voltage *less* negative than these values will permit conduction, they are known as critical values and the curves are called the critical curves. In most cases it is safe to consider that the triode thyratron will fire (conduct) when the grid potential is zero, or slightly negative, if the plate is positive. Alternating current is usually used for the plate voltage since the negative alternations of the sine wave cut off the tube and permit the grid to regain control.

Three basic ways are used to control the firing of the tube by adjusting the grid action. These are (1) changing the phase relations of the grid and plate voltages, (2) changing the value of the grid voltage about which the a-c signal alternates, and (3) using a waveshape other than a sine wave on the grid to act as a trigger.

Figure 2·5 shows grid and plate waveforms; in *A* they are in phase so the tube fires for one-half of each positive plate alternation. Shifting the

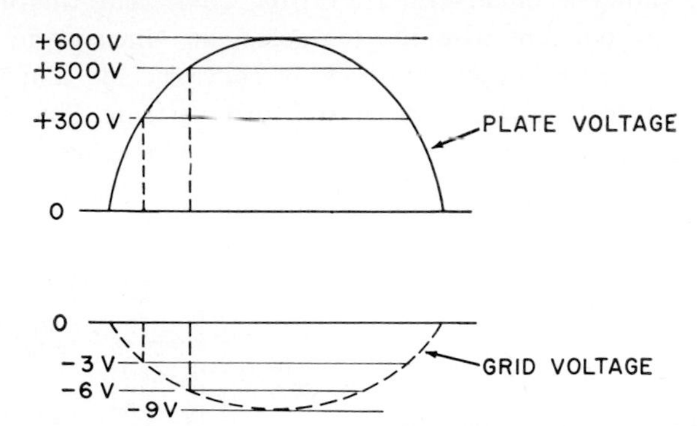

Fig. 2·4 Thyratron characteristics where plate (anode) voltage increases while grid voltage increases in a negative sense to prevent conduction.

grid voltage phase in one direction, as in *B*, increases the firing angle or conduction time. Shifting the phase of the grid voltage in the opposite direction, as in *C*, reduces the conduction time.

In Fig. 2·6 an a-c signal of variable amplitude is used. The grid a-c signal is superimposed on the steady-state d-c bias on the grid. For a small

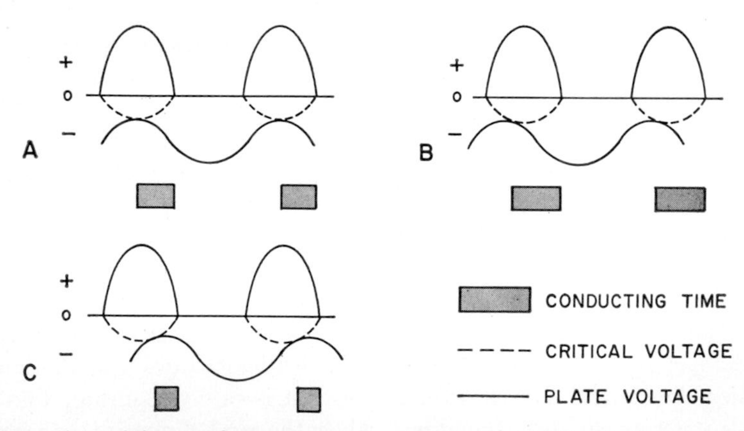

Fig. 2·5 Phase-shift control; negative grid voltage is phase-shifted, relative to the plate voltage, as a technique to vary the conduction time.

grid voltage the tube will fire at *A* and continue conducting until the anode goes negative at *B* because, once the tube fires, the grid loses control. A larger amplitude signal will cause the tube to fire at *C*, which is earlier than before. Again, conduction stops at *B*.

Many non-sine waves can be used to trigger thyratrons. A sharp wavefront is desirable for reliable triggering. Here a steady-state d-c bias keeps the thyratron from firing until the signal pulse allows the grid to become slightly more positive than the critical voltage and then the tube fires.

An example of a thyratron control is illustrated in Fig. 2·7. A phototube and a thyratron are used with an a-c power supply. The thyratron

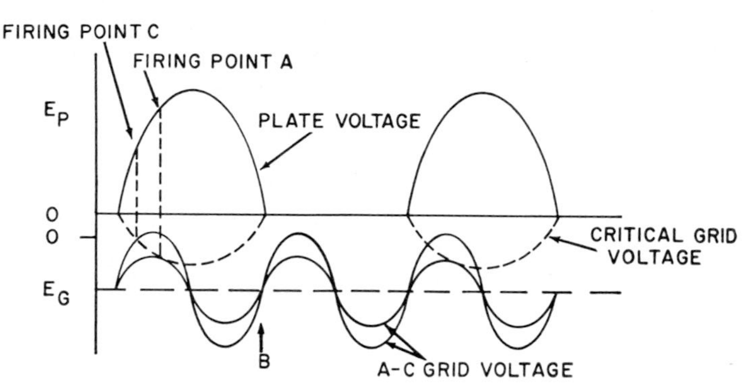

Fig. 2·6 Amplitude sine-wave control permits firing when the grid sine wave exceeds critical voltage. Note the shift in firing point..

plate supply is alternating current; hence the thyratron conducts when the plate is positive. The phototube cannot conduct while its cathode is positive (which is at the same time the thyratron is conducting).

When the phototube is dark, it will *not* conduct even when the cathode is negative. Hence the thyratron fires on the positive alternations because its grid is at the same potential as the cathode, and relay *R* is energized.

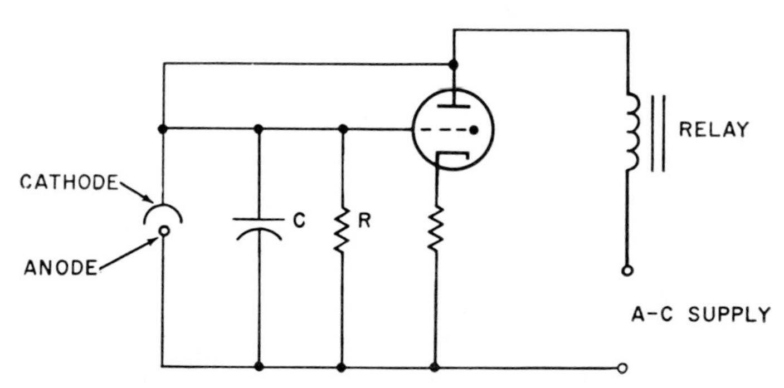

Fig. 2·7 Thyratron circuit controlled by a phototube in a circuit with alternating current as the power source.

When light hits the phototube cathode (and the cathode is negative), the phototube acts as a rectifier. The current through the phototube charges capacitor *C*, to the *peak* value of the alternating current applied to the phototube.

The phototube acts as a detector, and the charge on the capacitor keeps the grid of the thyratron at cutoff. As long as light shines on the cathode, the phototube keeps the thyratron grid at a negative potential. This prevents thyratron conduction and keeps the relay open. Interruption of the light removes the grid bias since the capacitor discharges through R. This fires the thyratron and closes the relay.

Thyratrons do not have an exact control by means of the grid voltage, for there is a range over which they can operate. In a typical range for

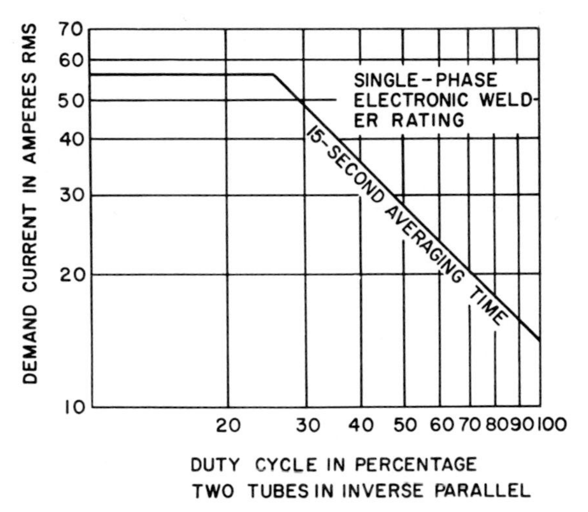

Fig. 2·8 Tube-type GL-6807 duty-cycle curve where maximum peak voltage (both forward and inverse) is 1,500 volts. (*General Electric*)

the GE-5663, at 250 volts on the anode the tube will fire at between 6 and 13 volts on grid 1 where grid 2 is -20 volts. Also, the current flow depends upon the duty cycle D where

$$D = \frac{\text{time on}}{\text{time off}}$$

A short duty cycle permits larger peak anode current than a long duty cycle.

As shown in Fig. 2·8, the GL-6807, GL-6808, or GL-6809 will have a current flow of 56 amp (rms) for two tubes in inverse parallel up to a duty cycle of 25 per cent, or 0.25. Above this the current drops off and, at a duty cycle of 0.65, current drops to 24 amp.

The RCA 3D22-A is a sensitive four-electrode thyratron of the indirectly heated cathode type. It is designed for use in relay and grid-controlled rectifier applications, particularly those involving motor-control service.

The 3D22-A has a negative-control characteristic which is essentially independent of ambient temperature over a wide range by virtue of the inert gas content. It also has small preconduction or gas-leakage currents, low control-grid-to-anode capacitance, and low control-grid current.

In power-control service, the 3D22-A is operated so that its output voltage to the load is varied by varying the time of firing during the a-c input cycle. When used for d-c voltage control, two 3D22-As in a full-wave circuit with resistive load are capable of handling up to 660 watts at a d-c output up to about 410 volts. When used for a-c voltage control, two 3D22-As in a full-wave circuit are capable of handling up to 800 watts.

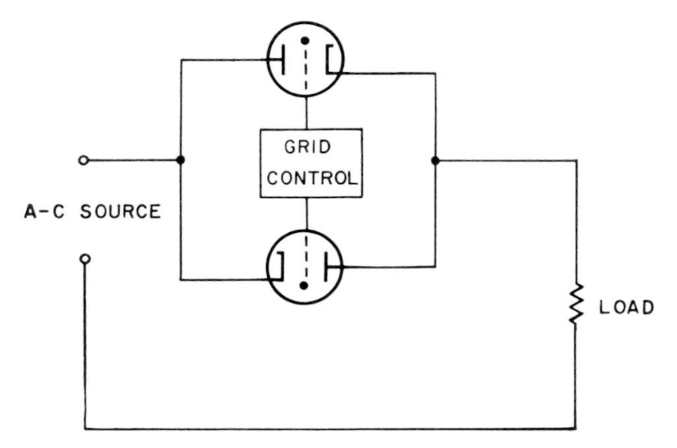

Fig. 2·9 A-c control using two thyratrons and an a-c power source.

A basic d-c voltage control using thyratrons is one in which the load and the tube are in series with the a-c line. Controlling the grid circuit allows a variation in that portion of the a-c positive alternation over which the thyratron conducts. When it conducts, it passes direct current through the load.

Figure 2·9 shows a-c power control in which each tube conducts on opposite alternations, allowing alternating current across the load. But here the amount of conduction on each half-cycle is adjusted by the grid-control circuit.

With a-c amplitude control, several degrees of control are available depending on the phase angle of grid-1 voltage relative to the anode voltage. If the grid-1 voltage lags the anode voltage by 180°, change in amplitude of the grid-1 voltage will initiate tube conduction early in the positive half-cycle of the anode voltage. This will effectively connect the load to the anode voltage supply for most of the half-cycle to provide maximum d-c voltage output. This type of operation gives abrupt change from zero to full d-c output voltage as the amplitude is decreased. If the grid-1 voltage lags the anode voltage by something less than 180°, change in amplitude of the grid-1 voltage can be used to cause the tube to conduct at only a few points in the positive half-cycle. This will mean that the a-c output voltage cannot be varied smoothly from the maximum value to zero.

The d-c bias method of control, in which the d-c level of the bias is variable, can initiate tube conduction only over the first 90° of the positive half-cycle of anode voltage. Initiation of conduction at the beginning of the half-cycle will connect the load to the anode voltage supply for most of the half-cycle to provide maximum d-c voltage output. If conduction is initiated at the 90° point, the d-c voltage output will be approximately one-half of the maximum value. To control the output voltage over a wider range, this method by itself is not practical.

With the phase-shift method, utilizing a sinusoidal grid-1 voltage, a wide range of control is available. If the grid-1 voltage lags the anode voltage by about 170°, conduction occurs near the end of the positive half-cycle, and consequently the load is connected to the anode voltage supply for a small part of the half-cycle to provide minimum d-c voltage output. If the phase of the grid-1 voltage is advanced so that the grid-1 voltage lags the anode voltage by less than 170°, tube conduction is initiated over a greater part of the positive half-cycle and the d-c output voltage will vary gradually from the minimum value to maximum. With this method, the use of adequate grid-1 voltage will ensure reliable control of the d-c output voltage.

Two figures illustrate a-c phase-shift control: Figure 2·10 shows d-c voltage control in which the d-c output (average) depends upon the portion of the a-c cycle over which the tube or tubes conduct. In Fig. 2·11, however, the a-c voltage across the load is a function of the conduction angle or firing time of the thyratrons. The greater the time of conduction, the greater the voltage across the load.

Ignitrons. These are special forms of diode power rectifiers. Figure 2·12 shows a group of these tubes. Small units, such as the two at the bottom, are air-cooled. All the others are water-cooled; the water inlets and outlets can be seen. Larger ignitrons have handles for ease of carrying. Ignitrons can carry currents up to thousands of amperes.

An ignitron is a gas-filled diode of special construction used to control large currents. There are many different ignitron types and most of them bear little physical resemblance to small vacuum tubes. An ignitron can rectify large alternating currents and can operate with steady loads of several hundred amperes. However, its greatest use is handling large currents of several thousand amperes for short times as an a-c switch. In effect, the ignitron replaces a mechanical switch, but it operates with little maintenance since it has no moving parts. A heated cathode is used, but rather than a filament there is a pool of liquid mercury. When this mercury is vaporized and ionized, very large currents can be handled.

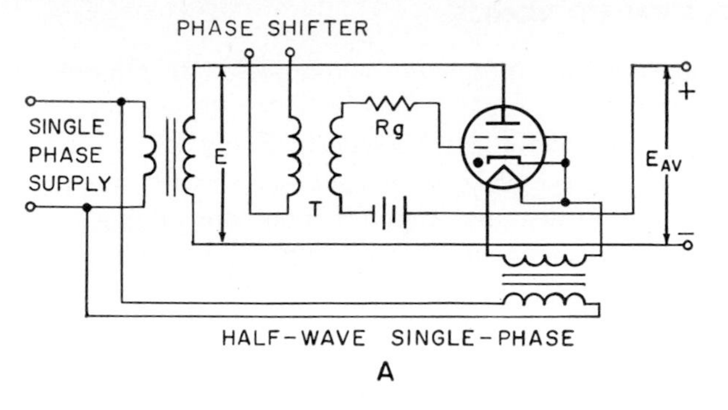

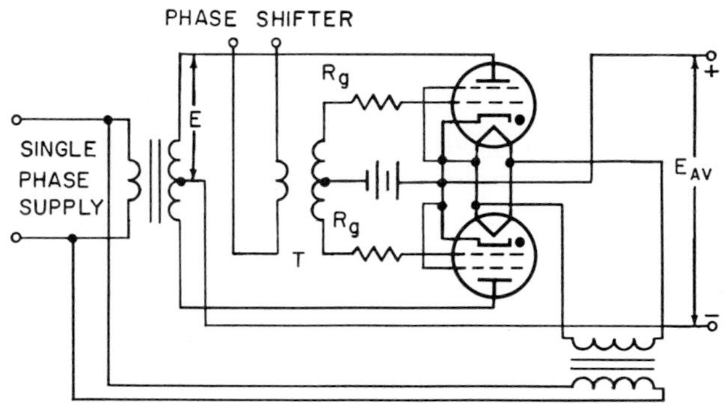

FULL-WAVE SINGLE-PHASE

B

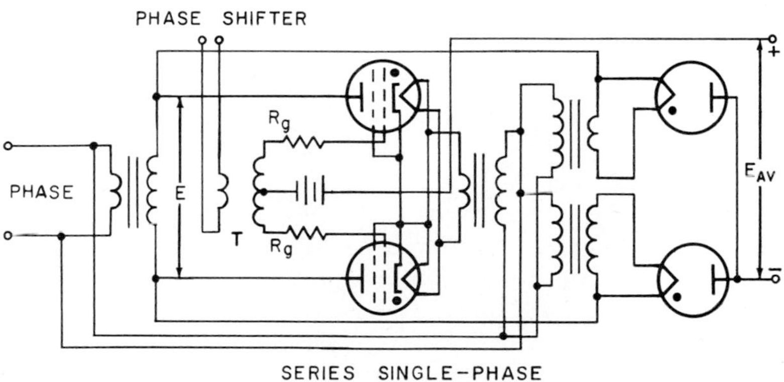

T = PEAKING TRANSFORMER
IN THE RECTIFIER TUBES MAY BE
3D22-A's USED AS DIODES. THE 3D22-A
IS USED AS A DIODE BY CONNECTING
GRIDS No2 AND No1 TO CATHODE (PIN 3)

C

Fig. 2·10 (*A*) D-c voltage control in a half-wave single-phase circuit; (*B*) d-c voltage control in a full-wave single-phase circuit; (C) d-c voltage control with series single-phase operation. (*RCA*)

Water cooling is used through the water jacket to remove the heat produced.

The parts of an ignitron can be seen in Fig. 2·13.

A tapered small piece of boron carbide is used as the starter or ignitor. The ignitor tip touches the mercury pool and (where it touches the mercury) forms a small depression. Although the boron carbide and mercury are in contact, there is a small resistance (100 to 500 ohms)

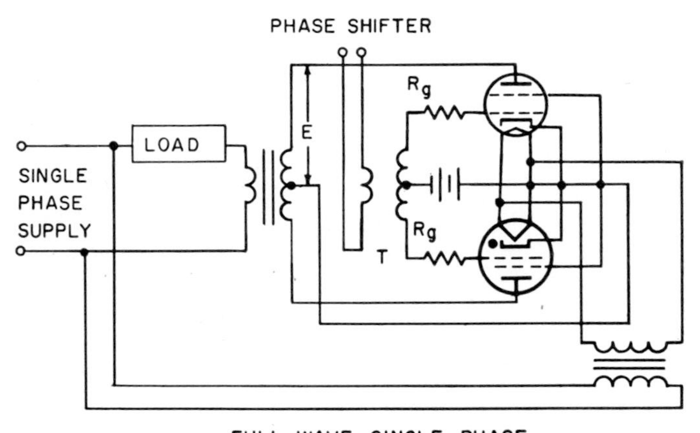

Fig. 2·11 A-c voltage control in a full-wave single-phase circuit. (*RCA*)

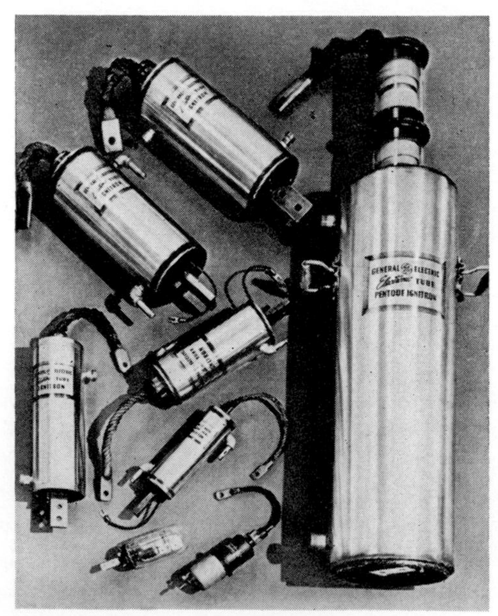

Fig. 2·12 Typical ignitrons covering a wide range of applications. (*General Electric*)

between them because of the nature of the materials. When a current of about 10 to 50 amp flows through this junction, an arc is developed which frees a large supply of electrons so that the tube can conduct from cathode to plate. A voltage of 100 to 200 volts is required to fire or start the ignitor action. Cathode spots are formed by the sparks from the cathode to the ignitor. These cause the cathode to emit large quantities of electrons.

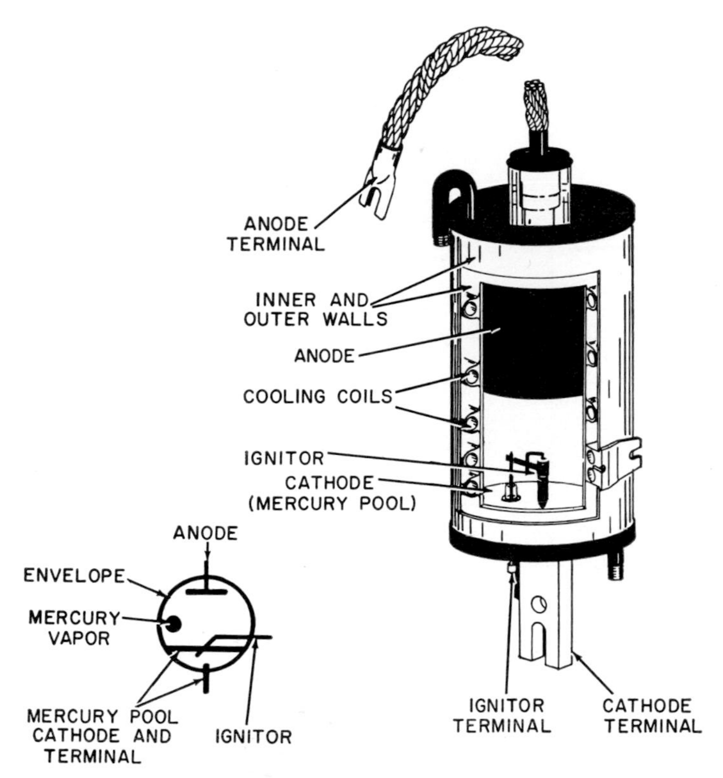

Fig. 2·13 Cutaway view showing ignitron construction; note the double metal walls with water-carrying cooling coils between them. (*National Corporation*)

After conduction starts, the ignitor circuit is turned off. The anode keeps the cathode in operation.

Electrons go from the cathode to the anode because the anode is positive and the electrons are negative. Graphite is the substance used for the anode. It does not normally emit electrons and it can radiate large amounts of heat to the cooling walls of the tube. Gas molecules fill the tube as the liquid mercury turns into mercury vapor. Electrons going from cathode to anode collide with gas molecules. This impact knocks electrons from the gas molecule. These electrons vastly increase the current flow. Also, because of the collision process, the gas molecule with one or more electrons missing becomes an ion; this process is called ionization.

In an ordinary tube there is a space charge, or cloud of electrons, near the plate. Ions tend to neutralize this space charge and increase the amount of current the tube can handle.

The tube terminals are of heavy construction because of the heavy current flow. There are three leads: cathode, anode, and ignitor. The envelope is a double-wall stainless-steel jacket in which water circulates for cooling between the two walls.

Ignitrons can pass very large currents for power control. The GL-5551-A can handle 56 amp in a-c control service. The GL-5553-B can control 355

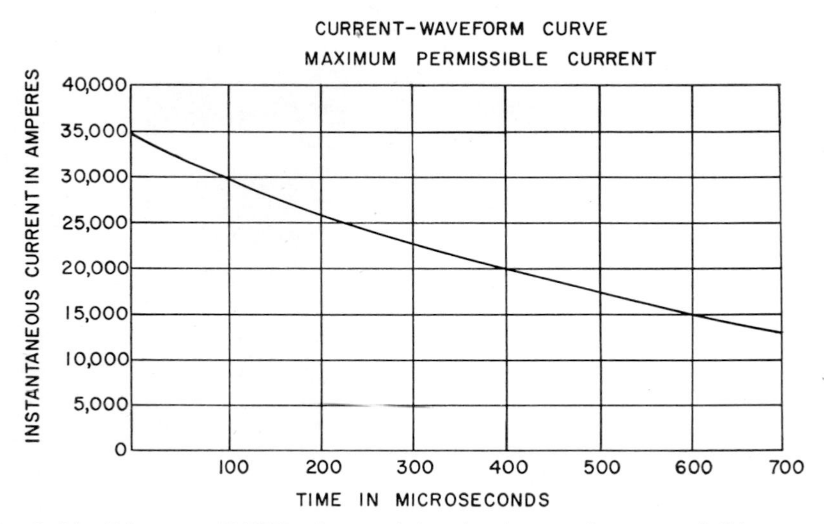

Fig. 2·14 Tube-type GL-7771 characteristics showing maximum permissible current. (*General Electric*)

amp. Even larger currents can be obtained, but the amount of current depends upon the time of the current flow. The GL-7171 has a peak current rating of 35,000 amp in circuits up to 10,000 volts. But the current flow decreases with the time the tube is on, as may be seen from Fig. 2·14. For 100 μsec the tube can carry 30,000 amp; for 225 μsec, 25,000 amp; and for 600 μsec, 15,000 amp.

Some ignitrons use grids for special functions. Basically the grid controls the ions so that the tube will not conduct when it is supposed to be off. Sometimes there are several grids. The one nearest the anode is the control grid; the other is the shield grid or baffle.

The shield grid prevents the drift of positive ions up through the control grid from the relatively large open volume in the lower end of the tube and thus further improves the ability of the tube to hold off voltage. This deionizing function of the grid is best obtained by applying a steady negative bias voltage to the grid with respect to the cathode and then superimposing a positive voltage peak at the point in the cycle when anode current is to be started.

The ignitron shown in Fig. 2·15 has a control grid, a shield grid, a gradient grid, and an auxiliary anode. As mentioned above, the shield grid is used to deionize the space near the anode and, in this tube, the gradient grid assists this action and reduces the voltage stresses. By means of the auxiliary anode an arc to the mercury cathode is maintained.

The auxiliary anode, or holding anode, has a low-voltage a-c power supply separate from the main anode. The current of the auxiliary anode is limited by series resistance to a value which is sufficient to maintain an arc to the mercury cathode. The arc is maintained from the time it is initiated by the ignitor to approximately the end of the positive half-cycle of the auxiliary-anode supply voltage.

This anode maintains the cathode spot when main-anode currents are very low. In some cases it is used to maintain the cathode spot until the main anode conducts after the ignitor fires. The current required to maintain the cathode spot depends upon the tube temperature and the residual ionization within the tube. This current may be as low as 3 amp if no anode current is carried, and up to 5 per cent of the main-anode current if the cathode spot is to be maintained after main-anode conduction. Just after main-anode conduction, the holding-anode current could be totally supplied by the residual ionization in the tube unless sufficient power is available in the holding-anode circuit to overcome this and retain the cathode spot.

Holding-anode ratings generally specify a minimum current of 7 amp rms (5 amp average) to ensure a continuous cathode spot.

The action of an ignitron as an a-c switch can be seen from Fig. 2·16. An elementary circuit of a welding control is shown; when the terminal *A* is positive (*AB* is the a-c source), the tube cannot conduct. When *A* is negative, the cathode is negative and the plate is positive. Conduction is now possible; the tube starts conducting (when the switch is closed) through the current flow from the cathode mercury pool to the ignitor. This starts the mercury arc and the tube is then a half-wave rectifier. Controlling the switch action will permit control over the current flow through the transformer to the welding arc through the step-down transformer.

There are two essential circuits in an ignitron used for welding control: the main, or cathode-to-anode, circuit and the firing, or ignitor, circuit.

In power rectification, a third circuit is used which is called the holding-anode circuit. In this, a small auxiliary anode is used to maintain a sufficiently high degree of ionization to carry the load current even at fractional loads.

In ordinary rectifier service, the ignitor starts the flow of current at the beginning of each positive half-cycle. The entire cycle of operation re-

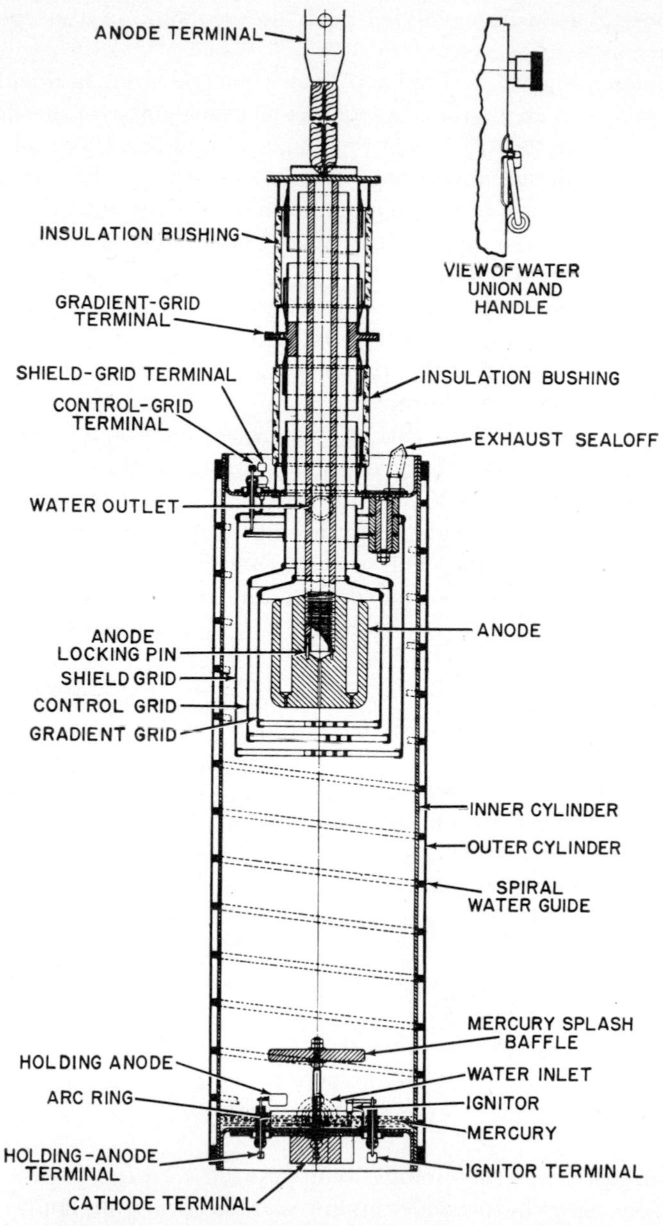

Fig. 2·15 General Electric–type GL-6228 showing construction details; note the three grids and two anodes.

peats continuously, with the ignitron tube allowing a surge of current through the circuit each time its anode is positive. The output is pulsating direct current.

For resistance welding, the fact that ignitron tubes change alternating

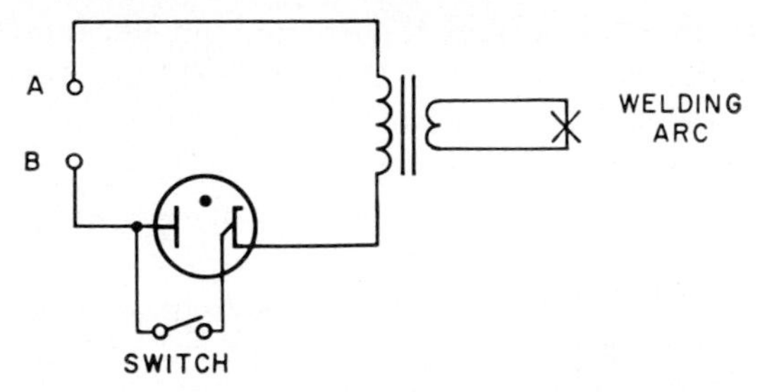

Fig. 2·16 Ignitron as a switch in a simplified welder circuit.

current into direct current is unimportant since resistance welding is usually done with alternating current. The rectifier effect is nullified by connecting two ignitrons in inverse parallel (the anode of one tube connected to the cathode of the other) as shown in Fig. 2·17. In this way

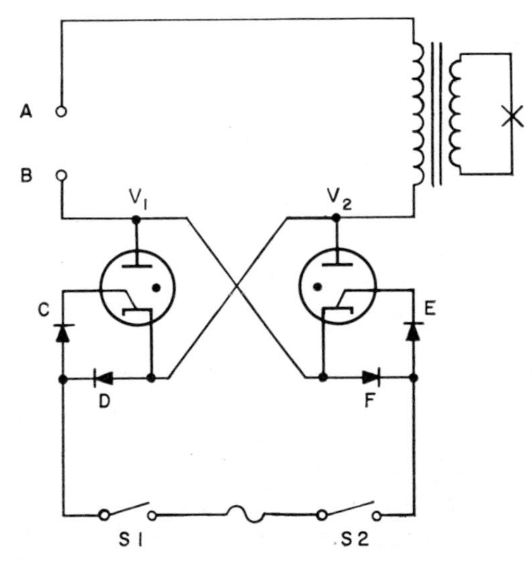

Fig. 2·17 Back-to-back ignitrons for welding with diodes (*C*, *D*, *E*, and *F*) for current limiting and control.

each tube conducts on opposite half-cycles and the output of both tubes together is still alternating current.

Ignitrons make possible precision control over each half-cycle. By controlling the point at which the ignitrons are caused to conduct current, it is possible to control the average amount of current flowing through the circuit.

In most welder circuits, the ignitrons are required to conduct in a complex sequence. For example, the material to be welded might require that the full amount of power be fed to the welding transformer for only three cycles and that the current then remain off for another four cycles to permit the metal to cool before the next weld begins.

In the same circuit, another welding job might require less heat (and therefore less current). If the job requires half as much current, it is possible to cause the ignitrons to pass current during only half of each half-cycle.

In Fig. 2·17 these tubes are connected back to back or anode to cathode. The rectifiers (*C*, *D*, *E*, and *F*) prevent harmful current flow from the

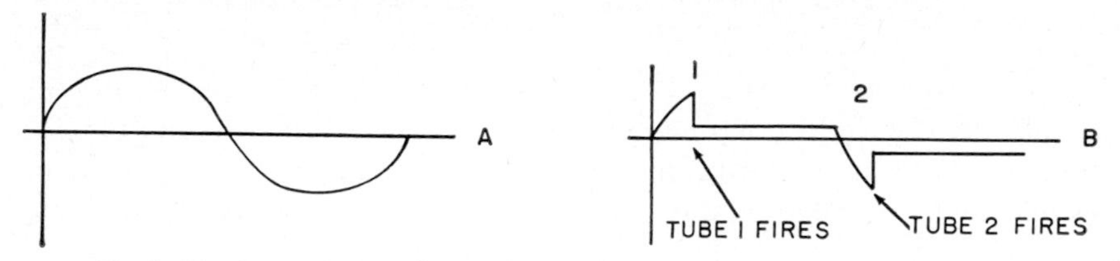

Fig. 2·18 Ignitron waveshapes showing a-c source (*A*) and firing points (*B*).

ignitor to the mercury pool. Normal ignitor current goes from the mercury pool to the ignitor. Reverse current could ruin the ignitor. When *B* is positive and *A* negative, electron flow is from *A*, through the transformer to the cathode of tube 1. But current goes *from* the mercury to the ignitor, through *C*, through *S*1 and *S*2, through *F*, and back to *B*. Rectifier *E* prevents reverse current in tube 2.

These tubes act as rectifiers, one tube for each alternation. By controlling the *portion* of the alternation over which each tube conducts, a control over the output current to the welding load is possible. Two switches are shown. *S*1 is the water-flow switch, which will stop the circuit if the water flow stops. The second switch (*S*2) is the actual control over the tubes. In some circuits the switch *S*2 is a thyratron, but the same action can be followed as with the switch.

In Fig. 2·18 the applied a-c voltage is shown in *A*, and in *B* the voltage across the ignitrons is plotted as it would appear on an oscilloscope. At point 1 conduction begins and the ignitron fires. The tube voltage falls to a very low value because of the low *IR* drop across the ignitron during conduction. Because of the ignitor voltage drop, required to start each tube for each conducting alternation, there are sharp pulses also at 2; this is conduction of tube 2.

In other commercial circuits the load, the control thyratrons, and the ignitrons are in series. By controlling the thyratron grid with a small signal, the thyratron firing cycle is adjusted, which in turn controls the ignitron firing time. This determines the average current per cycle through the load.

There are three main applications for which ignitrons are particularly suited: resistance welding, power rectification, and power conversion or transmission.

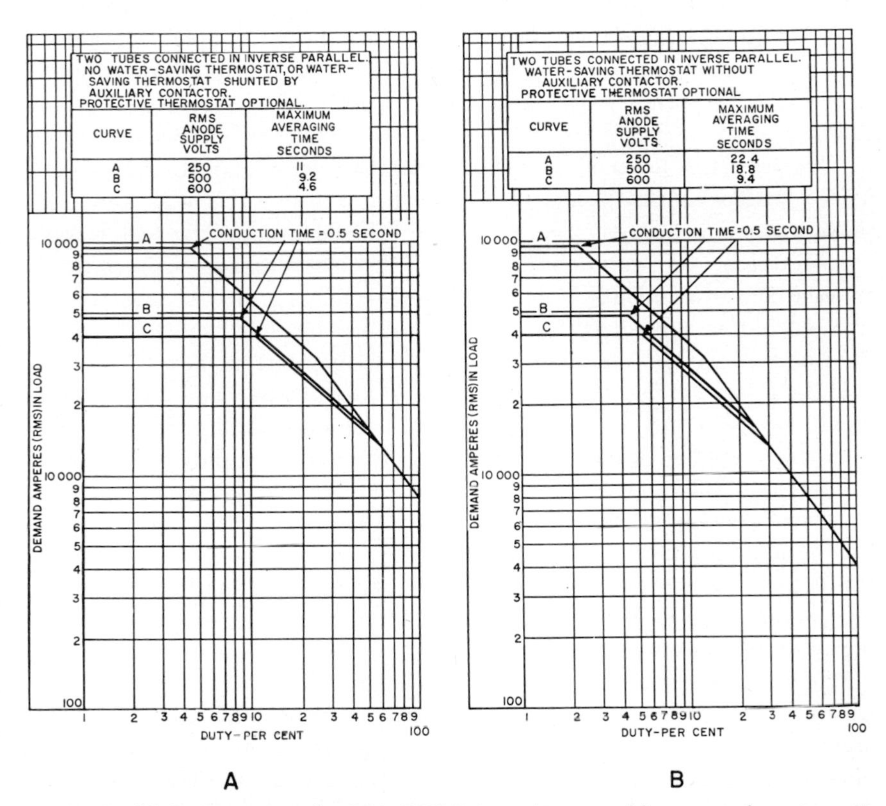

Fig. 2·19 (*A*) Rating curves for RCA 5553-B in resistance-welding control service with current as a function of duty cycle. No water-saving thermostat is used. (*B*) Same tube as in (*A*) except that a water-saving thermostat is used. (*RCA*)

Welding requires large amounts of current, and for this purpose ignitrons are usually used. The GL-5553-B is a water-cooled tube which can provide an rms current of almost 10,000 amp as shown in Fig. 2·19. This curve (*A*) is for resistance-welding service with the two tubes in inverse parallel. With a plate (anode) supply of 250 volts and a duty cycle of up to 4.5 per cent, 9,600 amp of current can be obtained. For a supply of 500 volts, current is about 4,800 amp; and at 600 volts it is 4,000 amp. At 100 per cent duty cycle, in all three cases, the current drops to 800 amp.

Figure 2·19*B* shows the same conditions but with the use of a water saver which allows circulation of water only while the flow is controlled. The GL-5553-B is equipped for mounting thermostatic controls which are calibrated either for controlling the flow of cooling water through the water jacket or for protection of the ignitron against overheating.

When the cooling water is circulated successively through the water jackets of two or more ignitrons, the water-saving thermostat, if used, should be mounted on the ignitron connected directly to the water supply.

The water-saving thermostat, which has normally open contacts, is calibrated to close a circuit energizing a solenoid valve in the water-supply line and thus permit water flow to start when the temperature of the thermostat mounting plate exceeds approximately 35°C. Because of the lag between the heating of the ignitron envelope and the functioning of the water-saving thermostat to start water flow through the water jackets, the ignitron may overheat before the flow of cooling water starts.

Such overheating can be prevented by the use of an auxiliary contactor shunted across the contacts of the water-saving thermostat and actuated by the welding-control switch. The contactor causes the solenoid valve in the water-supply line to open as soon as welding current flows.

If the water-saving thermostat is not shunted by an auxiliary contactor, it will be necessary to use a lower value of maximum average current than that specified when the auxiliary contactor is employed. The lower average current value is achieved by increasing the maximum averaging time and decreasing the maximum duty. Although the same maximum conduction time is permitted for both of these operating conditions, the use of the water-saving thermostat alone, without the auxiliary contactor, requires a longer interval between successive welds than when the thermostat is shunted by the contactor.

When a protective thermostat is used, it should be mounted on an ignitron from which the cooling water discharges into the drain. The protective thermostat is calibrated to open a set of normally closed contacts at a jacket temperature of approximately 52°C. The opening of these contacts causes a protective device to function. This device may be a relay opening the ignitor firing controls or, preferably, a circuit breaker which removes power from the ignitrons.

In welding applications ignitrons are used to control the primary current supplied to resistance welding transformers. They are used in voltage-supply circuits of 220, 440, 550, 1,100, and 2,300 volts (rms). The tubes function as switches and, by means of suitable electronic control, may be arranged to provide one, two, or a dozen cycles of current at given weld settings which may be repeated indefinitely without change in the number of cycles. As a result, uniformity in the welds is obtained and losses from poor welds are reduced.

Ignitrons for power rectification are available in sizes for d-c outputs from 40 to 1,000 kw with single units, depending on the operating voltage.

D-c voltages usually are 125, 250, 600, and 900 volts. These rectifiers are used to provide power for machine shops, elevators, mines, electrolytic reduction plants, arc welding, and similar types of service. Voltage-regulating equipment may be provided to give practically constant output voltage from zero to full load.

Ignitrons have a third class of application which is high-voltage d-c power transmission, or conversion of power at one frequency to power at another. Here the tubes are primarily for power conversion and are grouped to form units of 2,000 to 20,000 kw capacity; higher capacity may be obtained by additional units. These electronic power converters provide a nonsynchronous tie between two power systems and are able to transmit a constant amount of power independent of the usual variations in either the supply or receiving-system frequencies and voltages.

Ignitrons are rugged metal tubes, and they can carry thousands of amperes. But they do require care in storage and should be protected from vibration and shock. The basic stainless-steel construction of an ignitron protects the tube from physical damage during normal handling and storing; however, precaution must be taken to prevent the mercury in the tube from being deposited on the anode or the glass seal.

Any mercury droplets which are deposited on the glass seal or the anode can cause arcing, which can damage the tube. For this reason ignitrons should be stored in the vertical position. If, by accident, a tube is placed on its side, the mercury will run from its normal position.

To prevent operational delays, tubes should be kept ready as replacements for tube failures. Several tubes, depending on specific needs, may be kept on "replacement standby." This is done by keeping the tops of these tubes slightly above room temperature with a heat source such as a 100-watt lamp in a reflective enclosure.

Incoming power should be regulated as far as possible, since voltage surges in the lines may also cause arcing within the tube. Tube terminals should be clean and securely fastened to the mounting brackets. The tubes should be shielded from nearby high-frequency fields which sometimes initiate arcs within the tube. In addition, the tubes should be shielded from magnetic fields which may cause the arc to form on the tube sidewalls. Sufficient shielding of the tubes is usually made by the metal panel enclosures; however, high-frequency lines and conductors carrying large currents should still be placed away from the panels.

When replacing a faulty ignitron in the panel, be sure to check the rectifiers in the ignitor circuit. Their malfunction may have caused the original failure by applying negative voltages to the ignitor.

Because of the specific design-temperature limits and the heavy current flow, temperature control is important. Temperature control is obtained by regulating the amount of water flowing through the water jacket. Measurement of the water temperature at the water outlet provides an indication of the actual ignitron temperature. These tubes will often overheat if the water is turned off when the anode power is removed. To avoid this, the water flow should be maintained for the time specified in the instructions.

If an ignitron is operated for long periods near the rated minimum temperature, certain load conditions can cause high voltage surges in the tube. These surges will, in turn, cause breakdowns in the associated equipment unless it is adequately protected.

Ignitron failures or improper and sporadic tube operation can usually be traced to one of three basic conditions. These are:

1. Inadequate flow of water for cooling
2. Failure of the rectifiers which block ignitor current flow
3. Operation of the tubes outside of their published ratings

These conditions, where they exist, cause other conditions which lead to tube failure. Ignitron wetting is a special problem in these tubes. In proper or normal operation the mercury in an ignitron is uncontaminated. It does not become attached to the ignitor. The ignitor operates as though it were composed of a number of crystals in proximity to, but not touching, the mercury. When a voltage is impressed between the ignitor and the mercury, a cathode spot occurs at the juncture between the crystal and the mercury because of high voltage gradients. As long as this cathode spot forms on the mercury, the ignitor is unchanged and should last for years.

The mercury, however, may become contaminated. Ignitor wetting occurs when the mercury becomes impure and attaches itself to the ignitor. Under this condition, the crystal-to-mercury junctures become short-circuited. This will result in sporadic tube operation or complete failure of the ignitor to initiate the arc.

2·2 A-c power control Control of a-c power is the basic function of thyratrons and ignitrons. In this service they can act as both rectifiers and switches.

Methods of Power Control. Various methods of power control are used for different purposes. Figure 2·20 shows two basic applications. In welding (*A*), the a-c line power is rectified and controlled to send precise amounts of power through the welding transformer. For electric furnaces (*B*), the steady flow of current is adjusted to provide for the desired amount of heat.

Furnaces and welders are only two of the many devices requiring controlled power. Control of a-c power is also needed for operating many types of motors or other devices in industrial processes. Three methods of control are used.

One is the use of a switch which, when closed, permits the power to reach the load. With an a-c source of power the switch permits the load to operate for a portion of each cycle. The larger the portion of a cycle which is used, the greater the average power delivered to the load.

The second method is to use up some of the available power, as with a variable resistor in series with the load and the power source. A large value of resistance will, of course, consume a large portion of the available power and leave little for the load. A small series resistance will permit almost all the power to reach the load.

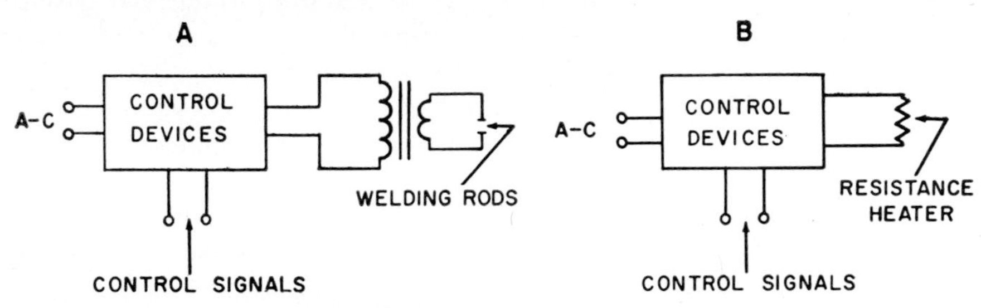

Fig. 2·20 (*A*) Power-control method as in welding with output (welding) transformer; (*B*) power control as in a furnace for electric heating.

The third method of control is the use of the transformer. An autotransformer with a variable tap may be used. Changing the position of the tap will vary the amount of power available to the load.

These three types of control are shown in Fig. 2·21. *A* shows the switch. Examples of switches are tubes such as thyratrons or ignitrons. *B* shows the power-consumption devices where *R* is the resistive element. Devices

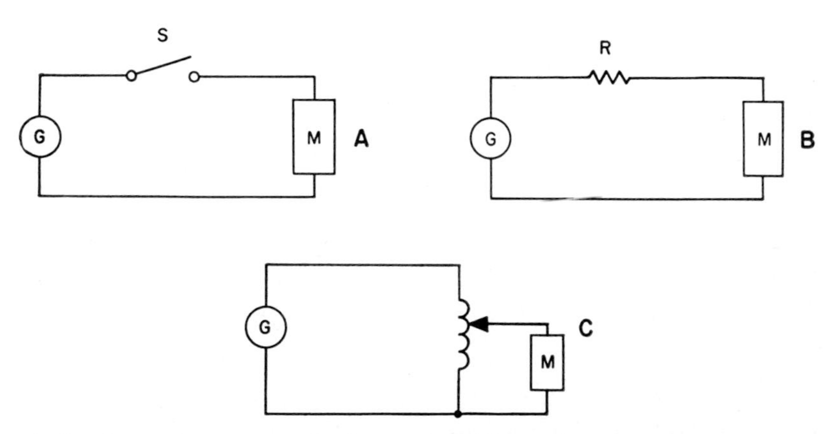

Fig. 2·21 Types of power control using a switch (*A*), a series resistance (*B*), or an autotransformer (*C*).

in this class include power resistors, tubes, and power transistors. The variable autotransformer is shown in *C*.

Switches for controlling power are of many different types, but they all have certain things in common. They all have the same output waveform; they all turn on at some time in the a-c cycle, which is the firing point or angle; they all turn off when the a-c supply goes negative; and they all are rectifiers as well as control devices and hence permit current flow in only one direction.

Switches control the a-c power by firing or turning on at a point in the positive half-cycle. If the switch fires at the positive peak, or 90°, one-half the available power in the half-cycle is applied to the load. Firing at 0° places all the power across the load. Thus the firing angle determines the amount of power at the load.

These switches include not only thyratrons and ignitrons but saturable reactors, magnetic amplifiers, and the controlled (semiconductor) rectifier as well.

The saturable reactor is a coil of wire wound around an iron core. When the iron is unsaturated, the reactor presents a high impedance between the power source and the load. A control winding is provided through which a direct current can saturate the iron core. Saturating the core is essentially magnetizing the iron in one direction so that no further change of flux can take place. The iron is no longer useful in making the coil a high impedance, and all that is left is the impedance of the wire, which is very low.

The saturable reactor is, in effect, a high impedance between the power source and the load. Saturating the iron core, which is equivalent to removing the iron suddenly, leaves nothing between the power source and the load except a coil of low-resistance wire. This occurs during every half-cycle. At the beginning of each half-cycle the iron is unsaturated and ready to be saturated again, with the firing time depending on the magnitude of the control current through the d-c winding. Varying this current controls the firing point and the power delivered.

The magnetic amplifier is similar to a saturable reactor, except that it has a rectifier in series with the iron core so that the bulk of the direct current needed to saturate the core is supplied by the load current. The control winding has only a very small requirement in determining the point in the cycle at which the iron becomes saturated. A magnetic amplifier can be considered a magnetically controlled rectifier in which the point at which the rectifier fires depends upon the current of the control winding.

The solid-state controlled rectifier is a solid-state device which acts as a thyratron. It is a rectifier, similar to the one used in the magnetic amplifier, but the forward firing point is determined by a d-c control signal in the gate. In the absence of a signal, the controlled rectifier will not fire and represents an open circuit. When the gate current exceeds a certain minimum value, the rectifier fires and conducts for the balance of that half-cycle.

Control Circuits. Thyratron control uses one of the techniques discussed in Sec. 2·1 to obtain direct currents over varying portions of the a-c line source. For welders, circuits such as shown in Fig. 2·22 are used. Phase shift for the thyratron grid is obtained, as in A, by the RC network of C and R_2. A grid waveform is taken from the adjustable resistance R_2. This changes the grid signal to permit variations in the firing angle of the tube and hence variations in the average direct current from the thyratron. This current is used in welding. A much wider phase shift can be obtained with the bridge network shown in B.

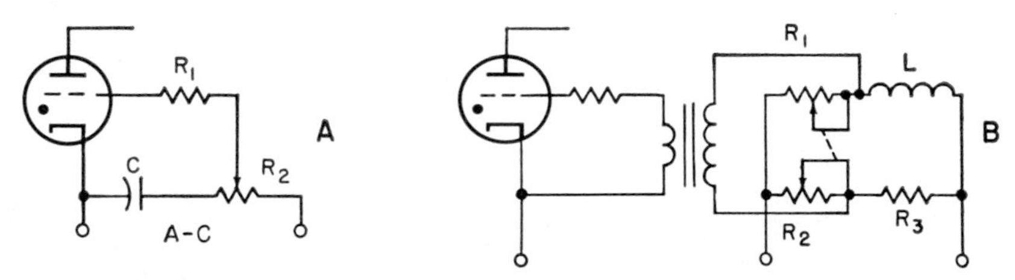

Fig. 2·22 Phase-shift circuits for thyratrons showing a simple *RC* phase shifter (*A*) and a dual phase shifter (*B*).

Motor control with thyratrons permits a smooth adjustment of the driving current. If the control voltage is positive, the motor will run at full speed since the tube acts only as a simple rectifier. By changing the d-c grid bias or by some other technique of grid control the average plate current will be varied, providing a method of speed control of the motor shown in the cathode circuit.

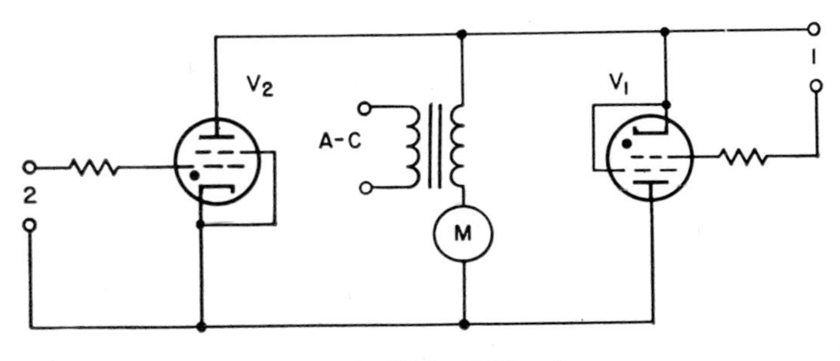

Fig. 2·23 Motor reversal for a d-c motor *M* driven by one thyratron for one direction and the other for the opposite direction.

Two of these circuits can be connected for motor reversal as shown in Fig. 2·23. For an a-c input (B) either rotation direction is possible. For clockwise rotation V_1 does *not* conduct (its grid voltage keeps the tube open) and V_2 conducts. This is shown in C, as a 180° firing angle. In counterclockwise rotation the situation is reversed; V_1 conducts and V_2 does not. With this circuit both the speed and the direction of the motor are controlled.

Table 2·1 Basic Power Rectifier Circuits *

Type of circuit →	Single-phase half-wave	Single-phase full-wave center-tap	Single-phase full-wave bridge	Three-phase star (half-wave)	Three-phase full-wave bridge	Six-phase star (three-phase diametric)	Three-phase double wye with interphase transformer	To determine actual value of parameter, multiply factor shown by value of:
Primary → Secondary →	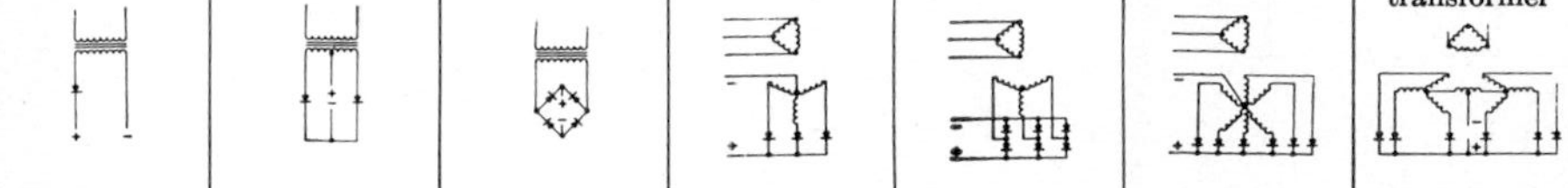							
One-cycle wave of rectifier output voltage (no overlap)								
Number of rectifier legs in circuit	1	2	4	3	6	6	6	
Average d-c volts output	1.00	1.00	1.00	1.00	1.00	1.00	1.00	Average d-c voltage output
Rms d-c volts output	1.57	1.11	1.11	1.02	1.00	1.00	1.00	Average d-c voltage output
Peak d-c volts output	3.14	1.57	1.57	1.21	1.05	1.05	1.05	Average d-c voltage output
Peak inverse volts per rectifier leg	3.14	3.14	1.57	2.09	1.05	2.09	2.42	Average d-c voltage output
	1.41	2.82	1.41	2.45	2.45	2.83	2.83	Rms secondary volts per transformer leg
Average d-c output current	1.00	1.00	1.00	1.00	1.00	1.00	1.00	Average d-c output current
Average d-c output current per rectifier leg	1.00	0.500	0.500	0.333	0.333	0.167	0.167	Average d-c output current
Rms current per rectifier leg:								
Resistive load	1.57	0.785	0.785	0.587	0.579	0.409	0.293	Average d-c output current
Inductive load	. . .	0.707	0.707	0.577	0.577	0.408	0.289	Average d-c output current

Peak current per rectifier leg:								
Resistive load	3.14	1.57	1.57	1.21	1.05	1.05	0.525	Average d-c output current
Inductive load	. . .	1.00	1.00	1.00	1.00	1.00	0.500	Average d-c output current
Peak to average ratio of current per leg:								
Resistive load	3.14	3.14	3.14	3.63	3.15	6.30	3.15	
Inductive load	. . .	2.00	2.00	3.00	3.00	6.00	3.00	
% ripple (Rms of ripple / Avg output voltage)	121%	48%	48%	18.3%	4.2%	4.2%	4.2%	
	Resistive Load	Inductive Load or Large Choke Input Filter						
Transformer secondary, rms volts per leg	2.22	1.11 †	1.11 ‡	0.855 §	0.428 §	0.740 §	0.855 §	Average d-c voltage output
Transformer secondary, volt-amp	3.49	1.57	1.11	1.48	1.05	1.81	1.48	D-c watts output
Transformer primary, rms amp per leg	1.57	1.00	1.00	0.471	0.816	0.577	0.408	Average d-c output current
Transformer primary, volt-amp	3.49	1.11	1.11	1.21	1.05	1.28	1.05	D-c watts output
Average of primary and secondary, volt-amp	3.49	1.34	1.11	1.35	1.05	1.55	1.26	D-c watts output
Primary line current	. . .	1.00	1.00	0.817	1.41	0.817	0.707	Average d-c output current
Line power factor	. . .	0.900	0.900	0.826	0.955	0.955	0.955	

* Assumes zero forward drop and zero reverse current in rectifiers and no a-c line or source reactance.
† To center tap.
‡ Total.
§ To neutral.
SOURCE: General Electric.

2·3 Power supplies A special form of power control is the d-c source known as the power supply. Taking alternating current from the line, these circuits rectify and filter the current to produce the required direct current for the use of other circuits.

Both vacuum-tube and gas-filled rectifiers are used, but their action is different from that of the tubes discussed earlier. These rectifiers are two-element tubes and are used in power supplies to rectify or convert alternating current to direct current. As with ignitrons, rectifiers conduct current only when the anode is positive with respect to the cathode. Unlike ignitrons, in which the point of conduction can be controlled by means of the ignitor firing time, rectifiers conduct over the entire positive half-cycle of the a-c voltage. The current delivered by the rectifier is a function of the load connected to it.

Low-voltage D-C Supplies. Power supplies of up to the 500- to 1,000-volt range are low-voltage. Their design and use is covered in many texts. The seven basic types found in industrial electronic equipment are those in Table 2·1. Only a short discussion of two types is included here and in the following section. One is the three-phase full-wave type commonly found in polyphase systems. It provides a voltage with a very small ripple. Very little filtering is needed to provide excellent d-c efficiency. It is found useful for moderate-power applications. This bridge has the lowest inverse potential applied to the tubes in comparison with any other type of three-phase circuit.

Six-phase operation provides still further reduction of ripple. This circuit is found useful in high-power applications where the six tubes can provide greater amounts of power.

High-voltage D-C Supplies. Many industrial devices require high voltage for their operation, such as the electronic air cleaner or X-ray equipment. Because of the current demands, r-f power supplies, with rectification after the oscillator signal is stepped up, are not satisfactory. The three main methods of obtaining high voltage are described below.

1. *High-voltage transformers and rectifiers.* A direct method of obtaining high voltage is the use of a step-up 60-cps transformer. Either single-phase or multiphase operation may be used.

Because of the high secondary (a-c) voltage, several rectifier tubes are used in series. This reduces the voltage requirement for each tube, but each still passes all the load current. For example, if the secondary voltage is 3,000 volts each tube can operate on one-third of this, or 1,000 volts, thus allowing the use of smaller tubes at lower cost. But separate filament windings are needed to prevent arc-over because of the voltage difference.

Voltage considerations for the rectifiers depend upon the peak inverse voltage (PIV), which is the voltage in the nonconducting direction. In the proper direction of current flow, from cathode to plate, there is a large voltage drop across the filter and load. Thus there is little voltage across the rectifier. If the a-c source is 1,000 volts (rms), the peak is 1,414 volts; and if the filter was charged to 1,300 volts (assuming no current drain), the voltage across the tube is now 2,714 volts peak or 1,414 volts (a-c) peak plus 1,300 volts d-c. This means that the tube in the above example would need a PIV rating of at least 3,000 volts d-c and more with a safety factor.

Use of a full secondary winding without a center tap provides for greatest use of the available voltage because a tap, as does a full-wave rectifier, divides the available voltage between the two rectifier tubes.

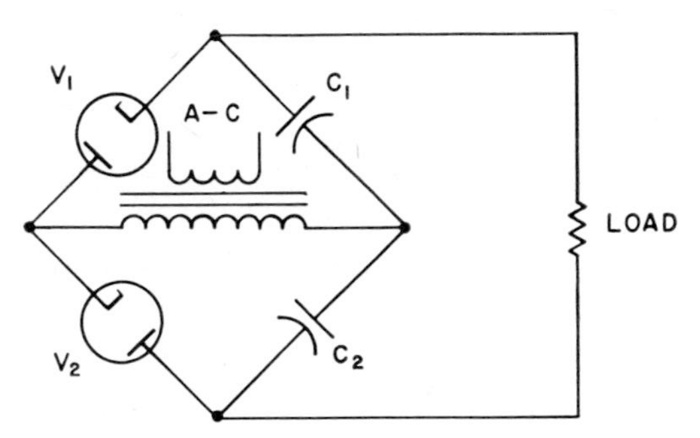

Fig. 2·24 Voltage doubler where the charge across both capacitors is across the load.

2. *Voltage multipliers.* Various circuits are used when a transformer voltage is required greater than that available from a given winding or when the load current requirements are not too great. Figure 2·24 illustrates such a circuit; on the first alternation when the tube V_1 conducts, C_1 is charged to almost the peak a-c value. Assume that this is 500 volts. On the other hand, alternation V_2 charges C_2. When the tube V_1 again conducts, with the polarity as shown, there is 1,000 volts available: 500 from the C_1 and 500 from the winding. R is the load on the circuit. This, then, is a half-wave voltage doubler. If another stage is added, the voltage available for the third rectifier is 1,000 and 500, or 1,500; this circuit is a voltage tripler.

Regulation, or the decrease in output voltage as output current increases, is not so good with voltage multipliers as with other rectifiers, so that they are used where small currents are required.

3. *Spark-gap rectifiers.* A mechanical rectification system is the spark-gap method shown in Fig. 2·25. A set of four spark gaps each is used: D, K, G, and F, which rotate on a shaft in synchronism with the applied a-c voltage. They are tied in pairs; D and F are electrically connected as are K and G.

Rectifier operation may be seen. Point B is negative and A is positive. Current flow is through the path $BKGC$, through the load and LFD, to A. On the other alternation the rotating gaps have moved, so G is now connected to the bottom of the secondary winding. Now the path is $AFDC$, through the load and *LKG*, to *B*. The next quarter-turn of the shaft repeats. In this way current flow through the load is always in the same direction and rectification has taken place.

Notice that this is more than just a mechanical device which can be replaced by an electronic device, because while it is full-wave, the *entire*

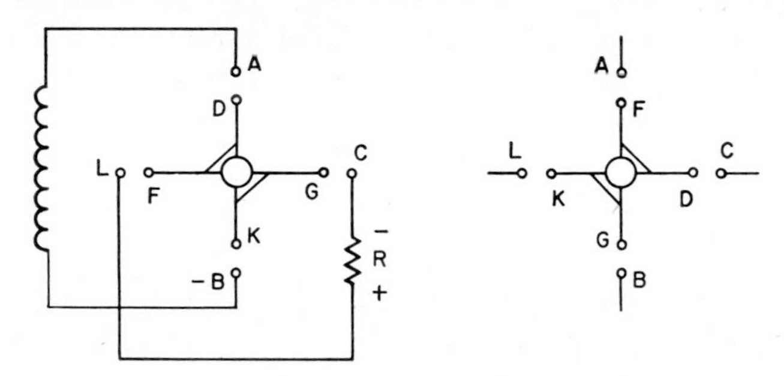

Fig. 2·25 Spark-gap rectifier or an electromechanical rectification system.

secondary voltage is used each time, not half the secondary voltage each alternation as with a full-wave tube rectifier.

REVIEW QUESTIONS

2·1 How does the thyratron grid control conduction?

2·2 Explain the difference between a thyratron and a vacuum-tube rectifier.

2·3 How can direct current be used to control a thyratron grid?

2·4 Why does the grid lose control when the tube fires?

2·5 How would a thyratron act if the grid were tied to the plate?

2·6 What is the function of the gas in a thyratron?

2·7 How can alternating current on a thyratron plate allow the grid to regain control?

2·8 How is alternating current used to control a thyratron grid?

2·9 What is meant by "duty cycle"?

2·10 If a tube is on for 3 sec and off for 4 sec, what is its duty cycle?

2·11 If a tube has a duty cycle of 0.4 and it is on for 10 cycles, how long is it off?

2·12 Why does a higher duty cycle raise the tube's temperature?

2·13 In a single-phase rectifier with a 10-volt drop across the tube, what is the PIV if 500 volts is applied across the secondary winding?

2·14 Define PIV. Why is it important?

2·15 Do two rectifiers in series each have less or more PIV than a circuit with one tube?

2·16 Why does a filter capacitor charge to peak voltage?

2·17 Why does a three-phase rectifier require less filtering than a single-phase rectifier?

2·18 What are three methods for the control of a-c power?

2·19 How does the ignitron differ from a thyratron?

2·20 How does the ignitor operate?

2·21 What is the cathode in an ignitron?

2·22 How is an ignitron cooled?

2·23 In addition to rectification, what else can an ignitron be used for?

2·24 Why are voltage regulators used?

CHAPTER

3

Solid-state Devices

3·1 Uses of solid-state devices

Active Circuit Elements. Solid-state devices cover a wide range of applications for industrial electronics, from transistor signal amplifiers to energy transducers of various types. Transistor circuits such as amplifiers are discussed in their specific applications in later chapters as are other specialized solid-state devices including transducers, detectors, and switches. It is the purpose of this chapter to highlight other less usual solid-state devices plus the power rectifiers and control elements.

Solid-state devices are important, for the future of small packaged circuits depends upon them. In these experimental integrated packages, several circuit functions such as inductance, capacitance, resistance, and amplification are combined into a single circuit element. A single piece of silicon or germanium has several circuit elements. Texas Instruments has demonstrated a complete solid-state multivibrator of match-head size. One circuit is a multivibrator with eight resistors, two capacitors, and two transistors. The other circuit is an oscillator with nine components.

A resistor, for example, is made by two ohmic contacts to the silicon or germanium material; transistors and diodes are made using the diffused-base technique. Capacitors are made by use of a large-area junction.

A subminiature circuit using transistors is shown in Fig. 3·1*A*. The largest unit is a complete four-stage amplifier. The RCA micromodules are another approach to microcircuits. Figure 3·1*B* shows a complete five-stage receiver using this solid-state technique.

Solid-state diode rectifiers, as examples of devices, are important in power conversion and control. But there are many other solid-state devices which serve new and unusual purposes.

Solid-state devices are performing circuit functions once only dreamed of. Ferrites, low-temperature amplifiers, controlled rectifiers, unusual transistorlike devices, and other solid-state materials are being developed at an increasing rate for use in electronics for industry.

One example is the Peltier or electronic cooling unit shown in Fig. 3·2. Using 10 amp of direct current at 0.25 volt, this refrigerator cools an infrared detector to about $-15°C$.

Peltier cooling uses electrons as a fluid to transport energy. Basically, a d-c potential is used to force electrons through dissimilar metal junctions. While the electrons are in one leg (or type of metal), they are at a given energy level. When the electrons move to another leg, they pass to a

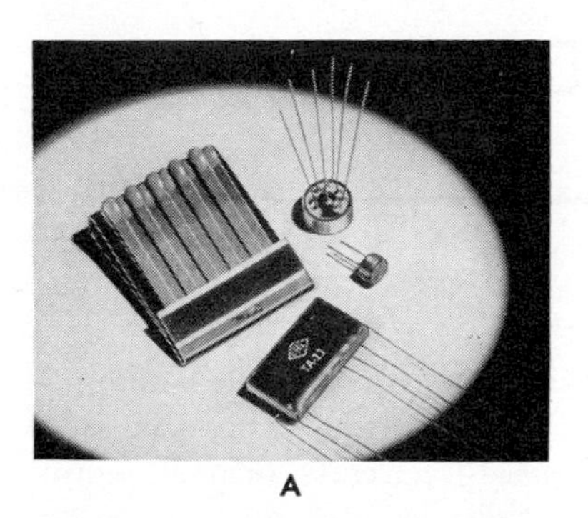

A

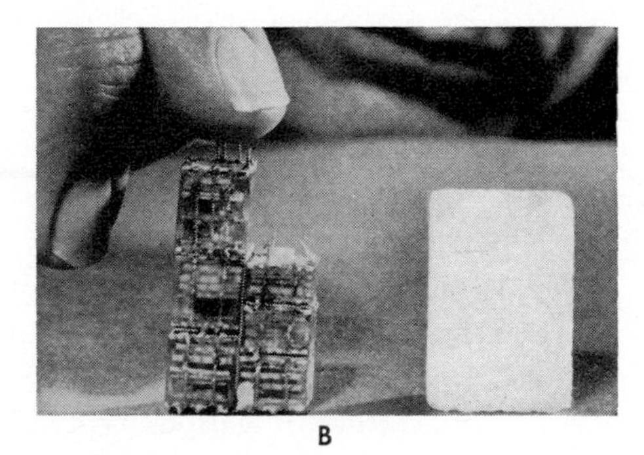

B

Fig. 3·1 (*A*) Transistor amplifiers by Centralab; each is self-contained with all parts except for the power supply, speaker, and controls. (*B*) A complete radio by RCA using the new micromodule technique.

different level of energy. If it is higher, the electrons must increase in energy; if lower, they must decrease.

In Peltier cooling they make up this difference by absorbing or rejecting this internal energy in the form of thermal energy, pumping it from the cool source to the heat sink.

Advantages of using the Peltier cooling unit include greater reliability, higher heat-sink temperatures, and a capability of accurately maintaining a static temperature at one particular level.

Some units, when coupled with a temperature controller, can either heat or cool by simply reversing the current. Peltier cooling is less expensive than Freon systems in applications with a heat rejection rate below 400 watts and with a temperature differential of less than 50°C.

Another simple, tiny low-temperature device, the cryotron, has been developed as a switching element. Figure 3·3 shows a cryotron circuit compared with a triode circuit. Because digital computers, for example, have thousands upon thousands of switching circuits, this cryotron could, if successful, reduce a computer now the size of a room to a size that would fit into a shoebox. Experimental devices operate at 4.2°K. At this temperature both the small tantalum rod and the niobium control winding are superconductive. Because there is no resistance in the tantalum rod, there is no voltage across the load resistance. No input power is required, since in the superconducting state the effective zero resistance of the niobium winding requires no voltage for the input current. As the control or input current increases, the magnetic field about the tantalum increases. At a critical value the tantalum suddenly develops its normal resistance and causes a voltage drop to appear across it and the load resistance. The

Fig. 3·2 Micro refrigerator for electronic cooling. (*Nortronics*)

cryotron circuit function is similar to that of the vacuum-tube circuit, but it works in a different way. In the tube, as the input or control voltage varies, so does plate current. In the cryotron, as the input current changes, the output voltage across the load changes.

Ferrites are another new material coming into use. Ferrites are ceramic magnetic materials made from oxides of metals such as iron, cobalt, nickel, aluminum, magnesium, and manganese. While they are metallic and magnetic, they are nonconductors and their hysteresis loss at microwave frequencies is very low. The electrons of the ferrite material are affected by external magnetic fields. Since the discovery of ferromagnetic resonance, in 1948, a whole new series of components has been developed for switching and for r-f use. Isolators or one-way transmission lines are important ferrite devices. A type of low-power isolator is shown in Fig.

3·4. This is a circular section of waveguide fed by a rectangular guide. The input signal has a voltage vector, at right angles to the first resistive vane, and is not affected by it. The ferrite-loaded section of the guide rotates the E (or voltage) vector by 45°. This rotation makes the input signal perpendicular to the second vane, and it passes through with only a small amount of loss, usually less than 1 db. If a signal is reflected from the load back into the output guide, its E vector is already at a 45° angle at the output end of the isolator. The reflected signal is rotated another 45° and becomes parallel to the first resistive vane, which absorbs the energy. This prevents the reflected energy from passing through.

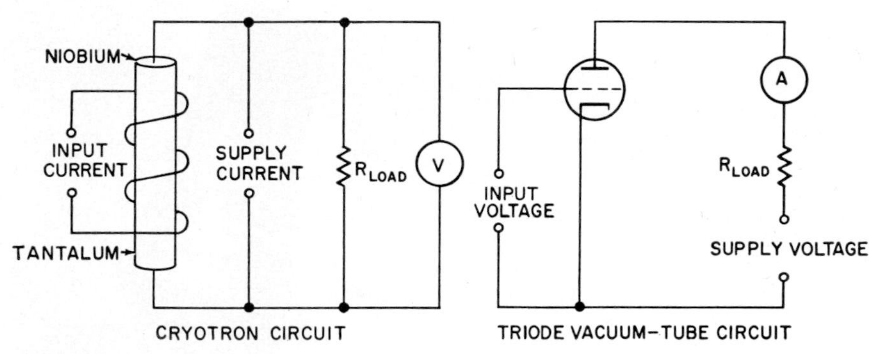

Fig. 3·3 The cryotron, which is a low-temperature switching device, compared with a similar vacuum-tube circuit.

Ferrites are also used in digital computers and control devices, where they act as a memory or storage for data. A large group of interconnected magnetic cores form the heart of the memory unit.

The memory has a large number of locations, or addresses, all of which can store information and retain it until needed. Each address can be described and located, so that either the information which it contains can be obtained or the location for new storages can be described. Reading out information from the memory does not destroy the data, in most cases, so that the same information can be referred to many times. In certain types of computer all the information needed for a solution to a problem may be stored in the memory, including all the steps, or program, required. In this way, once the proper information is placed in the memory, the computer is independent of all outside devices until a solution is reached. Then the final results are stored until they are needed by the output device.

Another solid-state device is the Magnetoresistor (Ohio Semiconductors), in which the electrical resistance is a function of the applied magnetic field density. Variations in the current in the control coil used with this device cause circuit changes used for control.

Printed Circuits. Printed circuits have advanced to a whole new series of materials and techniques to meet the needs of specialized industrial electronics. Glass and ceramics, for example, are two materials now being used for special applications.

Glass mat and cloth are used for high-quality laminated boards, but sheets of special glass also have characteristics which are applicable to printed-circuit use. Special types of glass known as Fotoform (Corning

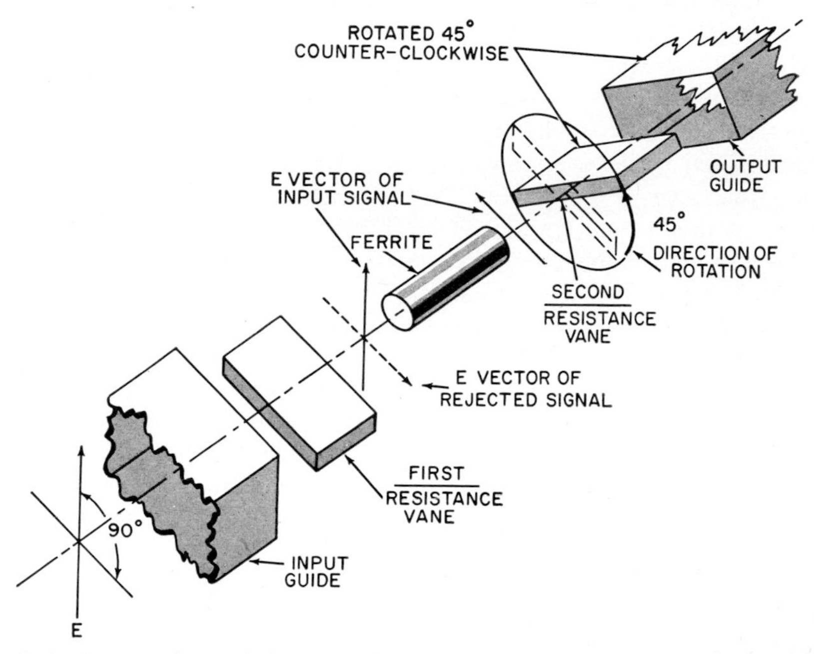

Fig. 3·4 Ferrite isolator which rotates the signal so that reflected signals are rejected. (*Electronic Technician*)

Glass Company) have good shock resistance and excellent surface resistance under humid conditions, and they can operate continuously at 500°C.

Etching is used to "machine" Fotoform. The glass plate is exposed to an ultraviolet light using a negative of the pattern to make a contact print on the glass. This is developed by heat treatment and etched to remove the undesired glass, such as in holes for mounting components. The resulting transparent glass may be processed and converted into other types.

Printed circuits include more than printed wiring. Some of the circuit components can also be printed by the same methods. A spiral of copper foil can serve as an inductance; a printed line of carbon ink can act as a resistor; copper pads on each side of the board can act as a capacitor with the base acting as the dielectric.

Active circuit elements such as transistors can also be made by these methods. By the use of photolithographic fabrication techniques, a transistor small enough to fit in a hole in a printed-circuit board has been developed by the Diamond Ordnance Fuze Laboratory. This development permits a transistor to become an actual part of a printed circuit.

Experimental transistors have been made this way and inserted into small holes on ceramic circuit boards. Extension of the technique may permit the manufacture of transistors directly on the baseboard. Present transistors made by this method are 0.05 in. wide by 0.01 in. high.

Hearing aids, pocket radios, miniature tape recorders, and other devices use printed-circuit amplifiers and circuit modules of different construction. New and smaller types with built-in transistors, resistors, capacitors, and wiring are shown in Fig. 3·1*A*. The smaller cylinder is the TA-6, the larger is the TA-7. Both are single-stage amplifiers. These are compared in size with the four-stage transistor amplifier.

The use of modules presupposes some system for mounting groups of circuits. In computer systems a large number of boards are connected into the electronic system; here modular designs eliminate a great deal of hand wiring. All active circuits are mounted on inserts, which are in turn attached to the modules. Modules are plugged into the subframes and the subframes are interconnected with cables. The module approach to design makes the addition of input-output facilities considerably easier and allows each module to be serviced independently to minimize computer downtime.

Not all assemblies use the plug-in technique. Packard-Bell has a different packaging approach, as shown in Fig. 3·5. The Multiverter has a single slide-out drawer which has a number of large boards. The power-supply board pivots so that all components are available while the circuit is still operative. Both the resistor matrix and the switching and control circuit boards are shown. These boards pivot into the box in the center. Then the box assembly pivots back into a slotted rest in the back of the unit.

Connectors are used in assemblies of systems using printed boards. Figure 3·6, as an example, shows a printed-circuit connector with the Bellows Action (Continental Connector) contacts for use with printed-circuit boards that can vary in thickness from $\frac{1}{16}$ to $\frac{1}{8}$ in. The spring-grip action is positive over the entire printed-circuit contact area, with no distortion or stress. Self-alignment of Bellows contacts allows for any residual warpage of the printed-circuit board. This series 600-93 printed-circuit connector is constructed with an extra rugged molding of glass-reinforced Plaskon alkyd 446 and is available in 15 contacts. Passivated stainless-steel mounting bushings allow for self-alignment of the connector.

Printed-circuit techniques extend beyond the circuit boards to the several types of printed cables which are now in use. Three different cables are Tape-Cable (Tape-Cable Corporation), Flexprint (Sanders Associates), and Polystrip (International Resistance Corporation).

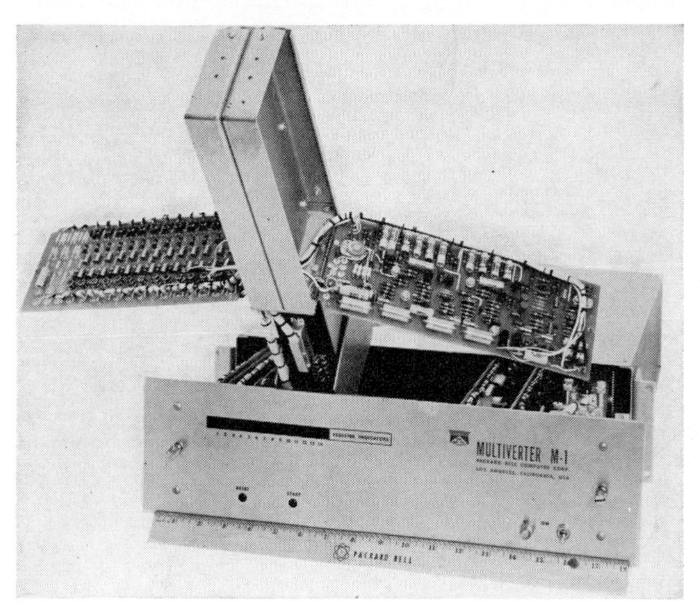

Fig. 3·5 Circuit packaging showing pivoted printed-circuit boards used by Packard-Bell.

Flexprint is a flexible printed wiring made of a conducting material, usually copper, bonded between two layers of insulation. It can be fabricated in the form of cables, wiring harnesses, and flexible printed cir-

Fig. 3·6 Printed-circuit connector. (*DeJur-Amsco Corporation*)

cuitry on which components may be mounted; the latter can be rolled into a compact cylindrical form or laminated, in part or completely, to other base materials to provide rigidity where required. Multilayer construction can be supplied to permit superimposing a number of layers of complex circuitry or wiring with suitable terminals exposed on one or both sides of the insulation. Terminal exposure on one side is preferred and always cheaper. Shorter routing of wires, reliable interconnection between circuitry on each layer, and a substantial reduction in the size of a wiring assembly may thus be achieved.

A complete circuit may be produced to bend or twist or fit into unusual spaces. Complete wiring harnesses tie a complex system together. Shielded wire can also be produced.

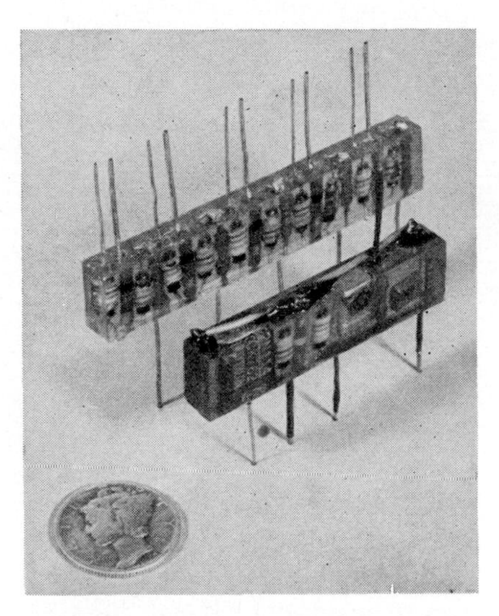

Fig. 3·7 Instrument-stick packages showing encapsulated components. (*Lind Corporation*)

One of the approaches to a rugged high-density group of standard components is the instrument-stick (Lind Corporation, Princeton, New Jersey) technique shown in Fig. 3·7. Packaging densities of from 40 to 74 per cent may be achieved. Instrument sticks are smaller and lighter than any other commercially available packaging method.

They are replaceable and repairable, and "three-dimensional" simplified wiring is possible when modules are stacked or used with printed-circuit boards. Aluminum sandwich construction may be used for internal thermal conductivity and for elimination of "hot spots" caused by dissipation of power transistors. Thermal-lag encapsulation may be used with low-power circuits for high-temperature environments lasting one or two hours. Vibration, shock, and acceleration resistance will meet most military specifications when sticks are strapped or cast together.

Protection against humidity, fungus, salt spray, etc., is provided through the use of double seals. All conducting surfaces are covered with casting epoxy. But this is only miniature-sized packaging of normal components; microminiaturization produces much smaller equipment from a different concept. In this new approach the solid-state materials are used to form the individual components such as resistors, capacitors, and transformers. They are combined in many ways to produce units with several circuit functions.

An example of new ultrasmall techniques is the micromodule (Fig. 3·1*B*) recently developed by RCA. The basic module is 0.3 in. square by 0.01 in. thick. It is a ceramic wafer, and each wafer is a part of an individual component. The materials on the wafer may be resistive, ca-

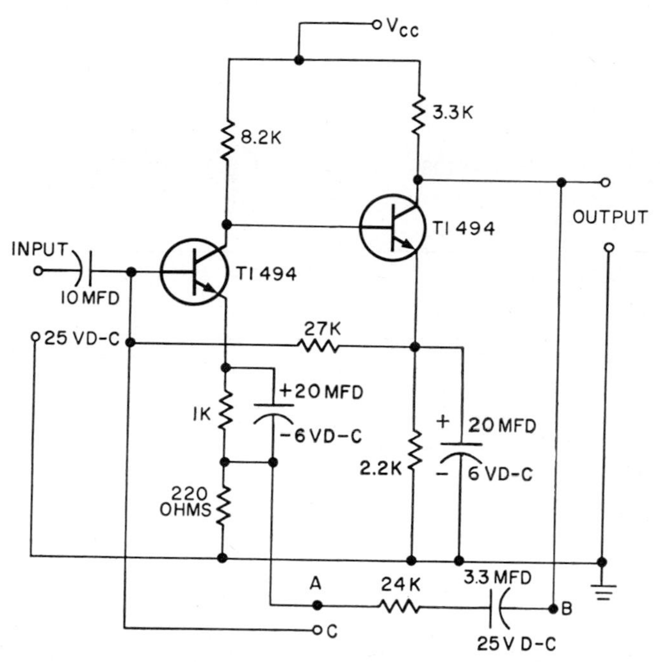

Fig. 3·8 Industrial amplifier for multipurpose use. (*Texas Instruments*)

pacitive, insulating, ferromagnetic, or semiconducting. Wafers, fused together and potted, form the circuit. In a demonstration, a five-module radio was shown with a fountain pen containing the entire radio including the loop antenna and variable capacitor for tuning.

3·2 Amplifiers Solid-state amplifiers are discussed in this book as a part of the equipment in which they are used. A typical transistor amplifier for industrial use is shown in Fig. 3·8. The circuit uses two silicon transistors, type TI 494, in a resistance-coupled amplifier suitable for any purpose where a signal level is to be increased. Gain is 40 db over a temperature range of -20 to $+100$°C with a d-c supply voltage from 10 to 25 volts.

The basic amplifier circuit also may be utilized as a sensitive a-c voltmeter, tuned amplifier, or tuned oscillator simply by employing different networks between terminals *A* and *B* or *C* and *B*.

Some basic transistor circuits are the common-emitter, common-base, and common-collector. In much the same way vacuum tubes have grounded-grid, grounded-plate, or grounded-cathode circuits. The positive voltage which is applied to the transistor collector roughly cor-

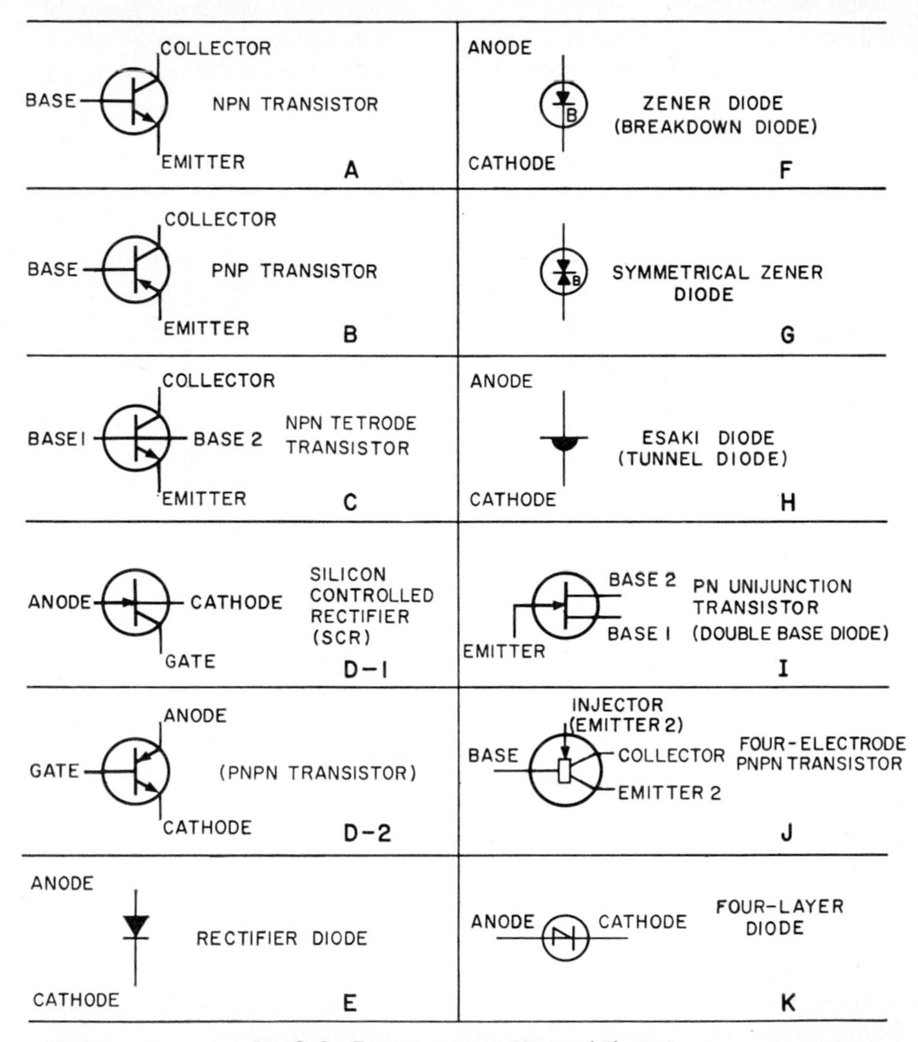

Fig. 3·9 Transistor types. (*General Electric*)

responds to the plate or anode of a vacuum tube. In the same way electrons are emitted by the cathode of a vacuum tube and collected by the plate of the vacuum tube. A control signal is applied to the grid of the vacuum tube, and this corresponds roughly to the base of the transistor. Transistors, because they are low-impedance devices, are current-operated; while vacuum tubes, being high-impedance devices, are voltage-operated.

The above discussion of the corresponding operation of a vacuum tube and a transistor is in terms of the grounded emitter or common emitter.

Some of the most widely used circuit symbols are shown in Fig. 3·9. Each of these is used and discussed in detail in several circuits in later sections of the book. The names given in this figure are common names; several manufacturers have special names for their own proprietary products. For example, the silicon controlled switch is known as a thyristor by RCA and Bell Laboratories, while Westinghouse calls it a trinistor.

3·3 Power diode rectifiers Diode rectifiers, both vacuum and gas-filled types, are used to produce direct current from alternating current as described in Chap. 2.

Semiconductors, which conduct electrons far better in one direction than the other, are not actually new, for copper-oxide rectifiers have been in use for many years. Because they are heavy and bulky they have been replaced by silicon, germanium, and selenium rectifiers for most power applications.

Selenium has been popular as a semiconductor rectifier because of the ease in manufacturing the sandwich of selenium which is the cell. A layer of selenium is placed between metal plates to obtain a unidirectional current flow.

Selenium rectifiers are very strong and withstand shock as well as vibration. They are capable of being constructed in stacks or groups to handle very high current applications. Selenium is found in power supplies having current characteristics of from 5 through 100 amp or more.

A germanium rectifier, made from a single crystal, has a high efficiency—about 98.5 per cent. This rectifier does not age or change its characteristics with time as a selenium rectifier does, but it is limited to low currents of up to the 100-ma range.

The number of silicon rectifier applications is growing. The silicon rectifier is smaller than the selenium rectifier but produces equal currents, it ages only slightly, and it operates at high temperatures. Currents of up to 5 amp or more can be obtained from silicon rectifiers. Characteristics of selenium and silicon can be compared from Fig. 3·10.

These considerations apply to power rectifiers. Circuits may be seen in Chap. 2. Semiconductor rectifiers which have a control element are discussed below.

3·4 Solid-state switches

Transistors. Transistor flip-flops are the most widely used semiconductor switches. There are two basic types. In the saturated circuit the transistors switch from saturation to cutoff. In the nonsaturating flip-flop the transistors do not reach saturation; hence these circuits are faster in operation. Figures 3·11 and 3·12 show examples of each type.

In a flip-flop one transistor is on while the other is off. Such saturated circuits are slower in operation (lower in frequency) than nonsaturated

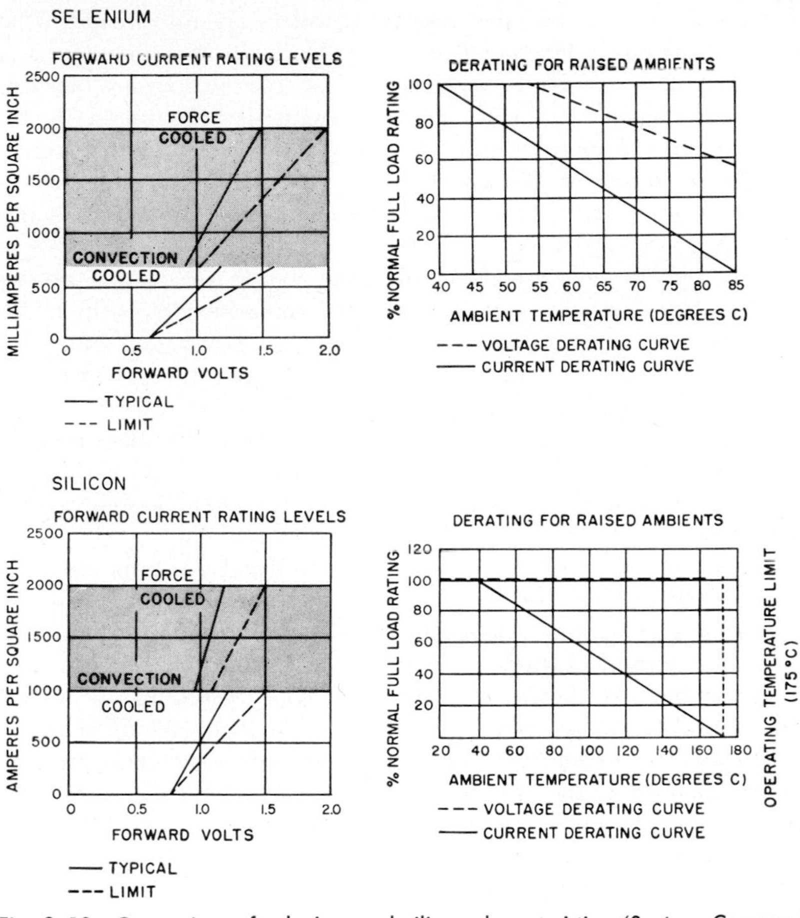

Fig. 3·10 Comparison of selenium and silicon characteristics. (*Syntron Company*)

circuits because of the storage time of minority carriers. Where the dissipation rating of a transistor is given, a saturated flip-flop can switch more current.

Steering circuits are used to permit a single pulse source to operate either transistor. The diodes D_S and resistors R_S form the steering circuits shown in Fig. 3·11*A*.

Capacitors C_K are used to speed up circuit operation. The diodes connected across R_S are required for the circuit to respond at the maximum repetition rate. If a number of flip-flop circuits are cascaded to form a

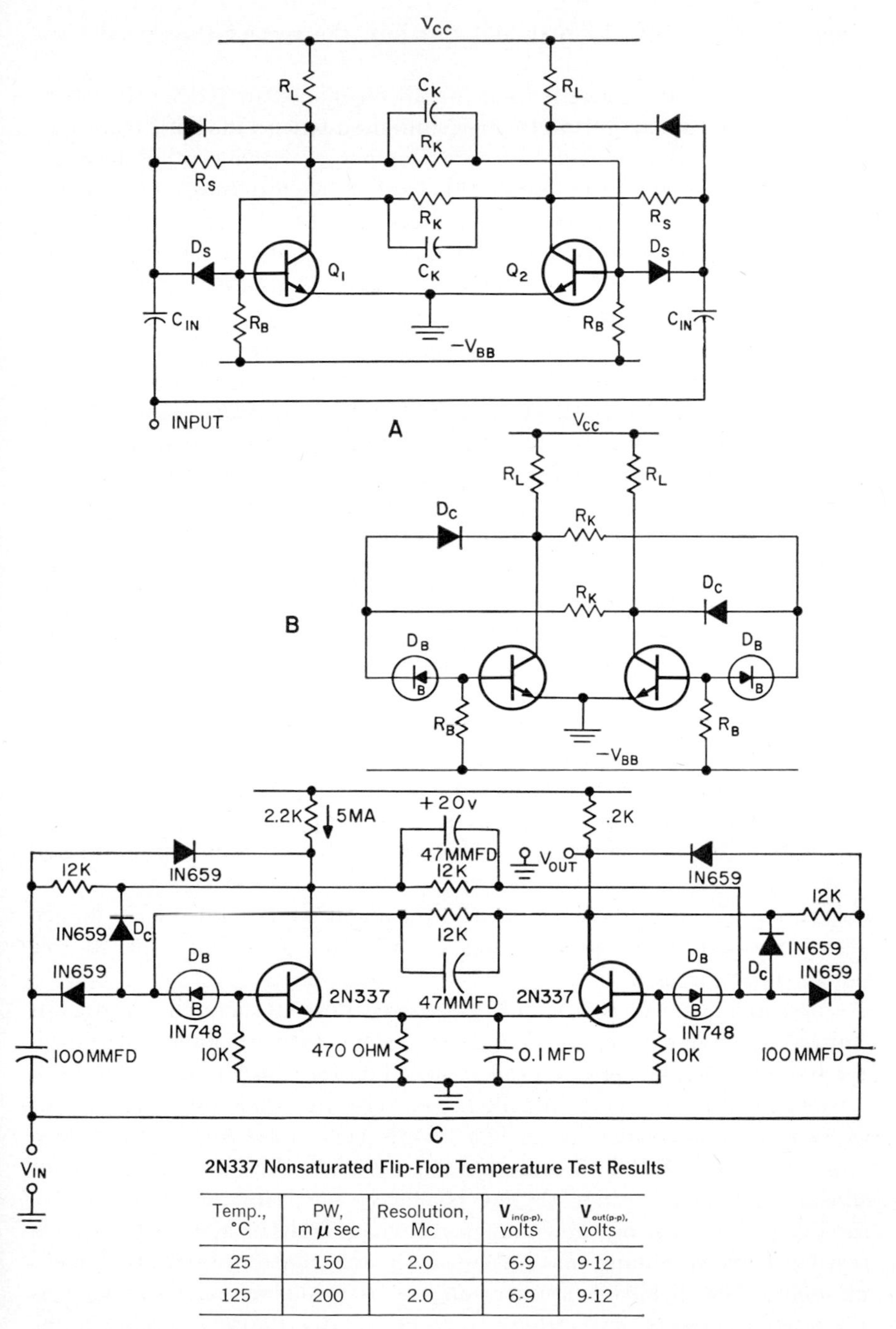

2N337 Nonsaturated Flip-Flop Temperature Test Results

Temp., °C	PW, m μ sec	Resolution, Mc	$V_{in(p\text{-}p)}$, volts	$V_{out(p\text{-}p)}$, volts
25	150	2.0	6-9	9-12
125	200	2.0	6-9	9-12

Fig. 3·11 (*A*) Basic flip-flop showing speed-up capacitors and steering circuit; (*B*) basic nonsaturated flip-flop; (C) complete nonsaturated flip-flop with test results. (*Texas Instruments*) binary counter, only the first and possibly the second stages need the diodes across R_S.

Negative base triggering is used for all flip-flops; for NPN transistors, negative pulses are applied to the base and used to turn the "on" transistor off. A flip-flop usually must drive a number of circuits; therefore, its output must be sufficient to handle this load. When flip-flops are cascaded, the output voltage or current must be greater than the input voltage or current amplitude required for triggering the next stage. The input pulse width and amplitude are also important and will usually determine the input and cross-coupling capacitor values.

One technique for designing a nonsaturated flip-flop circuit is to hold the transistors out of saturation with a reference diode and a computer diode. This technique is illustrated in Fig. 3·11*B*, which shows diodes added to the basic circuit. During the time that the transistor is on, diode D_C clamps the collector voltage above the base voltage by the reference voltage of D_B, thus keeping the transistor out of saturation. D_B is passing more than enough current to control the collector current and voltage. When the transistor is off, D_C does not conduct and D_B still maintains a reference voltage but now passes only a very small current.

Coupling capacitors, input capacitors, and steering circuits may be added to the circuit of Fig. 3·11*B*. Such a circuit is shown in Fig. 3·11*C*; here, however, the input steering diode connects to the junction of D_C and D_B rather than to the base of the transistor. Smaller trigger voltages are required with this connection.

Test results of the circuit in Fig. 3·11*C* are shown. In these results *PW* is pulse width of the input measured at the 10 per cent amplitude level; resolution is the maximum rate of the input pulses for full output voltage swing.

Figure 3·12 shows the component values and temperature performance data for a saturated flip-flop using 2N337 transistors. Only one 20-volt power supply is used.

Often in industrial circuits pulses are used for counting and control. An emitter follower can reproduce a pulse at the required impedance level. An inverter turns the pulse over as required by the next circuit it is to feed.

Controlled Rectifiers. Controlled rectifiers are solid-state switches or thyratrons. Semiconductor devices of this type are designed to replace ordinary relays, switches, and thyratrons. The solid-state thyratron or silicon controlled rectifier (SCR) is one of the latest additions to the rapidly growing list of semiconductor devices. Neither a transistor nor a rectifier, but combining features of both, the controlled rectifier opens up many new fields of application for semiconductors. Circuits now utilizing transistors or rectifiers may be greatly improved through the use of controlled rectifiers, and many new applications may now be converted to semiconductors because of the unique properties of the device.

It has the ability to change alternating current to direct current and, simultaneously, to control the power fed into a load.

Models about the size of a thimble are capable of handling loads varying from 200 to 1,000 watts at a stud temperature of 125°C. When switching at full rating, the controlled rectifier dissipates only 0.5 per cent of the controlled power. The controlled rectifier (Fig. 3·13) is a PNPN semiconductor consisting of three rectifying junctions. Breakdown of the

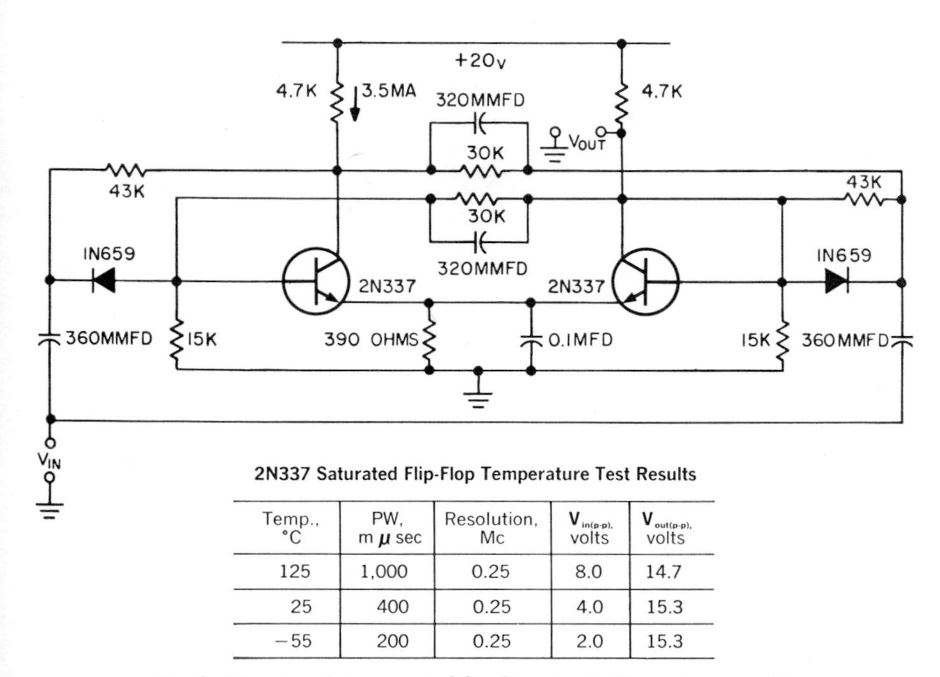

2N337 Saturated Flip-Flop Temperature Test Results

Temp., °C	PW, m μ sec	Resolution, Mc	$V_{in(p\text{-}p)}$, volts	$V_{out(p\text{-}p)}$, volts
125	1,000	0.25	8.0	14.7
25	400	0.25	4.0	15.3
−55	200	0.25	2.0	15.3

Fig. 3·12 Complete saturated flip-flop circuit. (*Texas Instruments*)

center junction can be achieved by applying an appropriate signal to the gate lead, which consists of an ohmic contact to the center P region. Breakdown occurs at speeds approaching 1 μsec. After breakdown, the voltage across the device is so low that the current through it is essentially determined by the load it is feeding. The controlled rectifier can switch 500 watts on the anode by the application of only 0.02 watt on the control gate. Peak current is as high as 150 amp. For industrial control applications it can replace thyratrons, circuit breakers, relays, power transistors, ignitrons, magnetic amplifiers, and other devices.

As shown in Fig. 3·14*A*, the source voltage is applied in series with the load and the rectifier. The rectifier will not conduct if there is no signal applied to the gate or if the peak voltage is 200 volts or less. The rectifier conducts when more than 200 volts is applied as shown in *B*. In *C* the peak applied a-c voltage is much greater than 200 volts.

Conduction can be controlled by the gate, which acts like the grid in a thyratron. The gate signal is generally applied from a high-impedance source which may be either alternating or direct current. For firing control, the gate voltage should be positive with respect to the cathode. The gate input impedance ranges from 10 to 100 ohms at the firing point. As the gate current is increased, a critical point will be reached and the device will break down at any positive anode-to-cathode voltage greater

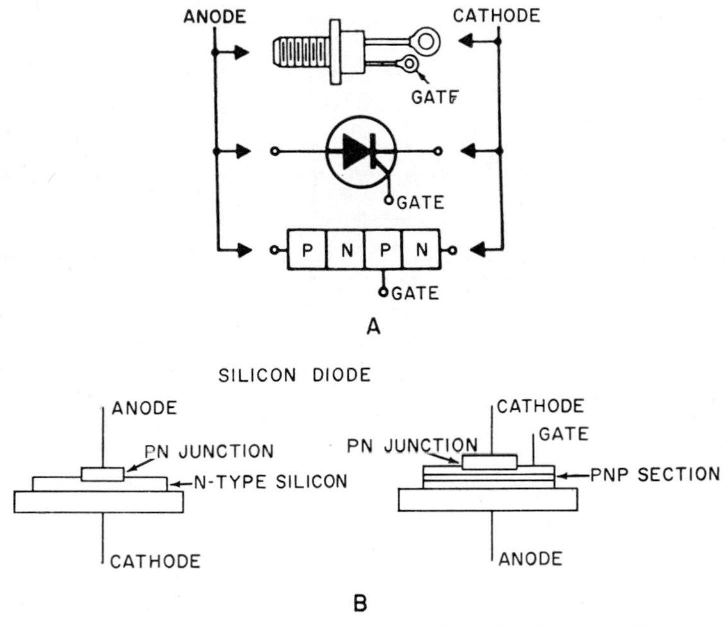

Fig. 3·13 (*A*) Parts of the semiconductor controlled rectifier (*General Electric*); (*B*) construction of the semiconductor controlled rectifier.

than a few volts. The gate loses control after breakdown, and cutoff can be obtained only by reducing the anode voltage to zero. This is identical to the loss of grid control in a thyratron.

The controlled rectifier has wide potential application in industry. Firing may be accomplished in as many ways as have been devised for firing thyratrons. The gate of the controlled rectifier has a much lower impedance than the grid of a thyratron and relies upon current rather than voltage for control. Figure 3·15*A* shows how to use this device to switch a d-c load. Momentarily closing switch *S*1 starts the conduction. The capacitor will charge after the switch is released. *R*2 determines the rate of charge. Closing switch *S*2 causes *C* to discharge and apply a reverse voltage which turns off the SCR. Figure 3·15*B* and *C* shows half-wave a-c circuits. Figure 3·15*D* shows a full-wave circuit.

The controlled switch (CS) is a PNPN device related to the controlled rectifier. It is similar to the SCR in both physical construction and theory of operation. The CS, however, has a much greater firing sensitivity. It is therefore useful in many low-level input applications that are not within the capability of the controlled rectifiers.

Consider two transistors connected together as shown in Fig. 3·16*A*. One is a PNP type and the other is an NPN type. Electrode *A* (anode)

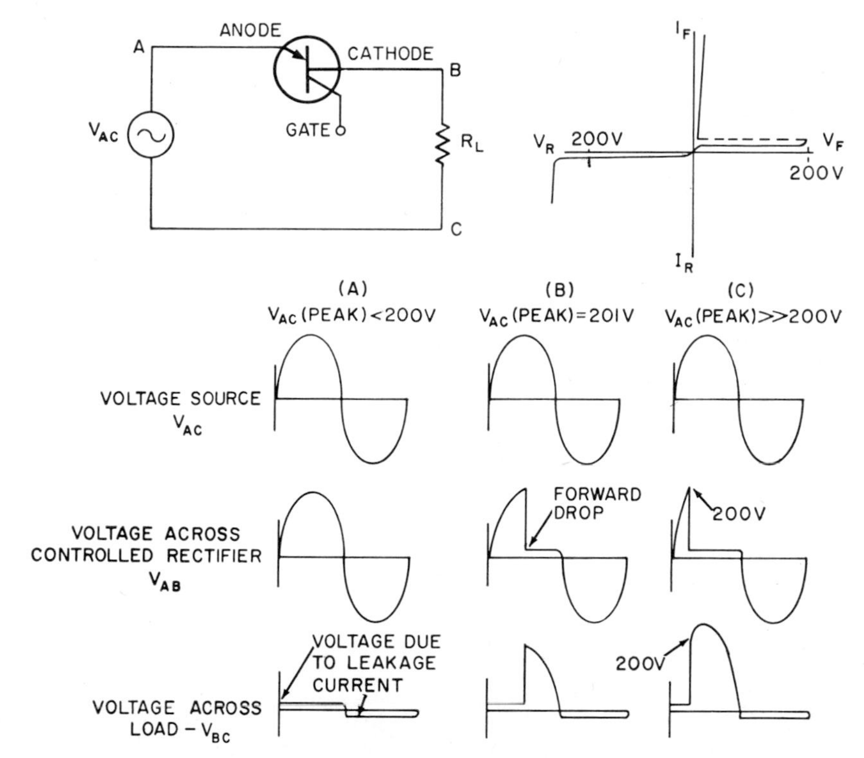

Fig. 3·14 Operation of the rectifier during conduction. (*General Electric*)

is attached to the upper P region; *C* (cathode) is part of the lower N region; and a third electrode, the gate, goes to the P region of the NPN triode. Actually, as shown in Fig. 3·16*B*, there are only four regions in a single wafer with three junctions as shown.

Schematically the transistor pair is shown in Fig. 3·17*A*, while *B* shows the circuit symbol for the CS. In operation the collector of Q_2 drives the base of Q_1 while the collector of Q_1 drives the base of Q_2. Where beta B_1 is the current gain of Q_1 and B_2 is the current gain of Q_2, the gain of this positive feedback loop is their product, $(B_1)(B_2) = B_3$. Where B_3 is less than 1, the circuit is stable. If B_3 is greater than unity, the circuit is regenerative. With a small negative current applied to terminal *G*, the NPN transistor is biased off and the loop gain is less than unity. The only current that can flow between output terminals A and C is the cutoff collector current of the two transistors. For this reason, the impedance between A and C is very high.

When a positive current is applied to terminal G, the NPN transistor is biased on, causing its collector current to rise. Since the current gain of

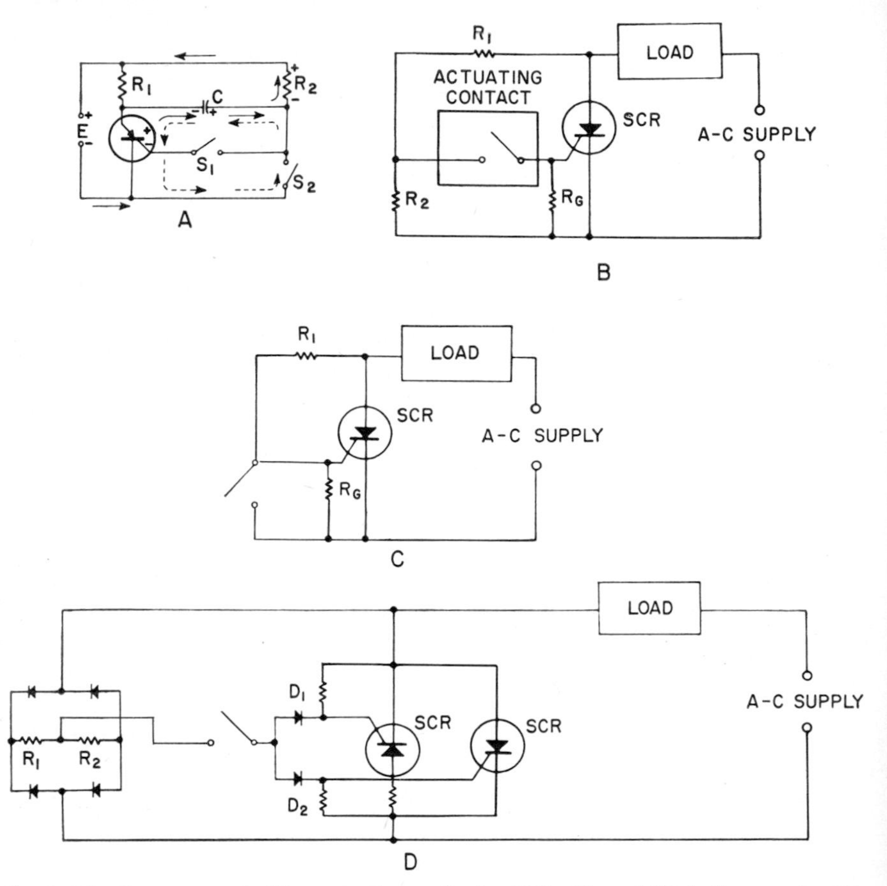

Fig. 3·15 Power control: (*A*) conduction paths for SCR; (*B*) and (*C*) half-wave circuits; (*D*) full-wave circuit. (*Solid State Products*)

the NPN, B_1, increases with increased collector current, a point is reached where the loop gain equals unity and the circuit becomes regenerative.

Collector current of the two transistors rapidly increases to a value limited only by the external circuit. Both transistors are driven into saturation and the impedance between A and C is very low. The positive current applied to terminal G, which served to trigger the self-regenerative action, is no longer required since the collector of the PNP transistor supplies more than enough current to drive the base of the NPN.

The circuit will remain in the on state until turned off. This is accomplished, for the CS, by reducing the collector current to a value below that necessary to maintain $B_3 = 1$.

Another device of similar construction is the silicon trigistor. As shown

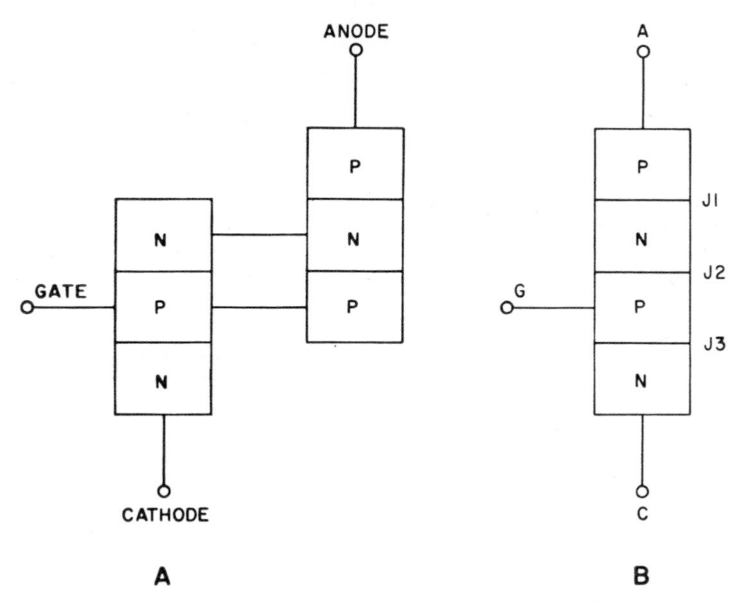

Fig. 3·16 PNPN transistors: (*A*) two transistors tied together; (*B*) three-junction and three-element transistor.

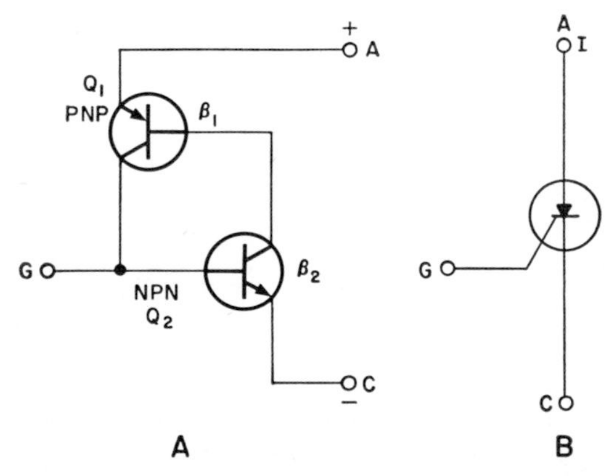

Fig. 3·17 (*A*) Interconnections in a two-transistor circuit; (*B*) circuit symbol for SCR.

in Fig. 3·18, it is a bistable semiconductor component with characteristics which approximate the circuit function of a flip-flop or bistable multivibrator. It is also a PNPN device but with the unique property of triggered turn-off as well as triggered turn-on control at its base.

The trigistor will turn on with the application of a low-level positive trigger pulse to its base. Once on, it will remain on without the need for sustaining base current. A negative trigger pulse applied to the base turns it off. It will then remain off until triggered on again. The circuit will

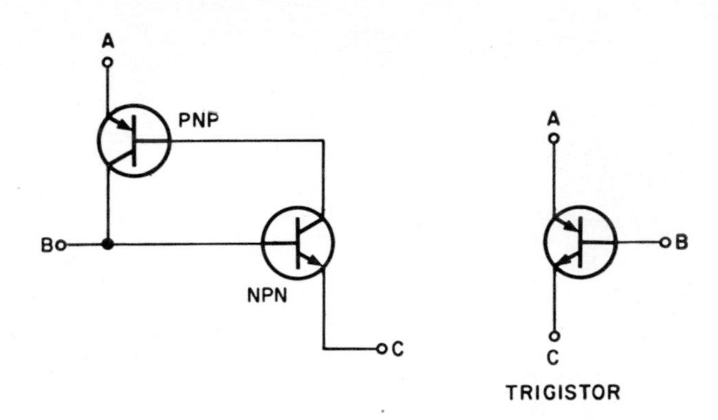

Device	A	B	C
Controlled switch	Anode or emitter 1	Gate or collector 1 and base 2	Cathode or emitter 2
Trigistor	Emitter 1	Base or gate	Emitter 2

Fig. 3·18 Trigistor and its two-transistor equivalent circuit.

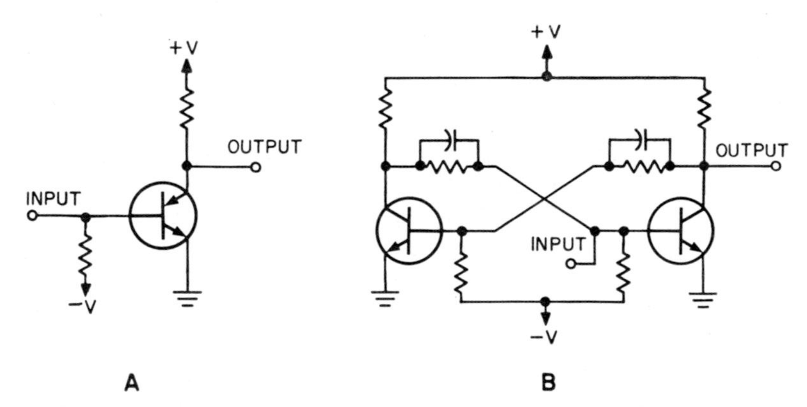

Fig. 3·19 (*A*) The PNPN trigistor basic flip-flop; (*B*) conventional flip-flop.

remain in the on state until it is triggered off. Turn-off is accomplished by a negative current pulse at terminal *B* which diverts the collector current of the PNP transistor from the base of the NPN transistor. Regenerative action is no longer sustained and the two transistors return to their stable cutoff condition.

Figure 3·19 illustrates the inherent simplicity of trigistor bistable circuits. A conventional transistor flip-flop is shown for comparison. Fewer components are required with trigistor switching circuits as compared with transistors or other switching elements. One trigistor will normally perform the same function as two transistors plus several associated capacitors and resistors.

Present units (Solid State Products) are designed for operation in the range of 1- to 8-ma collector current. When the trigistor is on, the voltage

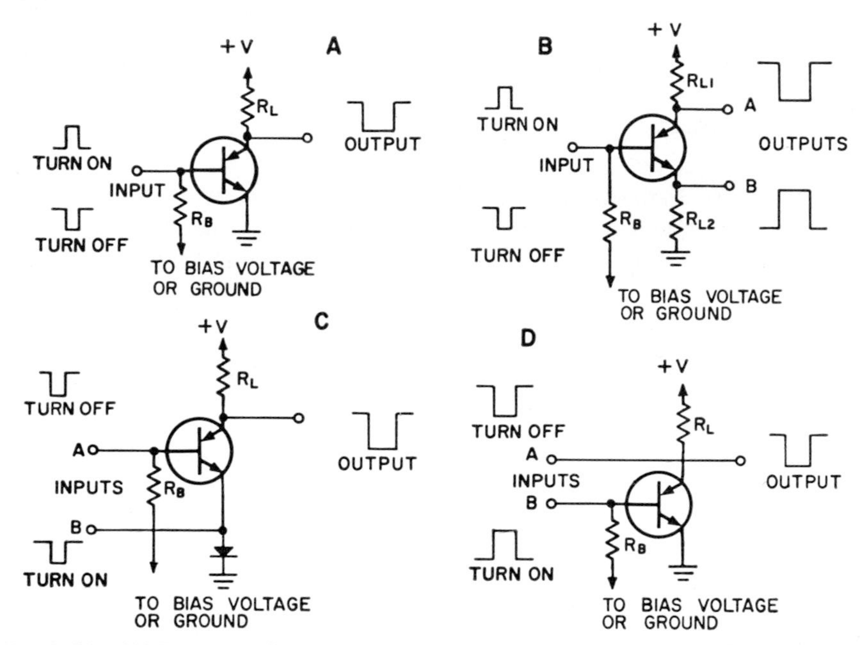

Fig. 3·20 (*A*) Turn-on and turn-off at the base connection for trigistor; (*B*) turn-on and turn-off at base with emitter and collector outputs; (*C*) turn-on at emitter, turn-off at base; (*D*) turn-on at base, turn-off at collector. (*Solid State Products*)

drop (collector-to-emitter) is approximately 0.8 volt. Its dynamic resistance is in the region of 10 ohms. When off, the trigistor has a very high impedance with leakage current normally less than 1 μa. Rise time is typically 0.4 μsec and fall time approximately 1.0 μsec. Repetition rates to 200 kc are possible. Peak pulse currents up to 10 amp for short-duration low-duty-cycle pulses are possible.

The trigistor is particularly suited to memory, counter, gating, timing, logic, and related pulse applications. These functions are usually achieved by combining one or more of the basic bistable circuits shown in Fig. 3·20 with appropriate coupling networks. These basic circuits show some of the possible variations to obtain a choice in triggering and output points. In all these circuits, the trigistor is turned on by making the base positive with respect to the emitter. It is turned off by making the base negative with respect to the emitter or by reducing the collector current to below the holding value. The trigistor has a preferred off state which it assumes before any base pulses have been applied.

Both turn-on and turn-off are accomplished at the base in the circuit shown in Fig. 3·20*A*. The output is taken directly across the trigistor. The output pulse width is determined by the time between turn-on and turn-off input pulses and is independent of input pulse widths.

The circuit shown in Fig. 3·20*B* is essentially the same except that output can be taken from both collector and emitter. Input voltage requirements for both turn-on and turn-off are increased by an amount equal to the output voltage at *B*.

Only negative triggering pulses are used in the circuit shown in Fig. 3·20*C*. Turn-on is accomplished by applying a negative pulse to the emitter across a silicon diode in the inverse direction. A diode is used in place of a resistor because its impedance drops to a very low value when the trigistor is on. Turn-off is accomplished by applying a negative pulse to the base.

The circuit shown in Fig. 3·20*D* is similar to that in *A* except that turn-off is accomplished by driving the trigistor collector negative. Essentially all the trigistor collector current must be bypassed to input *A* for turn-off to take place.

Determining component values for use in the basic trigistor circuits is relatively simple. R_L and the B+ voltage determine the on current level. For most circuits it is desirable to set this between 3 and 5 ma to ensure best possible trigistor performance throughout the operating temperature range. B+ voltages as low as +3 volts can be used. It is desirable, however, to set the B+ value well above this lower limit so the circuits are not sensitive to small voltage changes.

When the trigistor is off, collector cutoff current can act as a positive gate current signal, tending to turn it on. Resistor R_B is used to provide base bias current to ensure a stable off condition throughout the operating temperature range. For operation up to 125°C, the bias off current should be a minimum of 150 μa. Thus if a bias voltage source of −1.5 volts is used, R_B should be 10 kilohms or less.

A smaller value of R_B can also be connected directly to ground (or directly to emitter) at the expense of increased trigger-current requirements and a reduced operating-temperature limit. The criterion in this case is that the voltage (base-to-emitter) be held below +0.1 volt at 100°C. Since collector cutoff current can approach 50 μa at 100°C, the maximum value for R_B at 100°C is 2 kilohms when directly grounded. Higher values for R_B can be used if the operating-temperature limit is lower than 100°C. For operation above 100°C a negative bias voltage is recommended.

In the trigistor, turn-off is accomplished by a gate signal. In the CS, the gate turns the device on; lowering the collector voltage turns it off.

The CS is inherently a high-sensitivity device. This is because the input NPN transistor is designed to have high gain. It is therefore very useful for high-gain switching directly from low-level control signals. Usually the CS eliminates the need for intermediate stages of amplification in control circuits.

In those applications where the high sensitivity is not required, or not desired, any degree of reduced sensitivity can be achieved by gate

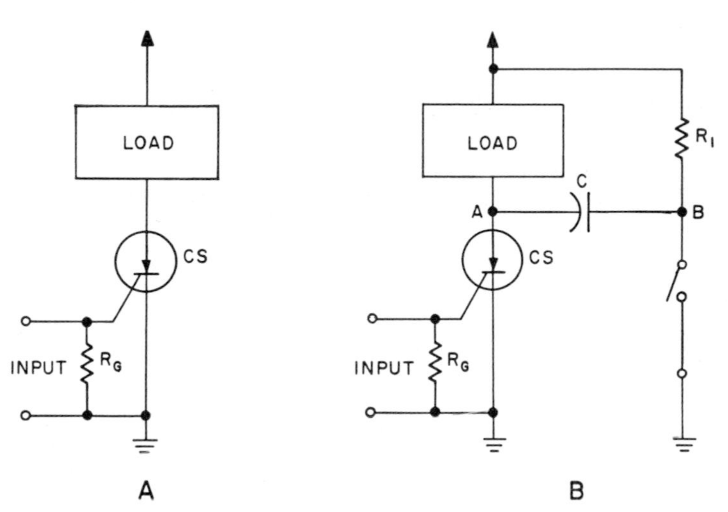

Fig. 3·21 *(A)* **Rectifier control by gate input;** *(B)* **gate turn-on with switch turn-off.** *(Solid State Products)*

biasing. In fact, some negative gate bias is recommended for all circuits to ensure absolute "no fire" stability. Since the CS has a very high gain, it should not be operated or tested with the gate open.

For many applications, stabilizing bias is easily achieved by a resistor between gate and cathode. The resistance value is determined by the maximum operating temperature for the application. Since biasing reduces firing sensitivity, the correct resistance value is a key part of the circuit design.

The input characteristics of the CS (gate to cathode) are similar to the base-emitter input of a conventional NPN silicon transistor. Firing occurs at a specific value of input current and voltage.

The CS can be used as a static switch, with either an a-c or a d-c source. Figure 3·21*A* shows the d-c source, where the CS acts as a latching switch. Once turned on by a control signal, it will remain on indefinitely. To turn it off, the anode current must be reduced to below the "dropout" level.

Figure 3·21*A* shows a simple latching-switch circuit. Resistor R_G provides a negative gate bias current and ensures a stable off condition. In this circuit, less than 20 μw input power (0.6 volt—20 μa) for a time duration of 1 μsec or longer will turn on load power up to 200 watts. By proper choice of R_G, any degree of reduced sensitivity can be designed for.

The CS will latch on at any load current above the dropout level. It will work as well with small loads (10 ma) as it does at higher load currents.

The circuit can be used as a single-contact latching switch for direct control of a given load. It is useful for driving relay coils or similar electromagnetic loads. With the CS, a conventional d-c relay can be converted to a high-sensitivity latching relay. For inductive loads a diode may be necessary across the load to eliminate voltage surge when the power is removed.

For the simple latching circuit, turn-off can be accomplished by removing the source voltage. The CS can also be turned off by the arrangement shown in Fig. 3·21*B*. As before, the CS is off until an input control signal turns it on. When the CS is on, the voltage at point *A* is approximately +1 volt. Capacitor *C* charges through R_1 until point *B* reaches the full positive supply voltage. When switch *S* is closed, point *B* is at ground and the capacitor drives point *A* negative with respect to ground. The current through the load is then diverted from the CS and begins to discharge the capacitor. Since point *A* is negative, the CS will turn off. This is called shunt-capacitor turn-off. The capacitor *C* must be sufficiently large to hold point *A* negative for the time required to turn off the CS.

If another CS is used in place of switch *S*, turn-off can be accomplished electronically as shown in Fig. 3·22*A*. Operation is identical to that described above except that turn-off is accomplished by a momentary low-level positive pulse at input 2. This circuit is actually a power flip-flop. When CS_1 is turned on, CS_2 is turned off by *C*. When CS_2 is turned on, CS_1 is turned off by *C*. A second load can take the place of R_1 so that double-pole single-throw switching between two loads is easily accomplished. This circuit is the static equivalent of the mechanical-contact arrangement shown in the inset as the set of contacts.

With an a-c voltage source, the CS acts as a controlled half-wave rectifier. It will block both the positive and the negative half-cycle until a positive control signal is applied to the gate. When this occurs, the CS will conduct during the positive half-cycle and block during the negative half-cycle for as long as the control signal is present. When the control signal is removed, the CS will block both half-cycles again, since it automatically turns off at the end of each positive half-cycle. By proper timing of the applied control signal, the CS can be made to conduct for all, or part of, the positive half-cycle. Thus proportioning control of the output is possible, as well as on-off switching.

Figure 3·22*B* shows a simple a-c static switch which supplies rectified half-wave direct current to the load. The input control signal can be a-c, d-c, or pulse. If the load is inductive, a diode placed across it gives a continuous current through the load during the negative half-cycle. The inductive field built up during the positive half-cycle returns stored energy during the negative alternation. The diode polarity permits this current to flow in the same direction as during the positive alternation.

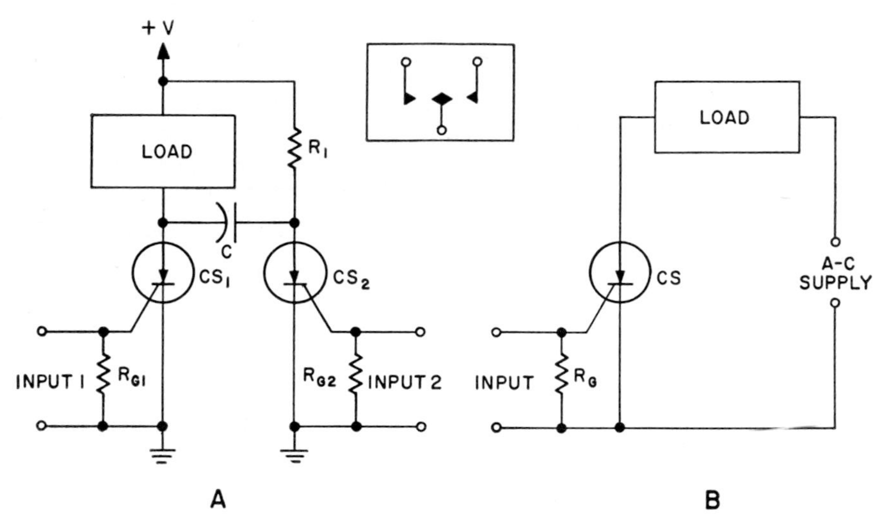

Fig. 3·22 (*A*) Dpst switch or power flip-flop; (*B*) a-c power as the source provides turn-off. (*Solid State Products*)

As used in a-c circuits, the positive control voltage applied to the CS gate must be kept at a low value during the negative half-cycle because of reverse leakage current. This current increases as the positive gate current increases, and, if this leakage current becomes large enough, it can cause thermal runaway.

Pulses of high power can easily be obtained from the CS. Outputs of 150 volts with up to 20 amp can be produced from inputs in the 20-μa and 0.6-volt range. This represents a power gain, in some cases, of 10 million.

A triggered *RC* pulse generator is shown in Fig. 3·23. Capacitor C is charged to the supply voltage through charging resistor R_C. When a trigger pulse is applied to the input, the CS fires, discharging C through load R_L. This circuit can be triggered at any desired repetition rate. The output pulse amplitude will be a function of the repetition rate, however, unless the time between pulses is greater than $3R_CC$. The output pulse voltage amplitude will equal the voltage across the capacitor less the drop across the CS. The available output current is limited by R_L. A positive pulse output can be obtained by placing R_L in the cathode leg and C from the anode to ground.

Figure 3·24 shows another type of pulse generator, which is also useful as a frequency divider. Frequency division to 1:10 is possible. In this circuit the charging resistor R_C is in the cathode circuit and the load R_L is in the anode. With no charge on the capacitor, the full supply voltage is across R_C. This provides a large negative bias between gate and cathode.

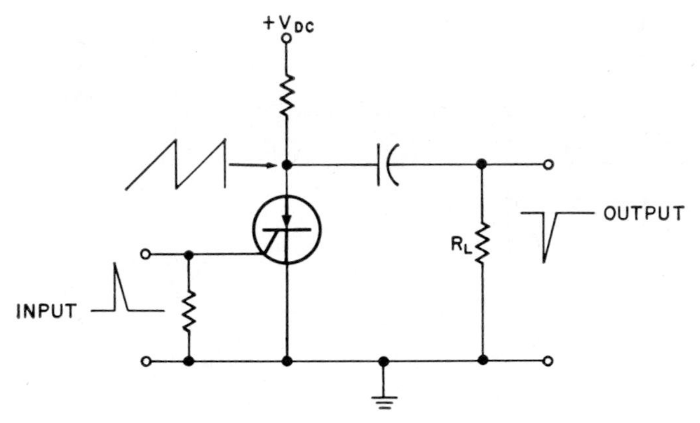

Fig. 3·23 Pulse generator using SCR. (*Solid State Products*)

Diode D_1 is in the reverse direction and prevents exceeding the gate-to-cathode voltage rating of the CS. As the capacitor charges, the voltage across R_C decreases toward zero. The CS will fire when the voltage at point A decreases below the gate voltage by an amount equal to the gate firing voltage of the CS plus the forward drop of D_1 (approximately 1 volt total). Assume an input pulse amplitude of +2.5 volts between gate and ground, an input pulse repetition rate of 20 kc, and a supply voltage of 20 volts. If the time for A to charge down to +1.5 volts is 500 μsec, then one output pulse will occur for each 10 input pulses. The output repetition rate will be 2 kc and frequency division by 10 has occurred.

In RC pulse circuits operating from a d-c source, which require the CS to turn off after each pulse output, R_C must be large enough to prevent the steady-state current through the CS from exceeding the dropout value. An upper limit also exists for R_C. It must supply enough current to permit the CS to fire. The requirement for a minimum value of R_C does not apply to circuits which are turned off by other methods. For example, the LC pulse generators turn off the CS by driving its anode negative.

The properties of the CS make it particularly useful in voltage-limit detectors and related voltage or current threshold actuating circuits. The high-sensitivity series with maximum gate firing current of 20 μa, and gate firing voltage of 0.52 ± 0.08 volt, have been specifically assigned for this type of application. The threshold firing point can be set at any desired value from 0.60 volt up by the use of an input voltage divider or Zener diodes.

The voltage-limit detectors can use either d-c or a-c load power. If d-c load power is used, the CS will "latch on" when the input exceeds the threshold voltage. If a-c load power is used, the CS will supply power to the load only when the input voltage exceeds the threshold voltage.

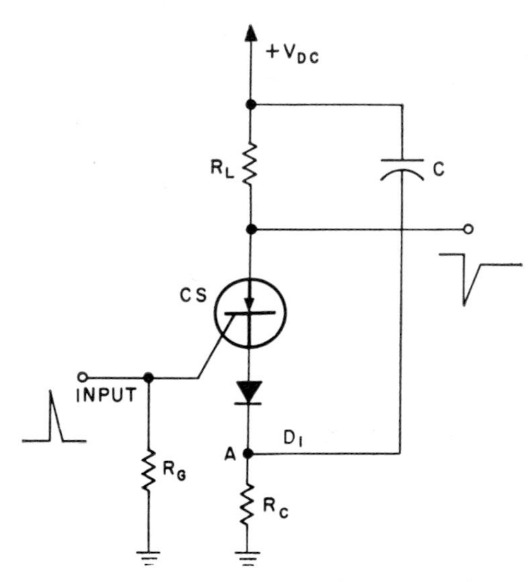

Fig. 3·24 Pulse generator as frequency divider. (*Solid State Products*)

Voltage-limit detectors, using the CS, are quite simple and are useful in a variety of timing, sensing, indicating, warning, and safety applications. Since the CS can handle high load power, it can directly actuate the controlled circuit in many cases. A low-power input from pressure, temperature, speed, flow, light, or similar transducers can be made to turn on the CS at a preset level. The CS can then actuate a control circuit, relay, solenoid, buzzer, indicator light, horn, or similar output.

Unijunction Transistors. The unijunction transistor is another new semiconductor component, originally called a double-base diode. It is different in both construction and operation from the conventional transistor, but the term transistor was used in the name of the component because it meets the industry-accepted definition for a transistor. A transistor is generally said to be any active semiconductor device having three or more terminals.

In contrast to the conventional junction transistor, the unijunction transistor is a device exhibiting open-circuit stable negative-resistance characteristics. Because of this, it is primarily useful in switching and oscillator circuits. In addition, it has the unique ability to sense voltage levels and temperature variations. Or, by various circuit modifications, it can be made insensitive to temperature and voltage variations. These devices are particularly useful for relaxation oscillators, sawtooth and pulse generators, pulse-rate modulators, pulse amplifiers, multivibrators, flip-flops, and time-delay circuits. Because of the high peak current rating (2 amp), the device is useful in medium-power switching and oscillator applications where one unijunction transistor can do the work of two

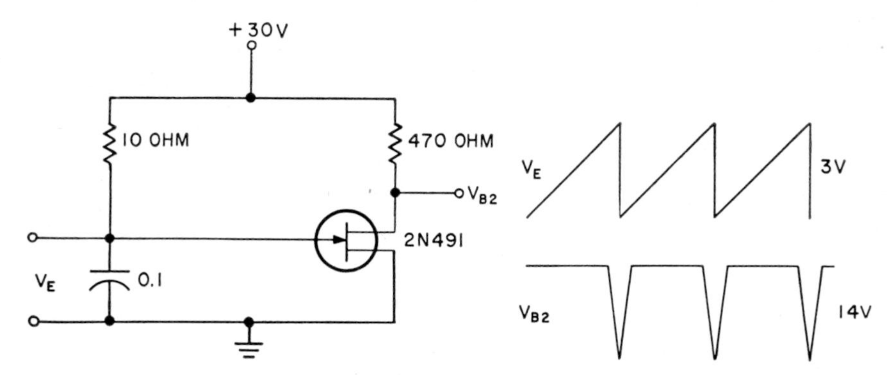

Fig. 3·25 Unijunction oscillator showing sawtooth and pulse outputs. (*General Electric*)

conventional transistors. In special voltage-sensing and locking circuits, one unijunction transistor has been used to replace as many as five conventional transistors.

The unijunction transistor can be operated in a number of different circuit configurations so that any of the three terminals can serve as a signal input or a load output. Externally it looks like an ordinary transistor. Internally it consists of a uniform doped N-type single-crystal silicon bar with ohmic contacts at each end and an aluminum wire attached to the silicon bar between them. The two ohmic contacts to the silicon bar are called base 1 and base 2. The room-temperature base-to-base resistance range is from 5,000 to 10,000 ohms. The aluminum wire connected to the silicon bar between the two ohmic contacts forms a PN junction, which is the emitter.

Figure 3·25 shows the unijunction transistor in a relaxation oscillator using two resistors and one capacitor. With the values indicated, the frequency is about 1 kc. One output (V_E) is a sawtooth, and the other (V_{B2}) is a pulse. Many circuit variations are possible. Figure 3·26 shows a frequency divider having a 100:1 countdown ratio.

Other Switches. Other solid-state devices are being developed as switches. One is the binistor (Transitron Electronic Corporation) shown in Fig. 3·27 as a four-layer switch. The two transistors *A* (an NPN) and *B* (a PNP) are combined to make the equivalent of the binistor as in *C*.

The binistor resembles a four-layer switch as shown in Fig. 3·28, but a difference exists in the design and use of this structure. The output current is taken from an intermediate layer and the upper junction serves

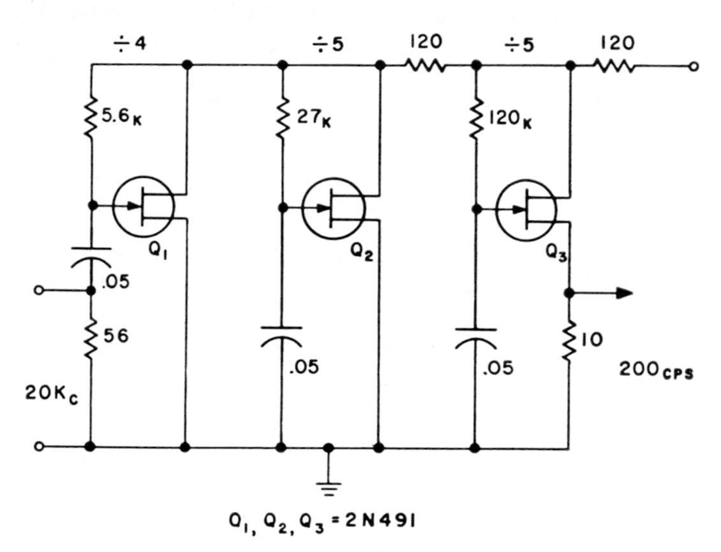

Fig. 3·26 General Electric unijunction frequency divider for division by 100.

only as a "latch" to hold the device on when in the conducting state. Figure 3·29 shows the characteristics.

If base current is applied, the collector voltage will fall because of normal transistor action. When the collector voltage falls below the injector clamp voltage, the upper PN junction becomes forward-biased and injector current begins to flow. This current is in such a direction as to aid

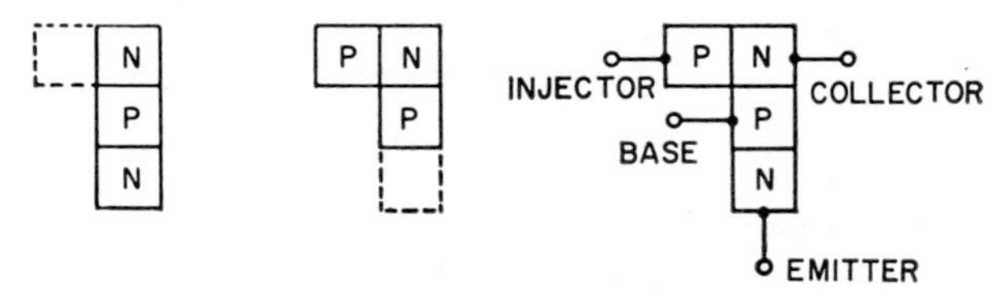

Fig. 3·27 NPN transistor as Transitron Electronic Corporation binistor equivalent.

switch-on of the NPN transistor. A regenerative action takes place, tending to drive both transistors into the saturated on state. No further external base current is required to maintain the binistor in this condition, the base current for the NPN transistor being supplied by the injector circuit.

Switch-off can be achieved either at the injector or, with particular binistor types, at the base. If the injector current is reduced so that the base current of the NPN transistor is insufficient to maintain the binistor in saturation, the collector voltage will rise. When the collector voltage rises above the injector clamp voltage, the upper PN junction becomes reverse-biased and regeneration occurs to complete the switch-off process. Therefore, the criterion for switch-on is that the collector voltage be taken

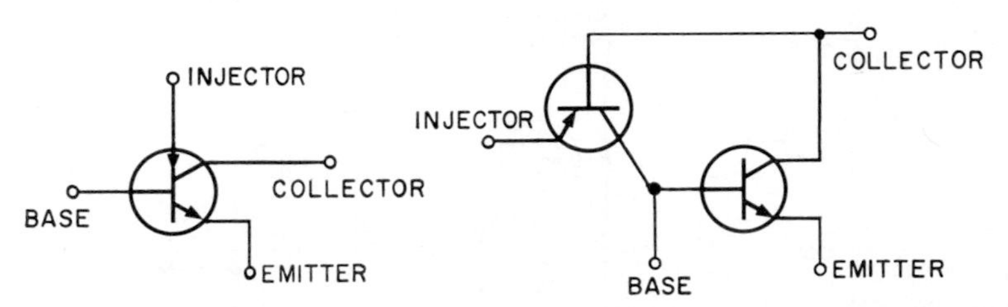

Fig. 3·28 Binistor diagram showing the four elements. (*Transitron Electronic Corporation*)

below the injector clamp voltage, and for switch-off that the collector voltage be taken higher than the injector clamp voltage. An example of a binistor is the ring counter shown in Fig. 3·30, which operates with a positive input pulse to the base of the transistor.

It is assumed that before the trigger pulse is applied one stage has been put into the on state and all other stages are off. When the transistor is switched on, injector current is diverted from the on binistor, thereby

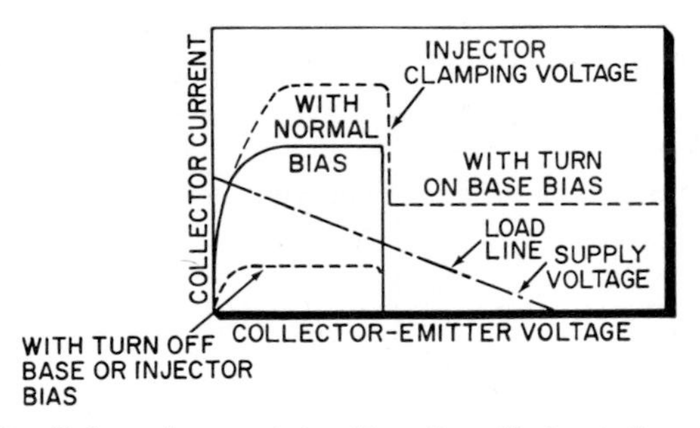

Fig. 3·29 Binistor characteristics. (*Transitron Electronic Corporation*)

switching it off. The positive-going transient at the collector of this stage is passed to the base of the next stage to switch it on when the transistor is switched off again. The on state has therefore been passed along from one stage to the next at the triggering frequency.

3·5 Applications for diodes Diodes are being used as active circuit elements in many types of electronic systems. In many ways the diode is used as a switch, but it can also function as an amplifier or a modulator.

Four-layer Diode. Because the switching properties of the four-layer diode closely approach those of the ideal switch, it is finding applications in many fields. In addition to its advantages as a semiconductor switch, the four-layer diode provides the unusual combination of power-handling ability and fast switching. Some of its present applications include pulse generators and amplifiers, oscillators, relay alarm circuits, ring counters, detonator firing circuits, magnetron and sonar pulsing, telephone switching, and computer applications such as magnetic-core driving. It has been used to replace relays, thyratrons, gas diodes, and switching transistors.

The PNPN four-layer diode is a two-terminal semiconductor switch (Fig. 3·31*A*) with a forward current as shown. This transistor diode is

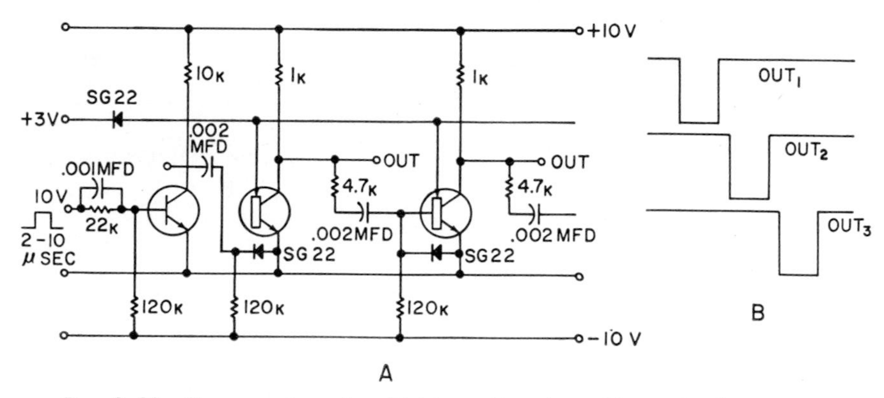

Fig. 3·30 Ring counter using binistors. (*Transitron Electronic Corporation*)

a self-actuated silicon switch with operating characteristics based on the principles of transistor action. It has two stable states: (1) an open or high-resistance state of more than 1 megohm and (2) a closed or low-resistance state of a few ohms. The device is switched from one state to the other by controlling the voltage across it and the current passing through it.

Because of this action, the transistor diode can act to switch a load as in Fig. 3·31*B*. A positive pulse on its anode will add to the d-c supply and cause conduction to begin and then continue. In the same manner, a negative cathode pulse can start conduction.

The voltage-current characteristic for the four-layer transistor diode shows the four essential operating regions as in Fig. 3·31*C*. These are:

I. Open or high-resistance state
II. Transition or negative-resistance state
III. Closed or low-resistance state
IV. High pulse current state

As shown, the voltage rises and reaches the switching voltage. The device starts to switch to the closed state. The current at this point is less than 200 μa. The device switches because of an internal feedback mechanism which allows the diode to pass a steadily increasing current if the voltage is held at a fixed value. Therefore, the transistor diode will pass whatever current the circuit requires.

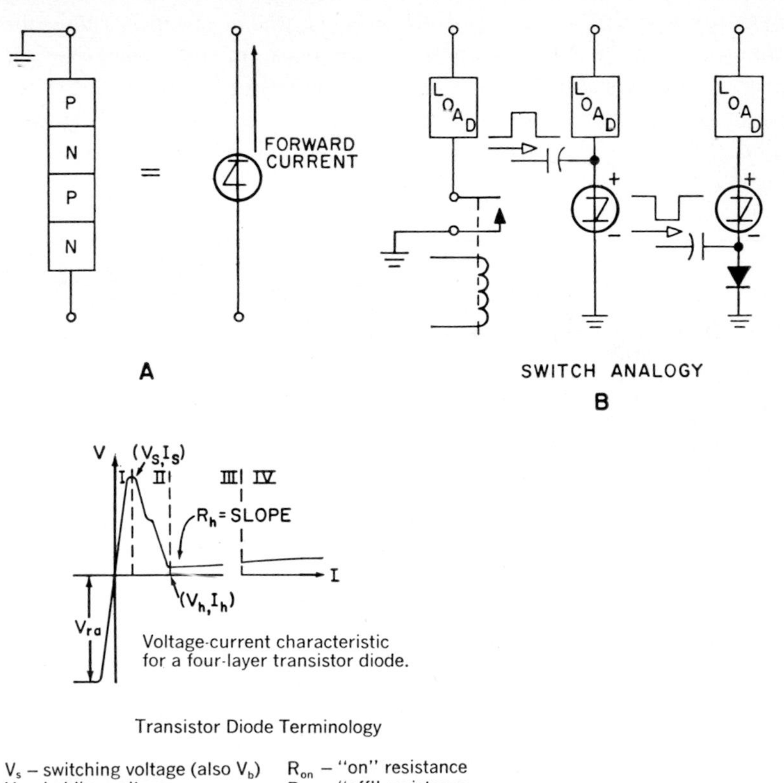

Fig. 3·31 **(*A*) Current flow in the four-layer diode; (*B*) switch analogy to diode operation; (C) four-layer diode terminology. (*Shockley Transistor*)**

The transistor diode will turn off when the current through it is reduced below holding current. The speed of this turn-off depends on circuit conditions. In a typical sawtooth oscillator, for example, the device must turn off with essentially holding voltage across it. This will take about 1 μsec under most circuit conditions.

In its off condition, the transistor diode may be considered a capacitance and a large resistance in parallel. This capacitance, which is similar to the collector capacitance of a normal transistor, has a value which depends on the actual voltage across the device. In its on condition, the transistor diode has such a low resistance that capacitive effects may be ignored.

In its off condition, the device will pass a capacitance current in response to a sharply rising voltage wave. If the rise rate of this voltage wave is large enough, switching will occur below the d-c switching voltage.

Since the resistance of the device decreases with increasing current (to substantially less than 1 ohm at high pulse currents), it can be destroyed unless the load current is limited. To switch the device on, the series impedance should pass sufficient current to exceed the holding current.

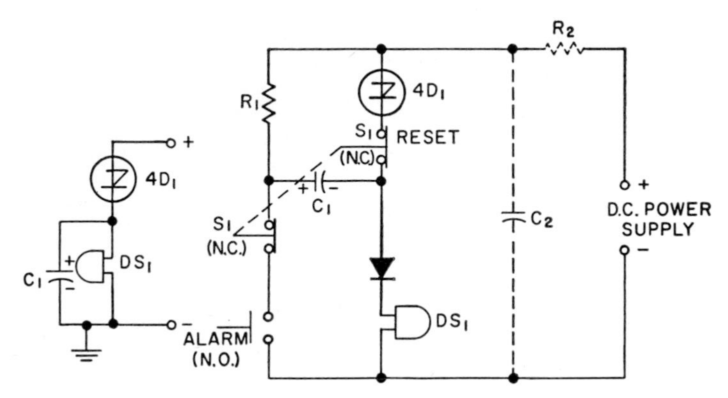

Fig. 3·32 Alarm circuit using Shockley transistor four-layer diode.

A basic alarm circuit is shown in Fig. 3·32. When the alarm contacts are closed for an instant or when they remain closed, C_1—which had charged to the d-c source voltage—will be in series with the source supply. The alarm closure grounds what was the positive side of the capacitor, placing its charge in series with the supply. This capacitance must be recharged between successive operations of the alarm contact.

Capacitance values between 0.005 and 0.2 μf are typical, and the capacitance must withstand full power-supply voltage continuously. R_1 is the charging resistor and must allow recharge of C_1 between operation of the reset and the next alarm. It must not cause excessive current drain from the power supply when the alarm contact is closed, nor must it cause burning of contacts.

A decoupling network (R_2 and C_2) may be required to retard rise of circuit voltage to prevent switching of $4D_1$ due to rate effect when power-supply voltage is first applied. Capacitance values of 0.01 to 1.0 μf and resistance values of 5 to 100 ohms are typical.

Either a momentary or a continuous closure of the alarm contact will turn $4D_1$ on and give a steady lamp indication. The lamp remains on regardless of alarm condition until reset. If the alarm condition has not been corrected when the reset button is operated, the signal lamp will return to its steady indication when the reset is released.

Several applications are shown in Fig. 3·33. *A* shows a sawtooth oscillator where *C* charges until the voltage across the diode is high enough to fire it and discharge the capacitor. *B* shows two pulse generators and amplifiers. The positive input is to the low-impedance circuit, while a negative pulse input is shown for a high-impedance circuit. *C* shows a flip-flop.

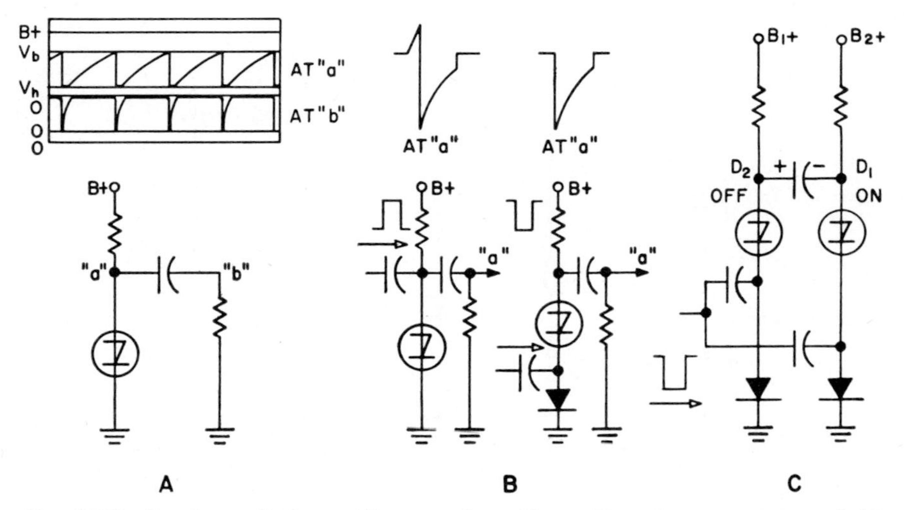

Fig. 3·33 Four-layer diode as (*A*) sawtooth oscillator, (*B*) pulse generator, and (C) flip-flop. (*Shockley Transistor*)

If diode $4D_1$ has just turned on, in the bistable circuit of Fig. 3·34*A*, point *A* will be at ground potential as long as the current passed by resistor R_1 is greater than the holding current of $4D_1$. As capacitor C_1 charges through resistor R_2, the voltage at point *B* increases until it reaches the switching voltage of diode $4D_2$. When this occurs, $4D_2$ switches on and point *B* is brought suddenly down to ground potential. At the instant of switching, there is a negative-to-positive potential across capacitor *C* between points *A* and *B*. Therefore, as point *B* is pulled down to ground potential, point *A* is driven to a negative potential before C_1 has a chance to discharge.

This reverse bias turns diode $4D_1$ off, and $4D_2$ now carries the load current through R_2. Capacitor C_1 then charges in the opposite direction through resistor R_1 until point *A* reaches the switching voltage of diode $4D_1$. As this occurs, diode $4D_1$ switches on, the voltage across C_1 turns diode $4D_2$ off, and load current once more flows through resistor R_1.

Figure 3·34*B* shows the monostable or one-shot configuration, where the circuit is arranged so that one leg is normally on, the other normally off. This is accomplished by choosing $4D_1$ to have a switching voltage less than the supply voltage and $4D_2$ to have a switching voltage greater than the supply voltage.

A negative-going trigger pulse injected by capacitor C_2 switches $4D_2$ on, which in turn switches $4D_1$ off through the coupling action of C_1. If R_2 passes current to $4D_2$ greater than its holding current, $4D_2$ will remain on until capacitor C_1 has been charged through resistor R_1 up to the switching potential of $4D_1$, at which time the circuit reverts to its initial

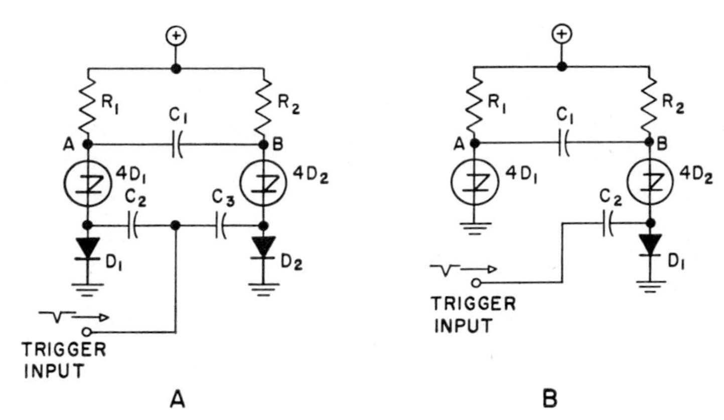

Fig. 3·34 (A) Bistable four-layer diode multivibrator; (B) monostable flip-flop. (*Shockley Transistor*)

stable condition with $4D_1$ on. In this operating mode, the time constant $R_1 C_1$ and the ratio between supply voltage and the switching voltage of $4D_1$ determine the duration of the period during which leg B will stay on.

In Fig. 3·34*A*, the bistable multivibrator circuit, the circuit is switched from A to B, and back to A, by successive negative trigger pulses inserted via C_2 and C_3. The four-layer diodes $4D_1$ and $4D_2$ are chosen to have switching voltages greater than the supply voltage. R_1 and R_2 are chosen small enough so that they provide holding current for diodes $4D_1$ and $4D_2$, respectively. Diodes D_1 and D_2 are employed to provide high impedances across which the negative trigger pulses are applied. They must have forward-current ratings adequate for the load currents determined by R_1 and R_2. C_1 must be large enough to ensure positive switching so that both sides are never on simultaneously.

Low-noise Amplifiers. In an attempt to reduce the noise which limits the signal that can be received, both low-temperature and diode amplifiers have been developed.

A new and different type of low-noise amplifier for uhf and microwaves is the variable-reactance diode amplifier. It is called a varactor or parametric amplifier.

Any nonlinear device can be used for a mixer, and some new types of experimental silicon diodes have a capacitance which varies with the applied voltage. This variable capacitance has been used to tune r-f circuits in receiver front ends. And, by the application of a pumping signal, the device can act as a converter with gain.

Bell Telephone Laboratories have developed a 6,000-mc converter with a noise figure between 5 and 6 db, and 8-mc bandwidth, and a pumping signal of 12,000 mc.

Converters can be of several types. There is a "down converter," in which the output is lower than the input frequency. There is also an "up converter," in which the sum of the input and the local oscillator (pump) is the output; hence the r-f signal is converted up. But there are also the types which are inverting and noninverting.

Parametric amplifiers, or paramps, use special junction diodes of either silicon or germanium. Because the capacitance of the junction varies with the applied voltage, the PN junction diode can be used as an amplifier. And, because of the low high-frequency losses, it can be used as an amplifier well up into and beyond the uhf region. It appears that varactors can operate to 10,000 mc or perhaps even higher. In application these devices can improve communication by radio between two points, for example, between pipeline pumping stations.

The varactor can also be used as a harmonic generator and as a modulator. Because of its nonlinear capacitance the varactor can also be used as a multiplier in harmonic-frequency generation. Modulation or up-conversion is possible with varactor diodes. Here the modulated subcarrier is amplified and converted to a new and higher frequency.

Diode Capacitors. Silicon voltage capacitors are diodes which can be used for many applications in tuning and control of radios as well as other electronic devices. They are directly usable as "signal seekers," for example, to replace manual tuning.

Since the capacitance of the device changes with the applied voltage, as with the varactor, tuning can be accomplished. Actually these solid-state diodes are not varactors since they are designed and produced for low-frequency tuning while the varactor is a high-frequency device. A better name for this voltage-dependent capacitor is the solid-state variable capacitor or the silicon capacitor, named after the material from which it is made.

Typical values as shown in Table 3·1 range between 6 and 88 $\mu\mu$f for the HC 7001 and 46 to 240 $\mu\mu$f for the HC 7005. The capacitance range, the Q at 5 Mc, and the d-c maximum voltage of these and others made by Hughes are given in the table.

A typical manual control circuit is shown in Fig. 3·35. Self-bias is reduced with this circuit. A d-c control voltage is applied across R_1; this voltage could come from the discriminator, as in an FM receiver,

Table 3·1 Hughes Silicon Capacitors

Model no.	Range, $\mu\mu$f	Q, 5 mc	Max d-c voltage
HC 7001	6–88	360	130
HC 7002	12–120	330	80
HC 7004	20–170	270	60
HC 7005	46–240	200	25
HC 7006	14–88	175	25
HC 7007	22–120	175	25
HC 7008	32–170	175	25

SOURCE: Semiconductor Division, Hughes Products.

or as part of an afc loop. R_2 is used to help isolate the tank circuit from the control circuit. Notice that, in the circuit shown, the tuned circuit has the coil L and the capacitance made up of two silicon devices. As the control voltage varies, the capacitance—and hence the frequency—also changes. R_1 can be adjusted for the amount of frequency change.

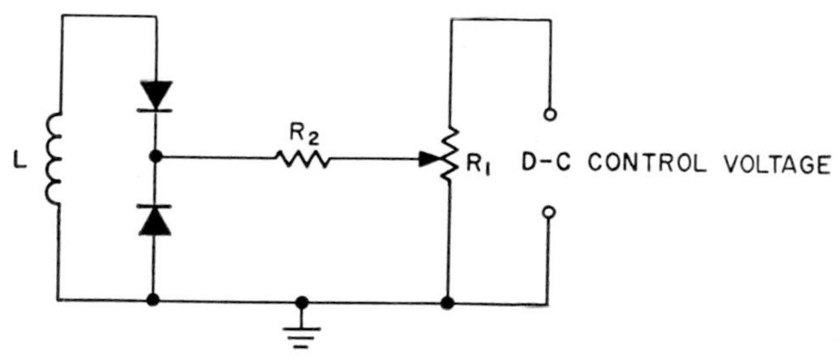

Fig. 3·35 Manual tuning using voltage-sensitive capacitors of Semiconductor Division, Hughes Products.

Automatic tuning or signal seeking may be illustrated by Fig. 3·36. Here C is much larger than C_1, the silicon capacitor. When the switch opens, C charges, through R, to the applied voltage. The rate of charge is determined by the time constant RC.

During this charging time the circuit frequency, as determined by L and C_1, sweeps through a range as determined by C_1 minimum and C_2 maximum. Since this tank circuit is part of a local oscillator, the oscillator signal sweeps a band of frequencies. Closing the switch S discharges C and starts the process again. When this circuit is used in a receiver, an intermediate frequency will be produced when the local oscillator hits the correct frequency to produce an intermediate frequency while beat against the incoming signal. This frequency is used to lock the oscillator in place when the switch is replaced by a complete circuit.

The application of the silicon capacitors to afc may be seen from Fig. 3·37, which shows the afc loop in a receiver. In this FM receiver, the d-c control voltage from the detector changes as the local oscillator drifts in frequency. As the control voltage is fed to the reactance tube, the tube output brings the local oscillator back to the correct frequency.

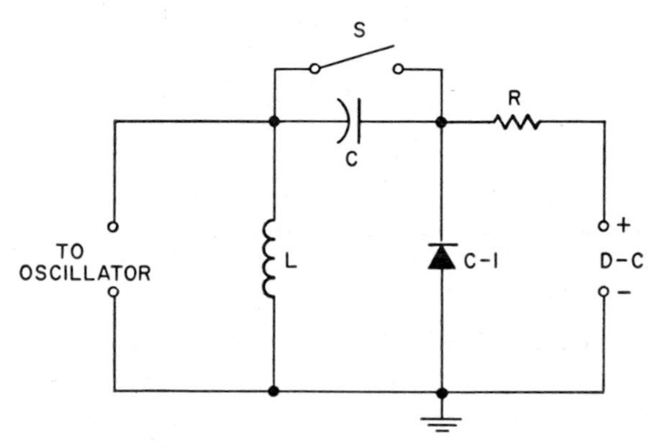

Fig. 3·36 Automatic tuning. (*Semiconductor Division, Hughes Products*)

This can also be done in a simpler circuit where two silicon capacitors are in series opposition across the oscillator tank circuit. Changing the control voltage varies the capacitance of the silicon capacitor, which, in turn, corrects the operating frequency of the oscillator as in Fig. 3·38.

Diode Regulators. Zener diodes are special silicon types used for voltage regulation. They are used just as voltage-regulator gas tubes are used,

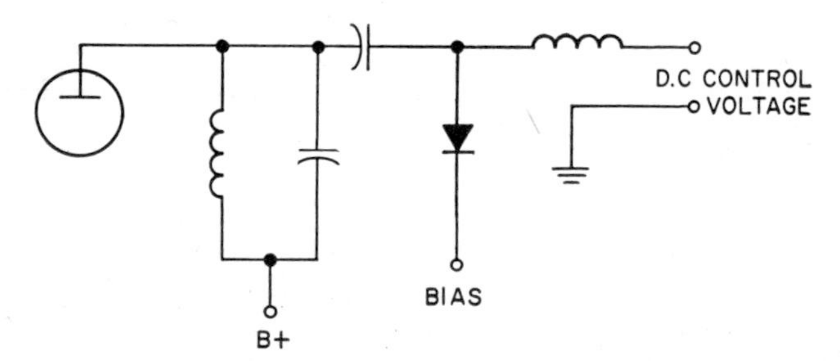

Fig. 3·37 Automatic frequency control. (*Semiconductor Division, Hughes Products*)

to provide for a constant voltage drop. Within limits, the output voltage stays constant as the current flow changes.

Silicon diodes are used for rectifiers where, with a positive anode or negative cathode, they conduct in the forward direction as shown by the positive voltage characteristics in Fig. 3·39. If a small reverse voltage is applied, the current flow is only in microamperes rather than in milliamperes as in the forward direction. The germanium diode, by contrast, does not have so sharp a front-to-back ratio although, as shown, it does conduct more in the forward direction than it does in the reverse direction.

But when the reverse voltage, which is negative on the anode, is applied to the silicon diode, it breaks down at a certain point. This sharply defined

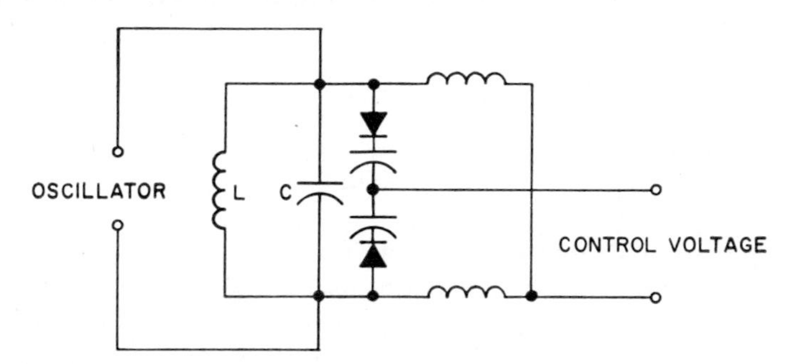

Fig. 3·38 D-c voltage control of an oscillator's frequency. (*Semiconductor Division, Hughes Products*)

voltage is the Zener or breakdown voltage, at which point the silicon diode conducts much as a gas tube does. This breakdown, in a silicon diode, is not damaging to the device and the diode recovers when the reverse voltage is removed.

The important feature is that, at breakdown, the voltage is almost independent of the amount of current flow. Again the diode acts as a

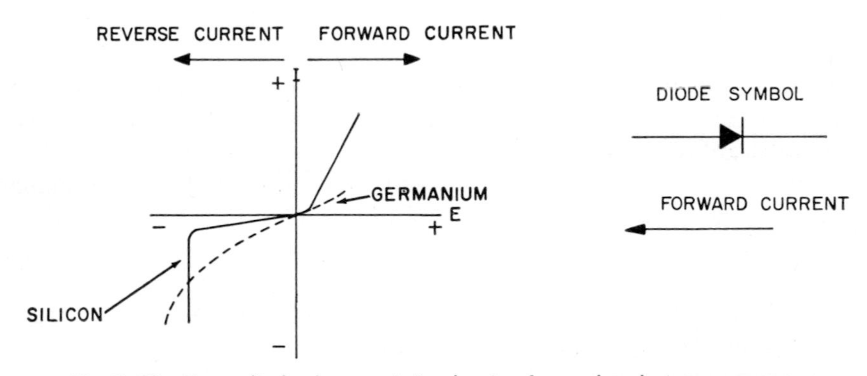

Fig. 3·39 Zener diode characteristics showing forward and reverse current.

voltage regulator whose voltage drop is constant over a wide current range.

Specifically designed for this use, Zener diodes are used for voltage regulation as well as voltage reference work. Silicon diodes for this application are designed for specific Zener voltages. Other silicon diodes, used as rectifiers, have their Zener voltage so high that this point is never reached in normal use and the diode does not break down under the application of the inverse voltage of the circuit.

Tunnel (Esaki) Diodes. Invented by Leo Esaki, the tunnel or Esaki diode is a new semiconductor device of wide potential application as an amplifier, a switch, and an oscillator.

Ordinary transistors, or even vacuum tubes, depend upon a charge carrier being influenced by the electric field between the emitter and collector electrodes. Their speed, therefore, is a function of the time required for the charge carrier to travel through this region. Tunnel diodes, however, operate on a different mechanism with a theoretical frequency limit of 10^7 mc, which is much higher than that of any other semiconductor device.

This diode, which gets its name from the tunnel effect, operates on the principle of quantum theory wherein a particle disappears from one side of a potential barrier and instantly appears on the other side. The particle does not have enough energy to go over this potential barrier, and scientists have therefore assumed that it tunnels through the barrier.

A characteristic curve is shown in Fig. 3·40*A*. As voltage V is applied to the diode, starting at zero, current rises sharply to a peak at point 1. But if the applied forward voltage is increased further, the current dips sharply into a valley, as shown, and then again rises to point 4 and even higher.

This negative-resistance region, from 1 to 2, makes the tunnel diode useful as an active circuit device.

Either germanium or silicon can be used for tunnel diodes. Peak currents can be anywhere from only a few microamperes to well into the ampere region.

A tunnel diode will also, because of its structure, conduct with an applied negative voltage, as shown in Fig. 3·40*B*. Such a diode is known as a backward diode.

Figure 3·40*C* shows a typical low-frequency circuit. If the incoming signal is a sine wave, its positive portion will add to or aid the battery voltage. Hence the current of the diode will drop (as in the region from 1 to 2) and there will be a decrease in the IR drop across R_1, the output resistance. In this way a positive-going signal produces a larger negative-going signal which is amplification.

The amplification can be used for an oscillator. Figure 3·40*D* shows a crystal-oscillator circuit at 27.8 mc. A 1N29 is used as the oscillator with bias stabilization by the backward diode HU-100. Note the symbol used here for the tunnel diode.

In Fig. 3·40*E* a different crystal oscillator is illustrated. Its frequency depends upon the crystal which is used. Here a different symbol is used for the diode.

An oscillator is the heart of the FM transmitter shown in Fig. 3·41*A*. Operation may be best explained by separating the circuit into two portions. Part 1 is a basic tunnel-diode oscillator whose frequency is primarily determined by the resonant circuit in the cathode. Resistors R_1 and R_2 provide a stable low-impedance voltage for the anode of approximately 150 mv. Capacitor C_1 is the r-f bypass for the anode.

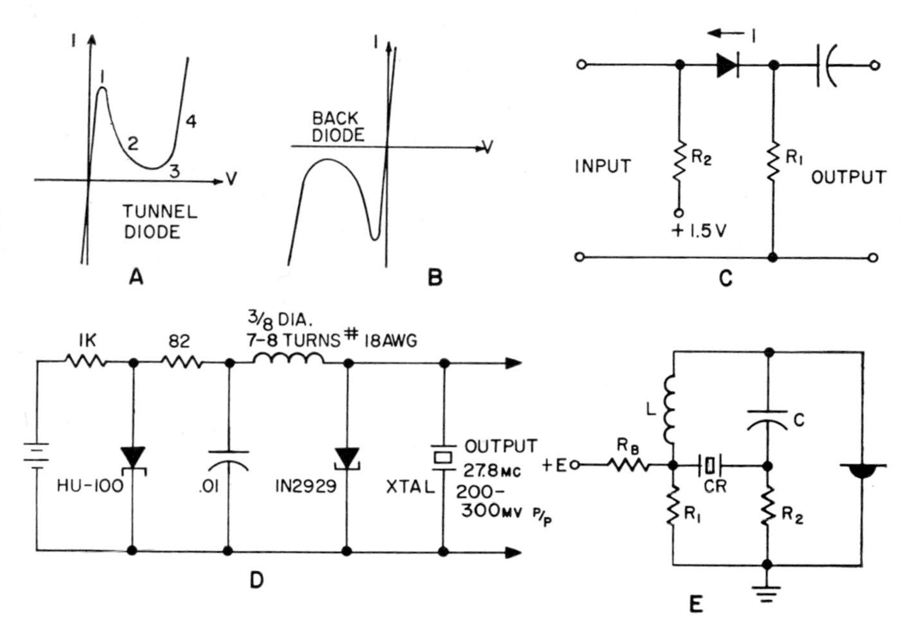

Fig. 3·40 (*A*) Tunnel-diode characteristics showing negative resistance; (*B*) backward-diode curve; (C) simple tunnel-diode amplifier; (*D*) tunnel-diode crystal oscillator at 27.8 mc using Hoffman Electronics Unitunnel or backward diode for bias stabilization; (*E*) General Electric crystal-controlled oscillator.

Part 2 is a transistor emitter-follower stage to amplify the audio signal from the microphone. The amplified audio is fed through capacitor C_2 to the anode of the tunnel diode. Frequency modulation is obtained by the instantaneous changing of the anode bias by the audio signal. Since the characteristic curve is not perfectly linear in the negative-resistance region, the negative conductance changes slightly with bias. FM deviations of ± 75 kc are readily obtainable with this type of circuit.

This transmitter has been successfully used as a wireless portable microphone which allows complete mobility on the part of the speaker. Of course, it has no wires or cords. When it was used with an average FM receiver having a sensitivity of 10 μv, an operating range in excess of 100 ft was obtained.

Tunnel diodes act as fast and effective switches. A simple bistable circuit or flip-flop is shown in Fig. 3·41*B*. Here the tunnel diode is biased

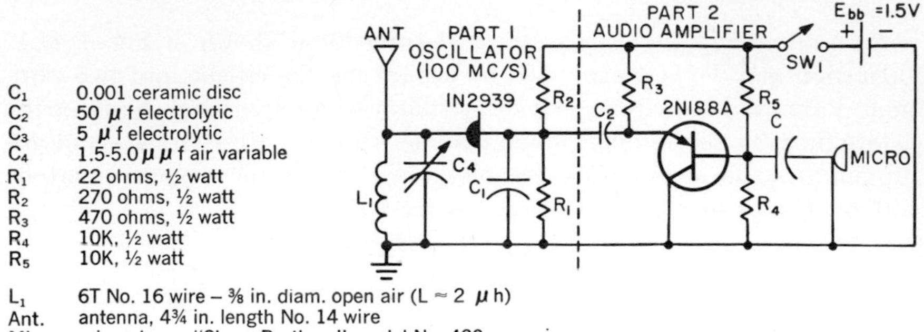

A

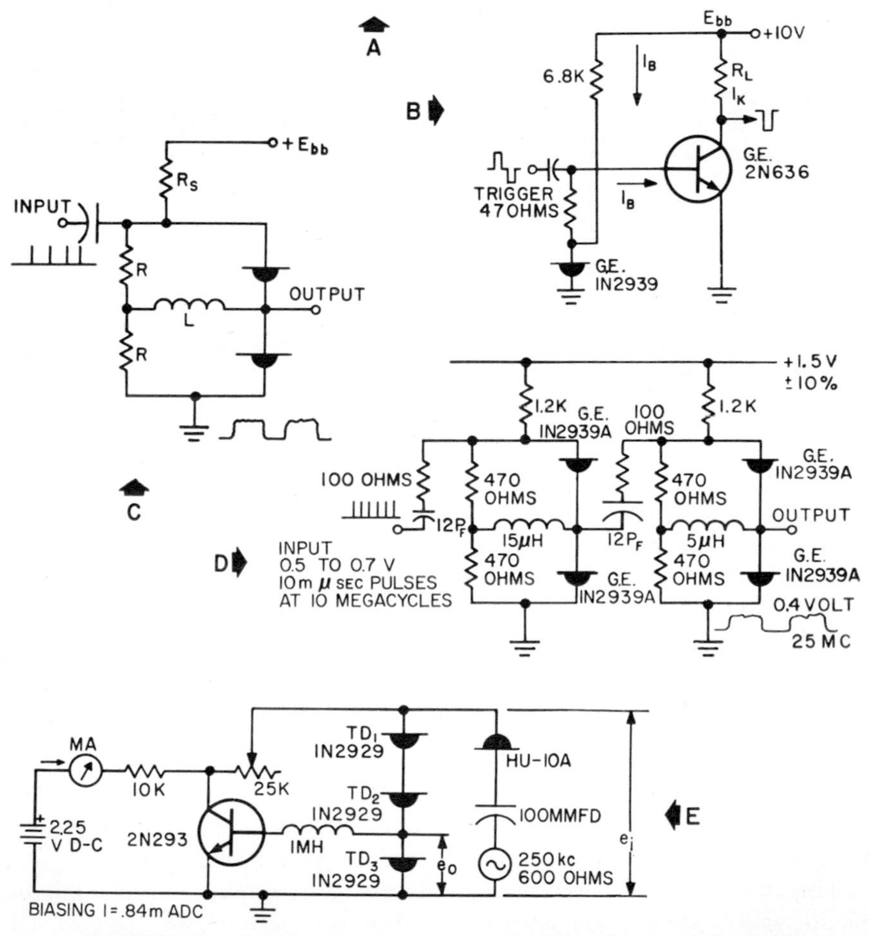

CONDITION: TD_3 must have peak current at least 50μa higher than that of TD_1 and TD_2. (I_P of TD_3 = 1.02 ma, of TD_2 = 0.95 ma, of TD_1 = 0.91 ma)

Fig. 3·41 (*A*) General Electric FM wireless microphone with tunnel diode; (*B*) bistable tunnel diode and NPN transistor circuit of General Electric; (*C*) tunnel-diode flip-flop; (*D*) General Electric tunnel-diode counter; (*E*) Hoffman Electronics tunnel-diode frequency divider.

in the low-voltage state by a current which is slightly less than the peak current. Since the transistor is in the off condition, the collector is at the supply voltage.

If a positive trigger pulse is supplied at the input such that the tunnel-diode current increases above the peak current, the tunnel diode switches to the high-voltage state. The tunnel diode will remain in the high-voltage state with a major portion of the bias current being diverted into the base of the transistor. A negative trigger returns the diode to its original state.

Figure 3·41*C* shows a tunnel-diode flip-flop circuit which requires trigger pulses of only one polarity. The supply voltage is of such a value that only one tunnel diode can be in the high-voltage state. The difference between the two tunnel-diode currents flows through the inductance. When a positive trigger pulse turns the diode which is in the low-voltage state to the high-voltage state, a voltage is induced in the inductance (because of the decreasing current through it). This voltage is of a polarity such as to reset the other tunnel diode to the low-voltage state. Each pair of trigger pulses completes one switching cycle.

The basic flip-flop circuit can be interconnected to form a counter as shown in Fig. 3·41*D*. With the values of components and input pulses shown, the counter will operate successfully up to 10 mc with supply-voltage tolerances of ± 10 per cent.

Figure 3·41*E* shows a tunnel-diode frequency divider with a 3:1 ratio.

REVIEW QUESTIONS

3·1 What are the main types of semiconductor rectifiers?

3·2 What are circuit modules?

3·3 How does the controlled rectifier operate?

3·4 What is the principle of the cryotron?

3·5 From what materials are ferrites made?

3·6 How does a magnetic memory use ferrites?

3·7 How is the four-layer diode like the solid-state capacitor?

3·8 How does the four-layer diode act as a switch?

3·9 What is a varactor?

3·10 What are its applications?

3·11 Give the advantages of the solid-state variable capacitor over the ordinary tuning capacitor.

3·12 How may a solid-state variable capacitor be used in afc?

3·13 What is the unijunction transistor?

3·14 What devices may the controlled rectifier replace?

3·15 What are the advantages in using a frequency above 60 cycles in a power supply?

3·16 How is voltage regulation used?

3·17 What is the principle of electronic cooling?

3·18 What are the advantages of printed circuits?

3·19 What is a steering circuit in a flip-flop?

3·20 How do saturating and nonsaturating flip-flops differ?

3·21 What is the resolution of a flip-flop in terms of frequency?

3·22 In Question 3·20 above, which type is faster?

3·23 Why is an emitter-follower used?

3·24 What are the major differences between silicon and selenium rectifiers?

3·25 What is a selenium rectifier's typical current capacity per square inch at 1 volt for convection and forced cooling?

3·26 In Question 3·25, what is the limit at 1.5 volts?

3·27 Under what conditions will a selenium rectifier carry 2 amp per sq in.?

3·28 What is this condition (as in Question 3·27) for silicon?

3·29 What can you tell about the current-versus-voltage relation of silicon as compared with that of selenium?

3·30 At 1 amp per sq in., what is typical and what is the limit for both types of cooling for silicon?

3·31 At 140°C, by what percentage is the load rating of a silicon rectifier decreased?

3·32 How is a silicon rectifier derated for voltage?

3·33 Explain the difference between the silicon and selenium current derating.

3·34 What is the principle of the semiconductor controlled rectifier?

3·35 How does this rectifier provide control?

3·36 How does the unijunction function as a relaxation oscillator?

3·37 What type of device is the binistor?

3·38 What is a solid-state capacitor?

3·39 How can silicon-diode capacitors act as tuning devices?

3·40 Explain the afc circuit in Fig. 3·37.

3·41 How does the control voltage shown in Fig. 3·38 change the frequency of the oscillator?

3·42 What is the function of a Zener diode?

CHAPTER

4

Switching Devices

In many ways switching devices are the heart of industrial electronic systems. All types of industrial electronic equipment use some form of relay or switch. In some complex equipment many relays are required. In other cases, only a few are needed. No industrial electronic equipment operates without some relays being used.

4·1 Types of devices Every industrial electronic system requires some type of switching device. From the simplest photoelectric relay to the most complex computer-controlled machine tool, switches and controls comprise an integral part of the electronic system.

The broad classes of switching devices are relays, the various types of electronic switches, and other types of switching devices, which include thermal and mercury devices.

Electromagnetic relays (Fig. 4·1*A*) are forms of electromagnets in which the current through the coil forms a magnetic field which attracts a clapper or actuator. Contacts are made, or broken, when the relay is energized. They are used to control a-c and d-c power and often to control the sequence of events in the operation of a system such as an electronic heater or welder. Solid-state circuit elements such as diodes, transistors, and magnetic cores have replaced ordinary relays in some applications, but almost every electronic system is turned on and off by a power relay.

Electronic switches (Fig. 4·1*B*) can be solid-state such as transistors or diodes, or they can be in tube form. Vacuum tubes and gas-filled tubes can both be used as switches. Other examples of switches are tubes such as thyratrons or ignitrons, magnetic devices including saturable reactors and magnetic amplifiers, and the solid-state controlled rectifier. A proximity switch is another type of electronic switch. Switches equipped with a mechanical trip lever can be used, but proximity switches have the advantage of greater life and are less subject to damage. In one type, an oscillator changes its frequency when a piece of metal approaches a pickup head or other sensing element. The change in frequency is amplified and sets other relays or machines into action. Another interesting type of electronic switch, which has no moving parts and no tubes, is a magnetic device with a sensing coil. When a ferromagnetic object moves past a magnetic field, the field is disturbed and the change is recorded. Photoelectric devices can also be classified as electronic switches. Light-sensitive elements enclosed in evacuated tubes or partially gas-filled containers

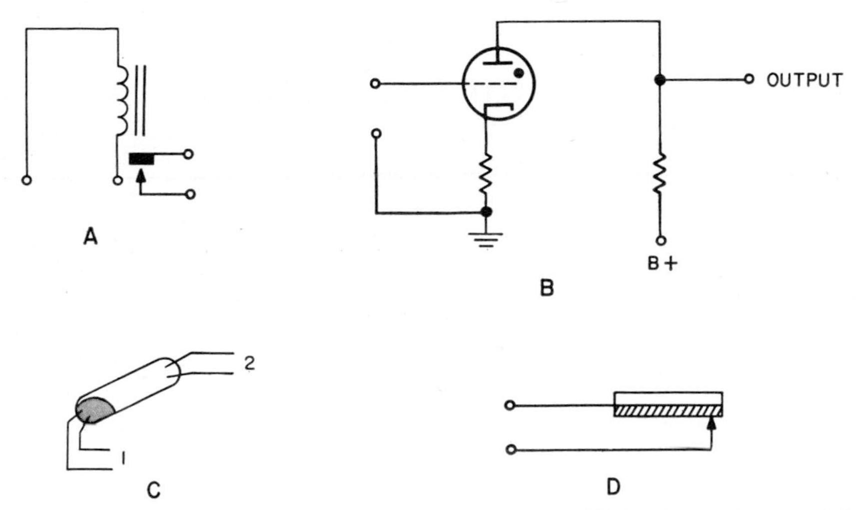

Fig. 4·1 (*A*) Electromagnetic relay as a switch; (*B*) a gas-filled tube used as a switch; (*C*) a mercury switch; (*D*) a bimetallic thermostat as a switch.

convert light energy into electric energy and become the sensing element of a switching system.

The third general classification of switching devices includes those that are thermally operated and mercury switches. Mercury switches are glass tubes into which stationary electrodes and a pool of loose mercury are hermetically sealed as shown in Fig. 4·1*C*. Tilting the tube causes the mercury to flow in a direction to open or close a gap between the electrodes to make or break the contact. Because of the hermetic seal, dust, dirt, moisture, and corrosive gases cannot enter the tube; hence they have a long life.

Thermal switches, or thermostats, make use of a metal strip and two contacts as in Fig. 4·1*D*. Two unlike metals are bonded to form the strip; when current passes through the bimetallic strip, the metals are heated because of their resistance. A heated metal expands and, since these are two different metals, they expand unequally, which causes the strip to bend as shown. The opening and closing of the contacts is the control.

Other more specialized devices include crossbar switches, used to make connections in a matrix of horizontal and vertical leads so that the proper interconnections can be quickly and easily made; stepping switches, used where a common or control point is to sample or be connected to a large group one at a time; and precision switches, used where the output is a precise function of the operating or actuating member.

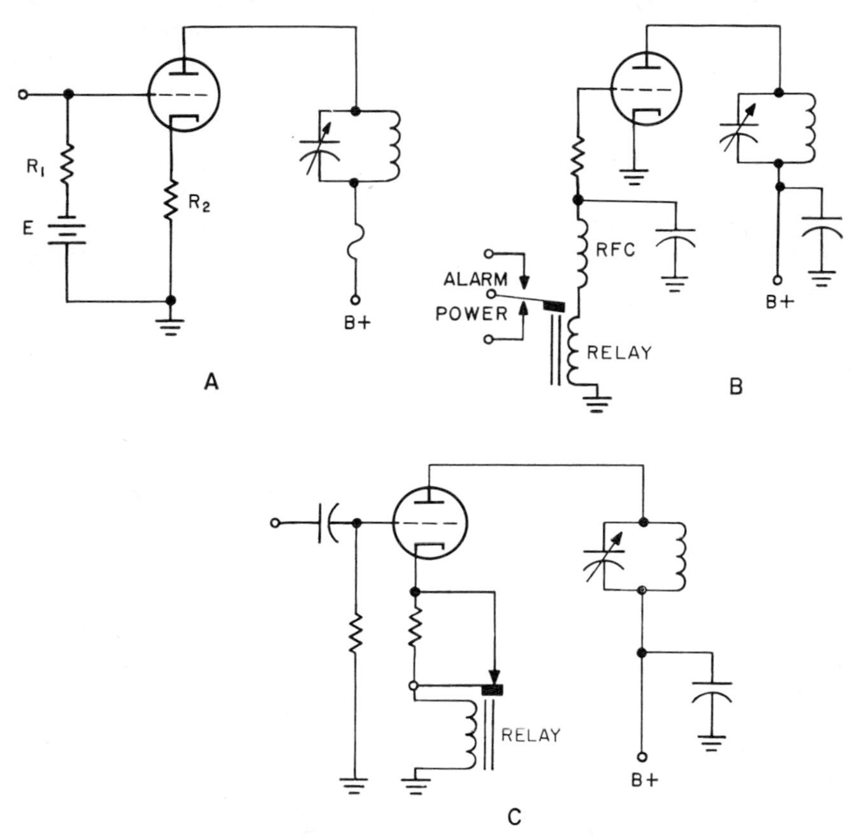

Fig. 4·2 (A) Protection of a vacuum tube showing a bias voltage, a series cathode resistor, and a fuse in the plate circuit; (B) grid-circuit control relay used to remove power if the grid-circuit signal fails; (C) cathode relay which adds circuit resistance when the plate current increases.

4·2 Relays One important use of relays is in an electronic circuit which would run away to the point of self-destruction if signal and other control voltages were lost or upset. The class C r-f power amplifier is a case in point, as shown in Fig. 4·2.

Amplifiers whose grid bias is developed as a result of a grid signal can make use of one or more protective devices. If for any reason the bias is lost, the plate current could increase until the tube and other components are damaged. Figure 4·2*A* shows a basic r-f class C amplifier.

Several things that do not require any special switching mechanisms can be done to protect this circuit from excessive current. A bias battery (E) could be placed in the grid circuit to furnish at least a minimum amount of bias in the event of a signal failure. A resistor (R_2) installed between cathode and ground tends to be self-regulatory. Briefly, its action is as follows: As tube current increases, the voltage drop across the resistor increases; the polarity is such as to make the cathode more positive than the grid; conversely, the grid becomes more negative than the cathode and reduces current flow. Still another protection method is to install a fuse in the plate circuit.

Relays could be used to advantage in some of these applications. Figure 4·2*B* shows a normally open relay in the grid circuit. When grid current flows, the relay is energized and closes the contacts. In the event of a loss of signal, with a resulting loss of grid current and bias, the relay armature could sound an alarm, shut down the equipment, or even switch in an emergency circuit.

A relay with normally closed contacts placed across a resistor which is in series with the cathode (Fig. 4·2*C*) affords protection by effectively inserting the resistor only when needed. During normal operation, tube current is insufficient to attract the armature. In this position the contacts are closed and the resistor is shorted out. Should the current increase to the danger point, or other preset value, the contacts will open and place the resistor in the circuit.

In a similar manner, a relay may be placed in series with the B+ lead going to the tube's plate. This relay is normally closed and wired in such a manner as to enable the contacts to open and close the plate current path. Should the plate current increase beyond a prescribed limit, the relay will open the circuit. A latch arrangement is required to hold the relay open; otherwise it will act like a buzzer. Except for the relay in the cathode circuit, it is necessary to reset the other relays. In some applications the buzzer action is desirable, because it does sound an alarm and does not have to be reset when the trouble is eliminated.

Relays are used to control power in almost every type of industrial electronic system, including dielectric and induction heating, X-ray equipment, counting and measuring equipment, and alarm systems.

Relay Construction. In the basic electromagnetic relay shown in Fig. 4·3*A*, the coil is wound on an iron core. When direct current passes through this coil it creates a magnetic field which attracts the iron armature from its support by the spring. Pivoting at the point as indicated, the armature moves and opens the contacts connected to the load. In a variation, the contacts are normally open rather than normally closed. Figure 4·3*B* shows a relay with three contacts. With no input current the spring holds the armature in the up position, completing the path from 1 to 2. Thus 1 and 2 are normally closed while 2 and 3 are normally open. Current flow through the relay coil, when heavy enough, creates a magnetic field which closes 2 and 3 while it opens 1 and 2.

The mechanical structure of a basic relay may be seen from Fig. 4·4. A winding is placed around the core, and when the current flows in the coil the magnetic field attracts the armature, which is hinged. As the

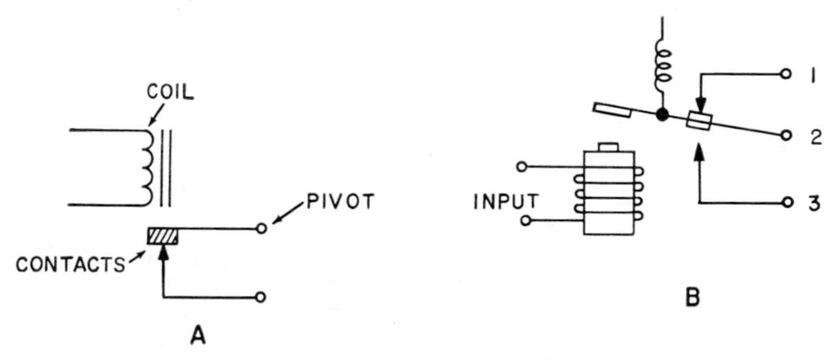

Fig. 4·3 (A) The basic electromagnetic relay; (B) relay operation showing normally closed contacts 1 and 2 and normally open contacts 2 and 3.

armature moves, because of the magnetic attraction, it closes (or opens) the contacts mounted on springs.

Relays may be classified in many ways, but there are two large groups defined by their use. General-purpose relays or telephone relays form the largest group of electromagnetic switching devices; they are usually d-c operated. Special-purpose relays are usually designed for specific

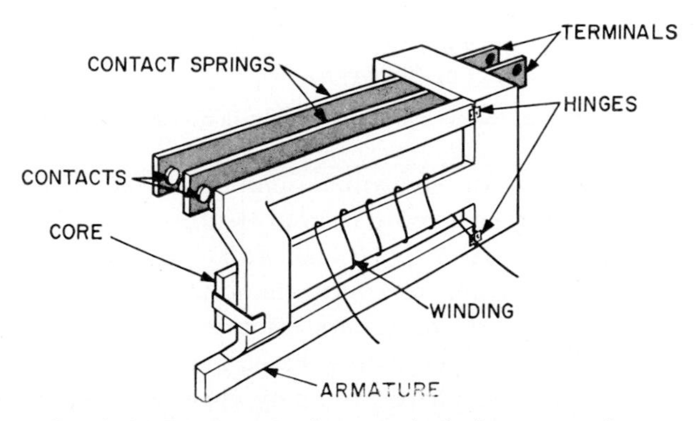

Fig. 4·4 Construction of a typical telephone-type relay.

applications with certain outstanding features such as high sensitivity, high speed, or the ability to carry large currents.

Thousands of different relay types, both general-purpose and special-purpose, are available for many applications. Many of these have certain features in common; as an example, there are several basic contact arrangements. In relay terminology the following expressions are used: spst, single-pole single-throw; spdt, single-pole double-throw; NO, normally open; NC normally closed; break, meaning relay action breaks or opens the circuit; and make, meaning relay action makes or closes the circuit.

The heart of the relay is the junction of the contact points. Several types are shown in Fig. 4·5. The flat contacts (*A*) require heavy pressure;

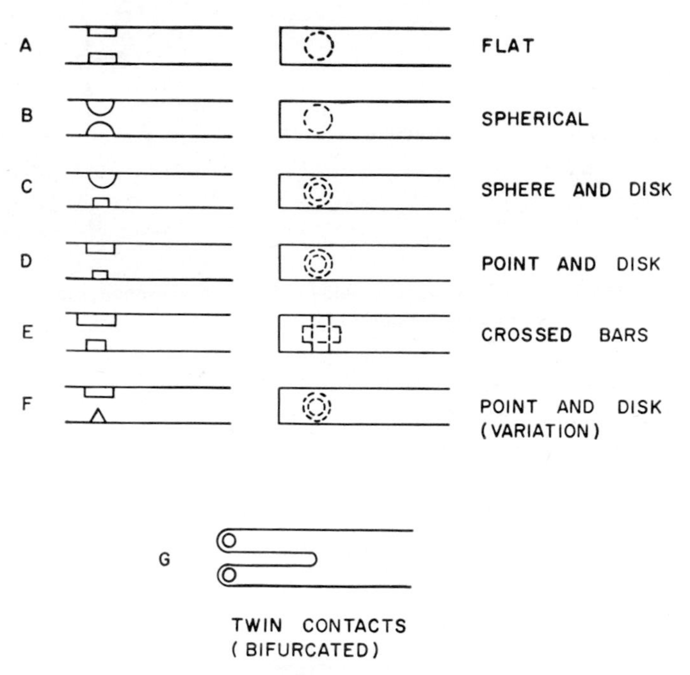

Fig. 4·5 Seven different types of relay contacts.

a half-round (*B*) reduces the contamination of the contact surfaces which hinder relay operation. The combination of the first two is in common use (*C*). Other types are shown in *D*, *E*, and *F*. The bifurcated or twin contacts shown in *G* give greater reliability than the others.

Relay contacts are made of several metals; copper is usually used only in the larger relay types, while silver and silver alloys are more common in smaller relays. There are many types of relay contacts with different characteristics. For example, code 0-18 is a palladium-silver type which is in wide use. These contacts are resistant to tarnish, will make or break a load up to 135 watts, and carry a load of 150 watts. Code 9-18, on the other hand, is platinum-ruthenium, which has the same ratings but with longer contact life. Figure 4·6 shows a relay with four sets of contacts.

There are actually hundreds of relay combinations considering all the possible variations in contact materials, contact types, coil design, and other factors. But the switching or contact arrangements are one feature which is important. Eighteen possibilities are indicated in Fig. 4·7. *A* is single-pole single-throw, normally open, and make when energized. *B* is the same except that it is normally closed. *C* is spdt, which breaks before make. Others can be understood from their titles. And even this list does not exhaust the possibilities.

Relay Trees. Relays can drive other relays, as shown in Fig. 4·8. Here there are eight possible circuits from three relays. Where N is the number of relays, the number of controlled circuits is 2^N for any N larger than 1.

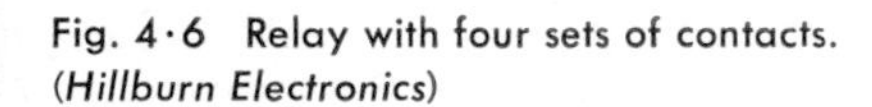

Fig. 4·6 Relay with four sets of contacts. (*Hillburn Electronics*)

Note that for different combinations of *A*, *B*, and *C*, one and only one circuit will be alive.

Latching Relays. When a set of relay contacts are to be held closed (or open) for a long period of time, latching methods are sometimes used. Figure 4·9*A* illustrates a simple latching mechanism. When the relay arm is attracted up, the latch moves as shown and locks the relay open. Figure 4·9*B* shows a complete latching arrangement. When the coil *B* is energized, the armature *E* is attracted toward the coil and this closes contacts *C*, which complete the circuit through the load. Because of the pivot action of this seesaw point, *F* moves down and is locked in place by the latch which is held by the spring. Now the contacts will remain closed when the coil is deenergized; a short current flow triggers this action and the relay stays locked. It can be opened manually by a button at *D* or electrically by attracting the latch bar in the other direction toward another coil *A*.

In Fig. 4·10 this is carried to another degree. Each time current passes through *A*, the armature moves to the left. It is restored by the spring action, and for each current pulse at *A* the contact bar moves down one notch. As shown here, two on-off sequences are required until the contacts complete the circuit through the load. This is a relay counter, using the latch principle. Other counts are quite possible, but as the number increases the device becomes more complex.

In some cases relays are driven by amplifiers. A transistor drive may be used, as shown in Fig. 4·11, where a dual-coil magnetic latching relay is shown. It has four contact pairs in a transistor-driven, magnetically biased relay-switching circuit. Applications of this circuit lie principally

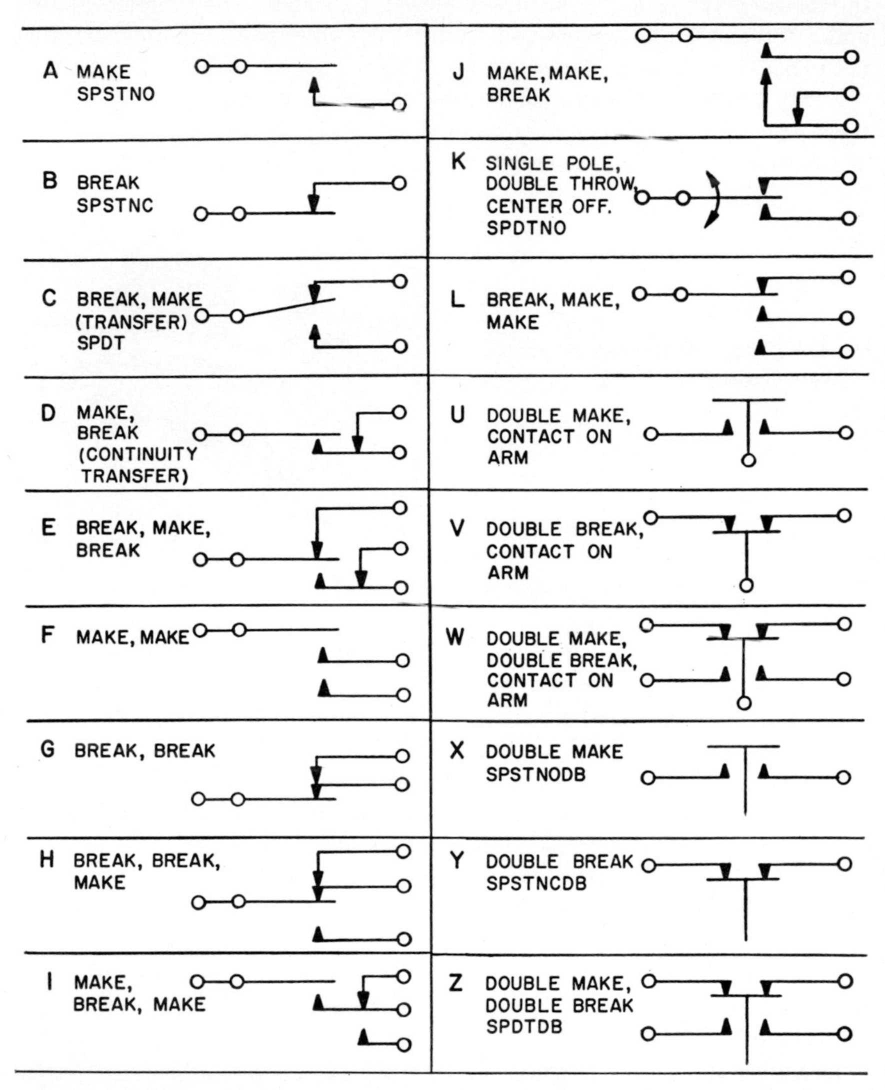

Fig. 4·7 Eighteen basic contact forms with their symbols and name identifications.

in the input and output devices of data-processing equipment where a frequency of operation which is intermediate in speed, between ultra fast electronic circuits and slow manual response, is required. The relay also provides higher power-handling capacity than low-current solid-state devices for driving electromechanical devices. A "set" signal switches the contacts in one direction where they stay until they are reset. In "reset" position the magnetic circuit holds the contacts until application of a "set" signal.

Meter Relays. Meters, of course, indicate a measured quantity such as current or voltage, and relays are used for control of circuits. A meter relay is a combination of the two used as in a bearing monitor. Originally

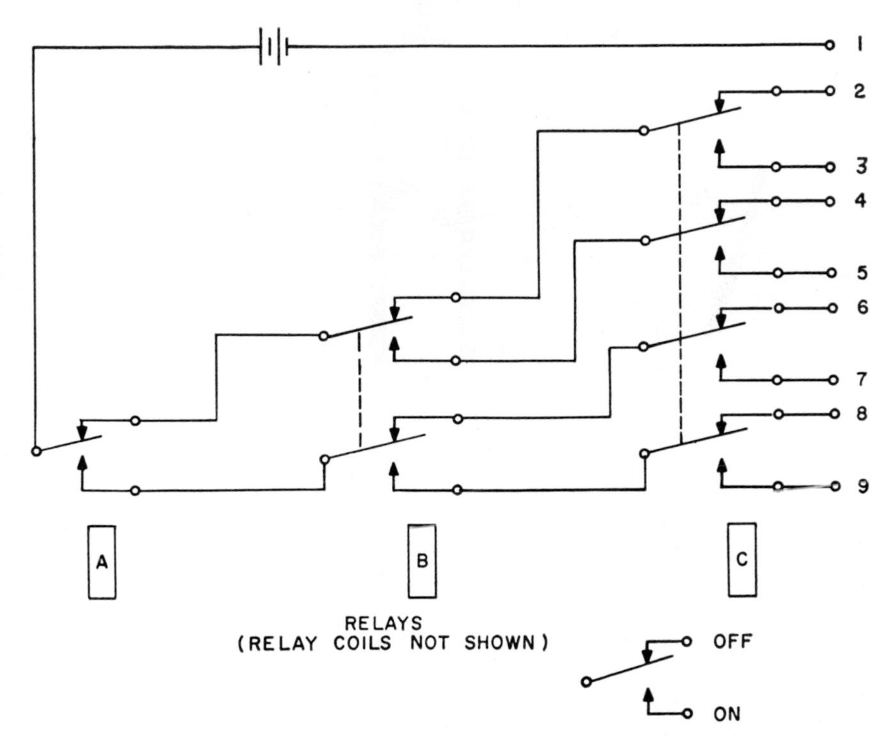

Fig. 4·8 Use of a relay tree showing three relays and four pairs of contacts.

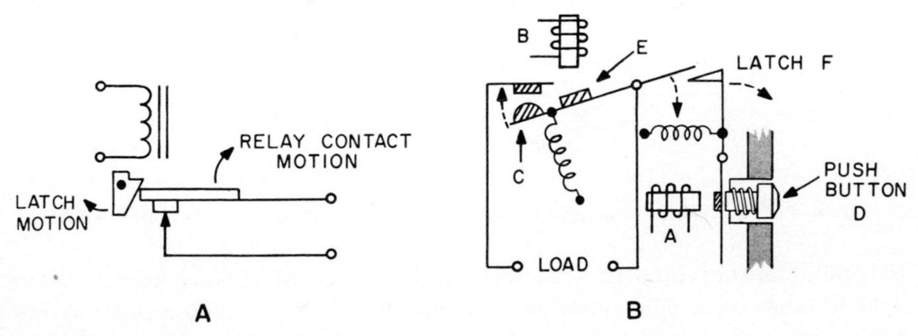

Fig. 4·9 (*A*) A simple latching relay; (*B*) a latching relay which is opened by use of the push button.

developed for safety equipment in pipeline pumping stations, bearing monitors give immediate warning or shut down when any bearing gets hot. Each bearing or other danger point has its own thermocouple and indicator. These show continuously the normal running temperature. The

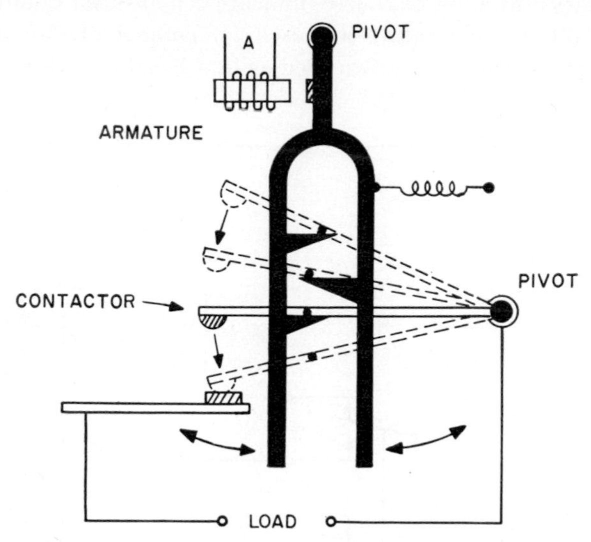

Fig. 4·10 A stepping relay showing the four different possible positions of the contact arm.

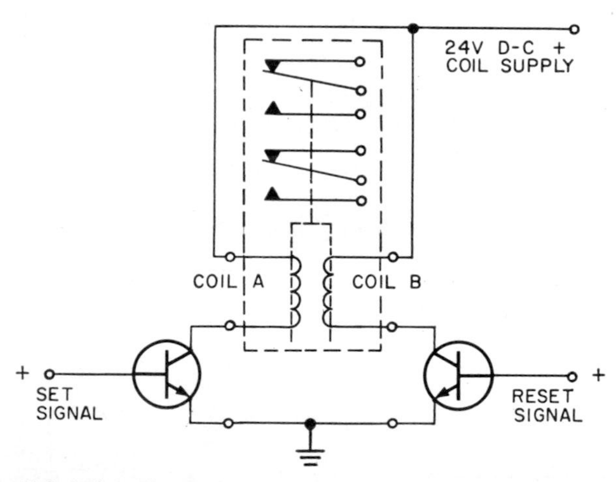

Fig. 4·11 U.S. Relay Electronics magnetic latching relay using two transistors.

trip point is individually adjustable from the front of each meter relay. Any number of danger points can be monitored. Each meter relay shows a bearing temperature and contains an adjustable trip point. A thermocouple is required for each bearing. The thermocouple attaches to its bearing. A pair of lead wires connects between monitor panel and thermocouple. The monitor is one example of the use of a meter relay.

Contact meter relays are electromechanical devices that are being used more and more in electronic circuits because they have inherent characteristics that stabilize and simplify such circuits. Meter relays will do certain things, with relative ease, that are difficult to achieve by purely electronic means.

Like any other relay, a meter relay is an amplifying device. It is the most sensitive of all relays, being actuated directly by a signal as small as $\frac{1}{5}$ μa. It is also a sensitive moving-coil d'Arsonval meter, continuously indicating a signal from any variable that can be measured electrically through a suitable transducer. Finally, a meter relay is a monitoring device with easily adjustable signal set points, where control action is triggered. But distinctive as this combination is, it still does not at first thought indicate all the useful properties meter relays offer electronics designers. For example, it does not imply that a meter relay can be used as an amplifier with a power gain of 10^6.

The meter relay, which was patented around the turn of the century as an offshoot of a d'Arsonval meter, was not developed to a useful stage until less than 10 years ago. The basic idea involves mounting a contact on the pointer (carried by the moving coil of the meter) with a mating contact positioned inside the meter. Most commonly, the second contact is fastened either to another pointer, which can be adjusted, or to a fixed stud or pin. When the signal current rises or falls to the point marked by a fixed contact, a circuit is closed across the contacts. Through slave relays, this circuit usually works into one carrying much greater current, and useful work is accomplished.

There are two general types of contacts in meter relays: locking and nonlocking. The locking category is further split into locking-coil and magnetic types. The locking-coil type (Fig. 4·12) is the most reliable and it ensures the most positive control action. The locking circuit imposes some limitations, but these restrictions may be easily surmounted with a few simple tricks. Another important advantage of the locking coil is longer contact life.

By itself, the moving coil of a meter relay exerts only a few milligrams of pressure between the contacts. Such a nonlocking arrangement is unreliable because it frequently leads to sticking of contacts when the signal is removed. Reasons for this problem include arcing, mechanical wear, and natural adhesion between two pieces of metal.

In general, therefore, nonlocking meter relays are usually restricted to applications where (1) the signal abruptly rises above and falls below the point of contact, (2) contact load is less than 10 μa, (3) open-circuit voltage is less than 10 volts, and (4) a high-torque (1 ma or better) meter movement is used. Even under these conditions, nonlocking contacts generally will have a limited life of not much more than 1 million operations.

In the meter relay, one contact of the relay is fixed but adjustable, and the other is carried by the moving element of the meter as shown in Fig. 4·12. The indicating part of the meter is conventional, but because of the very low torque of the moving coil it cannot reliably actuate the relay. A locking coil is used (wound on the indicator coil) so that, when the contacts touch, the current through the locking coil holds them together. Reset (opening the locked relay contacts) can be either automatic or manual.

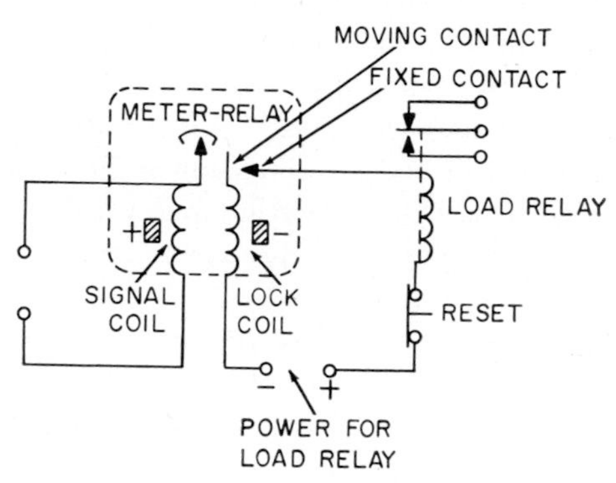

Fig. 4·12 Construction of the basic meter relay with a reset push button. (*Assembly Products*)

Signal current (to be measured) is applied to the input. The signal coil moves the pointer and the moving contact, until this contact touches the fixed contact. Direct current from the load-relay power source flows through the series circuit as shown. This current locks the meter relay in position and energizes the load relay. But because of this locking feature, the meter relay will remain closed after the signal current is removed. A reset button, in this case manually operated, breaks the lock-coil current when it is pushed. If the signal current has fallen below lock-in, the relay will stay open; if not, the relay will remain locked. Figure 4·13 shows a push-button reset.

There are many applications for a meter relay in industry. Virtually all industrial processes where meters or relays are used are potential sources for the meter relay. In thyratron triggering, as an example, the meter reads grid voltage and the relay contacts are set at the firing potential. In heat indicators the meter reads current from a thermocouple (a temperature-sensitive junction of two different metals) and the relay is a protective device to sound, or indicate, an alarm when the safe limit is reached. In pipeline equipment this is used to prevent running a bearing which is too hot. In vacuum-tube circuits a meter relay can indicate plate current and shut down the equipment when the plate current is excessive.

An automatic reset is shown in Fig. 4·14*A* where the cam rotation periodically resets the meter relay. The capacitor keeps the load relay closed while the cam tests the signal current and *R* limits the peak charging current. In another reset, shown in Fig. 4·14*B*, the load-relay contacts bypass the locking coil. When the meter relay closes it locks in, but the current flow through the load-relay contacts bypasses the lock-coil contacts. Because of this, the meter contacts open and the load relay stays closed until it is reset by the push button or the cam as in Fig. 4·14*C*.

There are cases where both upper and lower limits for a specific current are required, such as heater current for a tube. Figure 4·15 illustrates

Fig. 4·13 Assembly Products meter relay showing upper and lower limits for the indicator and a reset button.

a meter relay with two limits controlling the applied power to a piece of equipment whenever the meter reads above or below fixed limits. A manual reset is shown here together with a rectifier power supply for the direct current through the load relays. The upper and lower limits are shown each with a single load relay for alarm signals.

Several meter relays can be operated from a single power supply. Each may operate into its own load relay, or several may work into the same load relay. A single interrupter may be used for all, or some relays in the group may be operated on manual reset while the others are automatic. Both single- and double-contact meter relays as well as either high- or low-limit types may be used.

If the high contact on one meter relay and the low contact on another are apt to be made at the same time, it is necessary to break the separate locking circuits through each meter relay. Breaking just the common leg will not release the contacts.

Meter relays do not drop out as conventional relays do when current through the signal coil is reduced. To get on-off action, a sampling circuit with an automatic interrupter is generally used. The meter contacts are unlocked periodically to sample the signal current. If the signal is still up to the control point, the meter contacts reclose and lock again quickly.

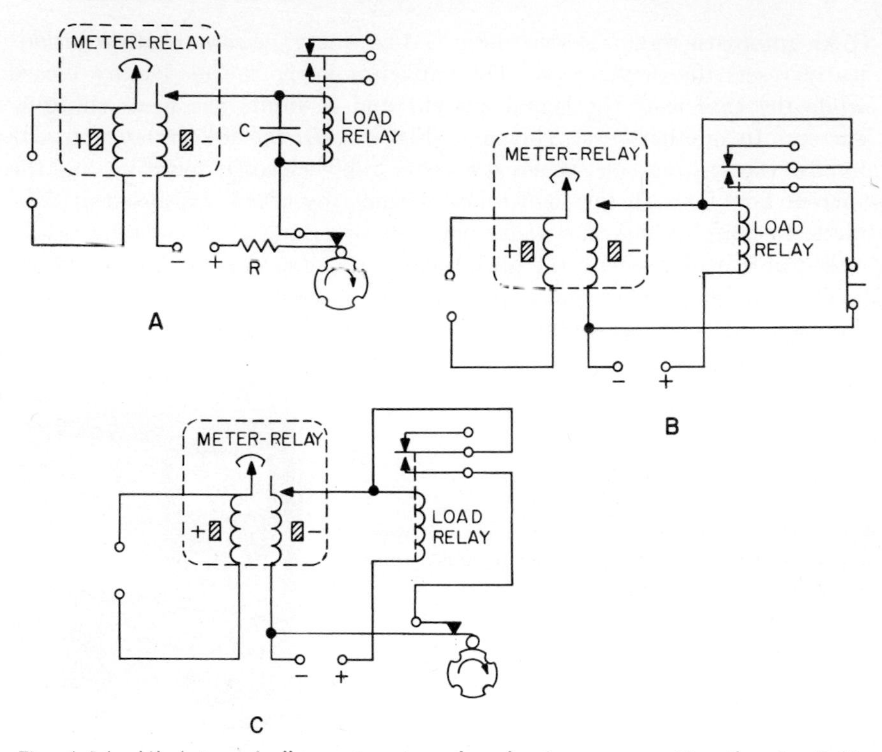

Fig. 4·14 (*A*) Automatically reset meter relay showing cam resetting the circuit (*Assembly Products*); (*B*) the bypass or shunt reset; (*C*) automatic bypass reset.

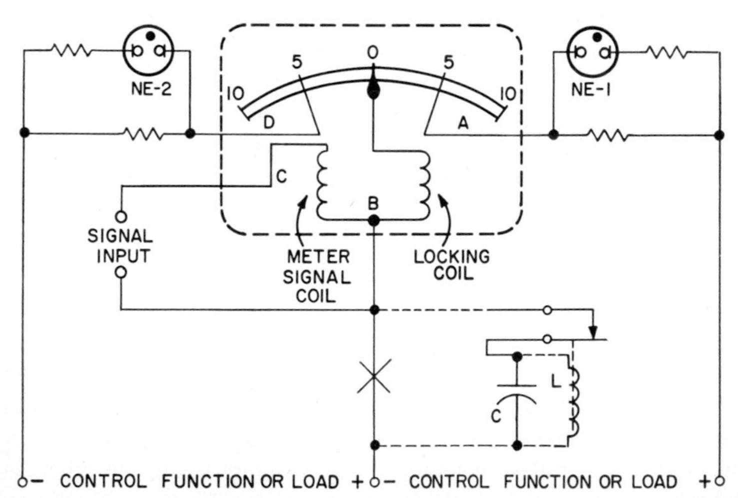

Fig. 4·15 Assembly Products double-limit meter relay with neon lights as alarm indicators.

However, if the signal has dropped, the meter contacts remain separated and the load relay releases.

One of the simplest automatic interrupters is that where the load relay does the interrupting; however, the load circuit will also be interrupted periodically. For alarms and some types of control this may be satisfactory. In some automatic controls the load relay holds steady unless the current in the signal coil drops.

One example of the use of meter relays is a Sim-ply-trol temperature control, which is an on-off control for furnaces, ovens, kilns, or industrial processes. Thermocouples are used as the sensing elements.

The heart of the Sim-ply-trol is a meter relay, usually calibrated in both Fahrenheit and centigrade. It serves as an indicator and initiates control action. All the other components in the control package are there to enable the contact meter relay to do its job.

Control action on a standard high-limit automatic control starts when the voltage generated by the thermocouple moves the d'Arsonval coil and causes a contact carried on the indicating pointer to touch one mounted on the control pointer. This causes a load relay to turn the heat off. A locking coil wound in series with the contacts locks them together to give greater contact pressure. Once contacts are locked together, they stay locked until automatically reset by breaking of the current flowing in the locking coil or by shorting of the locking coil. Spring mounting of one contact separates them forcefully to prevent sticking and arcing. This interruption occurs at a frequency of once per minute, usually, but can be varied to suit the control application. Only if the temperature has dropped below the control point will the load relay drop out to turn the heat on again.

Meter relays with pyrometer scales for indicating temperature usually are not made with isolated signal circuits. A bimetal attached to the bottom hairspring for pyrometer cold-junction compensation would interfere with the additional hairspring required for isolation. Voltage between the thermocouples and ground may cause an error in the pyrometer calibration. This will not occur if the thermocouples are insulated from ground. However, in some applications, it is desirable to attach the tips of the thermocouples directly to the metal frame of a machine for quick transfer of heat. In this case contact pyrometer relays with reversed polarity (positive input instead of negative to the common B terminal) should be used. The positive element of base-metal thermocouples has little if any thermoelectric effect with iron or other material usually found in machine frames.

There are many applications where a multicontact relay is required. For example, readings of pressure or liquid level, in steps, are needed in many industrial processes. In grading resistors, a multicontact relay can sort resistors according to value by steps.

Figure 4·16 shows a unique wattmeter, which both indicates and controls and which combines a meter relay and a magnetic circuit built around a Hall-effect solid-state device. The standard dynamometer wattmeter movement has been eliminated. This is possible, and power measurement is simplified, because of a special characteristic of the Hall-effect device that makes it uniquely suitable for indicating watts.

The Hall device puts out a d-c voltage that is the product of two inputs. This voltage is generated when a current is passed through a wafer

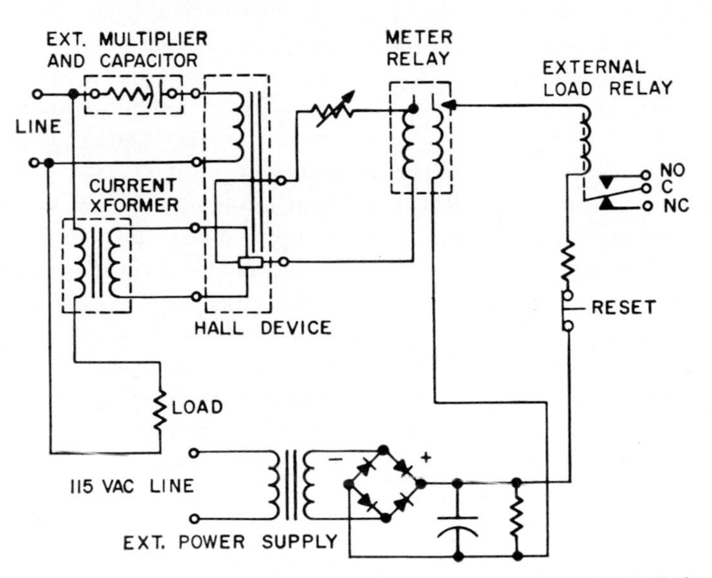

Fig. 4·16 Schematic of the Assembly Products wattmeter using the Hall device.

of indium arsenide which is placed in a magnetic field perpendicular to the direction of wafer current flow. The output voltage is developed across the wafer in a direction perpendicular to both the field and the current, and it is proportional to their product.

In this wattmeter, the magnetic field is created by current that is proportional to, and in phase with, the load voltage. The semiconductor wafer is excited by an in-phase analog current of the load current. The output voltage is then proportional to the power in the load circuit. The output is fed to a standard d'Arsonval meter relay that is calibrated for power measurement.

The meter relay has one or two adjustable pointers that set signal limits anywhere on the dial. Either a standard locking-contact meter relay or the new continuous-reading meter relay may be used. In either case, when the signal pointer reaches a limit, a locking coil is energized, providing firm closing of a circuit to an external slave relay. Control then takes any of a number of desired forms, with standard circuits.

In the case of the locking-contact type, when the signal returns within limits the locking circuit is first opened and the contacts are flicked apart by a spring. With the continuous-reading meter relay, the indicating pointer moves past the set point and causes toggle action that shorts out the circuits energizing the load relay and the booster coil.

Some meter relays use a taut-band movement as shown in Fig. 4·17. Because friction has been eliminated in the new meter relays, they are particularly suitable for control applications requiring full-scale sensitivities in the area of 0 to 5 μa or 0 to 2 mv.

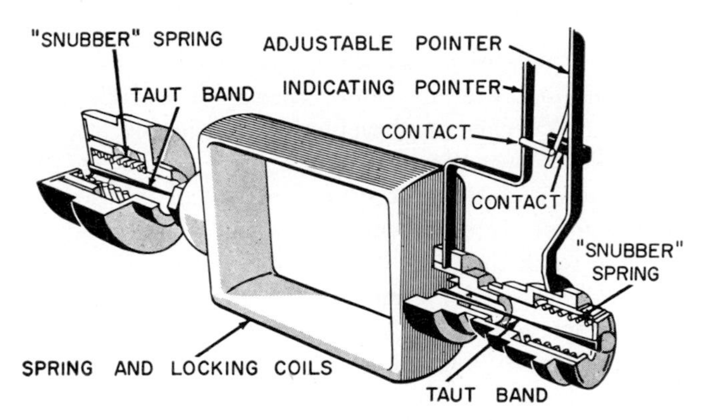

Fig. 4·17 Assembly Products taut-band meter relay.

For positive control operation, the new meter relays use the locking-contact action. Extra torque to lock pointer contacts, when the set point is reached, is supplied by a booster coil. Current to the locking coil passes through one of the two torsion bands that suspend the meter coil.

These meter relays use coil springs, coaxial with the taut bands, which are incorporated in snubber mechanisms at both ends of the meter movements. Under shock and vibration, the snubbers keep the taut-band system aligned.

4·3 Electronic switching devices Electronic switches form a large group of varied devices. They can be solid-state devices such as transistors or diodes, or they can be in tube form. Vacuum tubes and gas-filled tubes such as thyratrons or ignitrons can both be used as switches. Other examples are magnetic devices, including saturable reactors and magnetic amplifiers.

A proximity switch, for example, is a type of electronic switch. Switches equipped with a mechanical trip lever can be used, but proximity switches have the advantage of greater life and are less subject to damage. In one type, an oscillator changes its frequency when a piece of metal approaches a pickup head or other sensing element. The change in frequency is amplified and sets other relays or machines into action.

Another interesting type of electronic switch, which has no moving parts and no tubes, is a magnetic device with a sensing coil. When a ferromagnetic object moves past a magnetic field, the field is disturbed and the change is recorded.

Switching devices which are all-electronic in nature, as opposed to electromechanical devices, fall into four groups. These are:

Vacuum tubes
Gas-filled tubes
Semiconductor devices
Magnetic devices (covered in Chap. 11)

The operating principles of all these are different, but some of the functions they perform are the same. Electronic switches are used in many applications. Furnaces and welders, for example, are two devices requiring controlled power. Control of a-c power is also needed for operating many types of motors or other devices in industrial processes. Three types of control are used. One is a switch which, when closed, permits the power to reach the load. With an a-c source of power the switch permits the load to operate for a portion of each cycle. The larger the portion of a cycle which is used, the greater the average power delivered to the load. A second type of control is a variable resistor in series with the load and the power source; this uses up some of the available power. A third type is an autotransformer in which the tap is variable and changing its position will vary the amount of power available to the load.

Examples of switches in this application are tubes such as thyratrons or ignitrons, magnetic devices including saturable reactors and magnetic amplifiers, and the solid-state controlled rectifier.

Switches for controlling power are of many different types but they all have certain things in common. They all turn on at some time in the a-c cycle, which is the firing point or angle. They all turn off when the a-c supply goes negative, and they all are rectifiers as well as control devices; hence they permit current flow in only one direction.

Vacuum Tubes. An example of a vacuum tube as an electronic switch is the free-running multivibrator. By its own feedback between tubes the combination operates as a self-driven switch, and each tube has successive on-off periods. The tubes could be replaced by switches since they function in either of two states. The tubes are either at saturation, which is full conduction, or at cutoff so there is no current flow. A transistor multivibrator can perform the same function as the vacuum-tube type.

An electronic switch using the multivibrator finds an application in comparing two voltage readings on an oscilloscope. Figure 4·18 shows this switch. Signals *A* and *B* can be of any type and are the inputs. As the tubes turn on and off, one signal at a time is tied to the vertical input of the scope. Both signals appear to be presented at the same time.

Precision timing pulses for measurement may be generated using a vacuum-tube switch. A tuned circuit is in the cathode of the switched tube. As the pulse input permits the cutoff tube to conduct, the resonant circuit "rings" or creates a damped-wave output because of the flywheel effect. When this damped wave is clipped, it provides the timing-signal output.

Gas-filled Tubes. In many respects the thyratron is a near-perfect switch, for it has only two states—conducting and nonconducting—while

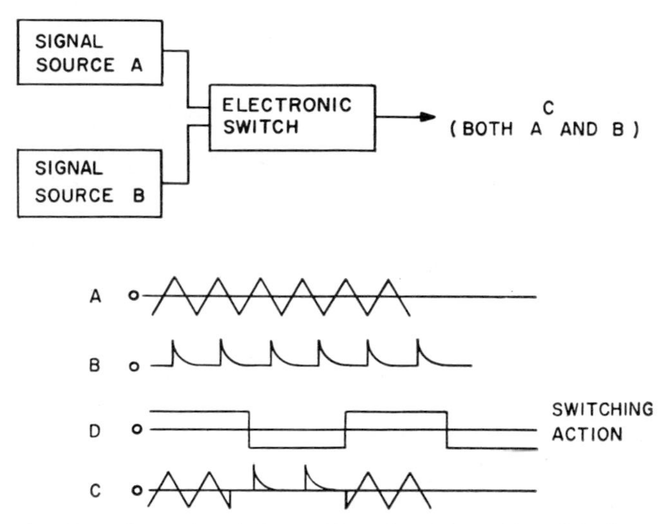

Fig. 4·18 Use of an electronic switch providing signal *A* and signal *B* alternately as an output.

a vacuum tube can assume many states between these extremes. Once the grid potential permits the tube to fire, only the reduced plate potential will turn the tube off.

In a time-base generator a thyratron has its anode voltage derived from the charge upon a capacitor. When this charge reaches the proper amount, the tube can fire, creating a sawtooth output for a time base. Usually the grid voltage, used to synchronize the output, causes the tube to fire just before it normally would.

Motor control also uses gas-filled thyratrons. Magnetic amplifiers provide the steep wavefront required for control of thyratrons and controlled rectifiers. A wide range of different inputs may be used as control inputs to the magnetic amplifier. A typical example of this control is given in Fig. 4·19. A pair of thyratrons are used to control the speed of the d-c motor using a magnetic amplifier. The speed-set control adjusts the desired operating speed of the motor. A reference voltage is compared with the armature voltage in one of the control windings. A different control winding measures the armature current and increases the motor drive as the load is increased. A resistance adjustment is provided for this winding which corrects for the *IR* drop in the armature. A third winding used for torque control turns off the thyratrons if the load current exceeds a predetermined value.

Semiconductor Devices. All semiconductor devices can be used as switches. Power transistors acting as switches, in pairs, are used to convert direct current to alternating current.

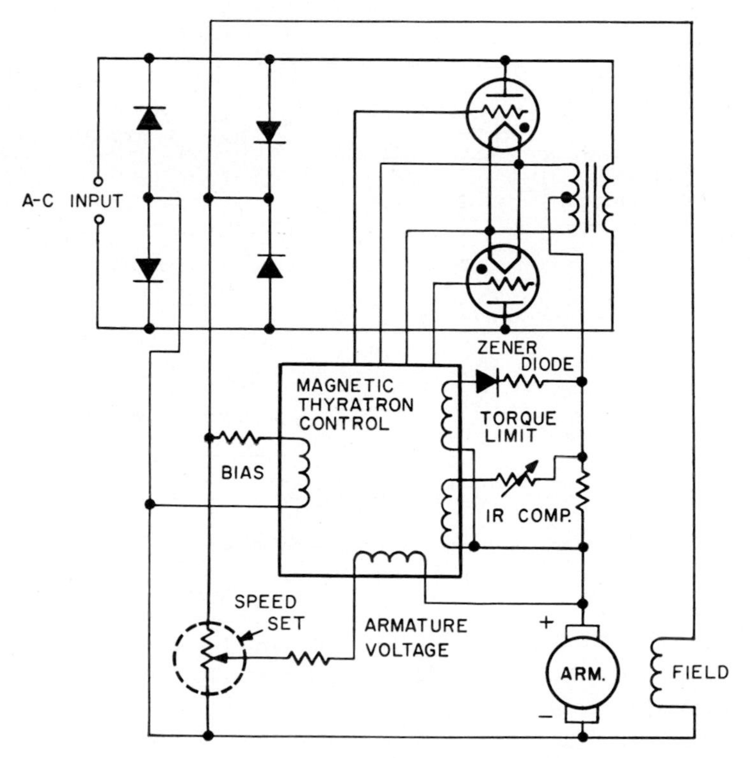

Fig. 4·19 Fairfield Engineering motor-speed control.

Less weight, longer life, greater efficiency, and fewer breakdowns are all possible with a transistorized power converter.

Power-transistor switches make effective devices for power conversion from d-c sources. They are more reliable than vibrators and are less complex and bulky than motor generators. And their efficiency is very high. High-voltage low-current direct current is used for mobile-radio plate supplies; low-voltage high-current alternating current is applied to servo systems and a-c motors.

Power conversion from low-voltage to high-voltage direct current uses a d-c to d-c converter. Conversion from low-voltage direct current to alternating current requires an inverter.

There are several forms of inverters in use. In 60 cu in. (3 × 4 × 5 in.) a 50-watt output at 5 amp and 110 volts a-c can be obtained from a pair of power transistors such as the 2N176 shown in Fig. 4·20. This is enough power to operate a small phonograph, an a-c/d-c radio, a dictating machine, a small soldering iron, or a trouble light. The unit weighs about

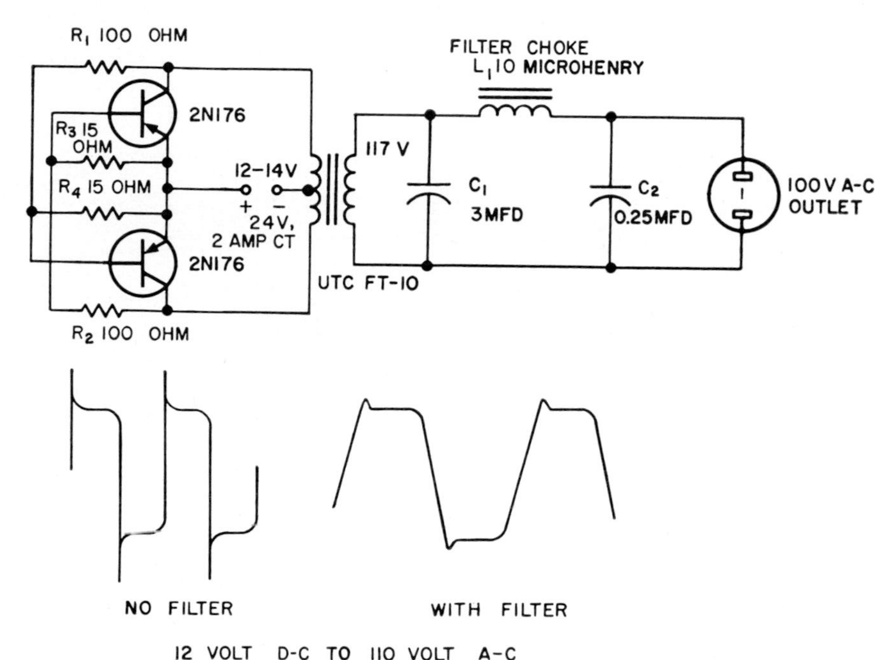

Fig. 4·20 Motorola inverter with 50-watt output.

4 lb and has an efficiency of about 75 per cent in converting direct current to alternating current.

As shown in Fig. 4·20, the base bias and collector-to-base coupling are provided through 100-ohm 2-watt resistors R_1 and R_2. Bias stability is provided by the 15-ohm 2-watt resistors R_3 and R_4.

This inverter, designed for use with a-c equipment, has a frequency output of approximately 60 cps, and the peak-to-peak voltage (between the flat tops) is 250 volts at 50 watts. For many applications this output is satisfactory even though the waveform is approximately a square wave without the filter.

The hash filter, composed of C_1, L_1, and C_2, removes most of the spikes, giving a trapezoidal waveform. With the filter the frequency is reduced to about 56 cps.

Much higher powers are also possible from inverters, for example, the circuit in Fig. 4·21, which shows a two-transistor inverter using the 2N1167 power triodes. They have a current gain of 25 at 25 amp and can thus switch 700 watts, which is 25 amp at 28 volts.

A 10-cfm blower is used to stabilize the case temperature to below 60°C. The transistors were insulated from the heat sink with an anodized aluminum washer, and silicone grease helped lower the thermal resistance from the transistor case to the heat sink.

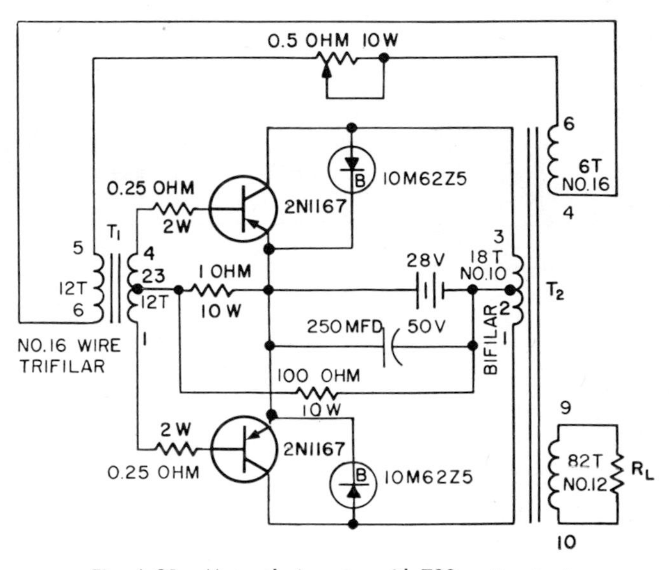

Fig. 4·21 Motorola inverter with 700-watt output.

The two-transformer design was used as shown, allowing the feedback transformer to do the saturating while reducing the collector-current spikes. Since the output transformer does not saturate, the saturation losses were considerably reduced. The driver transformer is designed to oscillate from 200 to 1,200 cycles with 2- to 12-volt drive. A multiple tap on the output transformer is wound on a double 4½-in. Hypersil C core, each half about ¾ sq in. in cross section. The driver transformer consists of three identical windings of No. 16 wirewound trifilar to provide various circuit configurations. The driver core is stack-laminated of Mumetal with a core area of about ⅜ sq in. and about 1- by 1- by ½-in. outside dimensions. The circuit operates common-emitter push-pull. A 0.5-ohm potentiometer was inserted between the feedback winding and the driver transformer to allow proper switching action and help control the frequency. Performance was improved by inserting small resistors in series with the base. These 0.25-ohm resistors equalize the drive and eliminate burnout problems.

From this circuit a power output of 575 watts was obtained with a power input of 700 watts, for an efficiency of about 82 per cent. Bias and feedback resistors account for about 25 watts while another 35 watts was lost in the transistors. The other 65 watts apparently was lost in the transformers.

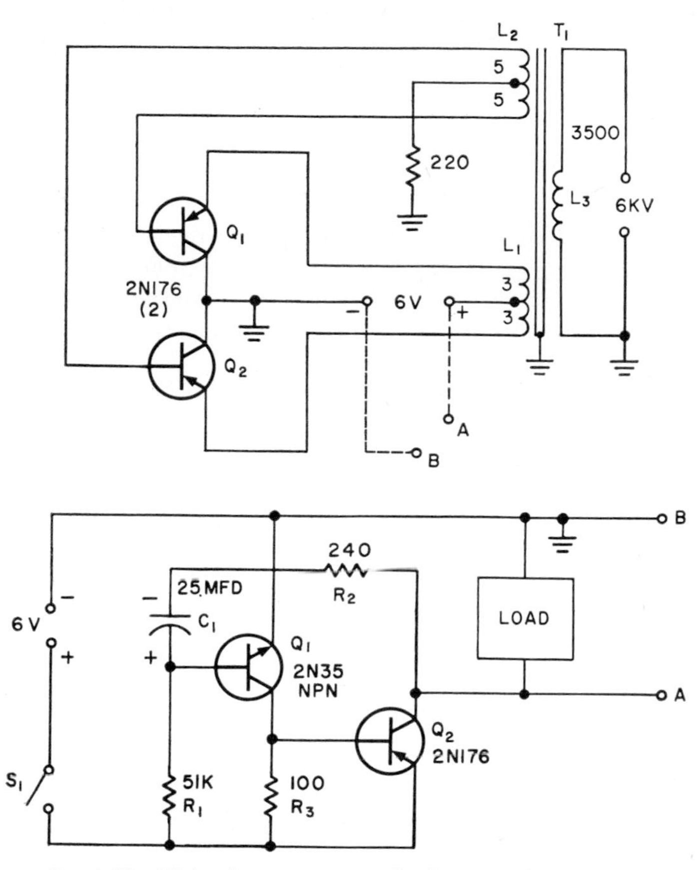

Fig. 4·22 High-voltage inverter with 6 kv output by Motorola.

An additional 20 watts was required to operate the blower, which was driven from the output. Currents of up to 34 amp were switched and power outputs of more than 700 watts were obtained on an intermittent basis.

Inverters can provide power for several applications. An inverter for 6 to 6,000 volts a-c is shown in Fig. 4·22. This 6 kv can be doubled and used for cathode ray tubes (CRTs). In the figure, on the left, is the inverter using a pair of 2N176s in a common-collector circuit. The transformer is special; the Ceramag U core, AP-11-264(2), is similar to the type used often in TV flyback transformers.

L_1 and L_2 are handwound, side by side on the cardboard bobbin with taps brought out on either side. The winding is covered with three thicknesses of bond paper saturated with coil dope. L_3 is universal-wound on this paper layer. The winding width is ¾ in. After every 250 turns a layer of doped paper is applied to the winding. This is to add strength to the coil, which builds up to a diameter of approximately 2¾ in. The coil is well doped, dried, and covered on the sides and top with masking tape painted with dope. Coil winding specifications are as follows:

L_1	6 turns, CT No. 18 Nylclad
L_2	10 turns, CT No. 18 Nylclad
L_3	3,500 turns, No. 36 Nylclad, silk

The inner lead of L_3 is grounded to any convenient point on the frame and the outer terminal is brought out by a well-insulated lead. Windings L_1 and L_2 must be correctly phased to obtain oscillation. Correct phasing is best established by trial and error. The base leads should be reversed if oscillation is not obtained on the first try.

This unit produces a 6-kv output (a-c) which can be combined with the unit on the right, as shown. The right-hand circuit is a time-base generator which can be used for a portable flasher for warning lights, as for construction hazards. They can also be used together.

Consider the time-base generator alone. Using a 6-volt supply, a 2N35, and a 2N176 power transistor, it is a relaxation oscillator. The network R_1C_1 is the time-constant circuit; R_2 controls the pulse duration.

With S_1 open, there is no output. When switch S_1 is closed, capacitor C_1 starts to charge, raising the voltage on the base of Q_1 positive with respect to the emitter. This charging rate depends on the time constant of R_1C_1. At some critical voltage, Q_1 begins to conduct, followed by Q_2. When Q_2 is driven into saturation, the full battery potential is across the load. Conduction of Q_2 causes capacitor C_1 to discharge until, at a low critical value, Q_1 is cut off, causing Q_2 to cut off also. This action is repetitive and establishes the time base.

The frequency of the circuit shown is approximately 90 pulses per minute. Other time rates can be obtained by using different values of R_1 and C_1. Frequency varies inversely with the value of R_1 or C_1. As the value of R_2 is increased, the pulse duration is also increased. In the circuit shown, the pulse duration is approximately 10 msec. If the value of R_2 were increased to 1,000 ohms, the duration would be approximately 50 msec.

Used as described above, the time-base generator can drive a 6-volt incandescent bulb for warning. Connecting A and B, as shown, allows the inverter to drive gas-tube lights.

Transistor d-c to d-c converters are used for equipment plate supplies because they are more efficient and reliable than other types of d-c power sources. Power outputs of up to 700 watts are possible. A rectifier filter added to an inverter makes a d-c to d-c converter.

Figure 4·23 shows a schematic of a transistor power supply designed to be added to, and used with, mobile radios. As shown, two triode power transistors operating as a flip-flop oscillator are used. This forms an oscillator which is a square-wave generator; the frequency is approximately 100 to 3,500 cps, which reduces the size and weight of transformers and filters.

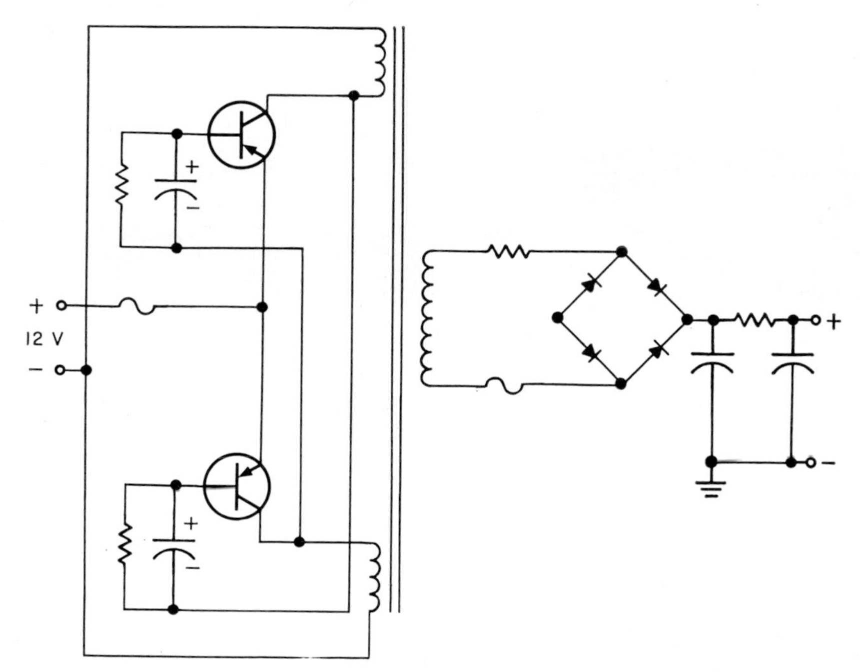

Fig. 4·23 D-c to d-c converter as used in a mobile-radio power supply.

The vibrator type of supply requires an iron-core transformer with materials and construction similar to a 60-cycle equipment; but this transformer, operating at 3,500 cps, becomes a small assembly weighing only a few ounces. Filtering and shielding can be reduced, because current switching in the transistors is electronic. Square waves are a source of harmonics into the r-f range, but the technique of suppressing these undesired components is not difficult.

Rectification of the stepped-up voltage appearing at the secondary of the transformer takes place in a bridge circuit using four silicon diodes. Ripple filtering is accomplished in a conventional RC network following the rectifiers. Output of the transistor supply shown is usually in the range of 200 volts at 100 ma, which is sufficient to power a 15-tube FM receiver. Overall efficiency of the supply is 70 to 75 per cent, which is approximately the efficiency of a vibrator supply employing a nonsynchronous vibrator and a selenium rectifier.

All components except the transistors usually are mounted on a small printed-circuit board. The transistors are mounted directly on the outside surface of the case, which acts as a heat sink. A protective bracket covers the transistors but still permits free air circulation around the heat sink to maintain the transistors within the recommended operating temperature.

There are some novel semiconductor switches such as the Raysistor (Fig. 4·24), which is a four-terminal electrooptical device. A voltage

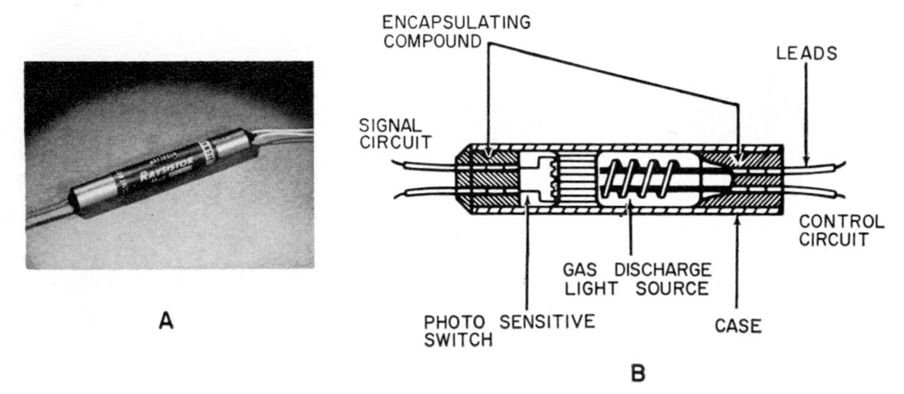

Fig. 4·24 (*A*) Raytheon Raysistor, which is an electrooptical relay; (*B*) construction of the Raysistor.

applied to two of the terminals controls the passage of a signal at the other two terminals. The control circuit consists of a light source which, when excited, lowers the resistance of a semiconductor in the signal circuit by a factor of about 1 million, thus allowing a signal (a-c or d-c) to pass. No electrical connection exists between the control and signal circuits.

The Raysistor controls signals through a low-noise resistive element with no switching transients or carriers. This pedestal-free analog switching provides an advantage not obtainable with semiconductor devices and electron tubes. It has no moving parts, is mechanically rugged, is nonmicrophonic, and is capable of 1 billion operations. Some applications are shown in Fig. 4·25, including an agc control, a suppressed-carrier modulator, and a relay. Agc varies the gain of the amplifier as a function of signal strength; the modulator cancels out the carrier and produces only sidebands, while the third circuit is the analog of a relay. Note that, if proper values are used, negative resistance can be obtained as shown.

4·4 Other switching devices Every system of industrial electronics requires some type of switching device. Electromechanical types and electronic switches are discussed above. They form two large classes of switching devices. Other more specialized devices covered in this section are mercury switches, thermal switches, and other specialized types.

Mercury Switches. Mercury switches are glass tubes into which stationary electrodes and a pool of loose mercury are hermetically sealed.

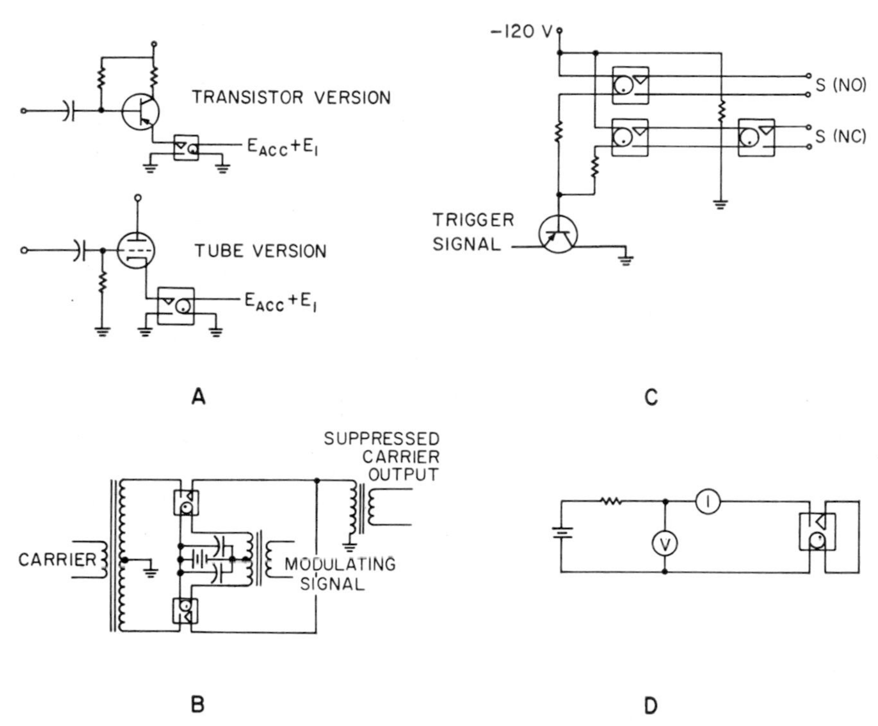

Fig. 4·25 (*A*) The Raysistor can provide a feedback path down to direct current and yet afford complete isolation of d-c levels. (*B*) Two Raysistors can form a suppressed-carrier modulator. Many related applications exist in which the Raysistor can be used as a chopper. (*C*) Three Raysistors can be combined to form an spdt relay. Various combinations can also be used for logic circuits in computer applications. (*D*) Raysistors with incandescent-type control circuits offer a negative resistance when the control and signal circuits are connected in series. Typical values are as follows: $V = 40$ volts; $1 = 9$ ma. $V = 17$ volts; $1 = 23$ ma.

Tilting the tube causes the mercury to flow in a direction to open or close a gap between the electrodes to make or break the contact. Because of the hermetic seal, dust, dirt, moisture, and corrosive gases cannot enter the tube. The switch is tilted so the mercury does not close the switch. Tilting in the other direction will cause the mercury to close the contacts.

Mercury switches can be arranged for fast or slow action as shown in Fig. 4·26. Device tilting action refers to the type of mechanism that drives the switch. For example, slow tilting action is provided by such devices as bimetal strips, thermal bellows, and floats. Fast tilting action is provided by foot pedals, rapid cams, freezer lids, etc., as in Fig. 4·26*B*. A slow-action type is shown in *A*. Some switches cannot be used for fast-acting mechanisms and others cannot be used on slow-moving devices.

For example, if a switch specified for fast action is used with slow tilting motion, a clean make or break will not result.

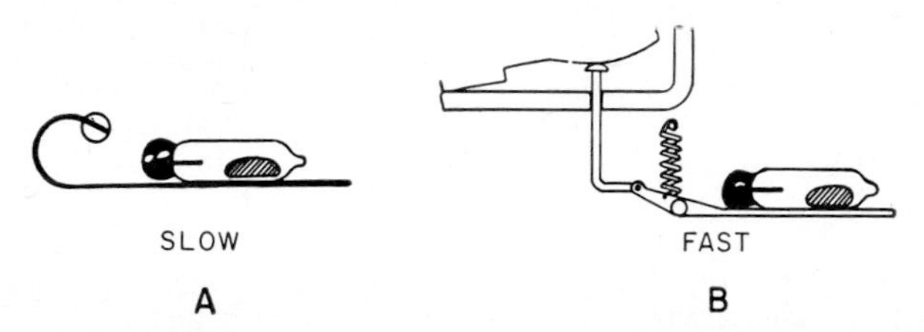

Fig. 4·26 (*A*) Slow mercury-switch action; (*B*) fast mercury-switch action.

Contact arrangements (Fig. 4·27) vary depending upon the application. Mercury switches are essentially single-pole devices. The most common are single-throw and depend on position for normally open and normally closed circuits. In two-circuit switches electrodes may be individually wired for opening or closing isolated circuits. A common lead, when provided, changes the switch to the conventional single-pole double-throw switching circuit.

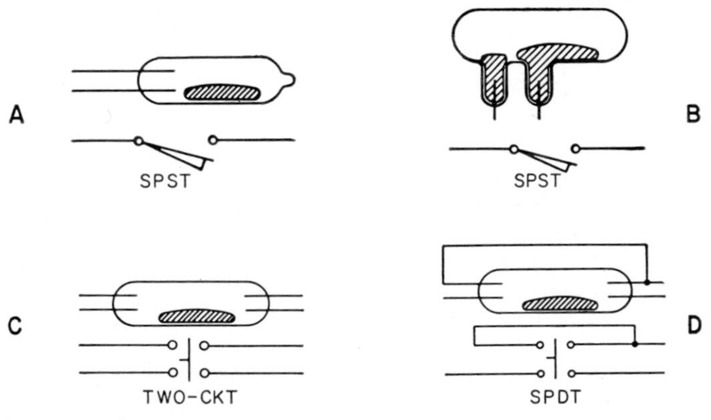

Fig. 4·27 (*A*) Single-pole single-throw mercury switch; (*B*) single-pole single-throw mercury switch with two mercury contacts; (*C*) two circuit switches making or breaking each of two circuits; (*D*) single-pole double-throw mercury switch. (*Minneapolis Honeywell*)

A is a simple single-pole single-throw mercury-to-metal switch. *B* is a mercury-to-mercury version of the same type for larger currents. *C* combines two single-pole switches. *D* is a single-pole double-throw mercury-to-metal contact.

The tilt and angle are important to the switch action. Figure 4·28 shows switch motion. In single-throw mercury switches differential angle is the motion in degrees from "just make" to "just break" of the circuit. In double-throw mercury switches differential angle is the motion in degrees from "just make" on one end to "just make" on the other end.

The differential angle differs with the type of switch. In selecting a switch for a particular application, choice of one that has a lower angle of operation than is actually needed provides a safety factor.

Also note that two switches having the same angle of operation may not have the same relationship to the horizontal. In design considerations, where space allows a greater movement than is necessary for operation, overtravel is desirable.

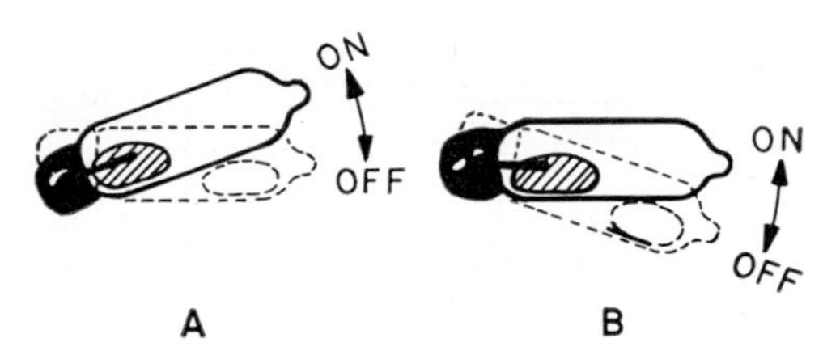

Fig. 4·28 Various mercury-switch tilt angles.

Most mercury switches offer design freedom with respect to overtravel. Overtravel after make or break is limited by the user's required lead-wire life and the electrode arrangement of the particular switch. Figure 4·29 compares two-contact and four-contact switches. They are shown together with the equivalent circuit of the switch.

The two-circuit switch operates like a double-pole switch with mechanically interlocked arms, because either circuit can be closed, but not both at the same time.

There is a wide variety of shapes and sizes available in mercury switches. Much modern high-speed production machinery does not receive maintenance until one of the component parts breaks down. Mercury switches can eliminate many of the switching troubles that cause prolonged interruptions of process operations and subsequent delay of production schedules.

There are over 1,000 different designs of these switches. The many sizes, kinds of glass, arrangement of electrodes, types of lead wires, and insulation can be varied to fill almost any need—from door interlock applications to the ultraprecise requirements of vertical gyros. Mercury switches potted in resilient material and enclosed in metal and nylon cases are available for locations subjected to shock and splashing chemicals.

Mercury switches are of three basic designs: (1) mercury-to-dry-metal contacts, (2) mercury-to-wet-metal contacts, and (3) mercury-to-mercury contacts.

Mercury-to-dry-metal contacts are used in applications with circuits ranging from tenths of an ampere to a few amperes and where economy is a factor. In this design, the mercury flows over and around the electrodes and fills the gap between them, thus completing the circuit. The

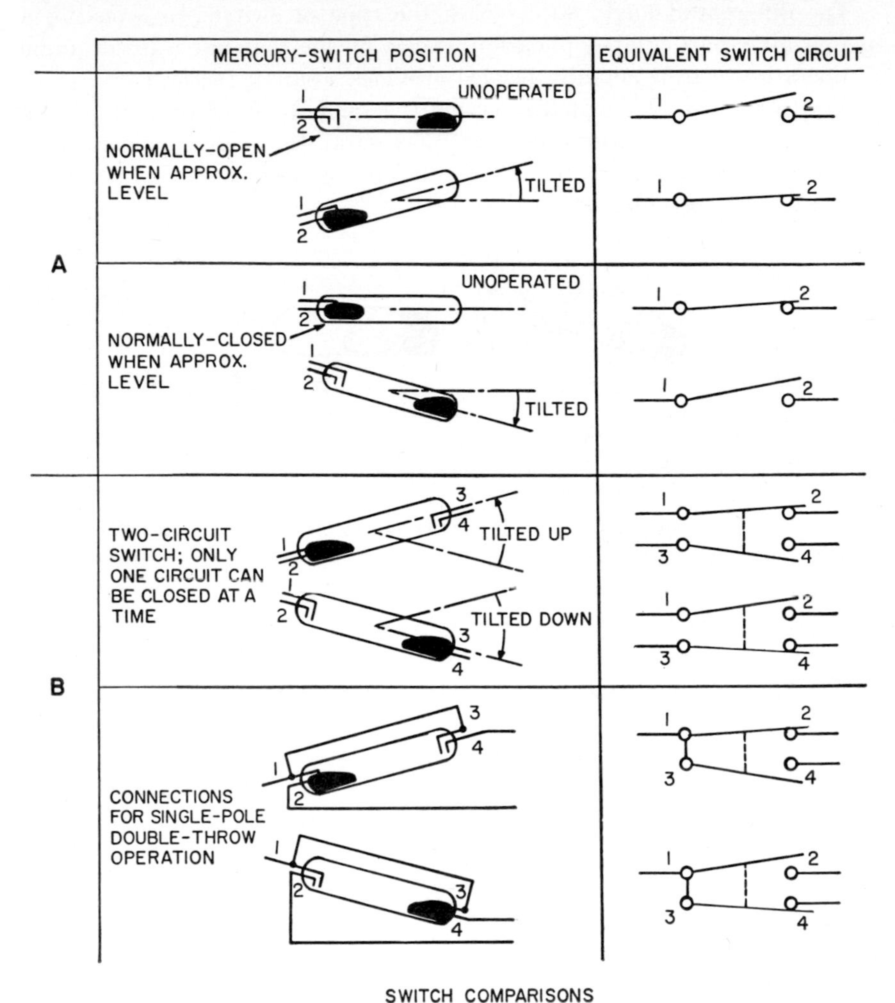

Fig. 4·29 (A) Two different types of two-contact mercury switches; (B) circuit arrangements available for four-contact switches. (*Minneapolis Honeywell*)

mercury breaks rapidly from the electrodes when the switch is tilted to "off," minimizing the possibility of arcing.

For circuits up to 10 amp at 115 volts a-c, the mercury-to-wet-metal contacts are generally used. In this type the electrodes have been specially processed in order to provide a snap-action contact and break.

For high-amperage heavy-duty loads, a mercury-to-mercury contact is employed. In this type the metal electrodes remain submerged in mercury and the switching action is accomplished through joining or dividing the two mercury pools. A ceramic or porcelain sleeve confines the electric arc, thus extending switch life.

Mounting of switches can be seen from Fig. 4·30, in which six types are shown. "Over center" mounting (1) improves the quick contact action

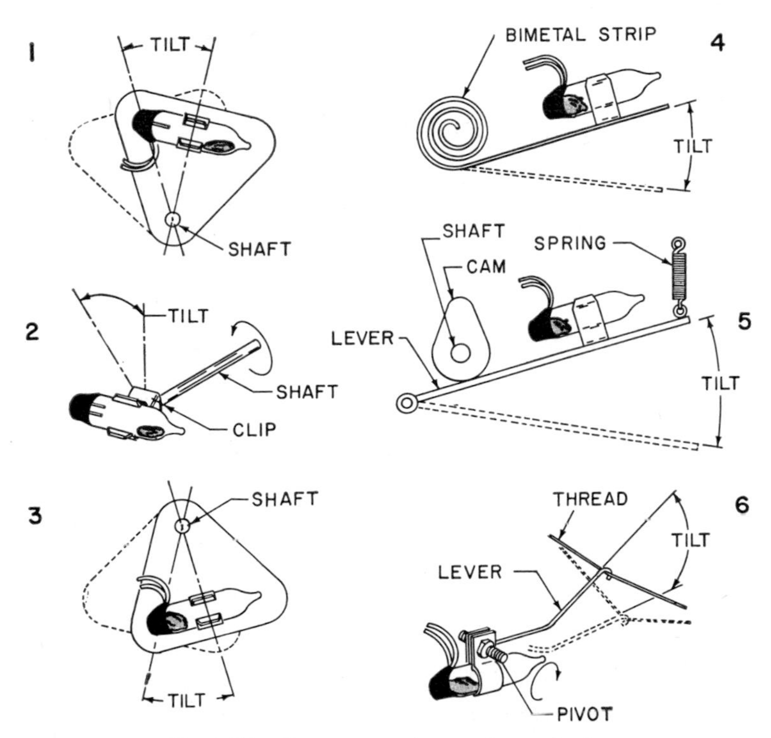

Fig. 4·30 Mercury-switch actuation methods.

of the mercury and is recommended when practical. If the support mechanism has a small free motion, an extra "over center" snap action can occur as the mercury will overbalance the device, picking up this free motion.

Where minimum operating space is a requirement, "on center" mounting results in a minimum displacement of each end of the tube as in 2.

If only minimum operating force is available, "below center" mounting is used as in 3. Lowest operating force can be secured by counterbalancing the switch and its support. The actuating mechanism and switch tube move while the mercury remains almost stationary.

For temperature control, the switch may be mounted on a bimetal strip, the angle varying with the temperature as in 4.

A mercury switch can be actuated by a cam-driven lever (5), or a pivoted actuator can be used with a mercury switch to detect a break in thread or wire tension as in 6. When the strand parts or slacks, the actuator arm drops and tilts the switch, interrupting current to the reel motor.

Thermal Switches. Large electronic devices such as industrial controls, transmitters (AM, FM, and TV), and digital computers use thermostats or thermal switches for temperature control.

A metal strip and two contacts, as shown in Fig. 4·31, make up the basic unit. Two unlike metals are bonded to form the strip; when current passes through the bimetallic strip, the metals are heated because of their resistance. A heated metal expands and, since these are two different metals, they expand unequally, which causes the strip to bend. The

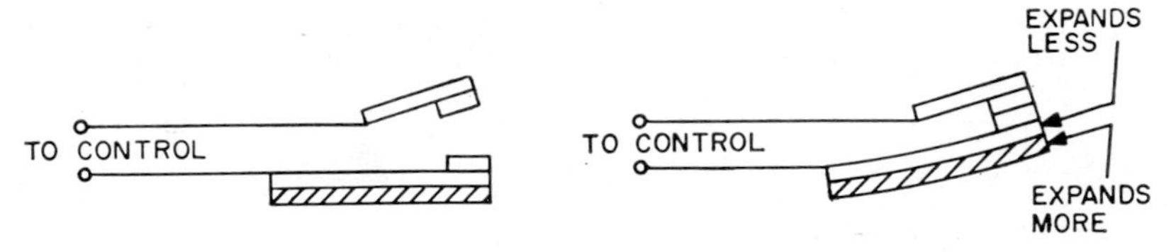

Fig. 4·31 Expansion of a bimetallic strip forming a thermostat.

opening and closing of the contacts is the control. In some thermostats expansion of the strip closes contacts rather than opening them.

There are many uses for thermal switches. A transmitting tube, for example, is cooled by a fan. Hot air rises to reach the thermostat. If, at a given point, the air is too hot, the thermostat strip expands, the contacts open, and the fan is turned on to cool the tube. In reverse, the cool air turns off the fan.

Or it is possible to control a heater and so the piezoelectric crystal in a precision oscillator. Here the contacts open when the proper temperature is reached; this turns the heater off. Again the natural-air cooling closes the contacts to start the heater.

This is the electric form of thermostat such as is found in an oil burner, where it turns off the ignition spark after the fire has started. In an automobile there is a nonelectric type; this opens when the water in the engine-block jacket is hot enough, which allows this water to circulate through the radiator and heater. Both types are used in electronics, but here we are concerned with the electric type.

A thermostat of any type is only an on-and-off or a go and no-go device. It is a switch which is either open or closed. Figure 4·32 illustrates this; *A* shows the thermostat position while *B* shows the result of the heating which is controlled by the thermostat. At the lowest or minimum temperature the contacts are closed and the strip is in normal (straight) position. But at time T_1 the current begins to flow through the contacts, the heater starts, and the temperature begins to rise.

Between T_1 and T_2 the contacts are closed and the heating continues. But at T_2 the bimetallic strip bends enough to break the contacts and the heating stops.

In this way the thermostat oscillates between maximum temperature (or contacts open) and minimum temperature (or contacts closed). Midway between these is the set point, so called because it is the point to which the thermostat is set.

But because of its thermal inertia, or the slow increase or decrease in temperature, every heated device lags behind the heater. When the thermostat in a transmitter signals that the tubes are too hot, it takes the fan a few minutes to cool the tubes off. In equipment design, allowances are made for this. The lag is illustrated by the exponential rise and decay of curve *B*.

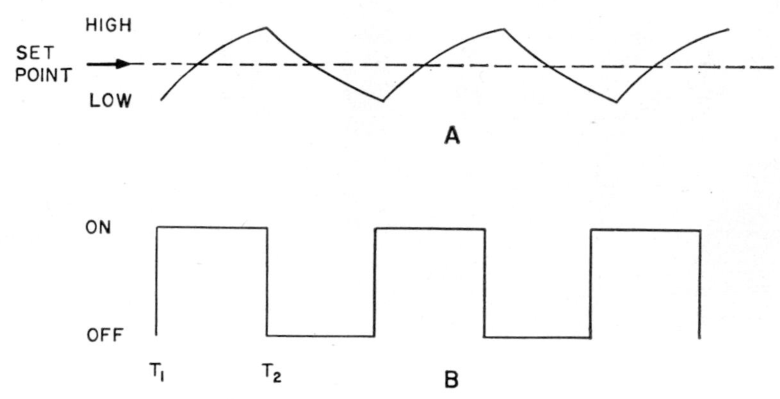

Fig. 4·32 (*A*) Operation of a thermostat above and below set point; (*B*) timing diagram showing heater being turned on and off.

A simple and inexpensive thermostat, of the type used for vaporizers, percolators, dryers, and other appliances, is shown in Fig. 4·33. It is designed for use where the heat generated by current flow through the bimetallic element is not objectionable. This type is quite like the tip-over switch and is mounted inside the base of the appliance. A pin or similar device (not supplied with switch) projecting through the appliance base presses against the lower contact spring to close the circuit. Immediately after the appliance is overturned or lifted from the floor, contact spring pressure opens the contacts to switch the appliance off. Stainless-steel contact springs have spot-welded silver contacts that last the life of the appliance.

Snap-acting units in which a bimetallic disk is used as the active element are shown in Fig. 4·34.

Typical applications include fire alarms, domestic appliances, television relay stations, communications equipment, refrigerator-freezer automatic defrosting protection, and motor protection. This type of thermostat can be made to either make or break an electric circuit in the event of a temperature change—either increase or decrease. Operating temperatures are from -20 to 300°F.

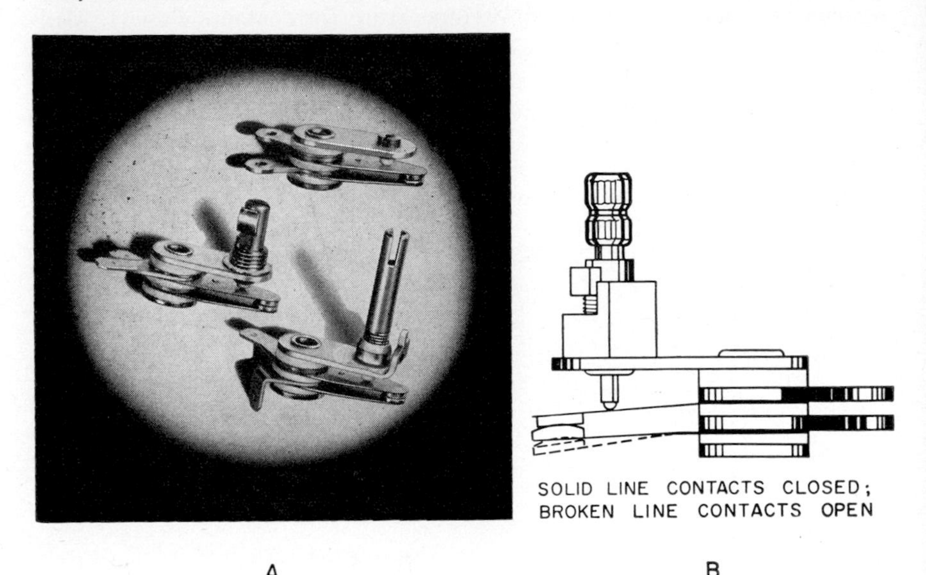

Fig. 4·33 (*A*) Stevens thermal relay or thermostat; (*B*) thermostat construction.

The schematic drawing in Fig. 4·34 shows the circuit breaking when temperature rises. Current flows from terminal 5 to terminal 6 on the silver side of a rigid nickel-silver contact disk (1). This disk is pressed firmly against the silver contacts (2) by a bimetallic disk (3). Since current follows along the path of least resistance, the relatively high re-

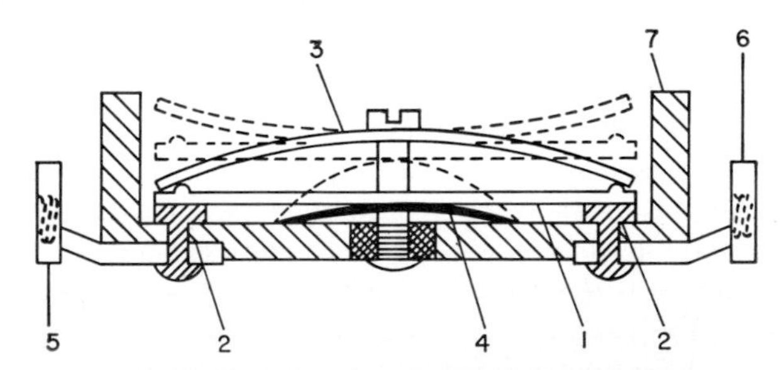

Fig. 4·34 Stevens snap-action thermal switch.

sistance of the bimetal eliminates large currents in the bimetal. When the bimetal reaches a predetermined temperature, it instantaneously snaps upward, releasing pressure on the contact disk which allows the return spring (4) to raise the contact disk off both contacts, thereby breaking the circuit with minimum arcing. As bimetal temperature then drops to a predetermined closing point, the bimetal again instantaneously snaps downward, applying pressure to the contact disk which forces it down onto the contacts, thus closing the circuit.

Crossbar Switches. Crossbar switches can be used in place of quite complex interconnections of relays. A three-dimensional arrangement of

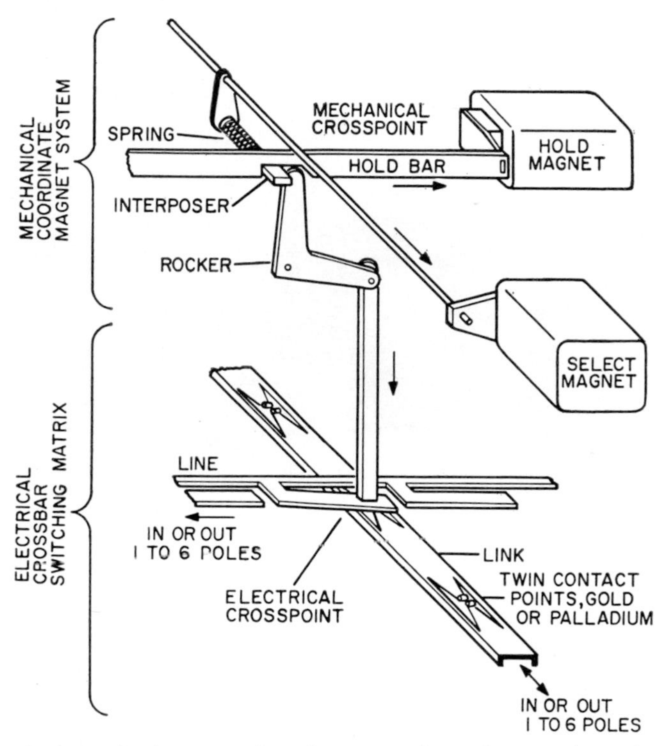

Fig. 4·35 A crosspoint in a coordinately operated crossbar switch made by James Cunningham, Son & Co., Inc., Rochester, New York.

electric and mechanical contacts makes up this switch. Mechanical connections are used to make the required electrical connections. These connections can be controlled in a variety of ways. In many applications the switch itself provides the logic and interwiring between the connected points.

A crossbar switch has a number of vertical relay units and a crossed series of horizontal bars actuated by selecting magnets. A holding magnet is used for each vertical unit.

A section of a crossbar switch is shown in Fig. 4·35. A mechanical cross point or junction is made by the mechanical coordinate magnet system. The select magnet operates before the hold magnet. Motion of the rocker arm brings the two contacts together at the electrical crosspoint.

This mechanical operation in turn makes electrical connection at the crosspoint, closing a circuit between one input and one output. In sequence, by similar operation the balance of the inputs are connected to outputs in any combination. Crossbar switches may be controlled by push buttons, dials, or punched cards; or they may be self-stepping.

Rotary Stepping Switches. Many control functions in industrial operations are accomplished by small, high-speed, multicontact, single-motion (rotary), magnet-driven ratchet stepping devices (switches) in response to electric contact pulses.

The switch mechanism consists of one or more wiping springs fixed to a shaft which is moved by a pawl and ratchet through the operation

Fig. 4·36 Automatic Electric Company rotary stepping switch.

of an electromagnet in response to momentary pulses of current. Each pulse causes the pawl to engage the ratchet; this moves the wipers, to which a circuit has been connected, one step ahead into contact with stationary terminals to which a selected circuit has been wired. These stationary terminals are assembled as a unit to form a semicircular bank and may contain up to 250 or more individual contacts. Thus this device permits the transfer of a common circuit to many secondary circuits.

It is usually necessary to extend circuits through the wipers and bank contacts over more than one conductor in order to perform the signaling, control, or transmission functions desired. Therefore, each terminal position consists of several rows or levels of bank contacts which are simultaneously contacted by separate wipers when the selection has been completed.

Rotary stepping switches such as the one shown in Fig. 4·36 are used in automatic telephone systems as linefinders, connectors, and allotter switches, as well as for trunk distribution, ring-code selection, registering the number dialed, sequence control, etc. In industrial signaling and control they are used in far greater numbers for applications requiring selection, distribution, totalizing, and counting. Typical applications include counting and routing of materials on conveyors, successive operations by machine tools, remote control of substations, control of radio transmitters and monitoring of programs, control of airport lighting and radio beacons, remote control of motor speed by varying field resistance, remote control of current-supply equipment for arc welding, automatic control of tear-gas release in burglar protection systems, remote control of aircraft signaling systems, and number selection for mobile radio units.

The operation of rotary stepping switches may be either manual, remote-controlled, or controlled by an automatic self-cycling or self-interrupted circuit. For automatic operation, the interrupter springs are

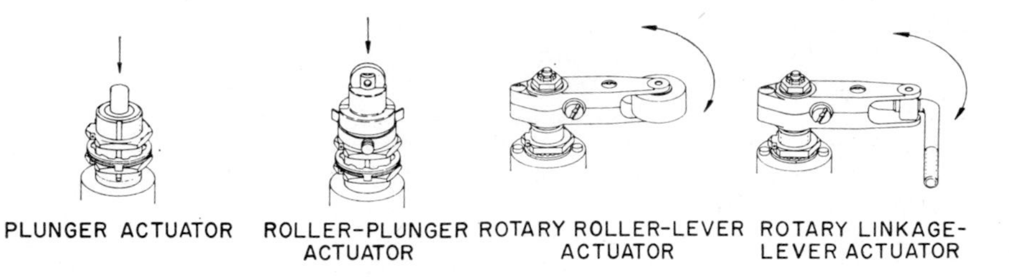

Fig. 4·37 (*A*) Plunger actuator; (*B*) roller-plunger actuator; (*C*) rotary roller-lever actuator; (*D*) rotary linkage-lever actuator. (*Microswitch*)

connected into the operating circuit and the switch acts as a nonnumerical or self-cycling device. It will automatically search its bank contacts for a special circuit condition such as the presence of ground or battery, or the absence of ground. For manual operation, control pulses may be transmitted to the magnet coil from a dial, push button, or other remote-control device.

Precision Switches. A precision switch is one in which the output is a direct function of the position of the actuator. Precision switches are widely used as limit switches to control a circuit when a moving member has reached a preset point and, in somewhat similar fashion, as safety switches to detect unsafe travel of such a moving member.

The operation of a precision switch is based upon snap-acting contacts actuated by the motion of a plunger. A typical precision switch is shown in Fig. 4·37. *A* is for use where the motion of the actuating force is in line with the plunger motion and overtravel is controlled. An ice-scraper ring on the actuator shaft prevents jamming or binding. Bushing mount provides for adjustment of the operating point of the switch. *B* is for actuation by cams and slides with a rise of less than 30°. The corrosion-resistant steel roller may be locked in increments of 45°. An ice-scraper ring on the actuator shaft prevents jamming or binding. The operating point is adjustable. *C* is for actuation by cams or slides with a rise of more than 30°. Actuation is in one direction—either clockwise or counterclockwise—with spring return to the neutral unoperated position. The roller-lever arm is adjustable to any position through 360°. *D* is for positive-drive actuation and release under most adverse environmental conditions of ice and dirt, with non-spring return. Actuation and release are controlled by an actuating device. The linkage-lever arm is adjustable to any position through 360°.

REVIEW QUESTIONS

4·1 How can a vacuum tube act as a switch?
4·2 Explain how a multivibrator is the basis for an electronic switch.
4·3 List the basic types of switches and tell how each works.
4·4 What is the purpose of R_2 in Fig. 4·2*A*?
4·5 What are the three protective devices shown in Fig. 4·2?
4·6 How does the grid relay shown in Fig. 4·2 protect the tube?
4·7 How would the relay in Question 4·6 act if grid excitation were reduced to zero?
4·8 Why is an expanded scale used on some meter relays?
4·9 What is the advantage of a multimeter using meter relays?
4·10 In what way is a relay like a thyratron?
4·11 Why are twin relay contacts sometimes used?
4·12 What are the following relay contacts: make, break, make; break, break; and make, make, break?
4·13 If *A* and *B* are on (down) and *C* is up in Fig. 4·8, which points are alive?
4·14 What conditions of *A*, *B*, and *C* in Fig. 4·18 are needed to produce a potential between 1 and 9?
4·15 What is the purpose of *A* in Fig. 4·9?
4·16 In what type of circuit would a simple latching relay have its greatest application?
4·17 How can a stepping relay be used as a counter?
4·18 What is a meter relay? How is it constructed?
4·19 What is the difference between the signal coil and the locking coil?
4·20 What is the function of the cam in a resetting meter relay?
4·21 Explain how a double-limit meter relay can be used.
4·22 What is the function of the Hall-effect device in a wattmeter?
4·23 Where would a very-high-sensitivity relay be used?
4·24 How does an electronic switch operate?
4·25 What is a d-c to d-c converter?
4·26 How is it used?
4·27 What type of rectifier circuit is shown in Fig. 4·20?
4·28 Agc voltage increases with an increasing signal. Explain how, in Fig. 4·25, this agc voltage changes the gain of either stage.
4·29 How is the carrier suppressed in *B* of Fig. 4·25?
4·30 Explain how the Raysistor can be a negative resistance.
4·31 What are the advantages of mercury switches?
4·32 To what is a mercury position sensitive?
4·33 How can a mercury switch act as an spdt switch?
4·34 What is the thermal-relay principle?
4·35 What is a bimetallic strip?
4·36 What is thermal lag?
4·37 How does a snap-action thermostat operate?

CHAPTER

5

Control Systems

Perhaps most of industrial electronics is concerned with control systems, but this chapter has four specific areas of interest. These are capacitance-operated devices, timers, temperature controls, and servomechanisms. These discussions are based largely upon the material in the first four chapters.

5·1 Capacitance controls Capacitance-operated relays are controls with many industrial applications. They can check or read the level of contents in a container, operate warning signals for overflow of containers, turn off power in a machine to prevent accidents (as when a punch-press operator has his hands in a dangerous position), or sense the locations of objects. The heart of the device is a relay whose operating point is indicated and triggered by a capacitance-sensitive device.

Such a system can be used to detect a change of 10 per cent or more in volume of material flowing in a chute, and the sensing element does not contact the material sensed. One short section of chute made of insulating material with the sensing element placed around it will do the job. Such a control can also detect a change of moisture content in material such as grain, flour, or gypsum. The volume of material must be constant, as all variables except moisture must be controlled in order to detect variation in moisture only. The distance from the material to the sensing element must be constant, and only the moisture content is measured.

Detection of change of flowing liquid is also possible. If, for example, there is gasoline flowing through pipe and then kerosene is put through, the unit detects when the gas ends and the kerosene starts because the dielectric constant of these two liquids is different.

In this application an insulative section of pipe is used with the sensing element around it, so that the pipe actually is not entered or broken, and the liquid sensed is not contacted by any sensing probe. Liquid level may be checked by inserting in the tank wall plastic-covered probes at desired levels of detection. The probes are watertight and may conform to the inside contour of the tank. The connection is made on an outside wall of the tank.

The dielectric character may change when a new or foreign substance is introduced into a liquid. The testing procedure is to "tune" control for a sample of the desired character, then replace it, in equal volume, with the sample and watch for a change in the control meter reading. More or less volume of the same solution, in a given container, will also be recognizable. It is important that containers themselves be uniform and that, when sensed, they always be at the same distance from the sensing element. Examples of this use are the detection of dirt in oil, water in oil, or salt in water.

In another application, a continuously running strip of material is marked at given intervals; and, later in the process, the marked points are to be detected. Small pieces of adhesive-backed aluminum foil are placed on the material at desired mark points for splices. Aluminum pieces are later detected, even when covered by other materials, and the spliced bits removed.

Relay Circuits. The principle of capacitance-operated relays is quite simple and is easy to understand. There is an oscillator operating at some convenient low radio frequency.

A relay is used in series with its plate circuit. The "sensor" or sensing probe is connected to the oscillator tank. Normally the tube oscillates and the current through the relay is not enough to close the contacts. When an object is brought near the probe, the oscillator is loaded down and it stops oscillating. This increases the plate current and closes the relay.

A schematic is shown in Fig. 5·1. In normal use the circuit oscillates and develops a small self-bias across R. C_1 is a small variable capacitor (padder) adjustable from 160 to 500 $\mu\mu$f. It is adjusted not for strong oscillations but so the tube just oscillates. Point A is the sensing probe, and when it is touched by hand the additional capacitance loads the circuit and stops oscillation. The lack of bias caused by the absence of oscillation causes a rapid rise in plate current, which closes contacts 2 and 3. In normal use 1 and 2 are closed.

Point B is the plate supply. If B is direct current, the circuit will stop oscillating for near contact as above; but when the object or hand is moved away from the sensing probe, the circuit will not start oscillating again. An a-c supply can be used at B to permit the tube to be self-starting. With alternating current supplied, the 6V6 is also a self-rectifier. A resistor, of about 1,000 ohms, in the cathode circuit will permit this oscillator to be self-starting on direct current.

More complex devices, based upon these principles, are available commercially. One is for industrial safety and involves an invisible radio-like field or screen of adjustable proportions set up around the dangerous work area. A person's hand—in fact, any part of his body—or even a tool entering the restricted field instantly actuates a relay. Connected to the clutch and brake, the control has the same effect as depressing an emergency stop button, bringing the machine to an immediate stop. Operation will not resume while the person or tool remains in the danger area, and because the field is r-f there is no sensation or harmful effect.

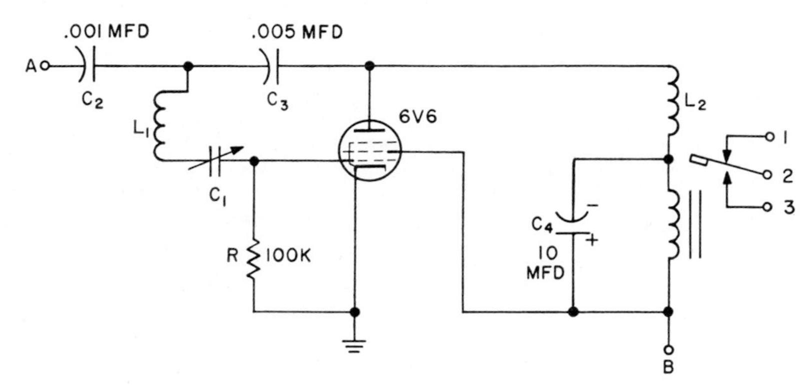

Fig. 5·1 Capacitance relay circuit with input A and relay output.

A field of radio waves surrounding the opening, as in a punch press, makes it impossible for the operator to activate the downstroke if his hands (or any part of his body) are in danger. There are several possible arrangements.

Figure 5·2 shows this capacitance-operated relay. It has a plus-minus feature where, after proper setup, any change of capacitance (added or subtracted) at the sensing element causes the control system to react. In normal operation the plus control relay is energized, and the minus de-energized. Any change of conditions or failure of any portion of the circuitry will cause the master relay to drop out. This stops the machine and protects the machine operator.

The unit controls machines by a shaped field of radio waves. The oscillator sets up protective radio waves in loops. To permit stock feeding automatically, a grounded shield can be placed in position to deflect waves from one loop, leaving an open area in the wavefield that is big enough for stock to pass through, but too small for hands.

The range and sensitivity of the radio field is adjustable by controls in the locked cabinet. The control is fail-safe under practical working conditions. In such case of failure a replacement chassis can be substituted in minutes, avoiding downtime while a new part is being placed in the regular unit.

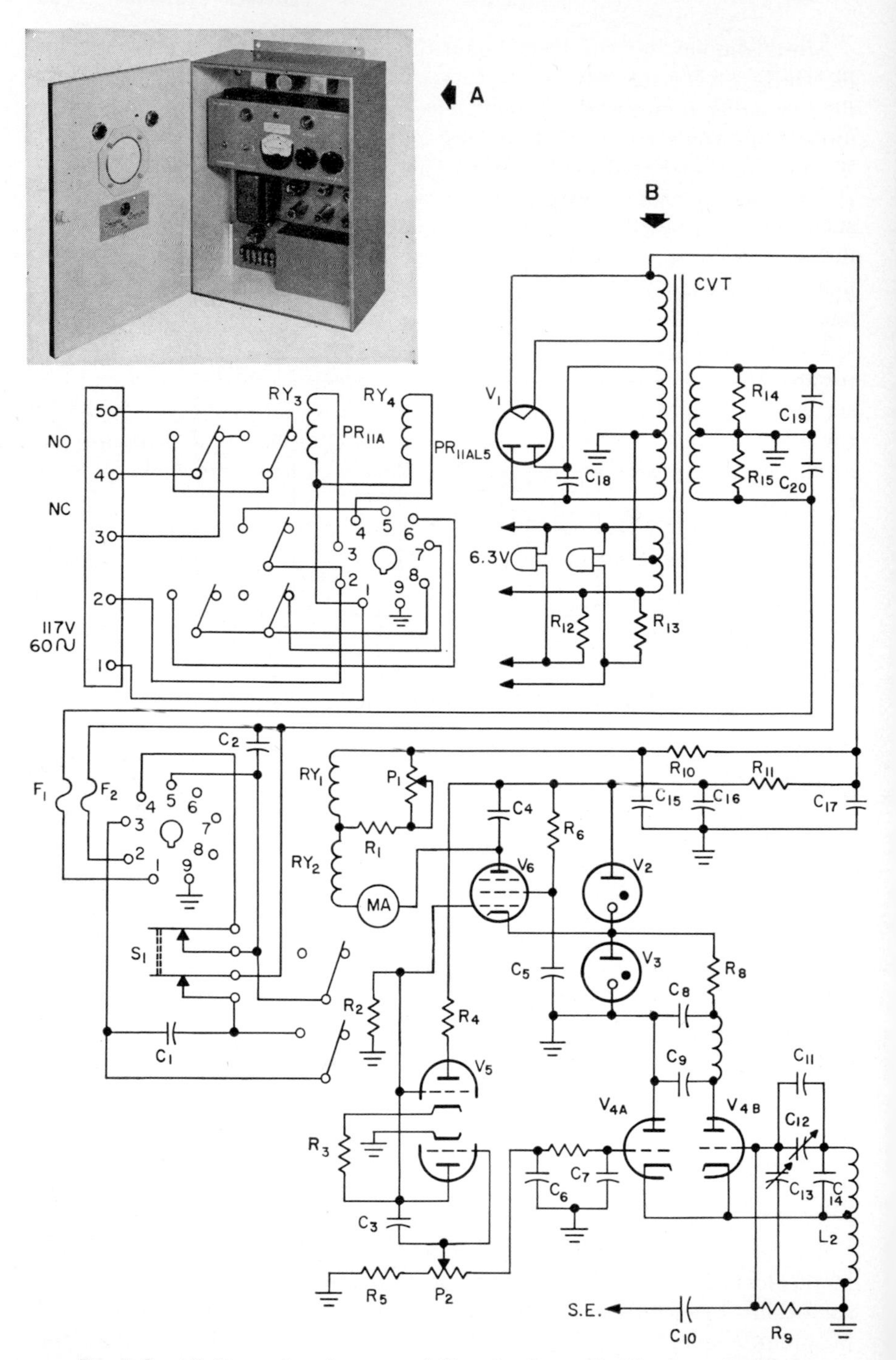

Fig. 5·2 (*A*) Electronic safety control (*Security Controls*); (*B*) schematic diagram.

The adjustments for range, sensitivity, and even substantial changes in sensing-guard size and form are easy. The only frequently used adjustment, for changes in operations, is tuning. This tuning control (C_{12}) adjusts the oscillator which feeds the protective r-f field, as shown in the schematic diagram of Fig. 5·2*B*.

This control gives dependable protection under difficult circumstances such as wide variations in power-line voltage. A unique feature of this device is that the predetermined range and the stability of the capacitance field is unaffected materially by wide variations in input voltage. A voltage regulator system includes V_2 and V_3.

Increased production is possible by using a cam switch to cut out the protective screen on the upstroke of the machine and allowing the insertion or removal of work during this portion of the press cycle. Thus, the operator has more time to place his work during the normal machine cycle, while still enjoying full protection on the dangerous, or downstroke, cycle phase.

V_{4A} is a low-frequency r-f oscillator with a loop antenna, from C_{10}, acting as the sensing element. Part of this r-f voltage is rectified by V_{4B} in the grid circuit. This voltage is fed to amplifiers V_5 and V_6, which control the relays (1 and 2). The relay action of 3 and 4 stops the controlled machine. Relays 3 and 4 are controlled by relays 1 and 2.

A modified device like that described above, available for other industrial use, is an electronic capacitance control designed to detect the correct position or size of metallic objects without contacting them.

The control is set up to any given constant, at a given time interval, and then recognizes deviations from the constant. The unit not only recognizes an error but also indicates how the sensed object is wrong—either too far from or too near to a predetermined position or setting. The control then either stops the machine, rejects the object, or actuates correcting mechanisms.

The sensing element contains no moving parts and does not touch the sensed object. It is surrounded by protective insulator material and is unaffected by vibration, dirt, and abuse. One control unit may use up to three sensing elements. The element will detect objects up to 2 in. away and can recognize deviations of 1/16 in. in nearby objects. It senses either ferrous or nonferrous objects. The control box may be mounted 15 ft away from the sensing element, a rugged coaxial cable connecting them. The sensitivity of the sensing elements is changeable by simple meter and distance control adjustments. The unit can be used to check proper positioning, check proper size and shape, detect moisture change, detect ionization change in liquids, etc.

The sensing element or probe is made of any metal, in any form or size conforming to the product detected (long, short, square, round, etc.) and can generally be easily fabricated. Any insulating material may be used as a covering to protect the probe against dirt, heat, etc. The field is shaped by the form and placement of the element. It can be expanded or contracted by a distance-adjustment knob and meter in the control cabinet. A coaxial cable connects the element to the control.

Capacitance Sensor Controls. Capacitance sensors are useful and sensitive devices for electronic control.

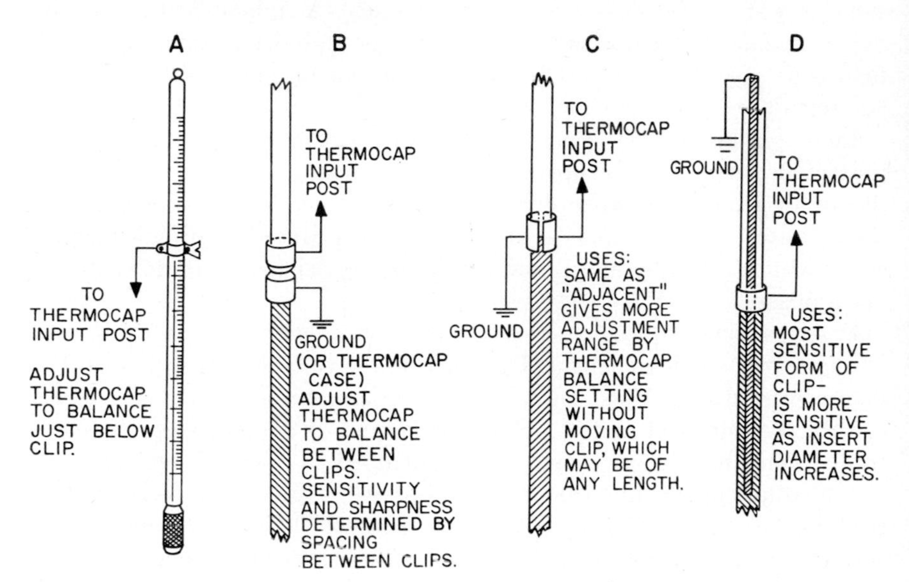

Fig. 5·3 (A) Capacitance control using a simple clip for thermometers or manometers; (B) adjacent clip for oil or organic liquids; (C) split clip for greater range adjustment; (D) concentric clip for metal rod inserted in tube.

Capacitance relays can also be used to monitor and control temperature. The sensing probe is a clip attached to the glass thermometer bulb. Variations in the height of the column cause changes in the relay output of the unit as shown in Fig. 5·3.

In *A* a simple clip is used with circuit-balance adjustment so the column of water or mercury is just below the clip. Any change in the column height will vary the capacitance fed to the control circuit. For oil or organic liquids in the glass tube a two-part clip (*B* or *C*) is used. More sensitivity can be obtained, as in *D*, by a metallic insert in the glass tube.

Capacitance controls can also be used to sense liquid levels as shown in Fig. 5·4. The level of the liquid is to remain constant in this tank as various amounts are withdrawn. A valve feeding a supply tank is under control of the capacitance relay. As the liquid rises, the capacitance increases. At a point within $\frac{1}{32}$ in. from the antenna (sensor) plate, the relay contacts open and the solenoid valve closes, stopping the flow. As the level drops, the relay contacts close and the container is filled through the valve. A timer is used which opens the valve for 2 sec and closes it for 2 sec until the proper level is reached. This prevents overflow.

An ice-accumulation control is operated by the difference in resistance between water and ice. Two probes are attached to one of the freezer coils, or to any freezer surface. The low-limit probe is usually set $1\frac{1}{4}$ in. from the pipe or surface and the high-limit probe $1\frac{1}{2}$ in.

When the compressor operates, ice forms on the pipes or the surface; this continues until the ice reaches the high-limit probe, when the control instantly shuts off the compressor. It remains off until the ice coating

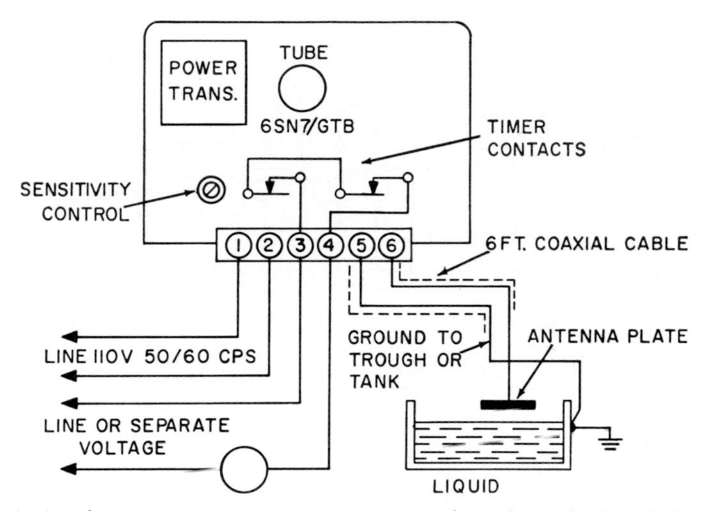

Fig. 5·4 Level sensing using a capacitor antenna plate above the liquid. (*Lumenite*)

melts enough to free the low-limit probe, when the control again starts the compressor.

Thus the ice on the coils or surface is constantly kept at an average of $1\frac{1}{4}$ to $1\frac{1}{2}$ in. This is sufficient to keep the cooling water always at an exact temperature with a minimum of compressor operation.

A boiler water control is shown in Fig. 5·5. There are three probes. If the water is used up so that the level drops below the medium-length electrode, the pump motors are started or valves opened and filling continues until the water again touches the high-level electrode, when it is stopped.

Should the water supply fail, for any reason, and the water level continue to drop until it leaves the low-level electrode, the relays controlling the electric controls of the firing (stoker, oil burner, or gas burner) would snap into action and cut off the fire. In addition, an alarm—either light or bell—will be put into operation, warning that the water supply has failed and the fire is out.

The burner will not start again until water reaches the high-level electrode or probe. Thus the automatic electronic safety watchman protects the boiler and prevents an explosion or burning out of the boiler or kettle.

The same probe technique can be used to control the filling of a tank truck so that the liquid stops at the desired level, and spillage is avoided.

In window advertising, a unit actuates a window display when a hand is placed against the disk of foil.

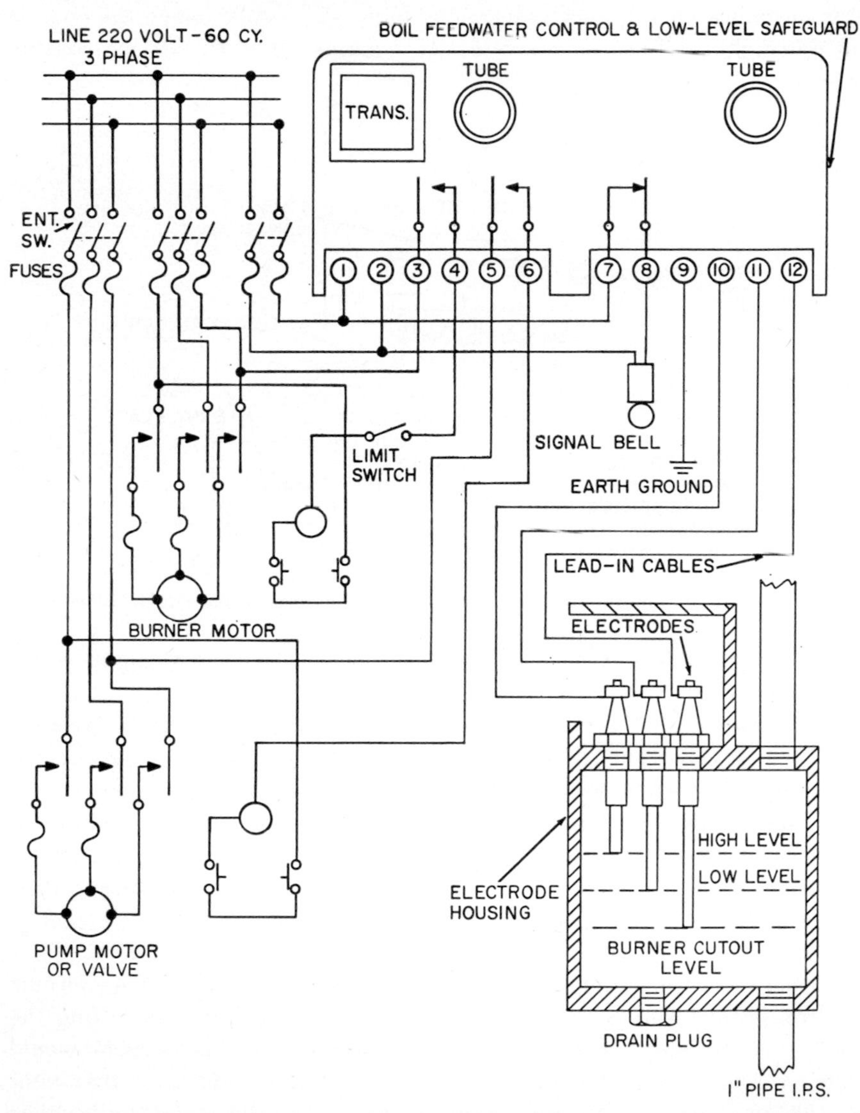

Fig. 5·5 Lumenite boiler water control where feed water pump motor stops at high-water level and starts at low-water level.

A capacitance relay can, of course, be used for protection against theft or burglary as shown in Fig. 5·6. The installation is quite direct.

1. Make external electrical connections as shown in the wiring diagram.

2. Be sure that the antenna, which may be a safe, metal plate, filing cabinet, metal chair, etc., is not grounded. The antenna should be mounted on insulators.

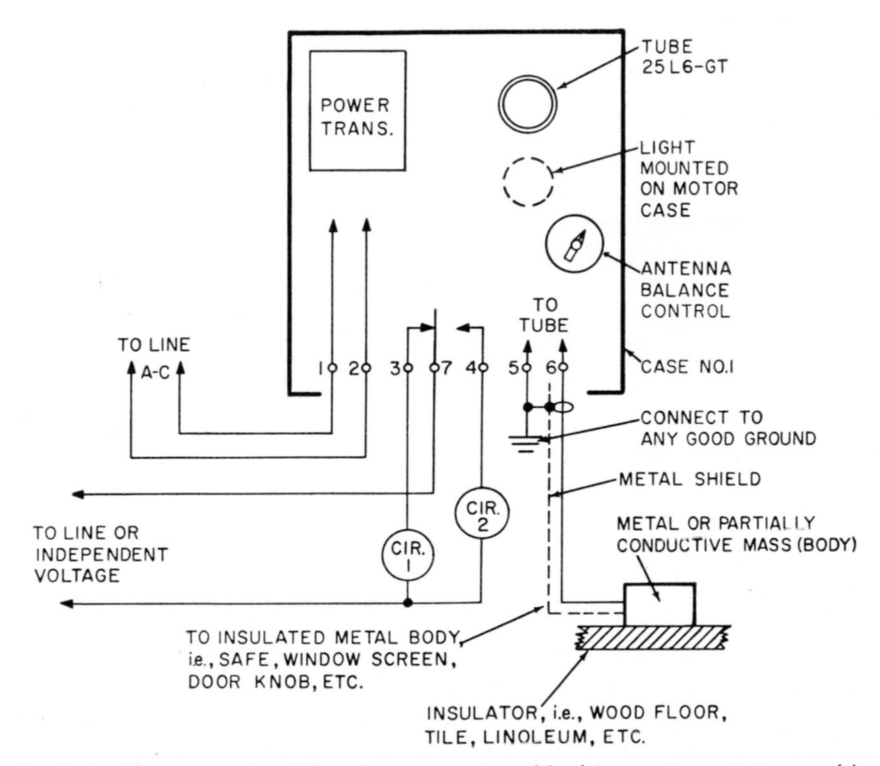

Fig. 5·6 Alarm operation: When the antenna (metal body) capacitance is increased by various methods (touched by hand, near a large metal mass, grounded by water, etc.), circuit 1 opens and circuit 2 closes. When the hand is removed or the large metal mass is taken away from the vicinity of the antenna, circuit 1 recloses and circuit 2 reopens.

3. Turn the sensitivity control knob right or left until the signal light on the case cover just turns off.

4. The electronic capacitance control is now balanced.

The sequence of operations is as follows:

1. When the antenna (metal body) capacitance is increased by various methods—i.e., the antenna is touched by hand, is near a large metal mass, is grounded by water, etc.—circuit 1 opens and circuit 2 closes.

2. When the hand is removed, or the large metal mass is taken away from the vicinity of the antenna, circuit 1 recloses and circuit 2 reopens.

5·2 Timers Timers are used in many types of industrial electronic controls. Application of the plate-supply voltage in a large power tube must be delayed until the heater is hot and the cathode is emitting electrons, or the tube will be damaged. Rectifiers for heavy-current industrial uses need time delays for the tubes to form the mercury vapor which conducts current.

Here are some examples of the use of time-delay circuits. In an electronic gluer using induction heating, delays are required to start the tubes in operation, to allow heater current before plate voltage, and to

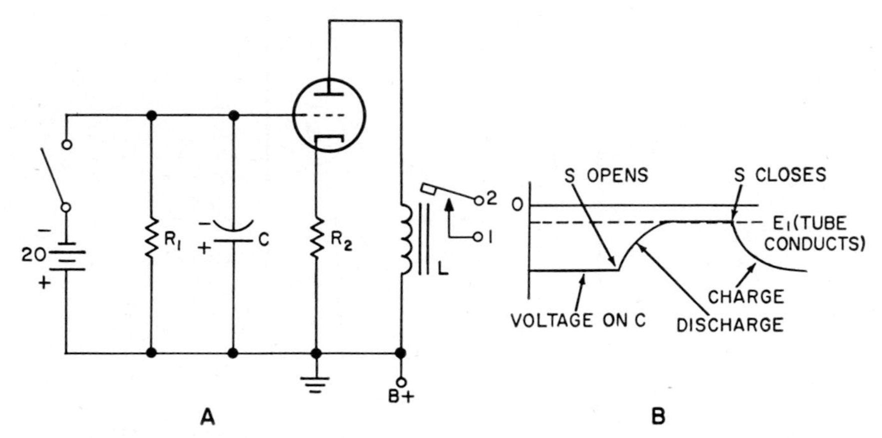

Fig. 5·7 (*A*) Electronic timing by capacitor charge; (*B*) charging curve.

control cooling. Another application of a type of timing circuit is in precision welding where the controls adjust the amount of welding energy, as measured in watt-seconds.

For these and other industrial applications there are two general uses for timer circuits and devices. A timer may delay the input from the circuit under control, as with a power amplifier where heater current is ahead of plate voltage; or it may be an internal device which adjusts the length of time a circuit is on.

Electronic Timer Circuits. Both electronic timer circuits and mechanical clock-driven devices may be used for timing.

RC circuits may be used to control amplifiers where the relay is in the plate circuit as shown in Fig. 5·7*A*. With switch *S* closed, the *C* bias of 20 volts is enough to cut the tube off. There is no current through relay *L*, and contacts 1 and 2 are open. Opening the switch leaves *C* charged to 20 volts, as shown, and the tube still does not conduct current. *C* discharges through R_1 and the grid voltage is reduced until it is small enough for the tube to conduct. This may be seen from Fig. 5·7*B*, where grid voltage is plotted. It remains at -20 volts until *S* is opened. As the capacitor discharges, the grid voltage becomes less negative until the point of conduction is reached. This is several volts negative, depending on the plate supply and the tube. When the plate current flows, the relay closes 1 and 2. The combined action of the cathode resistor R_2 and the remaining grid voltage limits the plate current.

The basic timing circuit is the RC network. Its action is familiar to all electronics technicians. When a voltage E is applied to C and R in series, the capacitor charges. At the first instant the voltage is applied, all the voltage is across R; this drop is given by curve B in Fig. 5·8. As

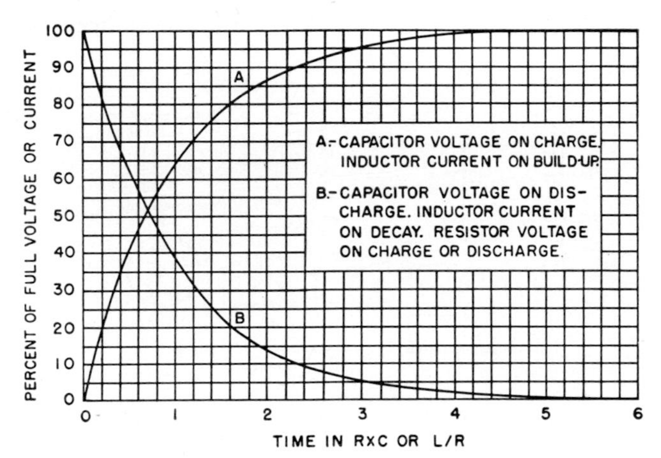

Fig. 5·8 Timing curve for resistance-capacitance circuit.

the capacitor-charge voltage increases, it follows curve A. The time constant then is the product RC, in seconds, where R is in ohms and C is in farads. After one time constant the capacitor has 63.3 per cent of final charge, and after $5RC$ sec it has almost 100 per cent of full charge. Some examples are given in Table 5·1.

Table 5·1

R	C, μf	RC	$5RC$
100 kilohms	5	0.5	2.5
1 megohm	10	10	50
2.5 megohms	2	5	25

The RC network has many different applications. A circuit with some modifications is shown in Fig. 5·9. Here contacts 1 and 2 are to be closed to start and to remain closed until switch S has been thrown from A to B. The relay is to keep open until the switch has been in position B for 4 sec.

With switch S in position A, as shown, the capacitor will charge up to the full applied B+ voltage, which is 50 volts. There will be current flow from cathode to grid and from cathode to plate which holds the relay closed as shown. Fifty volts across R_1 and C in parallel makes the grid and cathode potentials equal.

R_3, from the plate supply to ground through R_2, is adjusted so that there is +10 volts at B. When the switch is thrown to B, the 50 volts across C and the 10 volts from B to ground are in series, resulting in a

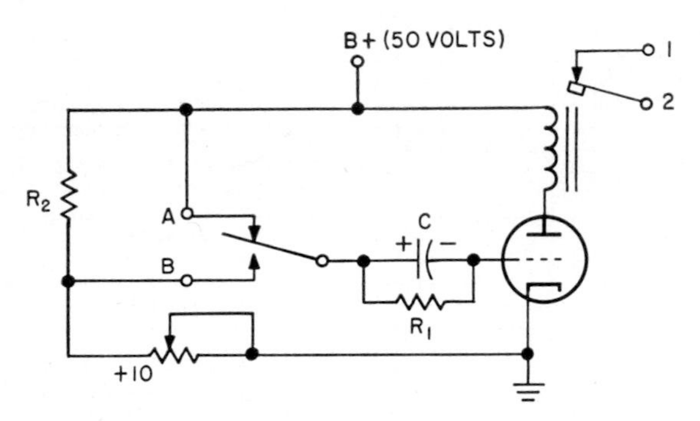

Fig. 5·9 *RC timing circuit using grid-leak timing.*

negative 40 volts at the grid. This cuts off the tube, which opens relay contacts 1 and 2. They will stay open until the charge on C leaks off through R_1. At a point where the grid is just slightly negative, the tube will conduct, thus closing the relay after 4 sec, until the circuit action repeats.

There is a limit to the time delays which can be obtained using a simple RC network but, together with an amplifier tube, the total time delay may be increased. In the circuit shown in Fig. 5·10, for example, the time delay of R_1, C, is multiplied by the gain of the tube. When the tube is placed in operation, there is a plate current which is small because C is not charged and the grid is at zero potential. Plate current is through relay L and R_3, which is about 50,000 ohms. The drop across R_3 is perhaps 10 volts or less, and C charges up to this value through R_1, which is 2 megohms. Because the grid voltage is equal to the charge across C, the potential on the grid increases and this increases the plate current.

The larger plate current causes a greater drop across R_3 with a higher voltage for C to charge to, and this regenerative action continues until the plate current is limited by the cathode resistance R_2, which is about 300 to 1,000 ohms. Thus the time constant for this circuit is greater than the product of R_1 and C. If R_1 is 2 megohms, C is 4 μf, and the gain of the tube is 20, the total time delay is 160 sec. It takes this long for the tube to reach its steady-state current and for the relay to pull in. At this current, contacts 1 and 3 are broken and 1 and 2 are made. The make of 1 and 2 permits the external circuit to function.

When the plate current is opened, by interrupting the plate supply, C discharges through the cathode-to-grid path and is then ready for the next cycle.

Pulse techniques are also used in other time or counter circuits. They are used, usually, where several different time intervals are required and each is needed to great precision.

Timing circuits can use the thyratron discussed in Chap. 2.

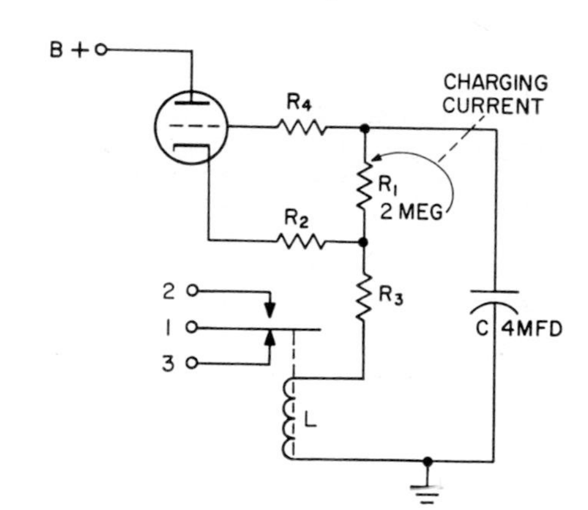

Fig. 5·10 Amplified time delay by grid-cathode charging current.

The sweep circuit is, of course, used for producing a time base in the electrostatically deflected cathode-ray oscilloscopes. A typical circuit for the 884 tube is shown in Fig. 5·11. The low range is from 20 to 60 cps using C_2; the highest range is from 3,600 to 11,400 cps using C_8. Variations within a range are obtained by changing R_6. To provide accurate synchronism, signals arc introduced into the grid circuit as triggers. Cathode bias is the same here as negative grid bias. Each of these synchronizing pulses causes the tube to fire (Fig. 5·12) just before it would normally. This locks the time base in synchronism with the trigger pulses for a steady oscilloscope picture. Where E_1 is zero, E_2 is the starting point for the sweep. E_4 represents the level at which the sweep would normally stop as set by the RC time constant. E_3 is the level caused by the trigger pulses.

Semiconductors, such as the controlled switch (CS) discussed in Chap. 3, can be used for timing as shown in Fig. 5·13. The CS will turn on after a preset threshold voltage at the input has been exceeded. This input voltage can be obtained from a capacitor being charged at a specified rate. A fixed time delay can be established by controlling the rate of charge on the input capacitor and the threshold point. This type of circuit will act as a time-delay static switch. Delay times up to 20

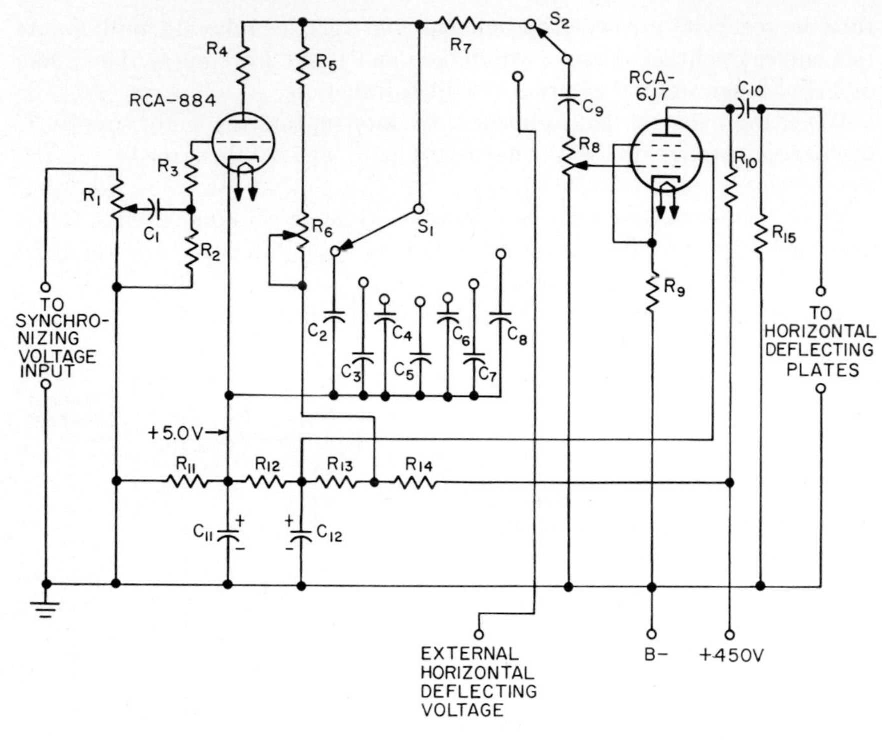

C_1 = 0.25 μf or greater
C_2 = 0.25 μf, 500 v
C_3 = 0.1 μf, 500 v
C_4 = 0.04 μf, 500 v
C_5 = 0.015 μf, 500 v
C_6 = 0.005 μf, 500 v
C_7 = 0.002 μf, 500 v
C_8 = 0.0008 μf, 500 v
C_9 = 0.5 μf, 250 v
C_{10} = 0.5 μf, 500 v
C_{11} = 25 μf, 15 v
C_{12} = 8 μf, 200 v
R_1 = 5,000-ohm (max.) potentiometer
R_2 = not greater than 50,000 ohms
R_3 = 2,000-3,000 ohms, 0.5 watt
R_4 = 350 500 ohms, 0.5 watt
R_5 = 0.3-0.5 meg, 0.5 watt
R_6 = 1-meg potentiometer
R_7 = 1 meg, 0.5 watt
R_8 = 0.5-meg potentiometer
R_9 = 850 ohms, 0.5 watt
R_{10} = 0.1 meg, 0.5 watt
R_{11} = 1,500 ohms, 0.5 watt
R_{12} = 25,000 ohms, 1.0 watt
R_{13} = 60,000 ohms, 1.0 watt
R_{14} = 60,000 ohms, 1.0 watt
R_{15} = 2.0 meg, 1.0 watt
S_1 = 7-contact sp switch
S_2 = spdt switch

Fig. 5·11 RCA linear sweep circuit and amplifier.

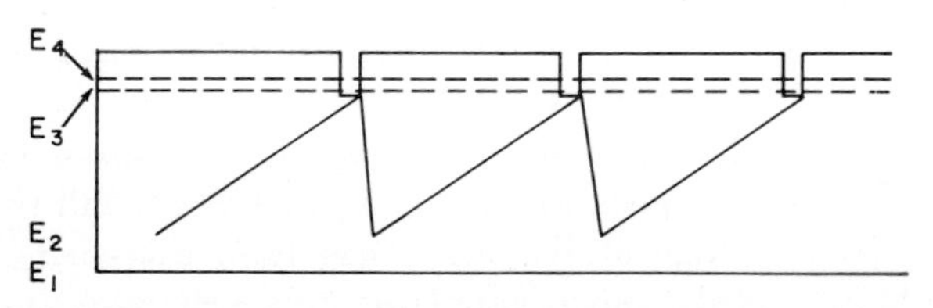

Fig. 5·12 Sweep waveforms showing sawtooth and trigger.

sec are possible and the circuits can be put in series for longer times. A time-delay static switch of this type can be very small in size yet control more than 100 watts of load power. Timing accuracy of better than ± 5 per cent can be achieved under varying temperature and supply volt-

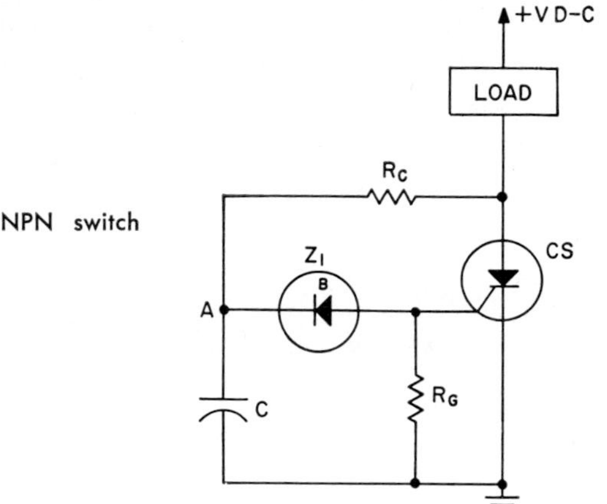

Fig. 5·13 Solid State Products PNPN switch timer.

age when the circuits have been properly compensated. Zener diode Z_1 establishes a reference voltage for the gate of the circuit.

A single-shot multivibrator, or an interval timer, is shown in Fig. 5·14. Here, when supply voltage is applied, CS_1 will turn on after the preset

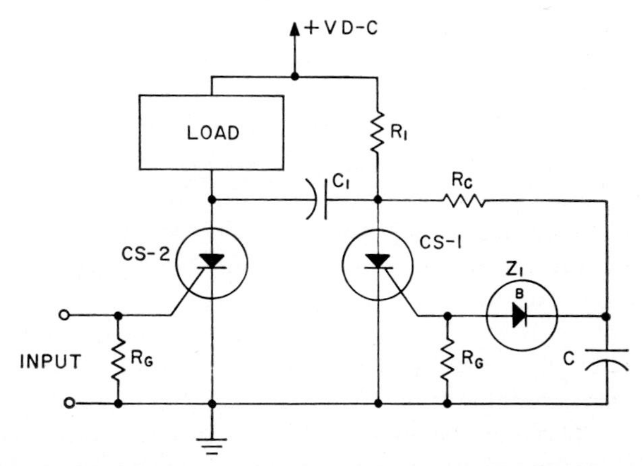

Fig. 5·14 One-shot multivibrator by Solid State Products; input causes power through the load and recycling.

time has elapsed. An input pulse will turn on CS_2. This turns off CS_1 by the action of the commutating capacitor C_1. When CS_1 has been turned off, C will commence charging and will turn on CS_1 after the preset time interval. This turns off CS_2 by the action of the commutating capacitor and removes load power. An input pulse should not be applied to CS_2 for a time period equal to two to three times the delay time after the supply voltage is first turned on. The time interval between input pulses should also equal at least two to three times the delay time. This is necessary to permit C to discharge after CS_2 has turned on and to allow C_1 to charge fully to the correct polarity.

Mechanical Timers. Both electronic systems and mechanical methods may be used for timing. Some units used for this purpose are synchronous motors similar to those used in electric clocks. These motors, through gear trains, cams, switches, and relays, then control the a-c power to

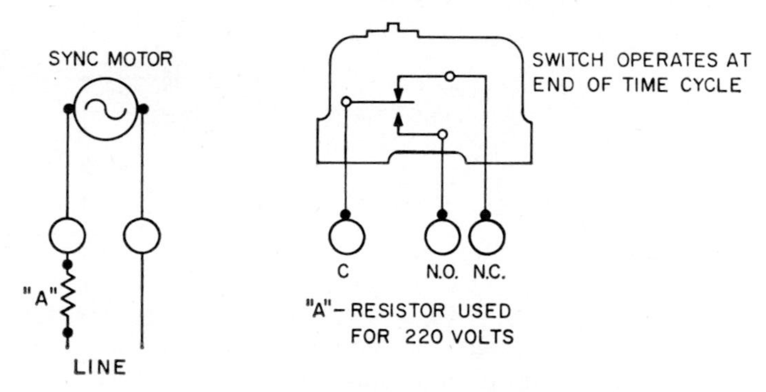

Fig. 5·15 Mechanical timer driven by a-c motor.

the various loads. There are a number of different types of mechanical timers; the first one to be discussed is the time delay.

Where an adjustable time delay is required, a timer like that shown in Fig. 5·15 controls the delay between the closing of one circuit and the predetermined opening, or closing, of another. As in the drawing, a synchronous motor is connected to the line. When the "start" switch (not shown) is thrown, the motor starts a gear train. At the proper time the switch operates, which closes the normally open (NO) and opens the normally closed (NC) contacts. Terminal C is common.

Examples of applications are radio transmitters or electronic induction heaters where the filaments are heated before the application of plate voltage. Delays are available from 15 sec to 5 min or more by a simple dial setting, or the delay may be an adjustment inside the equipment.

A percentage-time device is available in the range from 15 sec to 24 hr. For any given total time, the dial is calibrated in percentage and the on time is adjustable from 3 to 97 per cent of the total time. When the switch is closed, the timer (clock motor) starts. At the desired point the relay closes and energizes the load. This device is used for pumps, blowers, ovens, and other applications.

An interval timer is illustrated in Fig. 5·16. The time may be from 15 sec to 5 min. A mechanical push button is shown. When the button is closed (position *C*), the clock motor and the load are both across the a-c line. At the end of the preset time, the switch is opened by the clock and

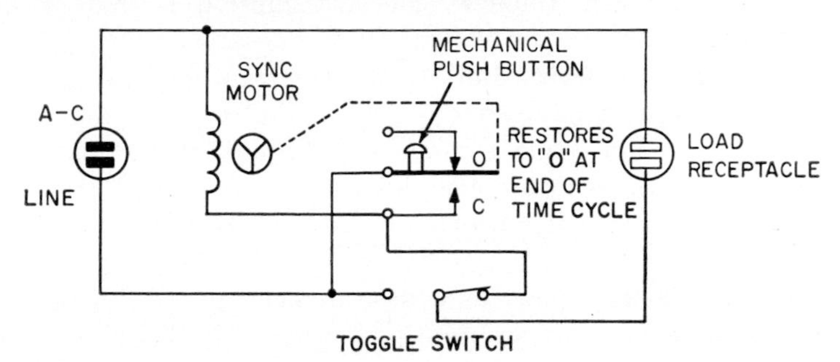

Fig. 5·16 Interval timer which disconnects load power and resets the button at the desired time.

the push button is restored to position 0. When the toggle switch is in the other position, the load is energized directly by the a-c line; the timer and the push button are out of the circuit.

In Fig. 5·17 two other parts are added. The remote control is in parallel with an electric push button as shown. When the push button is closed,

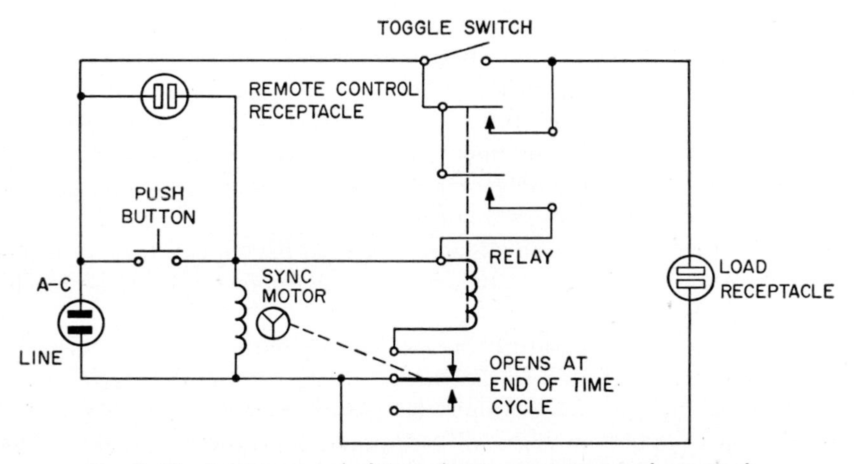

Fig. 5·17 Remote control of timer showing remote-control receptacle.

the motor starts and the relay closes. At the end of a timing cycle the motor-driven switch opens the relay. Again the toggle switch connects the load directly to the line. In plating, photography, or other chemical processes an interval timer is required. This can provide an alarm after a preset interval. After the pointer has been set for the desired time, a circuit is closed and a pilot light indicates that timing has started. As the moving indicator rotates toward zero, it shows the time remaining.

At the end of the interval an alarm sounds and the pilot light goes out. However, the alarm (buzzer) continues until it is shut off or the timer is started again.

5·3 Temperature controls Industrial processes using heat for various steps in the manufacturing cycle require accurate temperature measurement and control. For some simpler applications of the on-off type, thermostats can be used. More complex continuous control requires (1) a measuring device and (2) a control system or a recording system with manual control.

Three techniques for temperature measurement are the thermocouple, the resistance bulb, and a radiation-measuring unit. Each is available to cover different temperature spans. Thermostats are also used for temperature control and are covered in Chap. 4.

Thermocouples are bimetallic junctions such as copper-constantan, iron-constantan, or chromel-alumel which provide a current flow under the proper conditions. A thermocouple is made up of two unlike metals. They are joined at each end and an ammeter, or milliameter, is used at one end. If one junction is heated, a current flow will be detected through the meter at the other end. This current flow is a function of temperature, and the meter can be calibrated directly in degrees.

Materials whose resistance changes with temperature can be used to make a temperature-sensitive indicator. Variations in resistance are calibrated to read directly in temperature.

Any hot object radiates infrared rays which can be detected. The amount of radiation depends upon the temperature and, as with visible light, the phototube current increases as the amount of radiation increases, as when the object gets hotter.

Ordinary thermometers can, of course, be used for temperature measurement. With capacitance-operated controls thermometers can be mechanized for process control.

Thermocouples. Temperature ranges for thermocouples vary, depending upon the material used in the thermocouple wire and the size of the wire. Chromel-alumel, for example, is used up to 1200°C in No. 8 size but only up to 900°C in size No. 24. Copper-constantan in size No. 24 goes up to 225°C, but in No. 14 it goes up to 400°C.

The temperature ranges for different industrial processes vary widely. In the food industry some ranges are (in degrees F) 400 to 900 for bake ovens and 200 to 300 for cooking vegetables and fruit. For cement and lime kilns the temperature covers from 600 to 2600; ceramic kilns go from a low of about 1600 to a high of 2400. These are but a few of the many industrial uses of thermocouples.

Because of the number of uses there are, of course, many different techniques for mounting thermocouples. The actual circuit wires are covered and protected by tubes of metal or ceramic material.

Balancing Potentiometer. A temperature indication is but a part of the story. The readings are used for control and are recorded. Figure 5·18 shows a recording potentiometer used for this purpose. A potentiometer measures a voltage of unknown value by comparing it with a voltage of known value. The unknown is balanced electrically against the

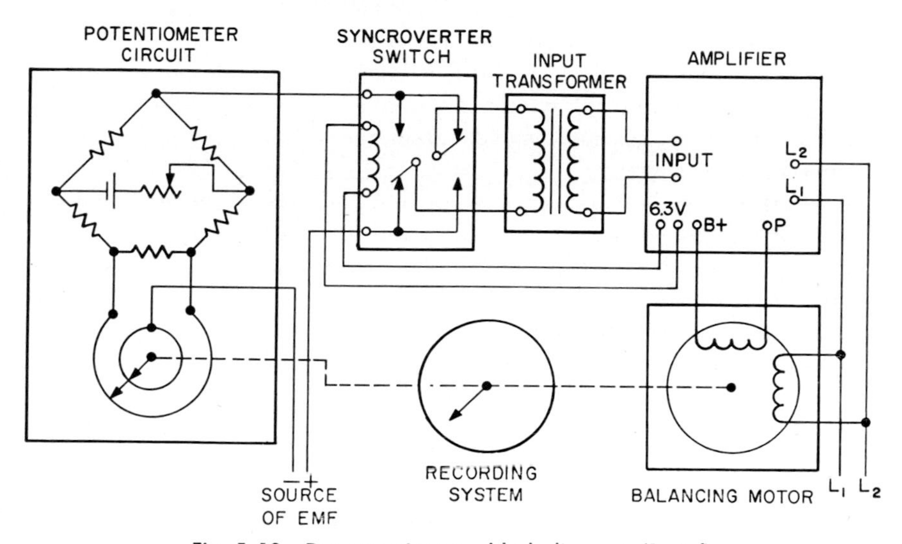

Fig. 5·18 D-c potentiometer block diagram. (*Bristol*)

known. When the two are brought to balance, no current flows in the detector circuit. In the Dynamaster potentiometer, balancing is automatically done by the motor-driven contact on the slide wire as shown in the figure.

The potentiometer circuit is supplied from a source of d-c voltage. This source may be either a dry-cell battery or a constant-voltage device. The potentiometer circuit consists of the sensing element, slide wire, and fixed resistors. The unknown voltage developed by the sensing element is connected into the potentiometer circuit so as to oppose a portion of the slide-wire voltage.

The balancing action is started by a change in d-c voltage output from the sensing element. The change in sensing-element voltage unbalances the potentiometer circuit and produces an error signal, which is converted to an a-c error signal. The a-c output voltage is stepped up by an input transformer and fed to the electronic amplifier.

The inverter converts the d-c unbalance signal of a potentiometer to an a-c signal which is amplified to operate the balancing mechanism. It is a nonresonant wide-frequency low-noise-level precision synchronous inverter, which effectively converts d-c signals as low as 0.05 μv.

The amplifier increases the a-c error signal to a power level sufficient to operate the balancing motor. The amplifier output voltage drives the balancing motor in the direction which will restore the potentiometer circuit to an electrical balance. The indicating pointer and recording mechanism are simultaneously positioned to the new value of the measured variable. This rebalancing action is continuous and begins immediately upon the slightest change in the sensing element.

The direction of the rebalancing action of the potentiometer is based on the phase relation between the a-c error signal and the a-c power-supply voltage fed into the reference winding of the two-phase balancing motor. This phase relation depends upon the direction of unbalance in the potentiometer circuit.

When it has been rebalanced, the entire potentiometer remains motionless until another change in measured variable occurs. This minimizes wear and ensures long instrument life. The basic function of the balancing motor is to drive the slide-wire contact as required to rebalance the potentiometer or bridge. The balancing motor also provides the power for moving the indicator, recording pen, and any control and alarm devices that may be in the instrument. The two-phase balancing motor is used. It has more than enough power to meet all requirements for slide-wire balancing and operation of pens, pointers, and associated control equipment.

Indicating Controllers. Thermocouples measure temperature; more accurately, a thermocouple provides a current flow which is temperature-dependent and this current is read on an indicating meter as temperature. The thermocouple and the indicator together make up a pyrometer.

The indicating pyrometer controller shown in Fig. 5·19 is used with thermocouples or radiation pyrometer heads to indicate and control the temperature of industrial process equipment. It can also be used for automatic control, signals, or alarms from measurement of millivolt signals from other variables such as voltage, current, power, speed, or smoke density.

The control action is electric two-position (on-off). A built-in load relay having a single-pole double-throw contact controls heating means, such as electric power to resistance heating elements or fuel to a burner.

The control system has four essential parts: (1) a photoresistive cell in the shape of a narrow strip; (2) a light source having a straight filament arranged parallel to the photocell strip; (3) a vane mounted on the temperature-indicating pointer, which passes between the photocell and the light source when the measured temperature increases to the control set point; and (4) a load relay.

The heart of this system is the temperature indicator and control vane.

The temperature-indicating mechanism shown in Fig. 5·20 is a d'Arsonval moving-coil system as used in electric meters. The two basic parts of this meter mechanism are the moving coil and a permanent magnetic field. The coil, which is rectangular in shape and wound with copper wire, is carried on a pair of pointed pivot shafts attached above and below it, which are carried in cup bearings. The coil also carries a pointer projecting at right angles from the center line of the pivot shaft. The tip of the

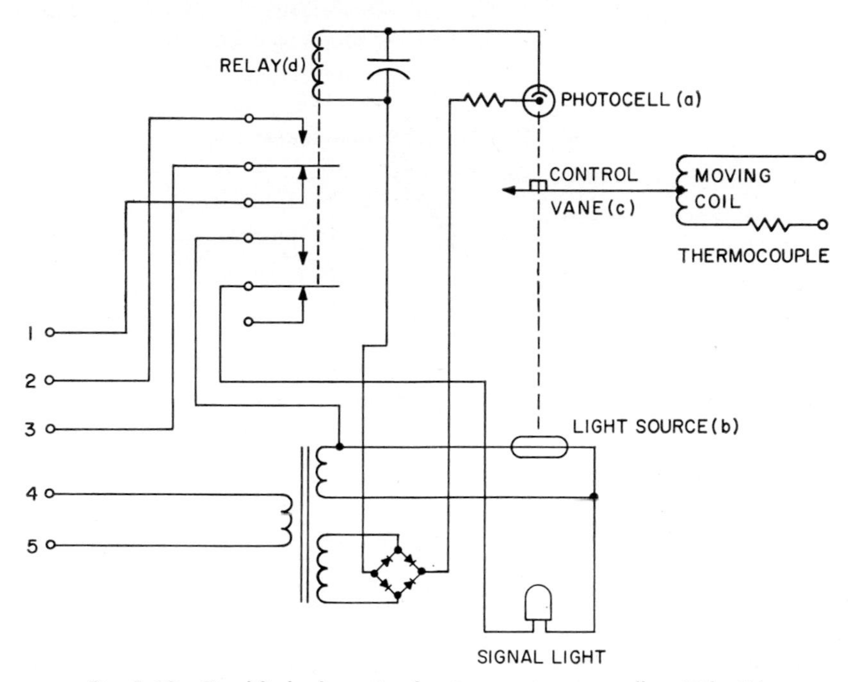

Fig. 5·19 Simplified schematic of a temperature controller. (*Atlantic*)

pointer travels across an indicating scale as the coil turns on its pivots. The drawing shows the essential parts of the moving-coil assembly.

The permanent magnetic field is developed by an alnico magnet whose poles enclose a cylindrical iron plug. The moving coil is positioned so that its two sides ride in the center of the air gap between the pole faces and the plug.

When an electric current passes through the coil, it generates a magnetic field, which reacts with the field of the permanent magnet to cause the coil to turn just as the armature of a motor turns. Two spiral springs are attached at their inner ends to the top and bottom pivot shafts, respectively, and their outer ends are anchored. These springs resist turning of the coil in proportion to the rotation of their inner ends. When current flows through the coil winding, the coil turns on its bearings until the resisting force of the springs is equal to the magnetic force. In this way, the rotation of the coil and the resulting pointer travel across the scale are made proportional to the amount of current flowing through the coil. The springs serve the additional function of providing a circuit to get electric current into the rotating coil.

Some means of adjusting the pointer to the scale zero when no current is flowing through the coil are necessary for two reasons. First, the moving coil and the pointer assembly are balanced carefully by adjustable weights but cannot be balanced perfectly and so are slightly position-sensitive. Second, the springs may relay or take a slight set with time. This zero adjustment is provided by attaching the outer end of the lower spring

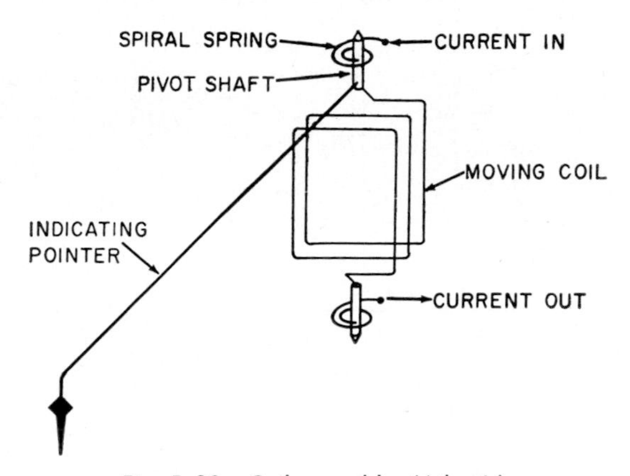

Fig. 5·20 Coil assembly. (*Atlantic*)

end not to a fixed point but rather to a small arm pivoted on the same center line as the coil bearings. This arm is friction-tight and can be turned through a limited angle by means of a lever or shaft mechanism generally operated from the outside of the instrument case. By thus moving the outside end of one spring, the pointer zero position can be adjusted slightly as necessary.

The mechanism used in the controller provides on-off control action with high sensitivity and high stability and without loading or restraining the indicating pointer.

The basic mechanism is a flag attached to the indicating pointer shown in Fig. 5·21. As the pointer moves upscale past the desired control point, the flag interrupts a light beam falling on a photocell, which controls a load relay.

A control arm is pivoted on the same center as the indicating pointer. The position of the control arm is adjustable by the simple cable-and-drum mechanism with its shaft projecting through the front cover for outside setting. The position of the arm is shown by an index pointer, which reads on the indicating scale. The photocell acts like a variable resistor controlled in value by the amount of light falling on it. When the cell is dark, it has a very high resistance. When the cell is illuminated, it has a low resistance. The cell is connected in series with a power source and relay coil so that its resistance controls the current through the relay coil.

When the temperature is low and the indicating pointer is downscale from the control-arm index, the light beam shines on the photocell, causing its resistance to drop and to allow a large current to flow through the relay coil. This pulls in the armature and so closes the relay load contacts to turn on a heat supply by operating a fuel valve or putting power on an electric heating element.

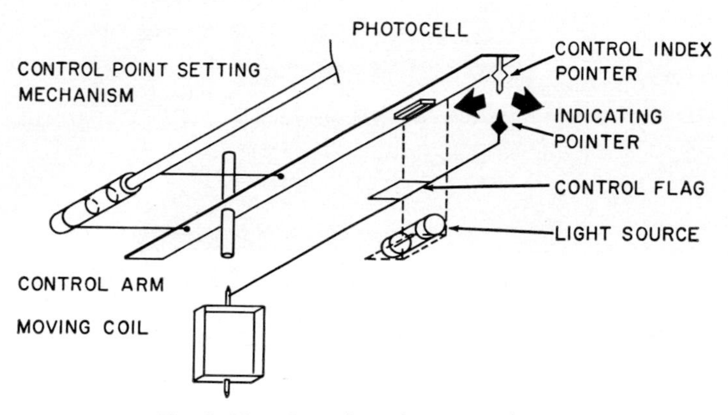

Fig. 5·21 Control mechanism. (*Atlantic*)

When the process temperature increases and the indicating pointer moves up to the control-index setting point, the pointer flag passes between the bulb and the photocell, cutting off the light beam. The photocell resistance then increases greatly, which cuts the relay-coil current almost to zero. This opens the relay load contacts and shuts off the heat supply.

Other Temperature Controls. Temperature controls can be used for alarm systems, as in cases where food in storage must be protected by low temperatures. A typical system, shown in Fig. 5·22, uses six temperature-control units and a master power supply. From left to right, the first unit has a thermostat which opens if the temperature goes up. An increase beyond the set point will cause an alarm, light the red alarm lamp, and indicate trouble from its location. The second unit covers two refrigerated units by connecting two thermostats in series. Either can cause an alarm. The last unit has both an upper and a lower limit. Any variation from the prescribed limits will cause an alarm.

While these units use thermostats, a thermistor (*therm*ostat-res*istor*) will perform the same function. A thermistor is a specially constructed resistor which has a negative temperature coefficient. This means that its resistance decreases as its temperature rises and increases as its temperature falls. A change in thermistor temperature can be caused by alternating or direct currents flowing through it. Thermistors are constructed from a mixture of metallic oxides; nickel oxide usually is the base, with manganese, uranium, or copper oxide and other materials.

Fig. 5·22 Temperature alarm showing six stations and settings for each. (*Walter Kidde*)

A bridge is a sensitive device for measuring resistance, inductance, or capacitance, and it forms the basis for many test devices. In the bridge shown in Fig. 5·23, R_2, R_3, and R_4 are 250 ohms each; and R_5, which is variable from 0 to 125 ohms, is called the balancing control. Resistor R_5 varies the current flowing through the circuit. This controls the initial current flowing through the thermistor. R_5 is varied until there is enough

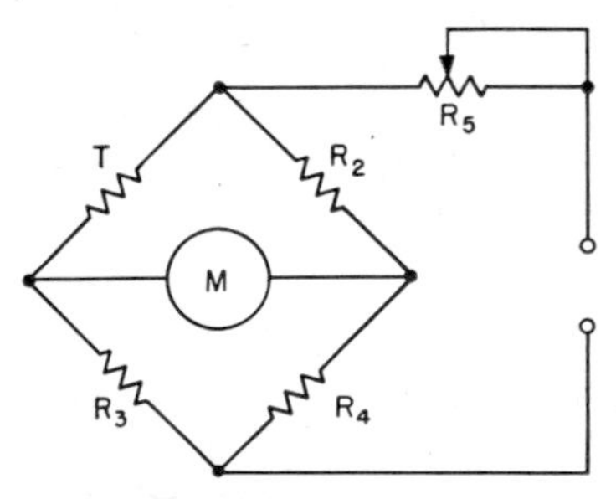

Fig. 5·23 Bridge circuit showing thermistor T and adjustment R_5. (*Veco*)

current flow through thermistor T to make its resistance 250 ohms. When all resistors including the thermistor are 250 ohms, the galvanometer reads zero. The bridge is now said to be balanced and is ready for making power measurements.

Should any one of the four resistors now be varied either up or down in value, the balanced condition of the bridge will be disturbed and a current will flow. The direction of the current will be determined by which resistor has changed and whether its resistance has been increased or decreased. Since the balanced bridge current is absolutely zero, any meter—regardless of its sensitivity—will read zero. Thus, if the meter movement is rated at, say 5 μa for full-scale deflection, even a tiny change of temperature will result in a large imbalance as indicated by the meter pointer.

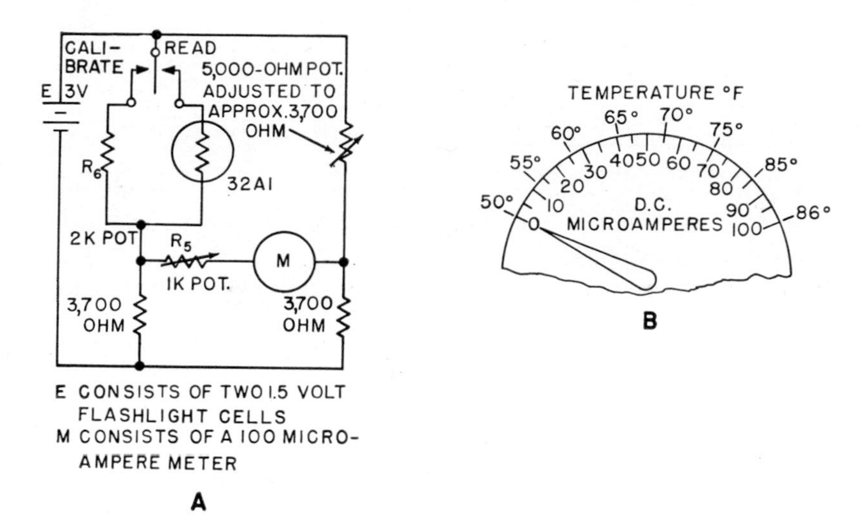

Fig. 5·24 *(A)* **Temperature-control circuit or electronic thermometer;** *(B)* **output linearity of circuit. (Veco)**

A bridge using a 32A1 thermistor is shown in Fig. 5·24. R_4 and R_5 are for balance and calibration. This thermistor has a sensitivity of about 20 mv per °C, which is measured by using the thermistor in a simple voltage-divider circuit with a d-c source at a reference temperature. As the thermometer changes the voltage drop across the thermistor is measured and plotted. The change is the number of volts, or millivolts, variation from normal for a temperature variation. Larger numbers mean greater sensitivity.

In the circuit this thermistor bridge is used as a temperature indicator; a typical scale is shown in Fig. 5·24*B*. This simple thermistor bridge has many uses. Since thermistors change value as the temperature changes and since the different temperatures cause the sensitivity of the bridge to change, the simple bridge requires frequent zero settings. Additional thermistors as well as compensating networks are added to balance out these effects and provide a more useful device.

5·4 Servomechanisms In a basic sense a servomechanism (from the Greek "servo" or slave) is a machine designed to carry out orders. In many industrial processes, for example, there is a requirement for a system in which an electric motor must turn a valve. Control signals are fed to an amplifier whose output drives a motor, which is coupled to the valve in the process. Heavy currents for quite large motors can be obtained, where the thyratron, motor, and a-c source are all in series. As the signals into the tube control its conduction, the tube's current, which is d-c, controls the motor's current and hence its speed. But this provides

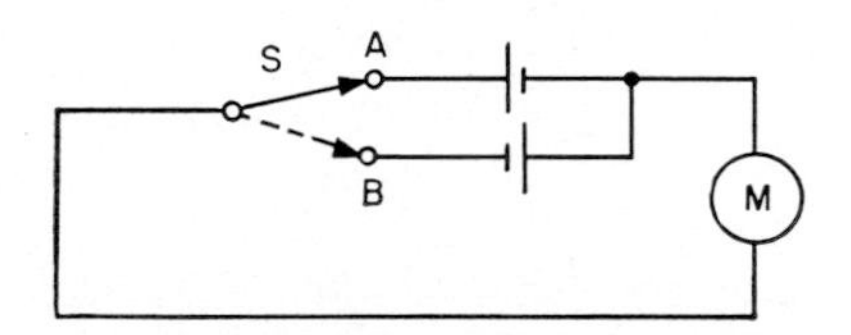

Fig. 5·25 Motor reversing by a simple switch.

only control of a single direction of the motor. A simple circuit such as that in Fig. 5·25 shows how reversing is possible by having two d-c sources and two switches.

The controlled rectifier can be used to provide motor control in both directions.

For simple a-c systems, the output circuit in Fig. 5·26 could be used. This provides a constant-amplitude a-c sine-wave output with reversible phase. It is therefore a reversing drive for a-c motors. The load voltage

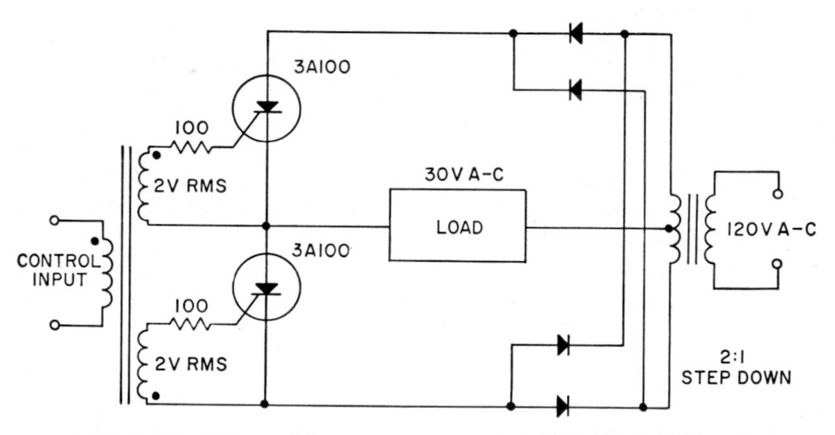

Fig. 5·26 SCRs used for a-c motor reversing. (*Solid State Products*)

will be either in phase or 180° out of phase with the supply voltage, depending on whether the control a-c input is in phase or 180° out of phase with the supply.

The Servo Principle. All the circuits discussed above are, in one sense, servos, for they respond to the signals which tell them to rotate in a certain direction. But this is not enough, for a true servo must not only control direction but also control the *amount* of rotation. This is done by use of the feedback principle; a control signal is fed into the control amplifier, which drives the load or motor.

But a feedback path is present by which the motor generates an error signal proportional to the difference between its actual position and the position it should be in as dictated by the control signal. Figure 5·27 shows a true servo. By means of a wheel or knob, the operator turns the signal potentiometer to a position marked as the required amount of rotation. The a-c signal across R_1 goes to the amplifier, which, in turn, drives the motor. It turns the motor in either direction by one of the techniques discussed above. Coupled to the motor is a second potentiom-

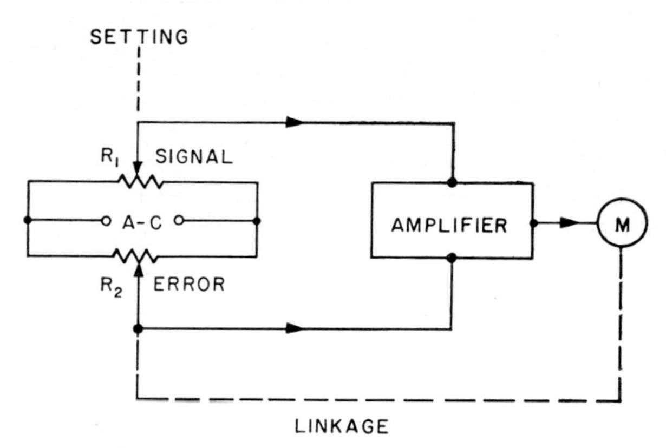

Fig. 5·27 Error and signal voltages in a servo system.

eter R_2. The signal across this resistance is the error signal, which differs from the control signal if, and only if, the motor has *not* turned the desired amount. In this way the error will be zero only if the two potentiometers are turned the same amount, which means the motor is in the proper position.

A servo system must accept an order or command, it evaluates the present position and compares it with the actual position, and it makes a correction until the error is zero. Hunting or oscillation, which sometimes occurs, comes about when the motor overshoots its desired position and returns but again overshoots. It thus, because of improper adjustment, oscillates about the correct point.

Servo Circuits. A simple servo amplifier for a d-c motor is shown in Fig. 5·28. An error signal, through transformer T_1, appears at the grids in push-pull, which is in two equal amplitudes, but they are 180° out of phase. Alternating current from T_2 appears at both plates in phase.

With no error signal, there is no grid signal for either tube. But because of the a-c plate supply, both tubes conduct equal amounts when their plates are positive. This causes equal but opposite voltage drops across the equal resistors R_2 and R_3. The d-c motor has a net zero drop across it, for the voltage across R_2 is equal to but opposite to the voltage across R_3.

But if an error signal exists, it appears at the grids in push-pull, so while one grid is positive the other is negative.

If the error signal makes the V_1 grid positive, V_2 will have a negative signal on its grid and there will be a larger drop across R_2 than across R_1. The motor will turn in one direction until the error voltage is zero.

If the error voltage is in the opposite sense, V_2 will conduct more than V_1 and the motor will rotate in the opposite direction until the error voltage is zero.

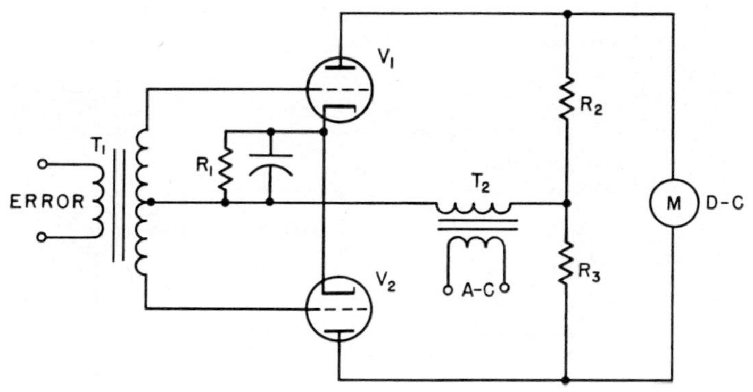

Fig. 5·28 Servo amplifier schematic showing error and line voltage inputs.

Often magnetic amplifiers or controls are used in servo circuits together with thyratrons. Magnetic thyratron controls (MTCs) are complete, packaged magnetic amplifiers designed to accurately control the firing time of thyratrons in control systems. Operating from low-level a-c or d-c control signals, they provide smooth power control.

The basic unit consists of a fast-response magnetic amplifier which generates a continuously phase-variable pulse. Featuring a rise time of less than 500 μsec and a pulse amplitude that varies with firing angle to a maximum of 140 volts, these drives permit positive, reliable firing of all types of thyratrons.

A built-in bias supply provides a negative voltage to hold off the thyratron grid in the absence of a control signal. A time-delay relay may be connected to protect the tube during the filament warmup. Any number of control windings or input circuits may be included to handle several independent input signals. These input windings are isolated from both the power supply and the thyratron.

REVIEW QUESTIONS

5·1 What is a servomotor?
5·2 Give two examples of its use.
5·3 How can a thermistor be used to measure temperature?

5·4 What sort of temperature coefficient does a thermistor have?

5·5 What is a capacitance relay?

5·6 How can a capacitance relay be used as an intruder alarm?

5·7 What is a transducer? Give three examples.

5·8 Upon what principle does the thermostat work?

5·9 Draw a simple circuit (not a bridge) showing how a thermistor can be used to measure temperature.

5·10 Show, using a block diagram, how a thermostat can protect a freezer against a high temperature.

5·11 How does an RC electronic timer operate?

5·12 How does an electronic timer differ from a mechanical timer?

5·13 What is a balancing potentiometer? Where is it used?

5·14 How does a thermocouple differ from a Peltier cooler (Chap. 3)?

5·15 Give three examples of the use of timers.

5·16 What is a servo's error voltage?

5·17 How could a servo and temperature sensor be used to control a furnace temperature?

5·18 What is an indicating controller?

CHAPTER

6

Photoelectric Devices and Controls

Photoelectric devices are used to count, measure, warn, and control for almost every industry. High-speed printing, as an example, is possible only by the use of photoelectric control by register marks to guide the printing, folding, and cutting. Protective devices use photoelectrics to shut off machines when the operator is in danger, and solar cells operate electronic equipment in remote areas and on faraway satellites.

6·1 Uses of photoelectric devices A few of these applications are shown in Fig. 6·1. *A* shows a loop control where the drive motors must keep a fixed loop, from the paper roll, so the printing equipment will not jam. If the loop falls below the lower light, the drive motor increases its speed.

If the loop is too short and does not interrupt the upper beam, the motor speeds up. In the same way computer-tape transports use two photoelectric devices. Because of the many rapid starts and stops, special measures are required to prevent breaking the tape. A typical system may be able to start or stop in 0.005 sec. Two columns are used, one for each reel. The switches are photoelectric cells which act to keep the tape loop above the lower switch as well as below the top switch. Each tape has its own drive motor which is connected to a servomechanism. When a tape reel starts, it is only required to draw from the loop while the other reel and its motor keep the loop the proper size by moving the tape. Tape breakage is thus avoided and quick stops and starts are made.

Figure 6·1*B* shows routine counting of objects on a production line. Each interruption is counted without touching the objects. *C* is an extension of *B*, but here a counter (discussed in the next chapter) stacks the precise quantities. *D* shows a type of cutoff control which automatically

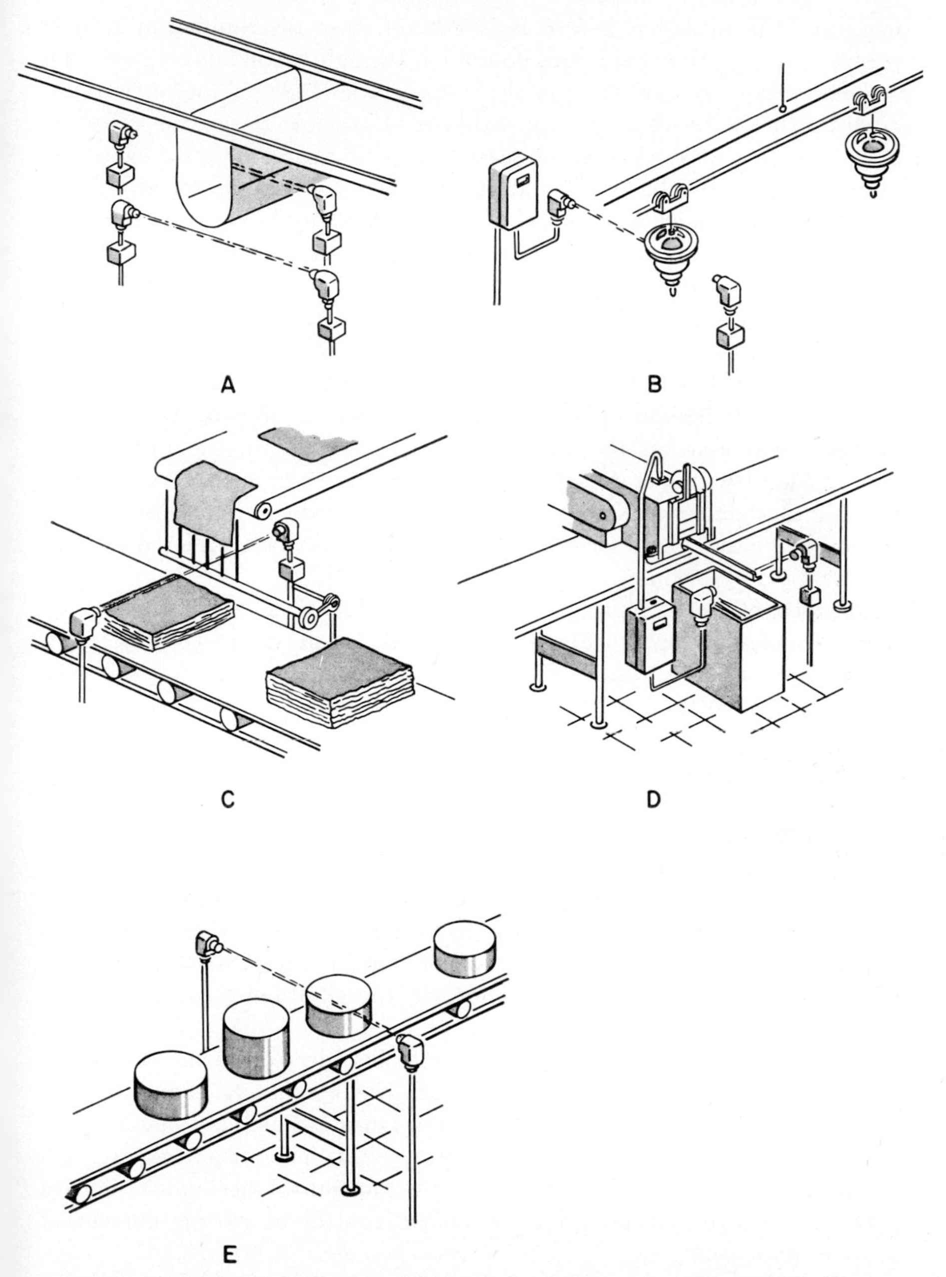

Fig. 6·1 Photoelectric applications: (*A*) loop control with upper and lower limit; (*B*) automatic counting; (*C*) stacking exact quantities; (*D*) cutoff control; (*E*) size control. (*General Electric*)

measures the extrusion before it is cut off. Size discrimination is also possible, as in E, where low cans do not cut the light beam but high ones do.

There are many variations of these applications for photoelectric relay circuits. In a sales-counter application it is possible to use the countermen to replenish stock bins or fill phone and mail orders during slack periods. A photoelectric relay and audible signal will automatically function when a customer approaches. Each counter may have its own signaling system, which is an advantage over the simple door buzzer.

In another system an automatic overhead door opener is controlled. Manually controlled doors may be converted to automatic doors. A manual system requires an operator when the door should be opened. An automatic system will open the doors as soon as the beam of light is interrupted. The doors will remain open as long as a beam is broken. As a person or vehicle approaches the door, from either direction, the interrupted beam of light will operate a photoelectric relay which controls the door-opening mechanism. A safety beam prevents the door from closing until the vehicle clears the doorway. The second beam, from either direction, acts as a safety beam.

One use of these devices is carton selection. One method using 31 possible codes is shown in Fig. 6·2, where the toggle-switch code selection is shown being set up in A. Five switches are located on the panel. By changing the position of these switches, an operator easily selects the carton type to be recognized. To facilitate this selection, an index is provided showing switch positions for each of 31 code variations.

As the carton travels along the conveyor line past a self-contained light source consisting of a bank of rugged light bulbs, the area of the carton surface bearing the code marks is illuminated. Simultaneously, the reflected light from these code marks is read by 10 photocells located immediately below the light source. A recognition circuit then determines whether the code read by the photocell units is identical with the one which has been preset into the unit by the positioning of the toggle switches.

If the code is that of the carton type to be selected, an internal relay is actuated. This relay has both make and break contacts, each set rated at 115 volts at 2 amp for a resistive load. Duration of relay operation is nominally 0.1 sec regardless of conveyor speed. The output signal provided by operation of this relay may be used to control mechanical devices such as conveyor switches for automatic control of carton movement or to actuate counters.

Figure 6·2B shows the front of the unit with the lights and photocells.

The carton selector will recognize cartons at any conveyor-line speed from 30 to 200 fpm without any internal adjustment. Without adjustment, the unit will recognize codes within a 2-in. variation in distance from the carton surface to the photocell reader. Provision for greater distance variation is available on special order. This wide latitude in spacing usually eliminates the need for conveyor-line guides to ensure orientation.

The code consists of a series of five parallel bars. In the carton area provided for this code, the presence and absence of the bars have special significance to the reading unit and allow it to determine whether the passing carton is to be selected.

Code markings are most economically printed simultaneously with other carton printing. In certain circumstances the use of conveyor-line printers is indicated. A large variety of such printers is available to place

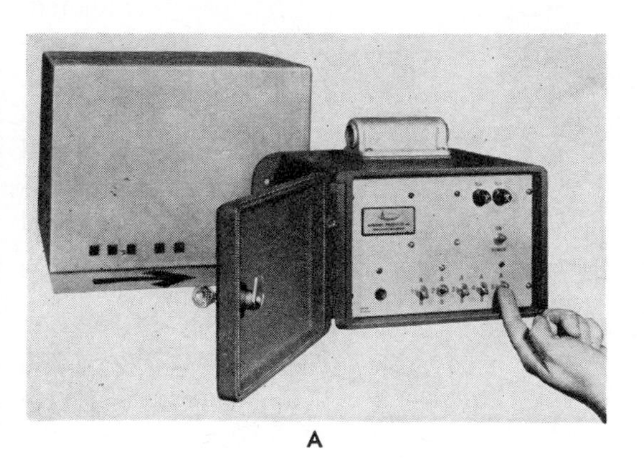

A

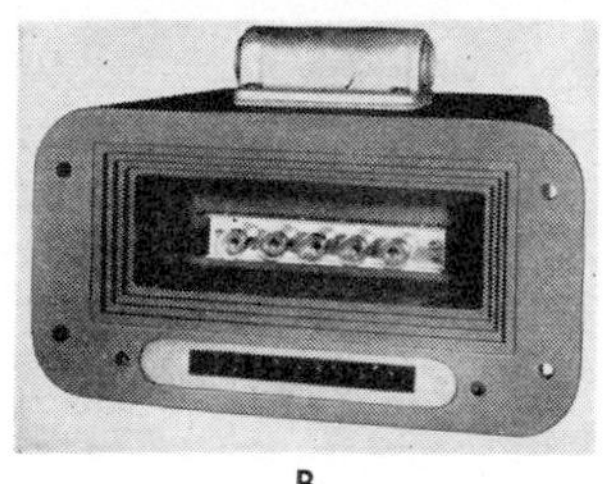

B

Fig. 6·2 (*A*) Setting five toggle switches for control of carton selector; (*B*) light sources and photodetectors for carton selection. (*Atronic Products*)

code markings automatically on cartons as they move over conveyor lines.

Code marks are read by 10 individual photocell viewers. A space must be provided on the carton for the five black bars which represent the code, since this is the total area viewed by these photocells. Because of variation in print registry and fold position of cartons, some tolerance should be allowed beyond the minimum height of the code bar to ensure positive reading by the photocell unit.

Figure 6·3*A* shows a hopper loaded with material for filling packages. Many industries such as food, cement, and sugar utilize hoppers and bins as reservoirs in a container-filling line. Although the feed to the hopper may have a steady rate of feed from an endless belt conveyor, the hopper can overflow if the filling line slows down or stops. This, of course, wastes material and results in increased cleanup expenses. A photoelectric control will automatically stop the belt feed when the bin level reaches a preset height and will prevent waste.

A window or hole is cut in the hopper at the level where the feed should be stopped. The sighting area should be clear of the incoming flow so as not to obstruct the light beam. When the solid level of the material breaks the light beam, it can either sound an alarm or stop the feed. Some conveyorized operations involve manual weighing of special pack cartons and masking. Because of the time it takes to perform these operations, it is impractical to have a steady flow of cartons to the weighing stations. This means an intermittent conveyor feed mechanism must be provided.

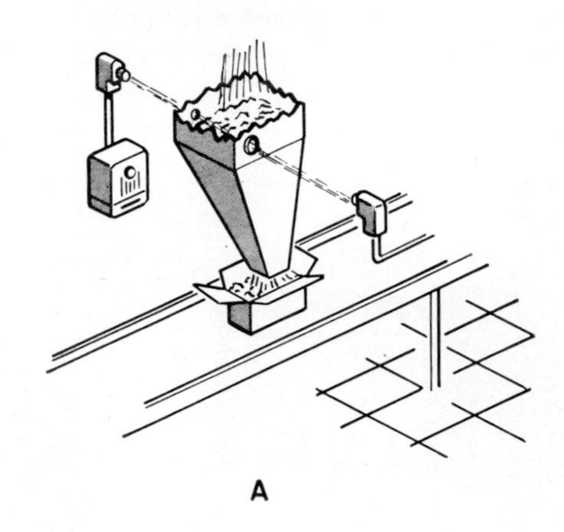

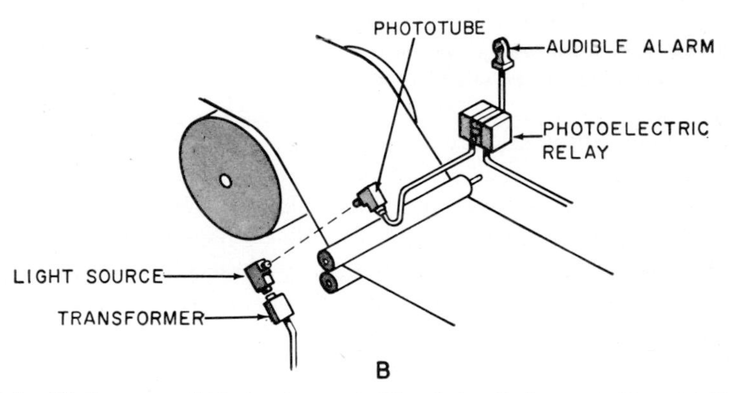

Fig. 6·3 (*A*) Hopper and bin level control; (*B*) web break detector. (*General Electric*)

Here is how the arrangement works:

1. The motor-driven conveyor delivers boxes to the weighing station.
2. The photoelectric relay is set to stop the conveyor when a box interrupts the light beam.
3. The electronic timer prevents constant interruptions by inserting a time delay equal to the average time it takes an operator to weigh and mark boxes coming from the conveyor.
4. If the box is not weighed and marked within a set time, the photoelectric relay stops the conveyor through a magnetic starter.

Other types of controls for photoelectric operation include register controls as shown in Fig. 6·3*B*. The one-way cutoff register control, as shown, is a special photoelectric relay designed to maintain register between the process and the material in a machine. This system is used on machines having material speed of from 5 to 400 fpm where control accuracy—of ±0.020 in. plus 0.005 in. times the register mark spacing in inches—is satisfactory. The control compares the position of a printed register mark, or salient position of the design, on the material with the position of the cutter or functional machine component. Material

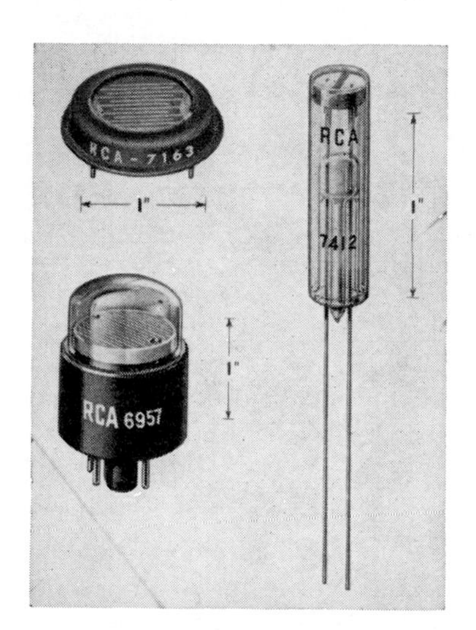

Fig. 6·4 Types 7163, 6957, and 7412 photoconductive cells. (*RCA*)

is usually overfed so that an error is accumulated in only one direction. The one-way register control energizes a solenoid-operated latching mechanism when a register mark is out of register to effect correction by momentarily causing the web to be underfed. With continuous-feed machines, correction is applied through a differential gear release or driving device.

6·2 Types of photoelectric devices There are three types of photoelectric devices: photoconductive, photoemissive, and photovoltaic. Devices which change their resistance are photoconductive or photoresistive; those which emit electrons are photoemissive; and those which generate power are photovoltaic. All these devices are used for inputs for industrial systems which count, control, regulate, and warn.

Photoconductive Devices. Photoconductive devices, as shown in Fig. 6·4, are thin layers of semiconductors such as selenium, silicon, cadmium sulfide, lead sulfide, lead selenide, and others mounted in sealed glass. A simple circuit arrangement, as shown in Fig. 6·5, has a voltage E impressed across the cell and resistor R in series. With no light the cell P has a high resistance which limits the current flow. When light falls on the semiconductor material in the cell, the resistance of the material decreases and the series current increases. Current flow through R, the load resistance, can be amplified for control. Or it is possible to replace R with a sensitive relay which closes when the current flow increases and thus controls a larger amount of power through the relay contacts.

Some photocells of this type have a good signal-to-noise ratio and sensitivity to red (and infrared) radiation. Their response is slow, however, they require an external voltage source, and cooling apparatus is required in some applications.

Light is measured in a very small unit known as the angstrom which has the symbol A. Visible light is from 3,800 to 7,600 A, ultraviolet is from 120 to 3,800 A, and infrared is from 7,600 to 100,000 A. Other colors

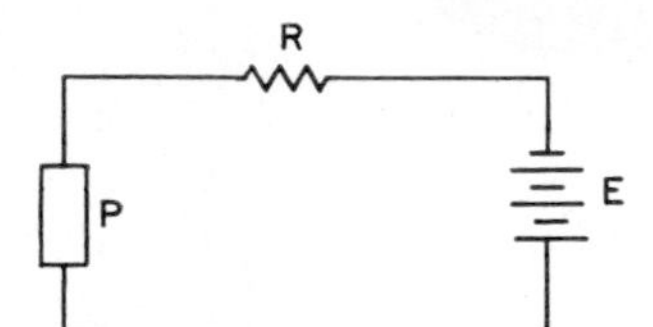

Fig. 6·5 Photoconductive circuit with cell *P*, load *R*, and power source *E*.

lie in between these extremes. There are a number of different color sensitivities available.

Figure 6·6 shows the 17 types of spectral-response curves. Each is a measure of relative sensitivity as a function of color (wavelength of light). Each has its special applications.

A response for a 7163 cell is S-15. This is a photoconductive type as in Fig. 6·7*A* and *B*. There is little response at either the ultraviolet end or the infrared end. Cells of this type have greatest sensitivity to yellow light at about 6,000 A. Figure 6·7*A* shows a simple on-off circuit while Fig. 6·7*B* shows a high-sensitivity circuit.

In *A* the a-c power is applied across R. By means of this sensitivity control, the voltage across the cell is varied. As the intensity of the light falling on the cell is increased, the resistance decreases; this actuates the relay. In *B* a rectifier and power supply provide direct current for the cell and relay. This permits faster operation of the circuit. Figure 6·7*C* illustrates the use of a triode amplifier with the relay in the cathode leg for a type 7412. With no light on the cell (the high-resistance case) the drop across the cathode resistor R_5 reduces the plate current and the relay drops out. When light hits the cell and reduces its resistance, the positive drop across R_4 appears as a positive grid signal which sharply increases the tube current. This current closes the relay. These are three typical circuit uses for photoconductive cells.

Photovoltaic Devices. Photovoltaic cells are also made from semiconductor materials, but they produce a voltage when light is applied. Because of this they require no external power; actually they convert light

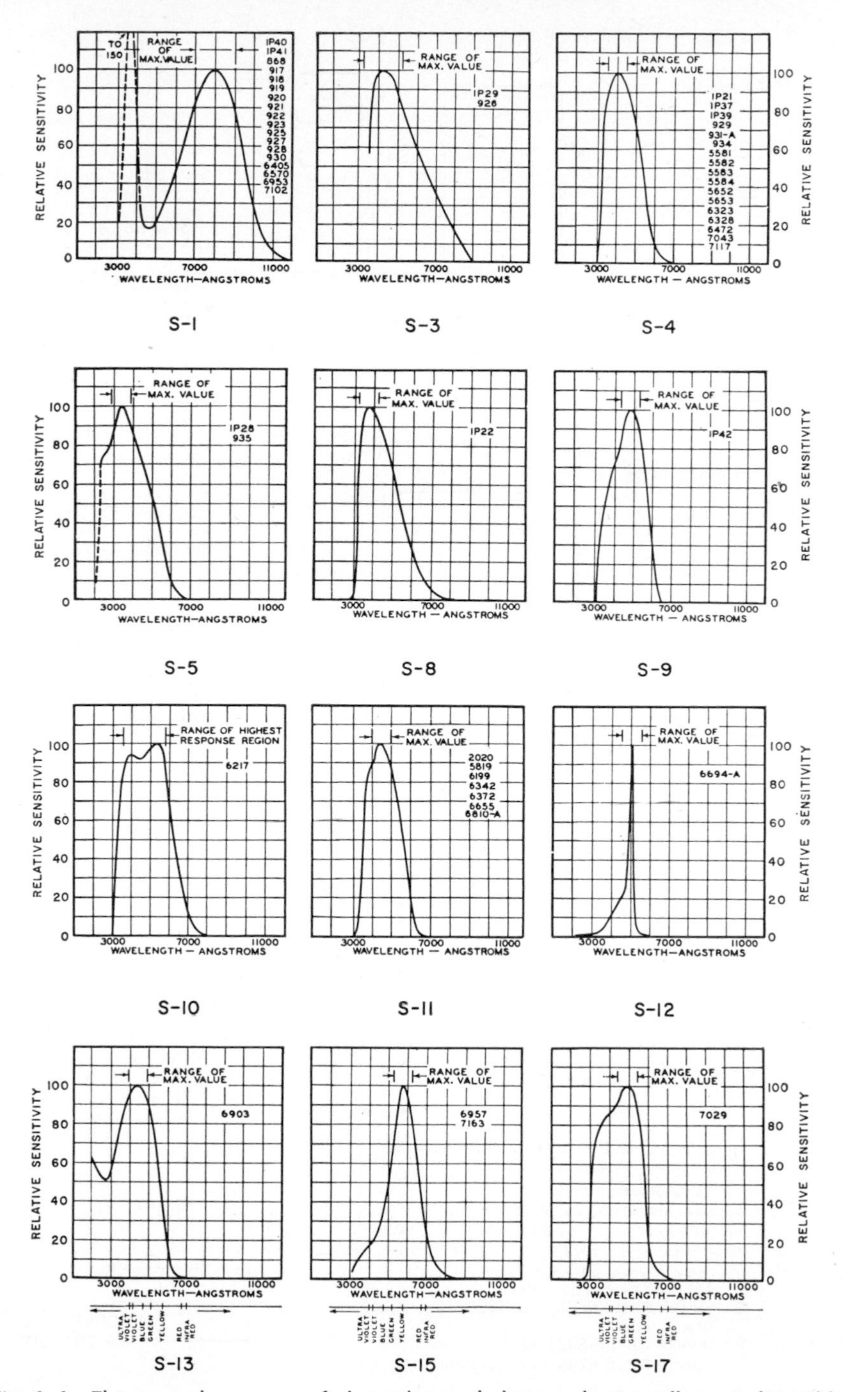

Fig. 6·6 The spectral responses of phototubes and photoconductive cells are indicated by S-designations and are shown in the above curves. These curves are for equal values of radiant flux at all wavelengths and give spectral sensitivity in relative units for the tube types listed on each curve. (*RCA*)

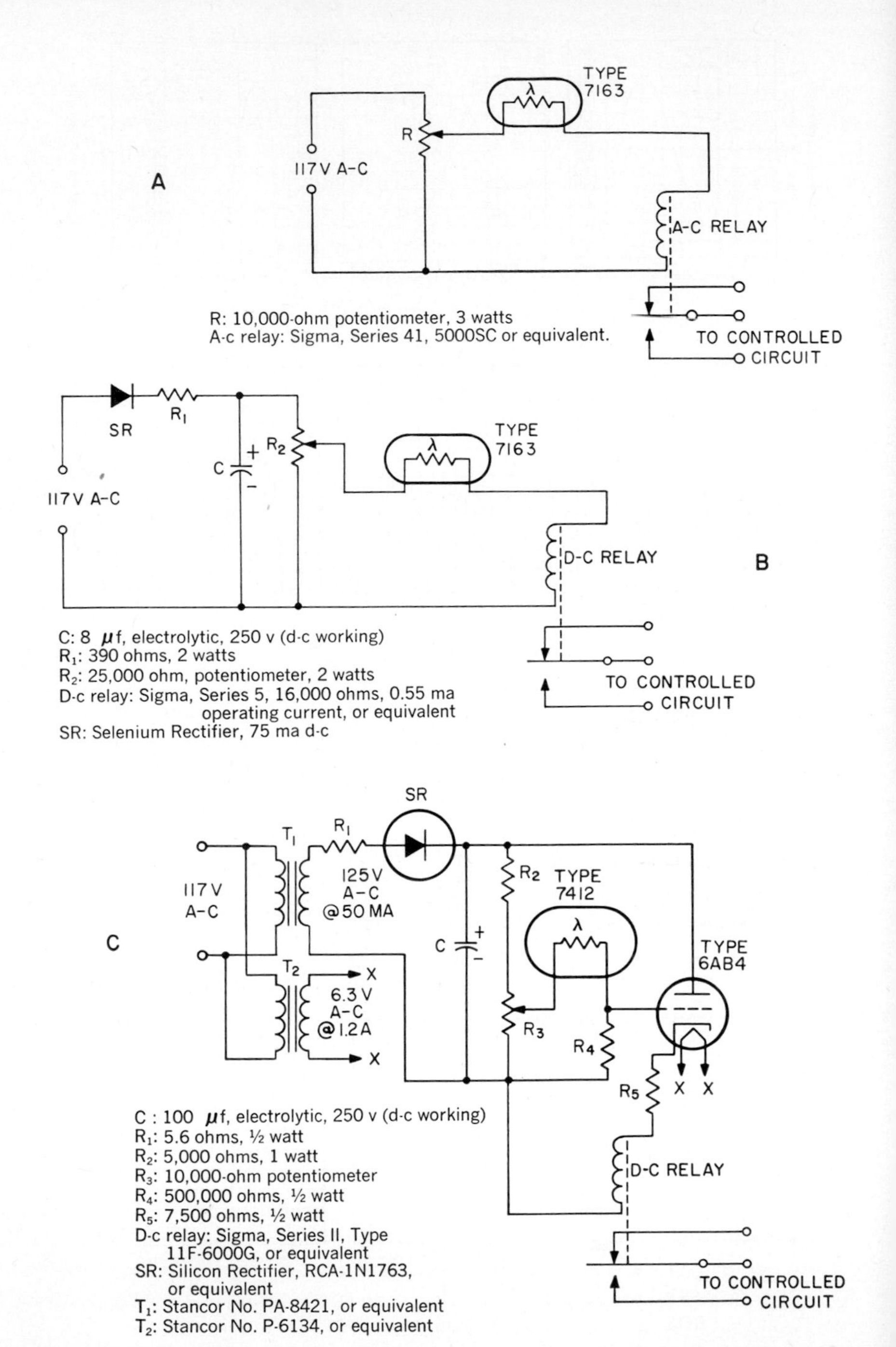

Fig. 6·7 (*A*) Typical on-off control circuit for type 7163; (*B*) typical high-sensitivity circuit for type 7163 (S–15); (C) typical circuit for type 7412. (*RCA*)

energy into electric energy. An example of the use of photovoltaic cells is in the production of electricity such as the experiments now going on by Bell Telephone Laboratories, who have installed large banks of these cells to provide electricity for rural telephones.

Several semiconductor compounds may be used for these cells, such as silicon and selenium.

A silicon photocell generates energy by use of a PN junction to produce electric energy from light. Pure silicon is a poor electric conductor. Adding small amounts of arsenic or phosphorus to the silicon crystal permits electron conduction since these materials have one more electron in their atomic structure than does silicon. N-type silicon, then, has conduction by electrons.

If boron or gallium is added to the pure crystal, P-type silicon results. This conducts by holes or positive charges, which are the absence of electrons.

When both N and P types are present in a single crystal, the NP junction is the boundary between the two types. In the presence of light, electron-hole pairs are formed as light energy falls upon the silicon. The electric field at the junction forces the electrons into the N side and the holes into the P side.

This potential difference (voltage) is the power output from the cell. In bright sunlight the voltage is about 0.6 volt, and cells can be placed in series for higher voltages. There is no chemical change, as in a battery; hence the solar cell has no limit to its lifetime of usefulness.

Only small amounts of power are generated by a single cell. An average cell will produce 30 mw per sq in. if the load resistance is about 4 ohms. As the load resistance is increased or decreased, the available power is decreased. Because of this, low-power cells are usually used in banks, to increase the voltage by series connections.

Some applications of photovoltaic devices use basic circuits utilizing the cell output without amplification. Balanced circuits employing two cells with compensating rheostats are used to eliminate light variations due to voltage fluctuations. Other versions of balanced circuits are available for use in colorimetric work such as vitamin determination, blood count, acidity measurement, turbidity determination, and other scientific applications.

Sensitive-relay applications utilize the cell output to actuate a relay. These circuits are so designed that, at a predetermined value of illumination, sufficient power output will be available to actuate the relay. Circuits such as these are used for counting, sorting, opening doors, turning lights on or off, burglar alarm systems, and other purposes.

Since the output of the cell is in the order of microwatts, amplification is essential in many applications. One of the simplest methods of increasing the sensitivity of a cell is to apply a small potential in the order of 2 volts in the reverse direction. For such a circuit the overall sensitivity, or microampere output per footcandle of illumination, is increased approximately 50 per cent. This is due to the fact that at increased illuminations the internal resistance of a cell decreases. The output of a cell in such a circuit is sufficient to operate a sensitive relay.

For applications requiring a still larger output, electronic amplification is required. Because the d-c amplification of such a low-impedance source

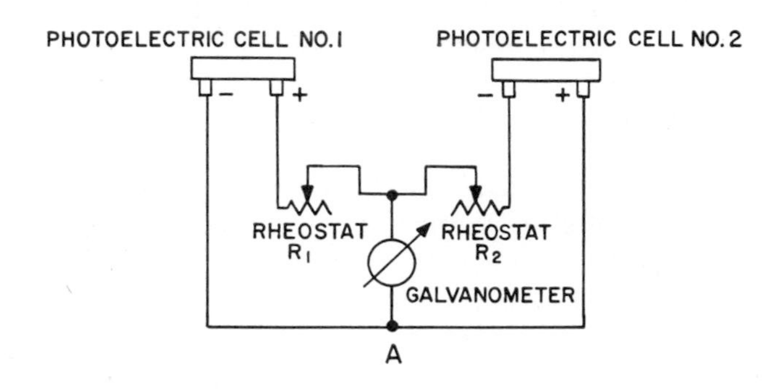

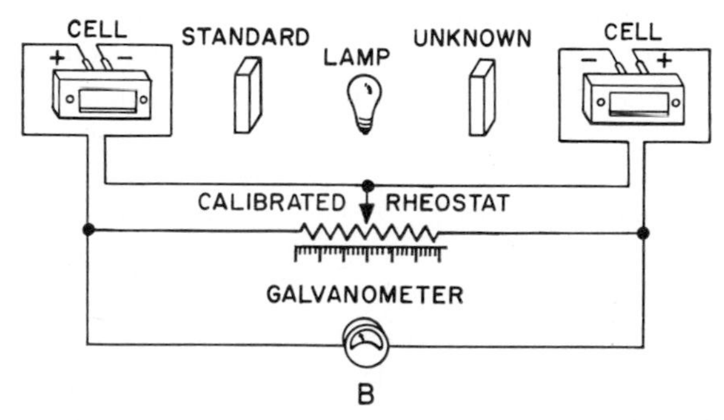

Fig. 6·8 (A) Balanced circuit for color; (B) circuit for comparing relative light-transmission properties of unknown samples with those of a standard. (*General Electric*)

as a self-generating photoelectric cell is rather difficult, the light falling upon the photoelectric cell may be interrupted by a sector disk and the resulting pulsating current generated by the cell amplified by an a-f amplifier. The field of applications requiring electronic amplification is limited to low-intensity light determinations and measurements.

Measurements are often made with these cells as shown in Fig. 6·8. In a balanced circuit (*A*) two cells are used in a bridge. R_1 and R_2 are adjusted for balance when the two colors are equal. In this way both cells get equal illumination. If a test color is measured against a sample, a reading on the meter will show any variation from the standard. Light transmission can be measured against a standard as in *B*.

Photodiodes are similar in action to photovoltaic cells, but they require a voltage source. Figure 6·9 illustrates their construction; light falling upon the PN junction decreases the resistance to current flow. A d-c voltage across the diode establishes the current flow, which varies with the incident light. Output is to the following control amplifier. Circuits such as these are designed for use in reading punched cards or punched tapes for computers or for sound pickup from motion-picture film.

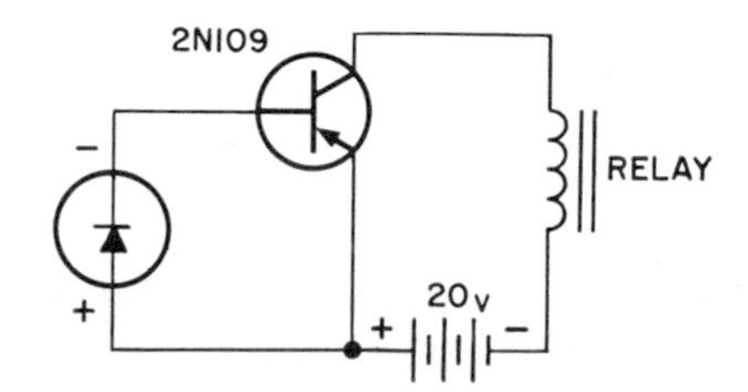

Fig. 6·9 Photodiode driving a transistor amplifier with a relay load. (*Hoffman Electronics*)

Photoemissive Devices. Photoemissive cells or phototubes are probably the most widely used in industry. The basic device (Fig. 6·10) has a large-area cathode coated with an active metal and has a rodlike anode to collect the electrons emitted by the cathode when light strikes it. The tube is either vacuum-type or gas-filled; both are in wide use. Multiplier tubes are used to provide a greater output signal for special applications.

Current flow, under the influence of incident light, is from cathode to anode, or plate. In Fig. 6·10*B* a simple vacuum-tube amplifier is illustrated. With no light and hence no phototube current the tube is cathode-biased to a small current flow. Conduction through the phototube, when light strikes, causes a positive grid signal and the plate current closes the relay.

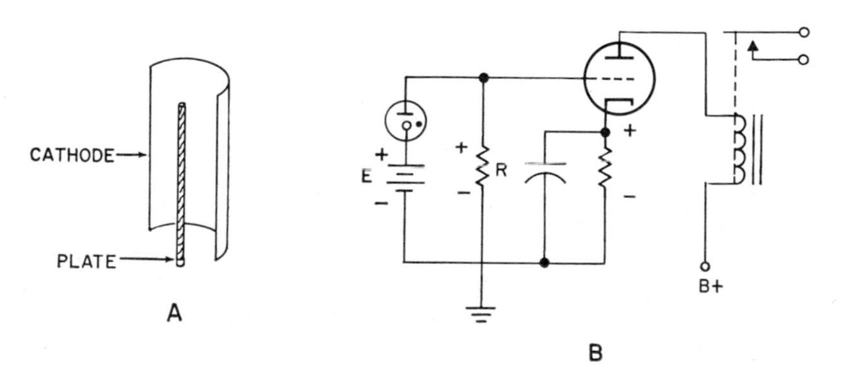

Fig. 6·10 (*A*) Photoemissive tube construction; (*B*) typical phototube relay circuit.

Sensitivity of a phototube is a measure of the amount of current output for a given light input. One important characteristic is the spectral sensitivity or the color (wavelength) to which a given tube is most sensitive. Spectral sensitivity may be considered similar to the inherent frequency response of a vacuum-tube amplifier since light is a form of electromagnetic radiation. Different colors are of different frequencies (or wavelengths) as shown in Fig. 6·6.

Cathodes of phototubes are made from the alkali metals such as lithium, cesium, sodium, potassium, or rubidium. All these metals are very active chemically, and when they are mounted in a tube, the light excitation causes electronic emission. A large-area cathode is used to increase the current flow to the anode.

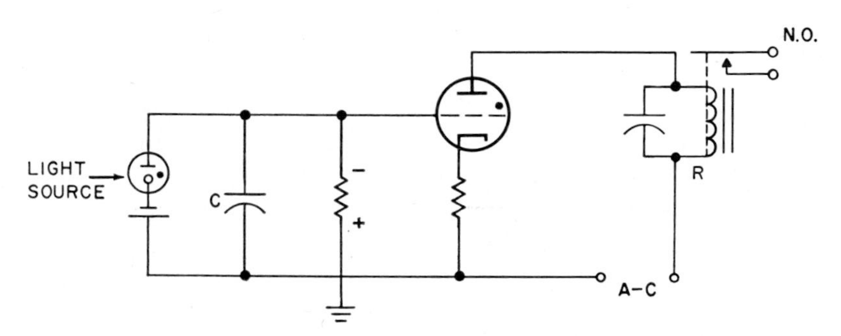

Fig. 6·11 Phototube relay circuit; contacts close when beam is broken.

A small amount of gas will increase the output by the ionization of the gas by the emitted electrons. This multiplication is, effectively, amplification, but it can only be carried to a certain point. If all the gas were to ionize, the tube would be damaged, so the gain of a gas-filled tube has limitations.

Gas-filled tubes are most usually of the S-1 type; vacuum phototubes are usually of the S-1 or S-4 type, but other spectral sensitivities are also available.

Gas-filled photoemissive tubes are used where high sensitivity is required. For example, the 6405/1640 gas tube has an S-1 response.

Gas-filled phototubes, having higher current outputs, can often be used to operate small thyratrons directly as shown in Fig. 6·11. These thyratrons can handle much more current than an ordinary vacuum tube and are better suited to switching and relay-type operations. A photo-operated relay using a gas phototube and a small thyratron is shown. This circuit requires only a filament transformer and takes its power directly from the line. Three resistors, a potentiometer, and the relay complete the circuit. The 2-μf capacitor bypasses the large inductance of the relay, thereby ensuring proper phase relationship between the plate of the thyratron and the anode of the phototube during the conducting part of the cycle.

Even greater output can be obtained by the photomultiplier as shown in Figs. 6·12 and 6·13. An average normal current for a phototube "diode" is about 10 μa. The multiplier tube has an average current of 1 ma and can provide a gain of over 200,000 or more. The 6903 multiplier tube has a gain or amplification of over 400,000 times.

A series of small anodes or dynodes are used, each with a higher voltage than the one before. Electrons reach dynode 1 from the photocathode, and secondary emission creates a greater number of electrons arriving at the higher potential of dynode 2. This continues for 10 dynodes. The signal output is from the final plate, which is 11 in the figure.

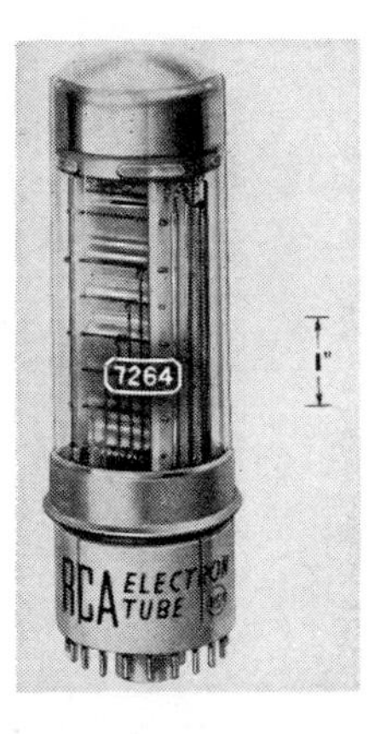

Fig. 6·12 RCA type 7264 multiplier phototube.

A typical photomultiplier circuit is shown in Fig. 6·13*B*. The use of individual bleeder resistors between each dynode ensures a constant voltage ratio between them and the anode at all times. This is essential for maximum gain and stable performance. High voltage can be obtained from the power supply as shown. Only minimum filtering circuits can be used since the current requirements of the multiplier phototube are only about 1.0 ma maximum. Signal output is taken off as indicated and may be further amplified by any suitable d-c amplifier, or it can be read directly on a microammeter.

Other multiplier tubes are available for use where large amplification is required for special applications. The 931-A is a multiplier phototube intended for general use in applications involving low light levels. It features a combination of high photosensitivity, high secondary-emission amplification, and very small d-c dark current.

The spectral response of the 931-A covers the range from about 3,000 to 6,200 A. Maximum response occurs at approximately 4,000 A. The 931-A, therefore, has high sensitivity to blue-rich light and negligible sensitivity to infrared radiation. Its sensitivity to incandescent light depends on the color temperature of the source.

The 931-A is capable of multiplying feeble photoelectric current produced at the cathode by a median value of 800,000 times when operating at 100 volts per stage. The output current of the 931-A is a linear function of the exciting illumination under normal operating conditions.

The frequency response of the 931-A is flat up to a frequency of about 100 mc per sec, above which the variation in electron transit time becomes the limiting factor.

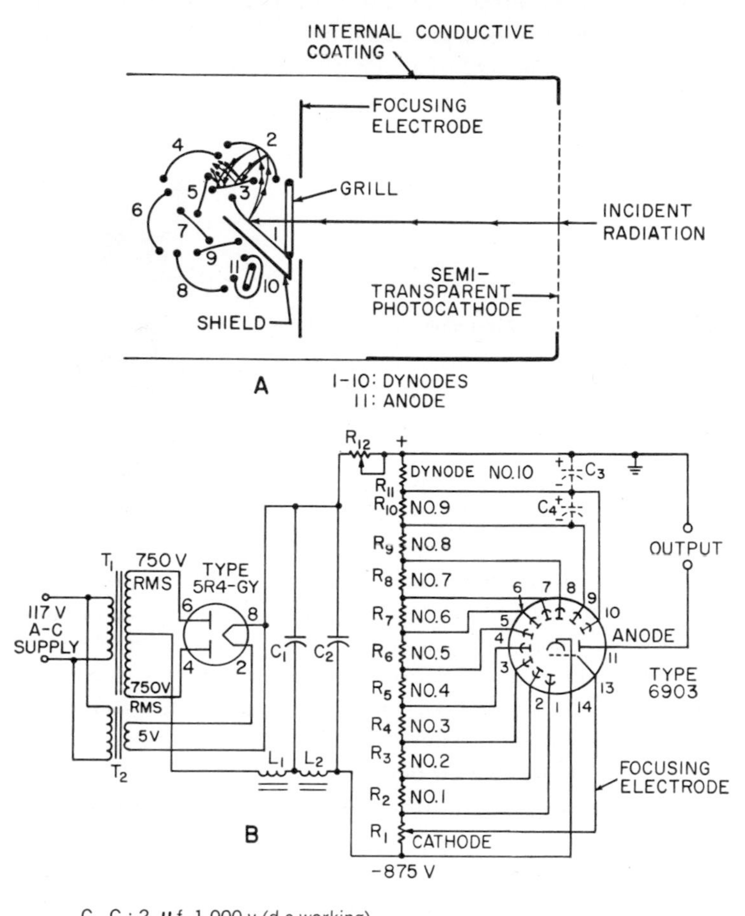

C_1, C_2: 2 μf, 1,000 v (d-c working)
C_3, C_4: 8 μf, electrolytic, 150 v (d-c working)
Required only if high peak currents are drawn.
L_1, L_2: United Transformer Corp. No. R-17, or equivalent
R_1: 50,000 ohms, 5 watts, adjustable (General Radio Co. Type 314-A or equivalent)
R_2, R_3, R_4, R_5, R_6, R_7, R_8, R_9, R_{10}: 18,000 ohms, 2 watts
R_{11}: 12,000 ohms, 2 watts
R_{12}: 100,000 ohms, 10 watts, adjustable (General Radio Type 471-A, or equivalent)
T_1: United Transformer Corp. No. S-45, or equivalent
T_2: United Transformer Corp. No. FT-6, or equivalent

Fig. 6·13 (*A*) Schematic arrangement of type 6903 structure; (*B*) full-wave rectifier power-supply circuit with voltage divider for supplying d-c voltages to type 6903 in applications critical as to hum modulation. (*RCA*)

With small size, rugged construction, high sensitivity, extremely low equivalent noise input, and freedom from distortion, the 931-A is for use in light-operated relays, in X-ray exposure control, in facsimile transmission, in light fluxmeters, in scintillation counters, in flying-spot video-signal generators, and in scientific research involving low light levels.

Other Photoelectric Devices. Semiconductor photoelectric switches form a separate class of devices.

The Photran (Solid-State Products, Inc.) is a miniature solid-state bistable switch with many properties that are similar to those of a gas thyratron. It is essentially a PNPN switch that is triggered by light energy instead of (or in addition to) electric energy. Before light strikes the Photran it is in a high-impedance (over 10 megohms) off state. When a light impulse is applied, it is switched to a low-impedance (under 10 ohms) on state. The device will then remain in the on state indefinitely until it is electrically turned off. The light energy used to trigger the device on need only be momentary since it is not required to sustain the device in the on state.

The Photran can often perform the dual role of light sensing and load actuating. Its power dissipation rating is 0.25 watt at 25°C ambient and 0.5 watt at 75°C case temperature. Its voltage drop when on is below 1.5 volts at all current levels under 200 ma.

The load power delivered by the Photran can be high as 40 watts continuous (200 volts to 200 ma) at an efficiency as high as 99 per cent. Peak pulse current up to 5 amp can be supplied in circuits which operate at sufficiently low duty cycles to stay within the average power rating of the device.

Figure 6·14 shows the trigger light intensity requirements as a function of the bias current. The time required to turn on the Photran is directly related to the light intensity used for triggering. When the light intensity is at the minimum required level, the turn-on time is typically 10 μsec. This time can be reduced to below 1 μsec by a sufficient increase in the light intensity. The turn-off time is primarily controlled by the recovery time of the device (typically 30 μsec).

Once light has triggered the device on, it will remain on indefinitely until electrically turned off by reducing its anode current to zero. In effect, the Photran remembers that a light pulse existed. This property can be used in a variety of applications to provide a new degree of electro-optical control and logic systems design using this binary memory property.

The light intensity required to trigger the unit can be set electrically by changing its gate bias current. This bias-control property allows simple electric gating. By setting a high bias level, the Photran can be made nonresponsive to light; and by subsequently reducing the bias, it can be made to react normally. The fact that electric triggering can be used in addition to optical triggering opens the possibility of an electrooptical temperature of the light source. The spectral response is similar to that of other silicon photodevices, ranging from 0.4 to 1.1 μ. Maximum sensitivity is at approximately 0.85 μ.

In counting, sorting, timing, indexing, and programming systems, it is frequently desirable to obtain a discrete pulse output for each inter-

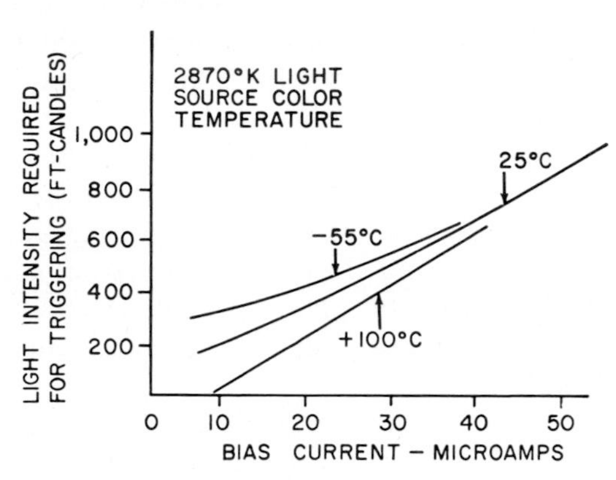

Fig. 6·14 Trigger current for a photoswitch where current is a function of light intensity. (*Solid State Products*)

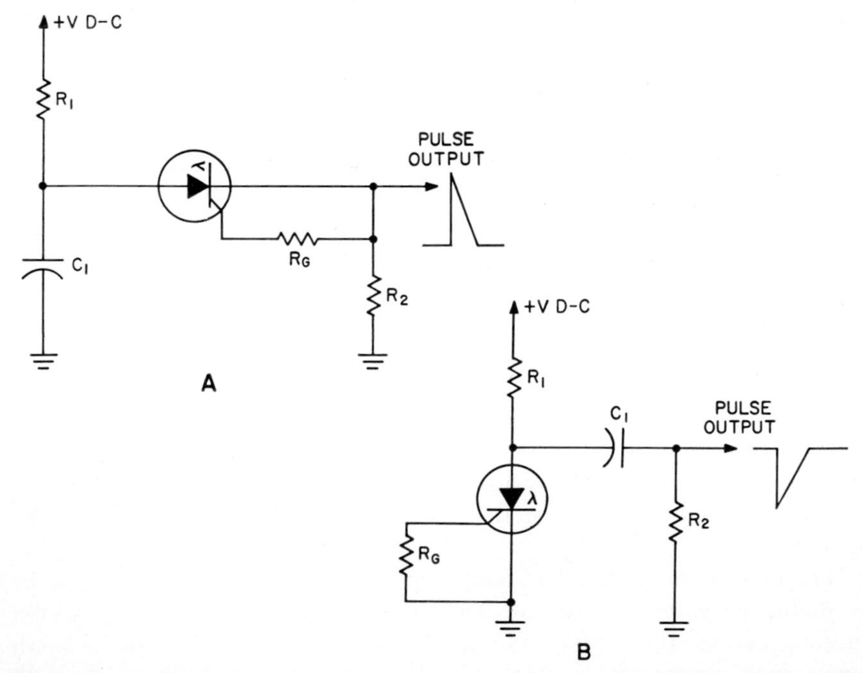

Fig. 6·15 (*A*) Pulse trigger output of positive pulse output; (*B*) negative pulse output. (*Solid State Products*)

ruption of light at the photocell input. Pulse-generator circuits shown in Fig. 6·15 provide a single pulse output each time a light beam has been removed and subsequently reapplied to the cell.

One mode of circuit operation is for light to be continuously on the cell. Under this condition the device conducts a small amount of current (under 100 μa) continuously through R_1, which prevents C_1 from charging. When the light is temporarily removed from the cell, it drops out of conduction and C charges to the suppply voltage through R_1. When light is subsequently applied, the Photran is triggered on and it discharges C_1 through R_2 and the load.

An alternative mode of operation is by light impulse. In this case, light is normally not present on the cell and C_1 is fully charged. When a light impulse occurs, C_1 is discharged through R_2 and the load. C_1 recharges after the light impulse terminates. For either of these modes of operation, the amplitude and width of the output pulse are set by circuit values and are independent of light intensity.

6·3 Photoelectric circuit applications

General-purpose Controls. In many cases a single photoelectric system—which is a light source, an amplifier, and an associated control circuit—can be used in several ways. A basic circuit is the self-rectifier amplifier.

In some photoelectric controls the operating speed is not important so that 60-cycle alternating current can be used for the phototube and amplifier plate supply. The conduction from cathode to anode as in either the phototube or the amplifier will take place only during the time the anode is positive, as with a rectifier. Actually the amplifier and phototubes act as self-rectifiers.

But although using the a-c line as the plate supply is incxpensive, it has drawbacks. A tube with a self-rectifier action is going from conduction to cutoff 60 times a second. For slow-speed operation, this is satisfactory; but for high-speed operation, d-c plate supplies are used for either the phototube, the amplifier, or both.

Figure 6·16 illustrates a circuit which is designed for use at speeds up to 1,200 operations per *minute*. This is not high-speed, in a radio sense; but if this device is counting packing cases on a conveyor belt, 1,200 per minute is very fast. *A* 6X4 is used as a half-wave rectifier. Note that the two primaries can be tied in parallel for 115-volt operation or in series for 230-volt operation.

Relay *A*, in series with the 5AQ5 plate, is energized when light is on the 1P40 phototube cathode. About 75 volts across the 1P40 and the 10-megohm load provide for a steady small current while light falls on the cathode. The drop across the load makes pin 8 positive; this is tied to the control grid of the pentode through a 1-megohm resistor.

A voltage divider (6, 3, and 1.2 kilohms) is also across the 1P40. A sensitivity control, the 3-kilohm variable resistor, is used to adjust the positive cathode voltages. When the light beam is interrupted, the 1P40 is opened and this removes the positive grid voltage. Because of the positive cathode voltage the pentode plate current is reduced and the plate relay *A* opens.

There are a number of safety features in this circuit. A failure in (1) the light source, (2) the rectifier tube or its associated circuit, or (3) the

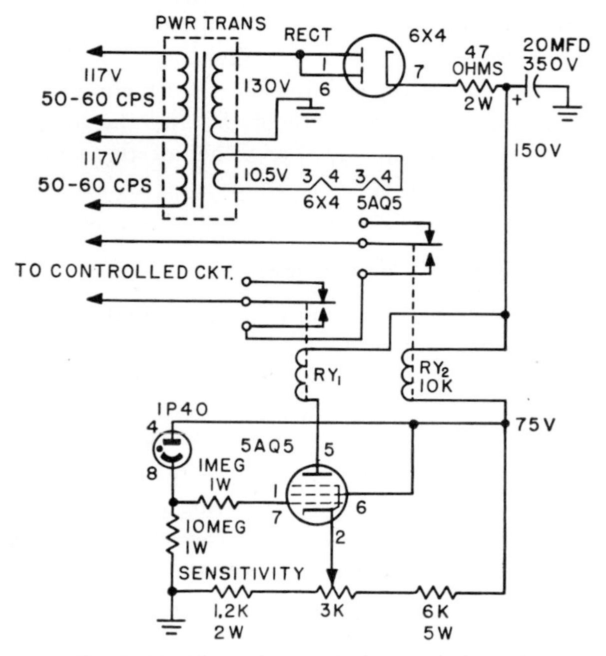

Fig. 6·16 Phototube circuit. (*General Electric*)

amplifier tube will each *alone* open the plate relay. In industrial applications where the control is being used as a safety device this "fail-safe" feature is important since the failure of the control will not go unnoticed.

The plate relay contacts are rated at 8 amp, 115 volts a-c. A second relay is a safety device, designed to energize after the circuit is warmed up. The phototube itself requires no warm up. It deenergizes when the phototube goes off or when the amplifier tube fails. There are many examples of uses for this relay. It may be used as a hand guard. If the operator's hands interrupt the light beam, the machine to which the relay is attached will not operate. In this application it is most important that failure of the phototube or the light source will also stop the machine. Photoelectric devices *alone* are not recommended for safety protection since they can be fooled or tricked.

A wide range of commercial units are available in controls for photoelectrics. Several of these are described below.

The 3S7505-GP105 is a tubeless low-cost relay using a cadmium sulfide cell for operations up to 200 a minute. This photoelectric relay is designed for indoor limit-switch-type applications. For reliable operation, the light beam must be completely interrupted. Nothing touches the object being moved, and no changes in conveyor design are necessary. The cadmium sulfide power cell operates the output relay directly. No tubes or transistors are used in the circuit. A separately mounted control panel is not required. External power supply comes from a transformer also supplying power to the light source used with the relay.

The CR 7505-B100 for operation up to 300 per minute is a long-distance photoelectric relay especially designed for outdoor application.

The semiconductor photoelectric switch (Sec. 6·2) can also be used for controls as shown in Fig. 6·17. The device can be operated from a-c as well as d-c power sources. When alternating current is used, the Photran will block during the positive half-cycle when it has not been triggered by light and will conduct when it has been triggered by light. During the negative half-cycle, the Photran acts as a blocking diode regardless of whether light is on it or not. If the device had been conducting during a positive half-cycle, it will automatically drop out of conduction during the negative half-cycle and must be retriggered with light before it will again conduct on another positive half-cycle.

It is frequently desirable to use a-c instead of d-c source power. This often eliminates the need for separate d-c power supplies and can simplify many control systems.

Several simple control circuits are shown where d-c source power is used and the Photran acts as a light-actuated latching switch. When light strikes the Photran, as in Fig. 6·17*A*, power is applied to the load. The Photran will continue to conduct when light has been removed. Load power is turned off by a reset switch or similar function.

The use of an a-c source (Fig. 6·17*B*, *C*, and *D*) will provide half-wave-rectified direct current to the load when light is applied and no output when light is removed. A diode D_1 is used when the load is inductive, such as in a relay or solenoid coil. The diode eliminates inductive voltage buildups and, equally important, allows the energy stored in the inductance during the positive half-cycle to be returned to the load during the negative half-cycle. The diode polarity is such as to allow the induced current to flow through the load in the same direction as during the positive half-cycle. This freewheeling action provides a continuous direct current through the load for both half-cycles.

A circuit which provides power to the load when no light is on the Photran, as in Fig. 6·17*D*, and removes load power when light is applied is also shown.

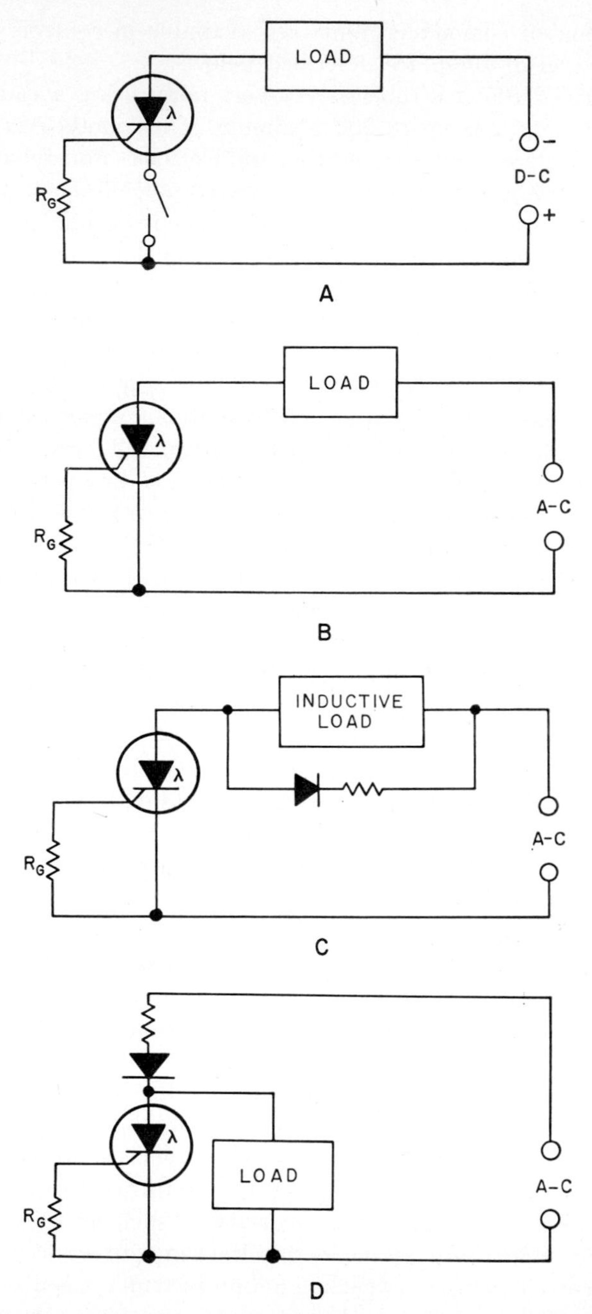

Fig. 6·17 (*A*) Photoswitch with d-c power and switch reset; (*B*) a-c power for reset; (C) circuit for inductive load; (*D*) shunt load. (*Solid State Products*)

Figure 6·18 shows other control circuits. In many control systems applications, it is necessary to supply relatively large amounts of power to a load from photocell inputs. In these cases a Photran could be used to actuate a power relay with one of the simple control circuits.

Four high-power output circuits are shown. The circuits *A* and *B* both provide full-wave d-c output to the load. In *A* full output is supplied when

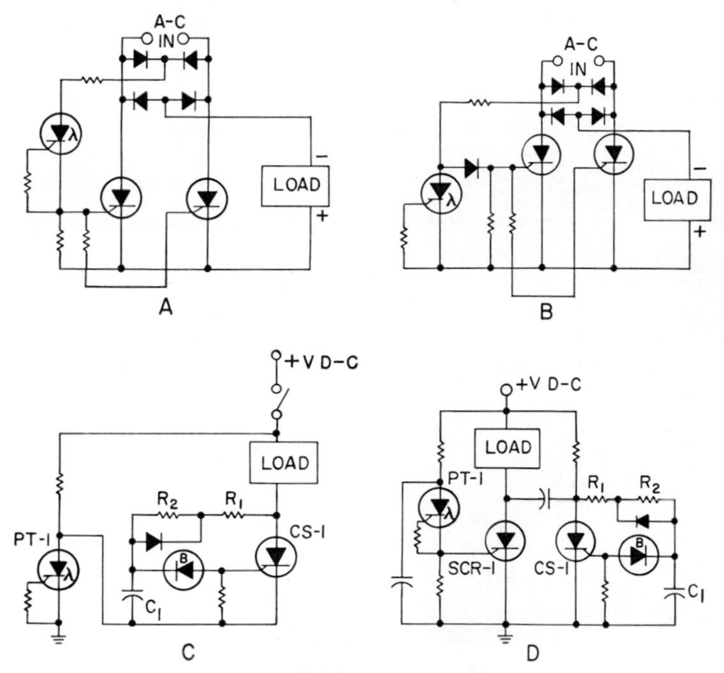

Fig. 6·18 (*A*) Output when light is present on the Photran; there is no output when light is removed. (*B*) The opposite action is achieved here which provides power to the load when light is removed and no power when light is present. (*C*) is a light-actuated time-delay circuit; when light is applied to the Photran, C_1 begins charging through R_1 and R_2. (*D*) is a light-actuated interval timer where the pulse generator is used to provide a gate firing pulse which applies power to the load for a predetermined time interval as established by CS_1 and its associated circuitry.

light is present on the Photran. There is no output when light is removed. The opposite action is achieved in the following drawing (Fig. 6·18*B*). This circuit provides power to the load when light is removed and no power when light is present. In both of these circuits, the Photran is used to control the gate voltage to the SCRs. The Photran is capable of supplying the necessary gate firing power to all currently available SCRs, including those with ratings up to 70 amp.

Two circuits for control of d-c load power are shown in Fig. 6·18*C* and *D*. One is a light-actuated time-delay circuit (*C*). When light is applied to the Photran, C_1 begins charging through R_1 and R_2. After a predetermined time interval, the controlled switch CS_1 is fired, applying power to the load. The circuit is reset by interrupting the d-c source power.

A light-actuated interval timer is illustrated in Fig. 6·18*D*. The pulse generator is used to provide a gate firing pulse to SCR_1. This applies power to the load for a predetermined time interval which is established by CS_1 and its associated circuitry. CS_1 will fire at the conclusion of this time interval, turning off SCR_1 and removing load power. This cycle will

Fig. 6·19 Silicon readout cells. *(International Rectifier Corporation)*

be repeated each time the light on the Photran PT_1 is interrupted and reapplied.

Data Processing. Photoelectrics have important uses in data processing such as, for example, in the reading of punched paper tape containing data stored for use in the computer. To read the holes and to convert them into electric signals, small phototubes are used. Light shines on the tape and as a hole passes the light it excites the photocell, which produces an output.

Figure 6·19 shows the silicon readout photocells which provide very fast response time (from 5 to 20 μsec) and are capable of reading 10,000 characters per second in perforated-tape and punched-card data-reading systems.

This application involves the reading of punched cards and punched tape for computer and control devices with preprogramming. The principle involved is very simple; however, until the advent of the solar cell, there were no photoelectric devices small enough and with sufficient power output to accomplish what was required.

In this application the assemblies of readout-cell segment positions are arranged in a mosaic pattern. Light, penetrating holes in a card or tape, impinges on the cell segment, which converts the light into an electric impulse. This impulse is then fed through a transistorized amplifying system which activates the computer or control mechanism. Only those cells which receive a light beam are activated. The other cells, of course, remain passive until they in turn are lit.

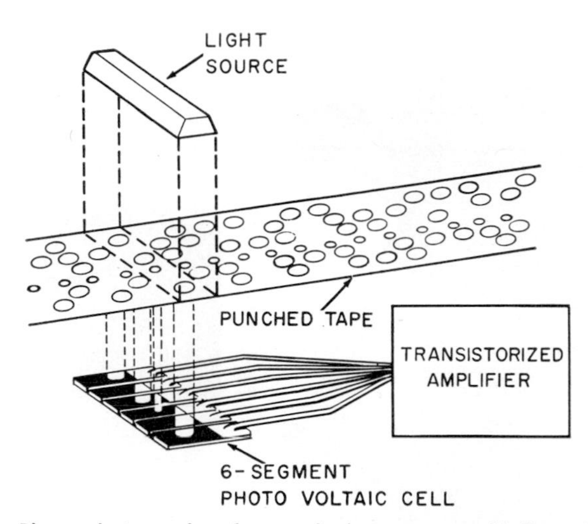

Fig. 6·20 Photovoltaic readout for punched paper tape. *(Hoffman Electronics)*

Figure 6·20 shows a six-element cell in this application. To perform these functions, an assembly (or assemblies) of readout cells is arranged in a preset pattern. Since the photovoltaic readout assembly and transistor are both low-impedance devices, optimum matching is readily accomplished. Readout cells convert light energy directly into a usable electric output, eliminating the need for external power supplies. These self-generating devices are also characterized by long life and a very short response time.

6·4 Industrial equipment Earlier parts of this chapter covered general principles and devices. This section describes some actual equipment.

Figure 6·21, for example, shows a hopper with a single light source and two sensing heads. When the material is below the bottom sensor, more fill is automatically added. The filling process stops when the material reaches the top sensor and breaks its light beam.

Photoelectric Alarm Systems. Light beams and photoelectric relays are also used in alarm systems. A modern version of such a system has transistors. The photoelectric alarm consists of a projector and a receiver. A low-visibility beam modulated at a specified frequency is projected across the protected area. An infrared filter makes the beam invisible even in the presence of smoke. Any interruption of this "black light" immediately transmits an alarm signal. Any attempt to bypass the system by substituting another light beam disturbs the frequency and triggers the alarm. Tampering with the projector, receiver, or housing, or breaking a line will also trip the alarm.

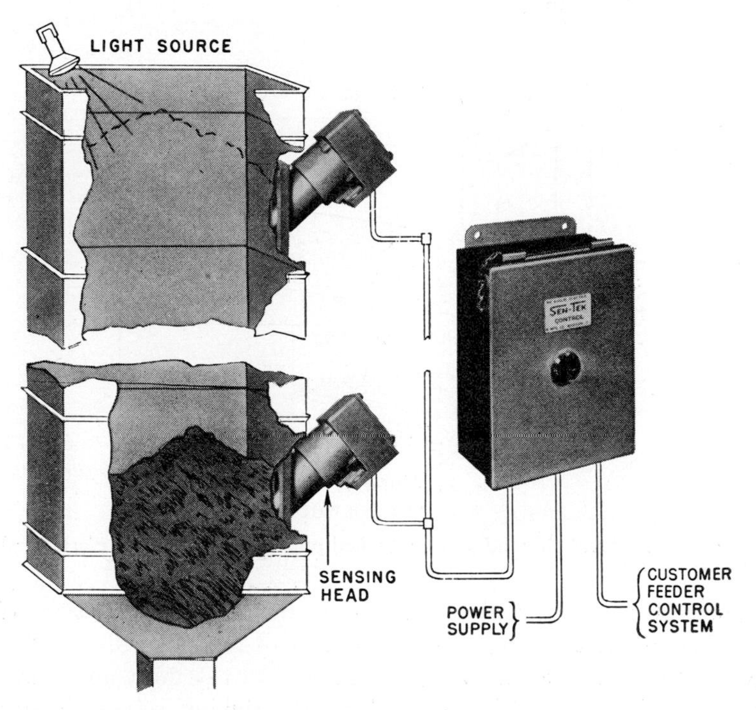

Fig. 6·21 Two-phototube sensor system as used in a hopper. (*Sen-Tek*)

The system operates at ranges up to 1,000 ft. Through the use of mirrors, the beam can be made to go around corners and completely encircle an area. An operating margin of at least 80 per cent reduction of the normal maximum light source intensity is provided to prevent accidental alarms due to gradual loss in output or efficiency.

A schematic of this type of alarm is illustrated in Fig. 6·22. The system may be operated from a 115-volt 60-cycle line or from a 12-volt battery. The a-c voltage is connected through SW-1A and fuse F (1.6 amp) to the primary of T_1. The transformer is a saturable type and together with capacitor C_1 forms a voltage regulator. This maintains a constant d-c supply of approximately 12 volts, which is independent of line-voltage variations. When a 12-volt battery is used, the positive lead is connected

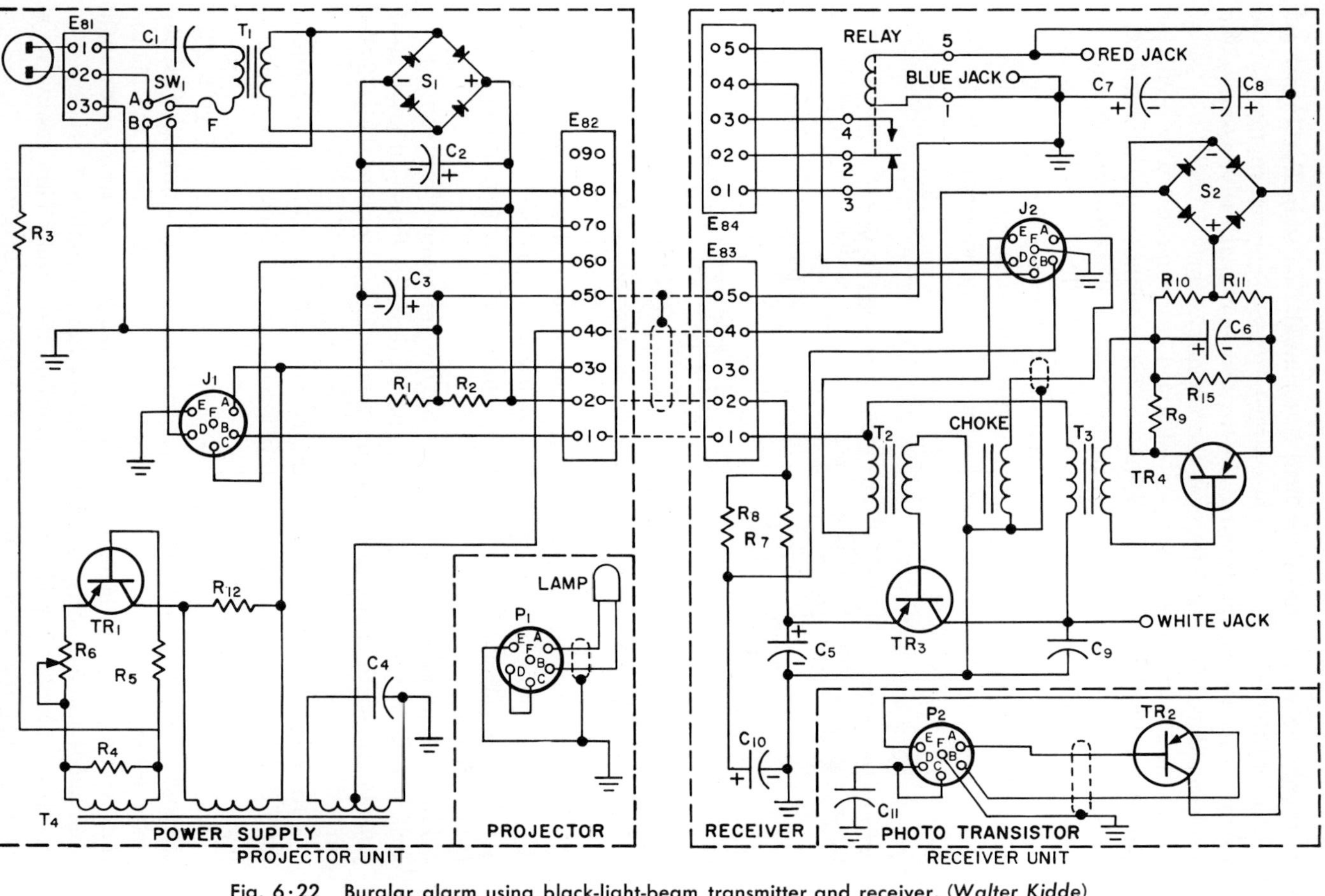

Fig. 6·22 Burglar alarm using black-light-beam transmitter and receiver. (*Walter Kidde*)

to terminal 8 and the negative lead to terminal 1 on strip E_{82}. In normal operation, the battery is trickle-charged from the power supply. In the event of an a-c power failure, the battery automatically assumes the load and thus prevents failure of the alarm system.

Modulation of the light beam is obtained electronically. The oscillator circuit consists of the power transistor TR_1; oscillator transformer T_4; resistors R_4, R_5, and R_6; and capacitor C_4. C_4 and its associated winding tune the oscillator to approximately 60 cycles. The center tap is connected to terminal 4 of strip E_{82} and supplies an a-c voltage to ground of about 20 volts. The lamp has a varying direct current passing through it which causes the intensity of the light to vary in step with the frequency of the oscillator.

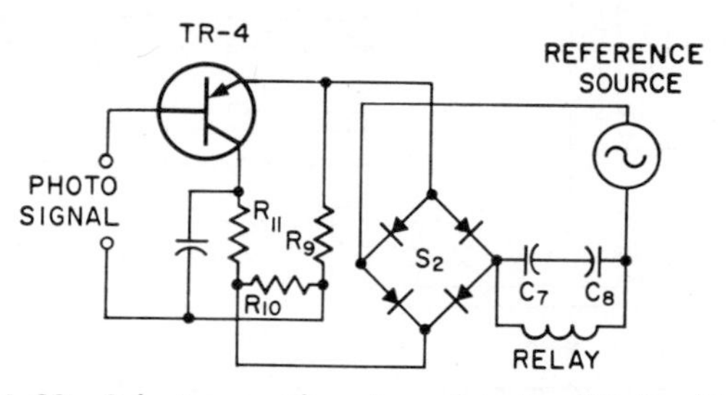

Fig. 6·23 Selenium-rectifier phase detector. (*Walter Kidde*)

Terminals 6 and 7 of strip E_{82} are connected to pins C and D of J_1. The corresponding pins on plug P_1 are shorted by a jumper, thus providing a means for the remote detection of tampering with the connection of P_1 and J_1.

Power is supplied to the receiver by the interconnecting cable from strip E_{82} in the projector to strip E_{83} in the receiver. Pulses of modulated light falling on the photosensitive junction of TR_2 modulate the collector current.

This 60-cycle current is transformer-coupled to the amplifier transistor TR_3. Capacitor C_9 corrects the phase of the amplified signal which is fed to the phase-detector stage TR_4.

The phase-detector circuit is the selenium rectifier S_2 (Fig. 6·23); transistor TR_4; bias resistors R_9, R_{10}, and R_{11}; and capacitor C_6. This detector compares the phase and amplitude of the signal from the phototransistor with the a-c signal, which is fed directly to the receiver through terminals 4 and 5 of strip E_{83}. Since both of these signals are developed in T_4, their phase (after slight correction) is the same. This results in a filtered d-c voltage across the relay which holds it energized.

Figure 6·24*A* shows the signal resulting from the modulated light striking the phototransistor. Figure 6·24*B* shows the directly coupled a-c signal or reference voltage. In Fig. 6·24*C* the reference voltage is shown after rectification. It is applied to the collector of TR_4 as a negative 120-cycle pulsating d-c signal. When a negative pulse of photovoltage is applied to the detector base and a negative pulse is simultaneously applied to the collector of TR_4, it will result in drawing more current through the collector circuit, as shown in Fig. 6·24*D*. By drawing more current in the collector circuit at alternate half-cycles, more current will be drawn through the relay in one direction than the other, as shown in Fig. 6·24*E*.

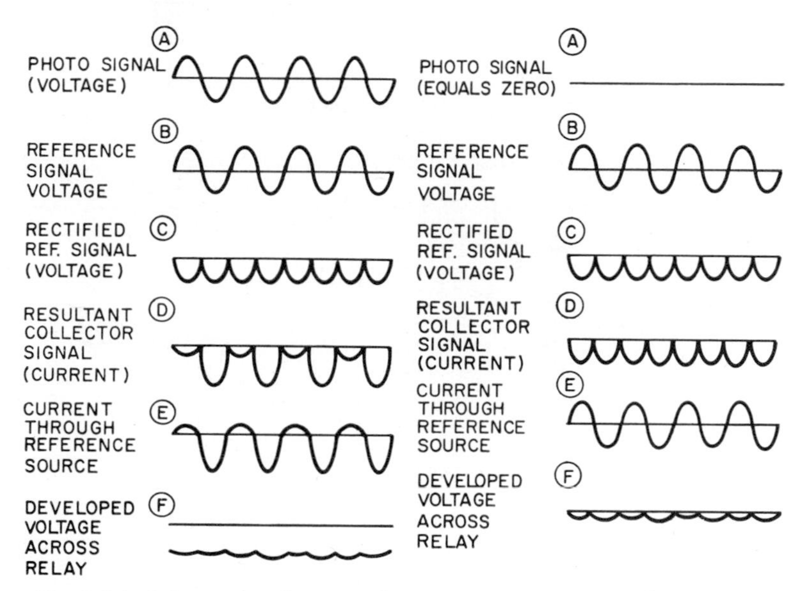

Fig. 6·24 Relay is closed in normal operation. (*Walter Kidde*)

Fig. 6·25 Alarm conditions for loss of light or improper frequency. (*Walter Kidde*)

Filtering the varying direct current (by means of C_7 and C_8) produces an average voltage which is applied to the relay; this holds it energized, as in Fig. 6·24*F*. The relay contains both a normally open and a normally closed circuit. The normally open contact is available at terminal 1. The normally closed contact is available at terminal 3, and the armature is connected to terminal 2 on strip E_{84}. Terminals 4 and 5 are connected to pins *C* and *D* of J_2. The corresponding pins of plug P_2 are shorted by a jumper, thus providing a means for remote detection of tampering with the connection between J_2 and P_2.

Figure 6·25 shows the operation of the system with a loss of light at the receiver, or the normal "alarm" condition. The reference voltage is applied in the same manner as shown in Fig. 6·25*B* and appears as a negative 120-cycle d-c pulse at the collector of TR_4 as in Fig. 6·25*C*. Since there is no signal applied to the base of transistor TR_4, the current through the reference source results in a net d-c value of zero across the relay; the negative current swing is equal in amplitude to the positive current swing.

If a light of different frequency is applied to the phototransistor, the polarity of the d-c voltage across the relay reverses at a rate equal to

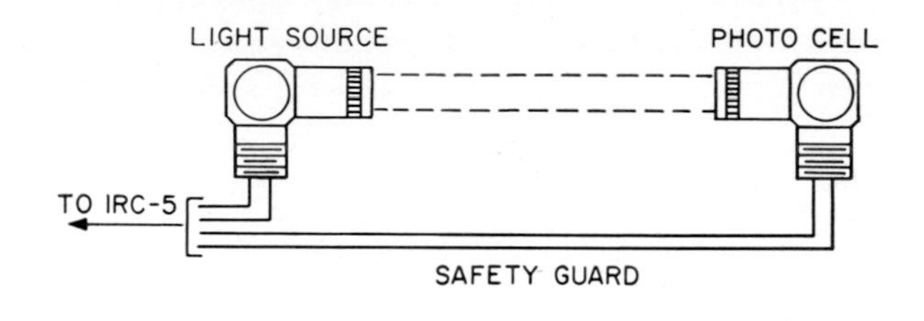

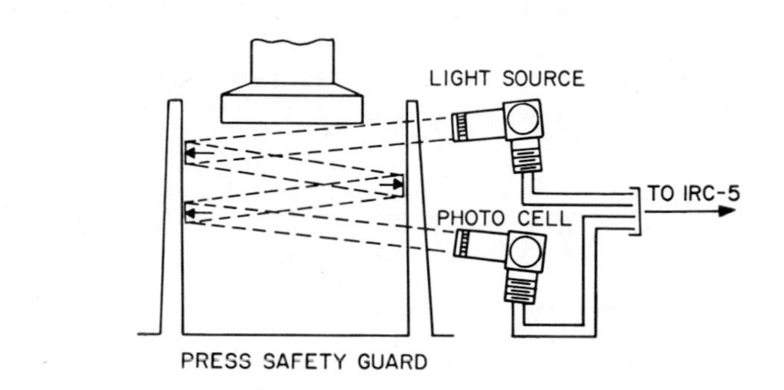

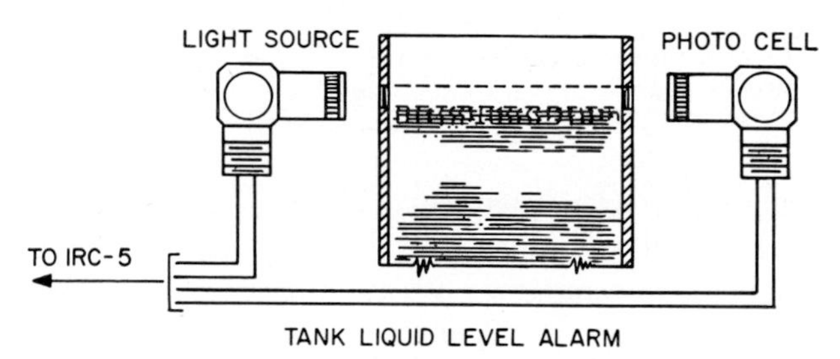

Fig. 6·26 Typical application of photocells showing safety guard, press safety guard, and tank-liquid-level alarm.

twice the difference between the light frequency and the reference voltage frequency. This results in a momentary dropout of the relay at each reversal. If the frequency difference is faster than the relay can follow, the relay will stay deenergized.

The system may be set up according to the requirements of a particular situation. The relay may be used as a normally open or a normally closed contactor.

General-purpose Equipment. Many photocells and their associated amplifiers are multipurpose. Some typical examples of the use of this equipment may be seen in Fig. 6·26.

Figure 6·27 shows the amplifier for this system. The Model IRC-6 control may be used in conjunction with any of the light sources and photo units. With a light change on the photocell of very short duration,

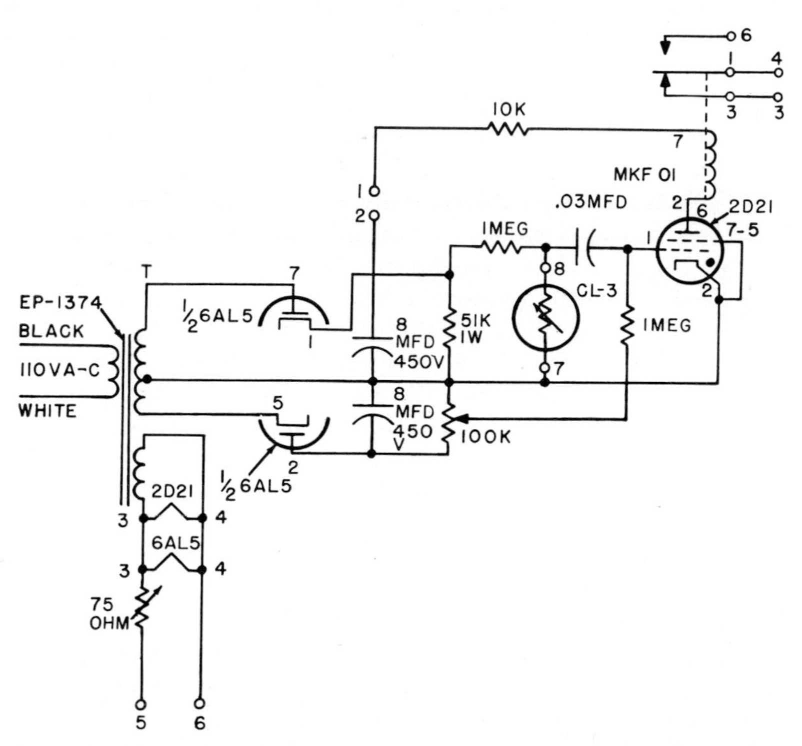

Fig. 6·27 Thyratron amplifier driven by photoconductive cell. (*Photomation*)

a power relay will operate and remain locked in. The duration of this light change may be as short as 1 msec. Once the relay locks in, the photocell loses control over the relay condition and relay can be released only by external means. The relay is controlled by a thyratron tube operating on direct current applied to its plate circuit. This direct current is provided by a built-in power supply. Because the relay locks in, it is necessary to release the circuit by external means. This is usually done in the form of a normally closed contact which is momentarily opened either manually or automatically. If it is to be opened automatically, this can possibly be accomplished by utilizing the rotation of a member of the machine on which the electric eye is applied. At a particular point in the operating cycle, a cam attached to the rotating member of the machine may be caused to trip the normally closed switch, thus providing a momentarily opened circuit.

The power relay is a plug-in type and is offered to energize its coil on either the make of light or the interruption of light to the photocell.

For relay response, the make or interruption in the light beam should persist for at least 1 msec. Once the relay operates, it locks in and further light changes on the photocell have no effect on the relay operation. After the relay is released by external means, the amplifier is ready for the next operation. Contact rating of the power relay is 10 amp at 115 volts a-c or 5 amp at 230 volts a-c, noninductive current. Supply voltage is 115 volts a-c and power consumption is approximately 10 watts.

Eight connections are provided on a terminal strip and are used as follows: two for the external normally closed relay release circuit; two for the power relay contacts (armature and *either* normally open or normally closed contact); two for the light source (no polarity); and two for the photo unit (no polarity). The 115 volts a-c single-phase power connection is made by splicing to two leads coming from the chassis, which are color-coded white and black, respectively. Only those wires from the light source and photo unit should be run in the same conduit.

A sensitivity control is included on the chassis, and rotation of same determines the amount of light necessary to cause the power relay to function. The lamp control varies the lamp intensity and makes it possible to select the minimum voltage consistent with satisfactory operation. Thus lamp life is greatly increased.

The systems in the drawings of Fig. 6·26 use transmitted light. Some others using this arrangement are shown in Fig. 6·28. *A* shows how falling objects break the light beam and, in so doing, trigger a counting device.

B illustrates various operations in monitoring edges, paper breaks, etc. Also, by using two electric eyes as shown in the center, a sheet can be automatically controlled whenever it deviates from the electric eyes. This is a rather common application in printing.

The system shown in *C* is for monitoring bottle caps. The problem was to ensure that the bottle was capped before it went to the next operation after the filler and capping machine. The thickness of the cap was about ⅛ in. Therefore, a beam set above the top of the bottle but broken by the cap would automatically reject any bottle which was not capped. However, since the bottles came at intermediate positions, it was necessary also to have a timer; in one method, if the bottles were continuously in touch with each other, a diagonal beam broken by two successive caps would immediately reject when either cap was not present. This is extensively used by pharmaceutical firms.

The system in Fig. 6·28*D* monitors a stop position which halts the work or tool as it strikes the light beam. *E* shows a water-column control as used in a boiler. Using two controls permits a "too high" or a "too low" warning.

Some uses, as shown in Fig. 6·29, require reflection.

A shows a very common application used not only in inspecting but also in automatic processing. These capacitors have one end black and

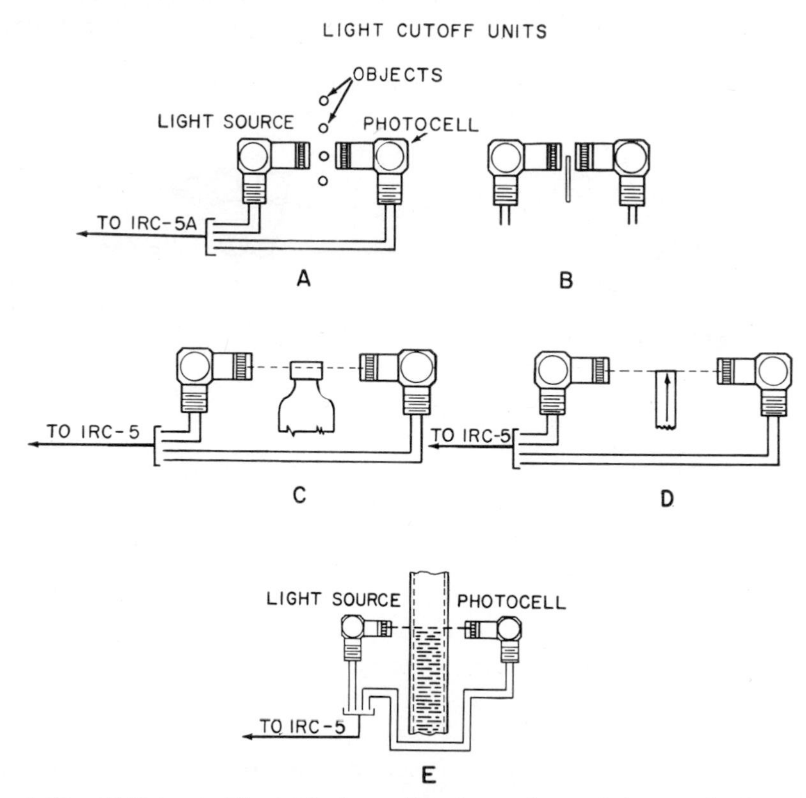

Fig. 6·28 (*A*) Light cutoff by small objects; (*B*) edge guide as a light control; (*C*) monitoring bottle caps; (*D*) monitoring a stop position; (*E*) gauge-glass-level alarm controls high or low alarm. (*Photomation*)

one tan. However, they must be presented properly to get the leads assembled. If one comes along in reverse position, it is automatically adjusted. The same system is used for inspecting the end caps, as in *B*, to ensure that the caps are properly placed. These are brown plastic caps, the inside having a rubberized surface. The reflection from the rubberized surface is quite different from that from the plain plastic, so that any cap coming along upside down is rejected.

Figure 6·29*C* shows the control of registration or other marks. This is a common application in printing, packaging, and other industries. The illustration shows the pickup of registration marks on an opaque surface by reflected light. A typical example is in packaging. The slippery outer covering is provided with a black mark which indicates that the packaging machine will fold it exactly on the proper edges without obscuring print. If the material lags a certain tolerance, the electric eye sends a signal to the adjusting mechanism, which feeds it back into registration. This same system is used in many printing applications, as well as on registration

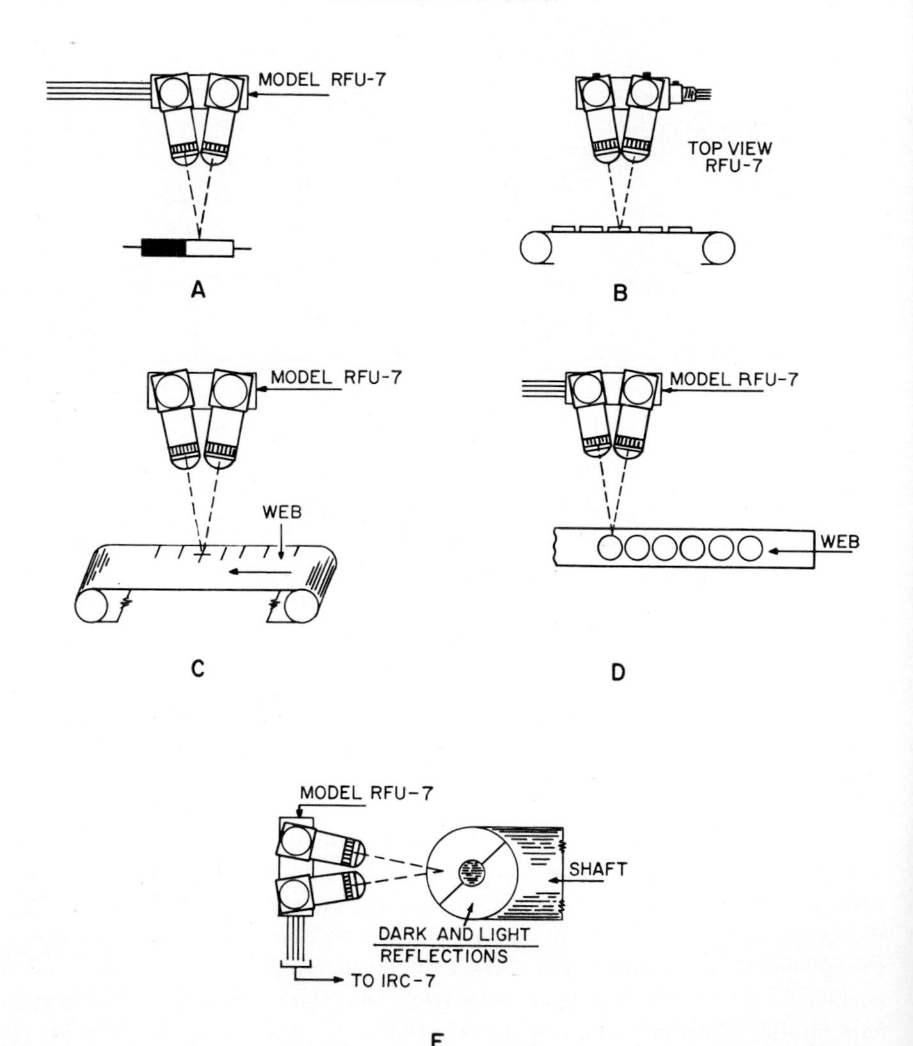

Fig. 6·29 (*A*) Reflected light from resistors; (*B*) surface discrimination; (*C*) packaging-machine pickup of register marks; (*D*) counting objects such as cans or rollers by reflection; (*E*) minimum speed safety alarm for turbines.

and line marks. It also is applicable to inspection of passing objects which should have a reflection at particular points. In the cigarette industry, some manufacturers use two such reflecting systems on the same pack, one to ensure that the revenue stamp is present and the other to ensure that the small tab end also is present. Absence of either one rejects the pack. Many large plants regularly use both the direct-action and the reflected-action beams for different operations in their works.

Figure 6·29*D* shows a simple counting system and *E* shows a speed monitor for turbines.

6·5 Solar-energy converters Energy from the sun can be converted into electric-current solar batteries which are photoelectric devices. Solar energy is a vast potential source of electric power. About 1,000 watts fall upon each square yard of the earth's surface in full sunlight. If this could be used, solar power could supplement other energy sources.

Our electricity now comes from coal and oil, which produce heat for steam turbines, and from vast hydroelectric power stations like those at Niagara Falls. Nuclear energy, of course, is also a potential source of electricity. There are nuclear-fission power plants now in operation and nuclear fusion, the controlled H-bomb, is a future possibility.

But solar energy is an important power source now in use for providing energy for rural telephone lines, eyeglass hearing aids, and solar-powered radios and satellites. The true solar cell is a photovoltaic device which produces a voltage when light hits the cell; hence it does not require a voltage source. It produces energy which can be used to run other electric equipment.

Solar cells are, of course, used in earth satellites such as the Tiros as shown in Fig. 6·30.

The photovoltaic effect was first observed in 1839; elements used in early experiments were capable of efficiencies of only about 1 per cent. They required a very large surface area to develop any real amount of power. But these field tests only proved the solar converters to be possible, not to be economically feasible for large-scale power generation.

The first feasibility study of a silicon solar converter was carried out by the Bell Laboratories during 1955 when a converter was installed on a telephone pole in Americus, Georgia. This operation proved that the relatively small area of silicon converters could provide power for operation of telephone lines. These new silicon solar cells were shown to be economically feasible devices which, once installed, will provide power for hundreds of years.

A silicon photocell which generates energy uses a PN junction to produce electric energy from light. Pure silicon is a poor electric conductor. Adding small amounts of arsenic or phosphorus to the silicon crystal permits electron conduction.

Silicon has four valence electrons. If an impurity such as arsenic, with five valence electrons, is placed in the silicon, one of the impurity atoms can replace a silicon atom in the crystal lattice. The material is left with an extra electron which is free to move if it is subjected to any force. Energy required to remove the electron from the five valence electrons is very small and can be furnished from the crystal thermal energy. This free electron contributes to the electric conductivity of the silicon.

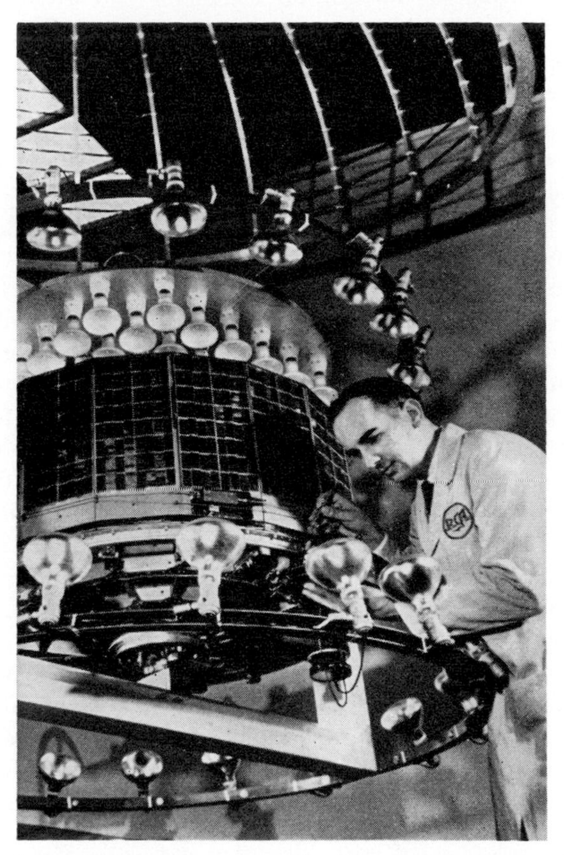

Fig. 6·30. Testing Tiros satellite solar cells. (*International Rectifier Corporation*)

As the conductivity is caused by electrons, which are negative charges, the material in this condition is called N-type semiconductor. If, instead of arsenic, an element such as boron containing three valence electrons is added, a condition will result in which the boron atom replaces a silicon atom. This material is left with a deficiency of one electron. This then is called a hole and can move around the semiconductor by being filled from adjacent valence bonds. Any bond which looses an electron to fill an adjacent hole then has a hole and in turn can be filled by a different electron.

In effect the hole is moving. As holes are actually a lack of negative charges, they are positive with respect to normal silicon. Since conductivity is caused by these positive holes here, this material is called P-type silicon. When N-type material and P-type material are in close contact with each other, some of the holes and the electrons intermingle with each other and leave behind a dipole layer due to the fixed charges of the impurities.

When both N and P types are present in a single crystal, the NP junction is the boundary between the two types. In the presence of light, electron-hole pairs are formed as light energy falls upon the silicon. The electric field at the junction forces the electrons into the N side and the holes into the P side.

This potential difference (voltage) is the power output from the cell. In bright sunlight the voltage is about 0.6 volt and cells can be placed in series for higher voltages. There is no chemical change, as in a battery; hence the solar cell has no limit to its lifetime of usefulness.

A typical solar cell produces about 0.3 volt. Current depends upon the light intensity, the cell area, and the load resistance. An average cell has a maximum power output of about 30 mw per sq in. of area with a 4- to 5-ohm load.

When light falls on the photocell, the energy-containing photons which make up the light penetrate the silicon sufficiently to be absorbed near the junction where the silicon contains boron on one side and arsenic on the other side. If the light is bright enough, some of the photons will have enough energy to break a crystal valence bond and create a hole-electron pair. The hole and the electron separate. The holes flow into the P-type material which is the material created by the boron, and the electrons flow into the N-type material which is the material containing the arsenic.

Although any one photon causes a very small amount of activity, there are millions of photons, so an external voltage is built up on the two connectors plated onto the silicon disk. If a load is connected across these two connectors, current will flow through the load. The light itself is acting as a current generator. When the current flows through the load, the hole-electron pairs that have been formed by the photons recombine and return to the same condition they were in before they were upset by the photon energy. Therefore, there has been nothing removed from the silicon and nothing added to the silicon during the process of creating electric energy from light. The process can continue as long as there is light energy available.

Figure 6·31 shows the characteristics of typical solar energy converters. In *A* the current as a function of light intensity is shown, while *B* shows the voltage variations. Since sunlight is the prime energy source, these cells must take maximum advantage of it. The sun's light is compared with both selenium and silicon cells, in terms of spectral sensitivity, in Fig. 6·32.

For industrial use, one manufacturer has power-supply modules fur-

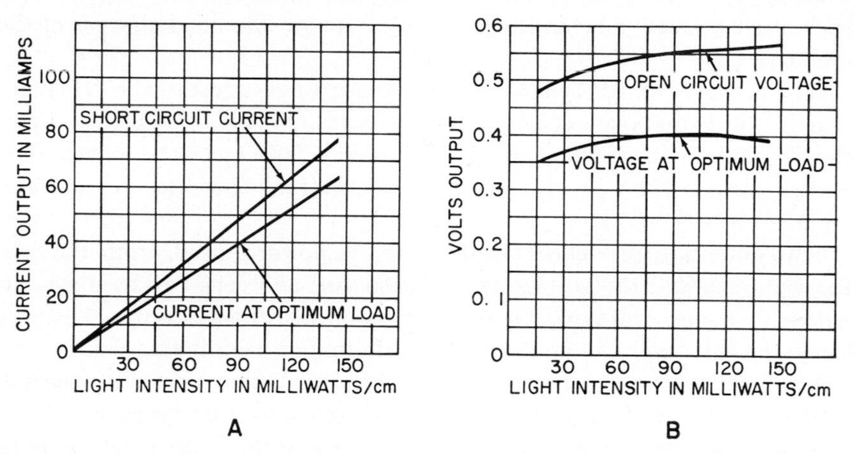

Fig. 6·31 (*A*) Current output of solar cells as a function of light intensity; typical characteristics based on cell of 1.8 sq cm active area; (*B*) voltage output of solar cells as a function of light intensity.

nished as 5-watt units. As many of these as necessary can be used in combination to furnish any amount of desired power.

The control elements are a relay and a back-current diode. The diode functions to prevent current developed by the storage battery from flowing back through the solar cells when they are not illuminated by sun-

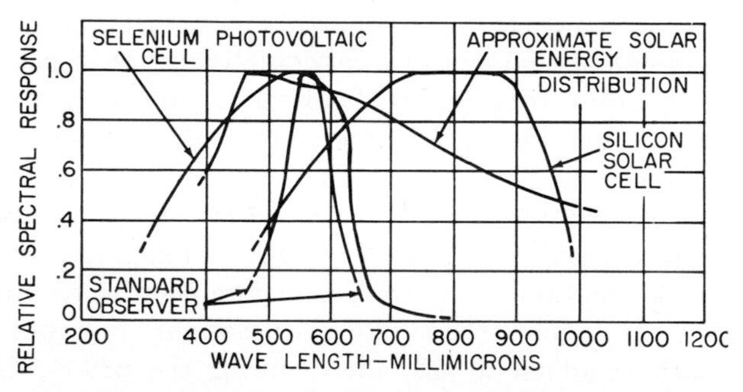

Fig. 6·32 Solar-cell spectral response. (*Hoffman Electronics*)

shine. As there is a voltage drop across any diode that can be put in the circuit, an appreciable power loss could be realized in the diode. In order to obtain as much power from the converter as possible, a relay is used to short out the diode during the time that the solar unit is providing a high charge. This relay is operated by separate solar cells connected directly to the relay coil.

Energy from the sun can be used directly as a source of heat by using a solar furnace made, in principle, from a parabolic mirror. The rays from the sun fall upon the mirror surface, where they are all reflected to meet at the focus. Spherical mirrors are not usually used since they do not bring the energy to so sharp a focus.

In a solar furnace the heat of the focused rays can be used, for example, to melt metals. It is also possible to build large devices such as this where the focused solar energy falls upon solar cells to produce electricity for the home. The Association for Applied Solar Energy has been studying this problem.

According to their figures, in the Temperate Zone about 580 kwhr falls daily on a roof of 1,075 sq ft. With an efficiency of only 10 per cent in the energy conversion process, this would produce 58 kwhr in a day. The average home uses only about 11.5 kwhr; hence it is possible—in theory, at least—to convert the sunshine on your roof into electricity for the house.

There are only two problems. The present cost is much greater than that of commercial electricity, and a storage system of some sort would be needed for rainy days. This storage system is the reason for the higher cost since it must have in reserve enough electric energy to carry over the longest span of cloudy days or days of snow and rain.

The control function of solar cells, in automation applications, will shortly outweigh their power-generating functions in importance. The list of these applications is growing. The automatic dollar-bill changer is in essence a "robot" cashier. It will accept a dollar bill, verify its authenticity through the use of light-to-electric-energy conversion principles, and make change in the desired coin combinations.

The optical guide-o-matic tractor is a unique operatorless tractor. Vehicle guidance is possible through the use of solar cells which register the light differentiation between a painted guideline and the surrounding floor or pavement area.

The cells control the guidance of the vehicle by following the line. They also act as a safety switch by instantly signaling the mechanism which turns off the power of the vehicle if it deviates from its prescribed path as little as ⅛ in.

A total of eight cells are used in the detection device. These cells are positioned in banks of two each on a U-shaped block. The arrangement is such that one set of cells "sees" the guideline while the other detects the adjacent areas. By balancing the various voltage outputs of the cell blocks, electric impulses can detect and correct the turning mechanism of the vehicles as it moves.

A unique application developed by a Texas oil company utilizes a special type of cell, approximately 2 in. in diameter, whose center is drilled out so that the contact strip is on the inside radius. The cell, with a rotating light, is placed in the drill bit. As the bit turns the light beam in a 360° azimuth, the light is reflected from the walls of the excavation. Variations in light reflection register on the solar cells, which send these fluctuations in electric impulses to the surface where they can be interpreted on instrumentation.

Oil prospectors can determine immediately the character of their drilling, without have to haul the drill bit up to inspect samples of the material being drilled. By using solar cells in this fashion the tedious, time-consuming, and extremely expensive older methods of inspection can be eliminated.

In hot-metal sensing, the cells are inserted into the heating furnace or near the hot metal. They convert the radiant energy given off by the hot metal into electric impulses which can be read on meters. Because the cells are extremely sensitive to the red portions of the spectrum, their output will be proportional to the color temperature of the radiation from the metal. The metal worker is thus able, by reading electric instrumentation, to control his processing with extreme accuracy.

In gas-fume detectors the cell acts as a triggering mechanism in a system which detects gas fumes and warns of dangerous concentration of these fumes in an enclosed space. This gas detector has a platinum filament that emits infrared rays when exposed to gas fumes. The radiations are picked up by a solar cell, which triggers a transistor amplifier that flashes a warning light on the boat control panel.

6·6 Infrared Of great potential use, infrared (IR) can detect any warm object. IR can locate railroad "hotboxes," which are overheated journal bearings, while a train is in motion. IR also can protect against intruders or be used for communications. An example of the use of IR is shown in Fig. 6·33 where the hot metal slab radiates heat (IR) which is detected and used to operate the control system for the rolling mill.

The physical basis for IR is simply that an object at a temperature above absolute zero (0°K or −273°C—every object is above this temperature) radiates invisible electromagnetic energy. This energy is propagated at the speed of light and occupies that portion of the electromagnetic spectrum between 1 and 100 μ (one micron equals 10^{-6} meter equals 10^{-4} centimeter equals 10^4 angstrom units).

This is the radiation that accounts for sunburn received at the beach. It is not the thermal effect due to conduction or convection. The comfortable ambient temperature of our planet earth is due to the thermal radiation received from the sun.

All objects are natural radiators of this invisible energy and can be detected by appropriate passive instrumentation which does not reveal the presence of the observer. For military applications, IR capabilities offer significant advantages.

Electromagnetic radiation incident upon a surface may be either reflected from the surface or transmitted through the surface of the body; some may be absorbed and converted into heat, while the remaining fraction may pass on through the body and escape. For bodies other than blackbodies, some radiation is absorbed and some is reflected; the ratio

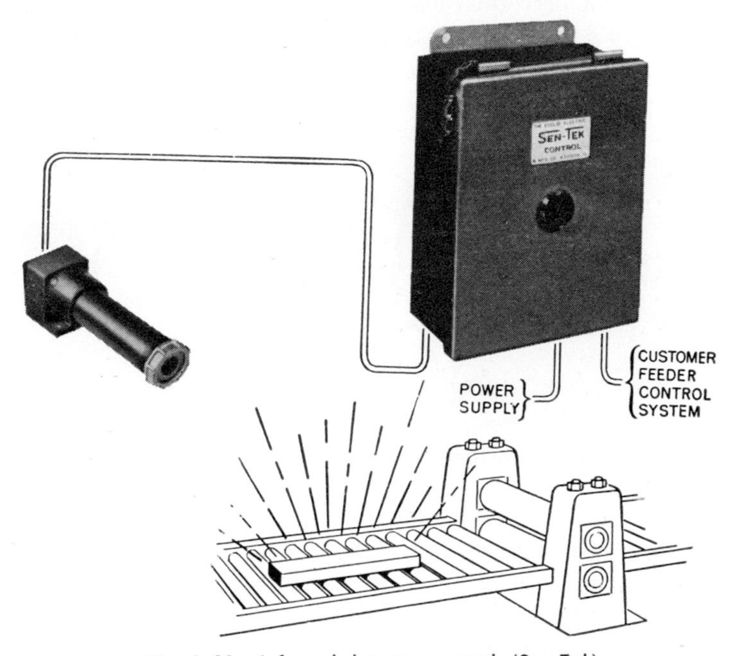

Fig. 6·33 Infrared detector control. (*Sen-Tek*)

of the absorbed energy to the total energy is termed emissivity. The emissivity, therefore, shows a comparison of grey bodies with blackbodies. The emissivity is also a function of the frequency. Much of the solar energy is concentrated in the visible spectrum, and any energy which is not reflected is efficiently radiated in the IR, with the maximum spectral emissive power occurring near 10 μ for surfaces at or near room temperatures. White painted surfaces are up to 19°C cooler in direct sunlight than an aluminum plate, for, although visual region reflectivities are similar, the metal has a low emissivity in the IR and thus cannot radiate away its excess heat at the same temperature.

There are two fundamental types of IR detectors. These are the thermal detectors and the photodetectors. Incident radiation changes the electrical properties of both types. Both thermal detectors and photodetectors are quantum detectors. Thermal detectors rely essentially on the heating effect produced by the absorbed radiation. Photodetectors exploit the phenomenon that photons of appropriate energy create free charge carriers in certain materials. The solid-state concept of a solid divides it into two thermodynamic systems: the lattice and the electronic systems. The electronic system consists of a valence band and a conduction band separated by an energy gap. The lattice system is composed of the atoms or molecules which constitute the solid and can be characterized by lattice vibrations. In photodetection, when the energy of the incident photon exceeds the energy gap, electrons are excited from the valence band to the conduction band directly across the energy gap. In thermal detectors,

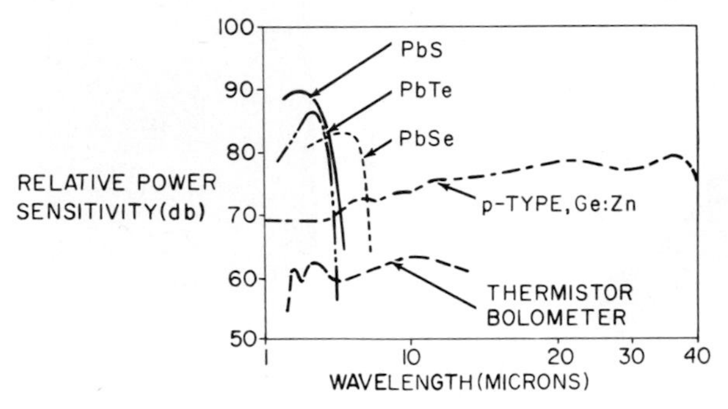

Fig. 6·34 Infrared sensors; power-versus-light wavelength.

the radiation is absorbed in the lattice, causing heating of the lattice and a subsequent change in the electrical properties of the detector.

Among the significant parameters associated with IR detectors, the most important are the detectivity or sensitivity, the time constant (which is an indication of the time required to respond to radiation), and the spectral sensitivity. Photodetectors have a larger detectivity and a smaller time constant than thermal detectors. However, photodetectors are sensitive over only the portion of the spectrum permitted by the energy gap. The sensitivity of the photodetectors may be improved at the expense of increasing the time constant. Thermal detectors exhibit a relatively flat response over the entire spectrum but have a lower sensitivity and a longer time constant. The characteristics of the photodetectors (lead sulfide, lead telluride, lead selenide, indium antimonide, zinc-doped germanium) and the thermal detector (thermistor bolometer) are shown in Fig. 6·34.

There are many potential applications for IR. Battlefield surveillance may be accomplished with either passive or active IR and achieves two distinct advantages simultaneously. First, it is not limited to the daylight hours of activity and, second, it affords a maximum degree of security to the observer. Active IR systems rely significantly upon the reflection of IR energy from the area under surveillance. It, therefore, requires a source of IR energy to illuminate the area. A passive system utilizes the inherent radiation emitted from the constituents of the area.

Equipment may be maneuvered into a tactical nighttime situation by equipping the drivers of the vehicles with IR binoculars and snapping IR filters over the headlights of the vehicles. The sniperscope, now called a "weapon sight," is an IR viewing device.

IR image tubes may be used to convert the periscopes used in tanks for night operation. A tank commander's IR observation system consists

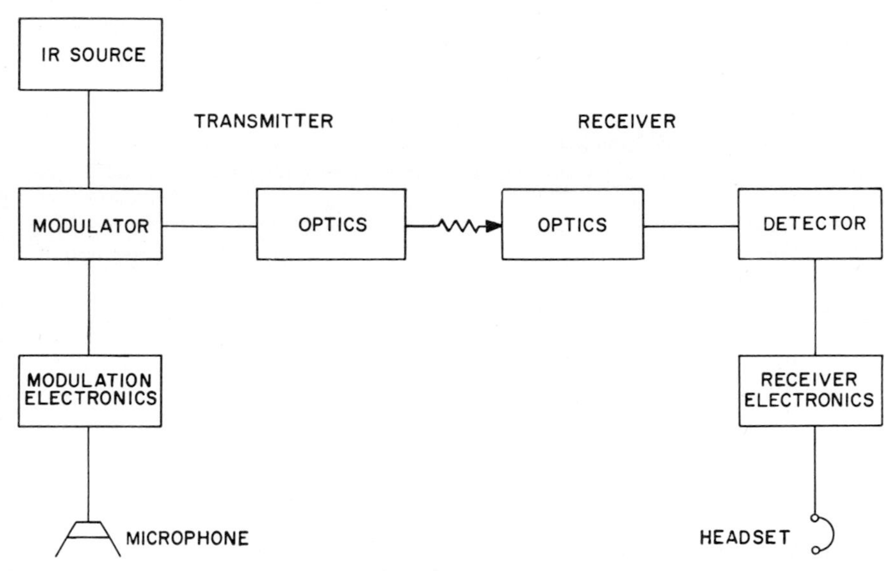

Fig. 6·35 Infrared voice communications.

of a dual-purpose IR-visible searchlight and a telescope. This tripod-mounted telescope enables a military commander to study a tactical situation, to deploy his forces effectively, and to observe the progress of a night encounter from some remote vantage point.

A block diagram of an IR communication system is shown in Fig. 6·35. It consists of an IR source, modulator, and optics which comprise the transmitter and a receiver that itself consists of optics, detector, and electrons. The principle advantage of such a communication system is privacy. This results from the narrow bandwidths, the lack of side lobes, and the ease with which the visible part of the radiation can be removed with filters. It affords a convenient private communication for distance as far as one can see.

REVIEW QUESTIONS

6·1 Why is a self-rectifying amplifier circuit slower than one with direct current applied?

6·2 On what principle does the photoemissive tube operate?

6·3 Why is the photocathode of large area in a photoemissive tube?

6·4 Draw a circuit where a phototube (emissive) closes a relay when light increases.

6·5 Why are amplifiers used with photoelectric devices?

6·6 List 10 applications of photoelectric devices.

6·7 What is a modulated light beam?

6·8 When is it used?

6·9 What is the difference between a photoconductive and a photovoltaic device?

6·10 What is a solar cell?

6·11 How does it operate?

6·12 Why is "black light" used in an alarm system?

6·13 What is meant by fail-safe?

6·14 How does a multiplier tube operate?

6·15 What is meant by "gas amplification"?

6·16 What is the purpose of the codes in Fig. 6·2?

6·17 Draw a photoconductive circuit. Why is a power source required?

6·18 From Fig. 6·6, which are the tubes most sensitive to the red end? To the blue end?

6·19 What is the difference between the S-12 and S-13 curves?

6·20 For what load resistance will the solar cell have the greatest power output?

6·21 How is the photovoltaic cell used for light measurement?

6·22 What is the current output at 2,000 ft-c in Fig. 6·13*C*?

6·23 Explain how the circuit in Fig. 6·11 operates.

6·24 How does the Photran operate?

6·25 How can it provide a pulse output?

6·26 What is the purpose of each relay in Fig. 6·16?

6·27 How does the circuit in Fig. 6·18*B* operate?

6·28 Explain how the Photran can trigger an SCR.

6·29 How are photoelectric devices used to read punched cards?

6·30 Explain the operation of the thyratron in Fig. 6·27.

6·31 Figure 6·31 shows the N2009 solar cell characteristics. As light increases from 60 to 120 mw (per sq cm), what is the current and power change?

6·32 What is the difference between an active and a passive IR system?

CHAPTER

7

Counters

From ball bearings to oil cans and from cigarettes to sliced bacon, almost every manufactured product is counted somewhere along the production line. Counters can measure liquids with a rotating flowmeter, they can measure length with a calibrated disk, or they can count individual items of all types.

7·1 Applications of counters There are many uses in industrial electronics for high-speed direct-reading electronic counters, which can be utilized to control any operation or to activate an alarm after a preselected total count has been reached.

A few of the many applications are batching and packaging pills, bottles, bottle caps, canned goods, pen points, machine parts, electronic components, and other parts in exact preselected quantities; controlling the exact length of stock in cutting operations; and controlling high-speed machinery. Any electrical, mechanical, or optical events which can be converted into electric impulses can be counted and controlled. Devices to effect this conversion may be photocells, magnetic coils, switches, and suitable transducers for pressure, temperature, velocity, acceleration, and displacement. A typical use of industrial counters is counting printed sheets, as shown in Fig. 7·1.

Many circuits used in industrial electronics are similar to those found in radio and television circuits and in test equipment. Counting and frequency measurement are closely related. For example, when individual cans on a conveyor pass a photoelectric cell they cause a pulsed signal output. These pulses are counted and an indicator provides a cumulative record of the number of pulses. This is a simple counting procedure. Frequency measurement, on the other hand, considers the number of pulses per unit of time.

A number of electronic and electromechanical configurations are possible for counters. Tubes, transistors, and relays are used in multivibrator, flip-flop, staircase, and blocking-oscillator circuits.

Just as it is possible to read out the number of pulses per second, it is also possible to determine time intervals by counting the number of pulses passed through a gate when the pulse frequency is known. Circuits can

Fig. 7·1 Electronic counters are used by Warner P. Simpson Company, Columbus, Ohio, on paper-folding machines to batch by predetermined counting. This eliminates measuring or weighing of stock prior to folding. (*Veeder-Root*)

be designed to operate in either the conventional decade (denary) or the binary numbering systems or in both. Readout may consist of individual or a combination of different forms, namely, direct reading, electric signals, punched tape, signal lights, alarms, or mechanical functions. Readout may be obtained in many cases without destroying the information in the machine in much the same manner as a subtotal is obtained from an ordinary adding machine. This permits the continued use of the previously stored data for additional operations.

There are several basic counter applications. In the production system illustrated in Fig. 7·2 finished items move along a conveyor as shown.

The pickup device is a transducer which changes energy from one form to another. Possible types are photoelectric pickups, magnetic pickups, capacitance relays, and induction coils. Different types of production require different pickup systems. All, however, produce electric pulses which are fed to the counter, which, when it reaches a predetermined amount, can trigger an alarm. It is also possible for the counter to read just the total number rather than to count in groups or batches. Control of the

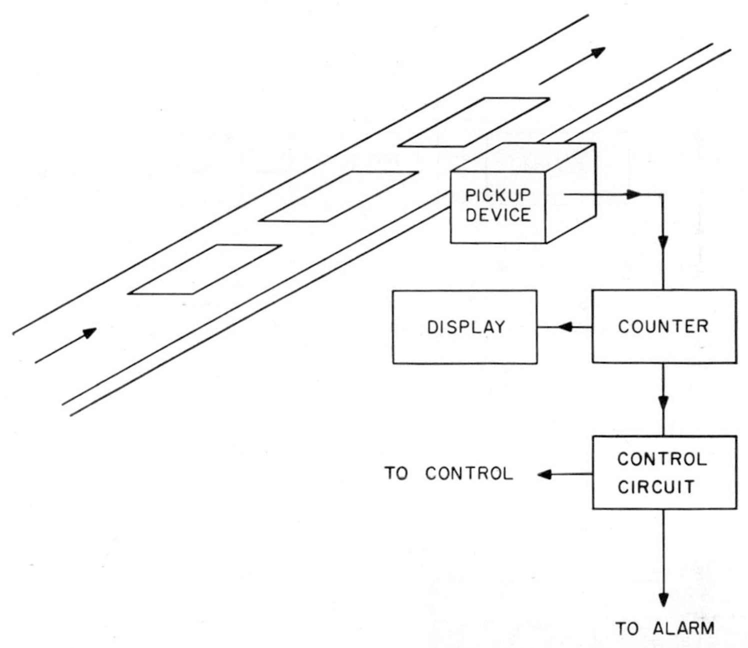

Fig. 7·2 Counter control and alarm in which the counter has a signal input and provides both display and control.

conveyor operation is also possible from the output of the counter circuit.

In a counter application for a speed indication in a jet-engine test, a sensing or pickup unit provides a signal to the counter for each rotation of the shaft. One possible sensor is a photocell reading one white spot on a dark area of the rotating part of the engine. Pulses are counted in time units, as revolutions per second, and they may be both displayed and automatically recorded.

A more detailed system is shown in Fig. 7·3. In this drawing, steel plates are moving down the line where they are stacked. Pulses from the pickup feed the preset counter which warns the operator of the approach of full count, then signals for a full stack. The number of plates in a stack depends upon the preset value. Motor control of the conveyor provides a constant rate of stacking and prevents accidents because of too great a stacking rate. Both an alarm and a printed record are part of this system.

Another use of a counter system is in the drilling, reaming, and honing of engine blocks. After these automatic operations, each block is scanned

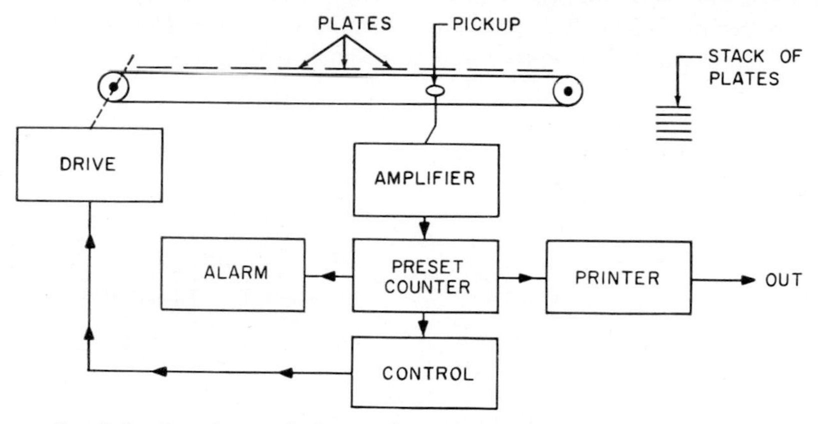

Fig. 7·3 Counting steel plates with counter printout, alarm, and control.

by photocells. A preset counter is adjusted for the proper number of holes. If the count shows a hole missing, perhaps because of a broken drill, the following production steps are not allowed to proceed on that block. This prevents damage to costly production equipment.

Fig. 7·4 Electronic counter.

A typical counter, which can be used for measuring time intervals, counting, and measuring frequency, is shown in Fig. 7·4. In measuring frequency, for example, a gate circuit is used at the input. A gate is an active circuit which can be opened (turned on) or closed (turned off) by a timing signal. The frequency to be measured is applied to the gate.

For a precise period of time the gate is left open and then it is closed. The frequency to be measured is counted during this time period.

Because the time period is controlled exactly, the counts per unit of time become a measure of the unknown frequency. The basic module of this unit is the decade counter.

7·2 Decade counters Electronic circuits which count by 10 form the basic unit of industrial counters. Any circuit which can act as a frequency divider can be made into a counter. Among the frequency dividers are blocking tubes, staircase counters, and multivibrators.

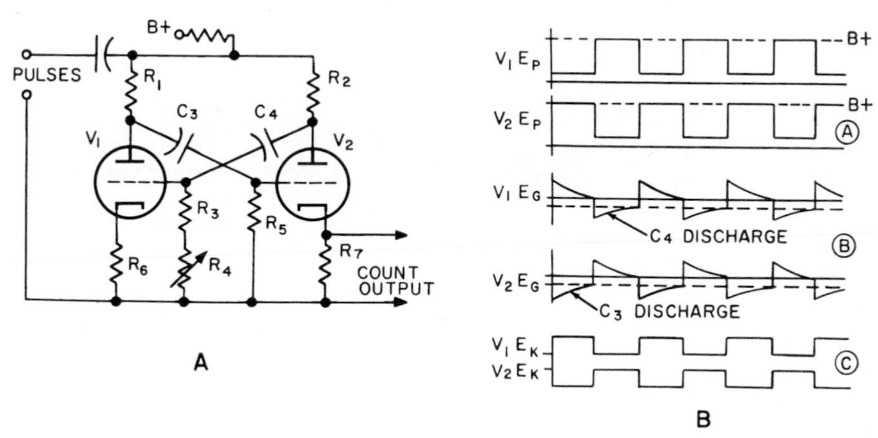

Fig. 7·5 (A) A typical plate-coupled multivibrator found in an all-electronic counting circuit. (B) Waveforms present at the plate, grid, and cathode of a multivibrator; the square-wave output is used as an on-off switch.

A typical multivibrator is shown in Fig. 7·5*A*. This circuit, shown in a free-running state, is well known to the technician with television experience and has its own natural frequency without any input. It is a two-stage cross-coupled amplifier using only resistors and capacitors. When it is turned on, V_1—it is assumed—conducts more than the other tube V_2. This reduction in the plate voltage of V_1 is coupled by C_3 to the grid of V_2. As the plate voltage of V_2 increases, this increases the grid voltage of V_1. This action continues until V_1 is at saturation and V_2 is cut off. At the point where the C_3 discharge across R_5 permits V_2 to conduct, the tubes reverse roles as shown in Fig. 7·5*B*. But if positive input pulses are applied as shown, they can lock the output so that there is effective frequency division or counting. Figure 7·6 shows one output for every eight inputs.

This is one way to use a multivibrator as a counter, but for industrial applications it is more often used to count by twos in a different circuit.

Redrawing this circuit with a common cathode load produces the bistable circuit shown in Fig. 7·7. As before, with no input one tube conducts more heavily and, because of feedback, goes to saturation. Heavy current flow through the cathode resistor R_1 is enough to keep the other tube cut off. A positive pulse through C_2 will appear at both grids and switch the state of both tubes. This circuit has two stable states, which are V_1 conducting, V_2 off; and V_1 off, V_2 conducting. It will remain in its last state until changed by an input.

In order to count, some designation is required for the on-off states. As shown in Table 7·1, V_1 off and V_2 on is a 0, while V_1 on and V_2 off

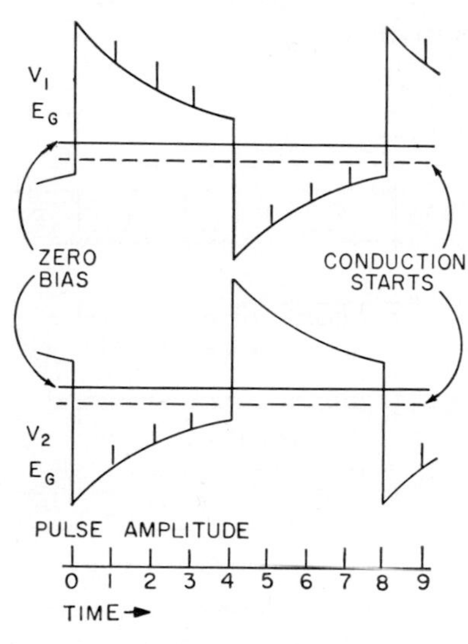

Fig. 7·6 Closeup of the grid-voltage curve in an 8:1 counter. Every eighth pulse on the grid of V_1 causes conduction.

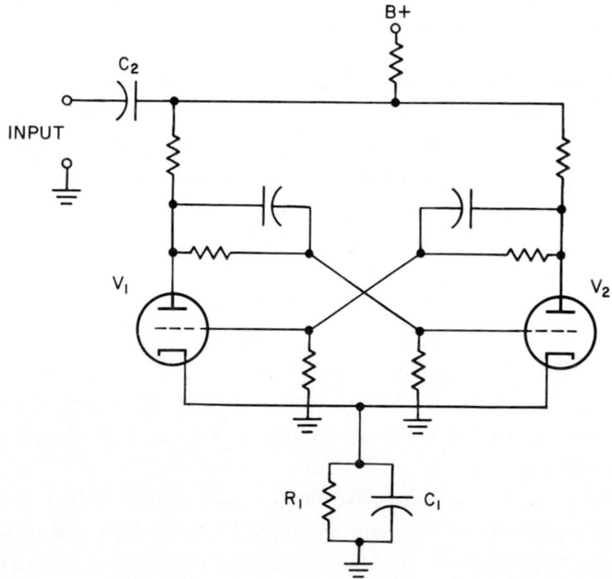

Fig. 7·7 Bistable multivibrator which has two stable states and which will stay at rest in either state.

Table 7·1 Flip-flop Indications for Changes of State

Pulse	V_1	V_2	Indicator
	Off	On	0
1	On	Off	1
2	Off	On	0
	On	Off	1
1	Off	On	0
2	On	Off	1

Table 7·2 Four-stage Counter

Pulse	Stage 1	Stage 2	Stage 3	Stage 4
In →	V_1 ¦ V_2	V_3 ¦ V_4	V_5 ¦ V_6	V_7 ¦ V_8 → Out
0	0	0	0	0
1	1	0	0	0
2	**0**	1	0	0
3	1	1	0	0
4	**0**	**0**	1	0
5	1	0	1	0
6	0	1	1	0
7	1	1	1	0
8	**0**	**0**	**0**	1
9	1	0	0	1
10	0	1	0	1
11	1	1	0	1
12	0	0	1	1
13	1	0	1	1
14	0	1	1	1
15	1	1	1	1
16	**0**	**0**	**0**	**0**
	$2^1 = 2$	$2^2 = 4$	$2^3 = 8$	$2^4 = 16$

is a 1. Thus it takes two pulses for the circuit to reach its original state. Starting with either a 1 or a 0, two pulses are required for a return to the condition the circuit was in without an input signal.

Each single (two-tube) multivibrator forms a stage in the counter. A four-stage flip-flop counter is shown in block form in Table 7·2. The stages

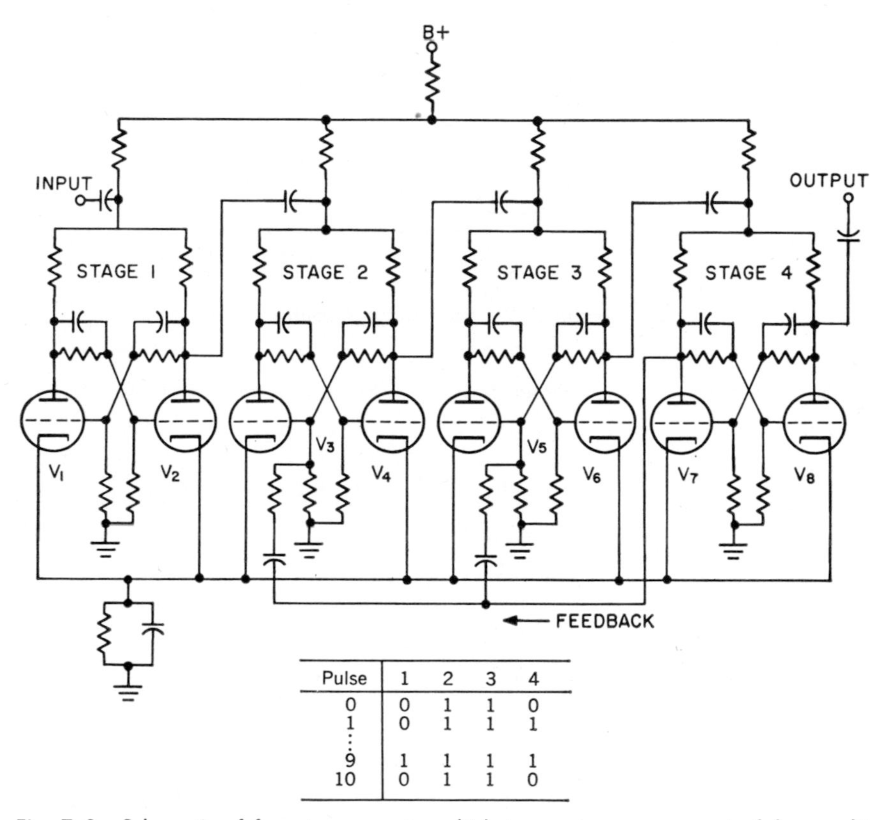

Pulse	1	2	3	4
0	0	1	1	0
1	0	1	1	1
⋮				
9	1	1	1	1
10	0	1	1	0

Fig. 7·8 Schematic of four-stage counter which is a series arrangement of four multivibrators, each coupled to the next stage.

are in series; hence, since stage 1 counts by 2, stage 2 counts by 4, stage 3 counts by 8, and stage 4 counts by 16. In summary, this means that after 16 pulses to the first stage input, the entire counter has returned to its all-zero state. But while a count by 16 is given by this four-stage unit, a count by 10 is usually needed. For a digital computer which operates on the binary system, a count by 16 is useful. (Binary numbers are discussed in Chap. 16.) For other industrial uses a count by 10 is desired.

A schematic of this four-stage counter is shown in Fig. 7·8. Without the feedback these eight tubes (in four stages) count by 16. Application of the feedback from the V_7 plate to the V_3 and V_5 grids makes this a count-by-10 or decade unit. This feedback causes the counter to start at a count of 0110 (rather than at 0000), so after nine pulses it has 1111. One more pulse, the tenth, causes V_8 to go from 1 to 0 (from off to on), and this provides a decrease in plate voltage. This decrease is the negative output pulse. V_7 is opposite to V_8; hence V_7 has a positive pulse which turns V_3 and V_5 on. Because of this, there is a 0110 condition after 10 pulses and an output. This same 0110 then becomes the starting condition for the next count of 10.

Neon lamps can be connected to light for each digit from 0 to 10 as a visual indication. The lamps are connected to the plates of the counter in such a way that only one light is on for each digit of the count.

7·3 Preset counters A number is set into preset counters before counting starts. When this number is reached, there is an output signal. In this way a predetermined number of counts triggers an output.

A decade counter, discussed above, may be thought of as a four-stage binary counter preset to read six counts. Ten counts, after the presetting, give an output. Normally a four-stage flip-flop counter would provide an output only after 16 input pulses.

Table 7·3 shows the preset counting sequence. *A* shows 16 pulses and the ordinary count without any circuit modification. If instead of starting

Table 7·3 Preset Counter States

Presets			Stages			
C	B	A	1	2	3	4
		0	0	0	0	0
		1	1	0	0	0
		2	0	1	0	0
		3	1	1	0	0
0		4	0	0	1	0
1		5	1	0	1	0
2		6	0	1	1	0
3		7	1	1	1	0
4		8	0	0	0	1
5		9	1	0	0	1
6	0	10	0	1	0	1
7	1	11	1	1	0	1
8	2	12	0	0	1	1
9	3	13	1	0	1	1
10	4	14	0	1	1	1
11	5	15	1	1	1	1
12	6	16	0	0	0	0

with 0000 the counter starts with 0101, as in column B, there will be an indication of a full count after six input pulses. Thus the counter is preset to read out at six. If, as in column C, the preset is 0010, there will be an output after 12 pulses. A preset binary counter reading out after 10 pulses is a decade counter, but any number may be used to preset.

Numbers are placed into the counter by electronically setting ones into the proper stages. Dials on the front of the counter are turned to the number desired. These readings are converted into 1s for insertion into the circuits. For a two-tube flip-flop a 1 means V_1 is on and V_2 is off.

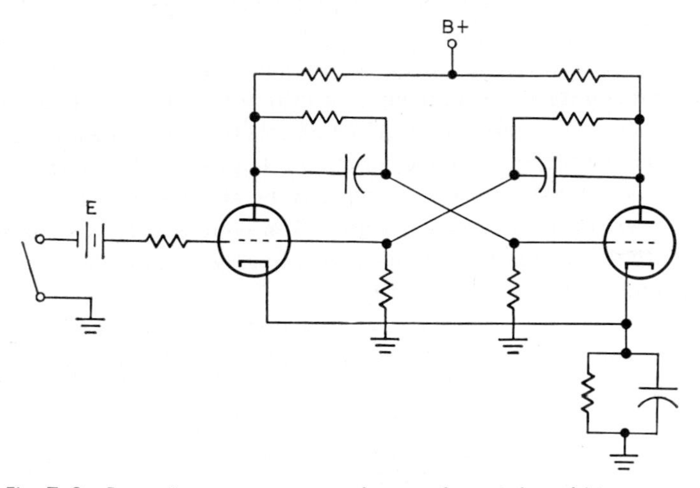

Fig. 7·9 Presetting a counter stage by use of a switch and bias source.

All stages which are preset to 1 require a positive voltage for the control grid. In a simple circuit a push button can be used to place a positive voltage on the grid of the stage which is to be conducting.

As shown in Fig. 7·9, a push button applies a momentary positive voltage to the grid of V_1 which causes it to conduct or to be set with a 1. Electronic circuits are used for presetting since push buttons or switches are too slow. As shown in Fig. 7·10, a thyratron tube V_1 is used as an electronic switch. V_2 is a tube in the counter which is to be set to a 1 (conducting). When a count is completed, the output relay closes contacts A. This fires the thyratron and places a positive potential on the grid of V_2, which places the tube in an on state.

Counters can do more than just count. In some applications their only function is to provide a continuous visual indication of a total number of items, but they can measure frequency, period, time, total events, and time ratios. The frequency can be periodic as with electrical frequencies or random as with nuclear particles. The period is the time necessary to complete one cycle of an unknown frequency. A panel switch is available in some units to permit measuring the average of 10 periods to obtain greater accuracy. Period measurements are especially useful for measuring low frequencies. But because the counter can measure accurate, predetermined time intervals, it may be used to obtain directly the frequency stability of oscillators or generators. This is useful in determining the effects of line-voltage variation, temperature change, and other factors.

Transducers which provide an output frequency related to the magnitude of weight, pressure, temperature, acceleration, and force make the electronic counter ideal for many industrial measuring applications. These techniques allow measurements to be made remotely, providing operator safety and economy in the amount of equipment required, and generally give many times better accuracy than that obtainable with ordinary gauges.

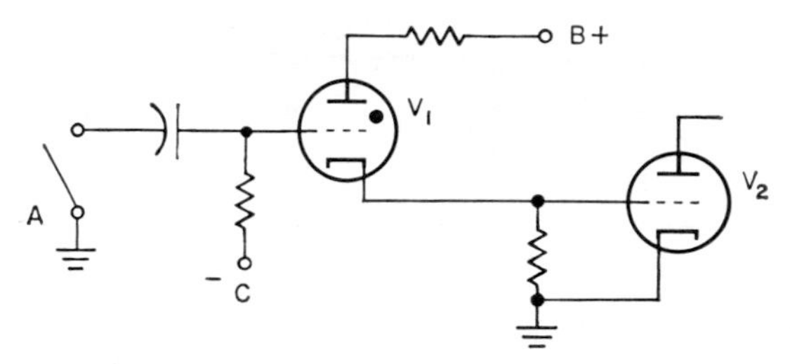

Fig. 7·10 Thyratron as a switch; when fired, it acts as a closed circuit.

For frequency measurement the unknown signal is shaped and then applied to an electronic gate. When the gate is open, the signal pulses are passed to the counter circuits. When the gate is closed, the counters display the counted value. The gate is opened for an accurate time interval based on the temperature-controlled quartz-crystal frequency standard and the count is actual frequency. For period or time-interval measurement the counter counts the output of the interval frequency standard during one or more cycles of the unknown frequency, or between two impulses, the interval between which is to be measured. The instrument is arranged so that the counted value can be displayed for a preselected period of time, adjustable up to 10 sec, or indefinitely by manual control.

7·4 Ring counters Ring counters have their stages connected together as a closed loop or ring. Only one stage is on at any one time. Input pulses are applied to all stages in parallel. For each input pulse the tube which was on is turned off and the next state is turned on. In a four-stage counter (Fig. 7·11) only stage 1 is on at the start. The first pulse turns 1 off and 2 on; the second turns 2 off and 3 on; the third turns 3 off and 4 on; while the fourth pulse turns 4 off and 1 on, which was the original state.

As a frequency divider a ring counter has a single output for each full series of pulses (N pulses for a counter of N stages), but the frequency-divider output can be taken from any stage.

A typical thyratron ring counter of four stages is shown in Fig. 7·12. In V_1 a negative voltage is applied to the grid, keeping the tube nonconducting. When this bias is overcome, V_1 fires and its neon lamp glows because of the drop across R_5. A series resistor R_6 limits the current flow through the neon lamp.

In sequence, the counter works this way:

1. The reset push button is closed for a moment. This fires V_1 and lights the zero neon since the plate voltage of V_1 drops sharply.
2. Bias voltage applied to all other stages keeps them cut off.

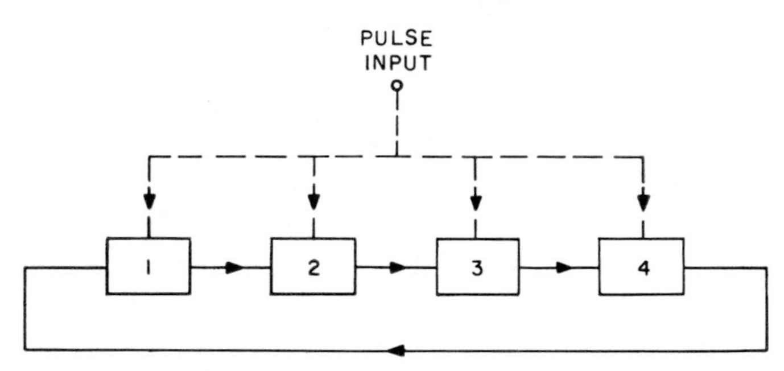

Fig. 7·11 Four-stage ring counter; input is to all stages but output from the last stage is fed back into the first stage as a closed loop.

3. The drop across R_2, caused by the V_1 cathode current, charges C_1. This voltage across R_2 is applied to the V_2 grid so that it is "primed" or almost ready to conduct.
4. At this point, with the neon indicator reading zero, a pulse is applied to all grids. It has no effect on V_1 because it is already conducting; it has no effect on V_3 or V_4 because of their bias, but it does cause V_2 to conduct. As V_2 draws current, there is a positive drop across R_3. This voltage, in series with the charge on C_1, is applied to the cathode of V_1. Together with the bias on the grid, this is enough to cut the tube off.
5. Now only V_2 is on and its neon lamp glows, indicating a count of 1. The next pulse turns V_3 on, and after three pulses V_4 is the only tube on.
6. The fourth pulse returns the counter to its original state.

Ring counters, by their design, count by ones just as a flip-flop normally counts by twos. Because of this, ring counters require more tubes for counting than do binary counters.

Ring counters have many applications. A decoder for selective calling for mobile radio is one example. In the block diagram in Fig. 7·13 an output is desired for a single code of three numbers. This allows one station to call another without disturbing the others. This system uses pulses and electronic counters; other pulse systems use counting relays.

Two main circuit blocks are used: the counter for counting the received code pulses and the digit register for code recognition.

Input signals which are groups of pulses are regenerated and sent to the counter. Assume that this code is 235. After two pulses have been received, the count of 2 is transferred to the first stage of the digit register. This stage has been preset to accept only a 2. Following the reception of the two pulses the ring counter is reset, it receives the 3 which is transferred to the second stage of the register, and the same thing occurs with the 5. After the code 235 is received and recognized by the digit

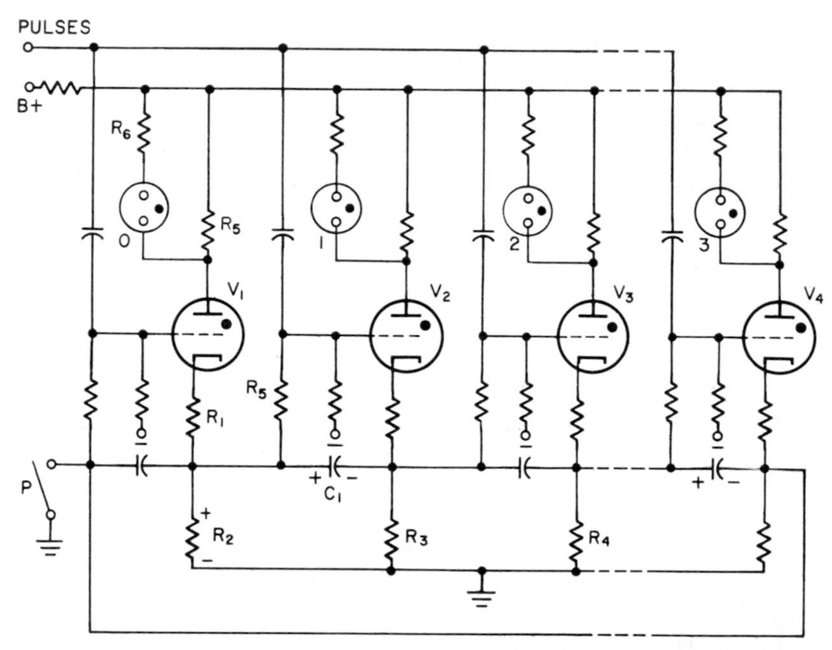

Fig. 7·12 Schematic of a ring counter using four thyratrons and four glow tubes.

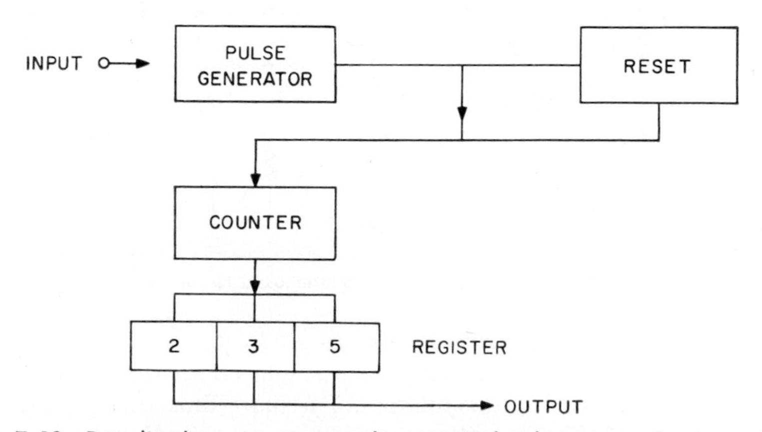

Fig. 7·13 Decoding by a ring counter where received code must equal preset code.

register, there is an output. This tells the radio operator he has a call on his receiver on the proper code. Each receiver has a code number to which it is preset and to which it will respond.

A second use for a ring counter is as a high-speed switch. In telemeter-

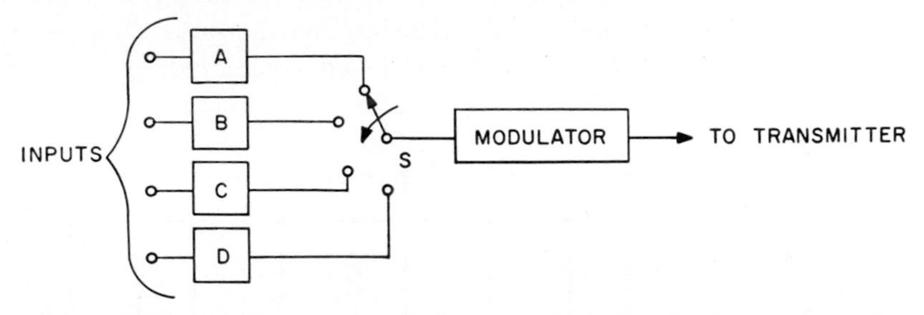

Fig. 7·14 Telemetering ring counter where samples of each channel are taken in sequence.

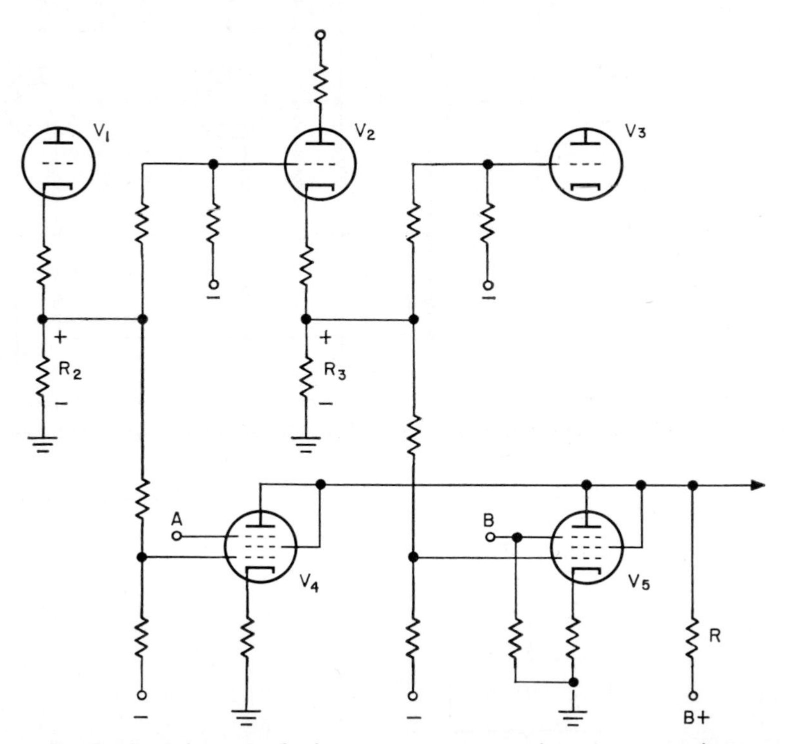

Fig. 7·15 Schematic of telemetering counter used to obtain sampling.

ing, for example, several channels such as A, B, C, and D are required (Fig. 7·14) on a time-shared basis on a single r-f carrier. Each of the four inputs is applied to a modulator for a short time through rotating switch S. Figure 7·15 shows three counter stages (V_1, V_2, and V_3) of a ring counter and two control tubes or gates below. Each gate has a bias voltage on its control grid and an audio signal on its grid 3. Each grid 2 is a screen grid. R is a common plate load for all gates.

When V_2 is turned on by the drop across the cathode resistor of V_1, V_4 also conducts and its audio signal appears across the common plate load R. None of the other gates can conduct. When the drop across R_3 causes V_3 to fire, V_5 conducts, allowing only the audio B to appear across R.

In a similar manner a ring counter acts as a switch to permit only one audio signal at a time to be sent to the transmitter modulator. A four-stage ring counter can be used for controlling four switching tubes or gates as a fast-acting electronic switch. The rate of switching depends upon the pulse source.

Ring counters described above use gas-filled tubes. But it is possible to use vacuum tubes as ring counters. Flip-flops can be arranged as ring counters by having all stages but one in the same state.

Four flip-flops can be series-connected to propagate a 0. With no input, stage 1 has a 0; all others have a 1. The counter reads 0111. After each successive pulse it reads 1011, 1101, 1110, and then 0111, or the original condition.

Ring counters can also be connected with binary counters. A combination of a five-stage ring counter and one binary stage makes a decade counter.

7·5 Other counting circuits In addition to decade and binary counters, there are other types for different applications. One of these is the electromechanical type shown in Fig. 7·16.

The counter shown in Fig. 7·17 has five rotating members capable of a total indication of 99,999. If necessary, more circuits can be added to provide higher totals. Printed wiring boards in contact with a rotating brush in each circuit indicate the position of each counting member.

By suitably interconnecting the circuits, a specific setup for each number can be obtained. These units can be made to remotely actuate control circuits, remote printing, card- or tape-punching operations, and other types of data-processing equipment. Transmission may be continuous or intermittent over common-carrier teletype lines, radio, or plain wire.

To operate, a counting pulse of suitable duration and voltage is applied to the coil or clapper-type solenoid. Each pulse or series of pulses, depending upon the type of triggering circuits used, causes the lowest-order counting wheel to move from one digit to the next. Transfer to the next higher order of indicating wheel is mechanical and occurs between the 9 and the 0 positions on the lower-order wheel. An indication of 45789 is shown.

In addition to the electromechanical counter, there are several types of electronic counter devices which are used. Perhaps two of the simplest circuits are the blocking tube and the step counter.

Figure 7·18 shows the blocking-tube circuit; when the circuit is turned on, the tube conducts. Plate current gradually increases, as the tube heater warms up, and this change is inductively coupled to the secondary. This winding is arranged so that the points, as marked, are of the same polarity. Thus, the increasing plate current causes the grid to go positive and draw current, which charges C and returns to ground through R. When the plate current reaches saturation, the magnetic field stops in-

Fig. 7·16 The high-speed predetermining counter is the basic counter in this complete line. It will count turns, strokes, pieces, lengths, or any other units to the sum set on the white wheels. When that run is completed, the counter acts to make or break a circuit, turn on a light to signal the operator, ring a bell, or actuate a stop motion to halt the machine.

creasing. At this moment there is no induced voltage and the negative charge on the grid starts to reduce the plate current. As the plate current is decreased, the plate field decreases and induces a negative voltage in the grid winding.

The tube is then cut off and stays cut off until the capacitor discharges through R. Because of the decreasing plate field, the plate voltage rises above the B voltage. When the grid voltage is small enough (near zero), the tube conducts and begins again. Counting is possible by applying positive pulses in series with the grid winding.

Stabilization is sometimes used by inserting an LC tank circuit in series with the grid as shown. This circuit is resonant at a somewhat higher frequency than the output frequency. At each pulse of grid current a sine-wave oscillation occurs in the resonant circuit. This voltage is arranged to aid the tube to conduct only on the proper count.

Another counter is the step, or staircase, counter shown in Fig. 7·19.

A series of positive pulses is applied to C_1. Each pulse causes current flow to charge C_2, through V_2 whose plate is positive, and C_1. During this time V_1 cannot conduct because its cathode is positive. But in the time between pulses C_1 discharges through V_1 because the negative charge on

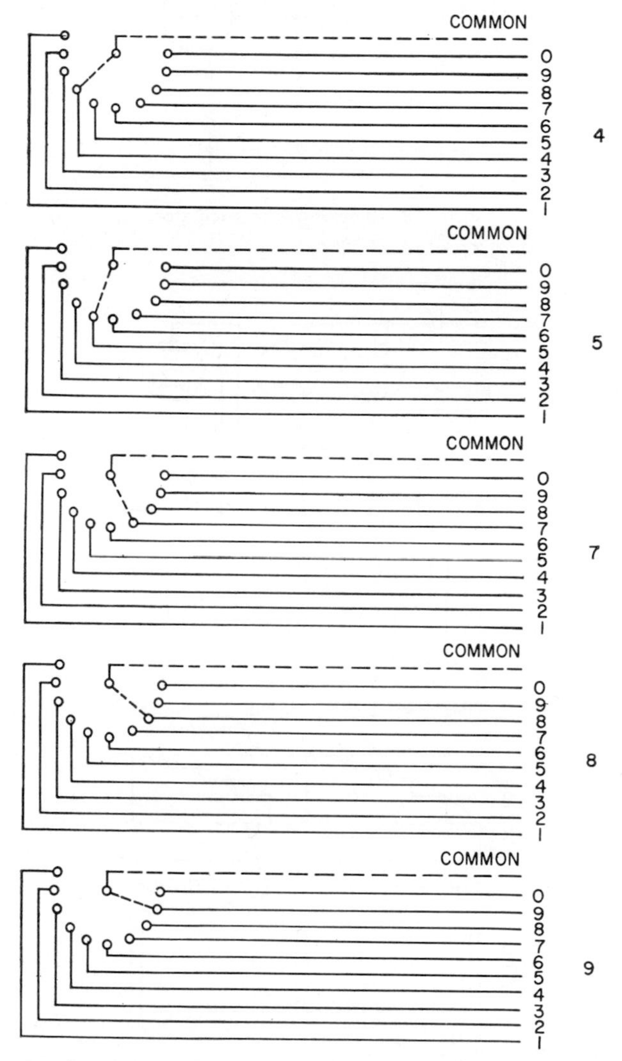

Fig. 7·17 Electromechanical counter diagram showing the number 45789.

C_1 makes the V_1 cathode negative. V_2 is negative and this tube cannot conduct, so C_2 is left charged. The next pulse charges C_2 still more (over and above the charge it has from the last pulse), and C_1 charges through C_2 and V_2. It discharges through V_1.

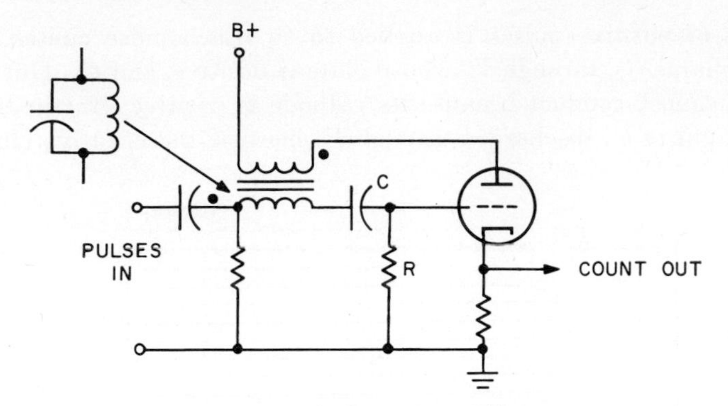

Fig. 7·18 Blocking-tube oscillator.

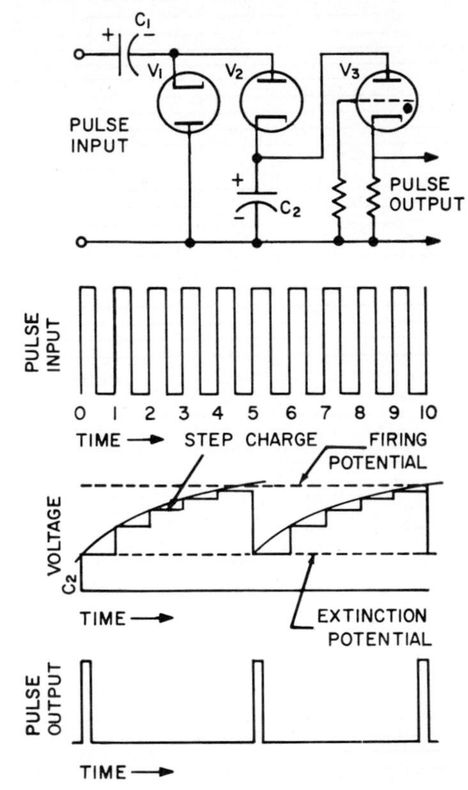

Fig. 7·19 Simplified version of a step-charge counter. The thyratron is fired every fifth pulse. Other countdown ratios are possible.

This continues and a step charge builds up close to the dashed capacitor charge curve. C_2 is discharged when it triggers the following circuit, such as a blocking-tube oscillator or a thyratron, as shown.

In the step counter with unequal steps there is a certain instability as the count increases. This is because each succeeding step is smaller and the difference between a trigger signal and no trigger signal is very small, which can lead to false triggering.

But this is overcome by the equal-increment step counter as in Fig. 7·20 where the discharge path for C_1 is now a triode. When a positive

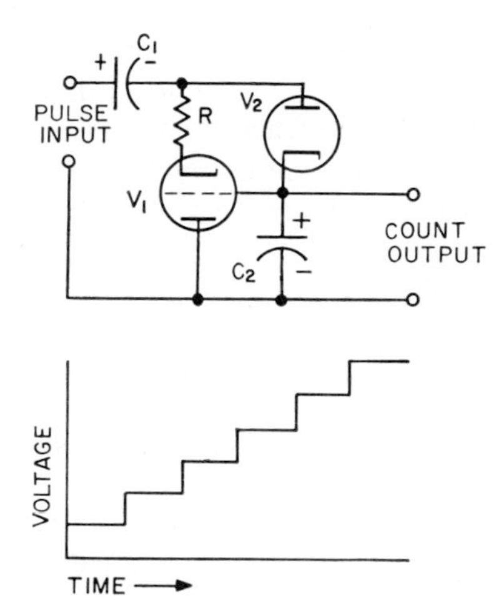

Fig. 7·20 For increased stability, this circuit provides a series of equal-sized steps.

pulse is applied, C_2 charges through diode V_2 as does C_1. When the first pulse is removed, C_1 discharges through V_1 as before, but now the resistance of the discharge path depends upon the triode grid bias which is the charge across C_2. The discharge path is through the V_1 triode and the cathode resistor. Across this resistor there is a cathode bias; grid voltage (positive) depends upon the charge on C_2. Thus the resistance of the C_1 discharge path varies with the charge on C_2, and C_1 is only partly discharged after each pulse.

Because of the remaining charge on C_1, C_2 is charged by the second pulse (and all succeeding pulses) by an equal amount. The problem of stability is thus overcome. Circuits like this can count to higher numbers than can the unequal-step counter.

Many industrial electronic systems count and measure. The results of counting or measuring must be displayed or recorded so that they can be used for quality control, for inventory, or for shipping records.

The process or production under control is measured by a transducer which changes the readings of temperature, pressure, humidity, or weight into electric signals. These signals are sent to the control unit, where they are converted into signals which can be counted; the results of measurement are also displayed and recorded.

7·6 Counting tubes Single tubes can be used for electronic counting. The two basic types are gas-filled and vacuum tubes. Each of these is different, but each is a complete counting unit when used with external components and the proper power supply.

Gas-filled Counters. A ring counter can be made by interconnecting a series of counter stages with the output fed back to the input. A glow tube can be considered as a complete ring counter in a single envelope. An example is the Sylvania decade counter or glow transfer tube, which is a gas-filled device employing 30 pinlike cathodes surrounding a disk-shaped anode. As shown in Fig. 7·21 the cathodes are identical physically but are electrically connected in three groups called guide-1 bus, guide-2 bus, and output cathodes. Conduction occurs in the tube between the anode and the most negative cathode pin, causing a glow on the tip of the pin.

The counter tube provides both visual and electrical readout information. Visual readout is obtained by observing the position of the glow through the face of the tube. Electrical readout occurs when a resistance is inserted in series with an output cathode, so that the ion glow on the cathode develops an output signal voltage across the resistance.

When the tube is counting, an ion glow moves around the cathodes spaced equally around a centrally located anode. Ten of the cathodes are output cathodes (K), and the remaining twenty are guide cathodes (G) whose function is to transfer the ion glow from one output cathode to the next.

The tube is operated by stepping the glow around the circular arrangement of cathodes. The transfer of the glow from cathode to cathode is accomplished by applying double pulses to the guide cathodes.

The counter tube can be visualized as containing 30 small neon glow lamps (cathodes) in a common envelope, sharing a common central anode.

If a sufficiently high voltage is applied to the anode through a suitable current-limiting resistor, ionization will occur between the anode and one of the cathodes. A glowing spot has now been formed on this cathode, which is known as the preferred cathode. As the ionization field builds up, the anode voltage drops to a level just sufficient to maintain the glow on the preferred cathode, but too low to cause any of the remaining cathodes to ionize. The glow remains on the preferred cathode until it is acted upon by some other force.

Since the positive ions in the field are attracted by a negative potential and the adjacent cathodes are close to the preferred cathode, the glow can be moved to another cathode by making the new cathode more negative than the preferred cathode. The glow can be transferred from cathode to cathode by making each succeeding cathode negative for an instant in the desired direction of travel.

An external d-c source is a simple means of causing the glow to advance by applying a negative transfer voltage as shown in Fig. 7·22. The cathodes K_1, K_2, K_3, and K_4 are equidistant from each other and spaced uniformly around the circular central anode. Voltage required to transfer the glow in either direction will be the same.

The glow could be moved from K_1 to K_2 to K_3 to K_4 by touching the test probe to each cathode in turn, and the process could be reversed by

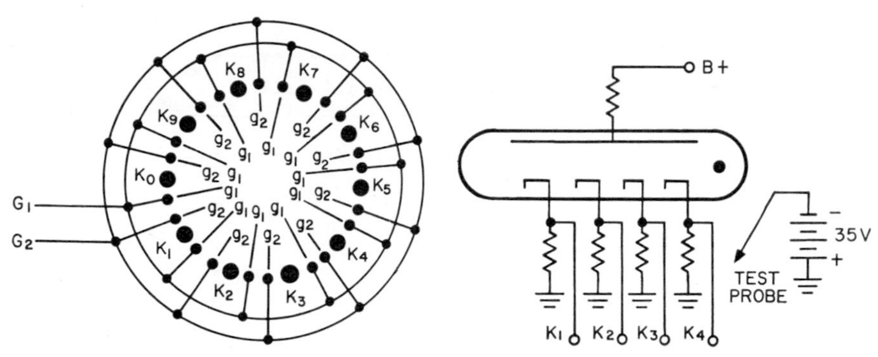

Fig. 7·21 Counting-tube structure showing grids and cathodes. (*Sylvania*)

Fig. 7·22 Starting the tube by the application of a test signal. (*Sylvania*)

going back in order from K_4 to K_1. When the cathodes are arranged in a circle, as in the counter tube, the glow can be made to complete a closed course repeatedly, as does a ring counter with four stages.

The 30 cathodes in the counter tube are arranged to permit the use of simple driver circuitry and to provide a definite direction of travel for the glow. The 10 output cathodes, represented by the larger dots and designated K_1, K_2, etc., are connected to pins on the base of the tube. The remaining 20 cathodes, g_1 and g_2, are divided into 10 pairs of guide cathodes, one pair following each output cathode so that a pair of guide cathodes lies between each two output cathodes. The guide cathodes are connected in two parallel circuits inside the tube, one containing all the g_1 cathodes and the other containing all the g_2 cathodes. These circuits are the G_1 and G_2 guides.

Two common G_1 and G_2 connections are provided on the base of the tube. Since there is no electrical difference between G_1, G_2, and K_1 through K_0, the tube can count in either direction.

A simplified counter-tube circuit showing d-c voltages is given in Fig. 7·23*A*. The glow is assumed to be concentrated on K_1, to begin, and the desired direction of transfer is from left to right. When the switch S_1

is closed, G_1 is made negative with respect to K_1 and the glow moves to G_1 at the point immediately to the right of K_1.

If the switch S_2 is closed, the glow is not affected, since G_2 is not made substantially negative with respect to G_1. However, if S_1 is opened while S_2 is kept closed, G_2 becomes the most negative point in the immediate vicinity of the glow and the glow moves to G_2. Opening SW_2

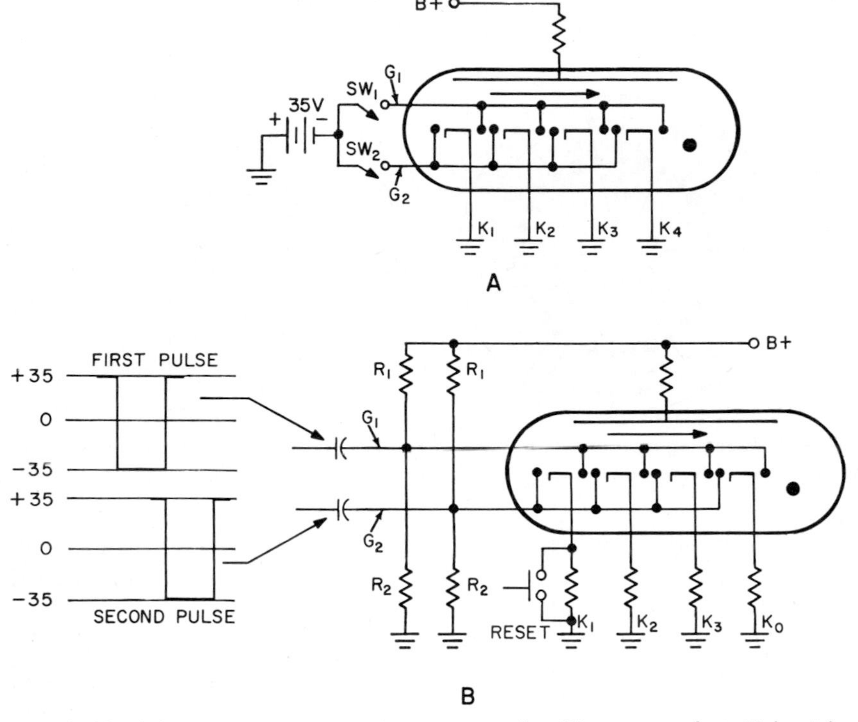

Fig. 7·23 (A) Propagating a count in a counter tube; (B) counter pulses. (*Sylvania*)

then allows the glow to move to K_2, which is now the most negative point near G_2, and the process continues on in this way; as the count is completed, the tube recycles.

In practice, two pulses are used to drive the counter tube as shown in Fig. 7·23*B*. In this circuit, the resistance R_2 terminates the impedance of the pulse generator. The ratio of R_1 to R_2 is such that both G_1 and G_2 are 35 volts positive with respect to ground during the no-pulse condition. This condition is known as back bias, and it is essential for the proper operation.

If a voltage is applied to the B+ line in the circuit, the glow will strike on one of the cathodes because the voltage differential between the cathodes and the anode will be larger than that between the guides and the anode. The cathodes are 35 volts negative with respect to the guides. A reset button starts the glow from K_1.

When the glow is on K_1 and the counting process can begin, a driver produces separate double pulses from single input pulses at the desired counting rate connected to the guide buses G_1 and G_2. As the glow rests on K_1, the voltage drop across the K_1 load resistor is in opposition to the 35-volt back bias. Assuming that the positive potential on K_1 is 25 volts, because of anode current flow, the effective back bias is only 10 volts. However, K_1 is still negative and the glow remains there.

As the first negative pulse is applied to the G_1 bus, the guide pin immediately to the right of K_1 receives the glow because it is the nearest G_1 pin to K_1. If the G_1 pulse is removed and a second pulse applied to the G_2 bus before the glow can return to K_1, the glow moves to the right again, to the G_2 pin. A slight overlap of the G_1 and G_2 pulses is necessary to provide a smooth transfer of the glow from one output cathode to the next.

When the pulses applied to G_1 and G_2 have decayed and the guide buses again have positive 35-volt potentials, the glow is attracted to K_2, which is nearing the G_2 pin and is 35 volts negative with respect to the guides. Then as anode current flows through the K_2 load resistor, back bias is again opposed by the voltage drop across the resistor and the glow can be transferred again by another input pulse to the driver stage. One cycle of operation has now been completed.

An output or signal pulse can be obtained from the cathode resistor.

Proper timing of the ignition and extinction of adjacent guides are required for reliable operation. Four steps must occur in order: ignition of G_1, ignition of G_2, extinction of G_1, and extinction of G_2. In order to satisfy the basic requirement, an overlap of guide pulses is necessary; the minimum overlap is 10 μsec for the medium-speed tube and 2 μsec for the high-speed tube.

A wide range of different circuits can be used with counter tubes. Figure 7·24 shows a drive circuit for 100 kc. This is particularly useful when counting down from a continuously running frequency standard, such as a crystal oscillator. The 270-kilohm anode resistor should be located directly at the tube socket. The 50-kilohm control should be adjusted for stable operation by adjustment of the counting-tube anode voltage. The 6C4 and its network provide the phase difference between the two guide signals.

Beam-switching Tubes. In addition to the glow type of counting tube, there is the beam-switching vacuum tube, which operates at higher rates. An example is the Beam-X (Burroughs Corporation) switching tube, which is a high-vacuum tube with a centrally located cathode surrounded by 10 identical arrays of electrodes as shown in Fig. 7·25*A*. Each array is composed of a plate (target), a grid (switching grid), and a beam-forming and locking element (spade). A small cylindrical permanent magnet was mounted outside the envelope on earlier tubes; new ones have internal magnets.

In normal operation the electron beam is formed between the central cathode and one of the 10 external arrays or "positions" as shown in Fig. 7·25*B*. The beam can flow to only one of the 10 positions at any given time. The greatest portion of the beam current flows to the target (plate),

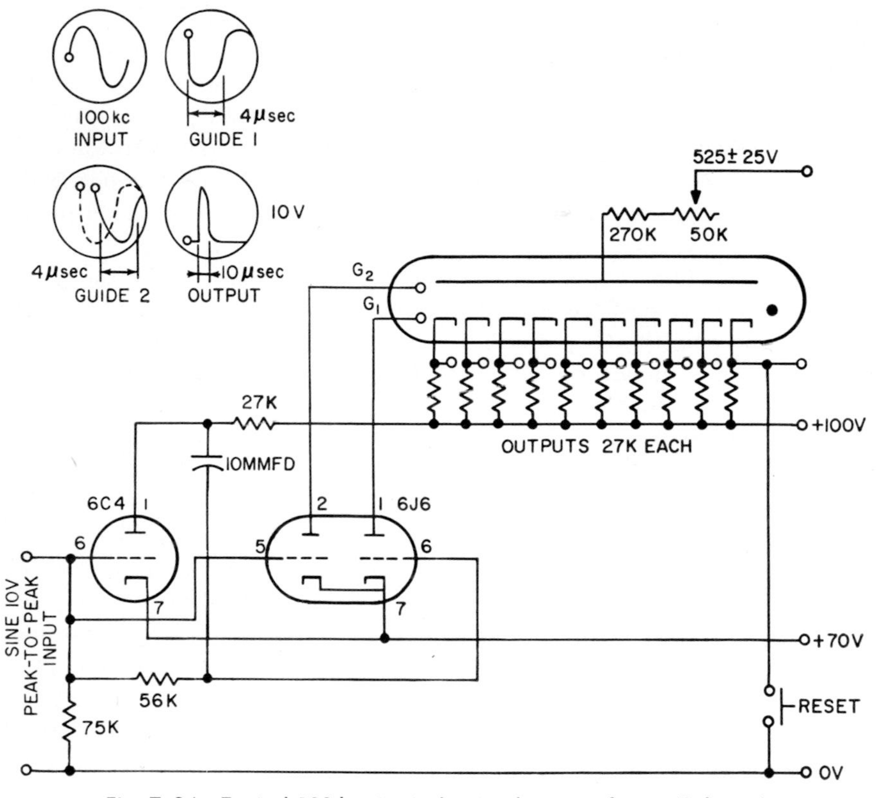

Fig. 7·24 Typical 100-kc circuit showing key waveforms. (*Sylvania*)

where it provides a useful output to operate a neon lamp or some other indicating device. However, the spade draws enough of the beam current to be maintained at approximately cathode potential and thus locks the beam on the target.

The beam is switched from position to position by applying a large negative pulse to the grid. This pulse frees the "on" target of the electron beam and allows the magnetic field to rotate the beam. The beam impinges upon the next spade, which conducts and falls rapidly to approximately cathode potential, thus forming and locking the electron beam on that position.

One of the 10 identical positions is shown in Fig. 7·25*B*. These three electrodes can (1) form an electron beam from the cathode to a single plate; (2) lock the beam in this position until it is switched; (3) switch the beam, either at random or in sequence; and (4) clear the tube or remove the beam and cut off the tube.

The characteristics of the beam are similar to those of a pentode plate current or constant current. Because this tube has both electric and magnetic fields, plate current will flow only when the positive potential is high enough to overcome the magnetic field. At low potentials this field is sufficient to cut the tube off.

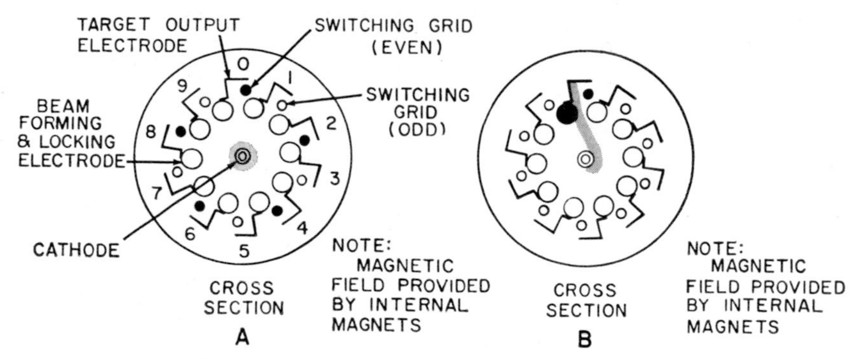

Fig. 7·25 (*A*) Cleared beam-switching tube; (*B*) beam-formed state. (*Burroughs*)

The tube is cut off when all three electrodes are positive. Lowering the spade voltage to a level too near that of the cathode voltage forms a beam to the associated target plate. The beam will remain on this spade until a further change in potential. Current flow across the spade resistor will keep the tube in this state.

Switching may be accomplished in many ways. The grid voltage may be lowered, which will divert current to the next spade and switch the beam. Because of the increased current, this next spade will have a voltage drop across its load which will keep the beam in this position.

Normally, switching is accomplished by the grids since they draw a very small current, and speeds of 0.1 μsec are possible. All the odd-numbered grids are connected together and all the even-numbered grids are connected. Because of this, direct current may be used as the switching voltage.

Several grid-switching sources are possible. A sine wave of frequency f may be used through a center-tapped transformer. With an applied a-c signal the tube is switched at a $2f$ rate. A flip-flop which produces a string of square waves can also be used for switching since two out-of-phase pulses are available, one from each tube. A single-pole double-throw switch, which lowers the grid voltage, is another switching source; or a single pulse input, as shown in Fig. 7·25*B*, will also provide switching.

If all the grids are lowered below the stable-switching point, the tube will provide its own output rate. The beam will rotate continuously at a rate determined by the time constant of the spade circuits.

When the d-c supply voltage is first applied to the switch circuit, the tube normally remains in its cleared or cutoff state. In order to initially form the beam to the zero position, the spade must be lowered. At the same time, the flip-flop must be set to the appropriate state so that the first input pulse that triggers the flip-flop produces a negative pulse

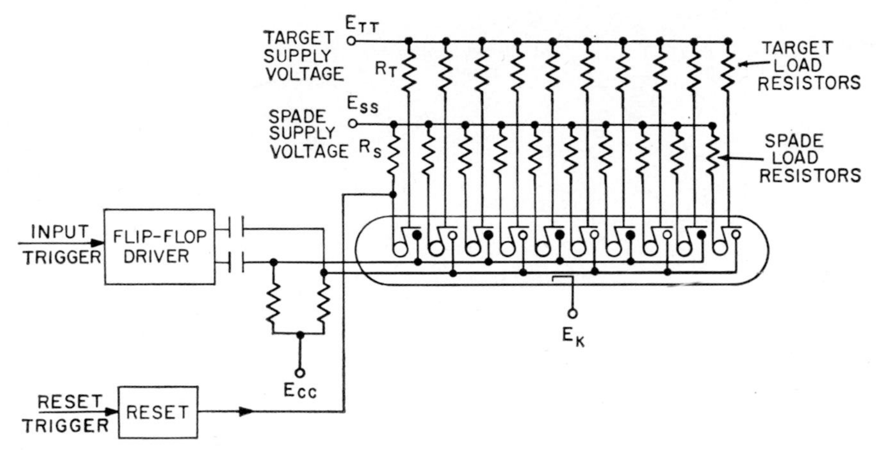

Fig. 7·26 Ten-position circuit showing input and reset trigger. (*Burroughs*)

from the flip-flop to the even grids to advance the beam from zero position. In order to reset the beam to zero from any position, the tube must first be cleared, as the beam will not step backward within the tube.

Figure 7·26 shows a typical 10-position circuit containing the voltages, components, and auxiliary circuitry required for decimal operation. A single B+ supply can normally be used.

One type of spade bus can be operated at any voltage from 15 to 75 volts by employing spade load resistors of appropriate values. The spade-bus voltage generally determines the magnitude of the constant-current output of the target. This output current can range from 300 μa to as much as 4 ma. Other types may be operated at higher voltages with correspondingly higher currents obtainable.

As in any constant-current device, the target supply voltage requirements are flexible. When the electron beam is formed to any one of the 10 target outputs, there is negligible current to any other target. The constant current output, in conjunction with the target supply voltage and load resistance, can be used to obtain output-voltage steps from 0 to 200 volts.

However, the grid-bias voltage is also a function of the spade-supply voltage and is somewhat positive with respect to the cathode, thereby ensuring a stable static beam. The grid-switching drive voltage requires a negative input whose magnitude is determined primarily by the spade-bus voltage and its load line, and secondarily by the target voltage and load line and the maximum frequency.

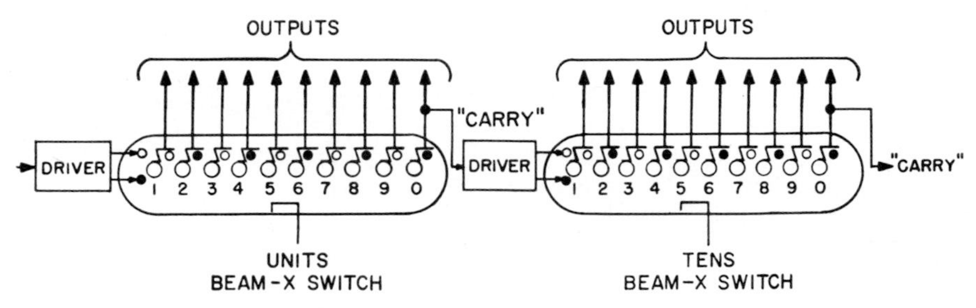

Fig. 7·27 Decade counting using two tubes in series. (*Burroughs*)

A rule of thumb is that the amplitude of the negative grid driving signal should be approximately equal to the spade-bus voltage. When the spade bus is operating at extremely high frequency or providing large target outputs, a larger input drive pulse is recommended.

The preferred method of driving the tube is by the use of a flip-flop.

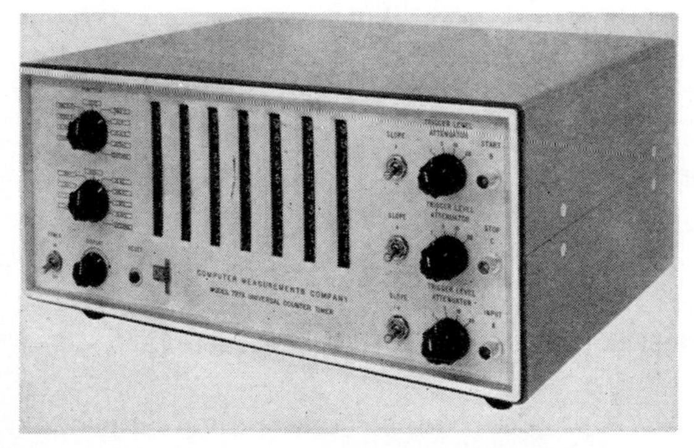

Fig. 7·28 Counter and timer with seven decades. (*Computer Measurements*)

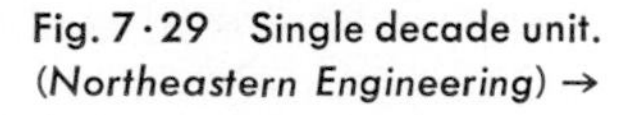

Fig. 7·29 Single decade unit. (*Northeastern Engineering*) →

The negative-going pulses from the alternate outputs of the flip-flop are coupled to the even and odd grids. Therefore, for each input pulse the beam is advanced one, and only one, position.

One of the basic functions of this tube type is counting (Fig. 7·27), in which pulses which can represent photocell outputs, mechanical switch closures, or some frequency are delivered to the switching grids of the first tube. The tenth output of this switch is used to advance the succeeding switch one position. Thus, with two switches, 99 counts can be accumulated; with three, 999; and so on indefinitely. In addition to the cascade or "carry" output, current is available at each of the 10 positions to

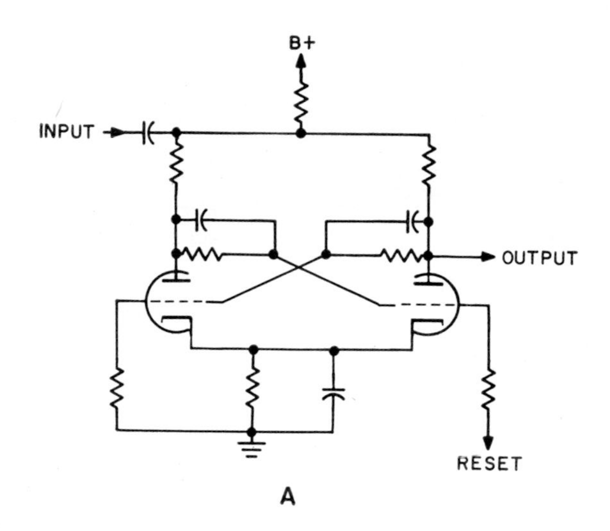

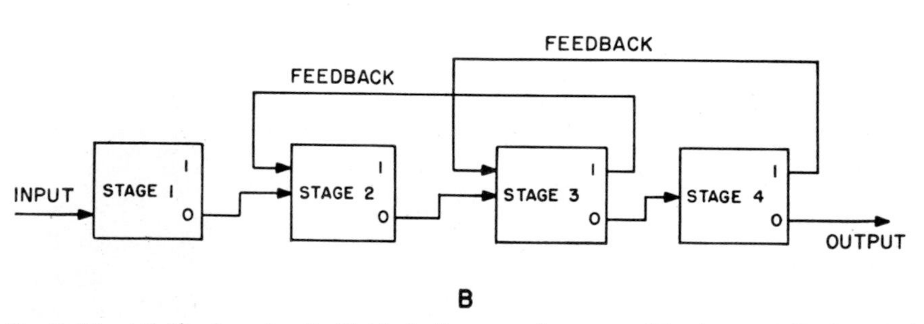

Fig. 7·30 (*A*) Flip-flop circuit; (*B*) block diagram of counter. (*Northeastern Engineering*)

activate visual readouts, show the progress of the count, activate a printer for recording information upon command, or perform useful work at the end of a preset number of events.

7·7 Counter equipment Many variations of counting equipment are available, based upon the use of decade, ring, and other counters. Figure 7·28 shows a typical commercial counter with seven decades and a visual indication of up to 9,999,999.

The basic unit of commercial counters is the decade counter. A typical one is shown in Fig. 7·29. This is a four-stage decade counter; each stage has a circuit like that shown in Fig. 7·30*A*. This is a flip-flop or a bistable multivibrator. Two feedback loops are used: one from stage 3 to stage 2 and one from stage 4 to stage 3, as shown in Fig. 7·30*B*. The resulting feedback causes a count of 0000, 0001, 0010, 0011, 0110, 0111, 1100, 1101, 1110, 1111, and back to 0000. Thus a series of binary (count-by-two) stages makes a decade counter.

REVIEW QUESTIONS

7·1 How can a counter be used to measure frequency?

7·2 What is the basic count of a single flip-flop? Of two flip-flops in series?

7·3 Draw a block diagram of a decade counter.

7·4 What is the purpose of a time base in a counting equipment?

7·5 What is the highest count which can be displayed in the counter shown in Fig. 7·4?

7·6 What determines the frequency of the multivibrator shown in Fig. 7·5*A* in its free-running state?

7·7 What determines the lowest plate voltage of V_1 in this figure? What determines the highest plate voltage?

7·8 Explain how the synchronizing pulses "lock" the multivibrator shown in Fig. 7·5*A* to their frequency.

7·9 What makes the multivibrator shown in Fig. 7·7 bistable?

7·10 What are its two states?

7·11 How many input pulses are required to return this circuit to its original state?

7·12 What is the purpose of the neon lamp in a counter?

7·13 *a*. Draw, in block-diagram form, a four-stage flip-flop counter.

b. With 0000 as the initial condition, what is its state after four pulses? After 10 pulses? After 16 pulses?

c. Why can this circuit, without feedback, not be used as a decade counter?

7·14 What is the purpose of the feedback shown in Fig. 7·8?

7·15 How does this differ from the feedback shown in Fig. 7·30*B*?

7·16 What is a preset counter?

7·17 Where is it used?

7·18 How is a counter preset?

7·19 Explain how a counter can measure time.

7·20 What is the purpose of the counter shown in Fig. 7·13?

7·21 How does a ring counter differ from a flip-flop counter?

7·22 Explain how a ring counter is like a rotating switch.

7·23 How can a blocking-tube oscillator be made to count?

7·24 What is a staircase counter?

7·25 Explain how a gas-filled counter tube is like a ring counter.

7·26 In a glow counter tube, how is the glow made to rotate?

7·27 What two functions does the glow perform?

7·28 Explain how a 10-position glow tube is a decade counter.

7·29 Does a glow-tube counter require feedback?

7·30 What is a beam-forming tube? How does it count?

CHAPTER

8

Data Display and Recording

Many industrial electronic systems count and measure production items as they are made. The results of counting or measuring must be displayed or recorded so that they can be used for quality control or for inventory and shipping records. Data display is also associated with remote indicators as in telemetering.

In the general system of industrial remote control the process or production under control is measured by a transducer which changes the readings of temperature, pressure, humidity, or weight into electric signals. These signals are sent to the control unit, where they are converted into control signals which are fed back to the original process. The results of measurement are also displayed and recorded locally.

Numerical values can be displayed so the operator can tell the operating conditions, such as temperature, or the total number of units produced. A continuous record of given readings in the form of a graph may be obtained with a recorder which provides a continuous permanent record of the meter readings.

Total counts of individual units passing down a line may be accomplished with various electronic counters with a visual display of the total count. This display is not a permanent record, however.

Data are transmitted, by radio, after passing through a code converter. At the receiving end, remote from the original measurement, the transmitted signals are converted into visual indications for observation.

There are many industrial uses for indicators. Three are shown in Fig. 8·1. Figure 8·1*A* shows data passing an observation point. A transfer relay drops the measured quantities to a series of indicators without impairing the flow of data from origin to destination. Often in the use of digital computers indicators are used as both a check upon the input and a technique for presenting the computed results as shown in Fig. 8·1*B*.

In some cases transducers which measure temperature, pressure, speed, or some other variable need both a translator and an indicator as shown in Fig. 8·1*C*. Here shaft rotation, an analog of the measured value, is converted into a digital reading.

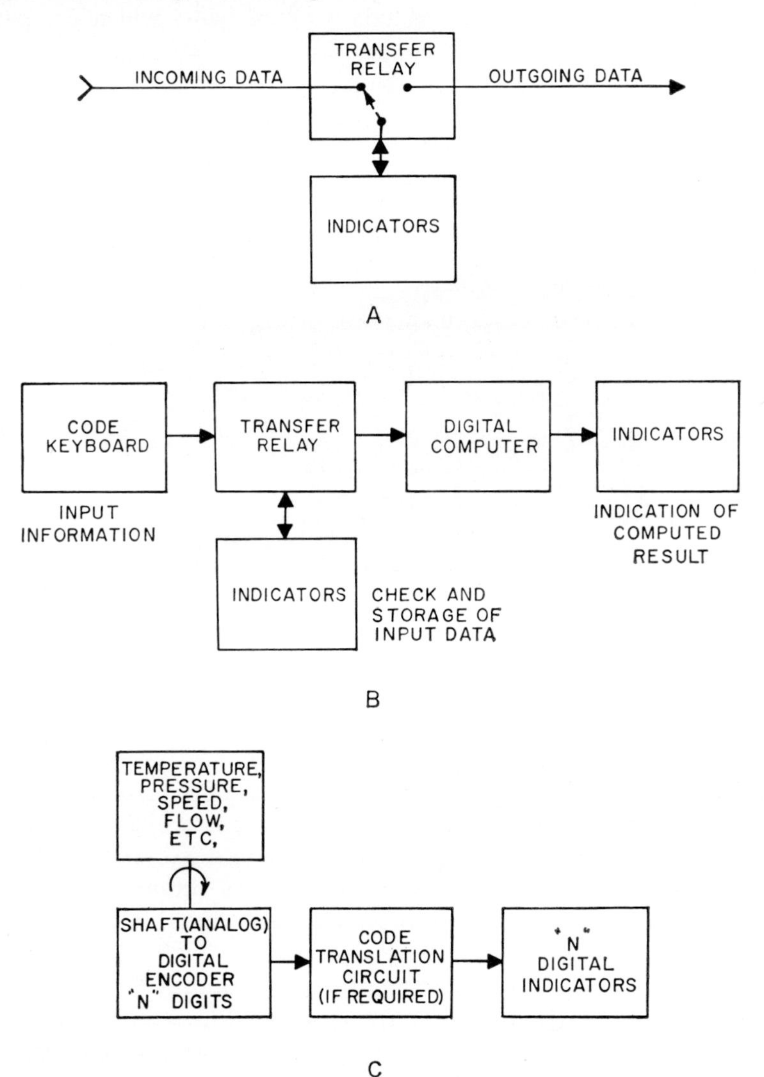

Fig. 8·1 Uses of indicators for (*A*) readout of transfer data, (*B*) check and storage of input data, (C) remote indication of analog data. (*Union Switch and Signal*)

Data which are the results of a measuring process may be in either digital form or analog form. Both types are displayed and recorded.

8·1 Data display Data are displayed in many industrial situations. An ordinary clock is an analog display since the hands rotate with time.

But a clock can also be a digital unit such as a highly accurate device for use in an industrial control system where data logging (recording) is required at specific times. The auxiliary digital time display is for the operator's use.

Other forms of data display include warning lights which are part of a control panel which includes caution and signal lighting.

There are a number of other methods and techniques for data display using lighted letters and numbers. One method, using a continuous belt with numbers printed on it, projects the proper number on a screen by illumination. Other methods use vacuum-tube indicators.

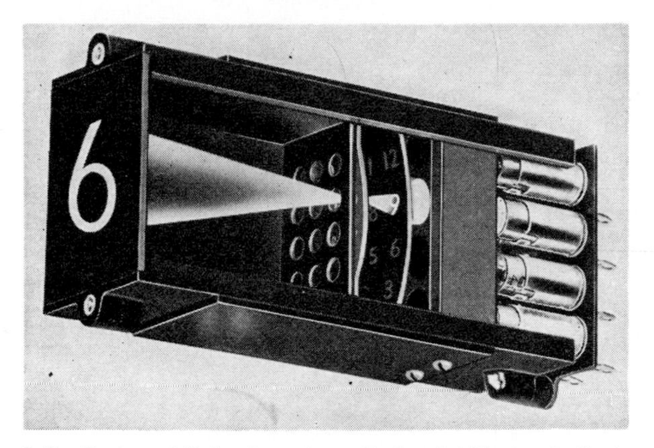

Fig. 8·2 Projected lighted numbers. (*Industrial Electronic Engineers*)

Figure 8·2 illustrates a technique using 12 lamps, each throwing a narrow beam of light through a transparent number. There are numbers 1 through 9, 0, a decimal point, and a colored light (usually red). One number at a time is switched on, and this number is projected on a translucent screen. As a different lead to the display unit is switched, the number changes. Stacks or groups of these units can be used to make any combination of digits or letters or letters and digits. All numbers are presented in the same plane for easy visibility.

Counters and Displays. Counters (Chap. 7) usually have some form of data display. Several counters are shown in Fig. 8·3. There are two different forms of data display here: the numbers on the counter wheels and on the counter tubes. Each of the three units in the center has tubes with numbers indicated around the tube face.

The Nixie tube (Burroughs Corporation) is a special indicator used as a decade-counter readout (Fig. 8·4). These tubes contain stacked elements in the form of metallic numerals. Application of a negative voltage to the selected character with respect to a common anode results in its becoming the cathode of a simple gas discharge diode. Only the selected information is visible in a common viewing area because the visual glow discharge is considerably larger than its metallic source.

Counters can use several other types of visual display. One type, which is large and easy to read, is made up of a matrix of 40 neon lamps for each numeral as shown in Fig. 8·5. It is designed to prevent visual errors

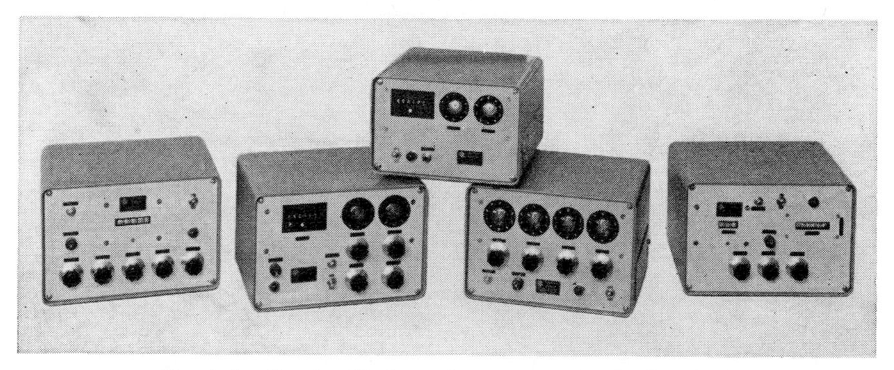

Fig. 8·3 Counter displays of various types. (*Veeder-Root*)

due to lamp failure and other potential sources of inaccuracy. Although many instruments use filament-type lamps for edge-lighted, masked, or projection-type illumination, this unit forms each 2½-in.-high numeral from 40 small, long-life neon lamps. The elimination of errors due to lamp-

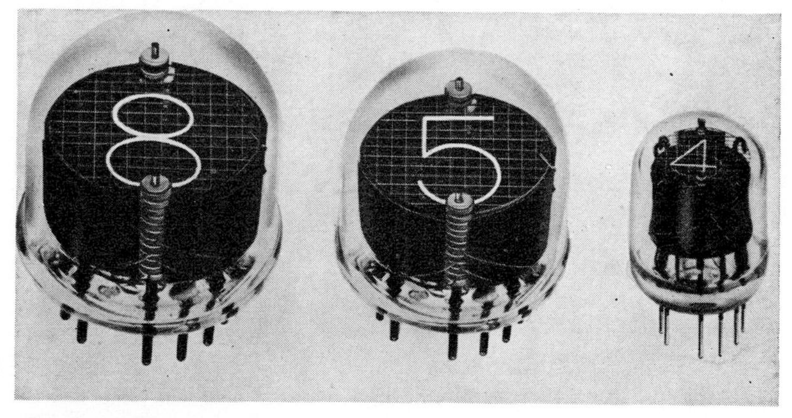

Fig. 8·4 Three sizes of Nixie tubes showing number indications. (*Burroughs*)

filament failure is an important safeguard to accuracy. Potential readout errors are reduced when information is taken from a straight-line readout, as compared with reading the necessarily small numerals on a digital counter, especially when they are staggered on 10 rapidly changing decade levels.

Individual 2½-in. numerals are formed on a flat plane by many small, high-intensity neon lamps. This type of presentation permits a wide-angle readability and is easily read even in strong ambient light.

A memory circuit retains the readout data until it receives a command from the counting instrument; this prevents confusion in reading while the counting instrument is cycling. The memory circuit may be over-

Fig. 8·5 Neon lights as indicators. (*Computer Measurements*)

ridden intentionally to observe low-speed cycling. Data may be retained indefinitely by preventing reception of the readout command from the counting instrument.

Digital Display Methods. Ordinary meter scales are subject to errors when test results are read, and for this reason digital voltmeters have come into use. A voltmeter reading is shown in Fig. 8·6. Note the sign and the decimal point. Operation of this unit is not complex. The in-

Fig. 8·6 Digital voltmeter display of numbers. (*Non-Linear Systems*)

coming voltage is sensed for polarity and then compared with a reference source. The result is displayed or printed for the output.

Other digital and alphameric indicators are electromechanical, d-c operated readout devices for displaying characters in accordance with a predetermined code. The character display of the indicators may be made in accordance with specific requirements.

The indicators are compact, self-contained devices and can be applied to the output of digital computers, teletype receiving equipment in conjunction with a buffer storage unit, telemetering systems, or whatever data need to be displayed. As shown in Fig. 8·7, they are designed for plug-in mounting in a row so that data or messages of any desired length can be stored, displayed, or transmitted. They operate from a conventional direct-wire closed circuit.

A feature of these indicators is their retentivity. They are so constructed that they store the positioned character electrically as well as visually; when desired and without any time limitation, the stored character can be transmitted into other indicators or to storage relays. While transmitting, the indicator does not move; hence the character displayed is retained

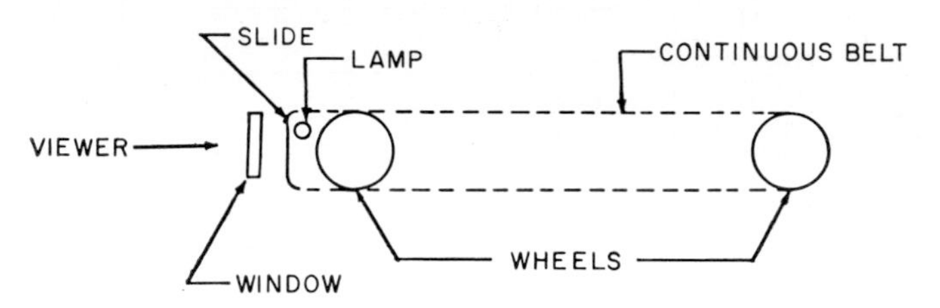

Fig. 8·7 Display unit using a continuous transparent belt.

and will remain until it is erased or a new character is received. The indicators receive and transmit the position code over the same set of control wires.

The indicators are controllable by binary code signals employing a minimum number of control wires, and they respond to simultaneous binary switching combinations. For each binary unit stored, an external relay can be eliminated.

The simplest way of displaying characters on these indicators is the use of push buttons or switches from a keyboard. If a switch or push button existed for each of the code wires operating the indicator, operating them in any combination would cause the indicator to seek a commutated position comparable to the selected switching combination, and the character displayed would correspond to that combination.

Data entry into the indicators from storage is simple and is accomplished by using storage-type relay contacts to replace push buttons or switches. Control in this case is exercised in some logical manner through external or automatic programming. For example, information received via teletype would be entered into a buffer storage and then transmitted to indicator storage. The buffer storage could serve many indicators, which in turn would store and display, freeing the buffer storage to accept other information.

An important feature of these indicators is their inherent storage and transmitting characteristics, which provide for data entry and retransmission. The indicators can be used to accept data from a source, free the source for other programs, and disseminate the data from one indicator to another as required.

The operation of the indicator is based on a positioning system utilizing a six-bit binary code. Seventy-two character positions are used. The relation of the visible character to its code for the standard digital indicator is the direct translation of binary-coded decimal notation to ordinary decimal notation. Other combinations can be developed.

The indicator is internally powered by a small d-c permanent magnet motor. A mismatch between a transmitted code and that commutated by the indicator will complete the circuit, causing the motor to operate. The

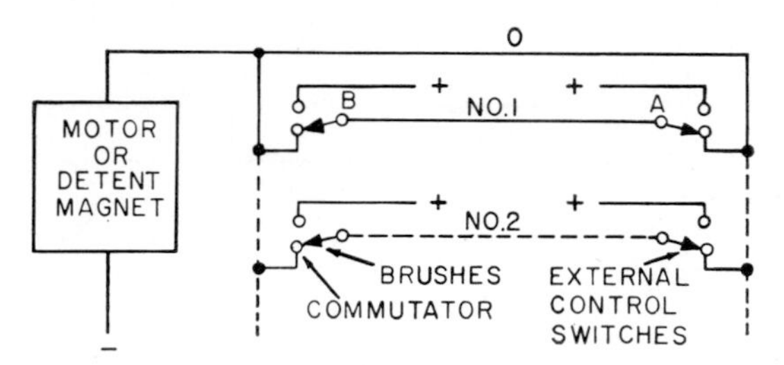

Fig. 8·8 Display-system operation. (*Union Switch and Signal*)

commutator is driven by the motor until the transmitted and commutated codes are energized and the indicator comes to rest, displaying the character whose code was transmitted. Once positioned, the indicator retains the information indefinitely, visually and electrically, until a new code is transmitted to it.

The basic principle of circuit operation of the digital indicator is shown in Fig. 8·8. Consider first only the wires shown by solid lines. *A* is a single-pole double-throw switch which will be used to control the indicator. *B* represents one commutator brush in the indicator. Brush *B* is shown in its down position, which represents making contact with an inner-area segment of the commutator. At certain other positions of the commutator the brush will make contact with an outer-area segment, represented by brush *B* in its up position.

With both *A* and *B* down, as shown, the circuit is open and the indicator cannot run. If switch *A* is moved to its up position, the circuit will be energized over wire 1 and will operate the indicator until *B* is also in its up position. If *A* is then moved down again, the indicator will again be energized by a current through wire 1, except this time the current will flow through wire 1 in the opposite direction and the indicator will be operated until *B* is in its down position. This might be termed an "agreement" circuit, since the motor will always operate until contacts *B* agree in position with contacts *A*.

Switch *A* can have only two positions; hence, a selection of only two characters could be made in the indicator. If circuit 2 (the dotted line) is now added, it can be seen that the indicator will operate until both the controls at B agree in position with both the switches at A.

As the two switches at A have four positions, more such control circuits can be added, each added circuit doubling the number of possible positions. There are 72 possible combinations.

The characters are assigned to the indicator belt in such a manner that standard five-bit teletype-code characters, signs, and other data can be displayed.

The operation of the alphameric indicators is based on a positioning system which uses a differential sprocket and belt arrangement. The belt

Fig. 8·9 In-line display of five-digit number. (*Industrial Electronic Engineers*)

carries 72 displayed characters and is supported by two sprockets. For each combination of sprocket positions, there is a distinct belt position, and there are 72 combinations.

Commutators, mounted within the two sprockets, are arranged so that signal brushes make contact selectively with the battery or with the detent magnet.

The indicator is internally powered with a small d-c motor with a permanent-magnet field. A detent magnet, when energized, withdraws the pawl from the detent wheel and closes the motor circuit. The detent magnet also closes a telltale circuit which may be used to determine whether an indicator is running.

In another system, different from but related to this one, digits are projected on a viewing screen, as shown in Fig. 8·9, where 28694 is indicated. The numbers are projected. By exciting the proper lamp, the desired number is projected upon the rear-viewing screen. Numbers 0 through 9 plus a decimal point are available in one case as shown in Fig. 8·10. S_1 is moved to the corresponding number, which is then seen on the screen. If a decimal point is to be shown in a given position, S_2 is used instead of S_1. For a red danger signal, S_3 is used. It is also possible to have words projected, rather than numbers, for certain applications. The basic module displays a single letter, number, sign, or decimal point.

A new and growing field is fiber optics, where bundles of solid material such as polystyrene are used to carry light.

The unique CRT in Fig. 8·11 employs a precision array of fiber optics ("light pipes") to facilitate the direct-contact exposure of light-sensitive materials. No auxiliary optics are required, and the light utilization of the CRT is enhanced by as much as a factor of 50. Because of low voltage

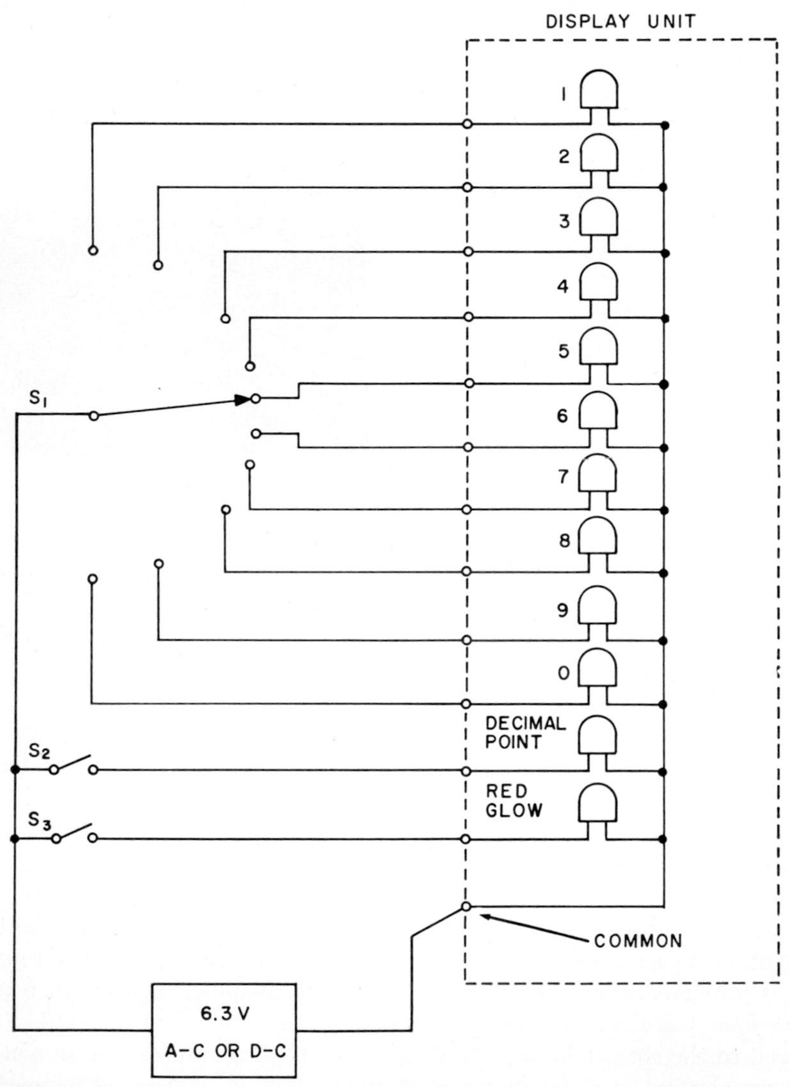

Fig. 8·10 Display unit connected to the customer's equipment. Display units suitable for connecting to the Coleman digitizer, and similar equipment, are available. Standard binary-to-decimal converters for use with the display unit are also available.

and electrostatic features of this CRT, it is unusually well adapted to a low-energy transistorized system.

The electron beam forms a dot pattern by exciting its CRT screen, whose light is carried to the face of the device by the fiber optics.

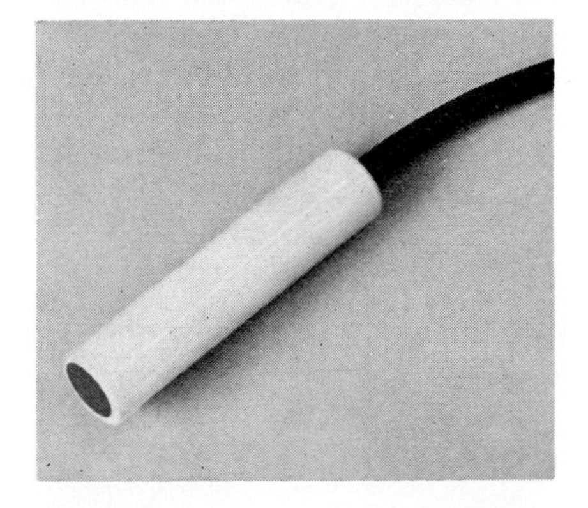

Fig. 8·11 Fiber-optics CRT. (*Litton Industries*)

Electroluminescent Display. Electroluminescence is a new and important display technique requiring low power. There are no filaments; hence the display lights almost at once with the application of power. A 10-segment lamp is shown in Fig. 8·12 as an example of this technique.

Electroluminescent displays may be used for letters and numbers.

Fig. 8·12 Letters are formed by energizing specific segments of the 14-segment alpha-type lamps or the 10-segment numeric-type lamps such as the one shown lighted above. Since all segments are on the same plane, the numerals and letters formed are distinct and easily readable even at wide-angle viewing and require a minimum of space. (*Westinghouse*)

Their construction is shown in Fig. 8·13. They employ no heated filaments, gas-filled tubes, or metallic vapors. The phenomenon of electroluminescence is produced by the passage of current through a thin layer of phosphor, and the lamp is not unlike a capacitor in its construction. A plate of specially coated glass is used as one conducting surface, then a thin layer of phosphor is placed next to the glass and the second conducting plate of metal is applied to this layer. A final moisture-barrier layer is placed on the back of the lamp. As alternating voltage is applied to the conductor plates, the phosphors are excited by the current passing through, and light is produced. The amount of light produced varies with voltage and frequency, but lamps must be designed for specific voltage and frequency ranges.

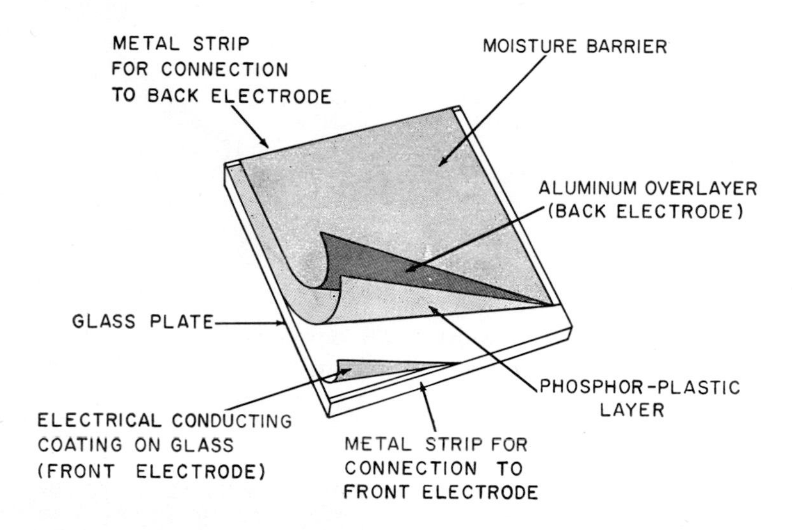

Fig. 8·13 Lamp construction. (*Westinghouse*)

Characters may be formed by two different types of lamps. One type has 10 segments and the other has 14. These lamps are excited by 240 or 460 volts at 60 or 400 cps.

For these lamps the brightness-versus-frequency curves are shown in Fig. 8·14*A*. For all lamps the brightness increases with increasing frequency. Brightness as a function of voltage, with a constant frequency, may be seen from Fig. 8·14*B*. The Society of Motion Picture Engineers gives a value of 7 to 12 ft-lamberts as acceptable for picture highlights, and a television kinescope has a highlight brightness equal to about 50 ft-lamberts.

Electric contact with the readout lamps is made through the pins molded into the back plate of the lamp. Specially designed sockets are of the flush-mounting type. The overall depth of lamp and socket is less than 1 in. Because of the arrangement of the segments and base pins, the same socket may be used for alphameric lamps. This plug-in arrangement makes it easier to replace the readout lamp than to change an ordinary light bulb.

Electroluminescent digital displays often use techniques for converting data from one form to another. For example, Fig. 8·15 shows a pattern of parallel X stripes and parallel Y stripes with the X and Y stripes at right angles. Between these stripes is the electroluminescent phosphor. If a potential is applied to one X stripe of conductive material and to one Y stripe, there will be a spot of light where they cross. Each stripe will

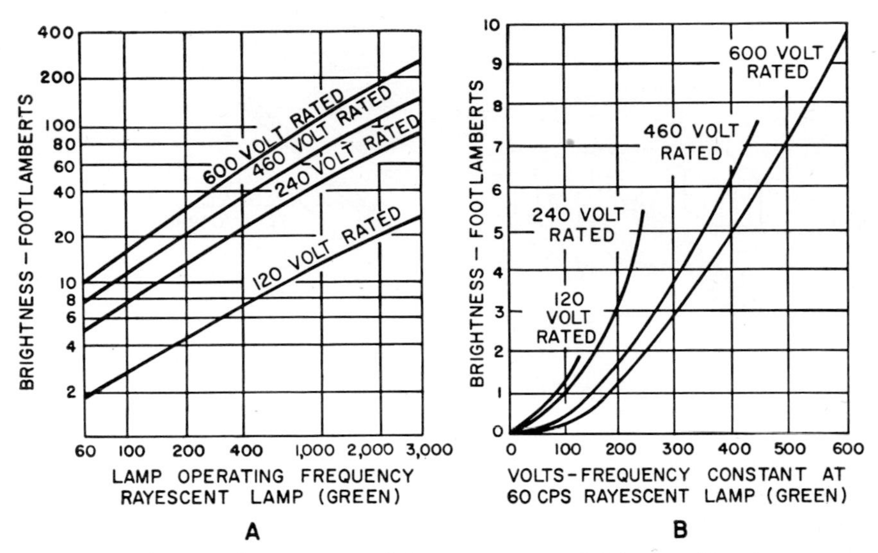

Fig. 8·14 (*A*) Lamp characteristics of frequency versus brightness; (*B*) voltage versus brightness at constant frequency. (*Westinghouse*)

emit some light along its entire length, but the crossover spot will be brighter. A resistive layer, added to the lamp, will inhibit all light except from the crossover spot if desired.

Maps and other displays can use this technique to light the points of specific interest. It can also be used for plotting boards.

Information conversion from decimals to visual displays can be easily accomplished by electroluminescence. Consider the switching action shown in Fig. 8·16. A lamp EL_2 is in series with an a-c voltage source and a photoconductor PC. When PC is dark, the voltage drop across it is high enough to prevent the a-c source from lighting EL_2. But when light falls upon PC, its resistance drops and this allows a large enough drop across EL_2 so that it lights.

EL_1 is the trigger lamp. When it is on, the resistance of PC falls and permits EL_2 to go on. This same technique is used in the translator shown in Fig. 8·17. A is the trigger lamp or EL_1, B is a mask or overlay for decoding, C is the photoconductor PC, and D is a nine-segment EL display. Each segment of the lamp is tied to a photoconductor and the a-c source in series to make a circuit. To make a 1, segment 1 is lit; to make a 2,

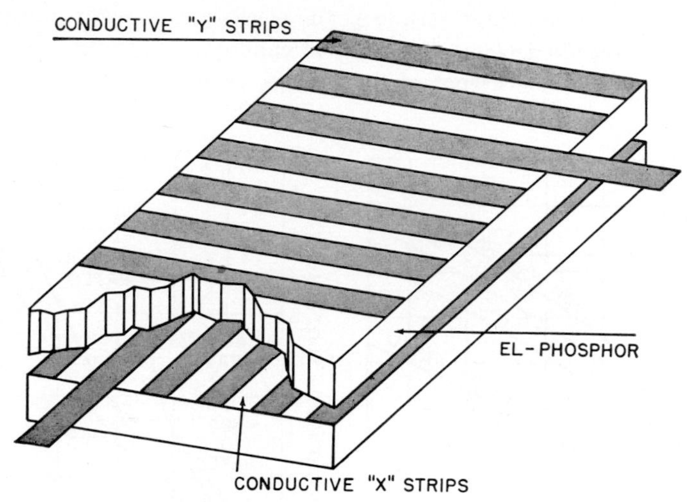

Fig. 8·15 Electroluminescent array. (*Sylvania*)

segments a, f, g, h, j, and p are required. Any digit from 0 through 9 can be illuminated as shown.

Suppose that a 7 is to be produced; from A the lamp marked 7 is lit. Through the mask this causes a, f, and n to light up and these segments produce a 7. Note that the mask allows three photoconductors to be illuminated by A through B. The effect is to create a visible 7.

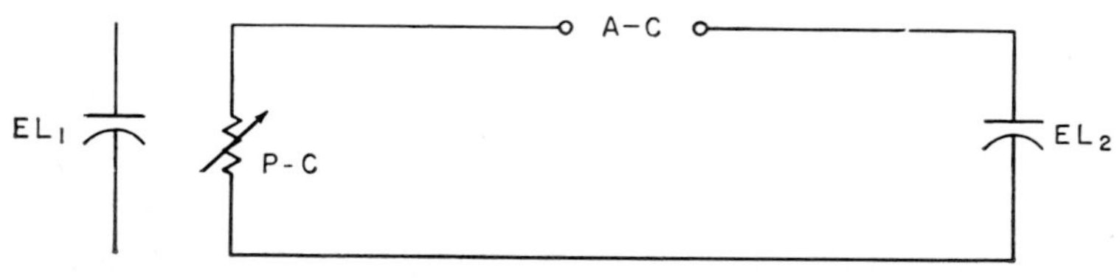

Fig. 8·16 Photoelectric circuit with electroluminescent elements and photoconductors. (*Sylvania*)

A thyratron can be an indicator with the glow of the gas as the visual indication. The 7401 (Tung-Sol) cold cathode fires at about 100 volts on the grid. It can be used as shown in Fig. 8·18A to keep the circuit closed, and to visually indicate this, once the switch S is closed. Because the grid loses control when the tube is fired, a push button is used for resetting.

Figure 8·18*B* shows this tube used with its own *RC* relaxation oscillator a-c generator to provide an output when light hits the cadmium sulfide photocell. By this means the indicator goes out when the photocell is in the dark.

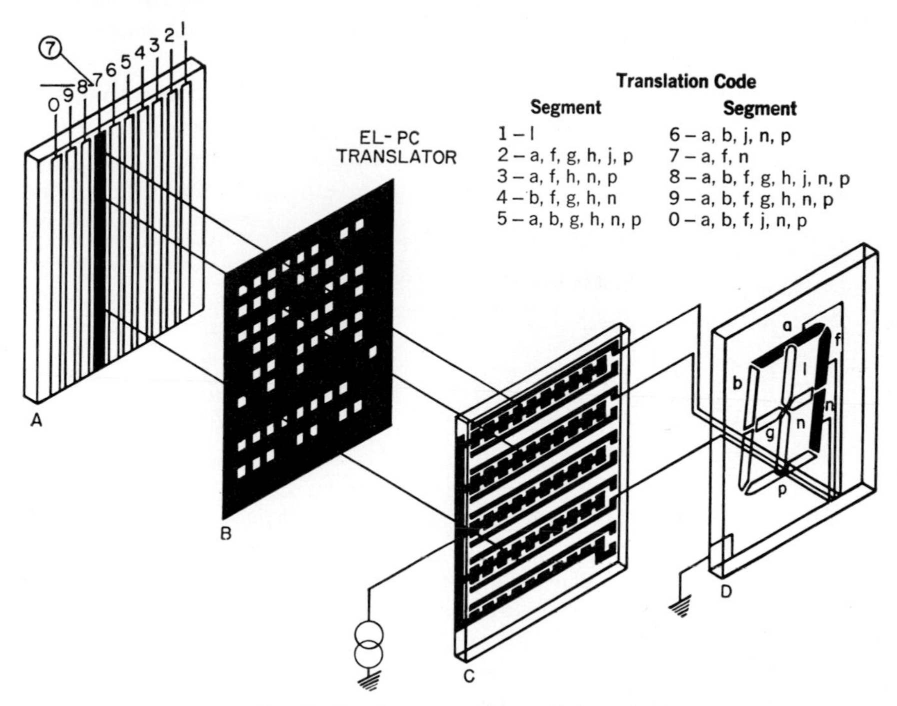

Fig. 8·17 Circuit translator. (*Sylvania*)

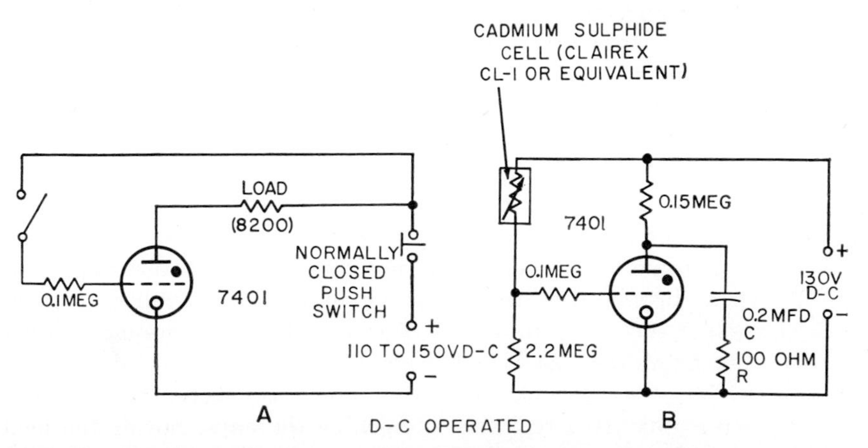

Fig. 8·18 (*A*) Indicator thyratron as relay; (*B*) relay for cadmium sulfide cell. (*Tung-Sol*)

8·2 Data recording For many industrial purposes data are both displayed and recorded.

Operation of a commercial recorder (Hewlett-Packard 560A) is based upon 11 units, one for each of 11 digits. The printer or recorder is driven by a counter circuit at a rate and accuracy determined by the counter. Printing rate is also controlled by the counter but cannot exceed five lines per second.

A voltage from the counter decade is fed to the comparison circuit as shown in Fig. 8·19. This is a staircase voltage as shown. The printer wheel is connected, by a commutator, to a voltage divider with taps. This produces a second staircase, taken from the printer wheel and also applied to the comparison circuit.

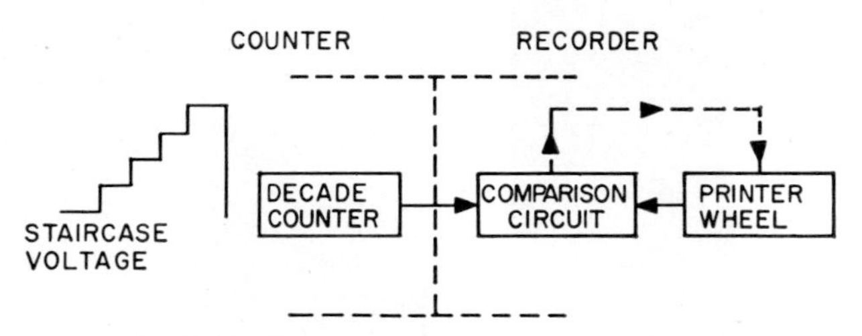

Fig. 8·19 Data recording on a single channel.

While the counter is in a counting period, the printer wheel is fixed and the comparison circuit is not operating. When a count is completed, the two steps are compared and the printer wheel rotates until the two steps on the staircase are the same. At this time both the printer wheel and the decade circuit in the counter are equal.

Because there are a number of parallel inputs from decade units in the counter, each decade output A, B, C, or D (in this example, where there are four) moves a printer wheel corresponding to each decade digit. When all printer wheels are correct, they mechanically print out on a strip of standard 3-in. paper or accordion-fold paper.

Analog Data Recording. Analog data recording is the formation of a printed chart or graph showing how one quantity varies in relation to another, usually time.

One technique of record keeping uses the recorder shown in Fig. 8·20. The reading, which in this case is in amperes, is given by the dial indicator and recorded on the roll of paper by a moving pen with its own supply of ink. This is a portable recorder with a windup motor. It will run for 60 days at a paper speed of 1 in. per hr or slower. Both alternating and direct current and voltage, as well as frequency, can be recorded.

For servicing ease, the inkwell is the throwaway type which is discarded when empty. It is replaced by opening the clips, raising the pen, and removing the entire inkwell. The chart mechanism is shown in Fig.

8·20 with the front plate removed to illustrate the clock-driving mechanism. The same device may be calibrated to record voltage.

Digital Data Recording. Most digital data are recorded either on magnetic tape or on punched paper tape. These forms also serve as storage media in computers. Information storage is a most important function in any data-processing system. Several types of information storage or memory are required in computer systems which may be classified accord-

Fig. 8·20 Analog recorder. (*General Electric*)

ing to the speed of retrieval of data after it is stored. A computer has a small high-speed memory for the rapid insertion and extraction of small amounts of data. A slower but larger storage is used for information not required in a very short time. A third type is bulk storage which is the slowest type.

The main high-speed computer memory is usually a magnetic-core matrix made up of thousands of tiny magnetic rings. The data can be retrieved in a short time, but the total capacity is limited.

A magnetic drum, which is a cylinder with a series of magnetic tracks, forms the backup or secondary memory. The access time, measured in terms of how long it takes to look up and find a certain piece of information, is longer than that for the core system. Drums, however, have a greater storage capacity.

Two types of input and output devices for the computer are magnetic tapes and punched cards. These are also a form of memory bulk storage; they have no limit to how much information they can store, but the access time is greater than for other memories.

Other forms of memory also exist in the computer in different circuits. A flip-flop, for example, remembers or stores its last state (either 1 or 0) until it is triggered and switches states.

There are three types of magnetic memory or storage. Magnetic cores are small toroids of magnetic material; magnetic-tape memories are reels containing a long strip of tape which can be magnetized; and magnetic drums may be considered as short pieces of tape, joined to make a circle and mounted on a cylinder. All storage forms accept input data and give output data.

All these devices store data in digital form. They are discussed in detail in Chaps. 11 and 16.

REVIEW QUESTIONS

8·1 What are four ways counters can indicate their output?

8·2 How does a digital clock differ from an ordinary clock?

8·3 What is an electroluminescent display? Upon what principle does it operate?

8·4 How are letters and numbers formed in an electroluminescent display?

8·5 Explain the electroluminescent matrix.

8·6 How does the output of this form of lamp vary with the applied voltage?

8·7 How does a change in frequency affect the light output?

8·8 Draw a simple system of projecting letters or numbers on a rear-viewing screen.

8·9 How would the system drawn for Question 8·8 work if the alphameric characters were part of a continuous belt?

8·10 How would this system work if the alphameric characters were all on one screen?

8·11 How does analog recording work?

8·12 Explain one digital recording method.

8·13 For what purpose are data displayed?

8·14 What are the three ways data can be presented?

8·15 What forms of display do counters use?

CHAPTER

9

Electronic Heaters

Electronic heaters are used to "sew" plastics, to produce fast-drying plywood, and for many other purposes where rapid and controlled heating is required. Induction heaters are applicable to heating of metallic materials, in which the r-f field induces the heating-current flow in the work. For nonconductive materials such as wood or plastics, dielectric heating is used, in which the wood molecules, acting as a capacitor dielectric, are stressed. This causes the wood to be heated throughout with a resultant rapid setting of the glue.

Industrial electronic heaters are of three types: motor generators, spark-gap generators, and vacuum-tube generators.

9·1 Types of heaters The three types of heaters are used for various applications. Motor generators are usually used at frequencies between 2 and 10 kc, with power ratings from 5 to 500 kw. Their prime use is for deep heating. Spark-gap converters have a frequency range from 25 to 250 kc, with power ratings ranging from 5 to 50 kw. This is power input, and because efficiency is only 50 per cent these units are comparable to other types of generators rated at 2.5 to 25 kw. Vacuum-tube oscillators have frequency capabilities ranging from 100 kc to 1 mc and higher, with power outputs from 1 to 400 kw or more. Their greatest use is in the range from 150 to 500 kc at 5 to 50 kw.

Vacuum-tube Generators. As a prime source of r-f power, vacuum tubes are the heart of most electronic heaters. In terms of electronics, the equipment is not at all complex. A typical unit is shown in Fig. 9·1*A*. Controls are on-off and power adjustment. A current meter can read either plate or grid circuits to check up on the operation. Figure 9·1*B* shows the basic features of the oscillator or power generator. Usually a power triode is used with three related circuits. One is the power supply, which can be adjusted for the desired plate voltage. Bias and grid control are provided by the grid circuit. Output is taken from the plate tank. Part of this output is fed back to make the oscillator function.

A commercial unit in which power is generated by the oscillator tube is shown in Fig. 9·2; cooling for the tube is supplied by the air blower. Together, the output inductance and tuning capacitor make up the tank circuit of the tube. Gas-filled rectifier tubes, a part of the power supply, may be seen in the lower right.

This generator can function as either a dielectric heater or an induction heater, as shown in Fig. 9·3. Alternating current from the power lines is

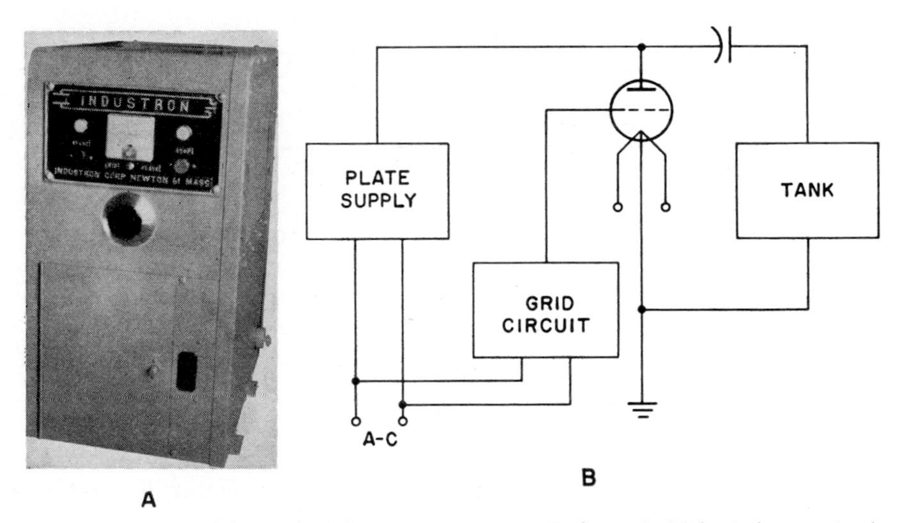

Fig. 9·1 (*A*) One-kilowatt high-frequency generator (*Industron*); (*B*) basic heater circuit showing major blocks.

fed to both the grid circuit and the plate circuit. In a low-power heater the grid circuit uses only grid-leak bias, and it has no requirement for a separate power source; but larger units use elaborate grid control circuits. The oscillator can have either of two tank circuits, one for each type of r-f heating. Usually only the feed from the tank to the load differs in these resonant output circuits.

Commercial equipment uses one of three basic oscillators: the Hartley, which uses a tap on the tank for feedback; the Colpitts, which uses the capacitance tap for this; or the tuned-plate untuned-grid, which uses a separate coil for the required feedback.

Motor Generators. A motor-generator set has a motor driving a generator whose output is of the desired frequency and voltage rating. Table 9·1 shows some typical ranges of power.

As an example, at an output frequency of 10,000 cycles and a power rating of 100 kw, a generator will produce either 220, 440, or 800 volts a-c. The motor for this set will be 175 hp at a rating of 220, 440, or 2,300 volts a-c.

Other motor-generator sets can produce 960 or 3,000 cycles with power outputs ranging from 50 to 250 kw.

Spark-gap Generators. A spark-gap oscillator or converter resembles a very early radio transmitter. The simplified schematic in Fig. 9·4

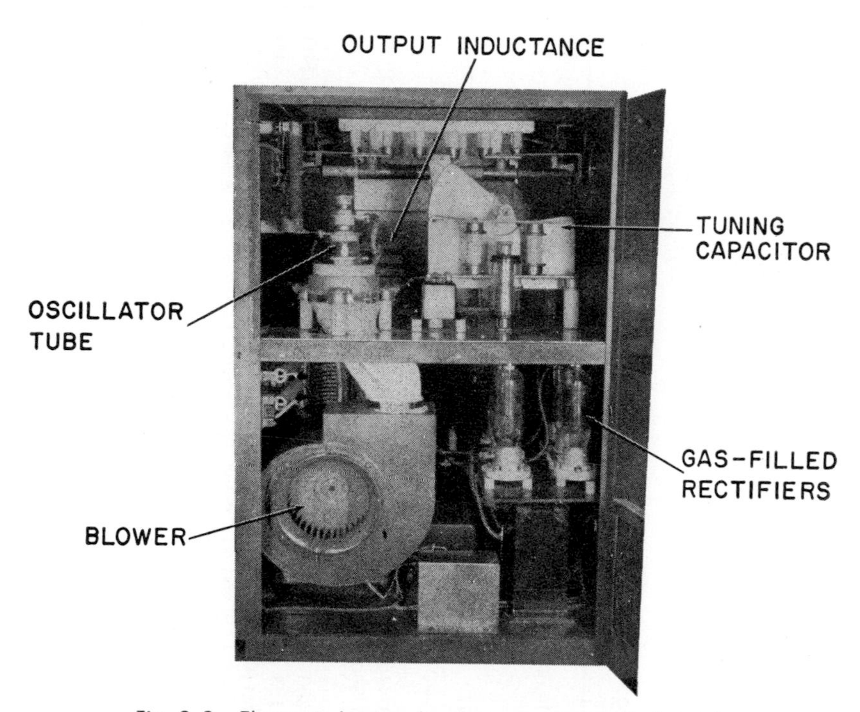

Fig. 9·2 Electronic heater showing components. (*Industron*)

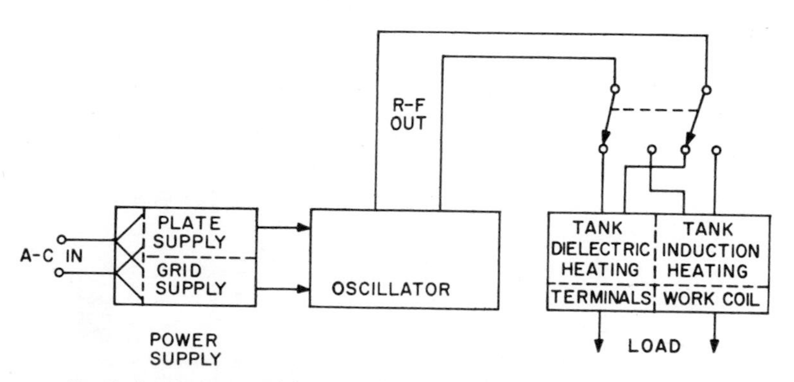

Fig. 9·3 Heater used for either dielectric or induction applications.

Table 9 · 1 Specifications for Motor-Generator Sets

	10,000 cycles				3,000 cycles				960 cycles	
Kw rating	50	75	100	150	50	100	150	200	175	250
Generator voltages available	220/440 400/800	220/440 400/800	220/440 800 †	400/800	400/800	400/800	400/800	400/800	400/800	400/800
Approximate excitation required, watts	300	300	360	600	300	300	600	600	600	600
Motor rating, hp	100	125	175	250	75	150	225	300	250	350
Motor voltages available*	220/440 2,300	220/440 2,300	220/440 2,300	440 2,300	220/440 2,300	220/440 2,300	440 2,300	440 2,300	440 2,300	440 2,300
Motor FL amps at 440 volts	120	160	218	300	104	190	282	385	315	450
Motor LR amps at 440 volts	546	641	1,370	1,530	425	845	1,800	2,250	1,540	2,250
Accelerating time, sec:										
At rated voltage	38	48	22	22	20	14	16	14	11	14
At reduced voltage	68	72	50	45	20	48	68	34	27	63
Approximate water flow required per minute:										
With 90°F water	5	7	8	10	5	5	8	10	8	10

With 70°F water (inlet water should never exceed 90° F)	5	5	5	6	5	5	5	6	5	6
Minimum water pressure at inlet, psi (water pressure at inlet should never exceed 100 psi)	18	18	18	18	18	18	18	18	18	18
Approximate water pressure drop through set, psi	15	15	15	15	15	15	15	15	15	15
Approximate net weight, lb:										
Open type	4,500	5,800	5,800	8,350	4,500	4,500	5,800	8,350	5,800	8,350
Enclosed type	4,800	6,100	6,100	8,850	4,800	4,800	6,100	8,850	6,100	8,850
Approximate outline dimensions, in.:										
Overall length, open type	49	58	58	67	49	49	58	67	58	67
Overall length, enclosed type	59	67	67	77	59	59	67	77	67	77
Width, including conduit box	37	37	37	46	37	37	37	46	37	46
Height, not including eye bolts	33	33	33	37	33	33	33	37	33	37

* Other voltages can be supplied for some ratings.
† 800 volts output requires larger frame unit.
SOURCE: Induction Heating Corporation.

shows the major components of this type of r-f generator. The input line voltage is stepped up by a transformer, resistors R limit the current flow, and S is a series of spark gaps and inductors. L_3 is a work coil which forms a parallel resonant tank circuit with capacitors C_2. This tank is in turn connected into a series resonant circuit which includes inductors L_1 and L_2 and blocking capacitors C_1.

The spark gaps are a series of air-cooled or water-cooled gaps, or combinations of both. Tungsten disks about 1 in. in diameter are used to form the gaps. Gap clearance must be carefully adjusted with a feeler gauge. Variable coils L_1 and tapped coil L_2 are adjusted for the desired frequency. An r-f ammeter is coupled to the output circuit to observe

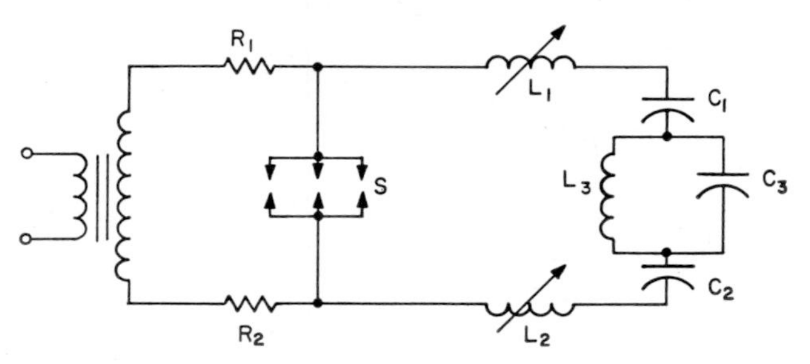

Fig. 9·4 Spark-gap heater, simplified diagram.

tuning adjustments for maximum r-f current in the work coil. The blocking capacitors are used to keep the high voltage from the output coil. There are several of these capacitors in parallel here as well as in the tank circuit. The work coils are at ground potential for safety. They are made of copper tubing to permit water to circulate through them, even while current is applied. In some cases currents as high as 1,000 amp can flow through the work coil. Greatest output is obtained when both circuits are resonant at the same frequency. If a different frequency is desired because of the depth-of-penetration requirements, the series tuning inductances may be adjusted.

Radio Interference. The Federal Communications Commission (FCC) requires a strict control to prevent excessive r-f radiation. All induction- and dielectric-heating generators and oscillators produce large amounts of r-f power which could cause interference. Commercial equipment is well shielded to prevent radiation interference with receivers. The servicing technician must be sure that this shielding remains intact after any repairs have been made.

All these heaters are, in effect, transmitters with small antennas. Most technicians know how easy it is to radiate energy even from well-shielded electronic devices. The problem with dielectric and induction heaters is severe since power ratings are up to 100 kw at frequencies up to 100 mc and even higher. Because of the potential interference problem, the FCC has ruled that no oscillator-generator may radiate more than 10μv per m as measured 1 mile from the unit. This maximum value includes all radiation, including harmonics, at any frequency.

9·2 Induction heating Induction heaters come in many sizes for various purposes. The size of the unit depends upon its power output rating and its application.

Special heaters are designed for specific requirements, such as the cold press in Fig. 9·5 and the wire stripper in Fig. 9·6 or the equipment for semiconductor manufacture in Fig. 9·7.

Fig. 9·5 Cold press for electronic gluing. (*Industron*)

Principles of Induction Heating. Induction heating is a method for obtaining localized and controlled heat by induced currents. Because the metallic workpiece has both eddy-current and hysteresis losses, the temperature increases in it in much the same manner as heat is produced in a transformer core. Typical advantages of induction heating over other methods are controlled areas of heating, rapid application of heat, precise control of the amount of heat, and uniformity of heat application. The principal disadvantage is high cost. This may limit applications to specialized cases which cannot be handled in other less convenient or less expensive ways. Applications include surface hardening and other types of metal heat treating, soldering, and brazing.

The intensity of a magnetic field depends upon current, number of turns, and core material. Magnetic materials heat up faster than non-

Fig. 9·6 Wire stripper. (*Reeves*)

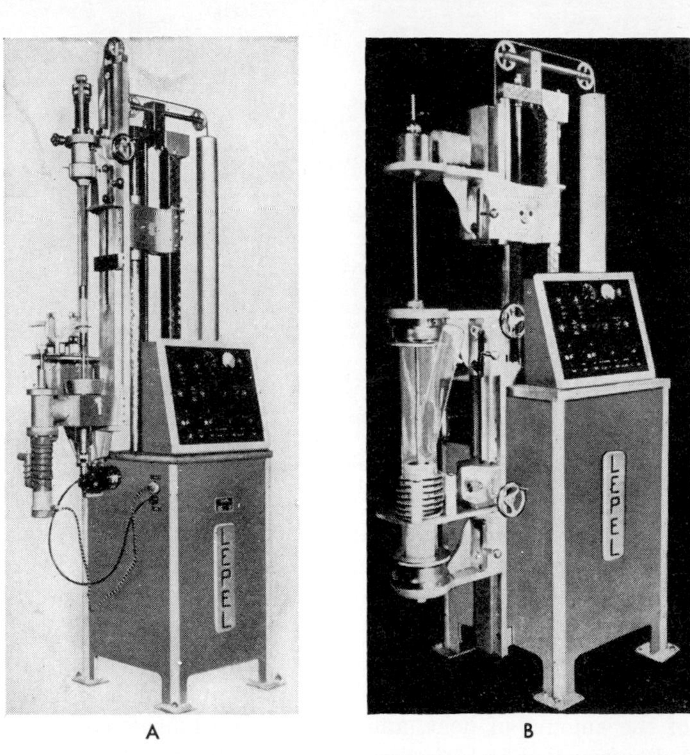

Fig. 9·7 Dual-purpose floating-zone and crystal-pulling fixture: (*A*) used as a floating-zone fixture; (*B*) used as a crystal-pulling fixture. (*Lepel*)

magnetic materials because they have both eddy-current and hysteresis loss; nonmagnetic materials have only eddy-current loss.

Induction heating encompasses a wide range of power and frequency. Frequencies extend from 60 cycles to 50 mc. Use of different frequencies provides for a variation in depth of electronic heating. The skin effect causes high-frequency current to travel on the surface. This same effect also manifests itself somewhat at lower frequencies, where some of the current flow is on the surface and some is below. This effect can be used to control the depth of induction heating.

For example, heating with a 2-kc current will cause a penetration of about 0.125 in., but a frequency of 200 kc will only heat a layer 0.020 in. deep. The thinner the layer desired, or the more shallow the penetration, the higher the operating frequency. The depth of penetration varies as the square root of the frequency. Heating depends upon several other factors, such as the characteristics of the metal work, the amount of power used, and the design of the shaped coil surrounding the work.

The work coil is inductively coupled to the oscillator tank circuit and acts like the primary of a transformer. The workpiece acts like the secondary. Cooling is used to prevent the work coil from melting. Single-turn coils are used to heat narrow areas. The size and shape of the work determines the number of turns and the shape of the coil.

Figure 9·8 shows some possible shapes for work coils. *A* is a regular coil for round objects; *B* is a pancake coil for flat objects; *C* is a coil for the internal heating of cylinders; *D* is a special shape used for brazing or soldering; and *E* is an example of a special shape for a specific purpose, namely, brazing carbide tips to cutting tools.

Work coils can be designed for individual applications. Waveguides for the electronics industry can be assembled using induction heating. In this case, as shown in Fig. 9·9, a combination pancake and internal-helix coil is used. The internal-helix portion of the coil heats the brass tube in the vicinity of the joint to be brazed, while the pancake portion of the coil heats the brass flange. Molten alloy drawn into the joint between the tube and the flange will appear at the bottom of the flange, signifying adequate flow and a sound joint. A uniform, small fillet remains at the top of the flange.

Many electric measuring instruments, such as ammeters or voltmeters, require permanent-magnet pole-piece assemblies. For this purpose high-permeability iron pole pieces are joined to strong permanent magnets, frequently alnico. Different techniques for joining have been used, including soft soldering and plastic bonding. The joining operation is accomplished rapidly and at low temperatures, thus minimizing the possibility of affecting the properties of the magnet. Figure 9·10 shows an alnico-magnet iron-pole-piece assembly for a d-c milliammeter joined by soft soldering. Sheets of soft solder are placed between the pole pieces and

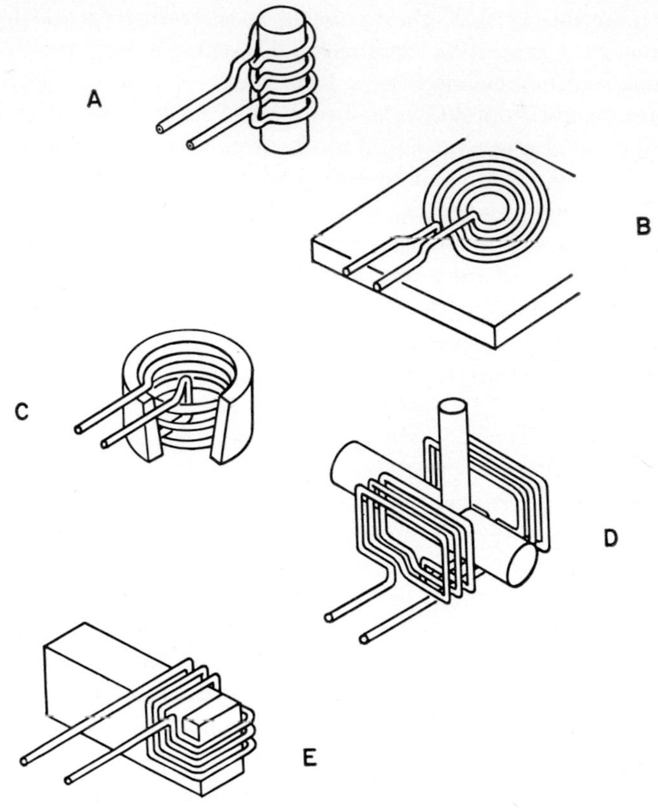

Fig. 9·8 (*A*) Coil shape around a rod; (*B*) pancake coil; (*C*) internal coil; (*D*) special shape for joining two rods; (*E*) corner hardening. (*Lepel*)

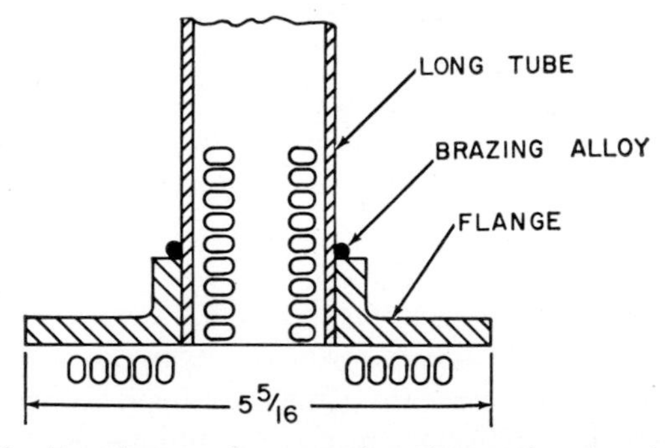

Fig. 9·9 Manufacture of waveguide with special work coil. (*Lepel*)

the alnico magnet, fluxed, and then heated locally by induction as shown. Rapid heating, confined to the vicinity of the joint, avoids overheating of the permanent magnet.

Power Generation. In every case the work coil is coupled to the tank circuit of the oscillator. Usually a step-down transformer is used. The

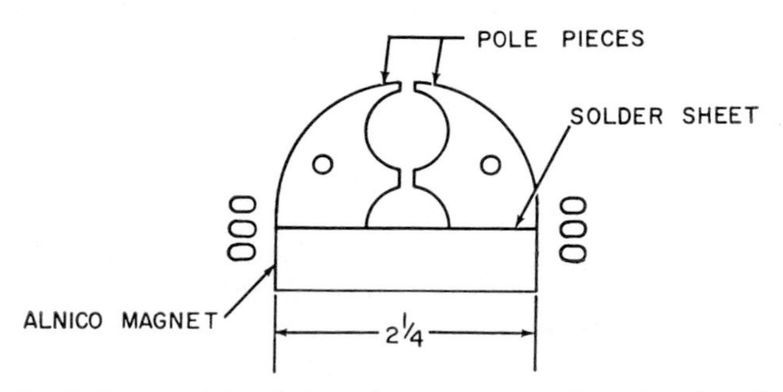

Fig. 9·10 Special coil shape for magnet manufacturing. (*Lepel*)

tank circuit obtains its power from the oscillator through the transformer.

A simplified vacuum-tube oscillator type of generator circuit is shown in Fig. 9·11. A power triode is connected in a plate-to-grid inductive feed-

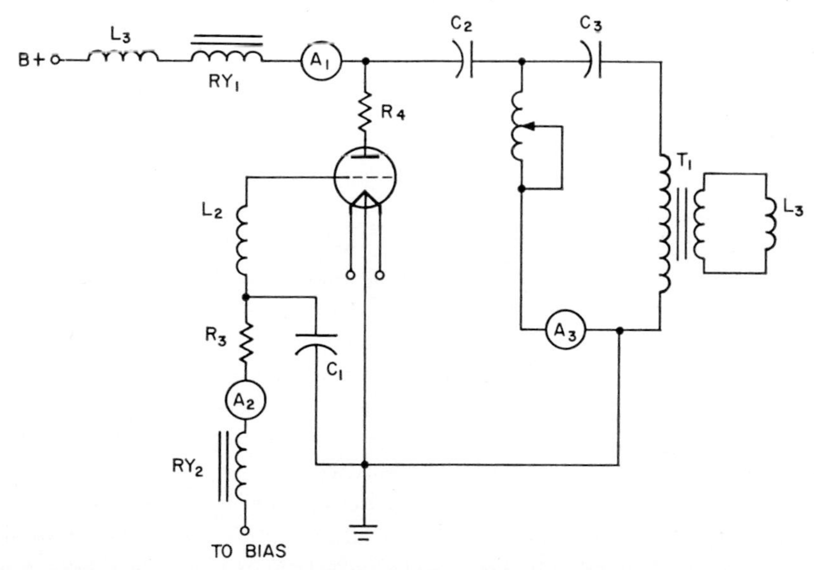

Fig. 9·11 Single-oscillator stage showing meters for adjustments.

back circuit. Resistor R_3 provides grid-leak bias, T_1 is the r-f output transformer, and the work coil is L_3. Blocking capacitor C_2 prevents the high plate voltage from reaching the output and ground. Capacitor C_3 and inductors L_1 and T_1 form a resonant tank circuit, and choke L_3

prevents radio frequency from getting into the plate power supply. Component values depend upon operating frequency.

Variations in the work-coil load are reflected back into the tank circuit. Adjustments of the tank inductance set the oscillator at the desired frequency. At resonance, grid current increases, plate current decreases, and r-f output current increases. Ammeters A_1, A_2, and A_3 are used to monitor circuit operation and also help troubleshoot when necessary.

Induction heaters, used for conductive loads, may be considered as three blocks. These are the oscillator or generator, the power supply, and the work coil and its coupling. Energy from the power supply is changed into r-f energy by the oscillator-generator, then applied to the load through the work coil. R-f power for induction heating uses class C oscillators in which the plate current flows for less than a cycle because of the heavy grid bias which is beyond cutoff.

Typical oscillators, discussed above, include the Hartley, the Colpitts, and the tuned-plate untuned-grid. All these circuits will give the same output power and the same efficiency with identical tubes.

Figure 9·12 shows a simplified circuit from a commercial unit, the General Electric 4HM-4OL1 or 4HM-4OL3. This is a shunt-fed Hartley circuit. R_{11} and C_{17} comprise the thyratron bias network. C_1, C_5, C_6, C_8, and C_9 are bypass capacitors. Plate voltage is applied to the tube (an ML 6426) through L_6, the filter choke. R_1, R_2, and R_3 form the grid leak. Plate current is read from M_1, grid current from M_3. L_3 and R_{10} are used to kill parasitic oscillations which can occur because of the lead and wiring inductances plus the stray and distributed capacitances. Both the transformer output and the series work coil with the tapped tank coil are shown as possible variations.

A-C Power Control. Induction heaters come in many sizes. A small unit may supply a power output of 20 kw with 250 to 275 amp of current in the output coil. A 50-kw unit may supply 400 amp. Water cooling of the parts with high current is common. Oscillator tubes are also cooled; both air and water are used. A modern air-cooled tube for this application is the Eimac 3CX10,000A3, which has a rating of 10 kw.

In an electronic heater, line power through a control system goes to both the plate supply and the grid supply. Plate voltage control is used to adjust the amount of r-f power output, while the grid supply prevents runaway conduction of the oscillator if the normal self-bias fails.

A-c power distribution (Fig. 9·13) in large heaters has several built-in safety features in common with much industrial equipment. These features make it impossible to start up the equipment in the wrong sequence, which would damage the oscillator tube.

The sequence of operation is given below:

1. Closing S_1 starts cooling by the fan (F) or, where water cooling is used, starts the flow of water.

2. Closing S_2, the filament power, lights indicator lamp L_1 and energizes relay RY_1, which closes the contacts of this relay.

3. Both T_1 (the filament transformer) and T_2 (the transformer for the grid-bias supply) are energized when the relay contacts close.

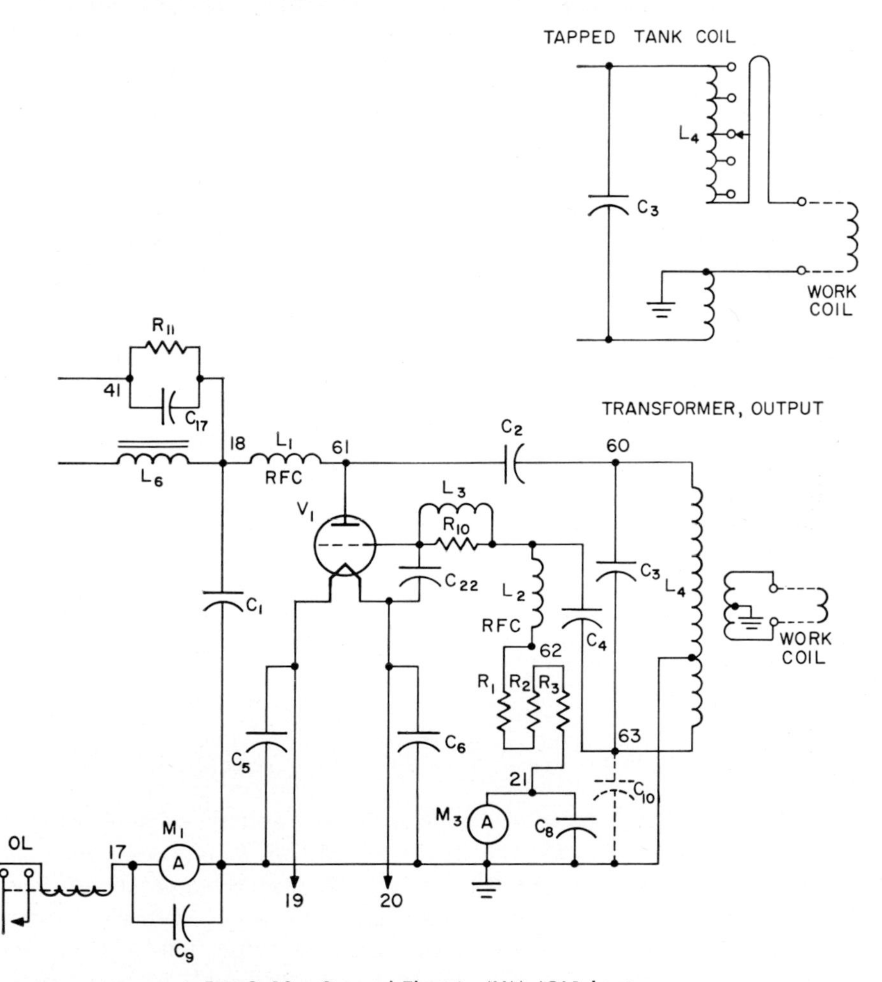

Fig. 9·12 General Electric 4HM-4OL1 heater.

4. One secondary on T_1 goes to a time delay RY_2 which, after about 5 min, closes the normally open contacts so S_3 (the plate switch) can be turned on.

5. If the interlocks (S_4 and S_5) are closed, S_3 energizes RY_3, which connects the plate transformer T_3 across the line. Lamp L_2 goes on.

6. If the plate current rises above a maximum value of 30 amp, RY_4 opens; in this way it protects the tube.

A diagram of the control wiring is required for troubleshooting. For example, if the oscillator and power-supply tubes are operative but there is no plate voltage, this diagram shows several possible causes of the trouble. The interlocks (S_4 and S_5) could be open, or the plate of the oscillator could have drawn so much current that the overload relay (RY_4) opened to prevent plate voltage. Other items which will open the plate circuit include fuses F_I and F_2, and S_2, which must be closed for the oscillator to operate. The control-wiring schematic is very useful in locating possible defects in induction-heating equipment.

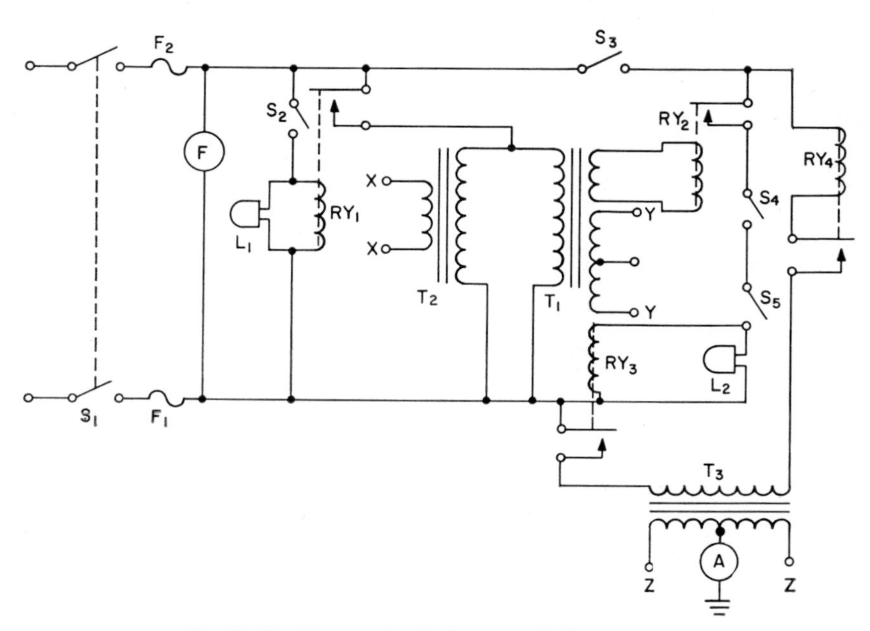

Fig. 9·13 Heater circuits for control of a-c power.

Figure 9·14 shows the control sections of the unit shown in Fig. 9·12. In the control section to the left, the secondary of T_1 is fused; when the filament contractor is closed, this means the water temperature (temperature interlock) is correct and the water pressure (waterflow interlock) is proper. After a suitable time delay, the start button turns the plate voltage on when M is energized and the contacts in each leg of the plate transformer (T_8) are closed. Plate voltage is interrupted if the overload relay OL opens because of excessive current or if the door interlocks are opened. The power supply uses six thyratrons.

Both diode rectifiers and thyratrons are used in induction-heater rectifiers. A simplified schematic (Fig. 9·15) shows a typical three-phase power supply using diodes. This, for example, could be from a 20-kw heater which has the 460-volt input stepped up to 7,900 volts and applied to the bank of six rectifiers, two for each phase. Each pair of diodes is in series; thus they split the applied voltage and their ratings are less. The filter output to the oscillator is 10 kv. Oscillator plate voltage is 6 kv at 500 kc.

Without power control, rectifier diodes provide high voltage which cannot be adjusted. In some cases taps can be provided on the transformer but, at best, this gives steps of control and not continuous adjustment.

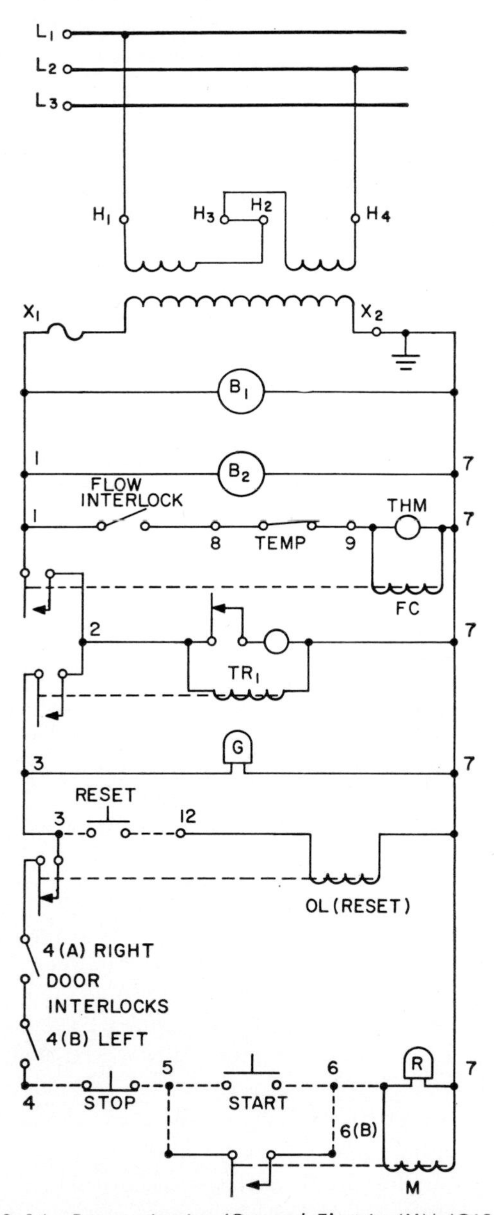

Fig. 9·14 Power circuits. (*General Electric 4HM-4OL2, L4*)

Cooling. Because of the heat generated in induction heaters, some method of cooling is required. Both water cooling and air cooling are used for oscillator tubes, while water is always used to cool other parts. In air cooling, small vanes are used, one at each air intake to each blower. A vane is connected to a switch which prevents the operation of the oscillator tubes unless there is a supply of cooling air to prevent overheating of the tubes. Water cooling includes the oscillator anode, which is inserted in a water jacket which has a continuous flow of water.

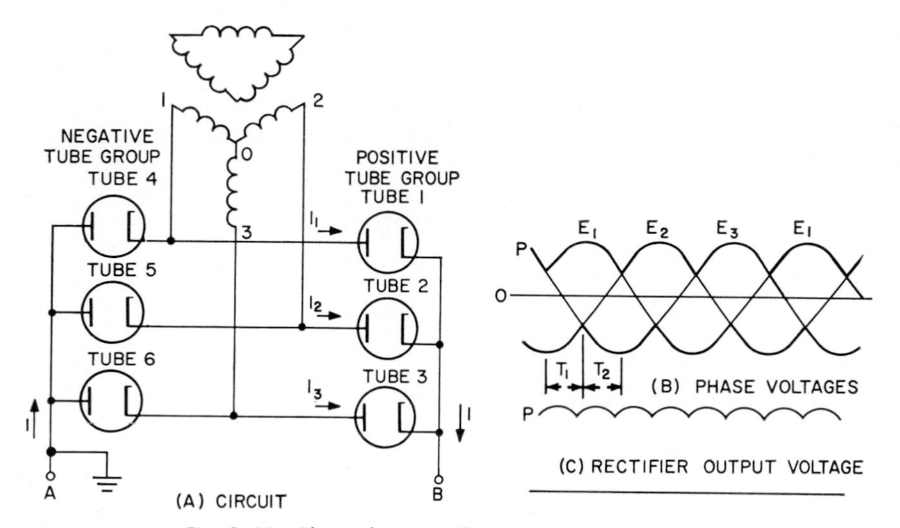

Fig. 9·15 Three-phase rectifier without power control.

Water is also required by other parts, as shown in Fig. 9·16, including the r-f output transformer, which converts the high-voltage r-f output current of an induction heater to lower-voltage high-amperage r-f current. It is then possible to couple the power into a very small load or into a very small section of a load by designing small inductor coils of very few turns. The output transformer should be as close as possible to the work, but it must not be on a metal plate since this would heat the metal and cause losses.

Primary turns for this transformer are cooled by water from the oscillator output terminals. The secondary turn must be cooled by an external source of water. The water should be connected first to the transformer secondary. After cooling the secondary, the same water may be used to cool the work coil. The temperature of the cooling water should not exceed 110°F as it leaves the transformer secondary.

Commercial Induction Heaters. A typical commercial unit is shown in Fig. 9·17. Power from the line goes to seven filament transformers. T_1, T_2, and T_3 supply current for the thyratrons, each of which is one leg of the three-phase power supply. Control of the grid circuits of these gas tubes allows adjustment of the plate voltage of the oscillator, which, in turn, varies the power output for heating. Three gas diodes, with current from T_4, T_5, and T_6, are used in the other legs of the power supply. No control over their conduction is available. Line A is the common positive bus of

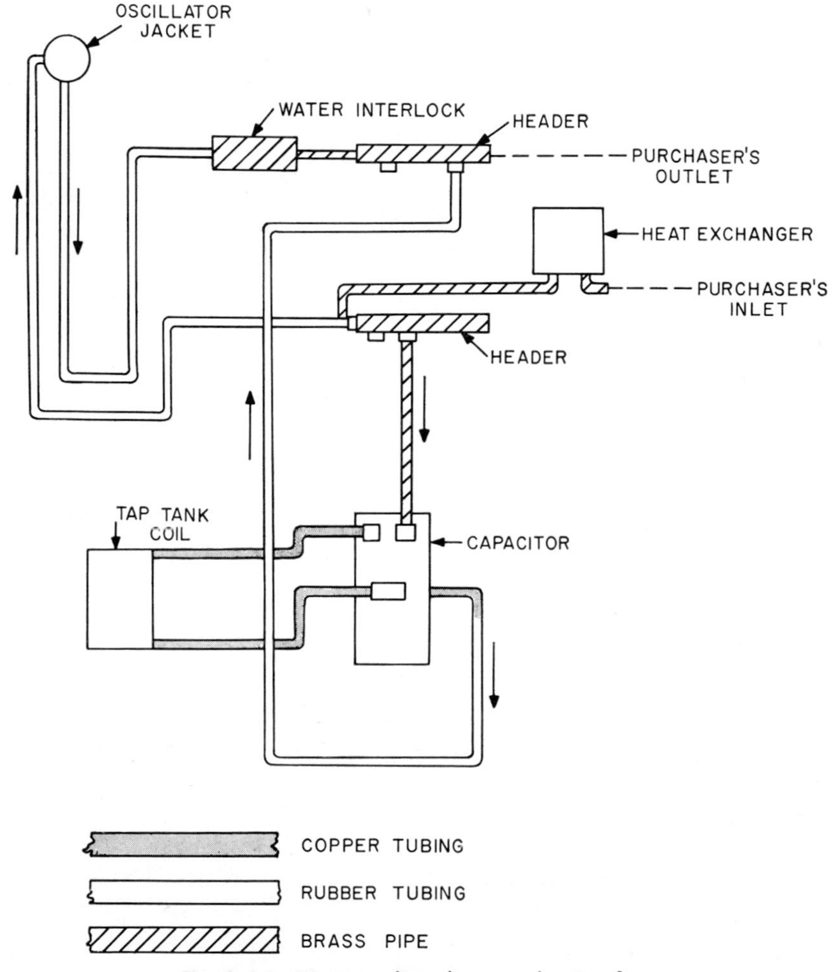

Fig. 9·16 Water-cooling diagram showing flow.

the power supply, and it goes to the plate of the oscillator through the LC filter. Line B, the negative end of the power supply, goes to ground through the plate-current meter. T_8 is for the heater of the oscillator, and the secondary has a filament voltage meter. A 6420 tube is used for 10 kw and a 6800 tube is used for higher power.

Control of any of these transformers is possible by changing primary taps as shown. This is most important for the oscillator filament since

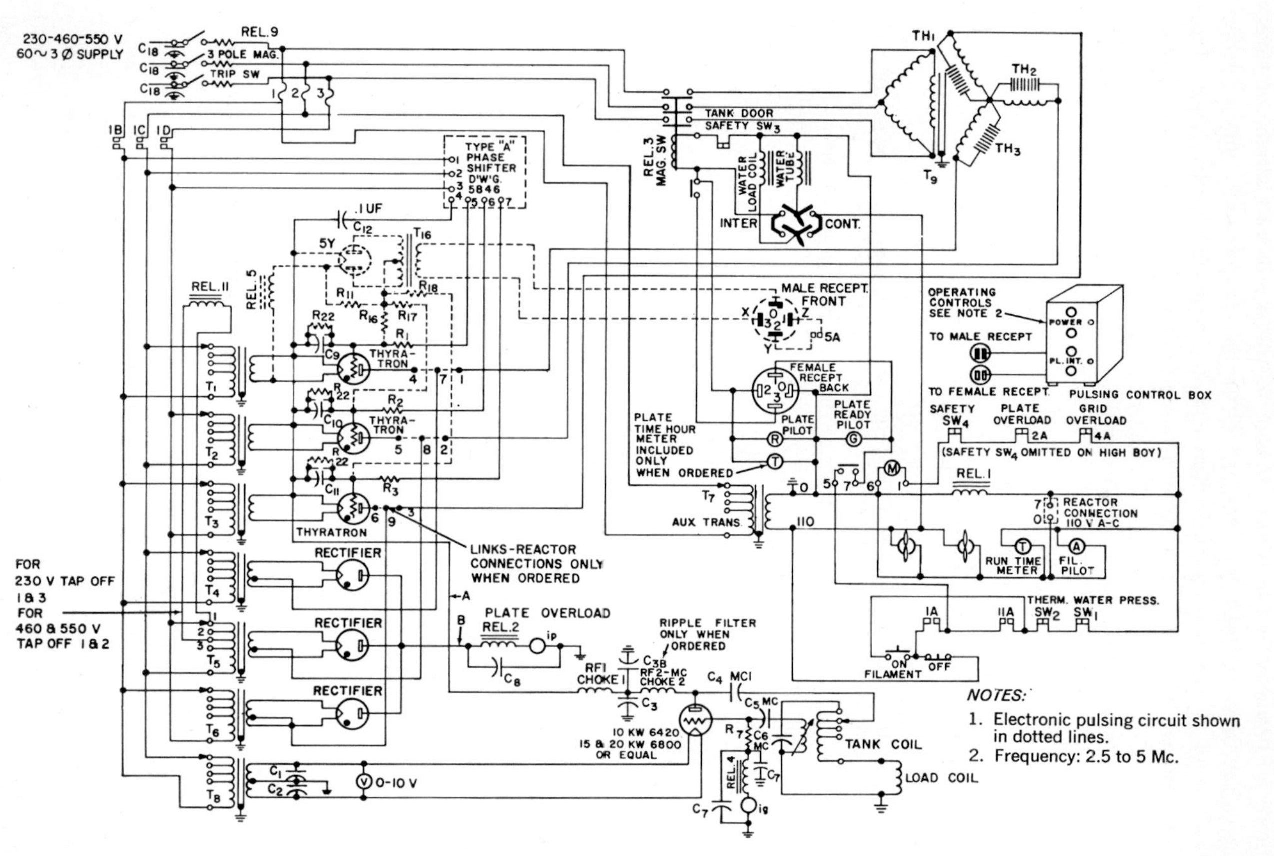

Fig. 9·17 Heater for 10, 15, or 20 kw. (*Lepel*)

tube life is a function of filament voltage as shown in Fig. 9·18. Increasing this voltage by 7.5 per cent will reduce the tube's life to 40 per cent of normal and, at the same time, increase the emission by 75 per cent. Reducing the voltage by 5 to 95 per cent of normal will reduce the normal emission by 30 per cent, but it will increase the tube's life by 180 per cent to almost three times the normal expectancy.

Three considerations in heater adjustments are work-load coupling, power control, and tube life. Adjustments of the filament voltage, as described above, can control the life of the tube. Power control is exercised by changing the direct current applied to the plate of the oscillator. As discussed in Chap. 2, there are several techniques used to vary the firing angle of thyratrons, and this is helpful in power control. In the circuit under discussion (Fig. 9·17), only the triode thyratrons are under control so that a pattern such as that in Fig. 9·19 results. Here, in *A*, a single phase is drawn. Diode conduction is 2 and 4; thyratron conduction is 1 and 3. Note that only the thyratron conduction changes in *B* and *C*.

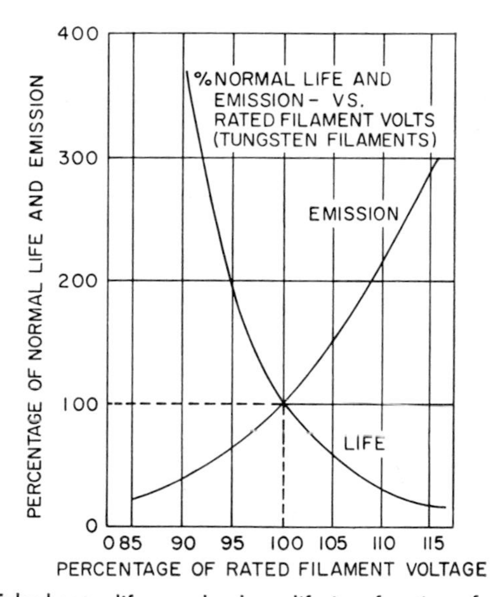

Fig. 9·18 **Tube-heater-life graph where life is a function of voltage. (*Lepel*)**

Figure 9·19*B* represents a condition where no control is used; hence there is no difference between the conduction of either tube type. But if the bias on the thyratron grids inhibits conduction for a portion of alternations 1 and 3, the total average d-c voltage output is reduced as in Fig. 9·19*C*. Thus, a change in control bias on the gas triodes varies the plate voltage for the oscillator. This adjustment is usually called power control.

Work-load coupling, the third variable, may be changed also. Variations in coupling reduce or increase the available power output. Because these changes are also reflected back to the tube as a change in its load, retuning may be required after a change in coupling. A load may be used in series; in this case a tap is used to adjust for the desired power.

9·3 Dielectric heaters Heaters used for nonconductive materials are called dielectric. The generators of this r-f power are quite like those of the induction heaters.

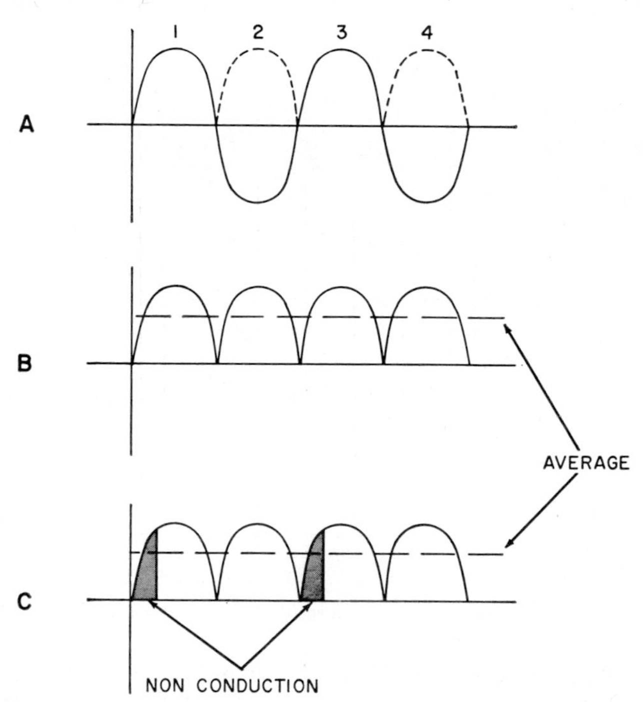

Fig. 9·19 (A) Thyratron conduction on a-c sine wave; (B) average conduction; (C) a-c alternations showing nonconducting times.

Principles of Dielectric Heaters. In dielectric heaters the work is placed between two electrodes which act like the plates of a capacitor, while the work acts like a dielectric. Charging and discharging the capacitor causes a rearrangement of the molecules in the dielectric, which results in heat. In dielectric heating this represents the useful output of the equipment.

The frequency of operation depends upon the power factor of the work material. Frequencies range from 5 to 100 mc and even higher. While dielectric equipment resembles induction-heating devices, it requires both a different type of output circuit and higher frequencies of operation. Installation and operation of induction and dielectric heaters should depend heavily on the manufacturer's directions. In most cases the equipment is installed by the equipment manufacturer, but troubleshooting and repair of these units is within the capabilities and test-equipment resources of most technicians.

Dielectric heating is a relatively new and different method of heating a variety of nonmetallic or nonconducting materials having suitable electrical and physical properties. Solids, liquids, and intermediate forms such as sheets, granules, or foams can be treated on either a batch or a continuous-flow basis.

Dielectric heating operates on the principle of converting high-frequency electric energy into heat energy. Since the energy conversion takes place throughout the mass of the material to be treated, transfer losses and unnecessary temperature gradients are reduced to an absolute minimum and the heating effect is instantaneous and quite uniform.

Development of heat in a dielectric material is brought about by molecular fraction. The molecules within the material are subjected to periodic stresses caused by an electric field alternating in polarity. The amount of heat developed in the material is directly proportional to the amount of r-f power applied to it. However, the voltage and frequency at which this power is applied depend on the form factor of the material and on the loss factor.

Most plastics and other insulators used in industry have loss factors high enough so that power can be applied at moderate voltage and frequency. Materials such as polystyrene with relatively small loss factor are impractical to heat by this method.

The physical requirements of the application will dictate the operational technique to be used. In all cases a generator or source of high-frequency power is needed, which must be the correct type, with power-output characteristics suitable for the particular application. This generator is connected by a suitable transmission line or feeder network to the electrodes (or dies) which apply the power to the material. These electrodes can be designed to assume almost any shape or form. Presses, jigs, or ovens contain the electrodes through which the current passes.

Applications of dielectric heating cover a wide range of industrial products. Uses include the welding of vinyl plastics and acetate, glue drying and edge bonding of wood, rubber vulcanization and softening, and curing of resin adhesives.

Uses of Dielectric Heating. Of the many applications, there are three which are most important. These are the r-f sewing of plastics, wood gluing, and material drying.

Plastic sewing or the bonding of thermoplastic sheets can be accomplished by an "r-f sewing machine." Thermoplastic materials melt when they are heated, and because of this they can be joined or welded where conditions of heat, feed, and pressure are controlled. Pliofilm, Koroseal, saran, and other materials are synthetics, produced in thin film forms, which lend themselves to these bonding techniques. About 150°C of heat is applied to the sheets to be joined by rotating electrodes which also apply proper pressure and maintain proper feed. Plastic products including raincoats, swimming pools, and large radomes are produced by using dielectric heating.

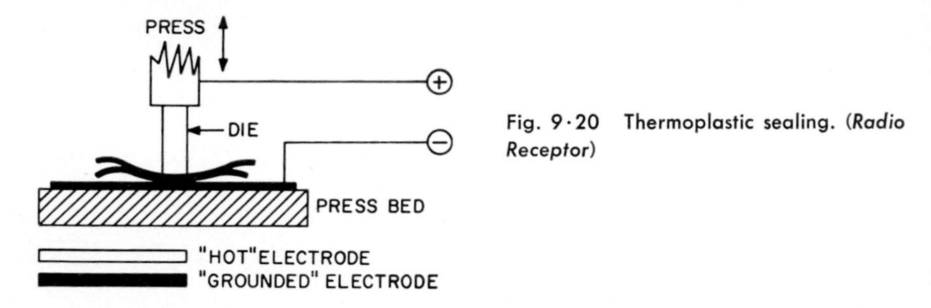

Fig. 9·20 Thermoplastic sealing. (*Radio Receptor*)

Figure 9·20 shows the technique by which the press forces the die against the bed. The two plastic sheets are joined by a combination of heat (from the r-f source) and pressure. In quilting plastic materials for upholstery and for automobiles a hydraulic press is used with dielectric

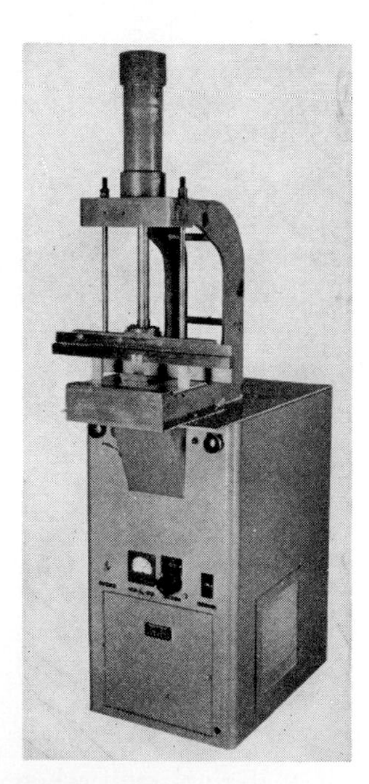

Fig. 9·21 Heat-sealing generator. (*Industron S-10*)

heating as shown in Fig. 9·21. Some plastic materials, as shown in Table 9·2 are better than others for this purpose. Vinyl materials have excellent heating properties. Others such as Teflon and the silicones have poor r-f heating qualities.

Table 9·2 R-F Heatability Chart

Material	Excel-lent	Good	Poor	Little effect
Acetate film		X		
Butyrate film		X		
Epoxy		X		
Kel-F			X	
Micalex			X	
Melamine formaldehyde resin		X		
Nylon, dacron, orlon, dynel (Mylar)		X		
Plexiglas, lucite		X		
Polyester resins		X		
Polyethylene				X
Polystyrene, styrene				X
Polyurethane foam		X		
Polyvinyl acetate		X		
Polyvinyl chloride (plastisol)	X			
Resorcinol formaldehyde resin		X		
Rubber	X			
Saran	X			
Silicone			X	
Teflon			X	
Urea formaldehyde resins		X		
Vinyl film	X			
Vinyl foam	X			
Water	X			
Wood	X			

SOURCE: Radio Receptor

Wood gluing is another important application of dielectric heating. The widespread use of plywood is due, in part, to the speed and economy of this type of heating. Sheets of thin wood are placed one on top of the other with the grains of the layers at right angles to each other, glue is spread between each sheet, and pressure is used to hold the sheets together until the glue sets. Dielectric heating is used to raise the temperature to hasten drying. Plywood manufacture is not limited to thin sheets. Blocks or strips of wood are also glued for flooring, blocks, seats, and other products.

Heating and drying of materials usually use a system by which the objects can pass through the r-f field in a continuous stream. Electrodes apply the r-f power from the generator to the work. Basic types (Fig. 9·22) are the flat-plate electrodes. As shown in Fig. 9·22*A*, the work is placed between the electrodes which form a part of (or are coupled to) the oscillator tank. For heating of bulk materials such as rubber or textiles an adjustable air gap (Fig. 9·22*B*) permits heating to the proper degree. Heating of certain materials for melting (Fig. 9·23*A*) requires the r-f field from the outer cylindrical electrode to raise the temperature of the material in the smaller inner container. If the material is moving, electrodes can be divided and staggered (Fig. 9·23*B*) across the work. For thin materials or coatings the electrodes are usually placed on one side as in Fig. 9·23*C*.

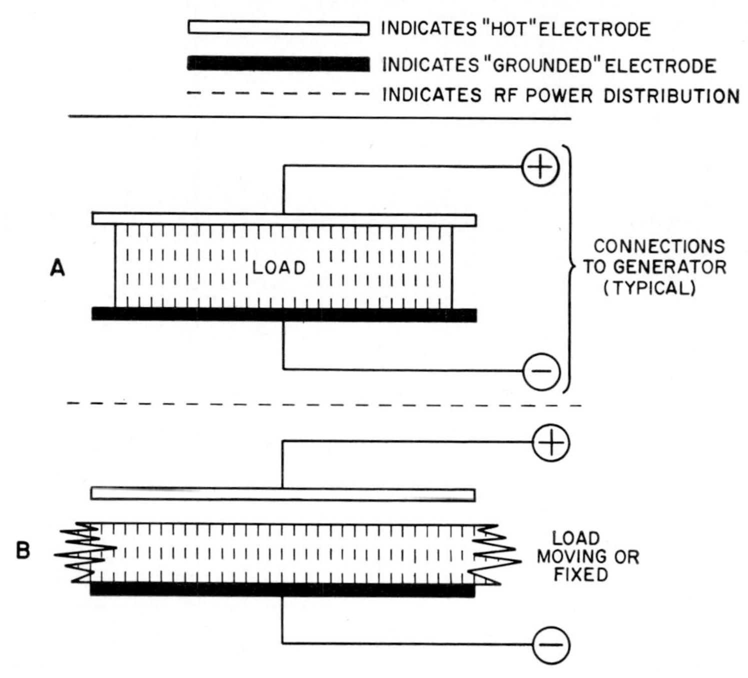

Fig. 9·22 (*A*) Flat-plate parallel electrodes; top and side pressure optional depending on application. (*B*) Flat-plate electrodes with fixed or adjustable air gap for heating bulk materials; batch-loading or continuous-flow. (*Radio Receptor*)

Heating of wood in the manufacture of assemblies requires special shapes. Figure 9·24*A* shows laminate curing or plywood forming with sheet-metal electrodes over wooden cauls. Gluing of subassemblies requires the application of heat at the glue line as in Fig. 9·24*B* while edge bonding, shown in Fig. 9·24*C*, requires pressure on the sides and top.

Commercial Dielectric Heaters. The electronic oscillators are the basic source of r-f energy for dielectric heaters just as they are for induction heaters. These are power oscillators, which have special circuit problems because of the high power. Line power through a control system goes to both the plate supply and the grid supply. Plate voltage control is used to adjust the amount of r-f power output while the grid supply prevents runaway conduction of the oscillator if the normal self-bias fails.

Two generator circuits are illustrated in Fig. 9·25. Both use two tubes for high-power applications. In *A* the plate tank is center-tapped, as is the grid coil. Because these tubes are in push-pull, twice the grid current flows through *R*, which is then one-half the value required for a single

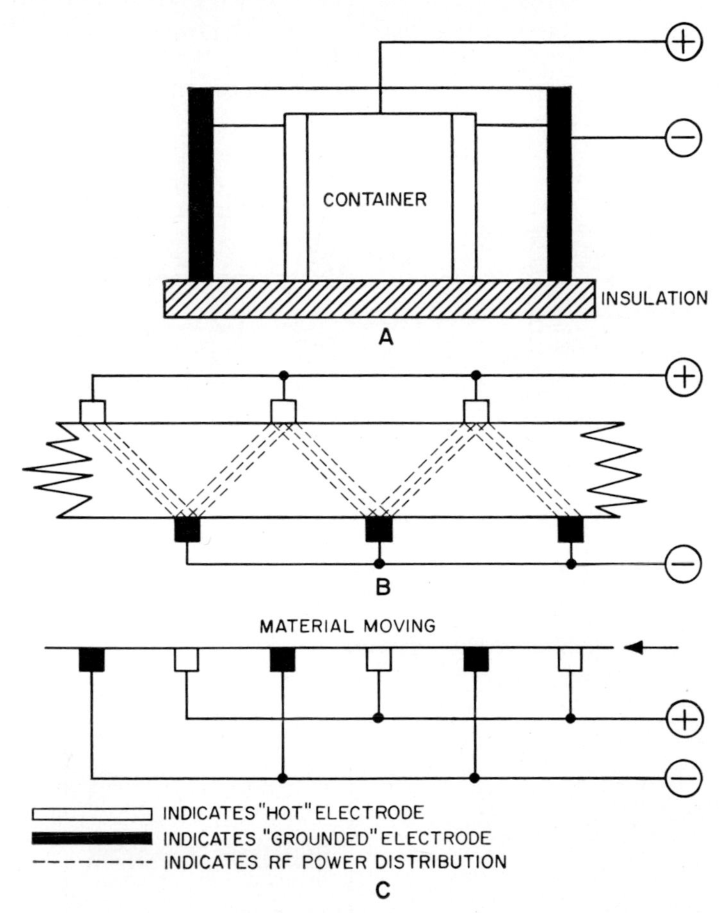

Fig. 9·23 (*A*) Curing liquid plastisol; melting or heating fats, chemicals, etc. (*B*) Staggered electrodes for partial cure. (*C*) Stray field electrode for thin materials or coatings in motion. (*Radio Receptor*)

tube. C_1 and C_3 are coupling capacitors designed to keep the d-c plate potentials from the work capacitor C_2. Work to be heated is placed between the plates of C_2 where it acts as the dielectric of the capacitor.

Two tubes provide twice the power output in push-pull (as in Fig. 9·25*A*) or in parallel (as in Fig. 9·25*B*). Here C_2 is again the work capacitor. C_1 and C_3 are variable to adjust the tank to resonance as the work material is changed.

In a tank circuit, the resonant frequency F is given by

$$F = \frac{1}{2\pi \sqrt{LC}}$$

Clearly, changing C will change the operating frequency. With no work being done, the capacitor has an air dielectric. Replacing the air

TYPICAL ELECTRODE CONFIGURATIONS FOR DIELECTRIC HEATING

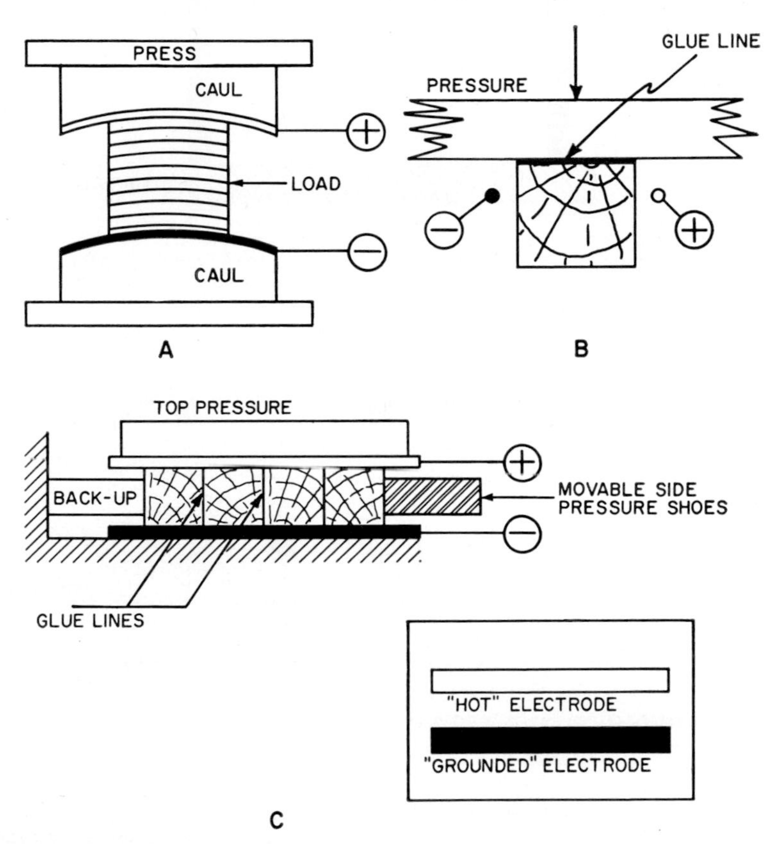

Fig. 9·24 (*A*) Laminate curing and plywood forming—sheet-metal electrodes over wooden cauls. (*B*) Selective glue-line curing; wood subassembly gluing. (*C*) Edge bonding. Side and top pressure provided by hydraulic or pneumatic power. (*Radio Receptor*)

with the work, as when a sheet of wood is to be heated, changes the dielectric. Wood is a poorer insulator than air; hence with wood as a dielectric, C has a reduced capacitance which, from the formula, changes the frequency of the oscillator. This change can be read on the current meters as they vary with the work. If necessary, the oscillator can be retuned.

To reduce this change in frequency, the coupling capacitors may be used as the work capacitors. Here no adjustment is usually needed when work is done in dielectric heating.

For a large power output to the work to be heated, two tubes are used

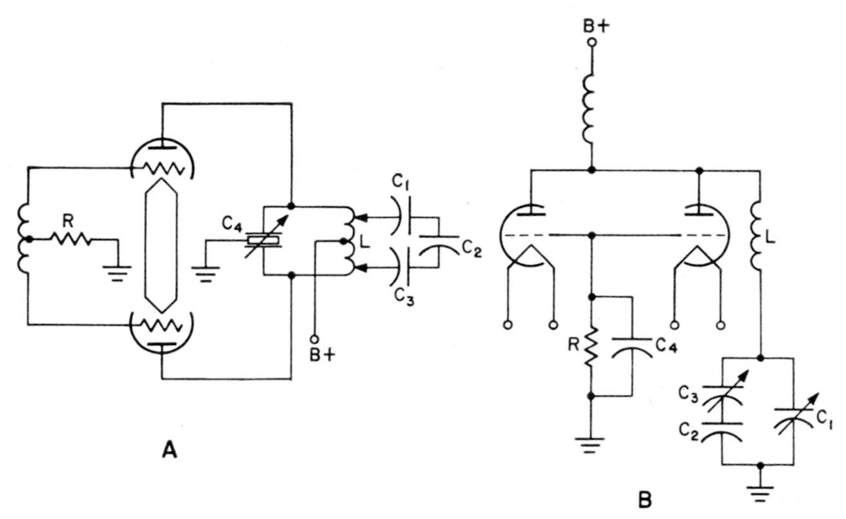

Fig. 9·25 Dielectric heater showing (*A*) shunt load and (*B*) series load.

Fig. 9·26 Air-cooled tubes. (*Industron*)

either in push-pull or in parallel as above. A view of an air-cooled pair is shown in Fig. 9·26. Note the vanes, at the bottom, used to protect the tubes, through interlocks, if the cooling-air flow stops.

REVIEW QUESTIONS

9·1 What is the difference between dielectric and induction heating?
9·2 Where is dielectric heating used?

9·3 In an ordinary non-heater amplifier, why is dielectric heating undesirable in a capacitor?

9·4 What is the advantage of wood gluing by dielectric heating?

9·5 What are the three types of heaters?

9·6 How does a spark-gap heater work?

9·7 Why is a motor generator limited to low frequencies?

9·8 How can a motor generator operate as a radio transmitter?

9·9 What is the function of the work coil?

9·10 Explain how a change in the size or shape of a workpiece affects the tuning of the oscillator.

9·11 What is skin effect?

9·12 What is the radiation limit to prevent interference?

9·13 In Fig. 9·12, what does meter M read? What is the function of C_4?

9·14 In Fig. 9·13, what protects the oscillator tube?

9·15 Explain the power supply shown in Fig. 9·15.

9·16 If the rms voltage between A and B in Fig. 9·15 is 800 volts, what is the peak inverse voltage across each diode?

9·17 Draw a water-cooled oscillator tube and explain its parts.

9·18 What type of power control is used in the heater shown in Fig. 9·17?

9·19 In Fig. 9·25, what is the function of these components: R_3, L_3, C_3, and R_4?

9·20 What happens to the emission and life if a tube is operated at 150 per cent of rated filament voltage? At 175 per cent? At 200 per cent?

9·21 A tube goes from 110 to 85 per cent of rated filament voltage. By what percentage are emission and life changed?

9·22 What is the function of the electrodes in dielectric heating?

CHAPTER

10

Welders

The process of welding involves heat to join two metals. Open flames can be used in some types of welding equipment, or heat created by electric energy can be used.

Welding equipment can be of several different types for various purposes. Automatic welding of small parts is a part of production-line manufacture of small parts such as diodes or transistors. Massive welders are found in heavy industry for all types of manufacturing from metal. Small precision welders are used in making missile assemblies and dental and medical equipment, as well as for other small precision manufacturing.

10·1 Methods of welding There are essentially three types of welding processes: forge welding, fusion welding (arc and gas), and resistance welding. Resistance welding is unique since welding heat is generated within the materials. It is not applied externally. No extra materials are involved in resistance welding, and pressure is applied to the materials during the weld process. A resistance weld is made by pressing two pieces of metal together while a heavy electric current is passed through them. The two metals are heated at their contacting surfaces to a welding temperature by the resistance offered to a flow of current.

Because pressure is used to force the heated parts together, the grain structure is refined, thus producing a weld with physical properties in most cases equal or superior to the parent metal. A feature of resistance welding is the short time required to produce the welding temperature. Many processes have been developed to utilize the basic theory of resistance welding. The most important of these are shown in Fig. 10·1.

Spot welding, as in Fig. 10·1*A*, is the standard, most versatile method of joining two pieces of metal together. Electrodes are held stationary while current is released to create a single weld spot directly between the electrodes.

Seam welding, shown in Fig. 10·1*B*, uses electrodes which are round or curved and are in motion while current is being released. Resulting welds are overlapping and form a continuous weld.

Projection welding may be seen in Fig. 10·1*C*. It is a process whereby current and pressure are localized by special shapes or projections on one

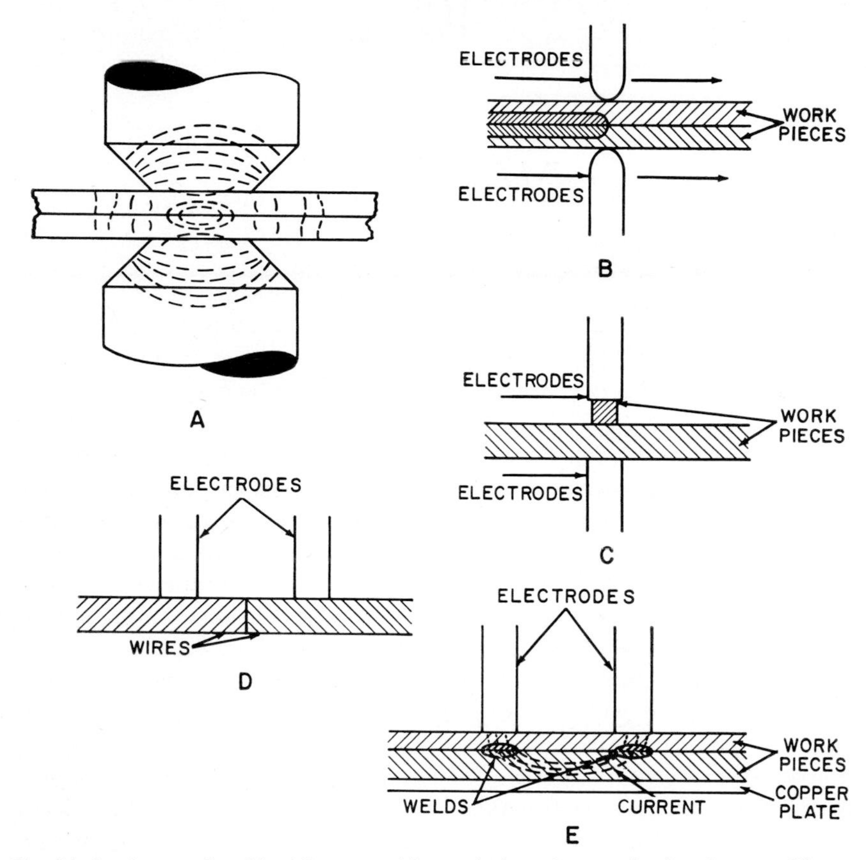

Fig. 10·1 Types of welds: (*A*) spot welding, which is the standard technique; (*B*) seam welding by electrodes in motion; (*C*) projection welding whereby pressure is localized; (*D*) butt welding for joining wires; (*E*) series welding, which is a special type of spot welding.

or more of the workpieces. Flat electrodes are used, as their shape is not important to the size of weld spot being formed.

Butt welding is most commonly used for joining wires, as shown in Fig. 10·1*D*. Materials are placed end to end and current is released. Total diameter of the wires becomes the size of the weld spot.

Series welding is illustrated in Fig. 10·1*E*. Technically, this type is an adaptation of spot welding. It differs from normal spot welding in that both electrodes contact the workpiece from the same side. Often a copper plate is used beneath the workpieces to aid in transmitting the current through the materials and back to the other electrode. Two or more weld spots are created, one under each electrode.

Both alternating current and energy storage are used for resistance welding. A-c welding equipment takes power directly from the supply line and, by means of some type of timing device, allows this line voltage to be released into the primary of a welding transformer. Usually, the windings of the welding transformer have various taps used to connect the welding circuit to the head or handpieces. The voltage input remains fixed, while output may be varied by as many taps as provided.

For effective, consistent results, the timing device must be more than just an "on-off" switch. It is usually designed with various controls used to determine the length of the current flow and to synchronize release time with the waveshape of the line input. Some units provide for control of the "upslope" and "downslope" of the discharge cycle (in addition to the actual welding-heat discharge time). A-c equipment is rated in kilovolt-amperes by multiplying the voltage times the amperage at the point of connecting the line current to the primaries of the welding transformer. Most kva ratings are based on half-cycle discharge times.

10·2 Types of welding equipment There are three basic types of stored-energy welding equipment: electrostatic, electromagnetic, and electrochemical. Electrostatic (capacitor discharge) is the type most commonly employed.

In the electromagnetic type, line current (usually three-phase alternating current) is rectified to direct current. When the direct current is connected to the primary of a special transformer, energy is stored until the circuit is interrupted.

The electrochemical type uses electric energy stored in a storage battery. The method of energy storage which is most versatile, and which occupies a minimum area, is the electrostatic capacitor-discharge type.

Capacitor-discharge welders draw energy from standard line sources (in medium- to small-sized units) at a low level. This voltage is stepped up and rectified, then stored in the capacitor bank. When the weld circuit is closed, a large quantity of energy flows through the primary of a pulse (welding) transformer and a secondary welding current of high peak is produced.

With stored-energy welding, the energy stored in capacitors may be varied over a wide range. Some units provide taps on the transformer to extend the duration of the weld time. Timing of the welding-current flow is essentially automatic, although it may be varied on some models.

A-C Welding. A-c welding uses one of the two techniques shown in Fig. 10·2. Both use a three-phase supply. The six ignitrons shown in *A* act as both switches and rectifiers. There are two tubes in each leg of the supply.

The effect in the secondary is a current pulse of one polarity followed by a pulse of the opposite polarity. In making a weld, either one pulse or a series of pulses which alternate in polarity are used.

In the system shown in Fig. 10·2*B*, separate rectifiers are used and the

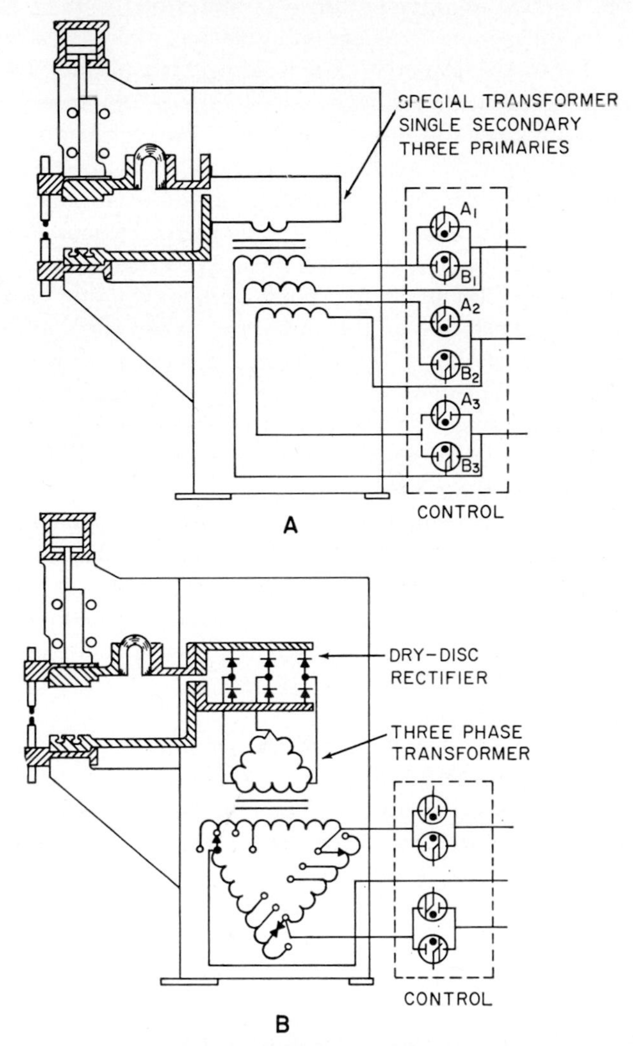

Fig. 10·2 (*A*) A-c welding in which ignitrons act as both rectifiers and switches to control the welding current; (*B*) a-c welder using separate rectifiers, in which the ignitrons act only as controlled switches. (*Taylor-Winfield*)

ignitrons act only as controlled switches. In both systems, control over the ignitron conduction varies the amount of welding current.

The a-c welder uses electric energy from the power line in a direct fashion, through power control, rather than store this energy until it is to be used. A single-phase welder has a timing diagram like that shown in Fig. 10·3. This welder control is for use with electric resistance spot and projection welders, which are air or hydraulically operated single-phase a-c machines. The welder may have a pressure switch or a firing switch which is actuated by the welder head movement, and this switch may be substituted for the squeeze timer of the control.

Heat control, by the phase-shift method, is included to allow finer control of the welding current than is possible by a welding-transformer

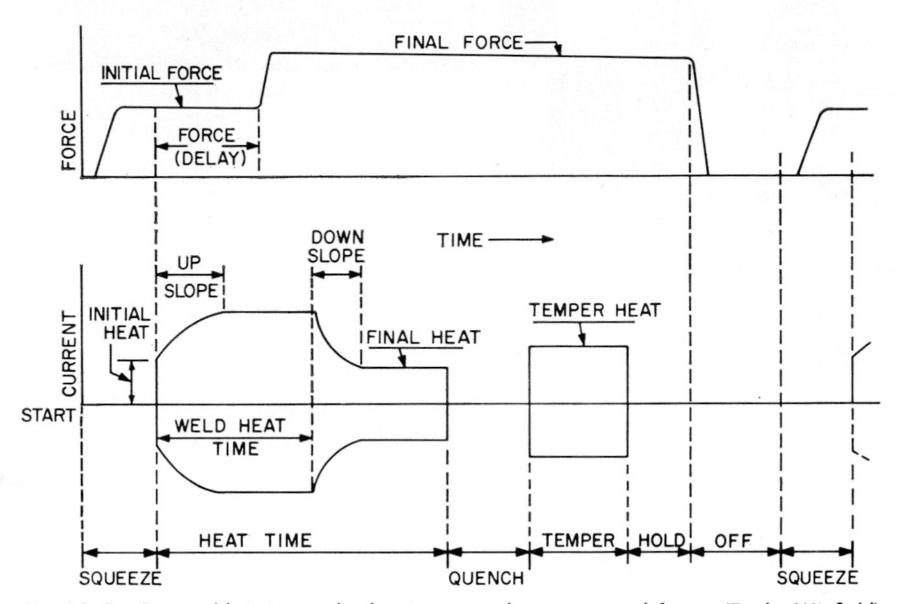

Fig. 10·3 Spot-weld timing cycle showing quench, temper, and forge. (*Taylor-Winfield*)

tap switch alone. Stepless adjustment of the welding current can be made with this control.

This basic control contains an electronic heat control, an ignitron contactor using two ignitron tubes, a sequence timer, and a synchronous-precision weld timer in a single enclosure.

As shown in Fig. 10·3, force is applied first, and then heat is applied to make the weld. Note that the heating current rises to maximum during upslope time and then decreases to final heat during downslope time. Maximum pressure or force on the work is delayed until after the start of the maximum heat. During quench there is force but no heat. The temper heat is less than maximum heat. This figure diagrams a single automatically timed weld.

A single-phase spot welder is shown in Fig. 10·4*A*. This is a control for electric resistance welders with air or hydraulically operated single-phase a-c spot and projection welders requiring a sequence of squeeze, weld, hold, and off times. It may be used with equipment which substitutes

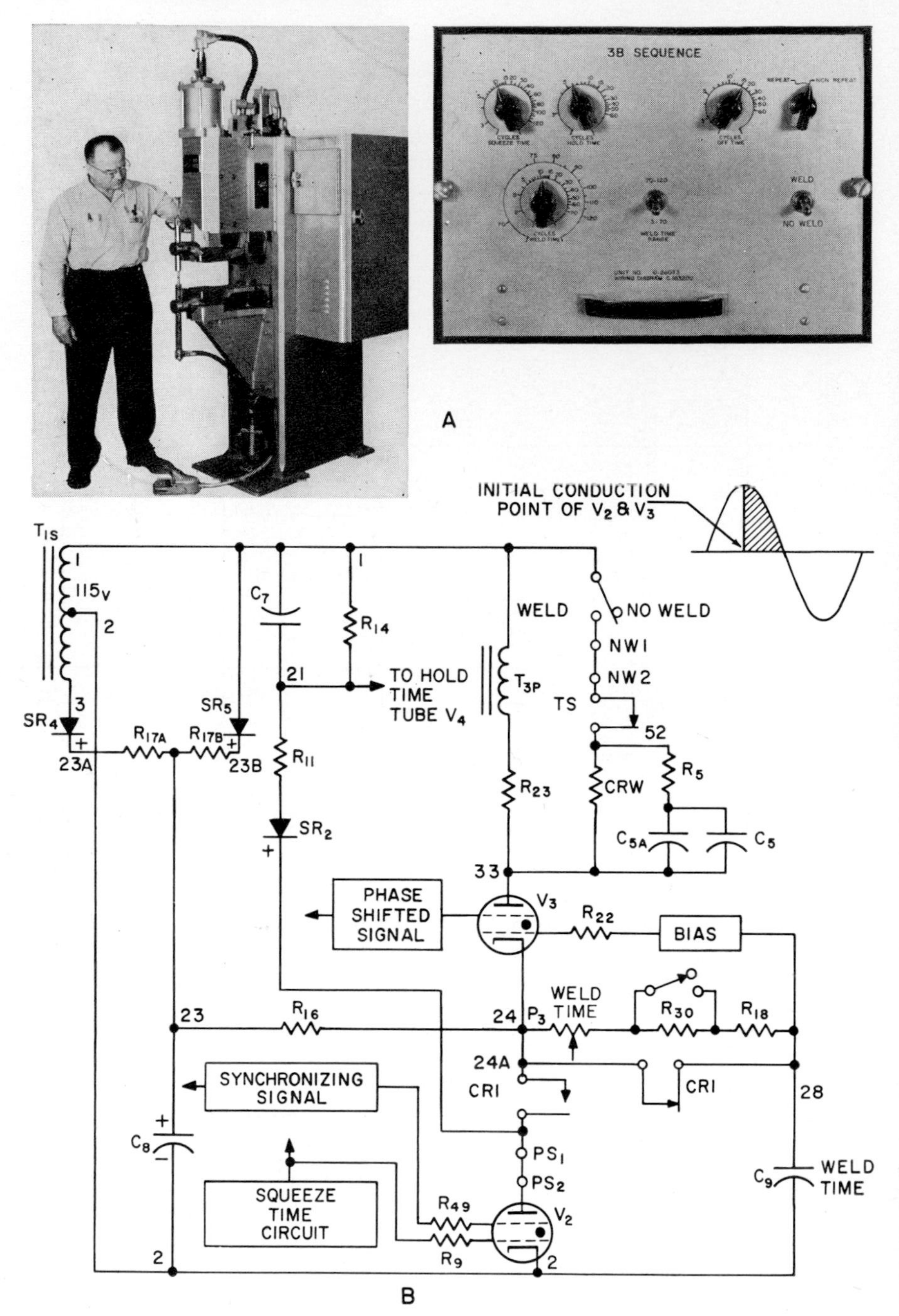

Fig. 10·4 (*A*) Taylor-Winfield N-2R-E-G1 welding control; (*B*) simplified schematic diagram showing series thyratrons.

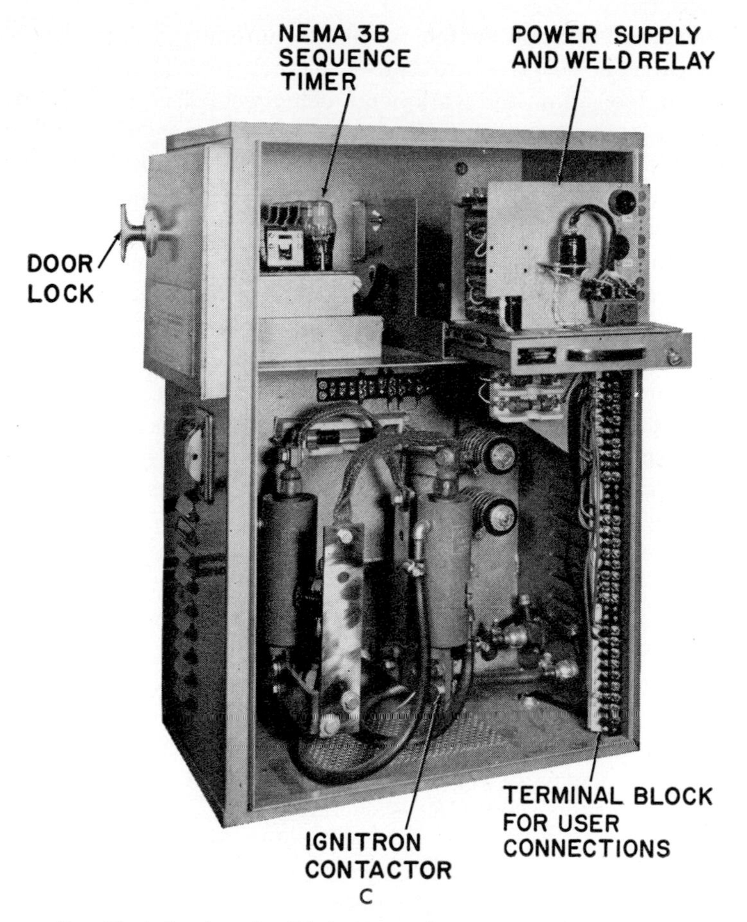

Fig. 10·4 *Continued.* (C) Ignitron-tube pair. (*Taylor-Winfield*)

a pressure switch, or a firing switch actuated by welder head movement, for the squeeze timer. The control contains one ignitron contactor, using two ignitron tubes, with associated timing circuits to supply power to the welder in accurately timed and controlled intervals.

Figure 10·4*A* shows the sequence timer; *B* is a simplified schematic of two of these tubes, V_2 and V_3; and *C* illustrates the ignitron pair.

The control consists of a sequence-weld-timer section, a power-supply section, and an ignitron contactor. The combination control is intended to properly coordinate and time the mechanical functions of an air-operated welder using a solenoid air valve with respect to the interval of welding-current flow.

The timer section controls the following functions in the following sequence:

1. When the initiation switch is closed (normally a foot switch), the solenoid valve is energized to cause the movable head and electrode assembly to descend on the workpieces being welded.

2. Squeeze time is the time interval after initiation which allows the welding force to build up to the desired value before welding current flows.

3. Weld time starts at the end of squeeze time and times the duration of welding-current flow through the work.

4. Hold time follows weld time and is the interval during which pressure is maintained on the work after welding-current flow ceases.

5. After hold time, the electrodes are released and the movable electrode assembly returns to the upward or starting position.

6. Off time is the time interval used only during automatic-repeat operation. It is the interval after the solenoid valve is deenergized until it is again energized. During this time the electrodes separate and close again, allowing the work to be repositioned for the next weld.

The contactor section contains the ignitron contactor, which allows the welding current to flow when the weld relay is energized.

During the weld time the operation of these two tubes (Fig. 10·4*C*) is as follows. Prior to tube V_1 conduction (not shown), weld-time capacitor C_9 is charged positively from a d-c source formed by transformer T_1S (1, 2, 3,), rectifiers SR_4 and SR_5, and storage capacitor C_8. This source acts through the *NC* contact of relay CR_1, and the contact shunts the weld-time potentiometer-resistor network of P_3, R_{30}, and R_{18}. After tube V_1 conduction, capacitor C_9 is trickle-charged through the weld-time resistance network as relay CR_1 is energized.

Tube V_2 now has a d-c plate voltage applied. A synchronizing (phase-shifted) voltage is applied to its screen grid to cause conduction at approximately a 90° point in the a-c plate voltage of tube V_3.

Tube V_3 is effectively in series with tube V_2. As tube V_2 conducts, tube V_3 conducts and capacitor C_9 voltage is applied to tube V_3 control grid. Tube V_2 conduction causes point 24 to approach point 2 potential, preventing further positive charging of capacitor C_9. Tube V_3 conduction energizes transformers T_3 and T_4 which indirectly cause the welding-current flow.

Weld-time capacitor C_9 discharges through the weld-time resistor network P_3, R_{30}, and R_{18} for weld time.

When the positive charge on capacitor C_9 is less than the negative voltage provided by the bias network of transformer T_6S_2, potentiometer P_4 (for calibration), R_{19}, and capacitor C_{11}, tube V_3 ceases conduction, ending weld time.

A synchronizing signal is applied to the screen grid of tube V_3, through transformer T_5, capacitor C_{14}, and resistor R_{24} to ensure full conduction of tube V_3 during its last conducting interval (half-cycle).

Counting tubes (Chap. 7) are used in some high-precision resistance-welder controls. The Dekatron cold-cathode gas tube is used to time the welds. A series of tubes is used and each counts to 10. In counting, the complement is initially preset and the output occurs at the zero cathode of the last Dekatron. For a count of 125 the complement is 1,000 minus 125, or 875. The initial glow is set on cathode 8 of the hundreds Dekatron, cathode 7 of the tens, and cathode 5 of the units. At the end of the count the output appears on the zero cathode of the hundreds Dekatron.

In a second method of counting, the Dekatron train, as shown in Fig. 10·5, is set initially to zero. The information is then picked off each

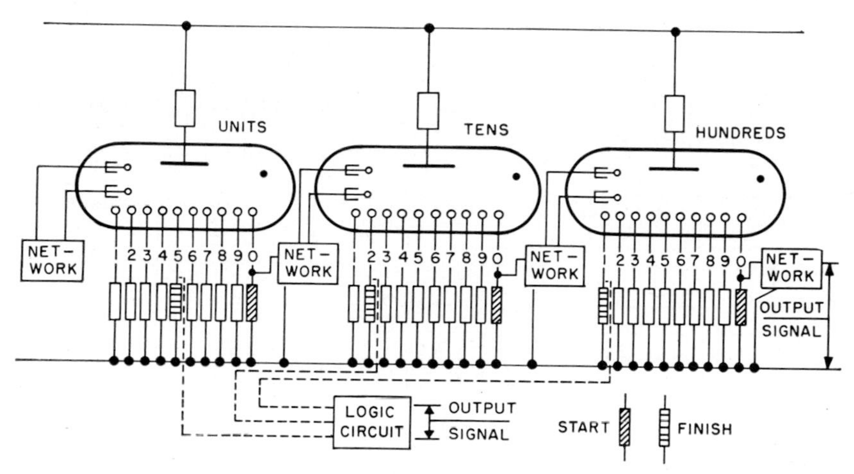

Fig. 10·5 Dekatron counter. (*Taylor-Winfield*)

cathode as needed and passed into a logic circuit which furnishes the output signal. For a count of 125 the initial glow is set on zero cathodes and the count appears at cathode 1 of the hundreds Dekatron, cathode 2 of the tens, and cathode 5 of the units.

The principal advantage in using a Dekatron or digital type of circuit is to allow accurate or precise timing of the desired periods of time. If the circuitry is designed properly and if it is not subjected to abnormal voltage transients caused by other apparatus (including arc- and resistance-welding equipment), the counting or time interval is precise.

Precise, short time intervals, below 15 cycles, on a 60-cps base, are no problem to achieve by other means. The most widely used is the *RC* timing network. Timing without deviation in this range is common if the same precautions are taken as are required with a Dekatron unit.

Greatest benefit from the use of the Dekatron unit will be achieved in timing intervals longer than 15 cycles. As the time intervals approach 1 sec or 60 cycles, it is practically impossible to have precise timing, that is, measurable variation in length of time, by using *RC* timing networks. An accuracy of ± 3 per cent would be good for such *RC* networks as used in resistance-welding controls. For these long time intervals the application of the digital counting system (of which the Dekatron is an example) is an advance in welding controls.

Equipment using this technique has controls used to preset each operation for the desired number of cycles. Using the 60-cps line input, the counters provide precise and accurate control of the welding.

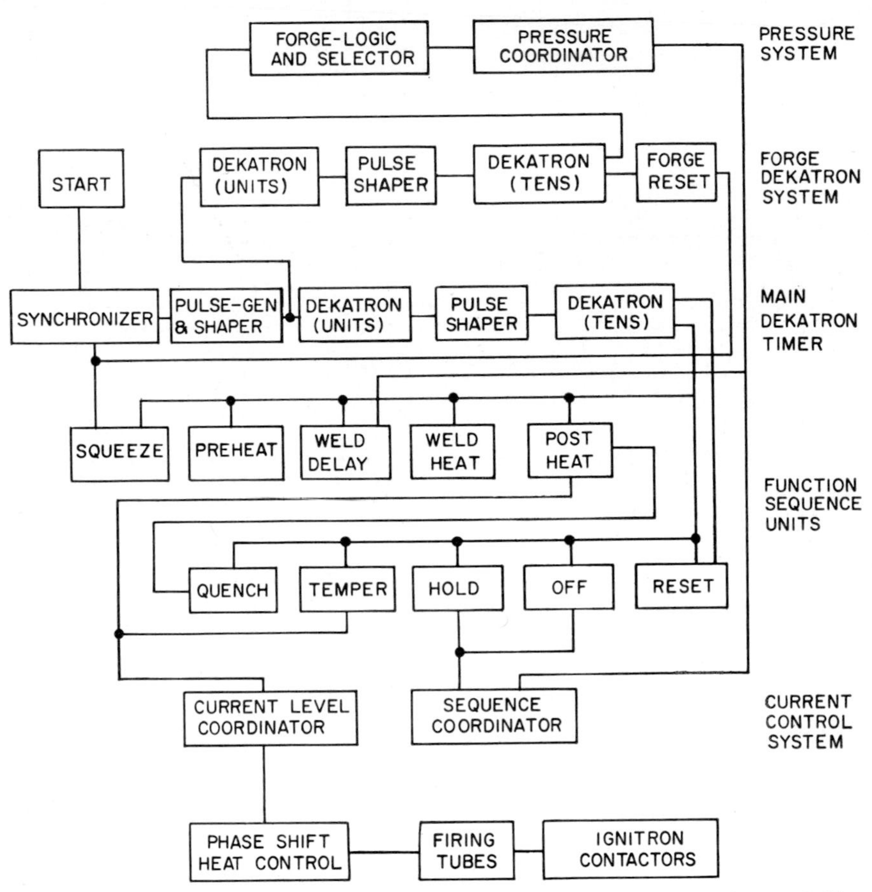

Fig. 10·6 Timing diagram showing three-phase Dekatron control. (*Taylor-Winfield*)

These decade counter tubes are used in a specialized form of counting. Each of the welding operations (Fig. 10·6) can be seen as a block-diagram function or as a sequence as in Fig. 10·7. In effect, the control programs the welder so that the results are uniform.

Stored-energy Welding. In other forms of welding, storage of energy is used to reduce the power-supply requirements. Since welding is not continuous in these welders, a reactance can build up a store of energy between welding times.

Stored-energy welding is used in place of slower, more costly, and less satisfactory methods of metal joining, such as soldering, silver brazing, riveting, and staking. Welding can be used for a wide range of materials

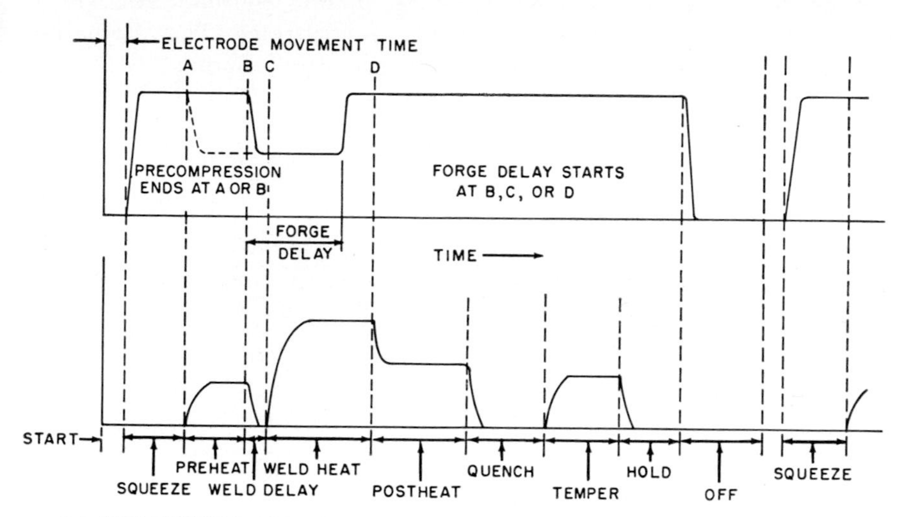

Fig. 10·7 Timing chart for three-phase Dekatron control. (*Taylor-Winfield*)

and thicknesses. This method allows many advanced designs by permitting efficient assembly techniques. It joins many materials formerly unsuited to resistance welding. Parts made of sensitive alloys such as precious metal or spring steel are undamaged and the welds are sound and clean.

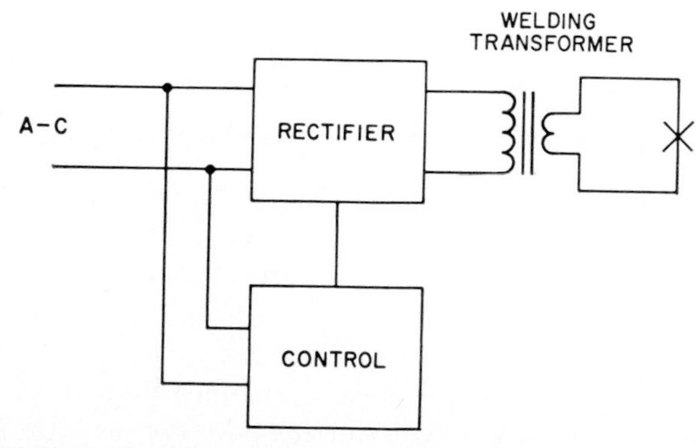

Fig. 10·8 Storage welding using the reactor-storage technique.

Figure 10·8 shows a basic form of storage welder. Line voltage is applied to both the control and the rectifier circuits. By means of an inductance or a capacitor, controlled-energy direct current from the rectifier is stored until required. Either the magnetic field of the coil or the electrostatic charge of the capacitor is used for this energy storage.

An example of the combined use of thyratrons and ignitrons is in the Sciaky or reactor-storage welding-control process shown in Fig. 10·9. Only one phase of a three-phase supply is illustrated. The alternating current is stepped down through T_1. A thyratron controls the ignitor

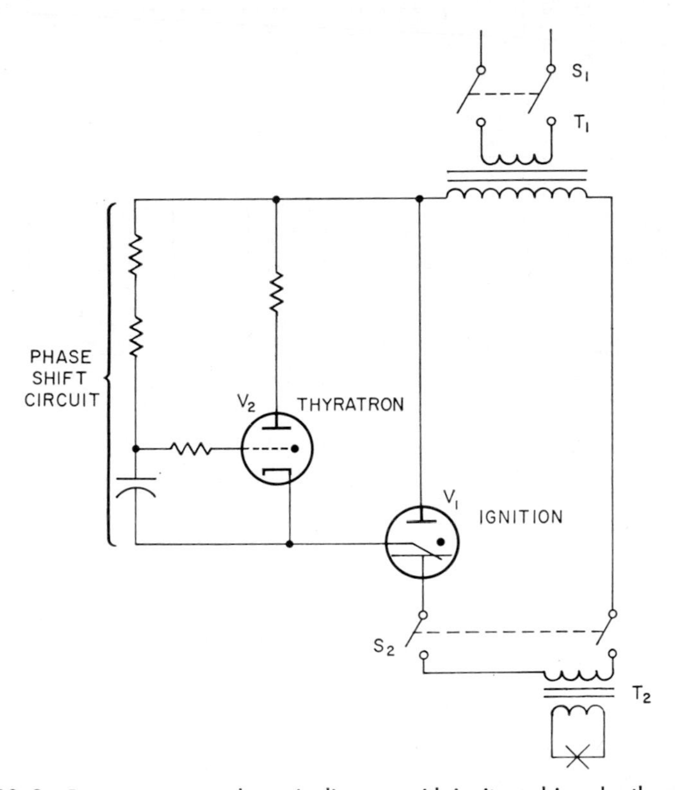

Fig. 10·9 Reactor-storage schematic diagram with ignitron driven by thyratron.

circuit as shown. A phase-shifted a-c signal controls the thyratron grid and the resulting thyratron conduction. When this tube conducts, it fires the ignitor circuit and starts the ignitron whose plate is connected across transformer T_1. The ignitron controls the direct current in the transformer T_2 and hence the welding current.

In operation, the work to be welded is clamped in place in this welder and S_1 is closed. S_2 is then closed, but this does *not* activate the welding current. As shown in Fig. 10·10, the primary current I slowly increases in an exponential curve because of the large inductance of the primary. There is a secondary current flow I_2 (in the opposite sense), which decreases as the primary current stops charging.

When S_2 is *opened*, primary current decreases but, because of the large magnetic field, there is a large surge of secondary current which does the welding. The peak current of the output (I_2) is four or five times the peak rectified current (I_1).

A modification of this principle, using a capacitor, is shown in Fig. 10·11. Here the rectifier charges C (a bank of capacitors) while S is open. When the switch S_2 is closed, the energy stored in C discharges through

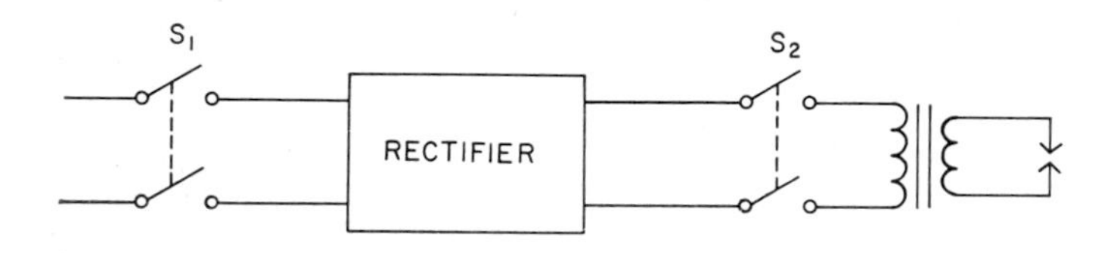

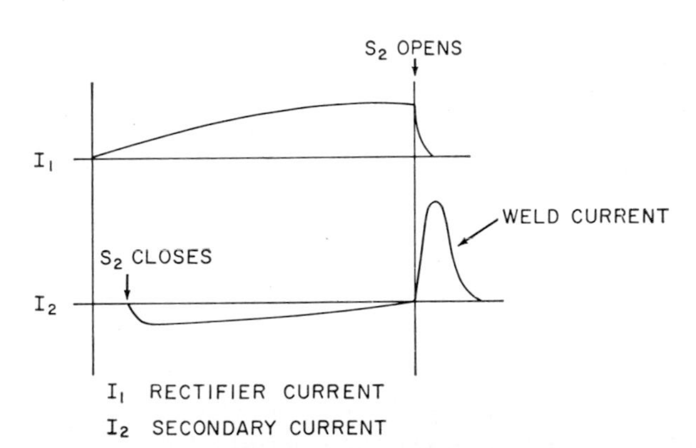

Fig. 10·10 Welding currents as they appear in the reactor-storage method.

the transformer to the load to make the weld. Actually, of course, the output *total* energy is equal to the *total* input energy (except for losses). But the output is delivered for a short period while the capacitance is charged over a long period. Thus a much smaller rectifier can be used than that which would be required to supply all the output energy at once.

A commercial example of a stored-energy welder is the Weldmatic Model 1027C shown in Fig. 10·12. This is a bench-mounted precision resistance welder, complete in itself. It is suited to joining small metal assemblies such as are used in electronics, instrument, and ordnance work. With the capacitor-discharge principle, it welds with a single short pulse of heavy current actuated by a foot-pedal control. Welds are made within milliseconds. Energy is accurately regulated by an electrode power supply with a single stepless control. Applied electrode pressure is directly indicated on the force scale and is adjustable over a wide range. Welds are uniform because firing pressure repeats exactly.

As in the schematic diagram, the 110- to 120-volt a-c input line voltage is stepped up and rectified to a predetermined d-c voltage. This voltage is then conducted to and stored in a bank of electrolytic capacitors. The amount of voltage maintained on the capacitor bank is determined by

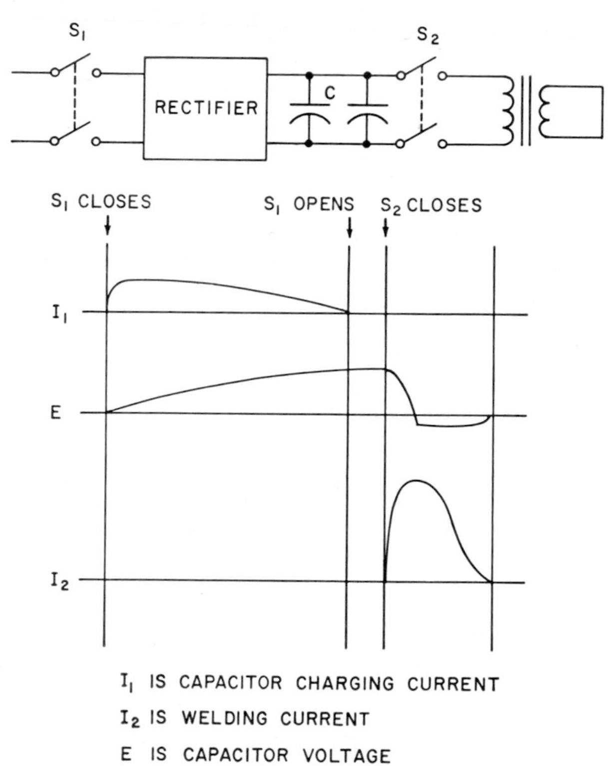

Fig. 10·11 **Capacitor storage; S_1 allows the capacitors to store energy which is discharged only when the second switch S_2 closes.**

the heat-control knob. The preset charge on the capacitors is retained until demanded at the welding electrodes incorporated in the welding head or handpiece. Actuation of the welding head or handpiece simultaneously applies the preset electrode force to the workpieces and trips the firing-circuit switch.

The energy stored in the capacitor bank is instantaneously discharged into the primary of a pulse transformer which produces a current flow of high magnitude to the secondary windings of the pulse transformer to which the electrodes are connected. The inherent resistivity of the parent metals to the flow of current at the point of electrode contact heats the metals to their welding temperature.

The preset electrode pressure, by action of the low-inertia upper electrode arm, then forges the two metals together while the weld nugget is in its molten state.

Immediately upon release of the pressure applied by the operator to the foot pedal, or to the hand lever on handpieces, the capacitor bank is recharged to its preset voltage.

A-c input is to TB_{401}; S_{501} is the on-off switch for power. Pilot light DS_1 glows, indicating current flow to the thyratrons V_{101} and V_{102}, but a time delay permits the filaments to heat before plate current is drawn to charge the capacitor bank C_{201} through C_{232}. R_{501} is the heat control whose adjustment changes the thyratron conduction. Charging current for the capacitor bank flows, through pin 5 of TB_{101}, through the thyratrons from their cathodes, which are tied to the capacitors through R_{13}.

The capacitor bank becomes charged to a voltage slightly higher than the reference voltage applied to the thyratron grids. This slight difference in voltage between the thyratron grids and cathodes prevents random firing of the thyratrons.

If the voltage tends to drop on the capacitors, the voltage difference between the plate cathode of each thyratron becomes sufficient to overcome the grid bias, and the tubes conduct until the capacitor voltage is brought up to the selected value.

The resistor R_{113}, in series with the thyratrons and the capacitors during the welding operations, stabilizes the changing rate of the capacitors when relatively low voltage is to be stored and permits reproducible welding currents of low values to be obtained.

The secondary winding of the power-supply transformer T_{101} is center-tapped to ground through the ground lead of the power plug and has each of its ends attached to the plates of the thyratrons.

The cathode of each thyratron is connected through stabilizing capacitors to ground and is also connected through the resistor to the contact of a relatively quick-acting double-throw relay K_{202}.

The relay armature (normally closed against the charging contacts) is connected to the positive side of the capacitors.

The built-in watt-second meter M is connected across the welding capacitors to register at all times the voltage (converted to watt-seconds) stored in the capacitor bank.

The two plates of rectifier tube V_{103} are connected, respectively, to the opposite ends of the secondary winding of the T_{102} power transformer. The cathode of the rectifier tube is connected through a filter circuit to the heat control R_{501}. The movable arm of the potentiometer is connected to apply a fixed, but adjustable, reference voltage to the control grids of the thyratrons.

The rectifier tube is isolated from the thyratron tubes by a time-delay relay to protect the thyratron cathode surfaces when the main control

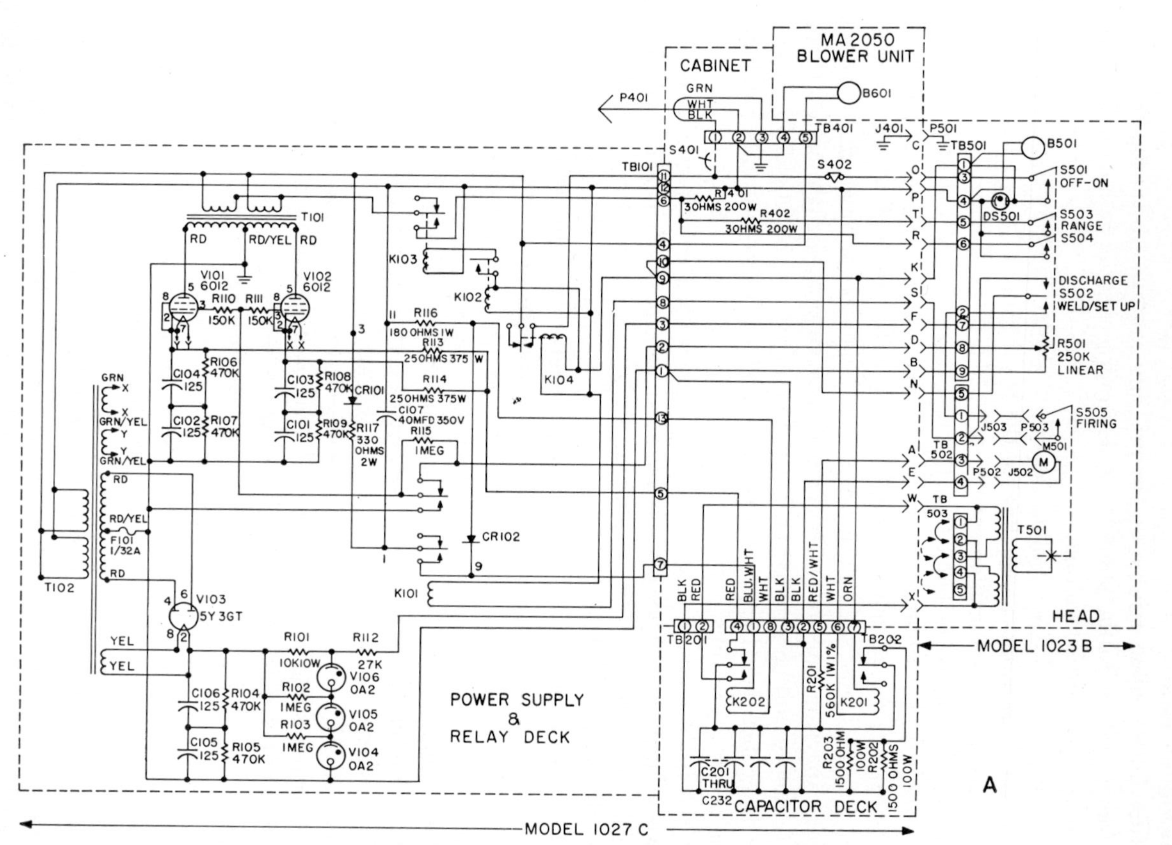

Fig. 10·12 (A) Schematic of a capacitor storage welder showing capacitor deck.

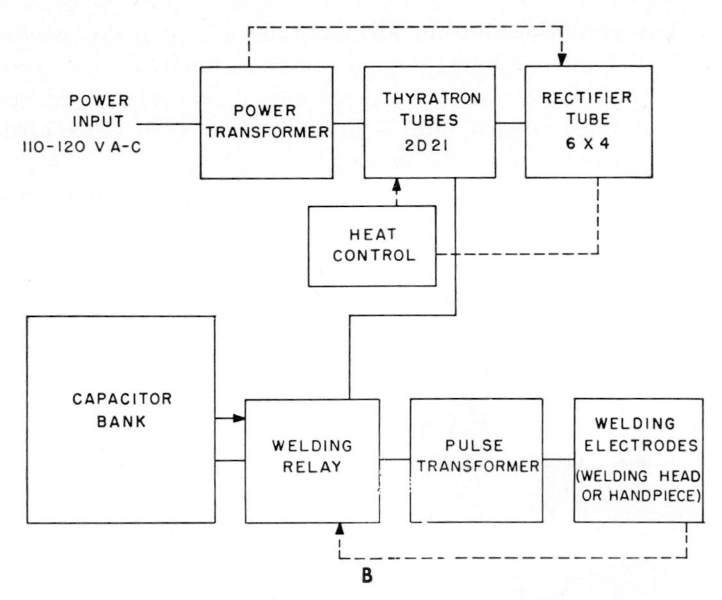

Fig. 10·12 *Continued.* (*B*) Welder block diagram. (*Weldmatic*)

switch is turned on. The time delay provides 30 to 45 sec of warmup time before relay K_{101} closes, completing the circuit. Resistors are provided in the grid circuits of each of the thyratrons to protect the grids. These are R_{110} and R_{111}.

The negative side of the capacitor bank is connected to one end of the primary coil of T_{501}, the welding transformer. The other end of the welding-transformer primary coil is connected to the normally open contacts of relay K_{202}.

K_{202} is the welding relay and K_{201} is a safety relay, while relays K_{101}, K_{102}, K_{103}, and K_{104} are used for proper starting of power to the circuits.

A different welder is shown in Fig. 10·13. The control of this welder employs the stored-energy principle to deliver a predetermined and precisely measured amount of welding energy in a minimum of time. The energy for each weld is accumulated between welds, and it greatly exceeds the demand upon the power source, thus making unnecessary expensive power lines with high current capacity and thereby reducing power charges. A very low line current for approximately 1 sec stores energy which is discharged in a few thousandths of a second.

The control automatically supplies the selected amount of energy to the weld, regardless of normal variations of line voltage. It contains a self-regulating charging voltage circuit which will ensure less than ± 2.5 per cent capacitor voltage variation for a ± 10.0 per cent line-voltage change.

The consistently precise short weld timing and the accurately metered pulse of energy obtained from an accurate welder make possible the successful production welding of material combinations not previously considered practical. Undesirable discoloration or oxidation at the weld due to duration of current flow is minimized or even eliminated when this type of control and welder is used.

The magnitude of the welding-current pulse depends on the value of capacitance, the capacitor charging voltage, and the welding-transformer turns ratio, all of which are adjustable to cover a wide range of welding applications.

Fig. 10·13 Table-model stored-energy welder. (*Taylor-Winfield HW-T-A-G*)

The basic control consists of a rectifying system to charge a group of capacitors to a predetermined voltage and then discharge the capacitors, upon initiation, through the primary of a welding transformer.

The control is initiated by a firing switch actuated by the welder ram movement after proper electrode force is supplied.

The control, with built-in welding transformer, provides the welding current, and the remainder of the welder sequence (such as squeeze and hold times) is provided by the welding machine.

Figure 10·14 shows the schematic diagram. Assuming that the power has been turned on, that the power switch SW_1 is in the "on" position, and that the door interlock switch (LS) is closed, the normally closed contacts open (37 to 5) and (37 to 38) close, placing a bias voltage on the grid of tube V_3 that prevents tube V_3 from conducting. The bias voltage source is from transformer T_1S_1 rectified through tube V_2, resistor R_5, and capacitor C_{10} (terminal 5 positive). After 1 min, time delay TD relay operates, deenergizing relay CR. Deenergized relay CR contacts

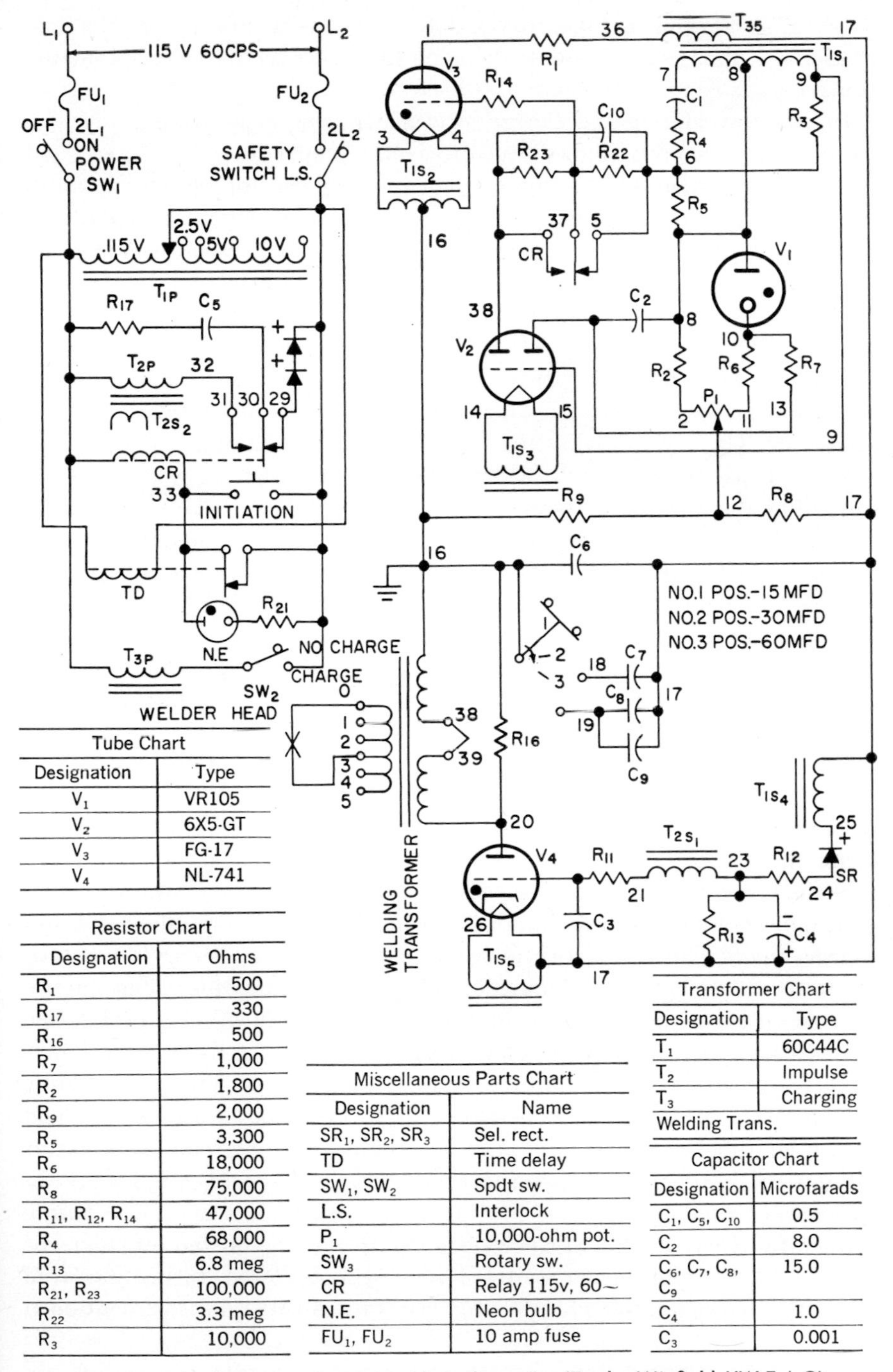

Tube Chart	
Designation	Type
V_1	VR105
V_2	6X5-GT
V_3	FG-17
V_4	NL-741

Resistor Chart	
Designation	Ohms
R_1	500
R_{17}	330
R_{16}	500
R_7	1,000
R_2	1,800
R_9	2,000
R_5	3,300
R_6	18,000
R_8	75,000
R_{11}, R_{12}, R_{14}	47,000
R_4	68,000
R_{13}	6.8 meg
R_{21}, R_{23}	100,000
R_{22}	3.3 meg
R_3	10,000

Miscellaneous Parts Chart	
Designation	Name
SR_1, SR_2, SR_3	Sel. rect.
TD	Time delay
SW_1, SW_2	Spdt sw.
L.S.	Interlock
P_1	10,000-ohm pot.
SW_3	Rotary sw.
CR	Relay 115v, 60~
N.E.	Neon bulb
FU_1, FU_2	10 amp fuse

Transformer Chart	
Designation	Type
T_1	60C44C
T_2	Impulse
T_3	Charging
Welding Trans.	

Capacitor Chart	
Designation	Microfarads
C_1, C_5, C_{10}	0.5
C_2	8.0
C_6, C_7, C_8, C_9	15.0
C_4	1.0
C_3	0.001

Fig. 10·14 Capacitor-storage-welder schematic. (*Taylor-Winfield HW-T-A-G*)

close (37 to 5), and the bias from capacitor C_{10} is removed from the grid of tube V_3. At the same instant, the neon light goes on, indicating that the control is ready for operation.

In standby, closing the charge switch SW_2 energizes transformer T_3. Tube V_3 will conduct immediately to charge capacitors C_6, C_7, C_8, and C_9, depending upon the position of the capacitor selector switch (point 16 positive and point 17 negative). Suppose C_6 is in the circuit (15 μf). The grid of tube V_3 is positive by whatever voltage is present between the arm of potentiometer P_1 (12) and point 8. This positive source comes from the power supply, consisting of T_1S_1 (8 to 9), tube V_2, and capacitor C_2, with tube V_1 regulating the voltage across P_1, R_6, and R_2.

In series with the bias voltage to the grid of tube V_3 is a phase-shifted a-c signal across resistor R_5. This signal lags the plate voltage of V_3 by about 90°. This signal makes for smoother trickle charging.

When capacitor C_6 is increasing in voltage because V_3 is conducting, the voltage across R_9 is rising to oppose the positive bias from the P_1 setting. When this voltage across R_9 reaches sufficient magnitude, the grid of tube V_3 will see a net voltage to the cathode which is negative, and V_3 will tend to block. Capacitor C_6 will then discharge through R_8 and R_9 until the a-c signal across R_5 fires tube V_3 enough to slightly recharge C_6. This trickling action continues until the firing switch is closed.

Capacitor C_5 is charged through rectifiers SR_2 and SR_3, limiting resistor R_{17}, and the normally closed contact of relay CR (29 to 30). Discharging tube V_4 is held nonconductive by a d-c negative bias supply in the grid circuit (terminal 23 negative).

For weld operation, closing the initiating contact causes a weld to be made in the following manner. Closing of the firing switch energizes relay CR. Energized relay CR disconnects capacitor C_5 from its charging source and connects it across transformer T_2 primary (30 to 31), producing a discharge signal which causes tube V_4 to conduct the energy stored in capacitor C_6 through the primary of the welding transformer. This energy is dissipated quickly into the weld nugget. The time duration is only a few milliseconds long and will depend on the turns ratio of the welder transformer, capacitance, and charging voltage used.

This relay also opens relay contacts 37 to 5, placing a bias in the grid circuit of tube V_3; therefore tube V_3 cannot conduct, and as a result capacitor C_6 cannot be charged during the time the weld is made.

To prevent unnecessary oscillation in the circuit of the welder transformer, the decaying flux of the welder transformer is shorted by means of resistor R_{16}. The initiating switch has to remain closed during the weld discharge. However, this closure time needs to be very short as mentioned above.

When the initiating switch is opened, CR will deenergize and will apply charging power to tube V_3. Tube V_3 will start conduction (half-wave) to charge capacitor C_6 as in standby.

REVIEW QUESTIONS

10·1 What are the three basic types of welding?
10·2 What is meant by resistance welding?
10·3 How does energy-storage welding operate?
10·4 What is the difference between capacitor storage and inductor storage?
10·5 What is spot welding? Where is it used?
10·6 What is the difference between butt and spot welding?
10·7 How do a-c and storage welding differ?
10·8 What is projection welding?
10·9 In a timing cycle what are heat, quench, and hold times?
10·10 Why is heat re-applied?
10·11 In Fig. 10·4, what is the function of C_9, V_3, and C_8?
10·12 Why are counter tubes used in welders? How are they used?
10·13 How many heat times are shown in Fig. 10·7? What is the function of each?
10·14 How does an inductance store energy?
10·15 How does V_2 control V_1 in Fig. 10·9?
10·16 How does a capacitor store energy?
10·17 Explain the block diagram in Fig. 10·13.
10·18 How do the two capacitor-storage welders (Figs. 10·13 and 10·14) differ?
10·19 How is storage welding controlled?

CHAPTER

11

Magnetics

The variety of applications for magnetics seems almost endless, for they extend from tape and wire recording to core memories for digital computers and magnetic amplifiers. There are new magnetic devices, one of which is the Magnetoresistor, which acts as a variable resistance. But perhaps data recording is the most important single use of magnetics.

Data are recorded for future use in many types of control systems. Magnetic recording is a method for storage of data until they are required by the system. A prime example is the use of prerecorded magnetic tapes which are used to guide automated machine tools.

11·1 Magnetic data storage Information which is to be saved or stored for later use is placed into some form of magnetic memory. The

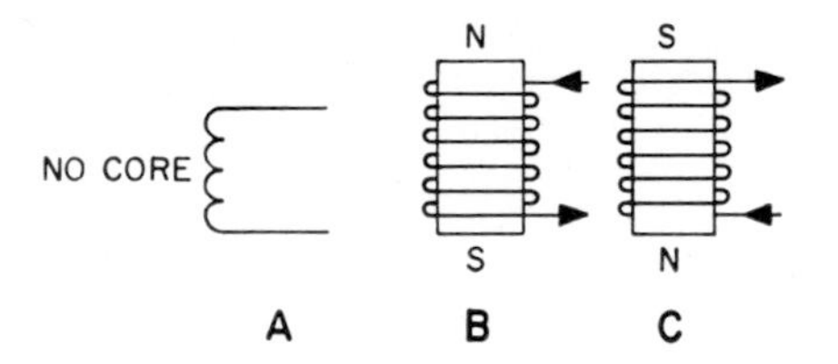

Fig. 11·1 Magnetic recording principle: (*A*) coil with an air core; (*B*) current flow creating north as upper end; (C) opposite direction of current flow.

three basic types of memory or storage are tape, core memories, and drum. Magnetic tape as a type of information storage is briefly discussed in Chap. 8, and a more complete coverage is included here.

All forms of magnetic memory depend upon a fundamental fact as illustrated in Fig. 11·1. A current flow through a coil, as in *A*, will create a magnetic field in a given direction. But if the core material has no retentivity when the current flow ceases, the magnetic field disappears. When a retentive core material such as that shown in *B* and *C* is used, the magnetic field is stored in the core material when the current flow stops. Current in one direction (B) creates a magnetic field in one direction while a current flow in the opposite direction (C) creates a magnetic field opposite C to the first one, in B. All types of magnetic memory systems use this basic principle.

Core Memories. Toroids of magnetic material can be used for data storage since they can be magnetized in either direction as shown in

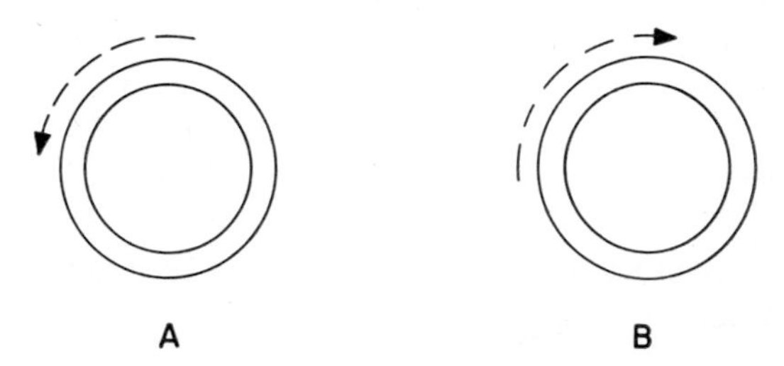

Fig. 11·2 Magnetism in either direction: (*A*) counterclockwise; (*B*) clockwise.

Fig. 11·2. Current flow in one direction will cause a clockwise magnetization, and current flow in the opposite direction will create a counterclockwise magnetization. An array or matrix of cores forms a magnetic memory for temporary retention of digital data.

Because of the bidirectional magnetic field, either a 1 or a 0 can be stored, depending on the direction of magnetization. (Chapter 16 covers binary numbers, which use a representation of 0 and 1.) The core will

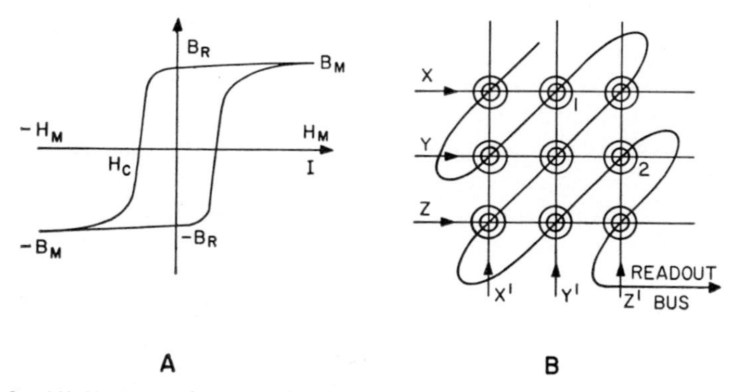

Fig. 11·3 (*A*) Hysteresis loop in which flux density is a function of magnetizing force; (*B*) magnetic-core-memory wiring in a digital-computer storage.

remain in a given state or direction of magnetization and thus store information by the direction of the magnetic flux density which remains after the current flow stops.

The hysteresis loop indicated in Fig. 11·3 for a magnetic core shows flux density B and magnetizing force H plotted. Suppose the initial state of the core is at $-B_R$. When a pulse of current is applied, equal to or greater than $+H_m$, the flux density will rise to $+B_m$. After the pulse the flux will decrease to $+B_R$. This is the flux at rest. This condition may be considered as the storage of a 1.

When a pulse of opposite polarity is applied, equal to or greater than $-H_m$, the flux density will move to $-B_m$. It will return at $-B_R$ when the current pulse is removed. Because of this, $-B_R$ can be said to represent the storage of a 0 or the opposite of a 1.

But look at the core starting from $-B_R$, which is the storage of a 0. If a *negative* pulse is applied, the core will *not* change its state. The core will go to $-B_m$ and then rest again at $-B_R$. Switching from one state to another requires a current pulse opposite to the present state of the core.

In the switching process the large rate of change in flux during t causes an induced voltage

$$e = -N \frac{d\theta}{dt}$$

where N is the number of turns in the output coil and θ is the flux in webers.

This output voltage occurs when there is a change of state for a core. From this a memory system may be developed. Each digit or "bit" of information is stored in one core. A pulse of current stores the proper information, either 1 or 0, in each core. When the information is desired, the required core is pulsed and the output is "read." Information is placed into this memory system by placing a 1 or a 0 in each core by a pulse on each of two "write" windings as shown in Fig. 11·3*B*. The readout winding also threads the core. When a negative pulse, or a pulse in the opposite direction, is applied to this winding, the core storing a 0 will not have an output. Those cores storing a 1 will read out a 1 in the output. Those storing a 1 will be switched and will produce an output; the others will not be switched.

In this way it is possible for the information retained, or remembered, by these cores to be read out at any time. Here in this system the stored information is lost when it is read out. The memory information may be restored or regenerated. In an actual memory a switch matrix system is used to pulse the cores with only a few tubes or transistors.

Magnetic cores are small, less than $\frac{1}{16}$ in. in diameter, and 10,000 can be used in a 10- by 10-in. area. Cores are mounted on a structure of vertical and horizontal wires, and each core can be identified by its intersecting wires. Another winding, this one through all the cores, is the readout winding. Two core memory planes are illustrated in Fig. 11·4.

The winding in a core memory is only one-half a turn, and a large current flows through the wires. A current of a certain size (I) is required to switch a core, and anything less than this will not switch a core. Figure 11·5 shows a part of a core matrix, with a driver tube for each line. V_1

fires a current pulse which goes through two cores; however, this is not enough to switch any core. V_2 is another tube whose plate current pulse also goes to two cores. But only one core, at the intersection of these

Fig. 11·4 Memory planes showing core wiring. (*Ferroxcube*)

two lines, will get the full current I. Because of this, only one core is switched.

One magnetic core can store one bit. Many cores are required for a large memory. Information can be placed in or read out of a memory of

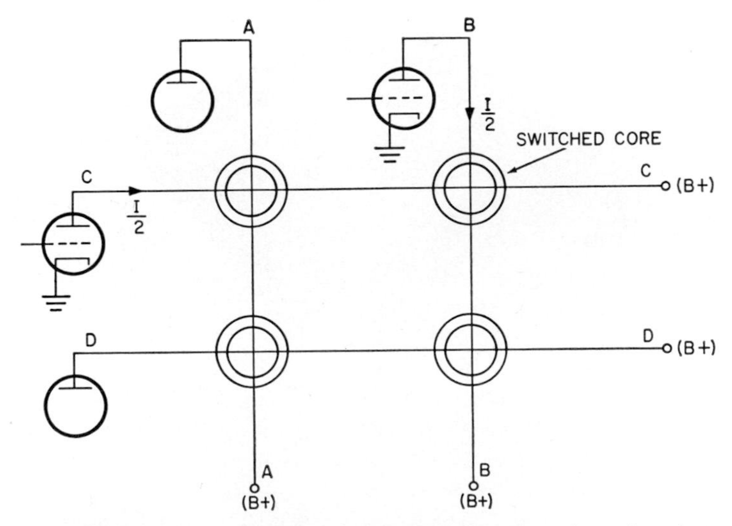

Fig. 11·5 Core switching in which the only switched core is where the wires carrying half-currents intersect.

cores very quickly. Each complete group of cores is a memory plane, and several planes are used. A core memory plane with 64 cores on a side has a total of 4,096 cores.

Fig. 11·6 Memory-plane set with planes, as in Fig. 11·14, plus associated electronics (*General Ceramics*)

Fig. 11·7 Memory stack. (*Ferroxcube*)

A memory-plane set is shown in Fig. 11·6. This consists of the memory planes, the circuits for driving the cores, and the power supply. When planes are stacked one on top of the other, a three-dimensional array (Fig. 11·7) results. This is used for mass data storage.

Magnetic-tape Storage. Magnetic tapes can store a great deal of information, but the access time is longer than that for core memories or drums. Because the tape is long—several hundred feet—it takes a long time to locate a specific piece of data. Figure 11·8 shows a bank of 10 tapes, each with two spools, as a computer input. This is the Philco 2000

Fig. 11·8 Magnetic-tape systems which are electronic "file drawers" in which to keep data. (*Philco* 2000)

system, which has tape storage as illustrated, and the number of tape stations varies with the needs of each installation.

A tape storage unit has two reels and the tape runs from one to the other, passing at least one magnetic head. The tape itself usually has three layers. The base, usually of a plastic such as Mylar, has a very thin layer of iron oxide which is the magnetic substance. A protective layer may be on top. This prevents the iron oxide from rubbing off. As shown, the tape moves past a head which may be considered as a core, as in a magnetic-core memory, except that there is a break or gap in the magnetic structure. In the structure itself the magnetic flux is contained. Across the gap the magnetic flux infringes onto the tape.

When current passes through the "write" winding, a magnetic field is impressed across the tape, recording a 1 or 0 depending on the direction of the current. If the tape had been passed through and information was recorded, the same head could be used to reproduce the recorded data. Each recorded section will induce a current flow in the "read" winding as the recorded flux cuts the read coil. Since reading and writing do not occur at the same time, a single winding on a single head can be used for both. In the write function, current flow in the winding will cause a magnetic field for recording; such a flow is induced by the passage of a recorded bit. The gap used for both purposes is very small, usually about 0.001 in., and is filled by a shim which is a small piece of nonmagnetic material. A single head with write and read amplifiers is shown in Fig. 11·9. Two write amplifiers or drivers are shown, one for each direction of magnetization. Since read or playback occurs at a different time, the amplifier for this purpose amplifies a signal picked up by the same coil winding.

Pulse packing is a measure of how dense the pulse recording is on the tape. One possible recording method is the return-to-zero (RZ) technique since the amount of magnetization returns to zero after each recorded

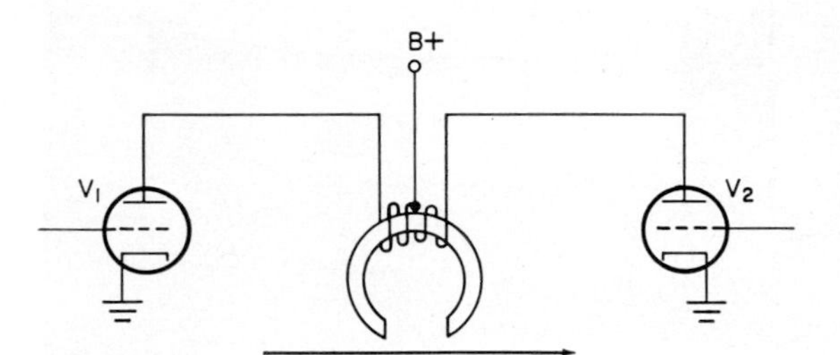

Fig. 11·9 Tube drivers which provide current for the magnetic "write" head.

bit whether it is a 0 or a 1. An increase of pulse density can be accomplished by use of a non-return-to-zero (NRZ) recording. In one NRZ method the direction of magnetization changes only when switching from one digit to another, as from a 0 to a 1 or from a 1 to a 0. It is also possible to have an NRZ system in which the direction of magnetization changes only to store a 1.

This increase in the density of the pulses is important, for tape is required to store huge quantities of data. At a density of 200 bits per inch of tape, a reel of tape 1,200 ft long and ½ in. wide can store more than 2,800,000 bits. Tape lengths of 2,400 and 3,600 ft are also used.

There is a difference between tape as used for audio reproduction, where the sound quality is important, and its use in computers, where the only important thing is the direction of the magnetization and not its amount. For this reason, currents used in the write head are made heavy enough to saturate the tape.

A computer tape moves at a speed of 100 in. per sec or more. Because the computer may be looking for specific data, the tape must be stopped quickly when the proper point is reached.

Because of the many rapid starts and stops, special measures are required to prevent breaking the tape. Each tape has its own drive motor which is connected to a servomechanism. When a tape reel starts, it is only required to draw from the loop while the other reel and its motor keep the loop the proper size by moving the tape. Tape breakage is thus avoided when quick stops and starts are made.

In some high-speed applications tape reels feed into bins, where the tape lies in loose folds. One disadvantage of tape is its long access time. In some cases the computer may have to search 3,600 ft of tape to locate specific data. A magnetic drum has a much shorter access time.

Magnetic-drum Storage. Magnetic tapes are very useful for data storage, but they have a limitation in the speed at which they can be run. The motion of the recording head or pickup past the recording medium can be speeded by using a magnetic drum, which has strips of magnetic ma-

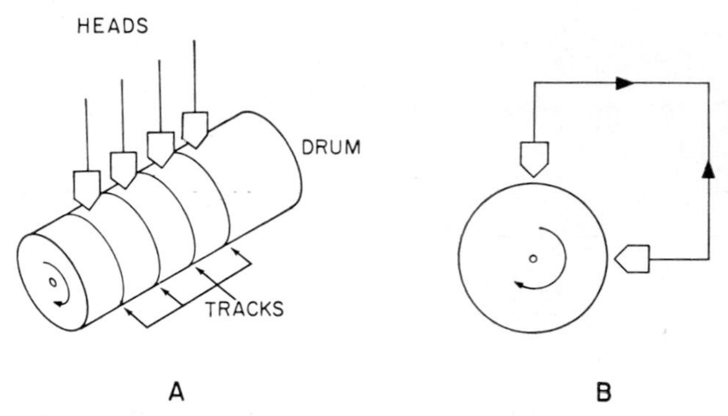

Fig. 11·10 (*A*) Magnetic drum showing tracks for data storage and the heads; (*B*) recirculating loop for data return.

terial as shown in Fig. 11·10. Each of these can be considered as a strip of magnetic tape with a head which is either recording or playing back.

A magnetic drum is a cylinder usually rotating at 1,200 to 15,000 rpm, with 3,600 rpm being an average value. Information may always be located, in a single track, in less than one revolution. If the drum is under a head at a certain spot, the next desired spot of data could be almost a full revolution away; average access time for a drum rotating at 3,600 rpm would be 0.0083 sec or 8.3 msec. This is the time required for the drum to make one revolution.

With parallel magnetic tracks at a density of 1,000 pulses per square inch, drums can store millions of bits of data. Drum diameters vary from about 4 to 15 in., and the drums may also vary in width, depending upon the required number of tracks. The number of tracks depends upon the width of the drum. The number of bits on each track depends upon the diameter of the drum. A typical drum is shown in Fig. 11·11.

The area beneath a head is a magnetic track made up of tiny bits of recorded data. One hundred recordings per inch may be used with more than 30 tracks per inch along the width of the drum. Even with small recording-playback heads it would be impossible to place 200 heads across a 12-in. drum length, for this would leave only 0.06 in. for each head. Instead, the heads are staggered so that their positions overlap, allowing a close spacing for the magnetic tracks. It is not necessary to have a separate read-write amplifier for each head, for switching can be used.

For identification, a separate track on the drum carries a series of prerecorded timing pulses. These are read by a head whose output gives a precise indication of the drum's rotation at all times.

Fig. 11·11 Magnetic storage drum showing head wiring. (*Bryant*)

Magnetic drums often receive information from the tape storage so that the data will be quickly available when required. Magnetic systems are often compared relative to their type of access. Core memories may be entered at any place; hence they are random-access. A tape requires reading from one end along the tape until the required information is found. For this reason these are serial-access storage forms. A drum permits random location of the track.

11·2 Magnetic amplifiers Coils with iron cores can be used for power control as saturable reactors or as magnetic amplifiers. They can control small amounts of power directly and also can be used to control other devices such as thyratrons which can carry larger currents.

An iron-core coil, in series with the load and an a-c power source, is a saturable reactor as shown in Fig. 11·12. In an unsaturated condition most of the voltage drop is across the coil L. When saturation is reached as at point 1, no further increase in the magnetic field is possible and the voltage drop then appears across R. This continues to the peak at 2 and until 180° is reached. Here, at zero voltage, the drop is across L

again until point 3. Thus, for the shaded portions there is power delivered to R as the load.

For saturable-reactor applications a square-loop characteristic for the core material is desired. This provides an abrupt transition from one state to another for precise control. Materials with curves like the one illustrated in Fig. 11·3A or even more nearly square in shape are used.

Rectifiers (see Fig. 11·13) allow current flow in only one direction

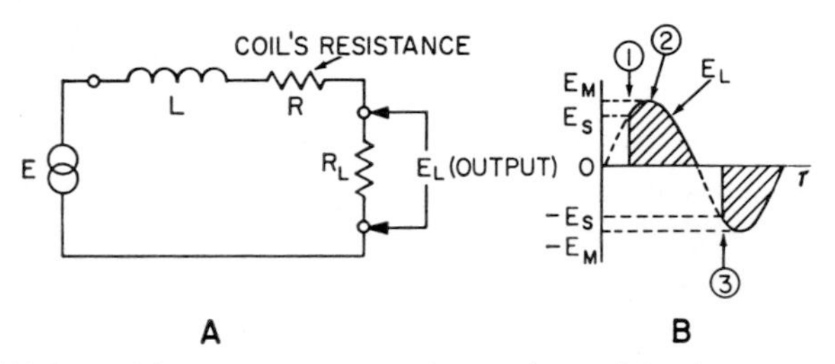

Fig. 11·12 (A) Saturable-reactor circuit with a coil L and load R_L; (B) output voltage for a-c input.

so the amplifier works on every other alternation. In a magnetic amplifier a rectifier is in series with the main winding so most of the current for saturation comes from the load current.

A magnetic amplifier, in a sense, is like other amplifiers since it has an input, it requires a power-supply source, and it produces an output signal larger than the input signal. In a simple vacuum-tube amplifier there is a grid signal, a grid bias for control of the grid voltage, and an output

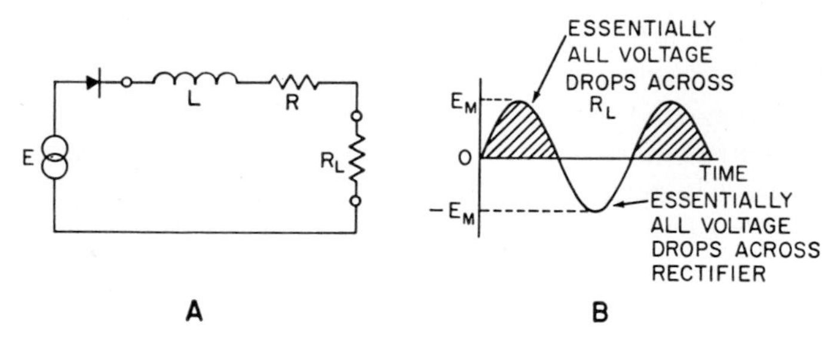

Fig. 11·13 (A) Saturable reactor with rectifier; (B) voltage drops for the output and across R_L.

voltage across the anode load of the tube. In a magnetic amplifier there is a magnetic core with several windings. Input is to a control winding, and output is from a load on a second and larger winding.

A basic magnetic amplifier with its two windings looks in many ways like an ordinary iron-core transformer. Input is to the control or smaller winding. Output is across the load, which is in series with the output winding and the a-c power source. With no current flow in the control winding, the secondary winding has a high impedance; hence most of the voltage drop is across the load winding and there is very little output across the load resistor. But when current flows in the control winding, the impedance of the output winding goes down, which allows a voltage drop across the load. Thus small variations of current in the control winding cause larger variations in the output and there is amplification.

Suppose that the control current is such as to make the load coil have an impedance of 10,000 ohms where the load itself is 1,000 ohms. A-c power is across the series circuit representing the secondary of the magnetic amplifier. With the high impedance and low value of load, almost all the applied alternating current will appear as a voltage drop across the impedance of the coil and very little or none will be available across the load as the actual circuit output.

But if the control current increases, the impedance of the load will drop to perhaps 10 ohms. Now the voltage drop is mostly across the load and very little is across the coil, whose impedance is now very low.

Magnetics are broken down into two basic circuits. The coil wrapped around a magnetic core which, by its nature, can be saturated by magnetic flux, is a *saturable reactor*. Used by itself it is a useful and used industrial circuit device. When this saturable reactor is used together with amplifier, rectifiers, and the like it becomes a *magnetic amplifier*.

Magnetic amplifiers go back at least to 1901 or earlier, but the first useful application appears to be the Alexanderson radio transmitters used during the First World War period. During the Second World War the German armed forces made extensive use of magnetic amplifiers; these devices now have many industrial applications.

When a voltage is applied to the circuit of a saturable reactor as shown in Fig. 11·14*A*, nearly all the voltage will drop across the coil up to the point where the core becomes saturated. Then the value of the reactance becomes essentially zero. During the remainder of the positive half-cycle, nearly all the voltage drops across the load. As the applied voltage starts into the negative half-cycle, the core goes out of saturation and once again the coil impedance becomes large; this action repeats itself on the negative half-cycle.

With an auxiliary winding on the core and a direct current of proper magnitude and direction passed through this winding, it is possible to select a bias point on the hysteresis loop as shown in Fig. 11·14*B*, which is similar to selecting an operating point on a tube characteristic by the use of a proper bias voltage. Figure 11·14*C* shows output, with variations.

The amount of voltage which can be absorbed by the coil on each positive half-cycle can be selected by adjustment of the bias current in the control winding. It is possible to control the amount of average power delivered to a load resistance by varying the magnitude of the bias current. Gain is achieved, since a small amount of bias current will result in a large change in load current.

The iron core now has two windings, an a-c or signal winding and a d-c or control winding. The direct current changes and this changes the magnetization of the iron core. Again the a-c output depends on the im-

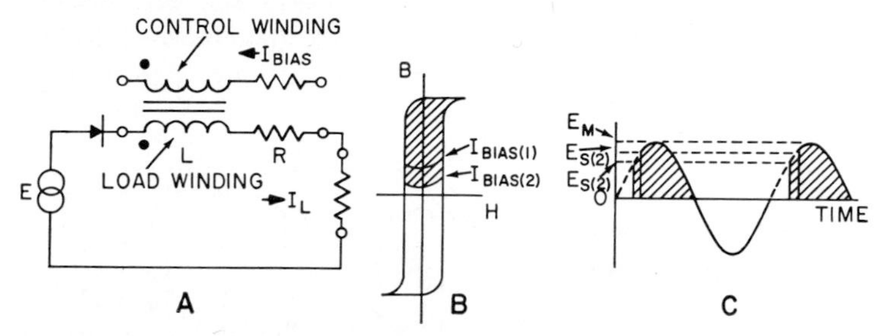

Fig. 11·14 (*A*) Load resistance, reactor, and rectifier in series; (*B*) portion of the hysteresis curve used by biasing; (C) typical output for (*B*).

pedance in the a-c circuit. If there is no d-c flow, there is a large impedance across the signal winding and a small output across the load. If there is a current flow through the control winding, the magnetic field is created partly by the direct current and only partly by the alternating current; the lowered impedance of the signal winding means a greater output voltage from the load.

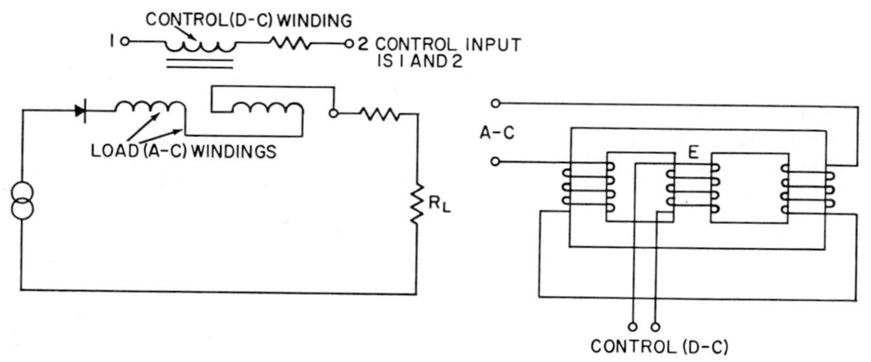

Fig. 11·15 Three-winding reactor showing split load (a-c) winding and three-legged core as the actual physical construction technique.

This circuit will not operate efficiently because there will be alternating as well as direct current in the control winding because of the induced field. A more effective circuit is shown in Fig. 11·15. Two a-c windings are now in series, and the a-c magnetic fields cancel in the center leg of the iron core because they are in opposite directions.

Figure 11·16 shows three modifications of the magnetic amplifier; A is a full-wave rectifier, B is a bridge rectifier, and C shows the use of a feedback winding used to increase the gain. Magnetic amplifiers of many different types are commercially available.

Magnetic Logic. Magnetic amplifiers may be used directly for control or they may be used in logic circuits. One type of logic circuit is the Ramey magnetic amplifier. Logic circuits are developed fully in Chap. 16 but, for this present discussion, we need to say only that an OR circuit

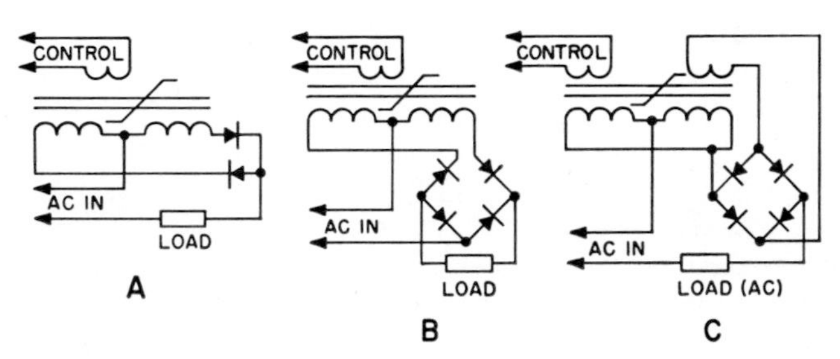

Fig. 11·16 (A) Full-wave-rectifier magnetic amplifier; (B) bridge-rectifier amplifier; (C) use of feedback to increase gain.

will operate if *any* input is applied and that an AND circuit will operate if, and only if, *all* inputs are applied.

In the basic Ramey magnetic amplifier (Fig. 11·17) the core would saturate if only the signal E_S were applied. A second winding is used, however, to apply a signal such that it opposes magnetization of the core by E_S. Being 180° out of phase with E_S, this second winding prevents saturation even if E_S is on; hence this second winding is called the reset or E_R. By blocking E_R, E_S can produce an output.

By blocking E_R with an appropriate input, as shown in Fig. 11·18, E_S will cause a half-wave output across the load R_L. Diodes D_2 and D_4

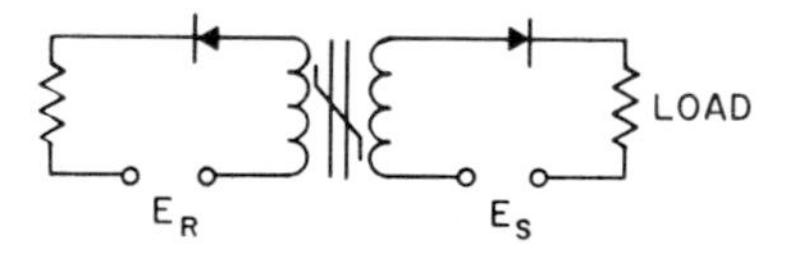

Fig. 11·17 Basic Ramey unit used for magnetic logic circuits in control.

permit the input to block the resetting signal. A d-c bias is required to obtain the proper impedance level. In Fig. 11·19, by connecting the points A and B of Fig. 11·18, an OR circuit is obtained for which an output will appear across the load if any single input is used. In a like manner the AND circuit in Fig. 11·20 can be created such that any input (1, 2, or 3) will cause E_R to be blocked so that there will be an output.

Other magnetic circuits are also used for logical switching. A in Fig. 11·21 shows a magnetic switch, which uses magnetic cores and silicon rectifiers. Three input windings are available for control. A square-wave source is employed for excitation. Use of the 2,000-cps carrier has distinct advantages in that core sizes may be smaller, achieving a considerable reduction in size and weight of the overall system. The high-fre-

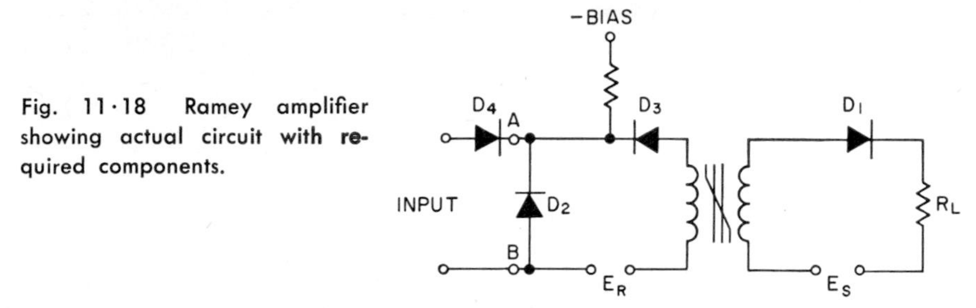

Fig. 11·18 Ramey amplifier showing actual circuit with required components.

quency carrier also makes possible fast switching times of 1 msec. Because of the small-sized cores, high sensitivities are achieved. Only 1 ma from a 5-volt source is required to switch 20 watts. The output of each mag-

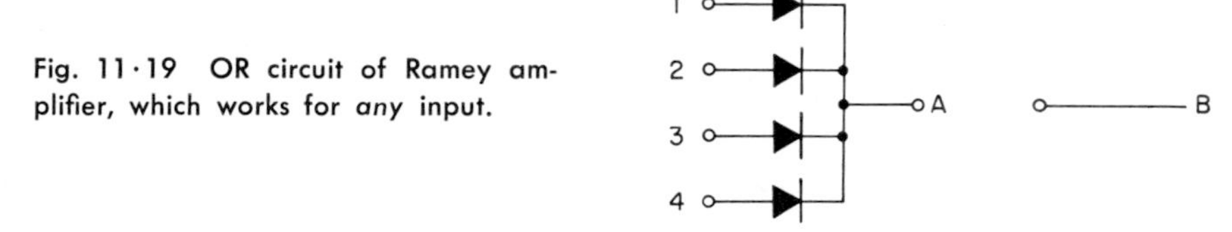

Fig. 11·19 OR circuit of Ramey amplifier, which works for *any* input.

netic switch is 28 volts d-c at 20 watts. Since the excitation frequency is a square wave, the switch output (which is a full-wave-rectified square wave) is practically pure direct current.

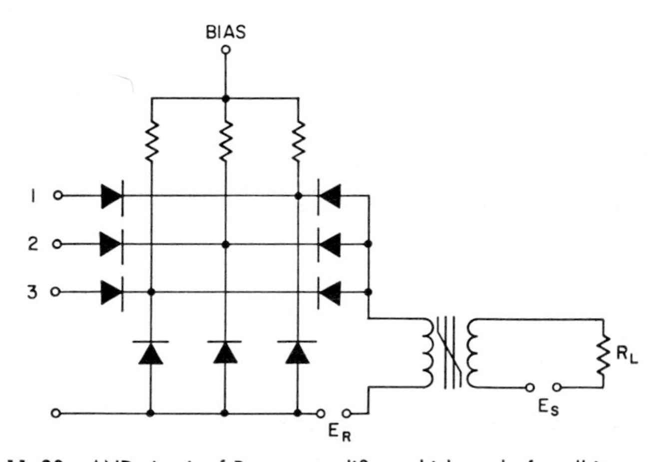

Fig. 11·20 AND circuit of Ramey amplifier, which works for *all* inputs only.

This magnetic switch has the ability to perform all the principal logic functions such as AND, OR, NOT, and memory, in addition to switching and power amplification. This is accomplished by utilizing two control

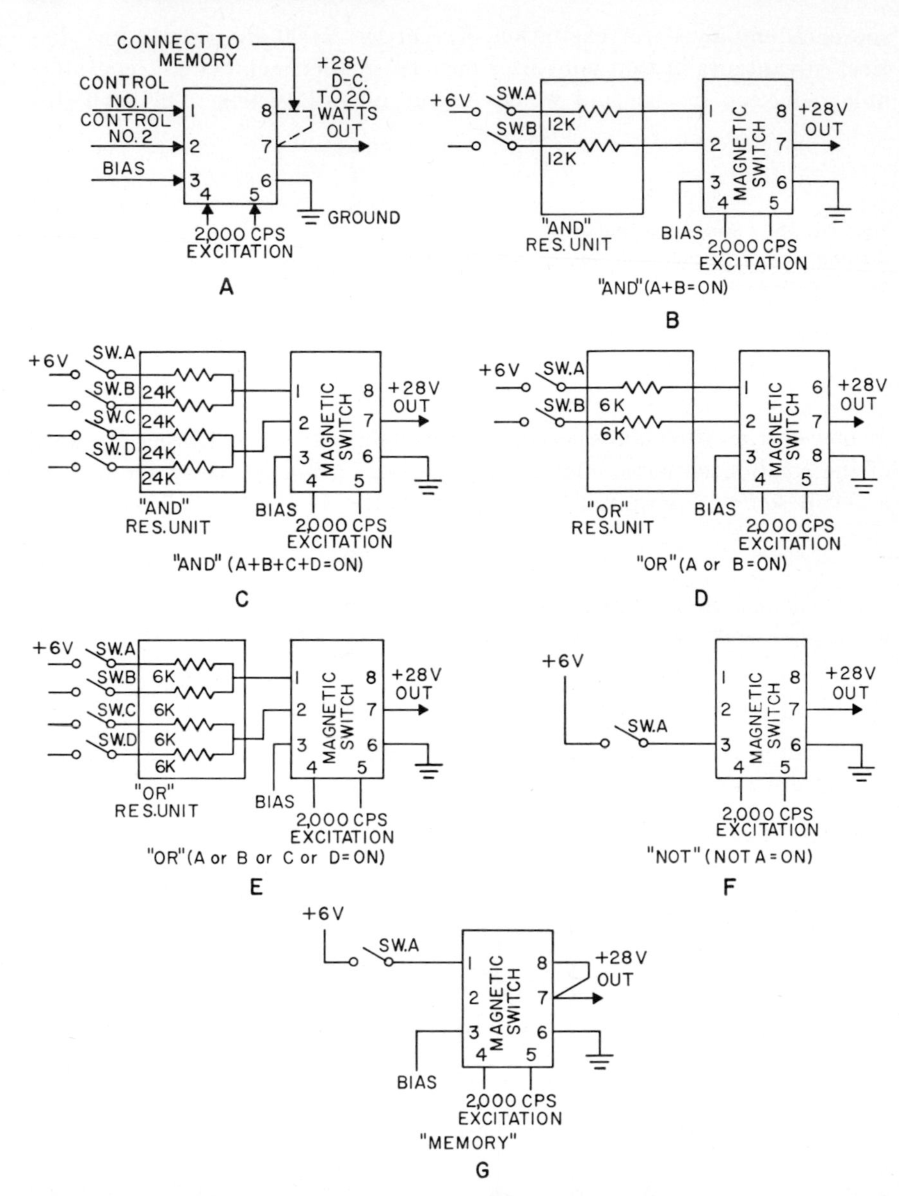

Fig. 11·21 (*A*) Basic magnetic-amplifier module; (*B*) AND circuit; (*C*) four-input AND circuit; (*D*) OR circuit; (*E*) four-input OR; (*F*) NOT circuit; (*G*) memory circuit. (*Magnetic Amplifiers, Inc.*)

windings: a bias winding and a feedback winding. The unit will switch on with 5 volts applied to the bias winding, and when the sum of the currents flowing through the control windings reaches 1 ma, a 6-volt d-c source is used for control-signal excitation.

B and *C* in Fig. 11·21 show the AND function; *D* and *E* are OR circuits. The resistance of each control winding is 130 ohms, small compared with values employed in the networks. The NOT function utilizes the bias winding to turn the switch off from a normally on condition, as shown in *F*.

The memory function is accomplished by employing a positive feedback winding, which maintains the switch in the on condition after the control signal has been removed, as shown in Fig. 11·21*G*. It is returned to its former condition (off) by the application of a negative control signal.

Uses of Magnetic Amplifiers. Magnetic amplifiers provide an output which is a sharp pulse with a fast rise time on the leading edge, and this can control other devices for passing large amounts of current, such as for motor speed adjustments.

Magnetic amplifiers provide the steep wavefront required for control of thyratrons and controlled rectifiers. One of these is the magnetic thyratron control (MTC). A wide range of different inputs may be used as control inputs to this magnetic amplifier. A typical example of this control is given in Fig. 11·22. *A* shows a pair of thyratrons (with a magnetic amplifier) used to control the speed of the d-c motor. The speed-set control adjusts the desired operating speed of the motor. A reference voltage is compared with the armature voltage in one of the control windings. A different control winding measures the armature current and increases the motor drive as the load is increased. A resistance adjustment is provided for this winding, which corrects for the *IR* drop in the armature. A third winding, used for torque control, turns off the thyratrons if the load current exceeds a predetermined value. *B* shows a regulated high-voltage supply, while *C* shows a three-phase bridge circuit for thyratrons.

Magnetic amplifiers also can be used with a controlled rectifier as shown in Fig. 11·23. The magnetic amplifier is tied between the anode and the gate. When the core of the amplifier is not saturated, the controlled rectifier is off. At some point, determined by the control-signal input to the magnetic amplifier, the amplifier fires (is saturated) and the gate has a current input. This, in turn, fires or turns on the controlled rectifier. Because the rectifier is on, it provides power to the load; but, at the same time, this conduction removes the supply voltage of the magnetic amplifier, which prevents a further increase in the gate signal of the controlled rectifier.

In another application an oven temperature is controlled. When thermostat contacts are closed, a heavy direct current flows in the control winding, there is a large inductive drop across the signal windings, and very little current flows through the heater. After the temperature is increased, because of the heaters, the thermostat contacts open and the direct current through the control winding stops. A large a-c drop across the saturable-reactor a-c windings effectively turns off the heater.

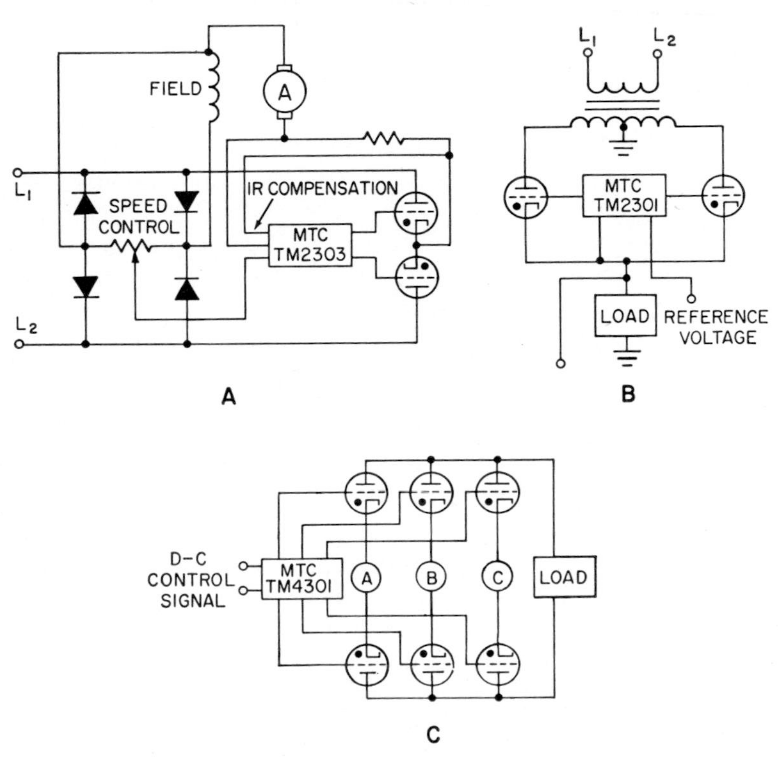

Fig. 11·22 (A) Motor speed control; (B) high-voltage regulated power supply; (C) three-phase full-wave bridge firing six thyratrons. (*Fairfield Engineering*)

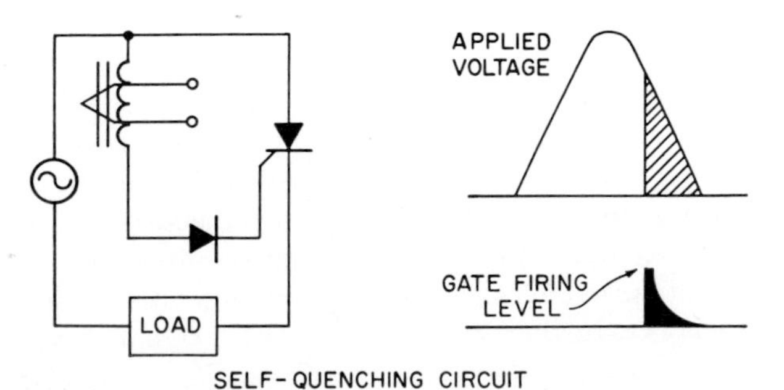

Fig. 11·23 Rectifier control with magnetic amplifier.

In a generator speed control a motor-driven a-c generator is to be regulated for frequency. Part of the generator output is taken off through a discriminator, to the saturable-reactor d-c winding. The discriminator may provide the required direct current or it may require a bridge rectifier because of the reactor circuit. In action this provides a current by the use of resonant circuits which varies as the frequency of the generator output changes. Because this current changes the reactance of the magnetic circuit, it effectively regulates the motor speed through the motor control block. Output frequency from the generator is a function of speed; hence the system controls the output frequency.

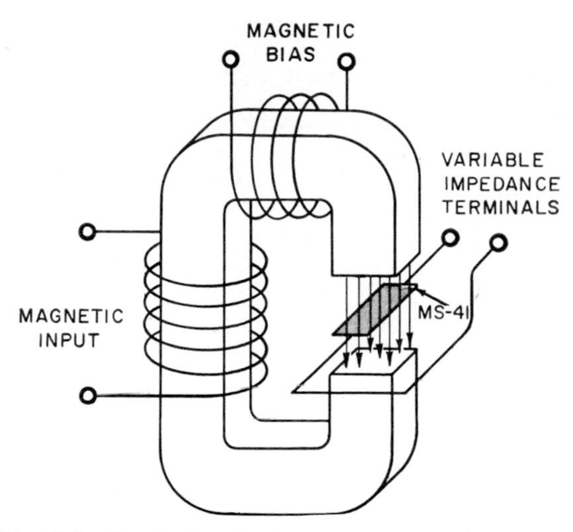

Fig. 11·24 Hall-effect device, the Magnetoresistor. (*Ohio Semiconductors*)

11·3 Other magnetic devices Magnetics are used in many ways. Some less usual applications are discussed in this section.

The Magnetoresistor. The Magnetoresistor (Ohio Semiconductors) is a new solid-state device in which the electrical resistance varies with the strength of the applied magnetic field. This unit, the MS-41, is shown in Fig. 11·24.

This unit has potential wide usage as a contactless variable resistor in control circuits. In addition to its use as a variable resistor in ordinary circuits, it will permit the development of circuits not easily achieved with conventional components. This device can also be used in such special applications as low-level choppers, power amplifiers, modulators, voltage and current regulators, and magnetic field measurements.

This magnetoresistive effect is based upon the interaction between certain semiconductors and the applied magnetic field. When a conducting material is placed in a magnetic field which is perpendicular to the direction of current flow, the resistance of the material is increased. For small fields the change in resistance is approximately proportional to the square of the magnetic induction.

The magnetoresistive effect is greatly enhanced at low temperatures. At a temperature of 80°K a change in resistance of the order of 1,000:1 should be obtainable using high-purity indium antimonide.

This unit can be either a nonlinear or a linear element. In the low-field range, the change in resistance is a power function of the applied magnetic field. Since the field is proportional to the current flowing in the magnetic-circuit coil, the change in resistance is a power function of the

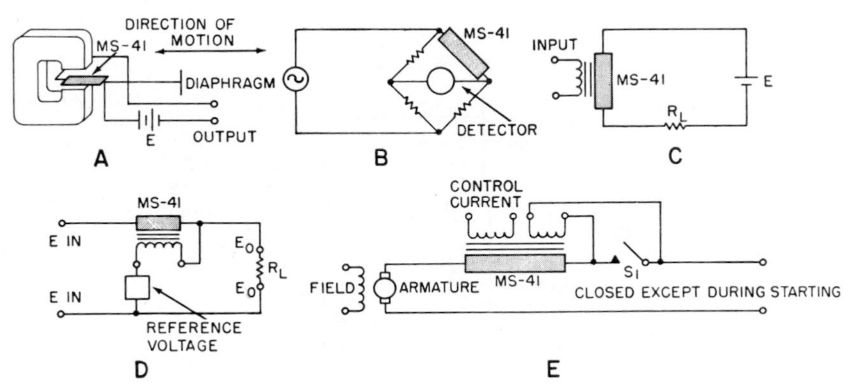

Fig. 11·25 Magnetoresistor used (*A*) as a transducer, (*B*) for measurement of small fields or incremental changes in a magnetic field, (*C*) as a circuit for a power amplifier, (*D*) as a voltage regulator, and (*E*) as a variable resistor. (*Ohio Semiconductors*)

coil current. A 10:1 increase in resistance is achieved with a 10-kilogauss field. In the region above 5 kilogauss the change in resistance tends to become a linear function of the applied field. Thus, by applying a magnetic bias, the properties can be changed from nonlinear to linear.

Circuit applications are shown in Fig. 11·25. *A* shows the use of the device as a transducer. Mechanically moving the unit in and out of a magnetic field will produce a resistance change that is a function of the distance moved. Such a device could be used as a microphone, phono pickup, barometer, or airspeed indicator.

B shows its use in a bridge circuit. By using the device in one of the arms of a conventional bridge circuit, incremental changes in magnetic field can be detected. A change in the resistance will result in a bridge unbalance.

C shows a power-amplifier circuit. A small amount of power may be used to control a larger amount of power in this circuit. The gain of such a system is limited only by the resistance of the input coil winding. The system has the advantage of complete electrical isolation between the input and output.

D illustrates a voltage-regulator application. Any change in the output voltage will tend to either increase or decrease the current through the current coil, thereby increasing or decreasing the resistance and stabilizing the voltage at a predetermined level.

E shows a variable resistor without a contact. This circuit illustrates variable control of a small motor. This same principle can be utilized to control the field circuit or a combination of both. The control of small motors is only one of many control applications.

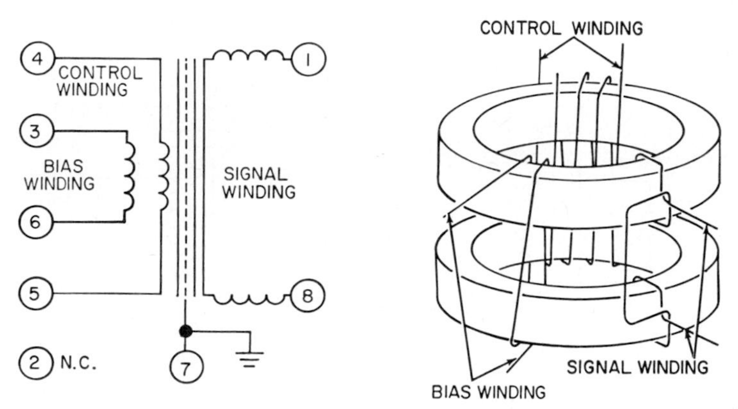

Fig. 11·26 Construction of one type of controllable inductor useful in r-f switches, in which two ferrite-ring cores carry a control winding and a signal (controlled) winding. When a direct current is applied to the control winding, the effective inductance of the controlled winding decreases as the ferrite core becomes saturated. A balanced signal winding and internal shielding reduce capacitive coupling and transformer action between the control and signal windings to a small value. (*CGS Electronics*)

Controlled Inductances. A special type of variable inductance is the Increductor (CGS Electronics, Stamford, Connecticut). There are two windings on the magnetic core: the signal winding and the control winding as shown in Fig. 11·26. As the current is changed in the control winding, the magnetic state of the core is varied, which, in turn, changes the inductance of the signal winding.

There are many possible applications for the device. Figure 11·27*A* shows the Increductor in a high-frequency Colpitts-oscillator circuit. The control-tube circuit for sweep frequency or FM use is also shown. Figure 11·27*B* illustrates an alternate oscillator circuit for lower frequencies. R_2 and R_3, together with C_3 and R_1, can be adjusted over an extended frequency range. Changing the control current causes the oscillator to vary in frequency because of the change in inductance.

Another common application of the Increductor is in tuned-amplifier stages. Figure 11·28 shows a wideband tuned circuit. Constant gain and bandwidth over the frequency range can be achieved. A number of similar stages can be cascaded into a tuned r-f receiver or used in conjunction with an Increductor-tuned oscillator stage for a superheterodyne receiver.

There are Increductors with a wide range of characteristics. Development has produced units with inductance changes of 400:1 in the frequency region to 10 mc, with a reduction in range over 10 mc. At 100 mc, frequency shifts of 3:1 in tuned circuits have been achieved. Oscillator units have been operated at signal frequencies in excess of 300 mc. *Q* values of over 200 have been achieved in the frequency region of several kilocycles to 10 mc. Above this range, *Q*s of 30 to over 100 have been achieved up to 250 mc.

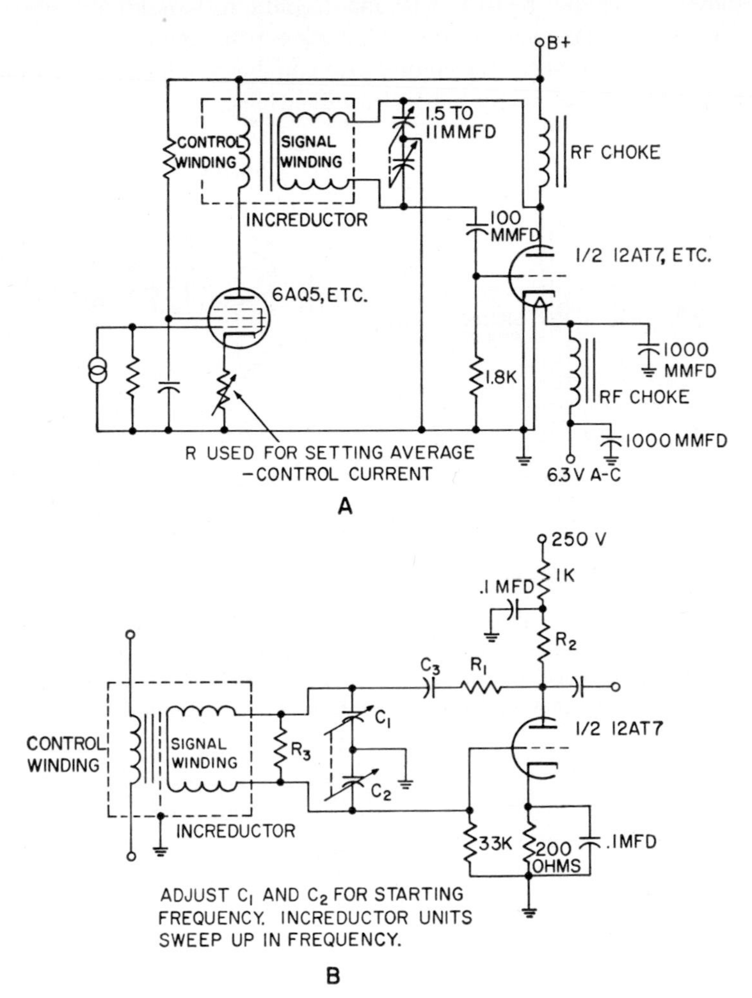

Fig. 11·27 (*A*) High-frequency magnetic-oscillator control; (*B*) low-frequency magnetic-oscillator control. (*CGS Electronics*)

Low-level Increductors require from 0.1 mw to 1 or 2 watts of control power for full inductance variation, with typical units requiring 50 to 250 mw. One high-power inductor delivers over 80 watts of r-f power with a 2:1 frequency swing and requires 10 watts of control power.

Proximity Switches. A proximity switch senses the nearness of objects for warning or control. It is a magnetic two-part device designed to deliver a d-c output of several watts when actuated by bringing a piece of magnetic material near its sensing head. It is self-contained. The only power source required is 117 volts, 60 cycles. The sensing head is a molded

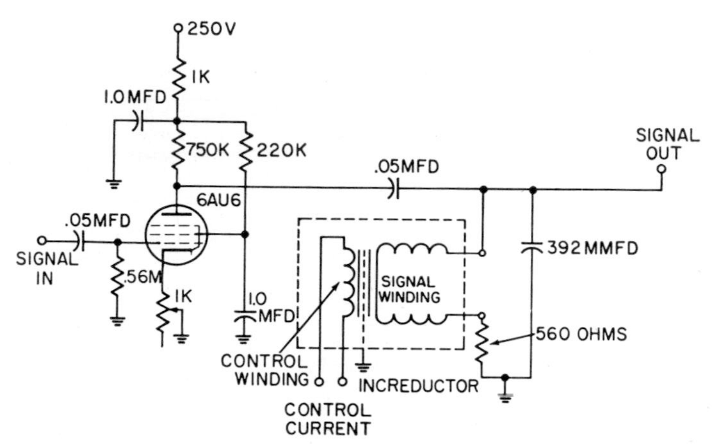

Fig. 11·28 Wideband amplifier control. (*CGS Electronics*)

unit containing half an inductance bridge. The rest of the bridge, power supply, transistor amplifiers, phase detector, and precision flip-flop are in a steel can. The two units are connected together by bringing three wires from the sensing element, usually mounted remotely on a machine, to the control element, usually mounted in the control cabinet. These wires are usually unshielded for runs of 50 ft or less. Shielded wire may be necessary when long runs are required or when switch wires are in proximity with high-current-carrying conductors.

The limit switch may be divided into eight parts as shown in Fig. 11·29. The power supply (1) delivers a-c and d-c voltages required for switch operation. These voltages are isolated from the line.

The balance element (6) and the sensing element (7) form an inductance bridge. Bringing the actuating member near the sensing element changes the output voltage from this bridge. An iron screw labeled Balance Adjust, mounted on the case of the control element, is used to set the bridge trip point. This makes it possible to vary the actuator trip distance which will produce a given bridge signal. A significant signal is developed only when the actuator is closer to the sensing head than the null point determined by the balance adjustment.

The a-c amplifier (2) raises the level of the bridge signal without significant distortion. This a-c signal is now rectified by a phase-sensitive detecting circuit (3). The filtered signal drives a precision flip-flop (4) which converts the analog signal to a true digital signal. Sufficient power is available from this circuit to permit driving a power-transistor amplifier (5). This amplifier produces a full-wave 24-volt d-c output with a maximum current drain of 335 ma. This maximum output current is reduced to

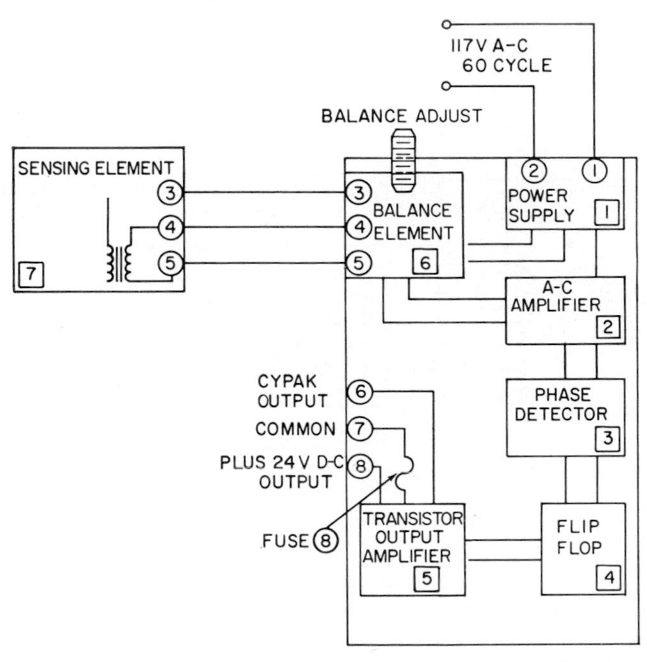

Fig. 11·29 Proximity-switch block diagram. (*Westinghouse*)

300 ma when light bulbs and other resistive loads are connected. The fuse (8), type MDL ⅜ amp, protects the power transistor from short circuits of all types.

The Magnistor. Magnetics technology has grown in expansion of its many applications and in the actual number of magnetic devices. One of these devices is the Magnistor (Potter Instrument Company), originally developed for a high-speed computer printer.

Basically the Magnistor (Fig. 11·30) is constructed of four separate coils (set, reset, interrogate, and signal output), wound around or through a toroidal ferrite-core structure as shown. By the use of special winding techniques for the coils plus high- and low-retentivity ferrite materials for the toroidal core, the Magnistor can, with associated circuitry, be made to function as a differentiating detector with a permanent or erase storage capability. This core has the set and reset coils wound around the high-

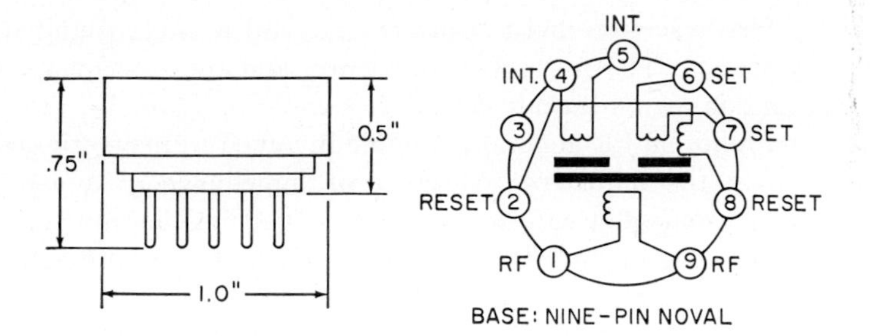

PT-1 Magnistor

Tentative specification	Resistance	Inductance	Current
Set	5 ohms	95 μh at 1 Mc*	100 ma ±20%
Reset	3 ohms	35 μh at 1 Mc*	120 ma ±20%
Interrogate	6½ ohms	700 μh at 1 Mc	120 ma ±20%
R-f signal	1 ohm	Set: 18 μh at 10 Mc Reset: 90 μh at 10 Mc	
Temperature: 100°C		Typical r-f signal voltage: 2 v rms	

*This value is for initial and final inductance; during the transition period the inductance may rise by a factor of ten.

Fig. 11·30 The Magnistor. (*Potter Instrument*)

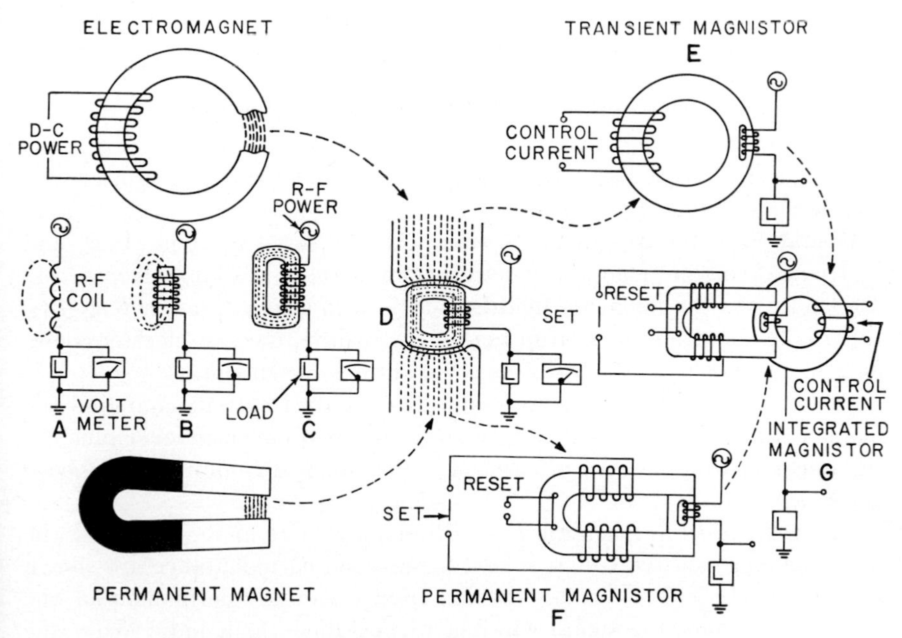

Fig. 11·31 Evolution of the Magnistor. (*Potter Instrument*)

retentivity ferrite section and the interrogate coil wound around the low-retentivity section. The set and reset section and the interrogate section share a common signal-output coil.

Operation relies on the general principle involved in magnetic-amplifier design wherein the inductive reactance or impedance of a coil wound around an electromagnet can be changed by varying the amount of flux in the core. The Magnistor differs from the magnetic amplifier, however, by using a 10-mc carrier frequency in place of the usual 60- or 400-cps carrier. Consequently, it has much greater speed of response and can be

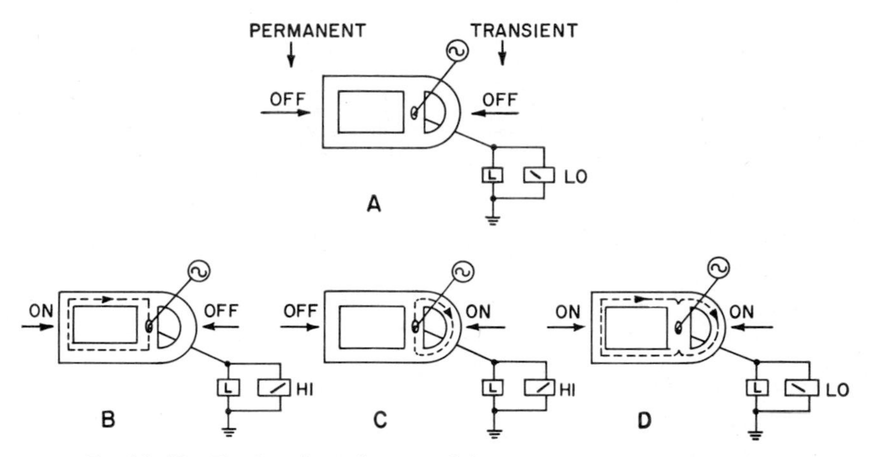

Fig. 11·32 The four logical states of the Magnistor. (*Potter Instrument*)

made in a smaller package. The magnetic amplifier and the Magnistor are compared in Fig. 11·31.

Operation of the Magnistor is shown by the series of steps *A*, *B*, and *C*. If an r-f coil and load are in series, as in *A*, there is a large drop across the load *L*. As the magnetic field is increased in strength, as in *B* and *C*, there is less and less of a drop across the load. Either an electromagnet or a permanent magnet can be used to form a Magnistor.

In *E* the transient unit is formed with a control current changing the magnetic field. A permanent Magnistor uses a permanent magnet plus the set and reset windings. An integrated Magnistor is a combination of both types, as in *G*.

The four possible magnetic states that may exist in the unit and the corresponding relative values of the signal-winding inductance are shown in Fig. 11·32. With no magnetic flux from either the permanent or the transient sections, the signal winding will exhibit a high inductive value as in *A*. With saturating flux from either of these two sections, as in *B* and *C*, the signal-winding inductance will be considerably reduced. With both sections energized as in *D*, the saturating flux through the signal-winding core is removed by the proper polarization and coercive effect of the energized interrogate winding.

It is evident from this that for equal input states such as off-off and on-on, a high inductance will appear; but for unequal inputs such as on-off and off-on, a low inductance will result. Thus this is a differentiating detector which distinguishes between equal and unequal states combined with storage for one state.

REVIEW QUESTIONS

11·1 Explain the basic magnetic recording principle.
11·2 Why is a square-loop material used for memory units?
11·3 How does a magnetic core record two magnetic states?
11·4 How does a magnetic amplifier operate?
11·5 Why do certain magnetic materials have a hysteresis loop?
11·6 How does the shape of this loop depend upon the material of the core?
11·7 What is a magnetic-core memory?
11·8 How is this memory used?
11·9 What is a magnetic-tape memory?
11·10 How is a memory plane constructed?
11·11 What technique is used to switch cores?
11·12 How is the switched core selected?
11·13 What is a memory stack?
11·14 For what purpose are the magnetic tapes in a computer used?
11·15 How does a magnetic drum differ from magnetic tape?
11·16 What is non-return-to-zero (NRZ) recording?
11·17 Where is it used?
11·18 Why is a special start-stop system used for magnetic tape?
11·19 What are the advantages of the magnetic amplifier?
11·20 What is a saturable reactor?
11·21 What is the diode rectifier shown in Fig. 11·13?
11·22 Why is bias added to a magnetic-amplifier circuit?
11·23 Why are the two parts of the main winding of the magnetic amplifier split?
11·24 How does a Ramey circuit operate?
11·25 How can it be used for an OR circuit or an AND circuit?
11·26 How can magnetics control thyratrons?
11·27 What is the Hall effect?
11·28 How can it be used for control?
11·29 What is the Increductor?
11·30 How can it control the frequency of an oscillator?
11·31 What is a magnetic proximity switch?
11·32 Explain the Magnistor.
11·33 How can the Magnistor be used for logic-circuit operation?

CHAPTER

12

Ultrasonics

Ultrasonics are sound waves above the range of human hearing. They are used for medical diagnosis, surgery, dentistry, testing materials, burglar alarms, soldering difficult metals, and storing information for radars and computers.

Navy-developed sonar (SOund Navigation And Ranging) is an underwater sound system for measuring the depth, range, and hearing of objects. Just as radio waves are transmitted through the air and reflected from targets, so ultrasonics are transmitted by water and reflected by underwater objects. In a commercial form, sonar can measure the depth of water under a ship or locate schools of fish; for the Navy, sonar can be used for communication as well as for detection.

Ultrasonic devices are used as flowmeters measuring the rate of flow in pipelines. Ultrasound is being used to turn television receivers off and on, to treat hospital patients, and to weld metals. It is used for material inspection and in alarm systems to detect intruders.

The ultrasonic gauge shown in Fig. 12·1 directly indicates the nosecone thickness of a Bomarc missile.

The electronic gauge offers several major advantages, one of them being the fact that it requires access to only one side of the object being measured. Thus, the thickness of the sheet at its very center is determined as easily as at the edge, which is impossible with conventional measuring devices. In fact, gauge of the metal may be found without lifting a sheet off the stack.

Using ultrasonic resonance to measure nondestructively, the electron gauge can determine thicknesses from about 0.005 to 2.7 in. to accuracies of 0.1 per cent, and even closer when an automatic recording accessory is used.

Inaudible sound waves, continuously varying within the range of 0.7 and 25 mc, are sent into the metal under test by a transducer, which converts the electric impulses it receives into mechanical vibrations. Returning waves in the metal are also picked up by the transducer and fed back into the electronic circuit, where resonant frequencies are amplified to show up as one or more traces on a cathode-ray tube (CRT). A calibrated scale in front of the tube face permits direct reading of thickness, which is inversely proportional to the resonant frequencies.

Fig. 12·1 Nose cone of Bomarc missile is checked to ± 0.001 in. thickness tolerance. Vidigage indicates thickness directly, without calculation, even when access from only one side is possible. (*Branson*)

The gauge is shown in use in Fig. 12·2, where strip and sheet steel are being measured to micrometer accuracies in a matter of seconds. All it takes is access to one side, and it is possible to determine metal thickness as accurately as a micrometer, yet more quickly.

Figure 12·3 shows an industrial cleaner for cam rollers. The unit uses a continuous chain conveyor to transport parts through the cleaning cycle. Parts are suspended from hooks on the chain.

The first stage in the system is a 30-sec spray with trichlorethylene. Here, gross soils are flushed from the parts to reduce contamination of subsequent stages. Liquid from this stage returns to stage 4 where it is distilled and reused. Insolubles sink to the bottom of the tanks out of the way.

Stage 2 is a 60-sec ultrasonic bath at 25 kc. Floating contamination, dislodged by cavitation, cascades into the spray tank and is eliminated.

Following ultrasonic cleaning, the parts pass through a second trichlorethylene spray for 15 sec, where they are rinsed of any residual contamination or loose soil clinging to the surface. From here they move to stage 4, a drain-dry process requiring about 30 sec.

The four stages are arranged in a cascade configuration, with each tank mounted higher than the one before. This allows the liquid to flow

Fig. 12·2 Testing metal thickness from one side by an ultrasonic meter. (*Branson*)

from the end of the line to the beginning, thus carrying soil to the least critical area.

Ultrasonics find many uses; some are rather unusual, such as the "meat finder." To operate it a small probe is held against the back of the steer, causing ultrasonic waves to be transmitted into the animal's body. These waves, although pitched so high that neither humans nor animals can hear them, do not cause hurt or damage any more than audible sound would. However, when these waves strike the boundaries between fat and lean, they literally bounce back and are picked up by the same probe.

The returning signals, as well as those that were sent, are changed to a visible pattern on a CRT. Depending on how deep the boundary is, the distance between these signals on the screen will vary, permitting direct and immediate interpretation of the fat layer.

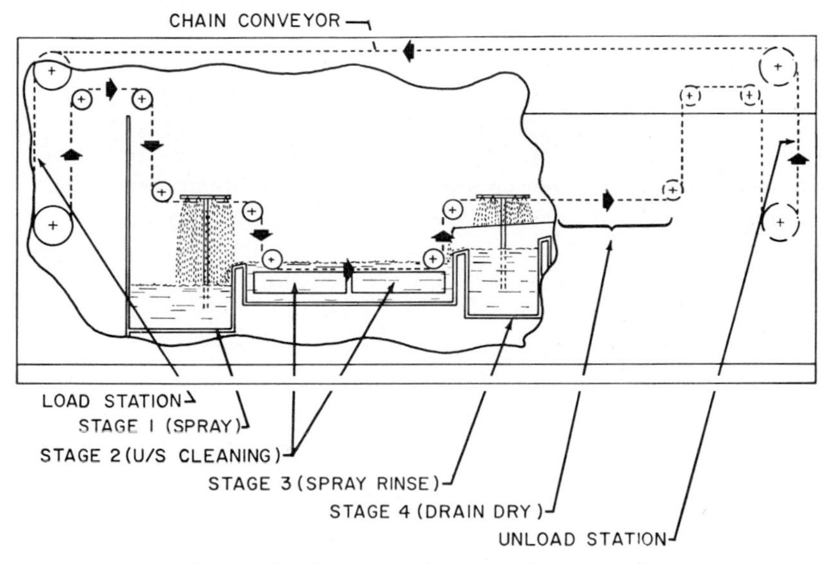

Fig. 12·3 Four-stage cleaning unit. (*Branson*)

By photographically tracing continuous measurements on the animal's back, a permanent record is obtained which produces a cross section of the loin eye and indicates the relative thicknesses of fat in the live animal.

The ultrasonic meat finder is, of course, an outgrowth of equipment originally designed and built for heavy industry. Used to detect flaws, laminations, or discontinuities in metals before they are fabricated into such critical structures as aircraft wings, the ultrasonic detector is a valuable device. Incipient cracks in railway-freight-car axles may be located by similar devices, and so may corrosion in fuel, oil, or storage tanks, even without emptying them.

Ultrasonics are also used in computers, where delay lines are used for data storage. An ultrasonic delay line is a rod of a transmitting medium which is either solid or hollow. Two crystals are used, one at each end. One is the transmitter and the other the receiver. A delay line is used to store information. In a computer this storage can be used as a memory element to retain information which is recirculated through the line until it is needed. In radar the delay line can be used in a moving-target indicator (MTI) system. Returning radar pulses are stored; since fixed objects always have an echo from the same position, these fixed radar returns can be stored from one sweep to the next. Electrically these signals are canceled out so that only moving targets are seen on the radar screen. Delay lines are useful for all types of electric timing and pulse formation. If a 10-μsec pulse is required, a line with this exact delay time will measure off a 10-μsec interval.

12·1 Principles If a loudspeaker cone vibrates at 1,000 cycles, we hear the musical note because the cone sends out sound waves through the air to our ears. But if the frequency is increased above about 20 kc, there are still sound vibrations in the ear which we cannot hear. Ultrasonics are these sound vibrations above the range of human hearing.

Low-power sound is used for inspection of manufactured products and for measurements. High-power sound is used for drilling, machining, and cleaning. A typical ultrasonic generator for cleaning small parts is a continuous commercial service unit which can deliver 500 watts into a resistive load over the frequency range from 12 to 50 kc.

It consists of a chassis on which are mounted high-voltage and low-voltage power supplies, an oscillator, and a power amplifier. A source of d-c polarization is also contained so that the generator may be used for either crystal or magnetostrictive transducer units.

12·2 Transducers A transducer converts electric energy into mechanical energy. Both magnetostrictive and piezoelectric transducers are used. A piezoelectric material, such as quartz, vibrates at its own resonant frequency when a voltage is applied across the quartz-crystal wafer. Electric energy is converted into mechanical energy; hence the crystal is a transducer or an energy converter. Such crystals are used in radio transmitters and receivers to provide a stable signal source for r-f energy in a crystal-oscillator circuit. Quartz, barium titanate, and other ceramics are used as ultrasonic transducers.

Acoustic power necessary varies considerably with different cleaning liquids and their characteristics. Under given conditions, the power requirement rises rapidly for frequencies above 50,000 cps.

Below 50,000 cps, the determining factor is audibility. Even if an operating frequency above human hearing is selected, during actual ultrasonic cleaning a broad band of waves is generated and the lower portion of this band may extend into the hearing range. For this reason, the frequency selected should be well above the audible range, but below the point where efficiency begins to fall off. In ultrasonic generators, vacuum-tube oscillators supply power to the transducers, which vibrate at the ultrasonic frequency to produce the useful output. The crystal is placed between two metal electrodes. Electric energy is converted into vibrations with efficiency of about 90 per cent.

Magnetostrictive transducers, the second type, are made of materials such as iron, cobalt, nickel, or alloys of these. When the transducer changes from one magnetic state to another, it also goes through a small change in size. This produces the vibrations.

Mechanical filters for electronics use this principle. The transducer shown in Fig. 12·4, which converts electric and mechanical energy, is a magnetostrictive device based on the principle that certain materials elongate or shorten when in the presence of a magnetic field. Therefore, if an electric signal is sent through a coil which contains the magnetostrictive material as the core, the electric oscillation will be converted into a mechanical oscillation. The mechanical oscillation can then be used to drive the mechanical elements of the filter. In addition to electrical and

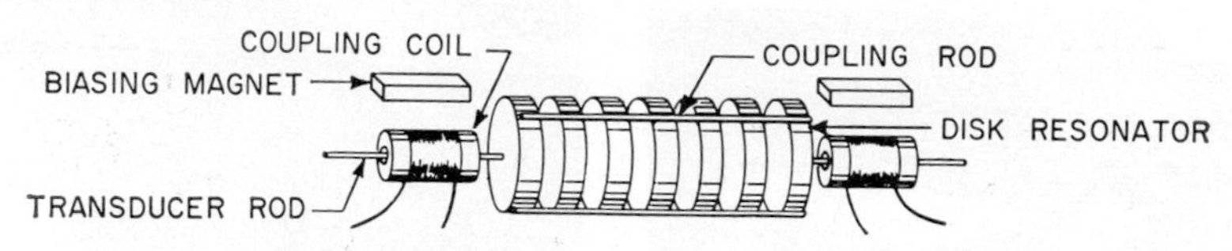

Fig. 12·4 Functional diagram of a typical mechanical filter. (*Collins*)

mechanical conversion, the transducer also provides proper termination for the mechanical network.

A transducer for ultrasonics (sometimes called "probe" or "searching unit") is a thin slice of piezoelectric material that reversibly converts energy between its electrical and mechanical states. It changes dimensions slightly in response to an alternating voltage and, conversely, generates a weak voltage when strained. It thus vibrates synchronously with the electric pulses sent to it and transmits mechanical pulses into the material. When the reflected acoustic energy returns from the interior of the test piece and acts on the transducer, it is converted back into electric pulses and returned.

The thin piezoelectric crystal is cemented into a supporting block of plastic and is protected against damage to its contact surface by a hard coating of aluminum oxide or other suitable material. The electrodes and leads are also encased in the plastic, and the entire assembly is mounted in a metal holder with a connector for the coaxial cable.

The crystals are cut to thicknesses for various frequencies between 0.4 and 10.0 mc. The lower frequencies generally have greater penetrating power, and the higher frequencies are preferred for the investigation of physical properties and the detection of minute flaws.

12·3 Applications Ultrasonics have five important applications. These are cleaning, inspection, detection, metal joining, and machining. Each is discussed below.

Cleaning. Cleaning with sound is one important application. Many intricate manufactured parts may be cleaned by ultrasonics, as on the aircraft shown in Fig. 12·5. Power from about 50 to 1,000 watts at frequencies between 40 and 50 kc is usually used. The parts (Fig. 12·6) to be scrubbed are placed in a cleaning solution. Ultrasound is applied through the transducer, which causes rapid motion of the solution in every crevice of the part. This motion aids the normal cleaning action, and all surface dirt, including oil, corrosion, and any foreign matter, is removed.

A small commercial model is illustrated in Fig. 12·7. The frequency is 40 kc. An oscillator in the unit supplies power to the transducer (barium titanate) in the bottom of the cleaning tank. The unit has an average power output of 125 watts.

Fig. 12·5 Ultrasonic cleaning is used for the parts of this aircraft. (*Gluton*)

Fig. 12·6 Glennite ultrasonic cleaner used by Lockheed. (*Gluton*)

Ultrasonic cleaning equipment consists of an electronic r-f generator and a transducer. The generator supplies the high-frequency electric energy which the transducer converts into high-frequency mechanical (ultrasonic) vibrations. The transducer radiating surface is vibrated at a high intensity to produce cavitation (formation of many small bubbles) in the cleaning solution, which provides an erosive action like scrubbing the surfaces of the part. This scrubbing action is effective even in small crevices, indentations, and holes where it would be very difficult to provide good cleaning by other methods.

Ultrasonics are normally used to remove loose chips, dust, and soils from slots, recesses, tapped holes, and similar locations which would otherwise require hand brushing. Parts requiring extreme cleanliness can be processed by using multiple ultrasonic stages with a continuous flow of clean solvent. Soluble soils can usually be removed in several seconds at low intensities. Slightly soluble soils and thin layers of insoluble matter usually require longer exposures. Printed circuits can be cleaned this way before they are dip-soldered. Figure 12·7*B* shows how the unit is connected to its cleaning tanks. A larger unit is illustrated in Fig. 12·8. Connections are the power-input plug, the interlocks (P_2, J_3), and cables to the two transducers from J_1 and J_2. T_1 has two windings, one for each filament of the two triodes in parallel. T_2 provides the high voltage, and both primary legs are fused. The schematic diagram is somewhat unusual and it is redrawn in B for clarity. The anodes are in parallel and are at almost ground potential through the output coil. One side of T_2 goes to one cathode and the other side goes to the cathode of the other tube. R_5

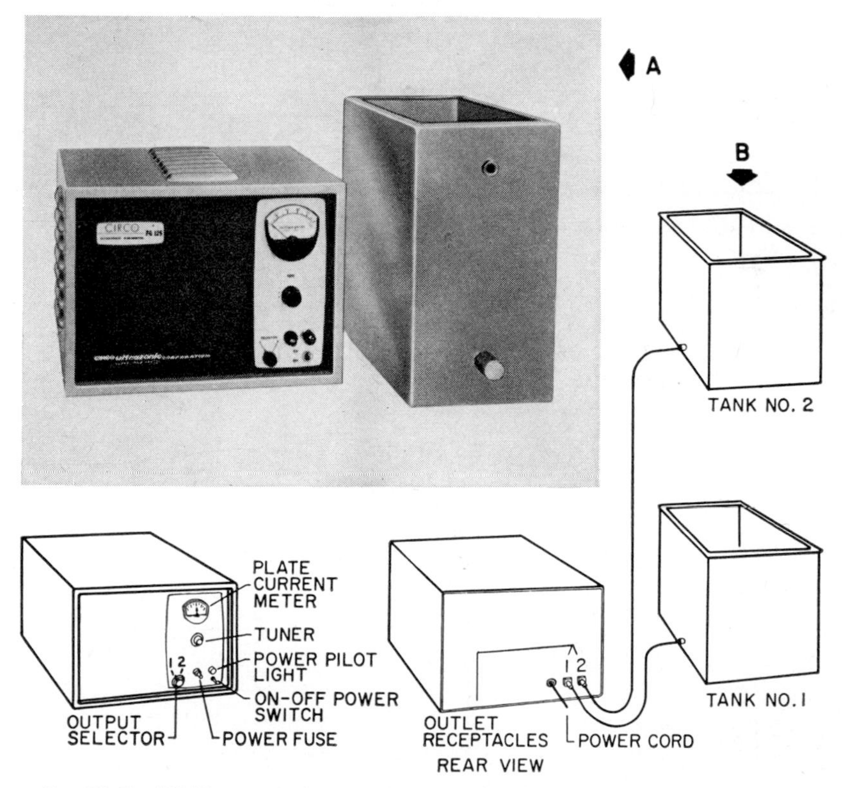

Fig. 12·7 (A) Ultrasonic cleaner PG-125 and tank; (B) wiring diagram. (Circo)

provides some cathode bias for these tubes. High-voltage alternating current is applied to the cathode, which causes the tubes to act as self-rectifiers. Of course, making the cathode negative causes exactly the same effect as making the plate positive. These cathodes conduct on successive alternations just as a full-wave rectifier does, but the tubes have their plates in parallel.

Grid bias is obtained from the drop across R_1 and R_2, from grid to cathode, together with the drop across R_5 in the cathode circuit. C_7 provides the feedback for this oscillator.

Ultrasonic cleaners are made for special purposes. A "white room" cleaner is used in special humidity-controlled, dust-free construction areas where critical assemblies such as missile parts are put together. This unit is designed for cleaning applications in such a room.

A degreaser is a unit especially suited for the cleaning of intricate precision parts which require the ultimate in cleanliness. Ultrasonic cavita-

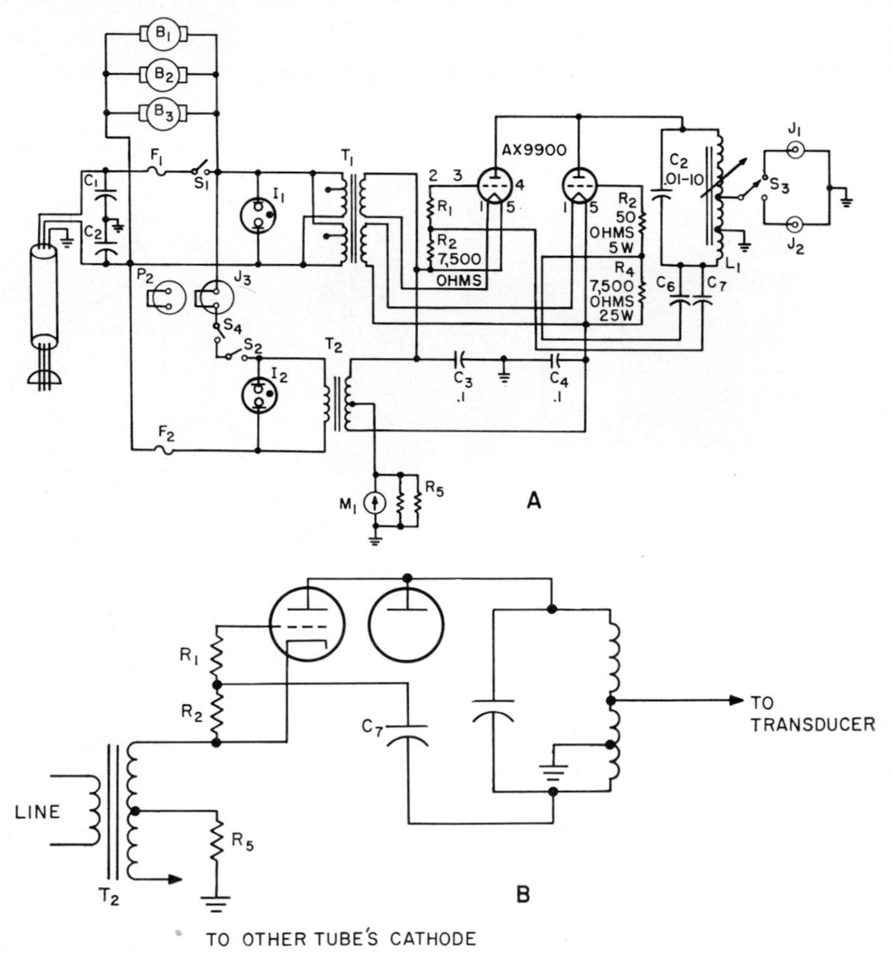

Fig. 12·8 (A) Ultrasonic-cleaner schematic diagram; (B) schematic detail showing oscillator. (Circo)

tion reaches deep into the most minute crevices with its effective cleaning action. In the cleaning of electronic components, gears, filters, nozzles, motors, and assembled units, degreasers are often used.

The degreaser has two chambers—a boiling-liquid chamber and an ultrasonic chamber—with the vapor area over both. The work to be cleaned is lowered into the boiling liquid, where most of the grease, oil, polishing compounds, and contaminants are removed. The parts are then transferred to the temperature-controlled ultrasonic chamber where the ultrasonic sound waves are directed against all surfaces of the work to remove all traces of dirt and contamination. The temperature of the ultrasonic chamber is held at approximately 140°F. A final period in the vapor area completes the cycle and the work is removed, clean and dry.

Inspection. Testing materials is another use of ultrasonics. In one form of nondestructive testing, flaws are located in metals and other materials. Using a frequency between 500 kc and 100 mc, the ultrasonic transducer applies a pulse to one end of the manufactured piece to be inspected. A

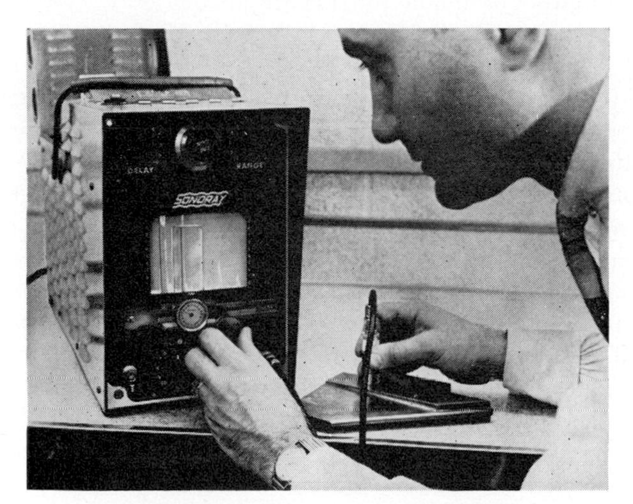

Fig. 12·9 Sonoray tester. (*Branson*)

receiving transducer picks up the energy sent through the work; a visual display (Fig. 12·9) is used to provide an indication of the received pulse. Any flaw such as a hidden crack in the metal workpiece will alter the displayed pulse and show the presence of the flaw. Small pieces to be inspected may be submerged in a suitable liquid.

Thickness of metals may also be measured by ultrasonics. Typical applications include aircraft propellers, large flat sheets of aluminum, hollow aluminum extrusions, milled and tapered aluminum sheets, castings, plate glass, seamless tubes, lead sheath, ship plate, bulkheads, and storage tanks.

Laminar discontinuities can be detected in clad metals, brazed assemblies, plate, strip, and tubing. For remote testing, extension cables up to 1,000 ft are used.

For nondestructive testing a beam of short pulses of acoustic energy at ultrasonic frequencies (above the audible range) is sent into the material under test. Each pulse normally completes a round trip back to the detector before its successor starts. Behavior within the material is shown by the trace on a CRT. Boundaries encountered by the sound beam reflect some of its energy. Any remainder is eventually reflected back to its starting point or picked up by a second (receiving) transducer, or it is lost by multiple internal reflections.

The trace is substantially linear in both time and amplitude. Pips in the trace indicate, in succession, reflections from the various discontinuities encountered; their relative heights show the strength of the returned portion of the beam. Distances to flaws and geometric boundaries can be determined accurately by reference to the scale over the tube face or to the pyramid marker available for superimposition on the basic line. With angle-beam tests, the marker is especially valuable for ease and accuracy in locating discontinuities.

The two techniques are *reflection testing*, where echoes or signal returns are detected to locate defects, and *transmission testing*, where the ultrasonic signals pass through the material and their attenuation is measured.

Figure 12·10 shows the reflection form of testing. In *A* the electric impulse is formed and passes through the cable to the transducer in contact with the surface of the test piece. The transducer converts the electric impulse to mechanical vibrations, and as they enter the material the first pip (initial pulse) appears on the CRT. In *B* the mechanical pulse has reached an internal flaw and a portion of the pulse energy is returned to the transducer by reflection, forming the second (or flaw-indication) pip. Remainder of the energy continues toward the back surface. *C* shows the remaining portion of the pulse energy, reflected from the opposite boundary, which has returned to its starting point at the transducer, been converted back to electric energy, and acted on the circuits to form the third (or back-surface) pip in the trace.

Since the pulses are generated in rapid succession and there is considerable persistence of vision and CRT phosphor, the baseline and pips appear simultaneously as a continuous, composite trace varying in shape only as the transducer is moved, as shown in *D*.

For transmission testing, as shown in *E*, two transducers are arranged coaxially, facing each other on opposite sides of the test piece. Pulses are formed and propagated as described above, but are picked up and returned by the second, or receiving, transducer. Ordinarily, only the first signal received is considered, and possible echoes are ignored. The reduction in amplitude of the received signal indicates the presence and size of discontinuities, but not their location in depth.

Detection. Burglar alarms use ultrasound. The room to be protected is filled with ultrasonic radiations from a transmitter; a receiver, which is the second transducer, picks up a portion of this sound reflected from objects in the room. In this system of detection a change in the frequency of the received sound, caused by the doppler effect, occurs whenever anything in the room moves.

If you ever stood at a crossing while a train passed with its whistle blowing, you probably noticed that as the locomotive passed you, the tone of the whistle dropped sharply. While the train is approaching, each sound wave from the whistle starts from a point slightly nearer to you than the previous one and reaches your ear sooner than it would have if the whistle were not moving. The frequency is higher when the whistle is approaching, and your ear hears this as a rising tone. As the train

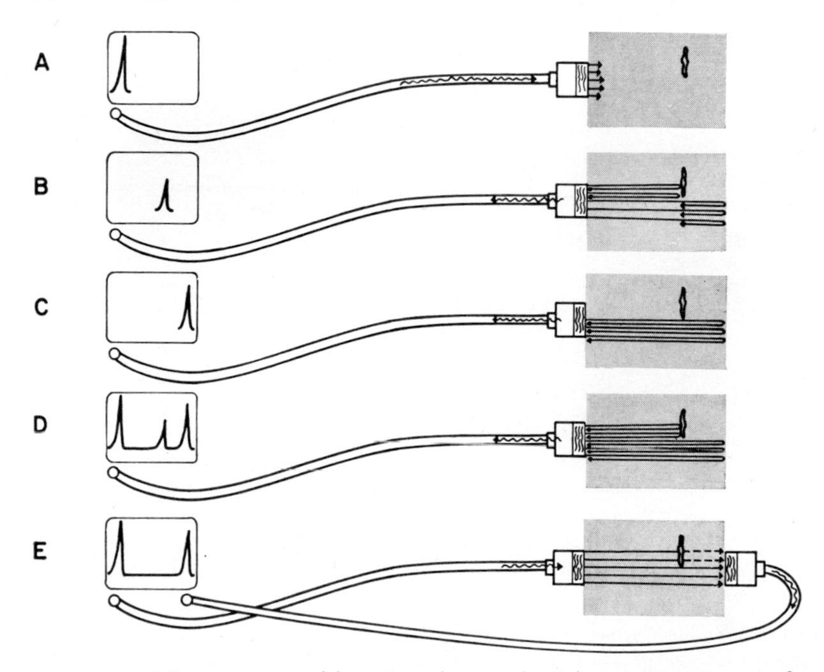

Fig. 12·10 (*A*) Pulse is transmitted from transducer to flawed part; (*B*) energy is reflected from flaw; (C) energy is reflected from boundary; (*D*) both reflections are received; (*E*) transmission of ultrasonic energy through flawed material. (*Branson*)

moves away from you, your ear receives fewer sound waves per second and you hear a falling tone. This peculiar effect which you get whenever a source and a receiver of waves are moving toward or away from each other is the doppler effect. It does not matter which of the two sources is moving; if the distance between them is changing, you will get a doppler effect.

When the source of sound and the receiver are both stationary, you can still get a doppler effect at the receiver if some object in their vicinity that reflects sound (and all objects do) is moving. This is the principle upon which the ultrasonic alarm operates.

Metal Joining. Ultrasonics have solved some of the most difficult problems in joining special metals. For example, two pieces of copper are easily joined together by the use of tin-lead solder. But there are many cases where flux, used to remove the oxide surface from the copper, is undesirable because of the post-soldering corrosion.

Ultrasonic soldering does not use flux; hence there is no cleaning problem. The energy from the transducer breaks up and disperses oxide and other films on the surface of the metal.

A typical small ultrasonic soldering unit is shown in Fig. 12·11. The electronic generator provides high-frequency alternating current to drive the transducer, which is inside the soldering head. The tip on the unit is

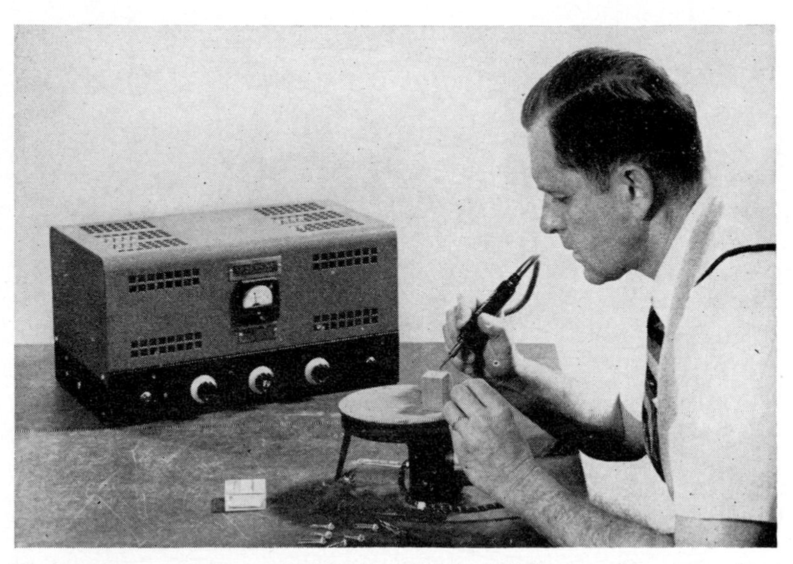

Fig. 12·11 Twelve-watt Sonosolder equipment being used to tin the surface of a brass contactor. *(Aeroprojects Inc., Sonobond Corporation, West Chester, Pennsylvania)*

the coupler for transmitting the energy to the work. The soldering head has a compressed-air supply for cooling. The workpiece and solder are heated by the platen and by a tip heater on the soldering unit itself.

Aluminum, becoming important as a substitute for copper conductors, can be coated with solder; the two pieces of tinned aluminum are then joined by heat and pressure. No flux is used.

Copper and aluminum wire can also be tinned in a continuous process by passing the wire through a solder pot which is receiving ultrasonic energy.

Brazing is used where the melting temperature of solder is too low for a given application. Metals with melting points between 850 and 1250°F have been used to braze copper and aluminum. A brazing unit resembles soldering equipment; however, a different type of tip is used for brazing and greater cooling is required because of the higher temperature. A small pool of brazing alloy is activated into which the joint is dipped.

Zirconium (an alloy used in atomic-energy equipment) and titanium (which is useful in jet aircraft) can both be brazed with ultrasonics. Welding of foil conductors and wires has also been demonstrated successfully. Both aluminum and copper have been welded to other metals. Ultrasonics are used for welding foils together and for welding foils to thicker pieces.

Machining. Ceramics and other hard, brittle materials are very difficult to drill and machine by ordinary power tools. But the increasing use of ceramics in many types of electronics has created the need for formed and machined ceramic parts. For this a drill bit is used with an action like a star drill used on masonry. The tool face is moved against the surface by rubbing, to produce a slicing action, or at right angles to the surface, as with a hammer to drill holes. Liquid abrasives are used between the work surface and the tool to carry away the bits of ceramic material as they are chipped off, to cool the vibrating tool, and to provide a good impedance match between the tool and the work face. Results are shown in Fig. 12·12.

High-temperature ceramic parts (such as printed-circuit boards) for missiles are drilled and shaped by ultrasonics.

All types of materials which are difficult to machine are being formed by ultrasonics. These include glass, quartz, ceramics, sapphire, tungsten, and semiconductors.

Ceramics may be machined ultrasonically after firing. Thus dimensional tolerances can be held closely. Multiple holes can be drilled at one time, thus reducing costs. The wide use of glass for mechanical components has been greatly helped by the development of ultrasonic machining. Ultrasonics permit much more rapid machining and almost limitless freedom in design when working on glass.

Ultrasonic slicing and dicing has replaced diamond sawing in the production of pellets for transistors and other electronic components. Again, many cuts can be made at one time with resulting machining economy. Ruby, cameo, jade, and many other precious stones can be machined ultrasonically. Many unusual effects can be achieved. Assymetrical shapes can be readily produced in carbide and hardened steel in much less time than is required by conventional methods. Typical applications are wire-drawing dies, extruding and blanking dies, and heading and stamping dies. Die repair work can often be done very economically by ultrasonics since the technique will work on the hardened tool and, because no heat is generated, no rehardening is necessary.

Other Applications. Ultrasound is found in other applications as well as those described above. Sonar, used to detect obstacles and determine depth for the Navy, has an industrial counterpart, for small boats, in the depth finder (Fig. 12·13). The indicator is on the left and the transducer, mounted through the boat's hull, is also shown. Ultrasonic pulses are sent down and returned. Their time for the round trip is a measure of the water's depth or the boat's distance from an obstacle underwater.

Another application is an ultrasonic caliper, which may be used as an ultrasonic flaw detector to measure thickness of metals and plastics, from one side, to accuracies within ± 0.010 in.

Operation of the caliper is simple. It consists essentially of a fixed, scribed, vertical line and a second line on a slider, movable across the face of the CRT. A synchronized dial indicator measures the amount of this motion, permitting the operator to determine thickness directly, without calculations or interpolation. Pulse-echo equipment is principally used

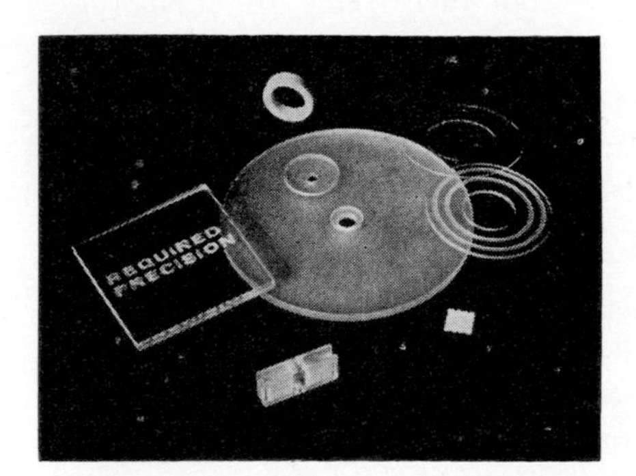

Fig. 12·12 Ultrasonic cutting of hard materials. (*Connecticut Instrument*)

Fig. 12·13 Depth finder for underwater sounding. (*Raytheon*)

to ascertain the presence of hidden discontinuities, voids, slags inclusions, and severe porosity, but the caliper permits accurate location of these and similar flaws. Knowing how deeply these imperfections penetrate into sound metal often makes the difference between scrapping the entire component or salvaging it by machining off the damaged sections.

A reference block of known thickness—and of the same material as the object to be tested—is used to calibrate the instrument. First the caliper is set to the known reading. Then, by means of the delay and range controls, both the ingoing and returning signal indications (traces) on the oscilloscope are adjusted until their leading edges coincide with the two vertical caliper lines. This provides a 1:1 ratio, so that any measured thickness will now be read directly by the indicator setting.

A flaw alarm for electronic monitoring is another device used with ultrasonic pulse-echo flaw detectors. Test signals which appear within preset limits are detected by the flaw alarm to actuate recorders or warning devices. Also, the alarm signal may be used to initiate corrective action automatically.

A flaw alarm lets the operator pay closer attention to materials under test, instead of constantly having to watch the CRT.

12·4 Industrial ultrasonic alarm system The doppler effect and ultrasonic frequencies may be combined to make possible a total-area protection system. The burglar alarm shown in Fig. 12·14 produces a sound wave of 19,200 cycles, which is well above the audible range for most people. This generated signal fills the area to be protected and is picked up by an ultrasonic receiver. Any intrusion or movement in the room, even a fire, could change the frequency of the received signal and trigger the alarm.

The doppler effect is usually explained by the train-whistle example described earlier. Any relative motion between the source and the receiver

Fig. 12·14 Ultrasonic alarm. (*Walter Kidde*)

will result in a changing frequency. Even if both are motionless, a moving object such as an intruder will reflect a changing frequency signal to the receiver, which can trigger an alarm. A moving intruder changes the echo note and causes the alarm to sound. A typical set of installations is shown in Fig. 12·15. Note the use of multiple transmitters and receivers. Security regulations for a drafting room having an area of about 4,000 sq ft required all drawings to be placed in storage cabinets when not in use. This resulted in a great loss of time; each evening the drawings were stored and each morning they were returned. The area had to be protected at all times, which was being done by guards around the clock.

An ultrasonic alarm system was used to protect the entire cubic content of the area. It can detect anyone who might have concealed himself within the area before closing time, as well as anyone attempting to enter. The drawings are now permitted to remain on the drawing boards at night. Lightweight fabric covers are used to make it impossible to photograph them through a window. Alarms are registered on a monitor unit which is located in a central guard station.

Another situation required protection of a multiengined classified aircraft. Because of the secret classification of materials and equipment installed in the airplane, it was necessary to protect it in such manner that the approach of anyone would be detected. Either an armed guard or an electronic system would have been required. However, as the airplane might be standing in any one of a number of locations, a permanently installed system could not be used, so transmitters and receivers were mounted on small wooden stands. All elements could then be positioned around the plane and the system put into operation in a few minutes.

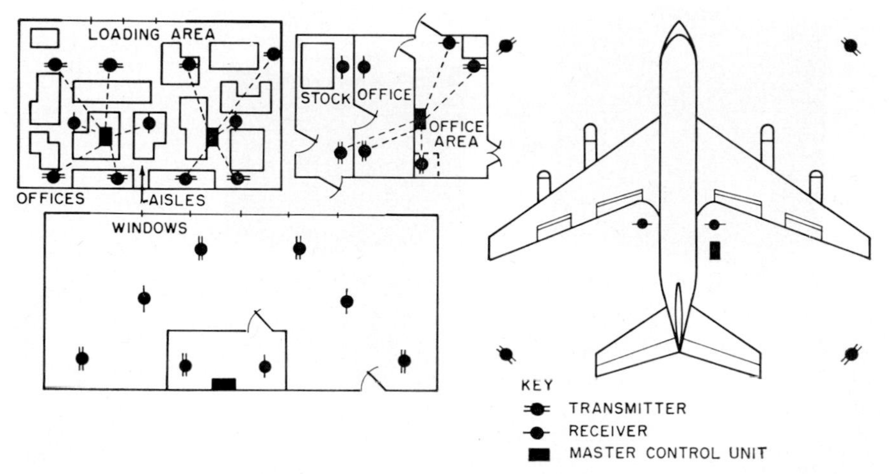

Fig. 12·15 Typical ultrasonic equipment layout. (*Walter Kidde*)

Any approach to or movement within the protected area could be detected and indicated on a distant monitor unit.

The complete schematic of the master control is shown in Fig. 12·16. The power supply uses a Sola constant-voltage power transformer which permits operation on line voltages from 100 to 130 volts at 60 cycles. The power transformer has three secondaries. Because the circuit is sensitive to 60-cycle hum, the filaments of V_2 and V_3 are supplied with direct current from a full-wave bridge rectifier S_1.

The high-voltage winding feeds a 6X5 full-wave rectifier in the conventional manner. The cathode is connected to a capacitor filter (C_2) and the voltage-divider network R_2 and R_3. This furnishes plate voltage for the relay tube V_3. The output of the rectifier also feeds a two-stage *RC* filter (R_4, C_3, R_5, and C_4) which supplies plate voltage to the amplifiers and the oscillator.

The first two amplifier stages (V_{2a} and V_{2b}) have a decoupling network R_6 and C_5 in their plate supply. A separate voltage-dividing network and filter (R_7, R_8, and C_6) is connected to the oscillator plate supply. The voltage from C_6 is also used to supply proper magnetic bias for the

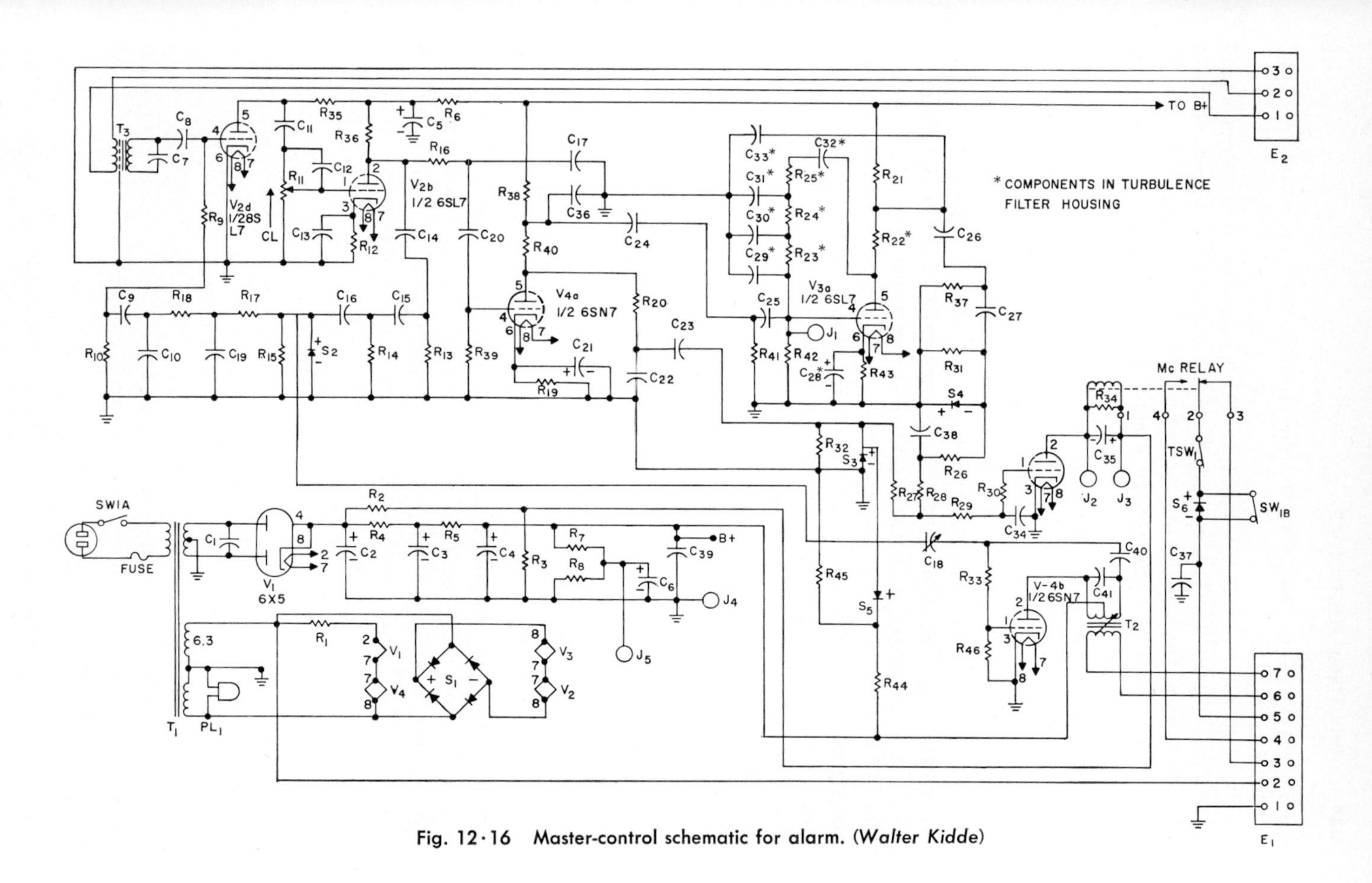

Fig. 12·16 Master-control schematic for alarm. (*Walter Kidde*)

operation of the magnetostriction transducers (microphones and speakers).

Tube section V_{4b} is the oscillator, which operates at a frequency of approximately 19 kc. The oscillator-output voltage is stepped down to 6 volts by transformer T_2 and is fed to terminals 6 and 7 on the E_1 strip. From there it is connected by cable to the windings of the transmitting transducers.

The 19-kc sound wave, radiated by the transducers, is picked up by the receiving transducer and converted to a 19-kc electric signal which is fed to the control at terminals 1 and 2 of E_2.

The signal is then fed into a step-up transformer T_3, the secondary of which is tuned to the oscillator frequency. Its output is coupled by C_8 to the grid of V_{2a}. The grid resistor R_9 is connected to the biasing resistor R_{10}. The 19-kc signal is again amplified by V_{2b}. Capacitors C_{14}, C_{15}, and C_{16} and resistors R_{13}, R_{14}, and R_{15} form a high-pass filter which couples the signal to the detector S_2. The network of R_{16} and C_{17} is a low-pass filter. The 19-kc signal does not appear at the grid of V_{4a}. The selenium diode is connected across R_{15}. At this point, the 19-kc signal from V_2 is beat against the original signal coupled back from the oscillator through capacitor C_{18}.

A moving object in the projected area causes a portion of the received signal to be of a different frequency from the transmitted signal. Any variations in frequency, phase, or amplitude will modulate the received 19-kc signal. The low-frequency modulation component appears across resistor R_{15}. The frequency of the variations in the received signal caused by an intruder will range from about 15 to 100 cps. Air turbulence will cause variations at lower frequencies, which can be filtered and compensated for, so as not to prematurely trigger the alarm.

The low-frequency signal resulting from mixing and detecting goes through the low-pass filter R_{17}, C_{19}, R_{18}, and C_{10}. It is then coupled to the grid of V_{2a} by capacitor C_9 and resistor R_9. This simultaneous amplifying of low and high frequencies by the same tube is known as "reflexing." V_2 is operated as a two-stage reflex amplifier. The amplified low-frequency signal then goes to the high end of the sensitivity control R_{11}. At these low frequencies, C_{12} appears as an open circuit and the setting of this control determines the amplitude of the low-frequency signal applied to the grid of the next stage. From the plate of this stage the low-frequency signal is fed through a low-pass filter R_{16} and C_{17} and coupled by C_{20} to the third stage V_{4a}. Since capacitor C_{14} is part of the high-pass filter, none of the amplified low-frequency signal can be coupled anywhere else but to the third stage.

The output from the third amplifier is divided into two portions. A low-frequency portion, about 5 cps, is fed through the low-pass filter

R_{20} and C_{22}, through C_{23} to the diode S_3 where it is rectified to give a signal that is positive with respect to ground. The higher-frequency component of the signal is coupled to the grid of a V_{3a} through C_{24} and C_{25}. The plate of a V_{3a} is coupled through R_{21}, C_{26}, and C_{27} to the diode S_4 where it is rectified in a direction that is negative with respect to ground. In order to improve the selectivity of V_{3a}, a negative-feedback filter (consisting of C_{28}, R_{22}, R_{23}, C_{29}, R_{24}, C_{30}, R_{25}, C_{31}, C_{32}, and C_{33} may be used. The filter is mounted in a plug-in case which fits into the V_3 socket; V_3 in turn plugs into the filter. Thus its use is optional, but it is advised wherever a condition of severe turbulence is encountered, such as that caused by space heaters.

These rectified negative and positive signals are added in the proper ratio, to overcome turbulence, by the resistance network R_{26}, R_{27}, and R_{28}. The resultant voltage is taken from the junction of R_{27} and R_{28} and fed to the grid of the relay tube V_{3b}. Resistor R_{29} and capacitor C_{34} serve as a filter circuit. Resistor R_{30} limits the grid current if the voltage across capacitor C_{34} becomes positive. In order to prevent the positive compensating signal from becoming excessively large, when the negative signal has reached a high enough value to saturate V_{3a}, a limiting circuit is inserted across the rectifier S_3. This circuit is a diode (S_5) used as a positive limiter. The voltage-divider network, composed of R_{44} and R_{45}, provides a positive bias on rectifier S_5. Rectifier S_5 will not conduct until the positive signal developed across S_3 exceeds the biasing voltage. When the voltage is exceeded, S_5 conducts and limits the positive potential that can be developed across S_3. In this manner, the positive signal can compensate but cannot overcompensate.

The negative resultant of the alarm signal is fed to the grid of V_{3b} and used to control the alarm-relay coil which is in the plate circuit. Capacitor C_{35} and resistor R_{34} are across the coil and act as a filter (to prevent chatter) and voltage limiter, respectively. Operation of the unit with the relay removed is to be avoided.

When not in the alarm condition, the relay coil is continuously energized. Jacks J_2 and J_3 are used to insert a high-resistance voltmeter to monitor the voltage across the relay coil. A voltage reading compared with the minimum energizing relay voltage will indicate how safely above the dropout point the relay is operating. Normally closed or normally open contacts are part of the relay and are available at the E_1 strip. An added safety factor is built into this unit by wiring the filaments of some of the tubes in series. If, for example, the oscillator tube V_4 had an open filament, the rectifier tube V_1 would be disabled and the alarm would sound.

This unit operates at 19 kc; this frequency may be compared with others for different ultrasonic applications in Table 12·1.

Table 12·1 Ultrasonic Frequencies and Their Uses

TRANSDUCER MATERIALS

Material	Frequency range	Max safe operating temperature, °C	Typical uses
Piezoelectric			
Quartz	100 kc–35 mc	550	Medical and nondestructive testing
Barium titanate	5 kc–10 mc	100	Most cleaning and processing applications
Lead zirconate or titanate	5 kc–10 mc	320	Most cleaning and processing applications and high-temperature uses
Rochelle salt	20 cps–1 mc	45	Sonar and depth finding
ADP	20 cps–1 mc	100	Sonar and electrooptics
Magnetostrictive			
Nickel	10 kc–100 kc		Ultrasonic cleaning, machining, drilling, soldering, melt treatment, and applications where transducer has pressure applied
Vanadium permendur	10 kc–100 kc		Same as nickel
Ferrite	10 kc–1 mc		

POWER APPLICATIONS

Application	Frequency
Brazing	15 kc–70 kc
Chemical processing	1 kc–1 mc
Cleaning	10 kc–1 mc
Coagulating	3 kc–40 kc
Degassing	10 kc–500 kc
Degreasing	10 kc–1 mc
Dispersing	1 kc–1 mc
Electroplating	10 kc–500 kc
Emulsifying	10 kc–500 kc
Impregnating	1 kc–1 mc
Machining	15 kc–40 kc
Medical therapy	100 kc–1 mc
Oil-well drilling	5 cps–30 cps
Precipitators	3 kc–40 kc
Pickling	10 kc–500 kc
Plastic welding	20 kc–90 kc
Shake table	15 kc–100 kc
Soldering	15 kc–70 kc
Welding	4 kc–40 kc

INSTRUMENTATION AND SENSING

Application	Frequency
Burglar alarms	20 kc
Flaw detection	40 kc–100 mc
Flow metering	10 kc–10 mc
Liquid gauging	81 kc–10 mc
Medical	1 mc–15 mc
Oral dentistry	25 kc
Remote control	15 kc–25 kc
Sonar	50 cps–100 kc
Thickness gauging	100 kc–50 mc

SOURCE: Ultrasonic Manufacturers Association.

REVIEW QUESTIONS

12·1 Name eight uses of ultrasonics.
12·2 How does sonar operate?
12·3 What is the principle of the thickness gauge?
12·4 Why is ultrasonic cleaning used?
12·5 Why are mechanical filters used?
12·6 What is a delay line? Where is it used?
12·7 How does an ultrasonic transducer operate?
12·8 What are some of the transducer materials?
12·9 What are the two types of ultrasonic transducers?
12·10 How does the piezoelectric transducer operate?
12·11 What is cavitation?
12·12 In Fig. 12·8, what are the functions of T_1, R_2, and C_7?
12·13 How does an ultrasonic degreaser operate?
12·14 What is reflection testing?
12·15 How is this used?
12·16 How can an alarm be made from ultrasonics?
12·17 How is transmision testing used?
12·18 What is the advantage of ultrasonic soldering?
12·19 Where is it used?
12·20 How can machining be done with ultrasound?
12·21 What is an ultrasonic caliper?
12·22 In Fig. 12·16, what are R_{43}, S_1, and R_{48} used for?

CHAPTER

13

Radiation Inspection and Detection

Warning IT IS IMPORTANT THAT EVERYONE HAVING ANYTHING TO DO WITH X-RAY RADIATION BE FULLY ACQUAINTED WITH THE RECOMMENDATIONS OF THE NATIONAL COMMITTEE ON RADIATION PROTECTION AS PUBLISHED IN THE NATIONAL BUREAU OF STANDARDS HANDBOOK, OF THE AMERICAN STANDARDS ASSOCIATION, AND OF THE INTERNATIONAL COMMISSION ON RADIATION PROTECTION, AND TAKE ADEQUATE STEPS TO ENSURE PROTECTION.

13·1 Uses of X rays X rays are used in many industrial applications, from inspection of opaque materials to chemical analysis. Three examples are welding inspection, thickness gauges, and specialized uses such as container inspection.

A noncontacting X-ray gauge can be used for measurement of pipe wall thickness. Specifically, this gauge can be used to indicate continuously the wall thickness of pierced, extruded, butt-welded, and stretch-reduced pipe and tubing, hot or cold; to detect eccentricity of pipes; to indicate weight per foot; to automatically mark the crop point; to automatically mark off-gauge areas; and to inspect for pulled-down wall, heavy ends, and defective welds. It is also useful for measuring walls of high-pressure cylinders. Any metal such as steel, brass, copper, aluminum, titanium, or nonmetallic material in diameters from 1 to 30 in. can be inspected.

A pencil of X-ray energy of fixed intensity is beamed through the wall or walls being measured, and the transmitted electromagnetic energy is collected by a receiver of rugged design which converts the received signal into an electric impulse which is in turn transmitted to the main electronic console. Here the amount of absorption provides signals to actuate either a marking device to define off-gauge areas or automatic alarms, if desired.

In weld inspection the tube, which is the source for the X rays, is inside the large-diameter cylinder. X rays passing through the welded seam will detect flaws before they cause a failure in operation.

The amount of material in a container after it has been filled can be determined within close limits by two X-ray beams through the container to detectors.

Some X-ray equipment is very large and used only in fixed installations, but other units are portable and can be carried to the work or even, in some cases, placed inside a large boiler or large casting.

Large X-ray equipment is shown in use in Fig. 13·1; *A* is an industrial installation where the operator, inside the large cylinder, is placing a film to the inner wall. X rays from the gunlike unit will penetrate the cylinder and provide (on the film when it is developed) a picture of the internal structure of this cylinder. Note that a rail-mounted crane permits the X-ray unit to be moved to the proper location as required. This unit has an X-ray tube.

With the advent of nuclear energy, certain radioactive substances have become available as radiation sources. Figure 13·1*B* shows a cobalt 60 radiographic unit as used to inspect a 10-in. stainless-steel casting for potential defects. This unit can be moved, inside the plant, by the wheeled dolly shown in Fig. 13·1*C*. Here the operator is positioning the film pack. After setting the proper time exposure, the operator moves to a safe position and the remote control takes over. After the exposure the radiation source, inside the head, moves to a safe position.

The units in Fig. 13·1*B* and *C* use gamma rays from isotopes. Radioactive isotopes (cobalt 60, iridium 192, and cesium 137) emit gamma radiation which, for all practical purposes, is equivalent in penetrating power to X radiation in the range of 1 to 3 Mv. Gamma rays are simply high-energy X rays given off by radioactive isotopes. Regardless of how radiation is generated, either electrically or spontaneously by nuclear disintegration, its penetrating power at similar energy levels is the same. The gamma radiation from isotopes is of such high energy that it is useful only for high-density materials—metals like steel, brass, bronze, and lead. X rays, on the other hand, can be generated at voltages as low as 30 kv for the radiography of materials of very low density.

Electromagnetic radiation in the form of radio waves is, of course, familiar from the lowest radio frequency through vhf and uhf television and up to microwaves as used for communications and radar. Light is radiation much higher in frequency than radio, but it is a special form

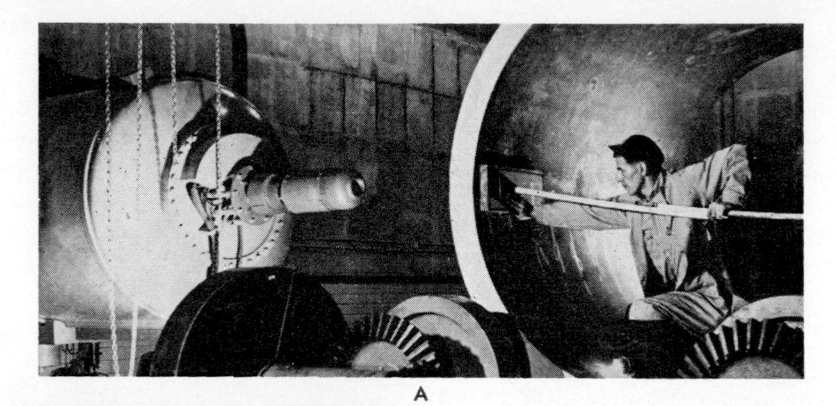

A

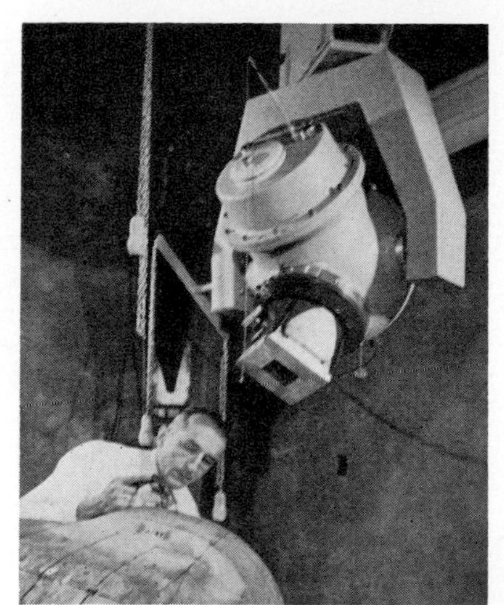

B

Fig. 13·1 (*A*) X-ray inspection of large welds (*General Electric*); (*B*) inspector setting up Cyclops cobalt radiography unit for radiography of 10-in.-thick stainless-steel casting (*Picker X-Ray*); (C) use of radioactive source for inspection (*Picker X-Ray*).

C

of electromagnetic radiation for detection. Our eyes detect this frequency of radiation.

Higher in frequency than light are the various types of nuclear radiation. Some elements such as radium and uranium are not stable in their natural state. Their actual physical structure constantly changes through decomposition; during this process energy is emitted by these materials. The particles emitted by radioactive materials are used as a means of detecting the presence of the materials. This radiation consists of at least three types of energy particles: alpha, beta, and gamma.

Alpha rays have high energies and a range which is quite short, only a few inches. As the radioactive materials disintegrate and change into new substances, the alpha particles are emitted. Their form has been determined to be that of a helium atom without the orbit electrons, or the same form as a helium nucleus. In helium there are two electrons in orbits around the nucleus. The nucleus has two protons, each with a positive charge, and two neutrons without any electric charge. Normally the two negative charges of the two external electrons are balanced by the two positive charges of the protons in the nucleus. However, the particles emitted by radioactive substances are helium nuclei and hence carry a double positive charge. They are easily absorbed, however, and a piece of ordinary paper will stop alpha particles.

Beta rays are high-energy electrons of greater force than alpha rays, and they travel far greater distances before they are stopped by collision with other materials, including air.

Gamma rays are electromagnetic radiations of extremely short wavelength. They are a type of X radiation which is very powerful and moves easily through most materials.

Portable X-ray units are available with capacities ranging from 90 to 270 kv. Such machines are suitable for a vast range of work in the plant and in the field. This is the sort of machine used for on-the-spot radiography of welds in cross-country pipelines. The apparatus is self-contained and is light enough to be handled by one man. Of course it needs a source of power, which in routine field work is supplied by a truck-mounted motor generator.

Over the 270-kv level, X-ray apparatus is generally of the stationary type installed in shielded radiographic rooms to which the work is brought either as single pieces or on a production-flow conveyor system.

Isotope gamma radiography machines, on the other hand, are very compact. They consist essentially of a storage pig of lead, thick enough to safely contain the radioactive source, and a mechanical means of moving the source out of the pig and into exposure position by remote control. They are completely self-contained: they need no power lines to operate them or water lines to cool them.

Because X rays are in wide use for inspection, fully portable units

(Fig. 13·2) have been developed. This two-package unit has the X-ray tube and all high-voltage components in one package and all the necessary controls in the other. Portable units similar to this are used in places where larger equipment cannot reach or for work which is so big that it cannot be brought to the equipment.

For inspection purposes the image of the X rays can be viewed in two basic ways as shown in Fig. 13·3. For some inspection of materials the X-ray source passes through the specimen, as shown in *A*, and falls upon a sheet of film. After development, any extra-dense sections or any voids in the castings will show up on the developed film. A fluorescent screen can

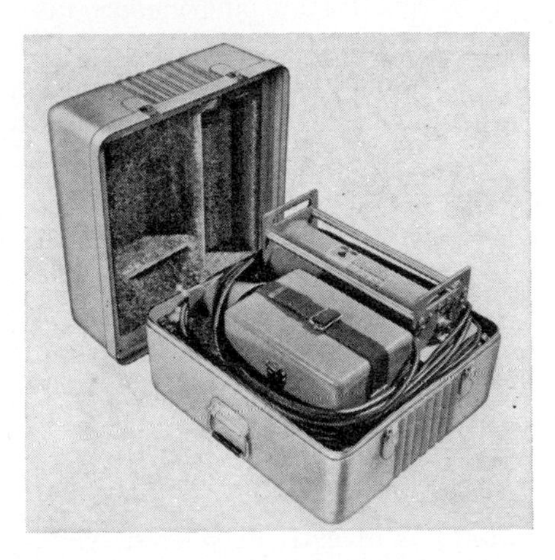

Fig. 13·2 Portable X-ray unit. (*General Electric*)

be used in place of the film if many objects are to be viewed in a short time and if no permanent record is required of each inspection.

In the second method an electronic image intensifier is used as shown in *B*. The X-ray beam penetrates the object to be inspected and the glass wall of the cylindrical evacuated tube, reaching the fluorescent screen mounted in one end of the tube. The fluorescent or input screen transforms the received X radiation into a conventional fluoroscopic image. The light of this image frees electrons from a sensitive photocathode which is in close contact with the input screen. On each point of the photocathode, the number of freed electrons per second is proportional to the brightness of the input screen at that point, and to the X-ray intensity. In this way the latent X-ray image with all its variations in intensity is converted into an electron image with corresponding variations in electron density just as in a TV camera tube.

The electrons released from the photocathode are accelerated by high voltage applied between the photocathode and the open anode positioned at the far end of the tube. The electron image is electrostatically brought in focus on a second fluorescent screen, or the viewing screen, which converts the energy of the impinging electrons into visible light. The viewing

A

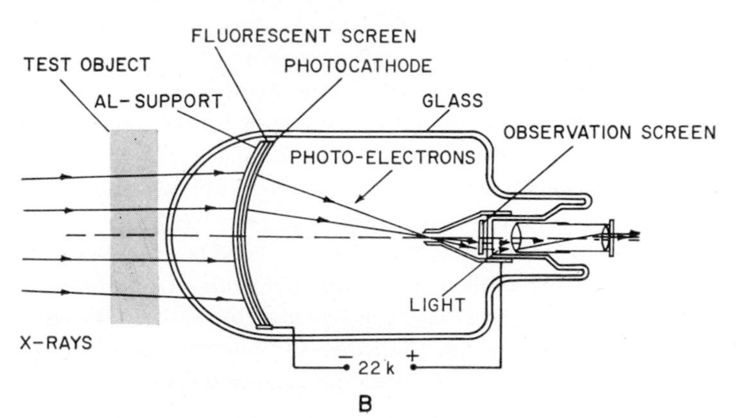

B

Fig. 13·3 (*A*) Mass exposure for inspection of many objects by one exposure (*Picker X-Ray*); (*B*) image intensifier (*Phillips*).

image is smaller than that of the input screen by a factor of 9, but it is far brighter.

Television also plays a part in X-ray inspection. A TV camera can be directed toward the observation screen of the image intensifier. The view is sent to the kinescope monitor where the brightness and contrast of the image can be improved. TV permits remote viewing at several locations without any potential radiation hazard to the viewer.

13·2 X-ray tubes X rays are an important means of inspection of objects such as filled metal containers or welding and castings. Roentgen or X rays are electromagnetic radiations just like radio waves but much shorter in wavelength. Their wavelength is between about 0.05×10^{-8} cm

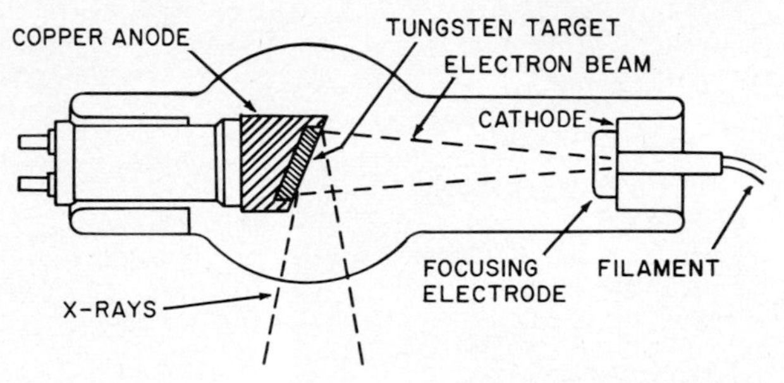

Fig. 13·4 Construction of an X-ray tube showing electrons generating X rays.

to 10×10^{-7} cm. Roentgen developed X-ray tubes as early as 1895, but the development of tubes used in modern equipment starts with the work of Coolidge, who in 1913 developed the tube shown in Fig. 13·4. Electrons from the heated cathode bombard the anode to produce the X radiation. Because of the tilt of the anode surface, the beam of radiation is directed outside the tube where it is used. Very high voltages (10,000 to 2 million) are required. Because of the high voltage and heating of the anode by the electrons, special design is required for the anodes. Some anodes are solid tungsten and others are made with a tungsten target mounted in a solid copper anode for better heat conduction. Other types have radiation fins on the anode or use cooling oil. Some tubes use rotating anodes to prevent a hot spot from developing in a certain place on the anode.

The pressure in a Coolidge tube must be very low. All tubes emit radiation over a band with a maximum determined by the accelerating voltage. Some tubes use several accelerating rings for very high voltage. Each ring has a higher voltage than the one before it, going from cathode toward the anode. Using this principle, electron beams of very high accelerations have been designed; some commercial tubes operate at 1 million volts.

Because a large percentage of the energy in the electron beam is converted into heat, some type of tube cooling is used in heavy-duty industrial X-ray equipment. Transformer oil or water may be pumped through the anode, gas such as sulfur hexafluoride may be used, or combinations of cooling are possible. In some commercial applications the tube and all the associated high-voltage components are mounted in an oil-filled case known as the tube head, which removes the problem of long high-voltage cables. A control unit contains the associated circuits for operation. An oil cooler uses a heat exchanger to dissipate the heat.

The X-ray beam from the tube is not homogeneous, for it has com-

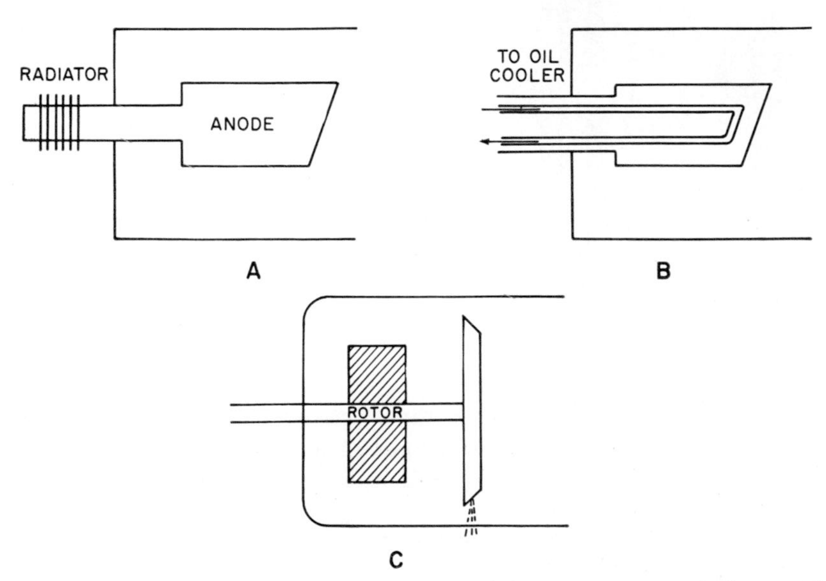

Fig. 13·5 (*A*) Cooling method using air; (*B*) oil cooling; (*C*) rotating anode for a cooling method.

ponents of different wavelengths. In a given application of a tube the shortest wavelength produced is a function of the applied high voltage, and as the high voltage increases the wavelength becomes shorter:

$$\lambda = \frac{12.34}{V}$$

where λ = wavelength, A
V = high voltage, kv

Thus for 24,000 volts λ is about 0.5.

X-ray tubes can be air-cooled (Fig. 13·5*A*) or oil-cooled (Fig. 13·5*B*). To prevent anode damage, often the anode rotates so no one spot is continuously used (Fig. 13·5*C*). Some typical tubes are shown in Fig. 13·6.

The tube is the heart of X-ray equipment, but it requires a power supply for its filament and high voltage for its anode.

This necessitates two transformers: a step-*down* transformer to reduce the voltage of the line supply to the 6 to 12 volts required for the filament and a step-*up* transformer to raise the line-supply voltage. This is raised 10,000 or more volts to drive the electrons from the filament to the anode

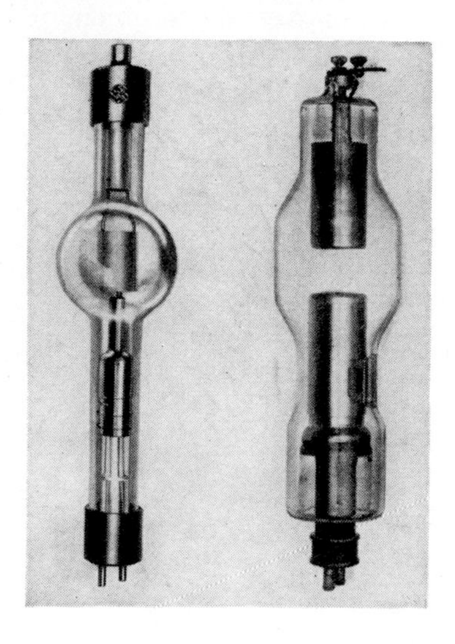

Fig. 13·6 Typical X-ray tubes. (*General Electric*)

with sufficient force to produce X rays. Two transformers are connected to the X-ray tube and to the line power as shown in Fig. 13·7. The "on-off" switch applies power to the filament, but the "exposure" switch turns

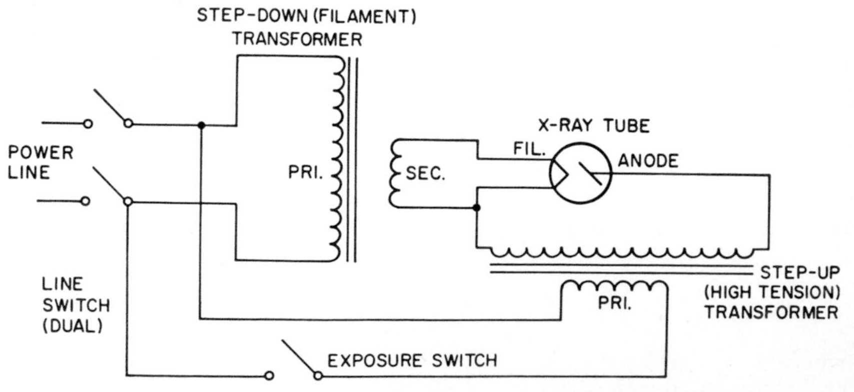

Fig. 13·7 Tube circuit showing two transformers, one for anode voltage and one for filament voltage.

on the high voltage for the exposure. Because "raw" alternating current is applied to the tube anode, this is a self-rectifying circuit.

Since the tube does not conduct when its anode is negative, there is a high inverse voltage across the tube during this time. By using rectifiers to provide the tube with direct current at a high voltage, the potential problem of the X-ray tube arc-over is avoided. These rectifiers can be half-wave, full-wave, and bridge circuits.

Exposures are timed by a clockwork mechanism or an electronic circuit, both of which are preset to the correct time.

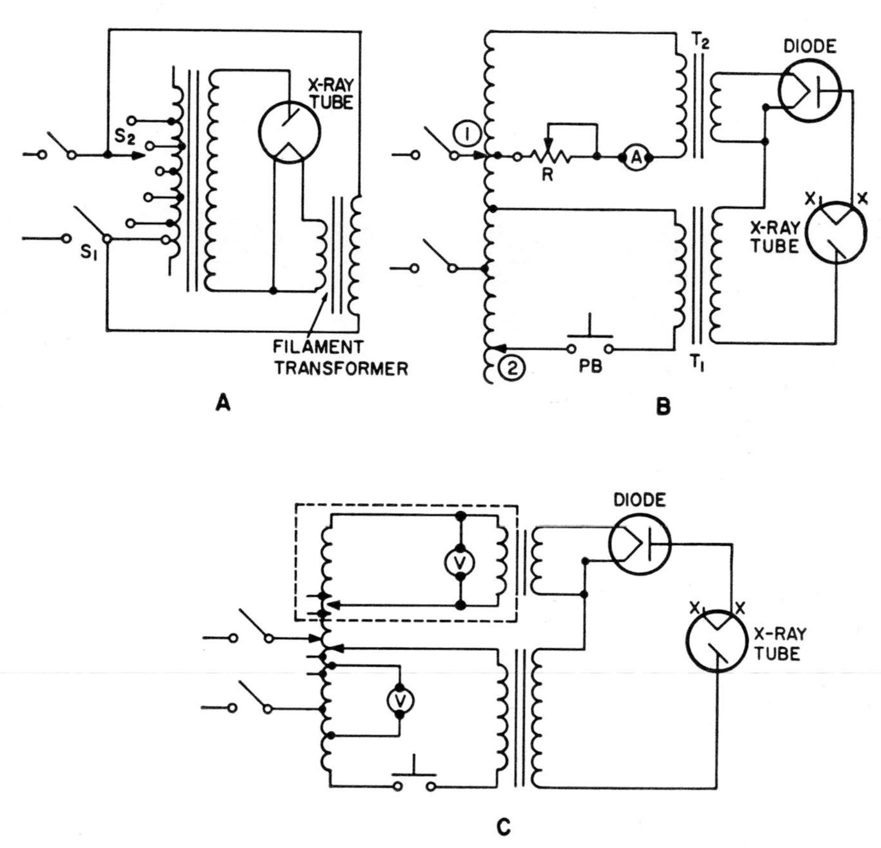

Fig. 13·8 (*A*) X-ray tube power control using primary taps; (*B*) use of a rectifier diode; (*C*) multitapped transformer.

13·3 Industrial X-ray equipment The X-ray tube and its power supply need two types of control. One is the adjustment of the applied voltages and the second is the exposure timing.

High-voltage adjustments are needed since the penetration of the beam is a function of the applied d-c voltage. A common method is to use an autotransformer, with taps, for various voltages. Figure 13·8*A* shows this for the anode supply only; the filament has no adjustment. *B* shows a circuit in which a half-wave rectifier is used for the anode supply. Here tap 1 is used to compensate for line-voltage variations which affect both filament and anode supplies. *R* is variable to permit accurate adjustment of the filament circuit to the tube. Meter A reads filament current. A step-down transformer T_2 provides the proper filament voltage. In the anode circuit a tap (2) is provided for high-voltage adjustment, a switch is used to turn the tube on for an exposure, and T_1 supplies an additional step-up for the anode potential. This is *not* a self-rectifying circuit since a diode rectifier is present.

Figure 13·8*C* illustrates a similar circuit arrangement except that there is a tap for the filament and two voltmeters are used. In some cases the high voltage for the anode has two taps, one for coarse and one for fine adjustments.

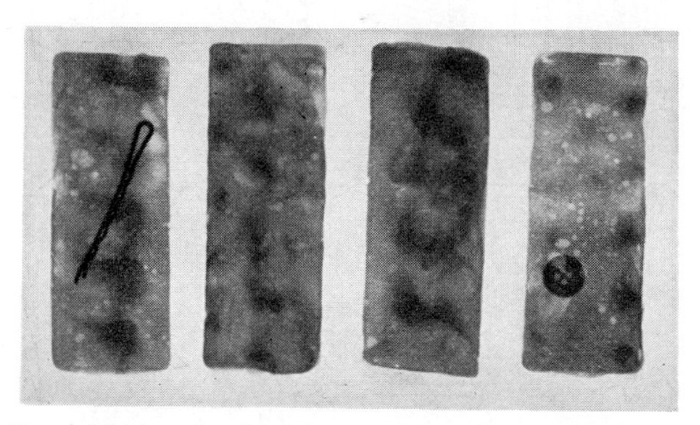

Fig. 13·9 X-ray photographs of candy bars showing imperfect bars. (*Picker X-Ray*)

X-ray Container Inspection. X rays are used for inspection in several ways. An X-ray picture of candy bars is shown in Fig. 13·9. Even with the greatest of care in manufacture and with modern equipment, foreign objects such as bobby pins or buttons can drop into a batch of candy. X rays catch these imperfect candy bars before they are responsible for a lawsuit. Wheat can be X-rayed to search out weevils. Dark spots on some of the kernels are weevils which can be located only by X rays. Ceramic resistors are sometimes ruined in production when solder from the pigtails runs down inside. By X-raying a batch at a time, the resistor manufacturer can quickly spot and remove the defective units.

All these provide for visual inspection by viewing of the X-ray picture, but there are also other forms of inspection.

An important application of industrial X rays is the continuous line inspection of the height of fill of closed containers such as canned milk, beer, or other liquids. A device for this use is the Hytafill (General Electric) shown in operation in Fig. 13·10. It protects containers against both overfill and underfill.

Figure 13·11 is a block diagram of this inspection system. The control circuits adjust the operating voltages for the high-voltage and filament transformers of the X-ray tube. Containers moving down the conveyor break the light beam (not shown), and the photoelectric relay opens the shutter for the X-ray beam. X rays from the tube pass through the container and strike the detector crystal. If the fill level is too low, the crystal output will be low.

The crystal signal is fed through the amplifier chain where it is shaped and amplified. When the container passes through the system, a pulse is generated which is the reject pulse. If the crystal has indicated a reject

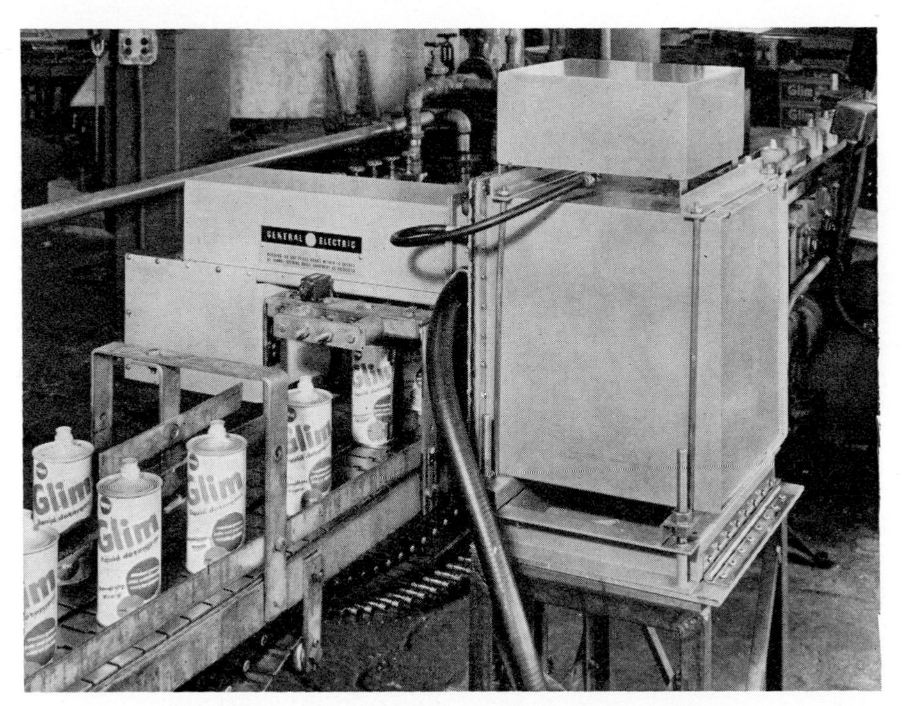

Fig. 13·10 X-ray container inspection. (*General Electric*)

container, the reject pulse from the pulse-forming stage combines with the amplified crystal signal and rejects the container by actuating the air blast as shown. Inspections up to 900 per minute with an accuracy within a few grams of container fill are possible with this rapid automatic system.

Cadmium selenide crystals are used to detect the X rays after they pass through the containers. Because the signals from the crystals can discriminate between fills of different heights, improperly filled cans can be rejected for either too high or too low levels, depending on the fill applications and the equipment used. All rejects are removed from the conveyor line.

The inspection scheme is shown in Fig. 13·12, with the beam at the correct level. In the case of underfills (*A*), the liquid surface in the container is just *below* the correct level indicated by the X-ray beam. In the case of overfills, the liquid surface is just *above* the correct level. In all cases the beam passes through the can parallel to the liquid level. The changing crystal can provide signals for fill plus container and for container alone.

For dual channels, two crystals are used; here one crystal detector is just below and the second is just above the correct limit, and the beam

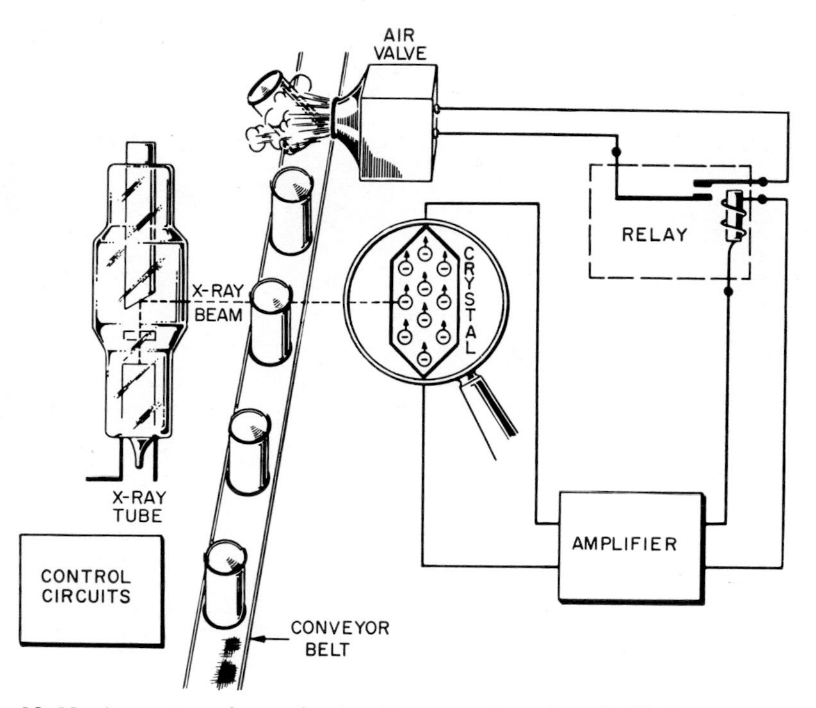

Fig. 13·11 Inspection scheme showing X rays passing through filled cans. (*General Electric*)

is set halfway between the two. The containers can pass down the conveyor in random sequence because a photoelectric relay senses the presence of a container and turns on the X radiation when the container is in position. The X-ray head contains all high-voltage components. An air-pressure reject occurs when a blast of air, from a slit in the copper tubing, pushes the rejected containers aside. The X-ray tube is water-cooled; controls are mounted in a cabinet.

Cadmium selenide crystals which are sensitive to both light and X radiation are used. By selection, the crystals (which are of match-head size) are sensitive to the X-ray emission. A crystal is mounted in a light-proof housing to eliminate the effect of light upon its electrical resistivity.

The self-rectifying (SPI-100) X-ray tube generates X rays only during the positive portion of the 60-cycle voltage impressed across the high-voltage transformer in the tube head, as shown in Fig. 13·13. For this reason, the X-ray pulses occur at a rate of 60 per second. The signal load resistor is connected in series with the detector crystal across a d-c supply voltage. The crystal signal consists of a-c and d-c components. The crystal and its load resistor act as a voltage divider. When the impinging X-ray intensities are of low magnitude, the crystal resistance is high and the voltage drop across the load resistor is a low value; when the impinging

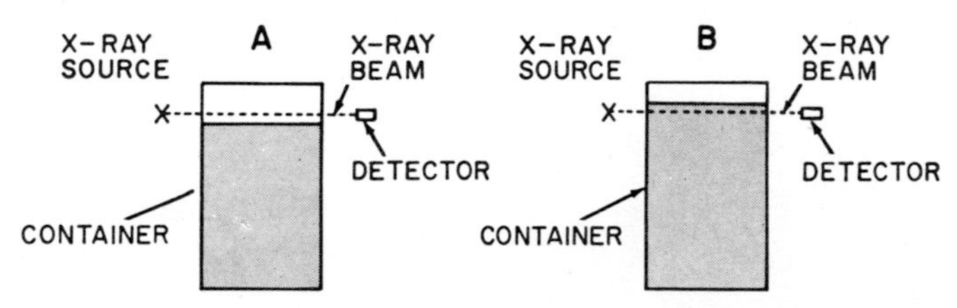

Fig. 13·12 (A) Underfill condition; (B) overfill condition. (*General Electric*)

X-ray intensities are of high magnitude, the crystal resistance is less and the voltage drop across the load resistor is a higher value.

The total current through the crystal is also a function of the supply voltage, the value of the series load resistor, the intensity of the X rays impinging on the crystal at a given time, and the history of the X-ray intensities just previously impinging on the crystal.

Most of the d-c signal component is removed by capacitor coupling in

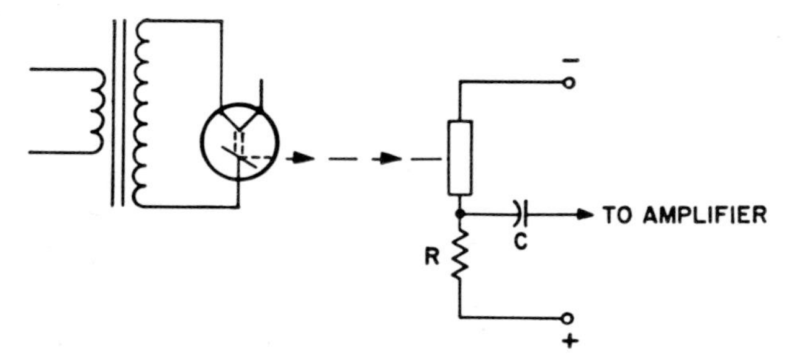

Fig. 13·13 Self-rectifier tube circuit.

the amplifier stages. The amplitude of the a-c component, after amplification and shaping, determines whether or not the unit will start a reject action. The a-c component amplitudes representing container plus fill material may be too large to represent accurately the height of fill, if the immediate previous history is that for no container and no shutter in the beam. For this reason, the unit uses a shutter to limit the X-ray intensities to magnitudes obtained with an empty container in the beam. The crystal signal must represent the filled container immediately after that container is inserted in the beam. If the previous history does not disappear immediately, an incorrect reject action will be initiated.

Other X-ray Equipment. Three different units are described below to show the range of available X-ray equipment.

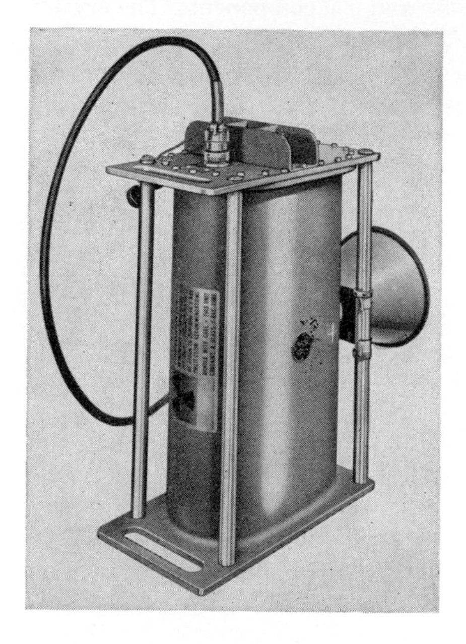

Fig. 13·14 Portable-unit tube head. (*General Electric*)

The General Electric LX-140 is a portable unit (Fig. 13·14) with a range from 70 to 140 kv. The power is a measure of the thickness of the material to be X-rayed (see Table 13·1).

Table 13·1

With	*You can radiograph*
50 kvp *	Very thin metallic sections or wood, plastic, etc.
150 kvp	Up to $4\frac{1}{2}$-in. aluminum or 1-in. steel
250 kvp	Up to 2-in. steel
400 kvp	Up to 3-in. steel
1,000 kvp	Up to 5-in. steel
2,000 kvp	Up to 8-in. steel
22 Mv	3 to 30-in. steel
Iridium 192	$\frac{3}{8}$ to 2-in. steel
Cobalt 60	1 to 11-in. steel

* Kilovolts peak

However, this is a very rough sort of guide; the thicknesses shown here are by no means the thinnest or thickest sections penetrable at each voltage. Calcium tungstate (salt) intensifying screens can reduce exposure time or increase permissible thickness.

This equipment, also shown in Fig. 13·15, uses sulfur hexafluoride gas for cooling the high-voltage self-rectifying tube. This figure illustrates the control box. Line "on-off" and X-ray "on-off" switches are shown. Each has its own indicator lamp. A dual spring-loaded switch ("line-load") allows the single meter to read either the incoming line voltage or the tube current. Various values of high voltage can be applied to the tube by setting the kilovolts peak dial, and exposures are timed by the dial calibrated in minutes. Three cable sockets are used for line-voltage input, load (which is the X-ray tube), and the third or interlock socket.

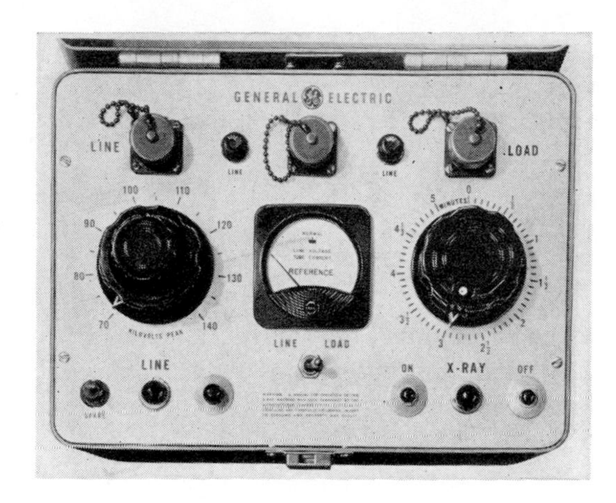

Fig. 13·15 Control box for portable unit. (*General Electric*)

Image Amplifiers. As discussed earlier in this chapter, an image intensifier is often used to provide images greater than are possible with ordinary fluoroscopic screens. The Picker 6171 is shown in Fig. 13·16. This is an electronic image amplifier that produces fluoroscopic images 1,000 to 1,500 times brighter than those of a conventional fluoroscope. The sharp detail of the amplified image shows up tiny defects in individual parts or assemblies (even moving ones). On a production inspection basis it is possible to find defects in potted electronic assemblies, fusing mechanisms, cord angles in tires, welds in 0.002-in. wires, and many others.

This unit (see Fig. 13·17) has a large-diameter image and uses geometric magnification of up to 20×. It has a brightness gain of 1,000 to 1,500 and employs lower voltage to provide better contrast and detail than conventional fluoroscopy.

As shown, the input element (photocathode phosphor) forms an image of the object through which the X rays have passed. The front layer emits light proportional to the X rays impinging on it. The second layer, in contact with the first, emits electrons proportional to the light.

These electrons are boosted in energy by high-potential acceleration (about 30 kv) inside the tube. The electron stream is then focused on the face of a small output phosphor, creating an intensely bright image.

Fig. 13·16 Image-intensifier system for X-ray use. (*Picker X-Ray*)

This super-bright image is magnified back to its original 9-in. size through an optical system producing a full-size amplified image that is 1,000 times brighter than that produced by fluoroscopy. Magnification is obtained by optical means and the image is presented upon a screen.

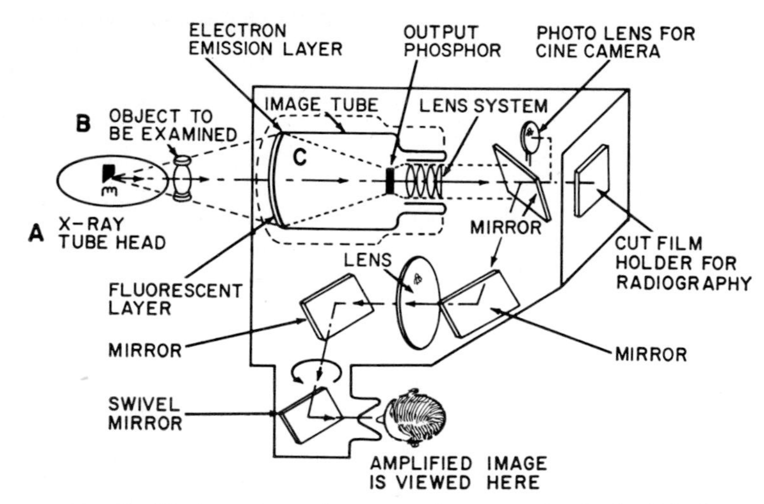

Fig. 13·17 Image-intensifier system showing optics. (*Picker X-Ray*)

Two very important uses of X rays are beyond the scope of this book. Medical treatment employs X rays of specified penetration and exposure. X-ray diffraction uses a beam to strike a sample substance whose scattering of the beam identifies the materials from which it is made.

13·4 Superhigh-voltage equipment X-ray equipment goes beyond that discussed in this chapter to superhigh-voltage units, such as the one shown in Fig. 13·18, which can penetrate casting, weldings, and forgings as thick as 10 in.

Particle accelerators (Fig. 13·19) such as the Van de Graaff generator can produce many types of radiation including positive ions, such as

Fig. 13·18 Superhigh-voltage X-ray unit. (*High-Voltage Engineering*)

Fig. 13·19 Van de Graaff generator. (*High-Voltage Engineering*)

protons, deuterons, and tritons, as well as extremely penetrating X rays.

Other accelerators are used for industrial radiation processing, which actually changes some materials in a chemical sense and sterilizes others for user protection. An accelerator of this type is the Dynamitron (Radiation Dynamics), which is a high-voltage generator with the cascaded rectifier tubes inside the rings (Fig. 13·20). The accelerator tube has many rings used to increase the applied potential which speeds up the electron stream through the tube. The tube fits into the center of the voltage generator and fits into a vessel.

Steam and gas (chemical) sterilization are being supplanted, in some cases, by the radiation process. Radiation sterilization holds promise of low-cost production-line processing of bulk materials. Objections of heat, moisture, and poor penetration are not factors. Time is reduced to fractions of a second, and goods can flow in an uninterrupted fashion down the production lines.

To sterilize a product, a radiation "dose" in the order of 3 million rad is required. This can be emitted from sources such as atomic reactors, radioisotopes, X-ray machines, or conventional particle accelerators.

Fig. 13·20 Dynamitron particle accelerator closeup showing cascaded rectifiers.

To penetrate to the depth of a product, sterilize within a package, or irradiate the opposite side of an object with a lethal dose, a radiation source of high voltage is required. The unit is made in the 1.5-Mev (million-electron-volt) sizes to penetrate packages of items such as cotton bandages, sutures, powders, drugs, plastic tubing, scalpel blades, or surgical needles.

Heat-sensitive items which would melt, discolor, or otherwise be destroyed in steam sterilization can, in most cases, be radiation-sterilized without damage. Almost instantaneous sterilization is obtained with a high-energy source by permitting extremely fast, uninterrupted flow of the product in front of the beam. Heat rise of only 3 or 4°C is encountered in the product.

The nature of the process eliminates any possibility of residual radioactivity in the product sterilized. The energy level of the beam is low enough to prevent reorientation of the atomic structures in the irradiated product. Only when sources in excess of 10 Mev are used does the production of radioisotopes become a problem.

13·5 Nuclear radiation detectors Radiation, as discussed above, has several forms. For protection and for prospecting, radiation detectors are used.

There are many possible methods to detect radiation; we are concerned only with those methods which are suitable for portable use. While some of the detectors in use are not so accurate as larger industrial detectors, they more than serve their purpose, which is to locate, with reasonable accuracy, radioactive ore. After location, samples are sent to laboratories for more complete and accurate testing including a chemical analysis.

Two somewhat crude detectors are the photographic method and the fluorescent method. In the photographic method, ore which is to be tested is placed on a small piece of metal on top of an unexposed photographic film. The film is protected from and never exposed to light. After about 24 hr, or even longer if the ore sample is a poor source of radioactivity, the film is developed. If the ore contains a radioactive substance, there will be an image of the metal plate on the film. The radiation will fog the film where it was not intercepted by the metal. This is an indication of radioactivity.

The fluorescent method is based on the fact that some materials will glow or fluoresce when excited by ultraviolet or "black" light. For example, some watch faces are painted with such materials. Ordinary sunlight contains a small amount of ultraviolet light. When the light source is removed, the luminous paint will continue to give off light. It does the same thing during exposure to sunlight, but the light given off is too faint to see.

This is a disadvantage to the method, because it cannot be used in daylight. A portable ultraviolet light source can be used as a detector for radioactivity in a dark room.

A more serious limitation is that some important radioactive ores do not fluoresce in their natural state while some of the materials which do are not radioactive ores. Some of the most important uranium-bearing ores cannot be located in this manner.

Fluorescence can be used, however, by heating a special compound and combining this, while still molten, with the ore to be tested. Uranium containing ore will fluoresce when excited by exposure to ultraviolet light.

The most effective portable radiation detectors now in general use are the Geiger counter and the scintillation counter. Both of these use different methods of detection from those outlined above.

The Geiger counter, built around the Geiger-Mueller (G-M) tube, is the most widely used low-cost instrument for radiation detection. A typical tube is shown in Fig. 13·21. A thin wire of tungsten is passed through the inner cylinder as indicated. This cylinder is thin metal or coated glass. This assembly is placed inside the outer cylinder, which is filled with a gas. In effect the tube is quite like a gas-filled voltage-regulator tube.

Both Geiger tubes and ionization chambers are used to detect radiation. As X rays pass through a gas they cause ionization; hence an electrode pair, mounted in a gas chamber, can act as a detector. A potential difference between these electrodes will cause a current flow as electrons, freed by ionization, travel to the positive electrode.

Both G-M tubes and ionization chambers are used. Their operation

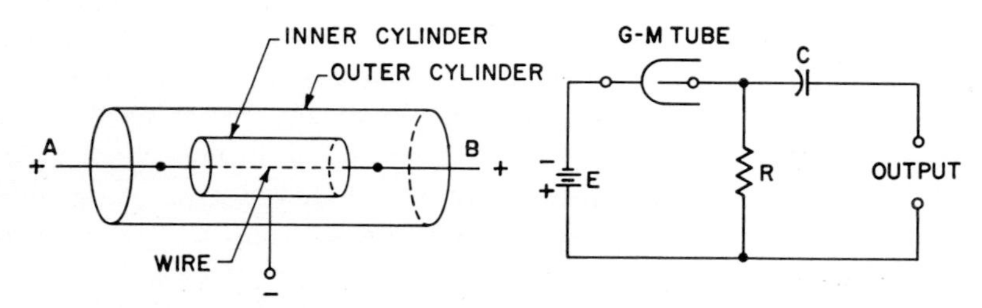

Fig. 13·21 Geiger-Mueller tube for radiation detection, showing basic circuit and construction of the tube.

may be seen from Fig. 13·22. With zero potential between the electrodes there is no current flow, for the ions and free electrons recombine after ionization. As voltage is applied and slowly increased, there is an electron flow from the attraction of the electrons to the positive electrode. This region from O to A is the ionization-chamber region. It is also the region of saturation.

Since all freed electrons are collected here, the chamber can be used to measure total radiation dosage.

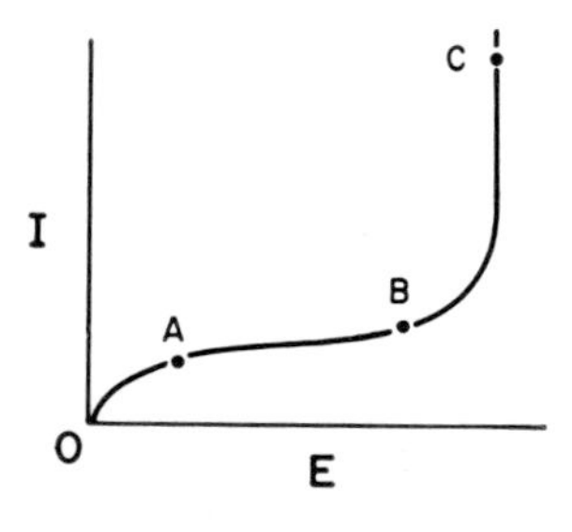

Fig. 13·22 Radiation-detector operation curve showing current as a function of the applied voltage.

If the applied voltage is raised, a single ionization creates an avalanche. Each freed electron now has enough velocity to free others; this process continues as gas multiplication does in a thyratron. Thus there is a very heavy current in the region B to C, which is the Geiger region. G-M tubes can count individual radioactive emissions.

A voltage is applied with the central wire positive and the inner cylinder negative. As a particle of radiation enters the tube from the outside, a gas molecule inside the tube loses an electron (a negative charge) by collision with the particle and becomes a positive ion (a positive charge). The freed electrons are attracted to the positive wire, but in their travel they collide with other gas molecules and produce other free electrons. Thus for one single particle of radiation many free electrons are produced. These are all attracted to the positive wire, which acts as a plate, or anode, in an amplifier tube. The total of the arriving free electrons acts as the output current. The positive ions, resulting from collisions which produced the free electrons, are attracted to the cylinder, which has a negative potential. Here they are neutralized by picking up electrons to make them neutral gas molecules again.

When a single particle of radiation reaches the G-M tube, a rapid multiplication of free electrons occurs which is the output current. In a very short time all the free electrons arrive at the wire and all gas ions arrive at the cylinder. The tube has provided an output and is now ready for the next particle.

In some cases external circuits are required to "quench" or turn off this action when it has started. There is also a great variation in G-M tube design and construction. Just as with a vacuum-tube amplifier, there are many commercial tube designs which perform the same function but have different operating characteristics.

Based upon the detection action of the G-M tube, many simple portable Geiger counters can be built. The design depends upon the nature of the particular tube and the final desired cost. Models are available from the simplest to quite complex units with complete amplifiers and several types of detectors.

Figure 13·21 shows a very simple unit; the d-c source is a battery with many small cells. This is possible because only a very small current is required, but the operating voltage for most tubes is between 300 and 2,000 volts. In this case the multiplication within the tube provides a series of current pulses through the load resistor R. Pulses are coupled through capacitor C to the output, which may be a pair of crystal earphones.

The tube is sometimes mounted in a hollow wand which is held facing the test ore or the direction of suspected radioactivity. By means of a cable the wand containing the G-M tube is connected to the body of the detector containing the amplifier, the power supply, and other circuit parts.

Radiation from radioactive materials can be converted into visible light by the use of the fluorescence effect of certain substances. Among these substances are zinc sulfide and calcium tungstate. When a radiation particle strikes a transparent crystal phosphor made of ore of these materials, a flash of light or a scintillation is produced. While these scintillations themselves may be used as indications, they are usually too faint and must be amplified.

Scintillation detectors may be used to record alpha, beta, or gamma radiations, depending on the type of materials used. These counters are more sensitive than Geiger counters, and they can also detect particles which are close together. It is possible by the use of counters to record in an accurate manner a large number of pulses in a very short time. In this way the scintillation detector is sensitive to both the rate (frequency per unit of time) and the strength of the radiation.

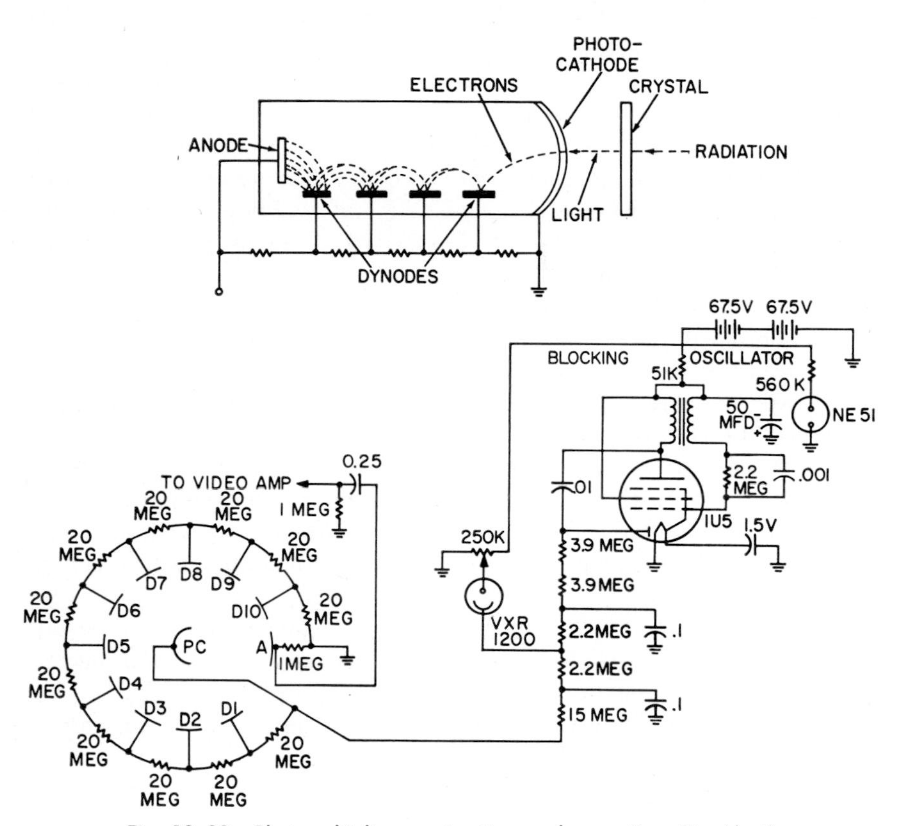

Fig. 13·23 Photomultiplier construction and operation. (*Du Mont*)

Because the phosphor screen converts the radiation into visible light, a different type of amplifier from an ordinary vacuum tube is required.

Radiation falls on the transparent crystal phosphor where it creates scintillations of visible light. This small signal must be converted into an electric signal and then amplified.

Any photocell sensitive to visible light can be used to convert light to a flow of current. Light falling onto the cathode causes the emission of electrons. The photocathode is transparent just as in television camera tubes. The light, created by the radiation reaching the crystal phosphor, falls upon the transparent photosensitive cathode. Electrons are emitted by the photocathode and this weak signal is amplified by electron multiplication. A series of small anodes (plates) are connected as shown in Fig. 13·23, so that each has a higher electrical potential as discussed in Chap. 6.

The basic structure of the photomultiplier tube is illustrated. Electron multiplication is used to deliver measurable output when the device is excited by few particles.

The actual circuit shows the Du Mont K1382 photomultiplier tube. Radiation particles strike a crystal phosphor, where they create scintilla-

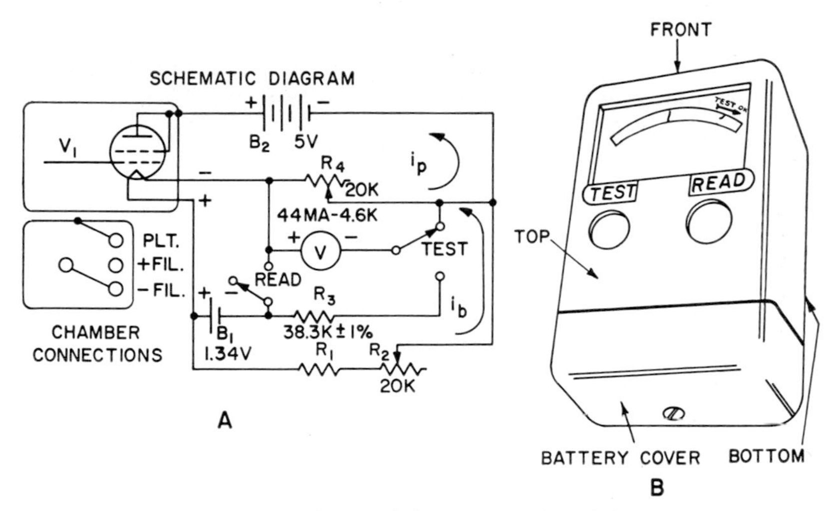

Fig. 13·24 Pocket-sized detector. (*Jordan Electronics*)

tions of visible light. The light is converted to an electric signal, which undergoes amplification.

Figure 13·24 shows the Minirad, a pocket-size detector which is a miniature gamma dose-rate meter for personnel protection use. The instrument indicates the rate at which nuclear radiation is being received and provides the operator with an estimate of the time he can remain in an area of radioactivity.

The operating controls (*B*) are two push-button switches. The one marked "test" connects the meter to measure the filament-battery condition. If the indication falls below the "test-ok" mark, the filament battery is near the end of its useful life. If there is no indication, the filament battery may be completely dead or the electrometer tube inside the ion chamber may be damaged.

When the "read" button is pressed, the meter indicates the gamma-ray dose rate. The indications are accurate for radiation being received from the front, the top (meter window side), and the bottom of the instrument. If the indication is below the bottom end of the scale, the dose rate is less than the rate to which the instrument is sensitive, but it is not necessarily zero.

Two mercury batteries power the instrument. The filament battery (Mallory RM-1) and plate battery (Mallory TR-114R) should both be replaced when the test indication is low. The Minirad uses a Neher-White ionization chamber. Radiation entering the chamber ionizes the argon gas, and the positively charged ions are repelled by the positive potential on the shell and are driven to the collector. The collector is connected to the grid of the electrometer tube, and increasing ion current flowing to the grid causes increased plate current. The meter is connected in the plate circuit to provide an indication of radiation dose rate. A "bucking" current flows from the filament battery B_1 through the resistor R_1 and potentiometer R_2 and through the meter in the direction opposite the plate current. This bucking current is adjusted by R_2 to cancel exactly the plate current when the chamber is subjected to the lowest dose rate on the scale. Under this condition, the meter will read at the bottom end of the scale.

The shunt potentiometer R_4 adjusts the sensitivity of the meter to provide the full-scale dose rate.

The test switch removes the shunt R_4 from the meter M and connects R_3 so that current flows from the filament battery B_1 through R_3, M, and the filament of the electrometer tube V_1. This permits a check not only of the battery but also of V_1.

REVIEW QUESTIONS

13·1 For 100,000 volts, what is λ for an X-ray beam?
13·2 Why are some tubes oil-cooled?
13·3 What are the three types of radiation?
13·4 What is the difference between a G-M tube and an ion chamber?
13·5 How does an X-ray thickness gauge operate?
13·6 How can X rays be used for inspection of welds?
13·7 What is the difference between an X-ray tube and a radioactive source?
13·8 What are alpha rays?
13·9 What are the three types of radiation?
13·10 What power source is required for radioactive inspection machines?
13·11 How does an image intensifier work?
13·12 What are the basic ways of viewing an X-ray image?
13·13 How does an ion chamber operate?
13·14 What are gamma rays?
13·15 How do beta rays differ from gamma rays?
13·16 Why is an image intensifier used?
13·17 Explain a photomultiplier tube.
13·18 How does an X-ray tube differ from a power diode?
13·19 Why are X-ray tubes cooled?
13·20 What are the methods of cooling?
13·21 How is television used for X-ray inspection?
13·22 Why are two transformers required for an X-ray circuit?
13·23 Where is the rectifier in Fig. 13·7?

13·24 What are three methods of controlling the power source of the X-ray tube?

13·25 Explain how X rays can inspect the contents of a filled can.

13·26 What is the detector in this inspection method?

13·27 What is the inspection rate?

13·28 Explain the controls in Fig. 13·15.

13·29 What two methods of viewing an image are shown in Fig. 13·17?

13·30 How are radiation detectors used?

13·31 Explain how a Geiger counter operates.

13·32 What is the difference between Geiger operation and ionization-chamber operation?

13·33 How does the power supply in the photomultiplier shown in Fig. 13·23 operate?

CHAPTER

14

Industrial Radio

Radio is an important tool for industry. This is an entire field in itself and requires techniques which differ from those covered in this book. (See Allan Lytel, "Two-way Radio," McGraw-Hill Book Company, Inc., New York, 1959.) It is the purpose of this chapter only to show how the several radio services tie into other industrial electronics. The Federal Communications Commission (FCC) has established different services for different types of radio users.

14·1 Radio services The two large dimensions of radio services are common carriers, such as telephone and telegraph, and the private business or industrial systems. In this book (and in this chapter) only certain aspects of radio are discussed, and these only as they relate to industry. Excluded are common carriers, since they are for public use, and broadcast, since this has no relation to industrial electronics.

Included in this chapter are the following: mobile-to-base or two-way radio, microwave radio, telemetry, and remote control.

14·2 Mobile radio (two-way)

Uses of Two-way Radio. Included under this topic are those forms of two-way radio most useful to industry.

Land-mobile radios permit rapid control of cars and trucks. Radios in vehicles (Fig. 14·1) allow plant protection to alert quickly emergency forces such as the fire truck shown. Repair and service trucks also can be routed with greater efficiency; shipping, loading, and unloading can be controlled; and lift trucks can be routed. Two-way radio equipment is small, lightweight, and easily installed.

By tying vehicles to a control point by radio, the plant has a tight control over widespread groups. This is useful for plant protection, warehouse or field work, and for all large organizations where the work force is spread out in a big building or in several buildings.

Base stations are the central control point for the two-way communication shown in Fig. 14·2. Messages usually originate from here to the mobile stations, most of which can reply via their own transmitters.

Personal receivers (Fig. 14·3) are used to supplement those for vehicles. Paging service, offered by Bell Telephone and others, allows one-way calling to alert the particular station to call in (using the telephone) to receive a message. For field work a portable two-way receiver-transmitter is very useful and time-saving.

Fig. 14·1 Fire truck with mobile two-way radio. (*Motorola*)

Industrial or Business. One of the uses of industrial radio is two-way voice communication between two stations at least one of which is mobile, or moving. Each installation usually has both a transmitter and a receiver. Frequency modulation (FM) or amplitude modulation (AM) may be used.

Every two-way communications system has one or more operating frequencies. The usual system has a base (fixed) station and one or more mobile stations such as those in automobiles, buses, or trucks. Radio is used for many industrial applications including delivery of fuel oil, construction and building, utilities, railroads, and automobile road service.

The FCC is responsible for licensing and control of mobile radio just as it controls the technical features of AM and FM radio and television broadcasting. The FCC approves commercial mobile equipment, issues the station licenses, and provides the licenses for operating and maintenance technicians.

Calls can also be made from a vehicle to a telephone by dialing a number in a local telephone system, in exactly the same way as when making a house-to-house call; only when placing a toll call must the long-distance operator be called first.

The system has a dial radiotelephone in a vehicle and a two-way radio system as a carrier to the local telephone company. A transmitter-receiver base station is connected to the local telephone system; termination and switching equipment at the base station automatically transfers

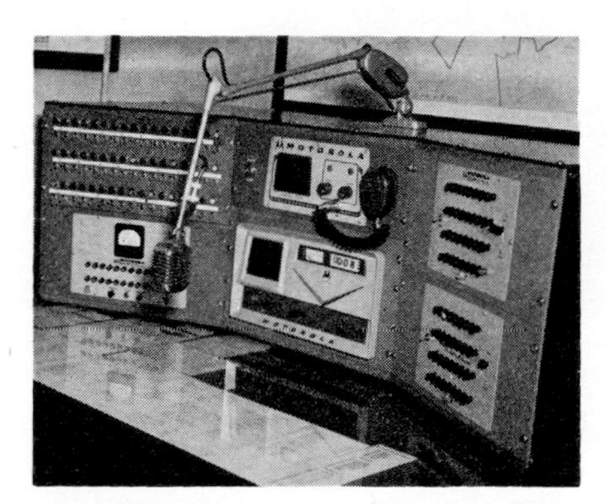

Fig. 14·2 Police department radio, Waco, Texas. (*Motorola*)

Fig. 14·3 Bellboy paging receiver. (*Bell Telephone*)

the radio calls to the telephone system and telephone calls to the radio system. Calls can also be made from vehicle to vehicle.

Most two-way radio, however, is used for voice communication from base to mobile station or from mobile to mobile station. In the vehicle a typical mobile-radio transmitter-receiver combination has an "on-off" switch, the volume control, and the squelch control. These are the only ones available to or required by the operator. A "push-to-talk" switch on the microphone is used during transmission.

The radio may be placed in the trunk, with the controls next to the driver; the entire unit may be placed next to the driver; or it may be installed on the seat of an industrial vehicle. A base station is a transmitter-receiver unit for use at fixed locations. It may be located at a desk; or, by use of a 117-volt a-c power supply, a vehicle can be converted into a base station; or remote control can be used.

Frequency bands for two-way radio cover several regions starting from 25 mc and extending through 470 mc. Frequency assignments in each band are usually on a nonexclusive basis. Several transmitters may share the same frequency.

Typical equipment is the small transmitter-receiver shown in Fig. 14·4. All operating controls are in the small box on the front of the unit. Cooling fins are provided on both sides of the case. A unit can be mounted in the front near the driver or it can be placed in the trunk. Modern two-way radios use solid-state devices and printed circuits for servicing ease.

Fig. 14·4 Complete call head and car two-way radio. (*RCA*)

The FCC is the government agency responsible for regulating all types of communication by radio. The technique is licensing of nongovernment radio stations and their operators. The Commission is not under any government department but is an independent Federal establishment reporting directly to Congress.

The FCC functions in the interest of all those who need radio communication. The limited amount of spectrum space is controlled and distributed among the several types of radio services. The radio spectrum is actually a natural resource. Because it is limited in size, maximum benefit can be realized only by careful use and conservation.

The FCC has two types of control over radio. One is the licensing of equipment and operators, and the second is the regulation of the technical operation of the equipment.

Repairs, adjustments, and tests made on transmitters during installation, servicing, or testing must be made only by, or under the supervision of, a licensed radio operator holding a First or Second Class Radiotelephone license. Operators of mobile radio in land-mobile service (above 30 mc) do not require licenses where radiotelephone (voice) is used. But only properly authorized operators can make adjustments which affect the operating frequency or make any repairs to the radio. Operators' licenses are issued after written tests are passed. There are four parts or elements to the FCC Commercial Radio Telephone examination: (1) Basic Law, (2) Basic Theory and Practice, (3) Radiotelephone, and (4) Advanced Radiotelephone. The First Class license requires all four elements, and Second Class requires the first three elements. Both classes require United States citizenship and the ability to read and write English. Examinations may be taken at any local FCC office.

The Commission's rules and their other publications may be obtained from Washington, D.C., office of the FCC. Lists of publications are available at any of its field offices.

There are specific items included in the FCC requirements for radio maintenance. The important requirements are frequency, modulation, and power measurements plus proper log entry.

The operating frequency must be measured when the transmitter is installed, when a change is made in the transmitter which affects the operating frequency, and at regular specific intervals. These must not exceed 1 month when the transmitter is not crystal-controlled and 12 months when the transmitter is crystal-controlled. Below an operating frequency of 50 mc the tolerance is ± 0.01 per cent and above 50 mc the tolerance is ± 0.005 per cent of the assigned frequency for most transmitters. Closer tolerances for transmitters will be required in the future for some services. These will be ± 0.002 per cent (25 to 50 mc) and ± 0.0005 per cent (50 to 1,000 mc).

A modulation check is required at the same time the frequency is measured. For FM transmitters the frequency of the carrier must not change more than a specified amount from the assigned center frequency. For AM transmitters modulation must be between 70 and 100 per cent for negative peaks of modulation.

The plate power input, which is plate current times plate voltage, must also be measured at the times specified above, and it must never exceed the licensed value.

The person making measurements and checks must sign his name and address in the station log, for he is responsible for the accuracy of his measurements. All radio transmitters must be licensed by the FCC before being put into operation. The license must be posted at the radio station. The method and form of license application depend upon the radio service to be used.

Conformance to FCC regulations is the direct responsibility of the licensed person making the measurements. He must know the proper procedures, and he also must know the accuracy of his instruments. The FCC monitors the operation of the transmitters and sends citations to all violators.

One fixed-base station and several mobile stations may be used to start a small two-way radio network. Other mobile stations can be added as they are required. The range and usefulness of the system of mobile communications can be expanded by using repeater stations. These are located at several strategic points; repeater stations can pick up all messages and rebroadcast them at greater power in both directions.

By definition, a repeater repeats. As applied to two-way radio, a repeater is a transmitter-receiver with special parts added to it. Its basic function is to receive a radio signal and automatically retransmit. In the mobile services the repeater is used three ways: as a repeater station, as a mobile relay, and as a combination base-mobile relay.

Radio signals at uhf, like television signals, travel in relatively straight lines; this is commonly called line-of-sight transmission. This makes base-to-mobile or mobile-to-base communications extremely difficult when they are over the horizon from each other, or separated by hills. To overcome this problem, repeater stations are used at some point of high elevation. By talking through the repeater station, base-to-mobile communications up to 100 miles are sometimes possible. So long as the repeater antenna can "see" the mobile antenna, then two-way communications are maintained.

The line-of-sight characteristic of radio signals is even more of an obstacle for mobile-to-mobile communications. Vehicle antennas are closer to the ground than base-station towers. This puts the horizon line much closer to vehicles, shortening the range of mobile-to-mobile coverage even more. But, with a repeater unit installed, each mobile unit can "talk" through the repeater to every other mobile unit. The only requirement, once again, is the line-of-sight clearance afforded by the high antenna at the repeater site. When the repeater is used in this function it is a mobile relay.

If no hills or tall buildings are available, the relay station is sometimes installed as a combined base and relay station in which the relay serves the function of a base station and saves the cost, also, of a separate base-station unit. Vehicles maintain their extended mobile-to-mobile coverage so long as they are in "sight" of the base-station antenna. This method of installation is sometimes preferred in small communities where range requirements are not so great.

New FCC rules and regulations effective Sept. 11, 1958, limit new applications of these repeaters and mobile relays to the uhf frequency bands above 450 Mc. This means that all such systems licensed for Business or Citizens Band applications must operate above 450 mc. These signals follow extremely straight lines and behave much like rays of light. In cities they bounce around buildings, just about like sunlight, and usually provide excellent coverage in built-up areas. But, like rays of light, once the horizon line or a hill does intervene, their coverage halts abruptly. This line-of-sight characteristic of 450-mc operation requires the broad application of repeater stations, mobile relays, and combination base-mobile relay stations.

Paging. One-way radio is a form of signaling or paging, as in a large warehouse or factory, using a small receiver as shown in Figs. 14·5 and 14·3. Coded signals are picked up by specific receivers only.

One-way city-wide radio paging falls under Part 21 of the rules and regulations of the FCC.

Frequencies are in the 30- to 40-mc range with transmitters of 250 watts. Since coverage of the primary city is desired first, with secondary

Fig. 14·5 Voice Director for in-plant paging. (*General Electric*)

coverage of the areas outside the city, the transmitter and antenna should be located in the downtown area. The antenna should be located on a tower or mast atop the tallest building. The transmitter should be located in a room on the top floor, or in a small housing built on the roof if the roof is flat. The transmission line between transmitter and antenna should be as short as is practicable to prevent loss of power in the transmission line.

A typical AM transmitter for this service uses remote control. The remote-control amplifier is required equipment and provides a means of connecting a microphone, tape recorder, repeater, or selective signaling encoder to the system. Although it serves as a preamplifier, it includes an audio compression circuit and an audio low-pass filter, which are circuit sections necessary to comply with the rules for amplitude-modulated transmitters.

The remote-control unit is connected to the transmitter over a single pair of telephone wires; the control circuits are so arranged that a short circuit between wires of the control pair will cause loss of modulation, but control of the carrier will be unimpaired. Grounding either side of the telephone pair will short-circuit the control voltage and will turn the carrier off.

The r-f final amplifier shown in Fig. 14·6 is typical of this equipment.

The type 660 power-amplifier circuit is shown in schematic diagram. The amplifier employs two Amperex type 6155/4-125A beam tetrodes in a conventional push-pull balanced circuit. Input is link-coupled from the r-f exciter, and output is link-coupled to the r-f transmission line with a small capacitor in series with the link to tune out any reactive component of the link circuit.

The amplifier is adjusted to tune to the intended frequency of operation while maintaining the correct loaded Q of the tuned circuit. Cross neutralization is employed to obtain complete stability at the fundamental frequency, and small parasitic chokes are inserted in each plate lead to prevent vhf parasitic oscillation.

The amplifier is bypassed and arranged in a symmetrical manner physically to preserve balance in the circuit and to return all circuit grounds to the sockets. An enclosure made of meshed screen is used to aid in the suppression of local radiation. The cabinet, of course, provides additional shielding in this respect. Forced-air cooling of the two 6155 tubes is accomplished by a small blower motor secured to the amplifier chassis.

The amplifier tubes are protected against plate-current overload by an overload relay which has its coil connected in the cathode return circuit and which opens the cathode circuit if the plate current reaches a preset, unsafe value. The overload control, which establishes the amount of cathode-current overload considered permissible, is located on the control panel. A reset button, located just below the adjusting potentiometer, is used for resetting the relay. A value of cathode current of 110 per cent is suggested as the overload trip current.

In addition to the above protective measure, a type 807 tube is used as a screen clamper which acts to reduce the plate current of the 6155 tubes if excitation, for any reason, fails or is cut off. The 807 simulates a low value of resistor connected between the 6155 screens and ground when excitation is absent, and the 807 conducts heavily. When excitation appears, however, the rectified grid current biases off the 807, and plate current in this tube ceases to flow. The 807, under this condition, is out of the circuit.

Using remote control, the operator receives a telephone message for a person with a given code. She sends out this code, by the use of push buttons as shown, to the transmitter, and the receiver for this code is alerted. The person wearing the receiver then calls in via telephone.

An antenna 70 ft above the ground in flat country will provide a primary service radius of about 7 miles. A 140-ft antenna will provide

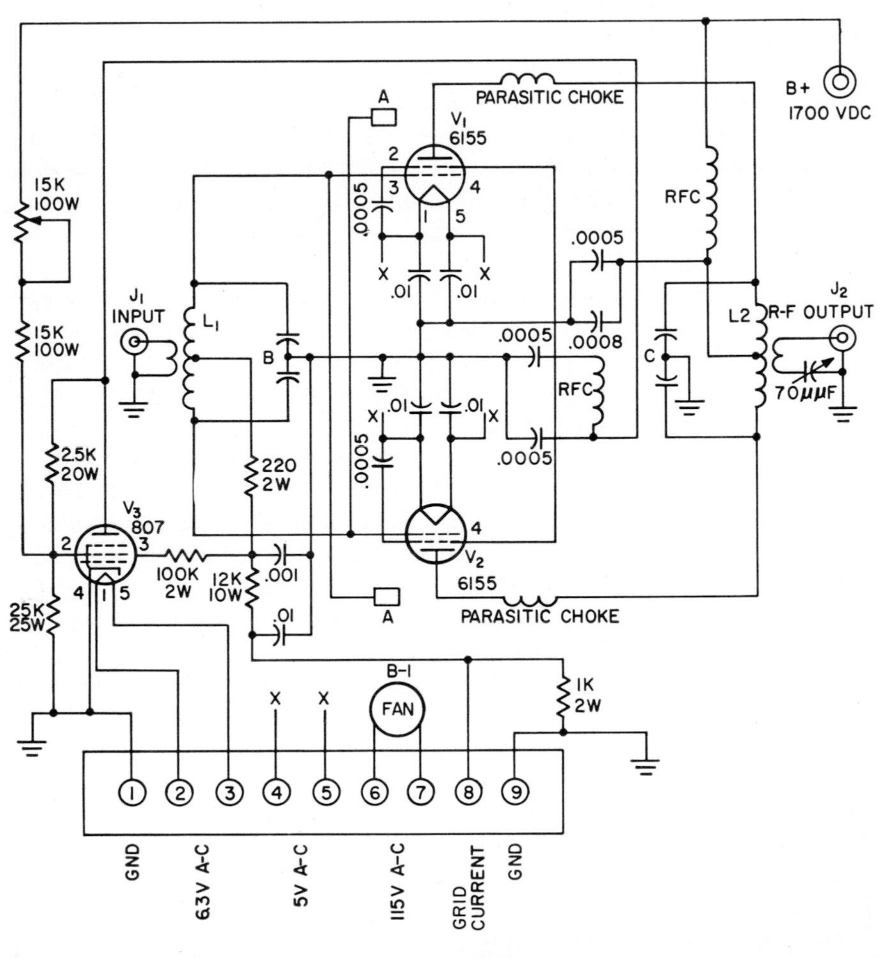

NOTE:
All resistors shown in ohms,
all capacitors shown in microfarads.
A – Neutralizing stud.
B – Dual 25-25 μμf Hammerlund HFAD-25B.
C – Dual 40-40μμf Cardwell X040DK (PL8056).

Fig. 14·6 Final r-f amplifier of paging transmitter. (*Tele-Communications*)

about 15 miles. The secondary area, beyond the radii given, is extremely useful to the subscriber.

The transmitter may be located in a building completely separate from the dispatch point. The control unit is installed at the dispatch point, and control of the carrier of the transmitter, as well as the audio transmission, is carried over a single-pair telephone line. This line is usually leased from the telephone company. The transmitter and control may be in the same building. In this case, of course, the control lines do not ordinarily have to be leased.

Calls are continuously broadcast during the hours of service of the station by means of a tape recorder of some type. Other methods of continuously transmitting voice are also used.

Marine Radio. For some industries radio communication with boats is important. In some ways this is an extension of other land-mobile services, but there are, of course, separate marine radio frequencies for this service.

Radiotelegraphy, which began with Marconi's original installation aboard the English lightship late in 1898, has grown to a world-wide system for commercial shipping. Two-way radiotelephone has expanded in recent years.

Radiotelephone permits voice communications from ship to shore, from ship to ship, and from shore to ship. Telephone companies operate land-based stations which tie in to the telephone system.

Marine radio telephone serves the three functions of safety, operations, and business. Safety is of prime importance, for marine radio and the instant two-way communication have saved many lives. Operations are the unique problems of navigation and movement of vessels. Business communications are related to commercial marine activities such as fishing, cargo carrying, or tugboat use.

There are several different frequency bands for marine radio. The 2- to 3-mc band, for example, is in wide use for both pleasure boats and commercial vessels. At 2.182 mc is the international distress and calling frequency. Radios are usually kept tuned to this frequency, and after establishing contact with the calling station both sender and receiver switch to another frequency for transmission and reception of the message.

Typical low-cost five-channel radiotelephones for the 2- to 3-mc band are shown in Fig. 14·7.

The 2- to 3-mc band is just a part of the radio spectrum for marine use. Other marine bands include portions of the 152- to 174-mc range, where 156.8 is the calling and safety frequency and the Business Radio service (formerly low-power industrial) has assignments in the 25- to 50-mc band. The 152- to 174-mc and 25- to 50-mc bands may be used for certain marine radiotelephone applications. Many land-mobile radios are also authorized in these bands. Connections to land telephone lines are possible using telephone-company-operated land-based stations.

Citizens Band Radio. Citizens Band Radio is a part of the communications band which has been given new life by the recent changes in the FCC rules of operation. The former 11-m band is now available for this use, for which no specific need must be established. There are four classes of stations which may operate on the frequencies. These are classes A, B, C, and D, and they may be used for business and industry.

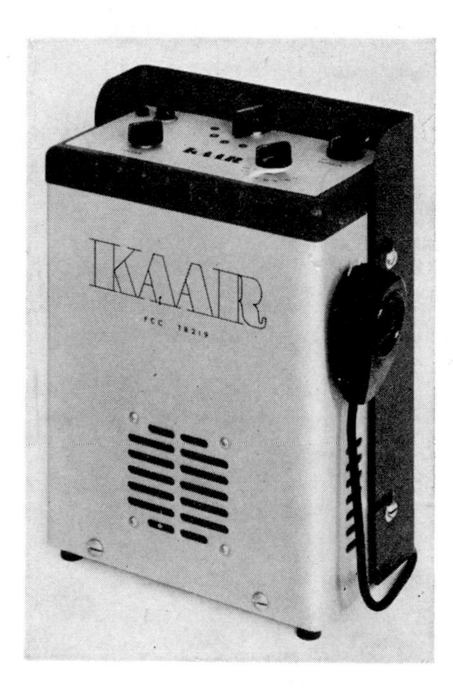

Fig. 14·7 Marine two-way radio. (*Kaar*)

Class A may operate from 462.55 to 466.45 mc and from 460.05 to 460.95 mc, with AM or FM, using up to 60 watts. Class B uses 465 mc, AM or FM, with 5 watts. Class C, which is for remote control only, uses 26.995, 27.045, 27.095, 27.145, 27.195, and 27.255 mc with crystal control and 5 watts of power (except for 27.255 mc, which may use 30 watts). Class D, with 22 channels, operates from 26.965 to 27.225 mc, with crystal control and 5 watts, using AM.

Class A Citizens Band Radio can use commercial FM two-way radio built for use in the 450- to 470-mc band. Usually this equipment is operated in commercial applications for industrial or land transportation services or for public safety, but exactly the same equipment can be used for class A. Selective calling equipment is available.

Class B radio equipment may be a superregenerative uhf band transceiver, or it may be ordinary uhf two-way radio used for the citizens' band.

The uhf radio signals behave quite like rays of light. This fact must be kept in mind at all times in order to obtain the best possible results with the equipment. Transceivers for this band are quite simple and give excellent short-range service.

Class C is used for remote control of garage doors, model aircraft, or model boats, and several types of quite simple equipment are available. This may use tone-modulation AM for control purposes. Before the change in FCC rules, only 27.255 mc was available; there are now five new channels for remote-control work.

Class D is the new band which is creating so much interest. There are no restrictions on the use of this equipment except operation within the

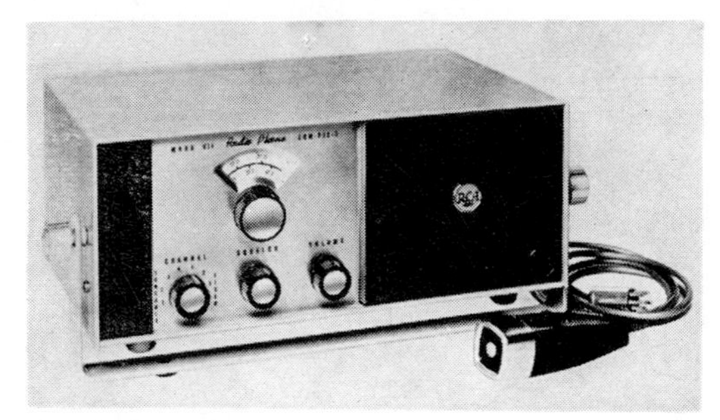

Fig. 14·8 Mark VII radiophone. (RCA)

law. The band is open for business or pleasure, but there is no guarantee of protection from interference. Twenty-two channels are available. Station licenses are available upon application, and no specific need is required. Typical equipment is shown in Fig. 14·8.

There is a very wide choice of citizens' band radio equipment, from do-it-yourself kits to complete gear that is ready to go. Hand-held portable sets (Fig. 14·9) are, of course, battery-operated while the others are meant for all types of mobile use from small boats to cars and trucks. They are used in all places from camping trips to everyday industry.

Most mobile equipment may be operated from 6 or 12 volts d-c and, for applications near power lines, from 115 volts a-c. Transmitters may be one or more channels; receivers may be crystal-controlled or tunable across the band.

Citizens' band equipment may be licensed as a two-way system to any citizen of the United States, 18 years of age or over. The citizen to whom the station or group of stations is licensed need have no particular technical qualifications, and no examination is necessary. The only requirement for station license and registration is Form 505. This is available from the local FCC office. Usually, a copy of this form is provided by manufacturers supplying citizens' band radio equipment.

It is not necessary to show any particular need for such communications in order to get a license. It is not necessary to prove that such a service is essential to a business. It may, in fact, be used for purely personal matters not related to business.

The 18-year age minimum does not necessarily apply to every operator of every station in a group licensed as a "system." The prime requirement is that the operation of every station in a particular system be under control of the licensee. This means, among other things, that the licensee is to make certain that stations in his system are not operated for illegal purposes, or in any manner contrary to law.

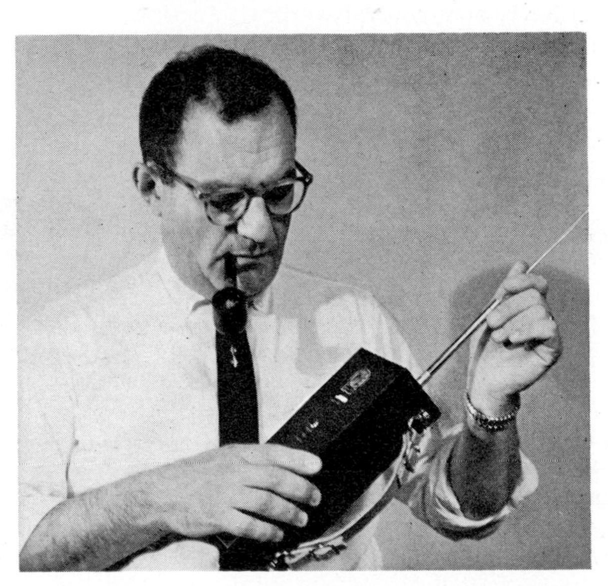

Fig. 14·9 Author with portable transceiver. (*Wightman Electronics*)

14·3 Microwave radio (two-way) Microwave two-way radio links, between two points, are used to transmit voice and telemetry signals for many industrial applications. Several frequency bands are available. Common carriers, of course, use microwave links for their special services, but special bands are available for private business or industrial use.

Two types of systems are "thin-route" and high-density. A thin-route system, with a bandwidth of 100 to 150 kc, can handle up to about 24 channels at once. A high-density system, with a bandwidth of several megacycles, can handle up to 720 channels at once. A very low power from the transmitter will enable proper service between hops of 30 to 50 miles because of the directive antennas.

A typical system is shown in Fig. 14·10. This is a two-way system and only terminals are shown. The terminating sets send the incoming speech or other signals to individual channel modulators, which convert the signals into a time-modulated pulse. The individual pulses from all channels are interleaved in the common multiplexing equipment and fed as a

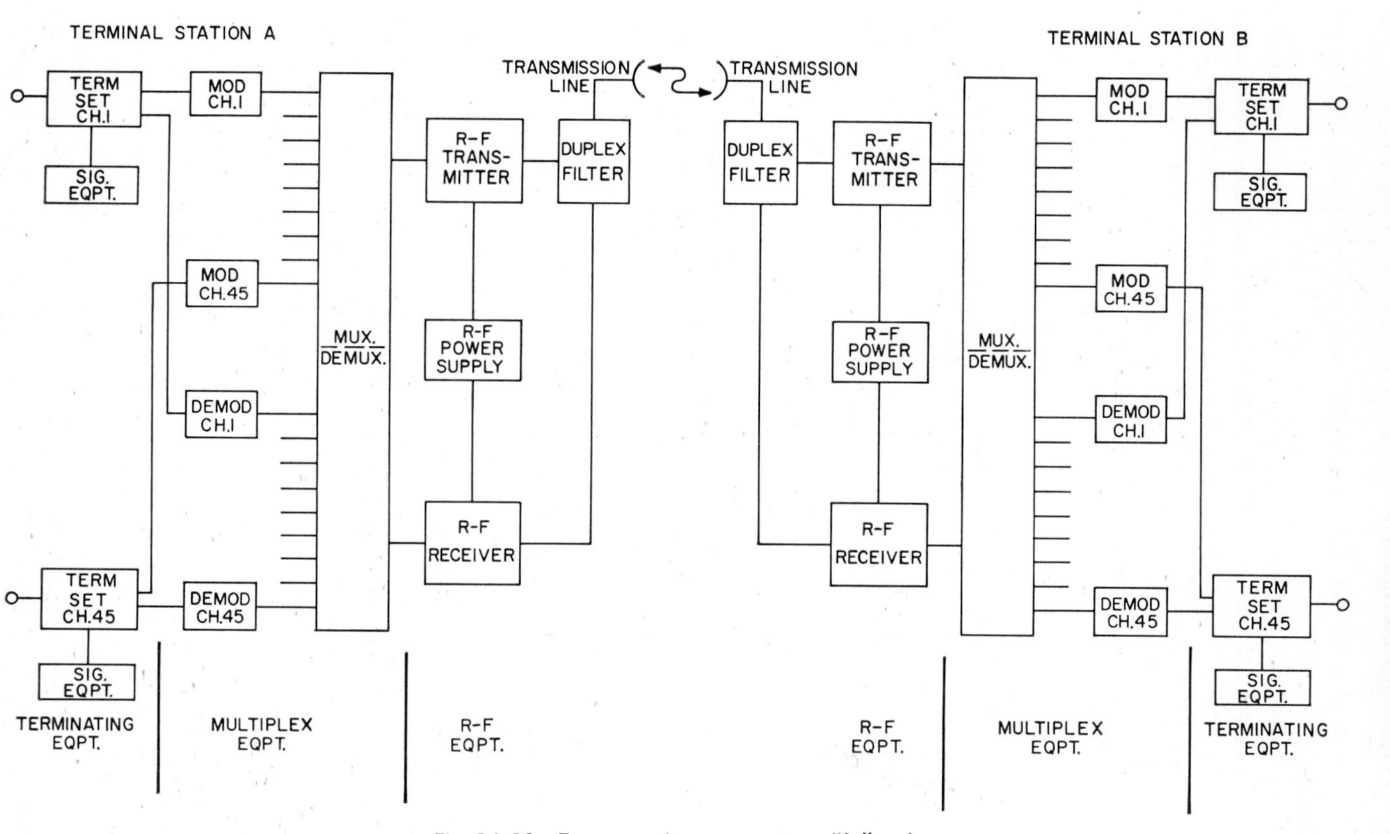

Fig. 14·10 Two-way microwave system. (*Kellogg*)

train of pulses to the r-f transmitter, which is pulsed on and off by this train.

The train of microwave pulses is then radiated into space through the antenna. The microwave signal is picked up by the receiving antenna and fed into the receiver at the distant terminal station. Here the pulse train is detected, amplified, and fed into the demultiplexing equipment to be finally separated as individual channel pulses at the individual channel demodulators, which also reconvert the pulse-time modulation (PTM) signal to audio.

Each repeater unit (Fig. 14·11) required to extend the range of the system consists of an r-f receiver and transmitter for each direction of transmission. Each receiver detects the microwave pulse train going in a particular direction and uses its output to modulate a transmitter facing the same direction.

For drop-channel operation a drop-and-insert unit is placed between the receiver and the transmitter; this unit converts only the desired dropped channel signal to audio. It also converts a locally generated audio signal into a pulse signal which it inserts into the proper position in the multiplex train that is fed to the outgoing transmitter.

Multiplex systems are used to send several audio or video signals on a single radio channel. As described above, the multiplex equipment contains the means for combining up to 45 separate voice-frequency channels into a time-modulated pulse train and for separating each of 45 channels from a time-modulated pulse train, thus restoring the original audio signal.

In the time-division system of multiplexing, each channel, up to the maximum of 45, is allocated $1/48$ of a frame interval for the transmission of its information. The remaining part of the frame interval is used for the transmission of a distinctive marker pulse to synchronize the multiplexing circuits. During the time allocated to a particular voice channel, a short pulse is transmitted. With no audio signal applied to the channel input, the pulse is positioned at the center of its time interval. An audio input (modulation) causes the pulse to deviate in time about its center position at an audio rate, the magnitude of the deviation being determined by the amplitude of the signal input at the instant of sampling. These pulses are combined into a pulse train and fed into the r-f transmitter. At the receiving station, the pulse train is detected in the r-f receiver and coupled to the receiving multiplex. The marker pulse is separated from the train and sequentially activates the individual channel demodulators. Although each demodulator has the complete pulse train at its input, the demodulator gate is open only at a specific time interval. The proper single pulse is then converted to the original audio signal and amplified to the correct level.

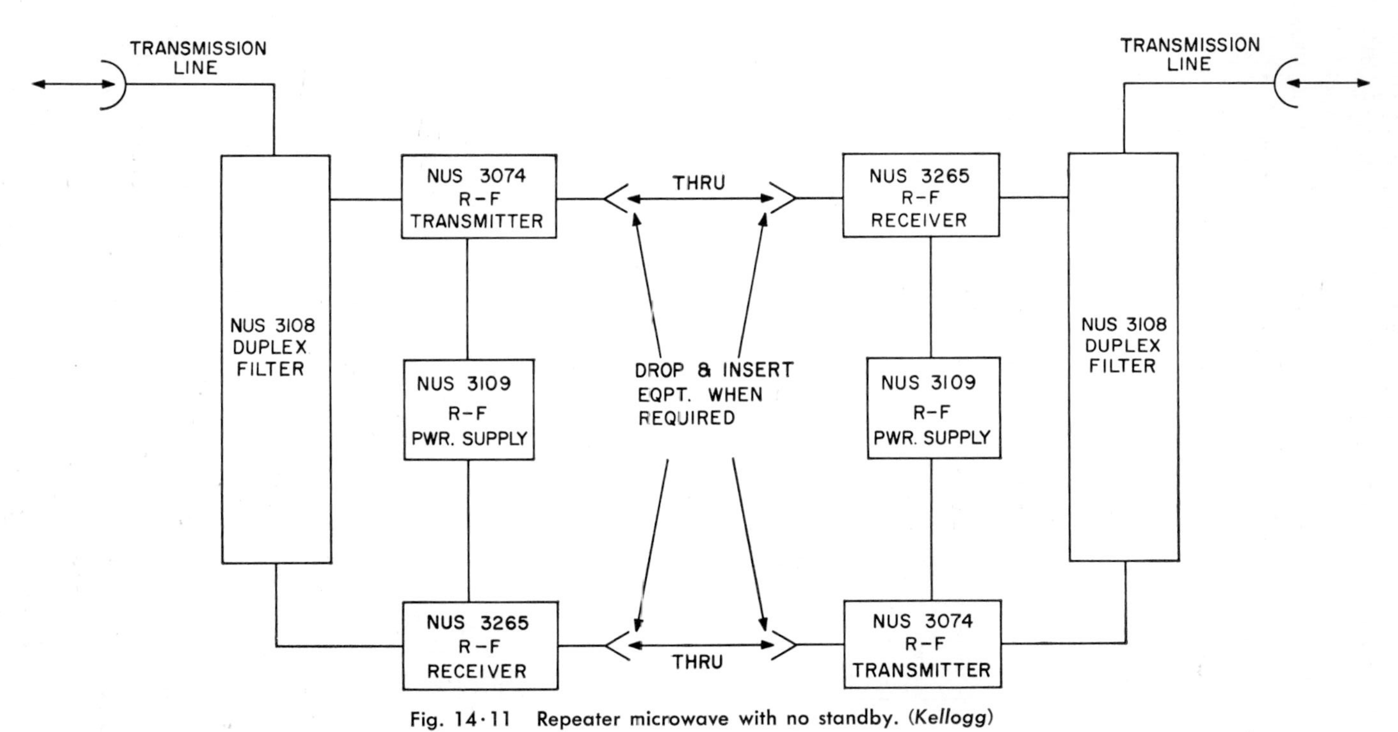

Fig. 14·11 Repeater microwave with no standby. (*Kellogg*)

14·4 Telemetry Television means to see at a distance and telemeter means to meter (measure) at a distance. Industrial control and missiles both use telemetry techniques to transfer actual readings great distances. Remote oil-pumping stations on pipelines have transducers which convert oil pressure, fuel level, and bearing temperatures of the pumps into electric signals. These signals are transmitted by radio to a master station where they are received, converted into meter readings, and displayed or recorded. Missiles and satellites also use telemetry to send information about their condition and their surroundings to ground stations.

Telemetering is the reproduction of a meter indication at a location remote from the original measurement location. The field is not new, for

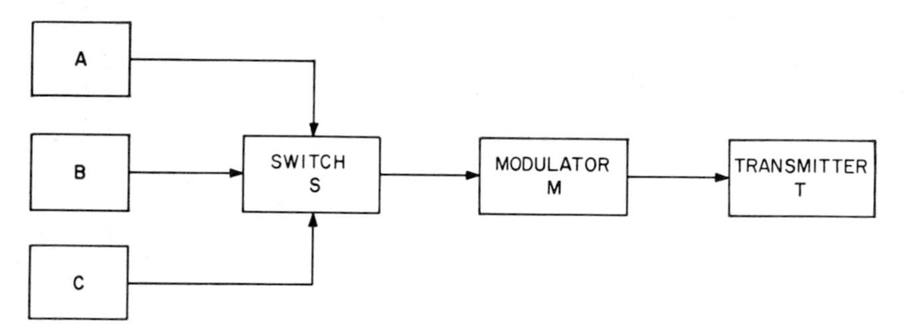

Fig. 14·12 Sampling method for three inputs *A*, *B*, and *C*.

current and voltage readings can, of course, be transmitted directly over wires. Carrier current is a method of wired radio where the measurements modulate on low radio frequency (below the AM band). This signal is then coupled to the transmission line which carries the a-c power from one power station to another.

Basic Principles. Radio is now widely used for telemetering because of its flexibility in situations where there is no existing wire link between two points. A basic system is one in which data from an electrical measurement modulate a transmitter. Radio is the link to the receiver, which demodulates the signal and presents the same data on the remote indicator. More than one channel of information or several meter readings are transmitted by a single carrier.

An industrial process may have several indicators (such as *A*, *B*, and *C* in Fig. 14·12) reading pressure, rate of flow, and temperature. These are devices which read the changes and translate them into electrical values. For example, temperature measurement by means of a thermocouple is recorded as a voltage so that the voltage amplitude is a function of temperature. This is indicator *A*.

A commutator switch *S* samples this voltage for a short period and then goes on to sample the next indicators in order before returning to *A*. In this way the indications are sampled and read one at a time, in sequence, into a modulator M. Thus the modulator sends out pulses of r-f energy through the transmitter T.

These pulses are received (R) and decoded (D) at a central control station where they provide a remote reading for the original indications as A, B, and C as shown in Fig. 14·13.

In the basic system, a quantity such as pressure is measured by the transducer, and this change is converted into a method of modulating the master oscillator. There can be several channels, each with its oscillator and modulator; these are known as subcarriers. All these subcarriers go to

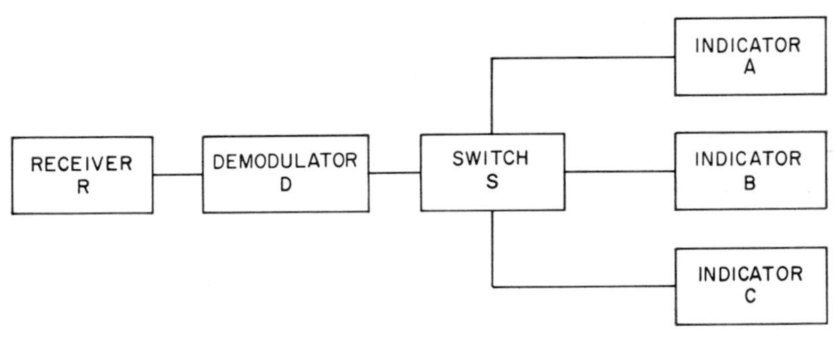

Fig. 14·13 Receiving end for three outputs.

a reactance-tube modulator which in turn modulates the main carrier oscillator. After amplification, this r-f signal is transmitted. At the receiver, each subcarrier is detected and converted into the original reading.

Sampling Methods. In Sec. 14·3 a pulse method for time-division multiplex is discussed. There are several possible sampling techniques.

The sampling process shown in Fig. 14·12 is one way of sending these readings on a single channel. Because the signals are taken one at a time, this is a sequential method of transmission.

Of the various methods of data transmission, one is illustrated in Fig. 14·14. A meter reads a quantity such as pressure, rate of flow, temperature, or humidity and has the meter shaft coupled to a variable capacitor. A change in meter reading changes the capacitance in the oscillator-1 circuit, which is low r-f. This variable frequency modulates the r-f carrier which is the transmitted signal. At the receiver this is demodulated and presented as the remote meter reading. Several remote stations can have readings on the same carrier.

Because there are usually many readings to be transmitted, an entire technology of telemetering sampling and modulation processes has been developed. If, for example, a continuous (not sequential) reading is required for quantities 1, 2, and 3, each reading first modulates a low-

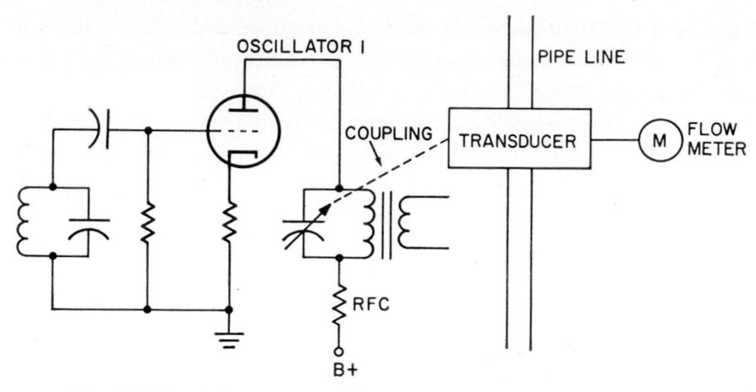

Fig. 14·14 Subcarrier modulation from transducer on pipeline.

frequency oscillator which is a subcarrier. Each of the subcarriers then modulates the higher-frequency carrier which is transmitted with all the required information to the receiver as in Table 14·1.

Table 14·1 Subcarrier Oscillator Frequencies

Oscillator	Frequency, mc	Modulation, ± 100 kc	Mixed with carrier
1	10	9.9–10.1	109.9–110.1
2	15	14.9–15.1	114.9–115.1
3	20	19.9–20.1	119.9–120.1

Combinations such as those in Table 14·2 are possible.

Table 14·2

Combination	Subcarrier	Carrier
a	AM	AM
b	AM	FM
c	FM	FM

In (*a*), the subcarrier channels are all AM and their modulation of the carrier is AM, but in (*b*) they modulate the carrier with FM. In (*c*), both the subcarrier and the carrier are FM. This technique allows a number of subcarriers which, by filters at the receiving end, are properly separated for their individual meter readings.

An example of the use of radio telemetering is shown in Fig. 14·15. A pipeline is installed cross-country to avoid digging up in built-up areas wherever possible. This means that the pumping stations, of which two are shown, are usually located in remote areas so that point-to-point microwave radio (voice) communication is used to connect these stations. By multiplexing several channels on one carrier as discussed above, there is no space for telemetering and voice with one frequency or single-carrier operation.

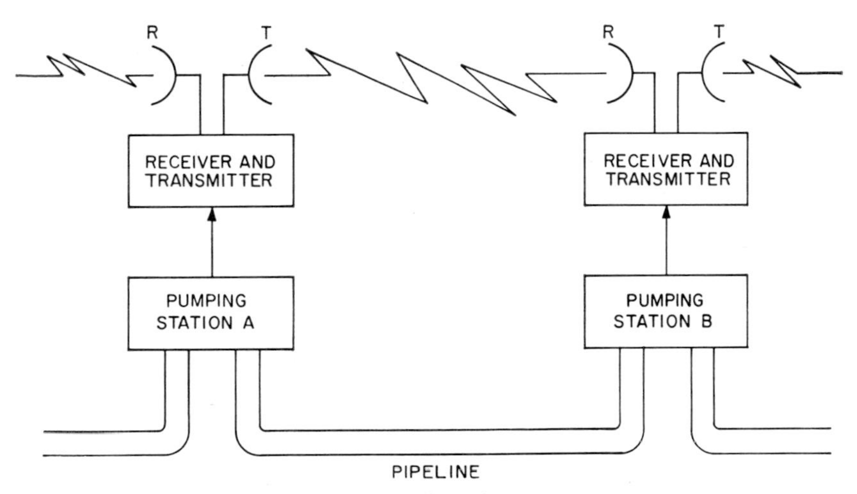

Fig. 14·15 Part of a telemetry system.

At *A* the received signal is amplified, sent to *B*, and so on down the line. But at *A* the required meter readings for pressure, rate of flow, temperature, and others are added to the r-f signal retransmitted to *B*. At *B* further additions are made for the readings at pumping station *B* and retransmitted.

The central control station receives all these signals and, because of telemetering, a complete picture of the operation of the pipeline at any moment is available.

If commercial a-c power is used, a telemetering channel measures the line current flow. If this drops to zero and the relay is still operating, then it has gone to emergency power—a diesel or gasoline engine. Then a telemetering channel reads the fuel level, which indicates how long the station can be expected to operate under these conditions. When a-c line power is restored, the emergency power generator is stopped automatically.

Telemetry can be divided into two types depending on the method of multiplexing, which is the technique for simultaneously sending many channels of information using a single r-f channel. In *time-division multiplex,* for example, 30 channels can be transmitted using a sampling technique so that one channel is sent at a time and all channels are sent in sequence. An example of this uses three channels where, on each of the three input channels, there is a varying voltage representing changing pressure, humidity, or temperature. These are sampled by a rotating switch, or commutator, so that only one signal at a time appears at the modulator.

Since each signal is sampled less than one-third of the time, the input is a series of pulses whose amplitude is a function of the size of the input signal. This is pulse-amplitude modulation (PAM). The system is a three-channel time-division multiplex using PAM.

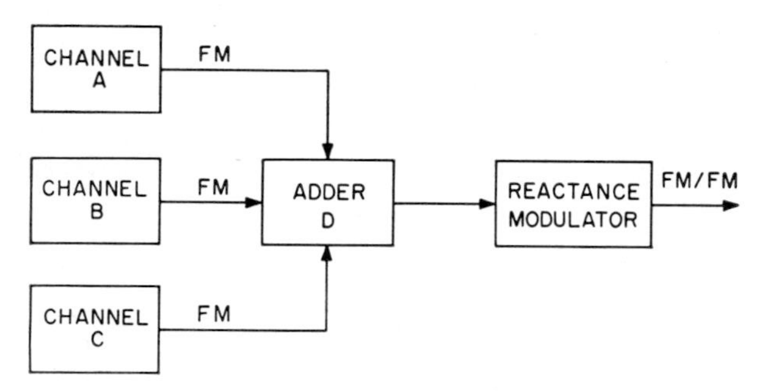

Fig. 14·16 Three-channel input to adder circuit.

If these three signals are varying at a slow rate, they are individually sampled and the average value of the signal is transmitted. The pulses from each sample contain information as to changing d-c values. With a high commutator rate the reconstructed signals at the receiver provide a continuous record of the original changing values.

In *frequency-division multiplex,* three signal channels each have their own subcarrier oscillator, which is FM as shown in Fig. 14·16. By means of the adder, these are brought together and fed to the reactance modulator D. Since each channel A, B, and C is FM and all three are again frequency-

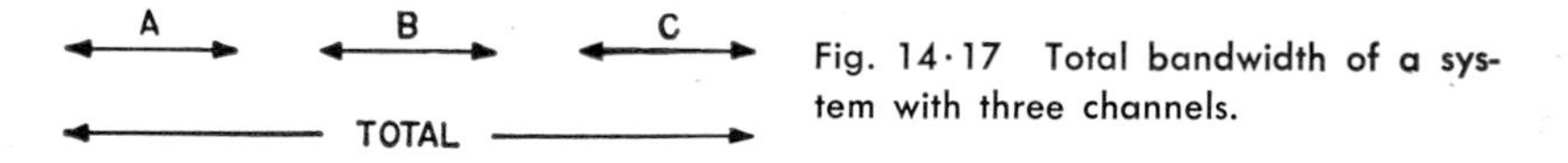

Fig. 14·17 Total bandwidth of a system with three channels.

modulated in D, this is an FM/FM system. Here all the carriers are of a different frequency and A, B, and C add to produce the total bandwidth. In each case, the actual bandwidth (Fig. 14·17) depends upon the deviation.

A more complete system is shown in Fig. 14·18 where a group of PAM signals are fed to modulator A and they modulate a subcarrier by FM. Other subcarriers such as B, C, and D also are modulated in the same manner. After passing through the adder, each again is frequency-modulated by E. Thus, this is a PAM/FM/FM system. The three pulse signals into modulator A are again transmitted in sequence as shown in the figure.

Pulse Techniques. Pulse modulation is often used. With a sine wave as the modulating signal it is, of course, possible to modulate a carrier with frequency modulation (FM), amplitude modulation (AM), or phase modulation (PM). A fourth method is pulse modulation, in which the original signal is changed into pulses.

In this method, the pulses can be modulated in one of several ways as shown in Fig. 14·19. In amplitude modulation (B) a series of slices or pulses sample the original signal, and the amplitude of each pulse sample

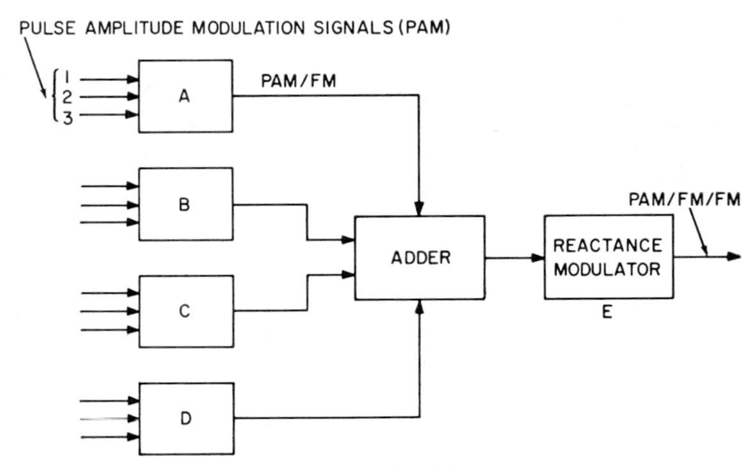

Fig. 14·18 PAM/FM/FM system.

corresponds to the original signal amplitude at that time. This is pulse-amplitude modulation (PAM). In pulse-frequency modulation (PFM), as shown in C, the pulse frequency varies as the original signal amplitude.

In addition to PAM and PFM, which are rather like ordinary AM and FM, there are other modulation techniques used with pulses. In pulse-position modulation (PPM), shown in D, marker pulses (M) are used and the information pulses vary in their position. They are farther apart during the positive signal peak and closer during the negative peak. In pulse-code modulation (PCM), shown in E, the number of pulses and their position form a binary code which is a measure of the original signal amplitude. Pulse-width modulation (PWM) varies the duration of the pulses to provide modulation.

These special types of pulse modulation are not difficult to implement. Consider the PWM system in F. The signal input is to be converted into pulses of varying amplitude.

While a large number of separate inputs are possible, only three are considered: A, B, and C. These can represent any three variables. A ring counter is a series string of tubes or transistors arranged with the output fed back to the input so that only one stage is on at any moment. This ring counter, which acts as a rotating switch, connects each input to the comparison circuit, in turn, for a short sampling period so that the sequence is *ABC, ABC*. Trigger signals for the ring counter come from the crystal clock. This same clock triggers a ramp generator and a flip-flop on. Voltage from the ramp generator rises until it is equal to the sampled voltage as determined by the comparison circuit. At this time the com-

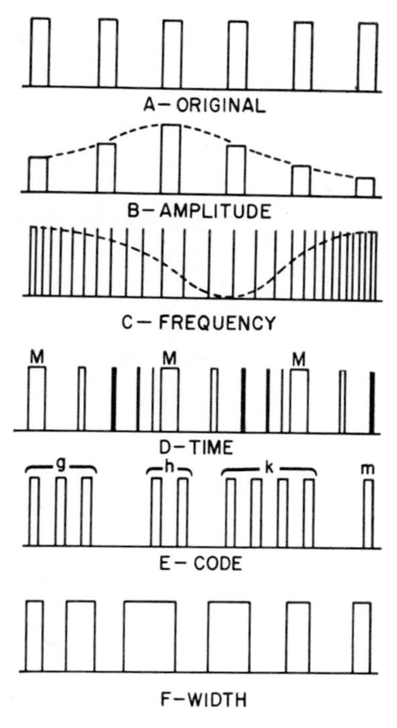

Fig. 14·19 (*A*) Pulse-modulation schemes start with original pulses; (*B*) amplitude modulation; (*C*) frequency modulation; (*D*) time or position modulation; (*E*) modulation by code.

parison circuit turns off the flip-flop, and by this technique the duration (or width) of the flip-flop output is a measure of the transducer input.

PCM reduces the original signal to binary pulse groups. As a first step, the signal to be transmitted is quantized. Each sampling is rounded off to the nearest whole number. This is required because these pulses are to be converted into binary code. When they are decoded at the receiving end, the original signal will be reconstructed.

In the binary code there is either a pulse, which is a 1, or no pulse, which is a 0. The count is by two rather than by ten, as in decimal, so the count from 1 to 7 in decimal is 001, 010, 011, 100, 101, 110, and 111. Each pulse sample is converted into a group of equal pulses; the number of these pulses corresponds to the amplitude of the original signal sample. These groups of pulses are then counted and changed to a binary code. The original transducer signal is changed to a PWM signal so the width varies as the original amplitude, as shown with PWM. A crystal clock provides accurate pulses which are fed to the gate. The number of pulses which pass through the gate is determined by how long the pulse-width modulator keeps it open. Thus, for a large-amplitude original sample there is a wide pulse from the pulse-width modulator, which permits a large number of pulses through the gate. For a small-amplitude original sample, there is a narrow PWM output and few pulses through the gate. A series of interconnected flip-flops at the binary counter produces a binary-code output from the clock pulse input.

Subcarrier modulation is sometimes used for each input signal, and these subcarriers are then combined. A number of different bands or channels are available as in Table 14·3. For the bands 1 through 18, the great-

Table 14·3 Telemetry Frequency Bands
(In cps)

(IRIG band) *	Center frequency	Lower band limit	Upper band limit	Band-width
1	400	370	430	60
2	560	518	602	84
3	730	675	785	110
4	960	888	1,032	144
5	1,300	1,202	1,398	195
6	1,700	1,572	1,828	256
7	2,300	2,127	2,473	346
8	3,000	2,775	3,225	450
9	3,900	3,607	4,193	586
10	5,400	4,995	5,805	810
11	7,350	6,799	7,901	1,102
12	10,500	9,712	11,288	1,576
13	14,500	13,412	15,588	2,176
14	22,000	20,350	23,650	3,300
15	30,000	27,750	32,250	4,500
16	40,000	37,000	43,000	6,000
17	52,500	48,560	56,440	7,880
18	70,000	64,750	75,250	10,500
A	22,000	18,700	25,300	6,600
B	30,000	25,500	34,500	9,000
C	40,000	34,000	46,000	12,000
D	52,500	44,620	60,380	15,760
E	70,000	59,500	80,500	21,000

* Bands 1 to 18, ± 7.5 per cent deviation; bands A to E, ±15 per cent deviation.

est deviation allowed is ±7.5 per cent, so for band 8, for example, the center frequency is 3,000 cps. With maximum deviation, the upper limit is 3,225 cps and the lower limit is 2,775 cps. In this way, the deviation never extends into the next upper band (9) or the next lower band (7). In addition to these 18 bands, there are five others (A through E) where the deviation allowed is twice the other, or ± 15 per cent. If these are used, there are certain restrictions as noted to prevent any interference or overlap. Where band A is used, bands 15 and B are not used.

Usually the frequency band from 216 to 235 mc is used. There are several names for these standards, including Inter-Range Instrumentation Group (IRIG), Research and Development Board (RDB), and Inter-Range Telemetry Working Group (IRTWG).

Industrial Systems. A basic telemetry system with 11 sets of tone oscillators and gates is shown in Fig. 14·20. The subcarrier oscillator is modulated by this unit's total output signal from the summing amplifier. When a gate is opened by a d-c voltage, the audio signal of the associated tone oscillator is fed to the output summing amplifier, which in turn can modulate a voltage-controlled subcarrier oscillator. As mentioned above, there can be a number of subcarrier oscillators feeding the master oscillator. Five of these used in this manner, each with 11 tone generators, would result in 55 information channels.

Oscillators for subcarriers can be inductance-controlled, capacitance-controlled, voltage-controlled, and so on. A technique for voltage control is a separate oscillator stage feeding a modulator (which also has an input), the transducer voltage. But in the interests of small, lightweight, and compact equipment a simple multivibrator circuit can be used as shown. A sensor or transducer is fed to a control tube whose output is a voltage varying as does the input. This voltage controls the frequency of operation of the multivibrator subcarrier oscillator.

If the transducer can produce a variable d-c voltage, this technique of control is used. Here the control tube acts as the modulator while the multivibrator is the subcarrier oscillator.

Many different devices are used for measurement of quantities to be used for telemetry. A simple mechanical bellows is used to measure atmospheric pressure. On the ground, the pressure inside the bellows is the same as that outside and it is in a state of rest. If the device is abroad a missile which is going up, there will be a decrease in external pressure from the atmosphere, which is the same as an increase in internal pressure, and the bellows expands. By coupling this bellows movement mechanically by linkage, the inductance or capacitance of the subcarrier oscillator can vary, thus creating a changing output frequency.

Another method of inductance control, where the transducer provides a current, is to use this current through a saturable reactor. As the control current changes, the tank inductance varies, which changes the subcarrier oscillator frequency.

Voltage-controlled oscillators require transducers which change their resistance with the quantity to be measured. In series with a voltage source, these variable resistances provide a variable current whose voltage drop is the transducer output. Two of these transducers measure temperature and humidity. A thermistor, which is the temperature-sensitive resistance discussed earlier, measures temperature changes. Humidity

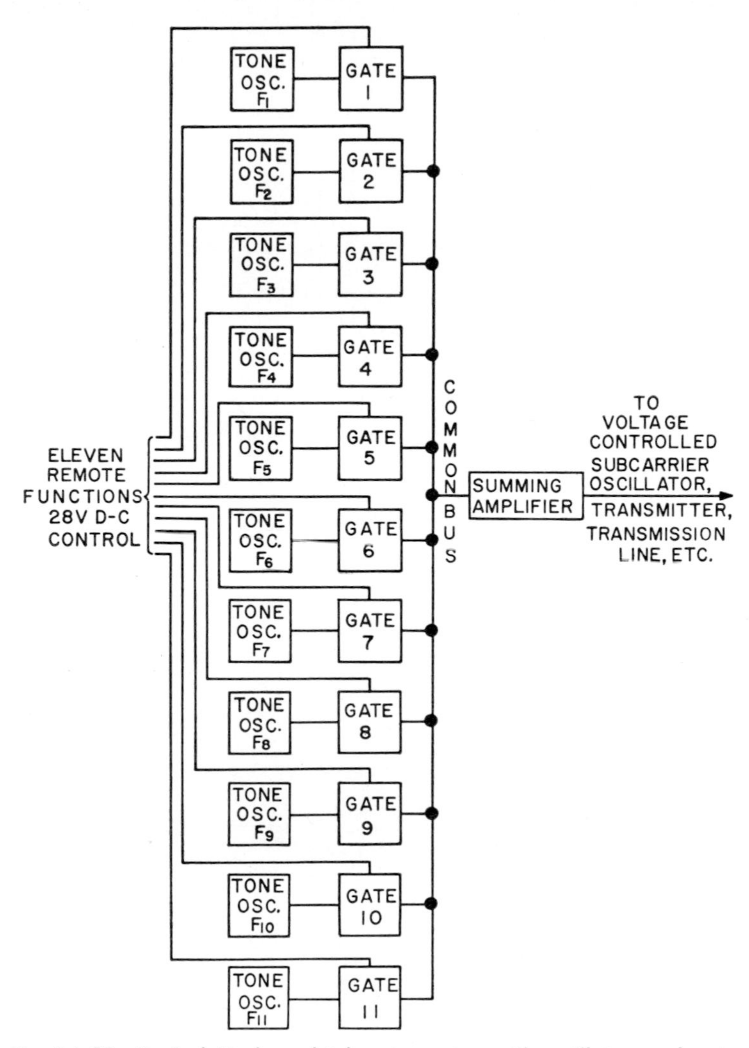

Fig. 14·20 Typical 11-channel telemetry system with oscillators and gates.

sensors are made of materials whose resistance changes with the amount of water vapor which is present.

A 30-channel pulse-width telemetry-system transmitter is shown in Fig. 14·21. Each of the multiple transducers is connected to an input relay. The relay string, as shown in the figure, is connected to the ring counter which closes each relay, in sequence, for 600 msec. Voltage from the transducer being sampled is compared with a rising voltage source or ramp voltage. This may be seen from Fig. 14·22, where three different ramps are shown. Each ramp is turned on by the clock-pulse leading edge.

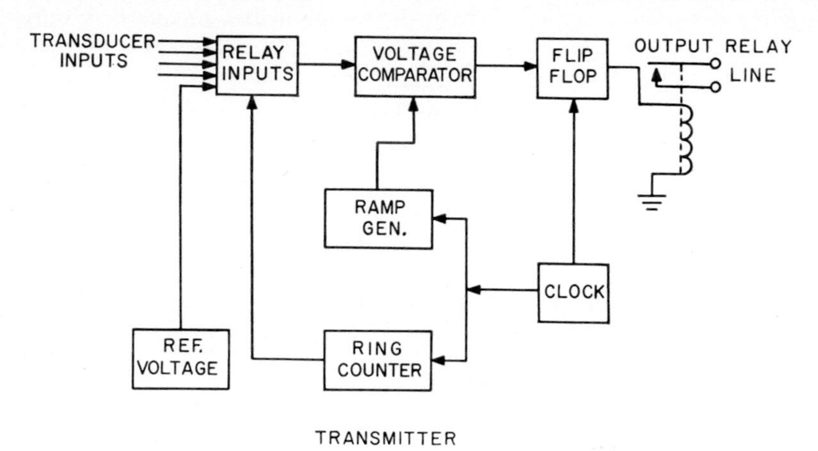

Fig. 14·21 Telemetry transmitter diagram. (*ASCOP*)

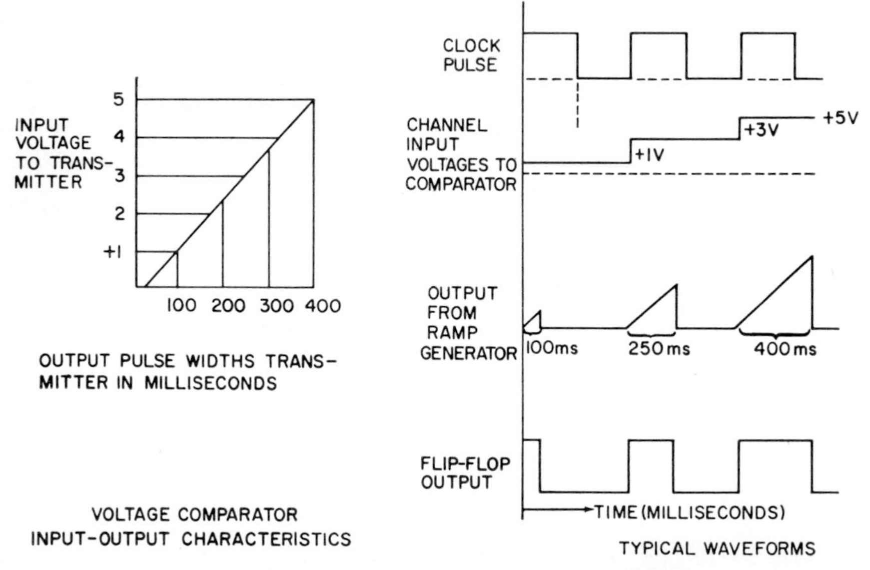

Fig. 14·22 PWM transmitter characteristics. (*ASCOP*)

At the same time the clock pulse turns on the flip-flop and closes the output relay for this channel.

As soon as the ramp voltage has risen to the same value as the transducer voltage, as seen by the comparator, the comparison circuit turns the flip-flop off and this opens the output relay. The conversion of d-c transducer level to flip-flop pulse width can be seen. Output pulse widths as a function of transducer voltage can be seen from the figure.

Receivers for this system can be seen from Fig. 14·23. A synchronizing pulse 1,200 msec wide, twice that of a clock pulse, is transmitted and received. At the receiver the sync detector recognizes this pulse and resets the counter to start a new count. In this way transmitter and receiver reach synchronism automatically when the units are first turned on. After the start, the counter advances for each closure of the input

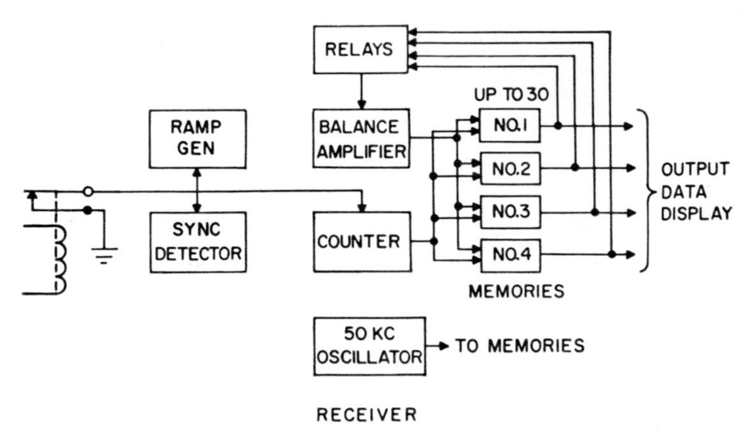

Fig. 14·23 Telemetry receiver characteristics to match the transmitter. (*ASCOP*)

relay so that the proper output display is connected for each transmitter pulse.

For each channel the counter starts an inverted ramp. The end of the input pulse terminates the ramp voltage and hence reconverts the data from pulse width to a d-c voltage. A balance circuit automatically adjusts the ramp voltage to start at zero, and a magnetic memory is used to retain the last state of each channel, in the event of power failure.

14·5 Remote control A discussion of remote control of devices by means of radio properly should be under Mobile Radio (Sec. 14·2), but it is included here because these techniques are useful for activating lights, gate locks, and other industrial devices.

The mobile radio can either receive or transmit codes for selective calling. In other industrial applications these same techniques are used for remote control; an example is the system used in Burbank, California, to control air-raid sirens (Fig. 14·24).

Selective calling permits the selection of mobile units by the base transmitter either individually, in groups, or all at once in response to different codes created by the operation of a telephone dial or push buttons.

At the mobile station, which is a vehicle or a boat, when the correct code is received for a particular vehicle, in a typical system, an alarm sounds a buzzer and a lamp is lighted. The decoder may open a normally muted loudspeaker, turn on a light, or sound an alarm to warn the operator of a call.

If the mobile units have a code sender and base stations have a decoder, vehicles can also call selectively. Base-station loudspeakers may then be muted until signaled the proper code of a mobile unit. In this way an operator can select a specific base station or one of the control points of

Fig. 14·24 Secode system in Burbank, California, for air-raid sirens.

a base station without alerting the others. Selective calling equipment can be installed on all mobile or base-station radios.

Other electric devices may also be remote-controlled from a mobile unit by selective calling. By equipping both mobile units and base stations with code senders and decoder units, any radio station can dial any other station in the same system without alerting the others just as if all stations were connected by a dial-telephone system, but the stations cannot (in this system) be connected by dial phones with the Bell Telephone system.

There are two types of equipment used for selective calling. The first type is digital, which uses a series of dual or single tones. These are transmitted and then decoded at the receiver. A second type also transmits a single tone or group of tones, but not in pulse form. These are detected at the receiver by a frequency-selective decoder.

Digital Pulses. This type of selective calling uses a series of pulses representing digits in the code. Each pulse represents the presence of tone modulation, if a single tone is used, or a change from one tone to another if two are used.

An audio oscillator and a telephone dial can be used at the transmitter. When a number is dialed, a number of tone pulses are transmitted so that the number of pulses is the dialed number. The equipment required for the mobile and the base station consists of a code sender, a decoder, and a call head.

The number dialed is often called a "code." In the telephone system, the first three digits dialed (the prefix) are called the code; the last four digits are called the number. In mobile radio, any numbers dialed to call a car or base station are called the code, or code number, of that station. Again in telephone language, the figures 0 through 9 are digits; combinations of these make a number. Thus the digits 4, 5, 6, and 7 make the number 4567.

By using a two-contact decoder, the dispatcher can also sound a vehicle's horn (or turn on its lights) to indicate an incoming call. This is especially desirable on utility or construction trucks where crews may be working in the vicinity of the vehicle.

Two codes (or call numbers) may be assigned to each vehicle. One code actuates the dash indicator light. If there is no response, the dispatcher dials the next code, which closes the second contact and turns on lights or horn for about 4 sec.

The code sender, made up of the dial and tone generator, is connected in parallel with the microphone; either input can modulate the transmitter.

When the dial is turned, contacts close and this closes a relay which applies plate voltage to the tone generator. At the same time it keys the transmitter on and sends the tone signal to the transmitter. Releasing the dial opens other contacts, which interrupts the tone signal.

As a number is dialed, the dial pulsing contacts alternately open and close the oscillator output circuit, thereby interrupting the tone signal and forming digital tone pulses to produce the particular digit dialed.

The dial return causes the code sender to produce "break" pulses or holes in the tone, the number of holes equaling the digit dialed. This system uses the principle of negative keying. It is the absence of tone after a gate tone has been sent that constitutes a pulse. A chief advantage of this system is high noise rejection. Selective signaling operates when the noise on the circuit is as much as five times the signal.

In operation, the base-station operator wishing to contact Mobile Unit 3, for example, simply dials the appropriate code. The base-station code sender produces the dialed pulse train which is transmitted to all mobile units. Only the decoder unit in Mobile Unit 3 will respond to activate the call head, which lights a call light and sounds a buzzer to indicate the incoming call.

A decoder is designed to count the pulses for the transmitted code and is connected to the receiver discriminator.

The decoder has a driven code wheel which has teeth and which is energized by an electromagnet. Each tooth contains a tiny hole into which a small wire "code pin" or a "contact pin" can be inserted by hand.

Electrically, the selector consists of a single electromagnet, equipped with two armatures. One of these armatures is fast-operating, and in some models it will respond to as many as 60 current pulses per second. This armature is coupled by a simple linkage to a driving arm, which turns the code wheel, one tooth at a time.

More than one code number is available to be used on a selector. This feature is often used in mobile radiotelephones, such as in police cars where each car may have its own individual code number, and each may be called by the central station, without signaling unwanted cars. However, each car in the fleet or group may have a second number which is the same for all cars. Thus if an "all cars" call is to be broadcast, the second or fleet number is called instead of the number for an individual car. This second number would then signal all cars at once. In addition, the use of additional contact pins can supply a "group call" system. That is, a number of cars, such as those in a certain class of service, or in a certain district, can have a group number. Calling the group number would signal all cars in the particular group.

The call head indicates the presence of a properly coded call. Then when the selector unit receives the correct sequence of pulses, it provides the momentarily closed circuit to the call indicator head or a similar external device. The call indicator head contains the buzzer, lamp, interposing relay, and other required circuitry for visual and audible indication of an incoming call.

An audible tone is heard from the speaker to announce a call. The "off" push button breaks the relay circuit, disconnects the speaker, and turns the lamp off. The "on" button allows the operator to monitor the channel before placing a call.

Tone Selective Calling. Tone signaling is a selective calling system using the frequency to distinguish between the different signal codes. The use of a-f signals, within the voice-frequency range, permits the signals to pass over wire lines (telephone) as well as radio transmission lines.

In some applications a fleet can be divided into several groups, each of which can be called by a different group code. At the transmitter the calling signal frequencies are produced by tone generators as shown in Fig. 14·25.

Tuned reeds are often used for tone generators. A tuned-reed relay is similar to a tuning fork. When energized by a magnetic field of the correct frequency, a tuned reed will vibrate and can be made to close an electric contact while vibrating. Some mobile radiotelephone systems use these reeds for selective calling of mobile units. The car radio is equipped with a combination of several reeds. The main transmitter is equipped to send out the different frequencies to match the many different reeds used.

When a certain combination of tones is transmitted, the reeds in the different cars will vibrate if their frequency is received. When the right combination of tones causes all the reeds in a particular car to vibrate, an electric circuit is closed which signals the driver by turning on his loudspeaker.

In operation, the desired code is selected from the appropriate tone generators by a telephone dial or by push buttons. The operator then presses the call button on the microphone; this connects the transmitter to

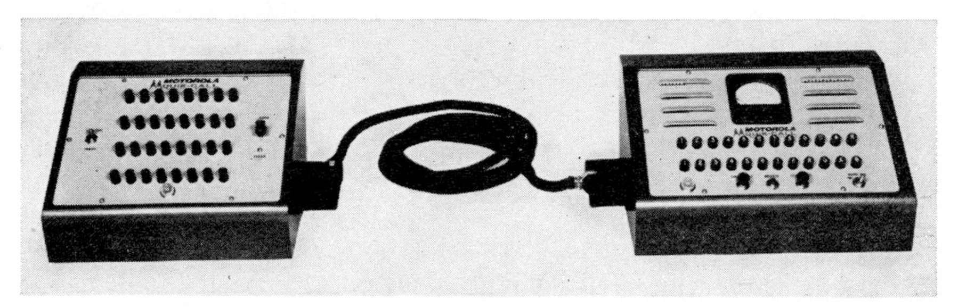

Fig. 14·25 Tone generators for selective calling. (*Motorola*)

the tone generator for a short pulse while the tone modulates the carrier. After a short period (about 0.5 sec) the switch connects the microphone for regular voice transmission and disconnects the tone generator from the transmitter.

All mobile units on the same carrier frequency receive this transmission and all receivers have a frequency-sensitive vacuum-tube switch. When the tone and the carrier are received, the switch unblocks or unmutes the audio stages. After the tone signal opens the receiver, the carrier signal holds the receiver in operating condition and the normal voice transmission is received. At the end of the transmission the switch again blocks the receiver until the next proper tone plus carrier. All receivers in a group receive the same message; all the other mobile receivers tuned to the same carrier frequency but having different tone codes are undisturbed.

Remote control can use either tones or pulses for turning lights on or off, for opening or closing gates, or for other purposes.

REVIEW QUESTIONS

14·1 What are some of the uses of two-way radio in industry?
14·2 How can materials handling use radio?
14·3 Explain the function and purpose of the FCC.
14·4 How does the FCC control radio?
14·5 What parameters does the FCC require to be measured?
14·6 What is paging radio? How is it used?

14·7 What is a radio repeater unit?
14·8 Where and how is it used?
14·9 How could an industrial plant use radio paging?
14·10 What type of circuit is shown in Fig. 14·6?
14·11 How is this amplifier driven?
14·12 What are some of the uses of marine radio?
14·13 What frequency bands are used?
14·14 What is Citizens Band Radio?
14·15 What frequency bands are used for two-way land-mobile radio?
14·16 What are the four types of Citizens Band Radio?
14·17 What is the purpose of class D Citizens Band Radio?
14·18 What are the class D license requirements?
14·19 What are microwaves? Where are they used?
14·20 How are the pulses in Fig. 14·19 modulated?
14·21 What is telemetry? Where is it used?
14·22 How is input sampling done?
14·23 Why are the inputs sampled?
14·24 What is a subcarrier and how is it used in telemetry?
14·25 Explain the functions of the subcarrier shown in Fig. 14·14.
14·26 If three subcarriers were 5, 7, and 9 mc and each was modulated ± 8 kc, show the output where the main carrier was 400 mc.
14·27 Do the same as above for twice the modulation limits.
14·28 What is an FM/FM system?
14·29 Explain PAM versus PPM modulation systems.
14·30 What is time-division multiplex?
14·31 How does it differ from frequency division?
14·32 In telemetry, what are the parameters of these bands: 5, 9, 11, 13, C, and E?
14·33 In Fig. 14·20, what is the function of the gates?
14·34 Explain the function of the ring counter shown in Fig. 14·21.
14·35 What is selective calling in two-way radio?
14·36 How can it be used for remote control?
14·37 What is the difference between push-button and dial calling?
14·38 How does two-way radio tie in to land telephones?
14·39 How do tone and pulse systems of selective calling differ?

CHAPTER

15

Industrial Television

Television is actually only beginning to show its vast potentialities for industrial uses. In telemetering for remote installations, such as isolated pumping stations on a cross-country pipeline, various conditions such as line pressure, fuel level for the power equipment, or outside temperatures are monitored automatically. The readings are sent by TV so that the operator at the control station can see the actual operating conditions as they exist at the remote location.

Television is often used to observe industrial operations and provide a visual presentation in a different location. A TV monitor can watch a railroad yard or a steel furnace or can protect an area against intruders. The use of industrial television (ITV) has grown very fast, and several organizations make specialized equipment for this purpose.

15·1 Applications of ITV One example, in telemetry, is the use of a camera for antenna boresighting (Fig. 15·1), a precision alignment so that the antenna will point at the missile. In Fig. 15·2 ITV is shown being used for inspection of missile parts to an accuracy of 0.001 in.

Underwater television, as another example, allows the remote observation of underwater objects and activities. It is used in constructing or evaluating the condition of pier and dock pilings and underwater moorings for buoys or ships, to evaluate damage and observe the repair of ship hulls, to monitor underwater drilling operations, for diver training, and for many other applications. The camera, including adjustment of the lens iris and focus, is completely controlled by the camera control unit. Housings are available for use by divers or for mounting in a permanent, fixed position for continuous monitoring. Special lighting equipment is available for use at night or at great depths where ambient light is low.

Figure 15·3 shows a boiler inspection-and-control system. In order to create a remote firing control system for the huge boilers of a power plant, the power engineer needs a system of continuously indicating the presence or lack of a flame within the furnaces. He must have this information available at all times at the central control panel. A remote system such as this greatly increases the operator's efficiency and brings

Fig. 15·1 ITV camera on missile-tracking antenna. (*Kin-Tel*)

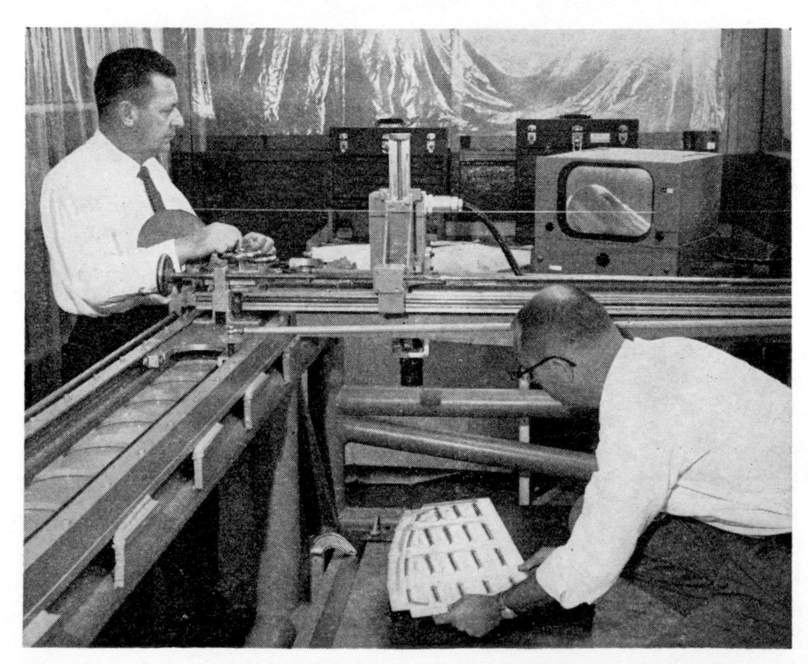

Fig. 15·2 Inspection of manufactured parts by television. (*Kin-Tel*)

the plant a step closer to the push-button power-generating station of the future.

Engineers of the Department of Water and Power, Los Angeles, California, designed the control room of the Scattergood steam plant, on the oceanfront at Playa del Rey, to operate with a remote firing control system.

Fig. 15·3 Scattergood steam plant using observation of furnaces. (*Kin-Tel*)

Each generating unit at Scattergood uses a divided furnace with a total of 16 burners; hence the problem of finding a sensing device showing the condition of all burners was difficult.

Installation of four closed-circuit television systems provides visual information for the operator that would otherwise require direct visual observation.

Each set of eight burners has a permanently fixed TV camera viewing its operation and sending the picture to a monitor in the control room. The operator can instantly view all 32 burners on the four monitors.

Both luminous and nonluminous flames can be seen on the monitors. Boilers in Los Angeles power plants must be equipped to burn both oil and gas. The red flame of fuel oil is easily picked up by the TV camera, but the blue flame is nonluminous and difficult to see. The use of filters and the high sensitivity and resolution of this TV system enable the gas flame to show just as strong as the flame of fuel oil. This quality becomes especially important during fuel switchovers, and it has reduced switching time from 30 to 2 or 3 min.

When observing furnaces, special protection is required by the lens because of the high temperature. A system for this protection is shown

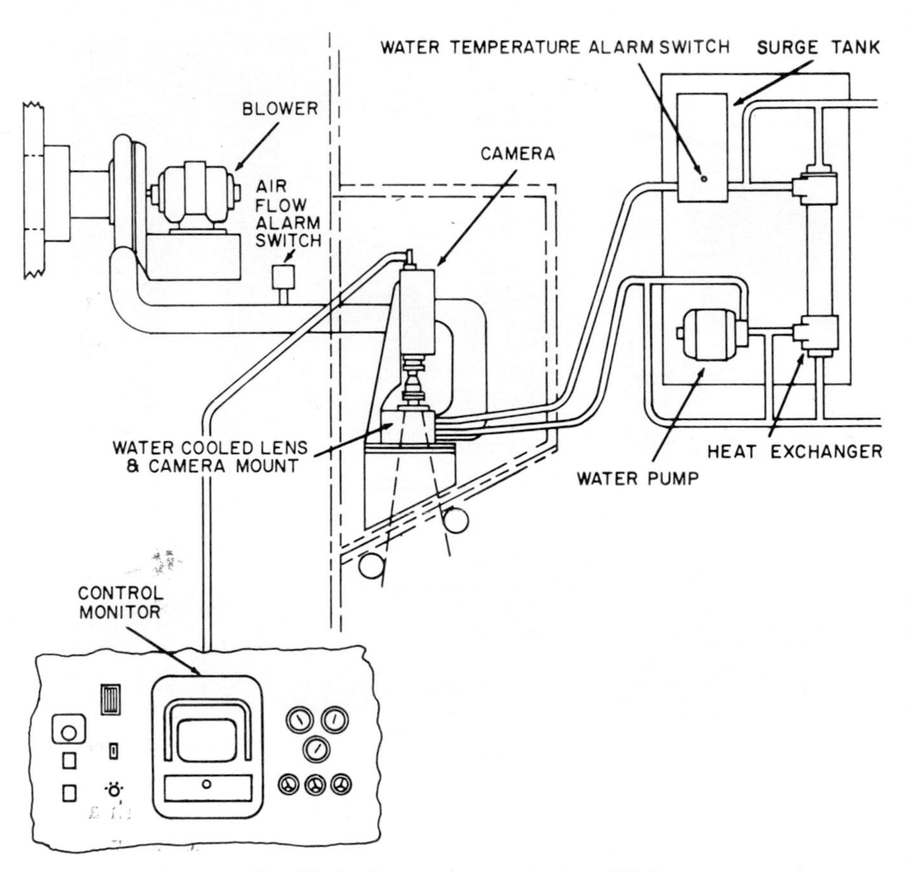

Fig. 15·4 Camera heat protection. (RCA)

in Fig. 15·4. A pump circulates the water in a closed system which includes the lens, the heat exchanger, and the water reservoir. Air from a blower is also used to cool the lens and to prevent an accumulation of combustion gases. Alarm switches are provided to guard against failure.

There are many applications of ITV in the Space Age. Closed-circuit television watches operational Atlas intercontinental ballistic missiles (ICBMs) in practice countdowns. Three explosion-proof TV cameras are equipped with remote-control zoom lenses and mounted on remote explosion-proof units at strategic spots around the Launch Operations building at Complex II of Warren Air Force Base, Cheyenne, Wyoming. Cables bring the pictures back to the above monitors in the Launch Control building.

The installation at Warren Air Force Base is virtually identical to closed-circuit TV setups at other operational Atlas sites either in existence or shortly to be completed at Vandenberg Air Force Base, California; Offutt Air Force Base, Nebraska; and Fairchild Air Force Base,

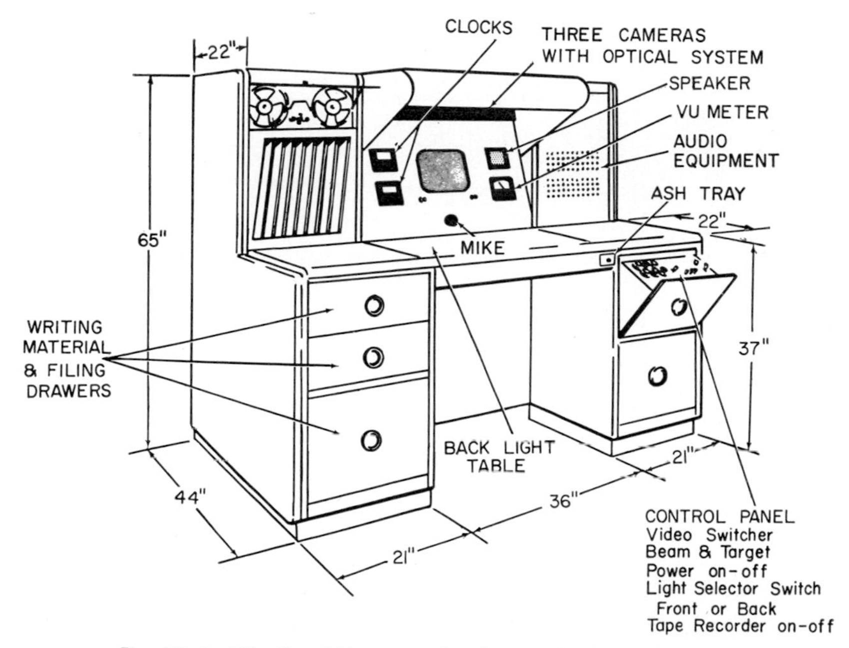

Fig. 15·5 Weather-Aide system for data transmission. (*Kin-Tel*)

Washington. The entire TV bird-watching program is known as Operation Lookout.

Industrial uses for TV are almost limitless. A Weather-Aide system (Fig. 15·5) is used at United States Air Force bases for transmission of weather information from the main weather office to briefing rooms. The system consists of a main console and a number of remote stations. The console contains three TV cameras to cover various-sized areas of the 36- by 22-in. working surface. Weather maps or charts being viewed can be either front- or back-lighted. Audio intercommunication is built in, and a tape recorder allows prerecording of an entire briefing. The monitor in the console shows placement of charts or maps as they are put in position.

Medical applications of ITV are also growing. In one of these a TV camera trained on a fluoroscope screen permits classroom viewing of details otherwise restricted to inspection in groups of three to seven students with consequent protracted exposure of radiology personnel to radiation. By use of TV and an associate intercommunication system, the radiologist is not required to wait until his eyes are accustomed to

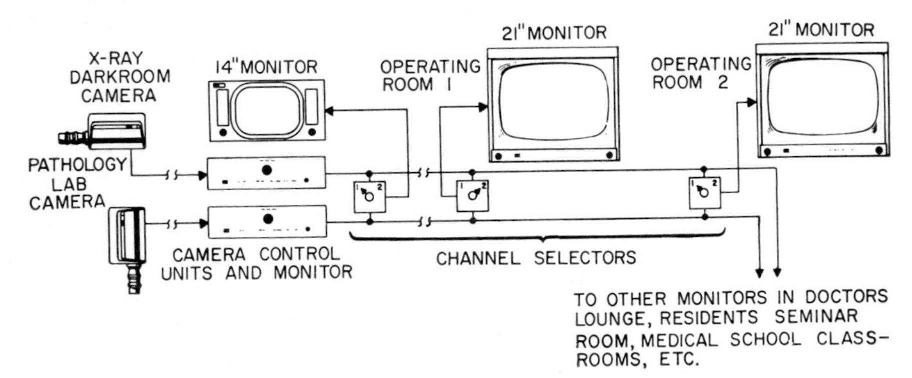

Fig. 15·6 Medical-school ITV system for use with two operating rooms. (*Kin-Tel*)

the dark, and he can direct the examination of 90 per cent of his patients without leaving his desk.

The block diagram in Fig. 15·6 shows a more comprehensive TV sys-

Fig. 15·7 ITV used with a microscope. (*Kin-Tel*)

tem with cameras in the X-ray darkroom and the pathology laboratory and with monitors behind glass panels in each of two operating rooms.

A simple channel-selector switch permits a surgeon to view a specimen or a still-wet X-ray film while scrubbed and with the patient on the table before him. Figure 15·7 shows a camera used with a microscope for a similar application.

Another application of ITV is the Datavision system shown in Fig. 15·8, designed to transmit printed data such as maps, charts, blueprints,

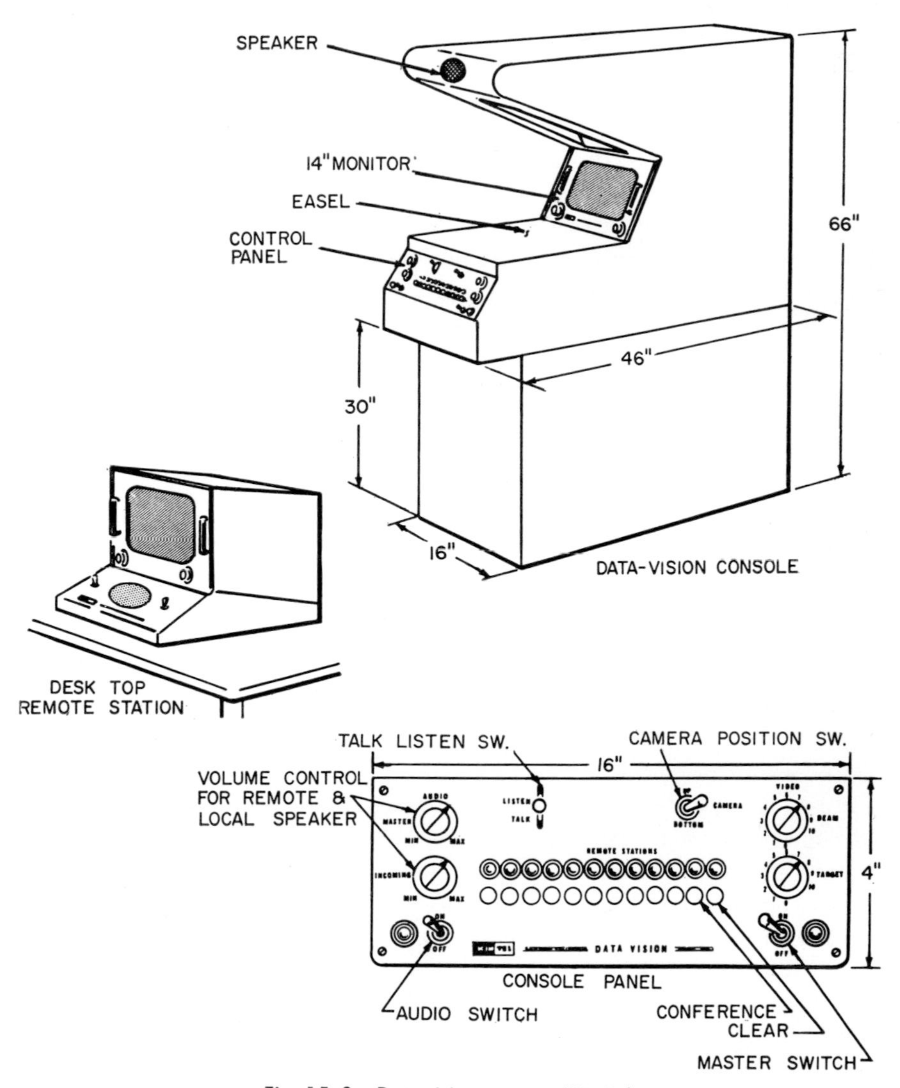

Fig. 15·8 Datavision system. (*Kin-Tel*)

parts lists, drawings, invoices, or other graphic material. The system consists of a transmission console and any number of desk-top remote stations. The transmission console houses the TV camera, monitor, and controls for adjusting the TV system to obtain the best possible picture.

The remote station can be either a 14- or a 21-in. monitor. Two-way audio intercommunication between remote stations and the transmission console can be obtained if desired.

This data transmission of all sorts of printed, drawn, or written records is an important aspect of ITV. Two stations, one for transmission and one for remote viewing, are shown.

The operation of the console is very simple. The operator turns on the equipment and adjusts the beam, target, and monitor controls for maximum picture quality. He may establish communication with the remote station by means of a standard telephone or the optional built-in intercommunications unit. He places the desired document on the lighted easel and determines the correct centering as shown on his video monitor. The view presented on the video monitor will show approximately an 8½-in.-high by 11-in.-wide portion of the lighted table. The operator can vary the camera position continuously, giving a total viewing area 11 in. wide and 20 in. long. A unit can be provided at the remote station which allows the operator there to remotely control the position of the camera.

The control panel for the Datavision console typically includes a buzzer to announce that a remote station is calling, lights to indicate which remote station is calling, volume controls for two-way intercommunication, a "talk-listen" switch, two a-c switches—one for all video equipment including lights and blowers and one for the audio equipment—and remote controls for the beam and target.

The Datavision remote station is complete with control panel, monitor, and control base. The panel on the control base will include an intercommunications unit and a remote camera position control, plus push buttons for selection of the proper originating console if more than one Datavision console is used.

When the remote-station operator pushes a button, the proper light on the selected console lights up. These lights remain on until the remote-station operator pushes his "clear" button. In this manner, the main console sees the number of remote stations desiring contact.

15·2 ITV systems Closed-circuit television instantaneously transmits visual information from point to point in an industrial plant.

Closed-circuit ITV consists of compact, rugged, low-cost television cameras and camera control units used with quality picture monitors or receivers. The monitors can be operated at a great distance from the camera location. Pictures of virtually any number of diverse operations can be brought to either widely separated locations or a centralized location. A three-unit (camera, camera control unit, and monitor) building-block approach, as shown in Fig. 15·9, allows extreme flexibility in system layout. Private, low-cost cable circuits connect the camera to the camera control unit and the camera control unit to the monitor. No government license of any kind is necessary for the operation of a closed-circuit television system.

A typical camera in a commercial TV system can be adjusted to a

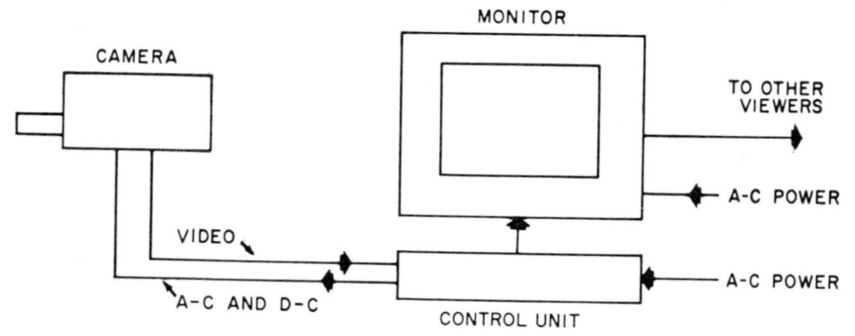

Fig. 15·9 Basic industrial television system showing camera, monitor, and control.

given picture coverage, but there are cameras (see Fig. 15·10) in which the movement and the lens may be changed by remote control.

This TV equipment features better-than-broadcast quality, automatic operation, and simple installation. A fully interlaced, 525-line scanning

Fig. 15·10 Zoom lens. (*Kin-Tel*)

system combined with an 8.5-mc bandwidth provides full 650-line resolution. Excellent pictures can be obtained under normal room lighting conditions.

The camera employs standard 16-mm lenses to focus the image of the scene being viewed onto the photosensitive face of a vidicon camera pickup tube. The vidicon converts the light image to an equivalent electric signal (video signal) which is amplified by a preamplifier in the camera and applied, through camera cable, to the camera control unit.

The camera control provides further amplification of the video signal from the camera and generates the necessary control and synchronizing voltages for the camera and monitor. The camera control can be set to provide fully automatic or manual operation of the TV system.

Figure 15·11 illustrates, in block-diagram form, the stages required for ITV. Since only horizontal pulses, vertical pulses, and video are needed and low power output is required, the equipment is less complex than that at a broadcast transmitter. The camera tube can be a vidicon, an icono-

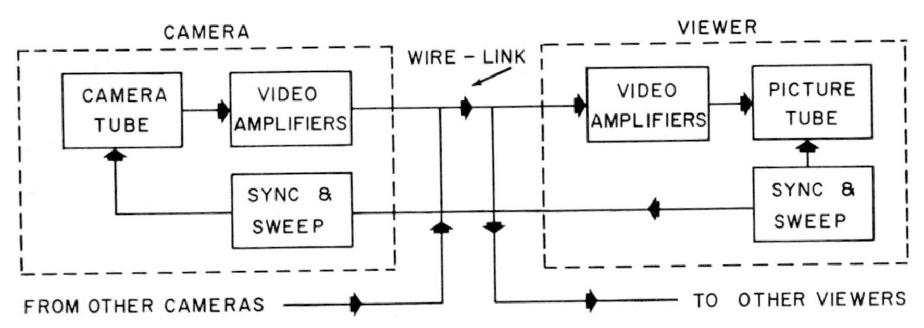

Fig. 15·11 System block diagram for the transmitter and receiver.

scope, or an image dissector which is provided with the required power and pulses from the sync and sweep blocks or camera control. Amplification of the camera output in the video amplifier is the third block.

A control unit does several things: it supplies heater and plate voltages for the camera tubes and other tubes; it has a video amplifier for the monitor; it has some form of camera-tube protection; and it provides synchronizing and sweep signals for the camera and the monitor. A typical control unit is shown in Fig. 15·12. This control is part of a system using a vidicon.

The vidicon is essentially an orthicon or low-velocity device. In the vidicon camera no electrical shading cancellation signals are necessary. Hence no shading controls are used nor is edge lighting or any other type of corrective lighting required for flat field.

Horizontal sweep is at 15,750 cps and vertical sweep is at 60 cps just as in commercial TV. The master oscillator is at 31.5 kc in synchronism with the 60-cps line. Frequency dividers (as in Chap. 7) count down from 31.5 kc to the vertical 60-cps timing.

A wired link, in this case, carries the signal to the viewer-receiver or monitor. A single coax can carry the pulses and video; or several cables can be used, in which case the video and pulses are not mixed. Again only a simplified receiver is needed; the r-f and i-f amplifiers are not required. A special receiver, built for wired ITV, will need only a video amplifier, a picture tube, and the sync and sweep blocks, in addition to a power supply. A monitor for ITV needs only these blocks, but a normal broadcast receiver can be used by bringing the video input directly to the video amplifier input.

Several viewers or monitors can share a single camera to allow viewing at several locations as in Fig. 15·13. By remote switching, several cameras can also be used to view different parts of an industrial operation. This is

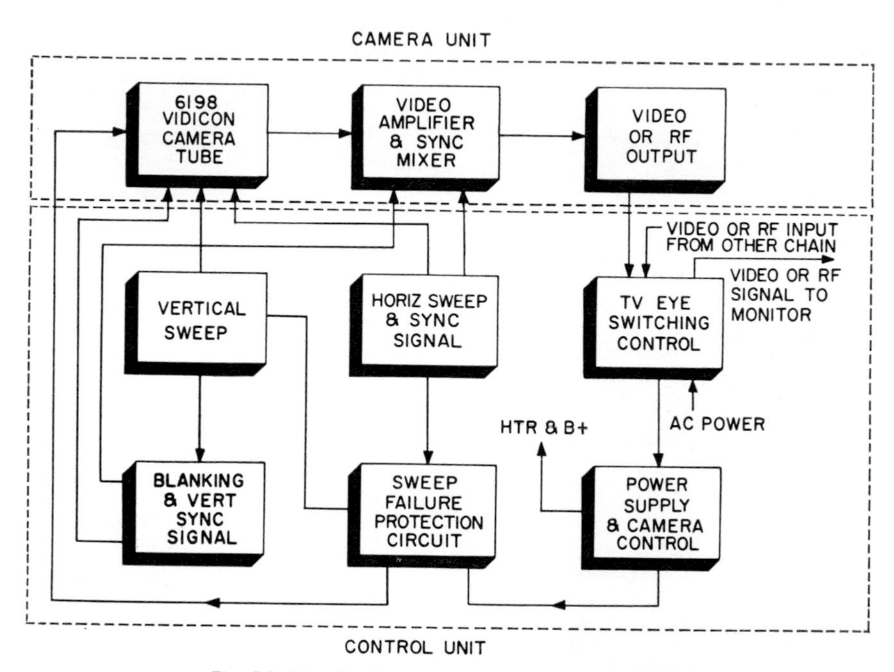

Fig. 15·12 TV Eye system block diagram. (*RCA*)

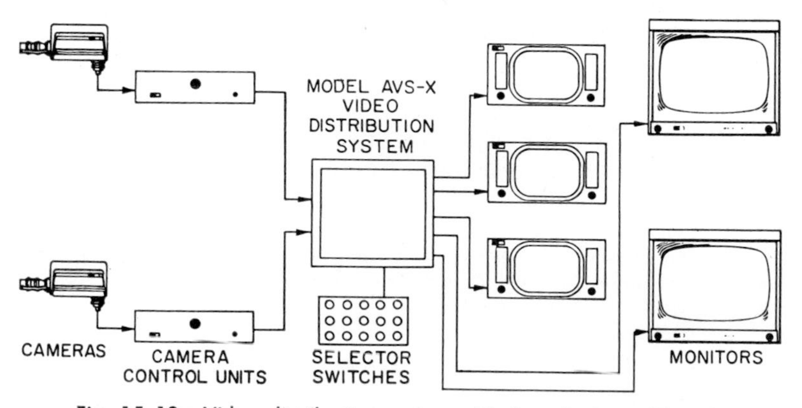

Fig. 15·13 Video distribution system with five displays. (*Kin-Tel*)

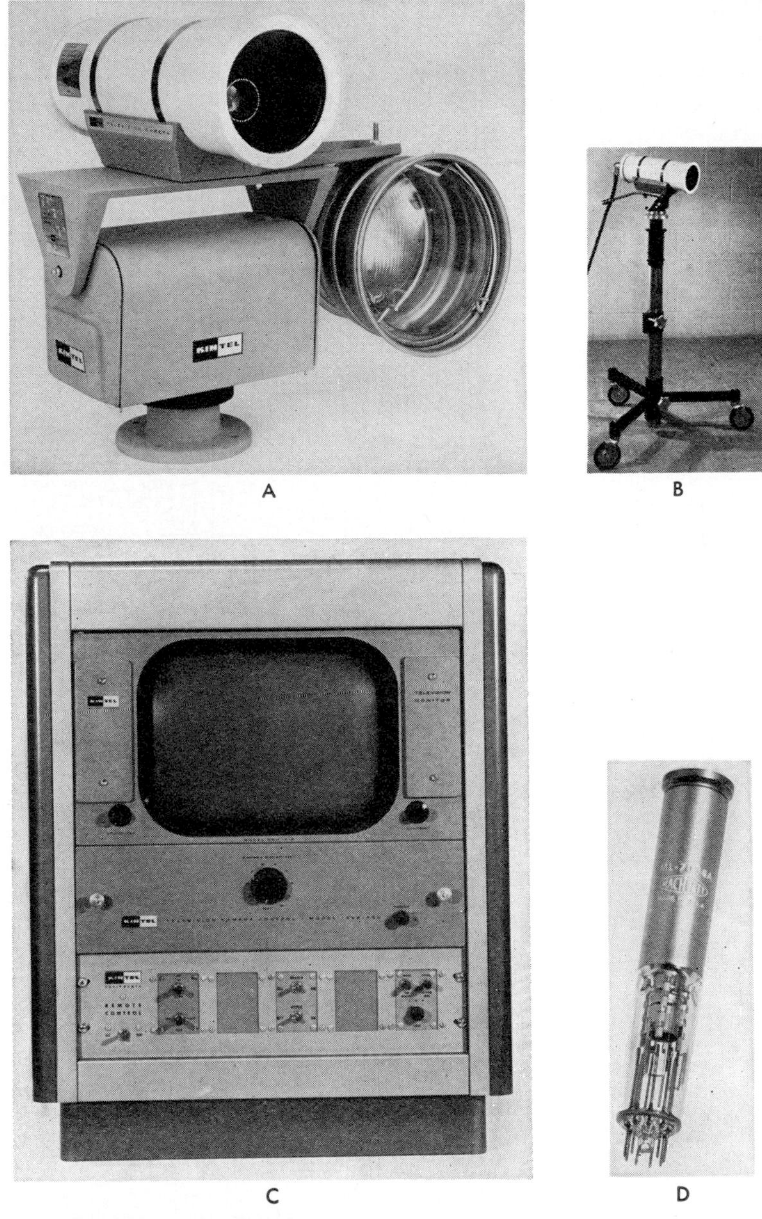

Fig. 15·14 (*A*) Enclosed ITV camera for protection from explosive hazards (*Kin-Tel*); (*B*) camera on a tripod (*Kin-Tel*); (C) camera control, monitor, and remote control (*Kin-Tel*); (*D*) camera tube ML-7038A (*Machlett*).

not the same as channel switching (no radio frequency is being used here), but a selector switch at each viewer permits the choice of any of several camera outputs by actually switching the input lines to the individual viewers.

A system for duty in extreme conditions of heat, vibration, or humidity is shown in Fig. 15·14. *A* is the camera on the mechanism which rotates the camera or turns it up and down. A light for extra illumination is shown on the right. This same camera on a tripod is illustrated in *B*. Figure 15·14*C* shows the monitor at the top, the camera control in the center rack, and the remote control on the bottom.

Standard 16-mm lenses are employed to focus the light image on the sensitive face of a vidicon-type pickup tube as shown in *D*. The electrical

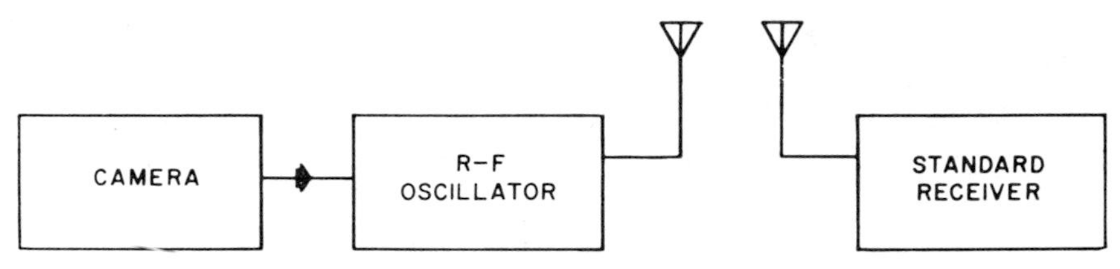

Fig. 15·15 Television pickup by standard TV receivers.

output of the vidicon is amplified by a high-gain, low-noise solid-state preamplifier. In addition to the vidicon and the preamplifier, the camera contains horizontal and vertical deflection coils with ring-magnet beam alignment, a constant-current electromagnetic focusing circuit, and a solid-state protective circuit which prevents damage to the expensive vidicon in the event of sweep-circuit or cable failure. The camera employs no electromechanical elements such as relays and no electrically variable components requiring adjustment. The vidicon assembly is protected against magnetic fields by means of a Mumetal shield. All necessary control voltages, deflection signals, focusing current, and operating power are supplied by the camera control unit through the interconnecting cable.

15·3 Broadcast ITV Another example of TV flexibility is shown in Fig. 15·15; here the camera is used to modulate a low-powered r-f oscillator. If the oscillator is on one of the commercial channels which is unused in the local area, a standard TV receiver can be used as the viewer without impairing the use of the receiver on existing local TV broadcast channels. This procedure is a simple and direct use of ITV at the lowest possible cost because of the use of existing receivers without modification. (FCC permission must be obtained for all TV transmitted by an r-f carrier.)

Elaborate systems can be made, as shown in Fig. 15·16, by using a number of units such as those shown in Fig. 15·15 with a microwave transmitter *T*. Several cameras (*A*, *B*, and others) can each modulate an r-f oscillator on one of the standard channels. These modulated signals are then used to modulate a single microwave transmitter. At the receiving end, the microwave receiver R heterodynes each of the single channels back to the original radio frequency, which may now be tuned in by any of the several standard receivers. This method is used for longer hops than are possible with a single r-f oscillator; remote links use this method.

15·4 The television camera lens Many of the techniques of ITV are similar to those of broadcast television. The receivers are probably most

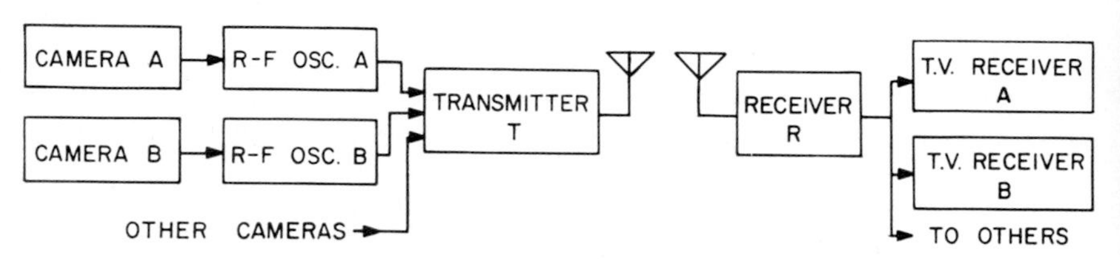

Fig. 15·16 Microwave transmission for closed-circuit television applications.

familiar to the technician, the camera probably least familiar. A camera with a built-in control is shown in Fig. 15·17. A shows the camera with the four-lens turret; B shows the internal construction.

Protective circuits are included to prevent vidicon damage in the event of sweep failure. With the exception of a single ruggedized preamplifier tube and the vidicon itself, transistor circuits are used for low power consumption and long trouble-free service. Adjustable phaseless aperture

Fig. 15·17 (A) Four-lens 20/20 camera; (B) inside the camera case. (*Kin-Tel*)

correction is available from -8 to $+8$ db. The packaging design involves swing-out circuit boards for ease of maintenance.

Figure 15·18 shows a close-up of a remote-control lens system. Both the iris or lens opening and the focus are adjustable.

Lenses are important to the application of closed-circuit television. If the ambient light level is low, a "fast" lens must be used, that is, a lens with a low "f" number or aperture.

If the physical size of the lens is important, remember that a fast lens has a larger diameter than a slow lens. The basic relationship is

$$f = \frac{\text{focal length } (FL)}{\text{diameter of lens } (d)}$$

Example: A 3-in. $f/2.5$ lens is 1.2 in. in diameter; OD of the lens barrel is $1\frac{13}{16}$ in.
A 3-in. $f/1.5$ lens is 2 in. in diameter; OD of the lens barrel is $2\frac{1}{4}$ in.

The depth of focus changes with aperture (iris setting) as shown on the facsimile of a typical calibrated focus control on a 16-mm lens barrel in Fig. 15·19.

Fig. 15·18 Close-up of lens mechanism. (*Kin-Tel*)

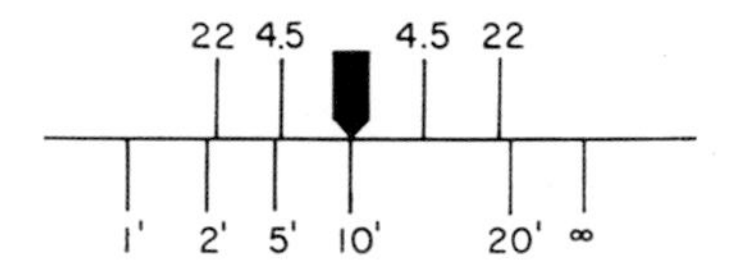

Fig. 15·19 Camera indicator (see text).

Focus is set at 10 ft. If the f-stop is at $f/4.5$, the depth of view will be from about 5 to 15 ft, but if the f-stop is at $f/22$, the depth of view will be from about 3 ft to some 20 ft for this hypothetical lens.

The resolution capability of a lens is sometimes, but not always, important. A television installation to be used for plant security requires high sensitivity when ambient light levels are low. Here resolution is not important. If the television system is used to transmit printed data or observe meter panels, resolution is important and sensitivity is not.

In most applications the camera position is predetermined. The proper lens can be selected, knowing any of the following:

1. The width of the object to be viewed
2. The angle of view required
3. The magnification desired

Figure 15·20 shows these relationships. The chart shows the lens-to-subject distance versus the width of the subject in feet. The table gives the angle of view of some standard lenses.

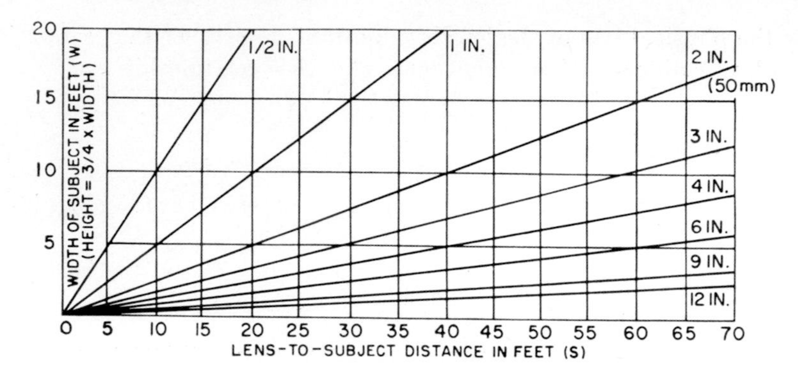

s lens to subject distance, ft
w width of vidicon target, in.
W width of subject, ft
FL focal length of lens, in.
FL $\frac{ws}{W}$

Angle of View for Some Standard Lenses

Lens	½ in.	1 in.	2 in.	3 in.	4 in.	6 in.	9 in.	12 in.
Horiz. angle of view	53°	28°	14°	9.6°	7.1°	4.8°	3.2°	2.2°
Vert. angle of view	41°	21°	10.7°	7.2°	5.4°	3.6°	2.4°	1.8°
Diagonal angle of view	64°	34.6°	17.8°	11.9°	9°	6°	4°	3°

Fig. 15·20 At top, lens charts relating lens distance to subject width; at bottom, angle of view for standard lenses. (*Kin-Tel*)

The angle of view for any lens can be determined using the formula

$$\tan\frac{\theta}{2} = \frac{\frac{1}{2}h}{FL}$$

where θ = total viewing angle
h = width, height, or diagonal of vidicon raster
FL = focal length of lens

If magnification is known,

$$FL = \frac{sm}{(m+1)^2}$$

where FL = focal length, in.
s = vidicon to subject distance, in.
m = magnification

Example: An object $\frac{1}{16}$ in. wide is to be viewed from a distance of 30 in. It is desired to have the object occupy the full width of the monitor screen. What lens is required?

Object width = $\frac{1}{16}$ in. = 0.0625 in. Vidicon raster width = $\frac{1}{2}$ in. = 0.5 in.

Therefore

$$\text{Magnification } (m) = \frac{0.5}{0.0625} = 8$$

$$FL = \frac{sm}{(m+1)^2} = \frac{30 \times 8}{(8+1)^2} = 2.96 \text{ in.}$$

Use a 3-in. lens.

It is important to know the inherent magnification of the ITV system. A 14-in. monitor as shown in Fig. 15·21 has a raster width of $10\frac{1}{2}$ in.

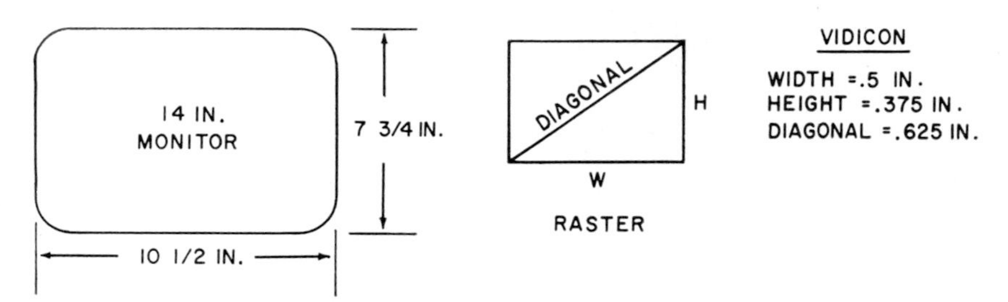

Fig. 15·21 Monitor and vidicon rasters. (*Kin-Tel*)

The width of the vidicon raster is 0.5 in. Hence, when a 14-in. monitor is used, the system has a magnification in the horizontal direction of

$$\frac{10.5}{0.5} = 21$$

REVIEW QUESTIONS

15·1 How can a camera be used as a plant protection device?

15·2 In a steam plant, what are the advantages of the use of ITV?

15·3 Why does a camera, as shown in Fig. 15·4, require protection from heat?

15·4 Give some examples of the use of ITV in medical schools.

15·5 How could a bank use a Datavision system?

15·6 What components make up an ITV system?

15·7 What is broadcast ITV? How does it differ from nonbroadcast ITV?

15·8 What are the functions of horizontal sweep, camera sweep-failure protection, and vertical sweep in Fig. 15·12?

15·9 What is a video distribution system?

15·10 How may a standard TV receiver be used in ITV?

15·11 What factors on a camera are varied?

15·12 If subject width is 15 ft, how far from a 2-in. lens can the subject be?

15·13 For a 4-in. lens, objects 40, 50, and 60 ft away from the camera must be how wide?

15·14 What is the relation between subject height and width?

15·15 What is the difference in angle of view between a 3-in. lens and a 9-in. lens?

15·16 What is the *f*-stop of a lens?

15·17 Is *f*/1.9 faster or slower than *f*/4.5? Why?

CHAPTER

16

Industrial Computers

Computers are fast becoming a part of every industry. Digital computers for control and data processing use logical circuits for their operation.

Logical design is employed in large-scale digital computers, but it is not limited to this area. The same principles have a direct bearing on, and increasing importance in, control circuits for other types of electronic devices. The automatic control of manufacturing processes is an area where logic plays an important role. Wherever relays, transistors, or vacuum tubes are used for control, logic is an aid to design and understanding of the circuit operations.

16·1 Principles of computers Relay circuits, as an example, make a convenient starting point, and their relationships are applicable to transistor or vacuum-tube circuits as well. The symbolism is taken from formal logic. As an example, there are two fundamental ideas of logic: a statement is true or it is false. In the same manner, a relay is open or it is closed. In logic there are the connectives used to express the relations between two statements. One is *OR*, which corresponds to relays in parallel, and another is *AND*, which corresponds to relays in series.

Symbolic logic traces its beginnings to the work of Gottfried W. Leibnitz (1646–1716), who tried to create "a general method in which all truths of reason would be reduced to a kind of calculation," and Augustus De Morgan (1806–1871). Both attempted to add logic to algebra. This was done by George Boole (1815–1864) when he published (in 1847) "The Mathematical Analysis of Logic" and his masterpiece (in 1854), "An Investigation of the Laws of Thought, on Which Are Founded the Mathematical Theories of Logic and Probabilities" (available in reprint form from Dover Publications, New York).

A number of works have appeared on the subject of mathematical logic.

But it was not until 1938 that the modern significance of this work was pointed out by Claude E. Shannon, then a student at Massachusetts Institute of Technology. His classic paper, A Symbolic Analysis of Relay and Switching Circuits, appeared in the *Transactions of the AIEE*, vol. 57, 1938.

Types of Computers. There are both digital and analog computers; in basic terms, an analog computer measures like a voltmeter and a digital computer counts like an adding machine.

Fig. 16·1 Data-processing center. (*National Cash Register*)

The typical large-scale digital computer shown in Fig. 16·1 is designed for use in data processing of large quantities of individual readings. An analog computer is shown in Fig. 16·2.

In an analog system quantities are measured (not counted) and fed as inputs to the computer. The computer acts on the inputs in such a way as to perform a number of mathematical operations or to solve an equation, and the results are usually plotted on a graph as output data.

To multiply two numbers using logarithms, the log of the first number is added to the log of the second and their sum is the log of the result. On a slide rule the same thing is done mechanically. A slide rule is only a table of logarithms.

An analog computer is more direct and simple than a digital computer, but it is not, ordinarily, a high-precision device. An analog computer is at its best when a precision of three or four digits is sufficient, no program branches are required, and there is one independent variable. Data which are available in analog form, such as varying voltage, can be directly fed to an analog computer.

The disadvantages of an analog computer are its lack of precision and its limited flexibility. Usually the computer has to be reconnected, by means of patch cords, to run a different problem.

A digital computer, on the other hand, counts rather than measures. Most large computers are digital in nature; they are versatile and, in structure, the electronic type has many small building blocks of basic circuits.

Fig. 16·2 Analog computer showing program patchboard. (*Philbrick Association*)

Analog Computers. These computers use basic relationships between two varying voltages in an analog representation.

Many electronic analog computers use amplifiers, which are high-gain stages which can be used for many operations. Examples of commercial amplifiers are shown in Fig. 16·3. By the use of simple networks a single stage (actually two tubes) can perform operations such as addition, subtraction (negative addition), multiplication (the input multipled by the gain of the amplifier), and division (multiplication with less than unity gain). All these can be accomplished quite directly.

Other mathematical operations can also be done in a single stage as shown in Fig. 16·4*A*, and this is the inherent value of the analog approach. Among the basic analog operations are integration and differentiation. Integration uses a capacitor C to store the charge received. The voltage across C is, at any instant, the integral of the charging rate. This circuit computes the time integral of the input. An amplifier is added to prevent the removal of the charge on C during integration.

The inverse of integration is, of course, differentiation or the determination of the rate of change of a quantity with respect to another quantity, usually time.

Digital Computers. There are many types of digital computers, including a cash register and an abacus. The digital computer *counts* quantities, whereas the analog computer *measures* them. Because numbers are used in the digital computer, any required degree of precision may be obtained by carrying more digits. Since digits are used, the computer components (transistors and vacuum tubes) respond to on or off states like a switch which is either open or closed.

In many respects a digital computer is a kind of automatic calculating device which follows directions. Many digital computers can be represented by the block diagram in Fig. 16·4*B*; there are, of course, variations of this fundamental computer. The basic units are (1) the arithmetic and logic section, which performs the actual arithmetic operations; (2) the memory, or storage, which retains the program, the problem, and the results; (3) the control unit, which directs the computer operation; (4) the input device for translating all input information to a form usable by the computer; and (5) the output device, which translates the computer output into a form which is usable.

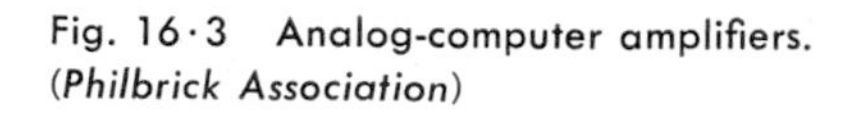

Fig. 16·3 Analog-computer amplifiers. (*Philbrick Association*)

The arithmetic unit is an accumulator which can store information and, under proper control, act on these stored data, which it obtains from the memory. A number from the memory can be placed in the accumulator; upon receipt of an order such as "add," a number will be added to that already present in the accumulator, which then contains the sum. When required, this can be transferred to the memory. It can also be retained by the accumulator for further operations. In the case of a negative number, the accumulator will recognize the negative sign. For a subtract operation the accumulator finds the difference and attaches the proper sign. To multiply, a series of additions is made in the accumulator. To divide, a series of subtractions is made.

The memory, or storage, has a large number of individual locations or addresses. Each of these can store information and retain it until needed. Each location can be described and located, so that either the informa-

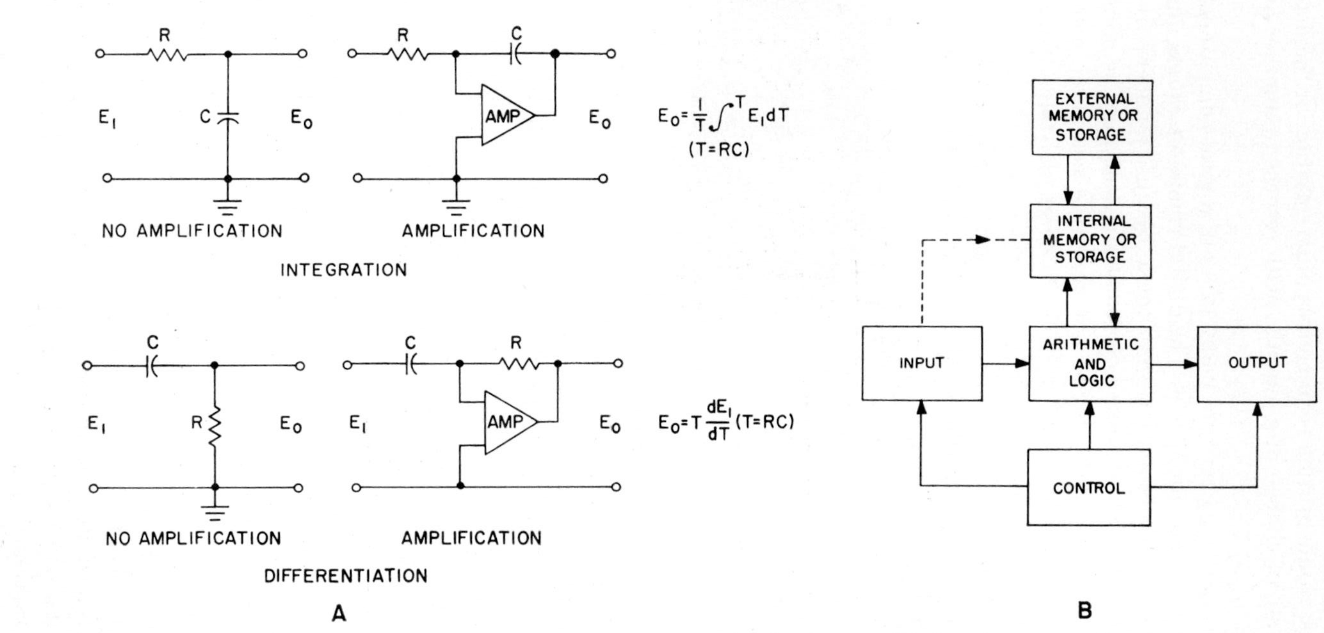

Fig. 16·4 (*A*) Analog operations of integration and differentiation; (*B*) digital-computer diagram.

tion which it contains can be obtained or the location for new storages can be described. Reading out information from the memory does not destroy the data, so that the same information can be referred to many times. In this type of computer all the information needed for a solution to a problem may be stored in the memory, including all the steps, or program, required for the solution. In this way, once the proper information is placed in the memory, the computer is independent of all outside devices until a solution is reached. Then the final results are stored until they are needed, when they are presented through the output device.

Magnetic drums are an important type of storage. This is a rotating metal drum with several tracks of recording material like loops of magnetic tape. Several heads may be used for each track; they record or read pulse information as the drum rotates. Clock pulses can be stored on a track to synchronize the entire operation of the computer.

Magnetic tapes (Fig. 16·5*A*) are also an important form of data storage.

A magnetic-core memory is a large group of tiny ring transformers. These are magnetized in one direction (clockwise) or the other (counterclockwise). The readout bus for taking information out passes through all cores. Other memories include the early use of relays (open and closed for 1 or 0), the Williams tube (a CRT with dots stored on the face for data), the capacitor store (in which the charge on a capacitor changes with the digital data), the transistor or vacuum-tube memory (in which the tube or transistor is either on or off for a 1 or a 0), and the delay line (which takes pulses and delays them until they are needed).

In *the control section* the implementation of the program is directed. The control unit observes the instructions and plans their proper execution by following the principles of operation for a given machine. This selection may translate a multiply order into a series of additions, which it actually is to the computer.

Input and output devices are similar, but they perform opposite functions. An input unit reads information from punched cards, magnetic tapes, or special keyboards. It also codes these data so that the computer can handle and use the information. The opposite function is performed by the output unit, which converts the results of computer operation to usable form such as typed sheets (Fig. 16·5*B*) or punched cards for the control of production machines.

A digital computer is a flexible and versatile instrument. It is most useful when high precision is required, when several independent variables are required, and when the program or instructions have several "forks" or possible paths. It is, of course, also useful when the input is inherently present in digital form.

Speed is a characteristic of digital computers; many can perform more than 10,000 operations in 1 sec. But this does *not* mean that a given

A

B

Fig. 16·5 (*A*) Magnetic-tape stations (*IBM*); (*B*) view of printer with case opened (*Analex*).

problem is solved faster than on an analog computer, for the faster digital machine (in terms of individual operations) may have to do thousands of operations to arrive at an answer.

16·2 Numbering systems Although we are most familiar with the decimal system based upon 10, there are other equally valid numbering systems using other bases. These include the binary, using the base 2; the biquinary, using 10 by pairs; and the Roman system of numerals. Our decimal system is used for most desk calculators, the Roman numeral system is still used for building dates, and the biquinary system is the basis of the abacus.

The decimal system bears closer examination. Take the digits 23, for example. In the decimal system we take this to mean 3 ones and 2 tens; to the base 5, however, this would mean decimal 13, for 23 is then 3 ones and 2 fives; using the base 8, 23 would be decimal 19, for it is 2 eights and 3 ones. This is shown in Table 16·1. In the first column we count by

Table 16·1 Counting from Decimal 1 to 10 in Three Numbering Systems

10	5	8
1	1	1
2	2	2
3	3	3
4	4	4
5	10	5
6	11	6
7	12	7
8	13	10
9	14	11
10	20	12

tens, in the second by fives, and in the third by eights. Counting in decimal is 0, 1, 2, 3, 4, 5, 6, 7, 8, 9, 10. To the base 5 the count is 0, 1, 2, 3, 4, 10; and to the base 8 it is 0, 1, 2, 3, 4, 5, 6, 7, 10.

The radix of a numbering system is its base; the radix of the decimal system is 10 and of the binary system, 2. Because we use numbers so very often we tend to forget that decimal numbers are not used in all applications. When we see 53 or VIII, what does it mean? The number 53 means 3 more than 5 tens; VIII is *not* 53 but the eighth in the series of Roman numerals. It is 1 five and 3 ones.

A numbering system is an ordered set of digits used for counting. When all numbers, in sequence, have been used in the first position, the second position is increased to the next symbol. For example, a system of ABC would count (as in Table 16·2) A, B, C, AA, AB, and so on. In terms of

Table 16·2 ABC Numbering System

1	A	11	CB	21	ACC	31	CAA
2	B	12	CC	22	BAA	32	CAB
3	C	13	AAA	23	BAB	33	CAC
4	AA	14	AAB	24	BAC	34	CBA
5	AB	15	AAC	25	BBA	35	CBB
6	AC	16	ABA	26	BBB	36	CBC
7	BA	17	ABB	27	BBC	37	CCA
8	BB	18	ABC	28	BCA	38	CCB
9	BC	19	ACA	29	BCB	39	CCC
10	CA	20	ACB	30	BCC		

decimal numbers A is 1, AC is 6, AAB is 14, BBC is 27, and CCB is 38.

Our normal numbering system is based upon tens. Any decimal number may be written as the sum of a series of numbers, each of which is a power of 10. This may be seen by making columns of tens and writing the digits in the proper places. The decimal numbers 352, 4,167, and 50,213 are written this way in Table 16·3. Each column, reading from right to left,

Table 16·3 Decimal Notation

10^4 10,000	10^3 1,000	10^2 100	10^1 10	10^0 1
. .	. .	3	5	2
. .	4	1	6	7
5	0	2	1	3

has an increase of 1 in the power of 10, so that 4,167 is, for example, 4(1,000) + 1(100) + 6(10) + 7(1). For mechanization, the decimal system of numbering (to the base 10) requires 10 different and recognizable states for each digit. In general, any numbering system requires m states, where m is the numbering base. Since a vacuum tube or a transistor usually has only two stable states which are dependable, 10 devices are needed for each digit. The necessary scheme is indicated below; each transistor is either on or off. A number $2{,}469_{10}$ is indicated in Table 16·4, where there are four sets of transistor indicators; each set has 10 indicators.

Those which are on read the number 2,469. The number $2{,}469_{10}$ can be written as a power series:

$$2{,}469 = 2{,}000 + 400 + 60 + 9$$

$$= 2(10^3) + 4(10^2) + 6(10^1) + 9(10^0)$$

Table 16·4 Indication of 2,469

Transistor	A 2	B 4	C 6	D 9
0				
1				
2	On			
3				
4	...	On		
5				
6	...	...	On	
7				
8				
9	...	...	...	On

Thus, for a four-digit number, there are 40 transistors needed for the proper indication of the number. Each digit has a string from 0 to 9, and only a single transistor can be on at any one time; this is wasteful.

Any number can be expressed as a power series as shown above for 2,469. However, any base can be used to express a given number; 10 is normally used for ordinary computation. But let the base be 5, as an example; a count from decimal 0 to 15 would be, to the base 5, as shown

Table 16·5 Counting to the Base 5

Decimal numbers	5^1	5^0
1	0	1
2	0	2
3	0	3
4	0	4
5	1	0
6	1	1
7	1	2
8	1	3
9	1	4
10	2	0
11	2	1
12	2	2
13	2	3
14	2	4
15	3	0

in Table 16·5. Representing $2{,}469_{10}$ to the base 5 would require fewer transistors.

$$2{,}469_{10} = 34{,}334_5$$

$$34{,}334_5 = 3(5^4) + 4(5^3) + 3(5^2) + 3(5^1) + 4(5^0)$$

$$2{,}469_{10} = 1{,}875 + 500 + 75 + 15 + 4$$

But other numbering systems are also possible and, for some applications, are more useful than the decimal system. One of these is the binary system.

Binary Notation. Switches for industrial control systems operate on a two-valued system such as on-off or open-closed. Vacuum tubes or transistors in modern digital computers also operate as simple switches which, as an example, either permit current flow or prevent current flow. These switching devices have but two states and cannot easily be used for representing the decimal system.

A numbering system most easily developed for use in industrial control systems or in digital computers is based on successive powers of 2. In this binary notation only two digits are used, 1 and 0. The first powers of 2 are 1, 2, 4, 8, 16, and 32. Just as 10^0 is 1, so 2^0 is also 1.

The binary number 101 is 4 + 0 + 1, or decimal 5, written 5_{10}. In the same way 10111 is 23_{10} or 16 + 0 + 4 + 2 + 1. A table for binary numbers is developed in exactly the same way the powers of 10 were shown for the decimal system. From 1 to 10 in decimal is, in binary, 01, 10, 11, 100, 101, 110, 111, 1000, 1001, and 1010. Other binary numbers are shown in Table 16·6.

Table 16·6 Other Binary Numbers

Decimal value	2^5 32	2^4 16	2^3 8	2^2 4	2^1 2	2^0 1
1	0	0	0	0	0	1
5	0	0	0	1	0	1
9	0	0	1	0	0	1
23	0	1	0	1	1	1
31	0	1	1	1	1	1
40	1	0	1	0	0	0
43	1	0	1	0	1	1
48	1	1	0	0	0	0
50	1	1	0	0	1	0
63	1	1	1	1	1	1

To represent $2{,}469_{10}$ in the binary (to the base 2) system, 12 indicators are needed as follows:

2^{11}	2^{10}	2^9	2^8	2^7	2^6	2^5	2^4	2^3	2^2	2^1	2^0
2,048	1,024	512	256	128	64	32	16	8	4	2	1
1	0	0	1	1	0	1	0	0	1	0	1

Thus, $2{,}048 + 256 + 128 + 32 + 4 + 1 = 2{,}469_{10}$ and $100110100101_2 = 2{,}469_{10}$.

To express a digit in the binary system, or a bit (from binary digit), only a single indicator is needed in either the on (or 1) or the off (or 0) state. Binary numbers are used in a great many digital machines for computation. The input data are converted to binary, usually in a coded system; the computations are made in the binary system; and the output data are converted from binary to decimal to be used external to the machine.

Binary numbers require very few rules and no tables for all arithmetic operations. One advantage of the binary notation is the simplicity of operations. *Addition* has four rules:

$$\begin{array}{r} 0 \\ +0 \\ \hline 0 \end{array} \qquad \begin{array}{r} 0 \\ +1 \\ \hline 1 \end{array} \qquad \begin{array}{r} 1 \\ +0 \\ \hline 1 \end{array} \qquad \begin{array}{r} 1 \\ +1 \\ \hline 10 \end{array}$$

If these rules are followed, any two binary numbers may be added directly. For example, to add decimal 12 and 5:

$$\begin{array}{rcr} 1100 & = & 12 \\ +101 & = & +5 \\ \hline 10001 & = & 17 \end{array}$$

To add 01011 and 00110:

$$\begin{array}{rcr} 01011 & = & 11 \\ +00110 & = & +6 \\ \hline 10001 & = & 17 \end{array}$$

For *multiplication* there are four rules which reduce to two: $1 \times 1 = 1$ and all other combinations are zero. For example, to multiply 1011 by 0101:

$$\begin{array}{rcr} 1011 & = & 11 \\ 0101 & = & 5 \\ \hline 1011 & & \\ & & \\ 1011 & & \\ \hline 110111 & = & 55 \end{array}$$

To multiply decimal 12 and 5:

```
   1100 = 12
    101 =  5
   ----
   1100
  11000
 ------   --
 111100 = 60
```

With binary multiplication there are no tables and no carries (except in adding partial products), and every product is equal to 0 except for 1 × 1, which is 1. Multiplication is a series of additions, and multiplying 14 by 8 in decimal is the same as adding 14 eight times. In binary, 10101 can be multiplied by 101 as follows:

```
  10101
    101
  -----
  10101
      0
10101
-------
1101001
```

This multiplication is a series of shifts and additions. With 101 as a multiplier, as above, a 1 is an addition and a shift left. The 0 is a shift but no addition.

Shifting right or left is just a form of multiplication. Starting with 101100, a shift left is 1011000 or two times 101100. Dividing by 2 is a shift right, or 10110:

$$1011000 = 88$$

$$101100 = 44$$

$$10110 = 22$$

Division is the inverse of multiplication; if the divisor can be subtracted, a 1 is placed in the quotient. If not, a 0 is placed there; for example,

```
      10101
101/1101010
    101
    ---
      110
      101
      ---
        110
        101
        ---
          1
```

Binary division is no different from long division with the decimal system.

For example,

```
          1101
1010/10001001
       1010
        1110
        1010
         10001
          1010
           111
```

Subtraction of binary numbers may be done in two ways. Direct subtraction is

$$\begin{array}{r} 101101 \\ -1011 \\ \hline 100010 \end{array}$$

Complements can also be used as a means of subtraction. In decimal notation the tens complement is the difference between 10 and a given number so that a number and its complement add up to ten. The nines complement is the difference between 9 and a given number. The tens complement of 7 is 3; the nines complement of 7 is 2. In decimal notation, complements can be added. Take 427; this may be subtracted from 564 directly, as follows:

$$\begin{array}{r} 564 \\ -427 \\ \hline 137 \end{array}$$

The tens complement of 427 is 573 and, adding,

$$\begin{array}{r} 564 \\ +573 \\ \hline 1{,}137 \end{array}$$

Thus, the answer is 137 by dropping the extra first digit. The nines complement of 427 is 572 or just 1 less than the tens complement 573. Again adding,

$$\begin{array}{r} 564 \\ +572 \\ \hline 1{,}136 \end{array}$$

But here the extra 1 (the end-around carry) is added to the 136, which is the sum without the first 1 digit. In binary the ones complement of 1011 is 0100; this is obtained by interchanging ones and zeros. To subtract 1011 from 101101 directly:

$$\begin{array}{r} 101101 \\ -001011 \\ \hline 100010 \end{array}$$

And, since the complement of 1011 is 110100, subtraction is also possible by addition:

$$\begin{array}{r l} 101101 & \\ +110100 & \\ \hline 1100001 & \\ 1 & \text{(end-around carry)} \\ \hline 100010 & \end{array}$$

Octal Numbering System. The switching device in a computer is required to indicate only an on or an off condition. However, in many cases computations are also made by other methods in addition to computer calculation. For example, to check out, or test, a program for a computer, the step-by-step procedure which the computer will follow must be checked by a human operator using a desk calculator.

For this purpose the decimal numbering system itself is not sufficient, for the machine (the computer) will use the binary system. But the binary system is a very tedious method for human computation, and there are no desk calculators which operate directly in this system. For checking a machine calculation, in binary, the octal system of numbers is used.

Table 16·7 Binary Triplets

Octal equivalent	Binary triplet	Octal equivalent	Binary triplet
0	000	4	100
1	001	5	101
2	010	6	110
3	011	7	111

The octal numbering system, to the base 8, is quite closely related to the binary system. Table 16·7 indicates the eight possibilities using three binary places. Binary groups of three expressed in octal are as follows:

$$\begin{array}{ll} 000 = 0 & 100 = 4 \\ 001 = 1 & 101 = 5 \\ 010 = 2 & 110 = 6 \\ 011 = 3 & 111 = 7 \end{array}$$

Binary, octal, and decimal may be compared as follows:

Binary	*Decimal*	*Octal*
010	2	2
+111	+7	+7
001001	9	11
101	5	5
+110	+6	+6
001011	11	13

To illustrate this principle, tables for addition and multiplication are indicated in Tables 16·8 and 16·9. For example, to add 321_8 and 405_8:

$$\begin{array}{r} 321 \\ +405 \\ \hline 726 \end{array}$$

Table 16·8 Octal Addition

	0	1	2	3	4	5	6	7
0	0	1	2	3	4	5	6	7
1	1	2	3	4	5	6	7	10
2	2	3	4	5	6	7	10	11
3	3	4	5	6	7	10	11	12
4	4	5	6	7	10	11	12	13
5	5	6	7	10	11	12	13	14
6	6	7	10	11	12	13	14	15
7	7	10	11	12	13	14	15	16

Table 16·9 Octal Multiplication

	0	1	2	3	4	5	6	7
0	0	0	0	0	0	0	0	0
1	0	1	2	3	4	5	6	7
2	0	2	4	6	10	12	14	16
3	0	3	6	11	14	17	22	25
4	0	4	10	14	20	24	30	34
5	0	5	12	17	24	31	36	43
6	0	6	14	22	30	36	44	52
7	0	7	16	25	34	43	52	61

This octal system can be used on special desk calculators; this last problem, in binary, would be as follows:

$$
\begin{array}{r} 011 \\ +100 \\ \hline 111 \end{array}
\qquad
\begin{array}{r} 010 \\ +000 \\ \hline 010 \end{array}
\qquad
\begin{array}{r} 001 \\ +101 \\ \hline 110 \end{array}
\begin{array}{l} = 321 \\ = 405 \\ = 726 \end{array}
$$

Thus, while the digital computer itself uses the binary system of numbering, the octal system can be used by a human operator (with or without a desk calculator) to check on the operation of the computer during the testing of a new program. Conversion from binary (computer language) to octal, which can be used by the human operator, is direct and simple.

Note the simplicity of the translation from binary to octal, for 111 is 7_8, 010 is 2_8, and 110 is 6_8. Or, to add 767 and 654:

$$
\begin{array}{r} 767 \\ 654 \\ \hline 1{,}653 \end{array}
$$

Perhaps an example can best be seen by steps:

457 multiplied by 263:

$$
\begin{array}{r} 457 \\ 3 \\ \hline 25 \\ 17 \\ 14 \\ \hline 1615 \end{array}
\qquad
\begin{array}{r} 457 \\ 6 \\ \hline 52 \\ 36 \\ 30 \\ \hline 3432 \end{array}
\qquad
\begin{array}{r} 457 \\ 2 \\ \hline 16 \\ 12 \\ 10 \\ \hline 1136 \end{array}
$$

By addition:

$$
\begin{array}{r} 1615 \\ 3432 \\ 1136 \\ \hline 151735 \end{array}
$$

Conversion from decimal to binary is the last remaining step needed for going in any direction between decimal, binary, and octal. Take the decimal number 765; referring to the binary numbers, the numbers are 1, 2, 4, 8, 16, 32, 64, 128, 256, 512, ... Each binary value is subtracted from the decimal number for the largest which can be subtracted. The ones and zeros are written as the binary value. Take decimal 765, for example:

765	
−512	1
253	0 (for 256, which is larger than 253)
−128	1
125	
−64	1
61	
−32	1
29	
−16	1
13	
−8	1
5	
−4	1
1	0
−1	1
0	

Thus, the binary number is 1 011 111 101, or

512	256	128	64	43	16	8	4	2	1
1	0	1	1	1	1	1	1	0	1

In this way decimal numbers can be converted to binary directly.

Numbering Codes. Conversion from decimal to binary, in either direction, is possible but not convenient. Combinations of decimal and binary are often used; one is the binary-coded decimal. Here, as shown in Table 16·10, each decimal digit is given a code. The number 18 is, in pure binary, 10010; and in binary-coded decimal it is 00001000. This system enables digits such as 6, 9, 3, or 4 to be converted separately rather than as parts of numbers such as 64, 91, or 37.

Table 16·10

Decimal	*Binary-coded decimal*	*Decimal*	*Binary-coded decimal*
0	0000	5	0101
1	0001	6	0110
2	0010	7	0111
3	0011	8	1000
4	0100	9	1001

Another code in frequent use is the "excess-3" code (Table 16·11), in which each number is represented as three more than its actual value. This permits direct subtraction using complements. To subtract 3 (0110)

Table 16·11

Decimal	*Excess-3 code*	Decimal	*Excess-3 code*
0	0011	5	1000
1	0100	6	1001
2	0101	7	1010
3	0110	8	1011
4	0111	9	1100

from 9 (1100), the 0110 is changed to 1001, which is 6. In the same manner 2 (0101) from 9 is 7, or 1010. This is a simplification of computer arithmetic.

Other codes are the gray or reflected binary (in which each change from one number to the next in sequence requires only a change in one bit) and the alphameric code of Table 16·12 (in which letters are represented

Table 16·12 Alphameric Coding

	Zone bits			
	00	01	10	11
0000				
0001			=	π
0010			+	÷
0011	0	A	B	C
0100	1	D	E	F
0101	2	G	H	I
0110	3	J	K	L
0111	4	M	N	O
1000	5	P	Q	R
1001	6	S	T	U
1010	7	V	W	X
1011	8	Y	Z	,
1100	9	.	?	:
1101		#	%	-
1110		!	"	
1111				

by binary numbers). A is 010011, C is 110011, Z is 101011, and both letters and decimal digits are coded. Zones are areas of the recording mechanism, as on magnetic tape, where these data are recorded.

16·3 Logic circuits Many types of switching complexes use logic for their circuit planning and layout. The two basic circuits are the AND and the OR. Both are forms of gates.

A gate allows current flow if and only if certain conditions of the input signal are met. The AND gate will provide an output only under conditions where *each* and every input signal is present. The OR gate (buffer) will provide an output if *any* of the input signals are present.

Any circuit which operates to either pass or not pass a signal flow (depending on its control of the signal path) is a gate. The fundamental

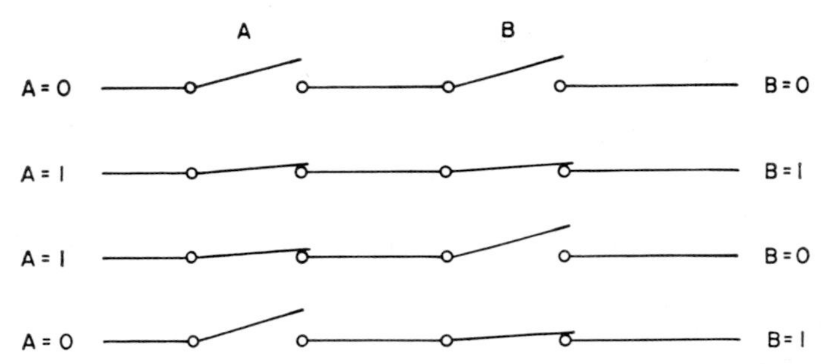

Fig. 16·6 AND-circuit switches. Circuit is closed only if $A = 1$ and $B = 1$ for $1 \times 1 = 1$.

circuit is a simple switch from input to output. When this switch is closed, signals (which arc in this case pulses) pass from input to output. And if the switch is open, signals do not pass from input to output. While this seems quite basic, it is, in truth, the action of all gate circuits, however complex.

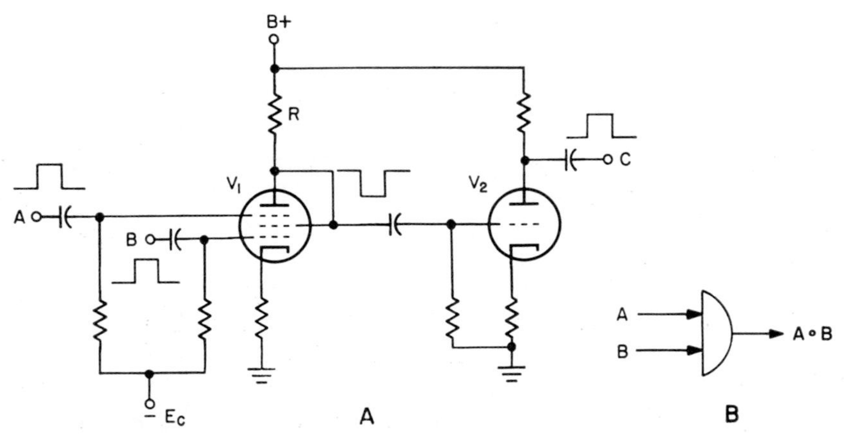

Fig. 16·7 AND circuit and symbol.

The AND gate is shown in Fig. 16·6. Where 0 is open and 1 is closed, the four possible combinations are shown.

Figure 16·7 illustrates, as an example of this type of circuit function, a sharp-cutoff pentode tube. With a positive voltage on the screen there is a negative voltage on both the suppressor and control grids. But either grid, if negative, is enough to cut the tube off. If a positive pulse is applied to input A alone, the tube will not conduct; if a positive voltage is applied to input B alone, the tube will not conduct. Only if *both* inputs are positive will the tube conduct. Thus if there is a positive pulse at A and a positive pulse at B, there will be an output pulse across R. An inverter amplifier can change this to a positive pulse so that a positive pulse representing a 1 will result if there is a 1 at input A and a 1 at input B. An inverter amplifier is shown following the AND tube. In A the grid bias cuts the tube off with no pulse input. If a positive pulse is a 1 and no

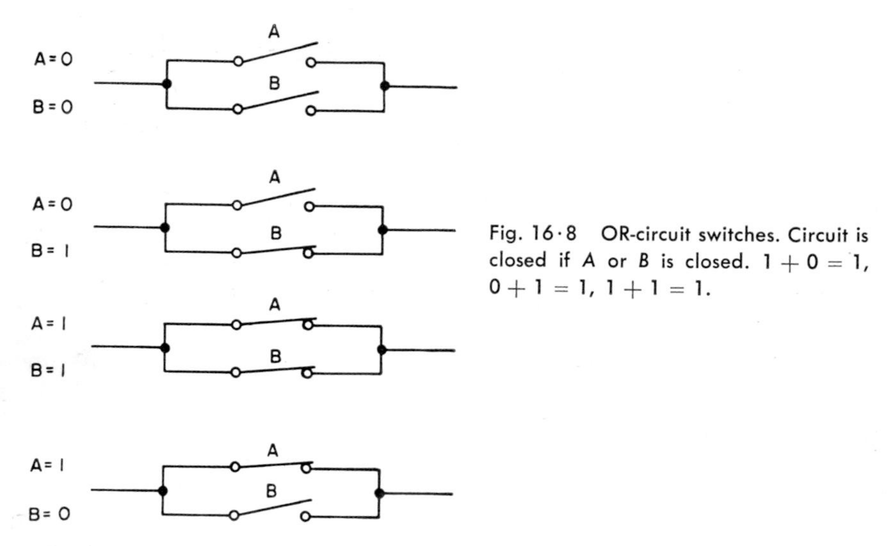

Fig. 16·8 OR-circuit switches. Circuit is closed if A or B is closed. $1 + 0 = 1$, $0 + 1 = 1$, $1 + 1 = 1$.

pulse is a 0, this circuit produces a pulse at C if there is a pulse at A and at B at the same time.

V_1 amplifies and inverts to produce a negative pulse which is reinverted by V_2 to produce a positive pulse output at C. The AND-circuit symbol as shown in Fig. 16·7B stands for a circuit which requires an input from A and B to provide an output.

OR circuits (Fig. 16·8) are parallel switches. Again 0 is open and 1 is closed. The four possible combinations are illustrated. The OR circuit operates if *any* of the input signals is applied. More than one input will, of course, also provide an output.

Transistor Logic. Transistor circuits can be used for any and all logic circuits. A typical circuit (Fig. 16·9) can be used for either an AND or an OR. This is a circuit building block, or module, from which control systems and computers are built.

The drawing shows the power-supply voltages as well as the 0 and 1 values. Note the truth table, which shows how the circuit behaves for various values or conditions of input signals.

The controlled switch (CS) is a semiconductor device which can function as a logic-circuit element. As an example, Fig. 16·10 shows an AND, which means that the CS will conduct if, and only if, there is a voltage at A and at B. In the circuit shown in Fig. 16·11 the anode is positive at

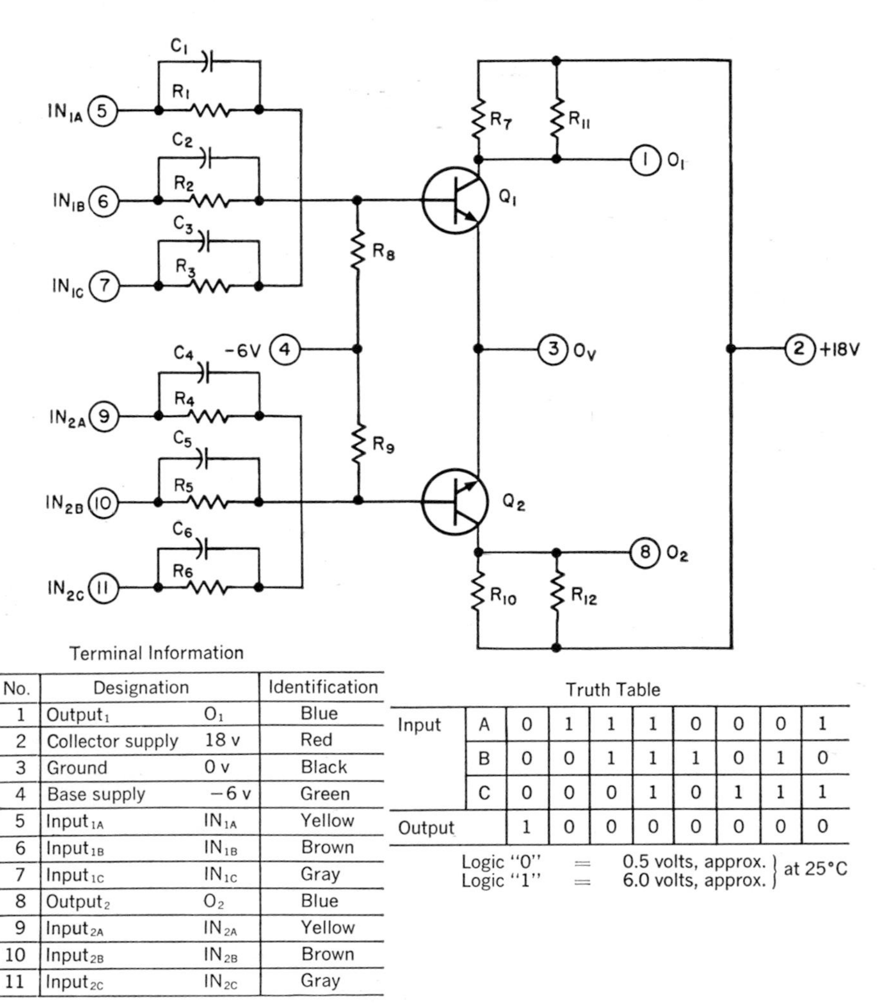

Terminal Information

No.	Designation		Identification
1	Output$_1$	O_1	Blue
2	Collector supply	18 v	Red
3	Ground	0 v	Black
4	Base supply	−6 v	Green
5	Input$_{1A}$	IN_{1A}	Yellow
6	Input$_{1B}$	IN_{1B}	Brown
7	Input$_{1C}$	IN_{1C}	Gray
8	Output$_2$	O_2	Blue
9	Input$_{2A}$	IN_{2A}	Yellow
10	Input$_{2B}$	IN_{2B}	Brown
11	Input$_{2C}$	IN_{2C}	Gray

Truth Table

Input	A	0	1	1	1	0	0	0	1
	B	0	0	1	1	1	0	1	0
	C	0	0	0	1	0	1	1	1
Output		1	0	0	0	0	0	0	0

Logic "0" = 0.5 volts, approx. } at 25°C
Logic "1" = 6.0 volts, approx. }

Fig. 16·9 Transistor logic circuit. (*General Electric*)

all times, and negative bias is applied to the gate. If either A or B is allowed to fire the CS by limiting the level of these signals, this becomes an AND circuit because both signals are needed to fire the CS. But if these signals are large enough, the circuit becomes an OR since either alone can fire the switch.

Similarly, Fig. 16·12 shows a negative input AND circuit as well as an

OR circuit. The circuit shown in Fig. 16·13 is another OR since either a positive input at A or a negative input at B will cause conduction.

If several inputs are combined, a single circuit with four inputs is achieved. A variety of combinations can then be established. Additional

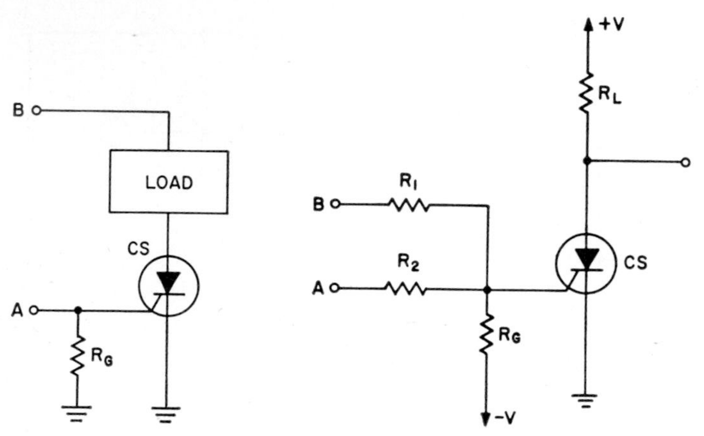

Fig. 16·10 Simple AND circuit. (*Solid State Products*)

Fig. 16·11 Positive AND or OR. (*Solid State Products*)

inputs can be added to such a combination as long as biasing is properly set.

Figure 16·14 shows a sequential AND circuit which is useful for safety,

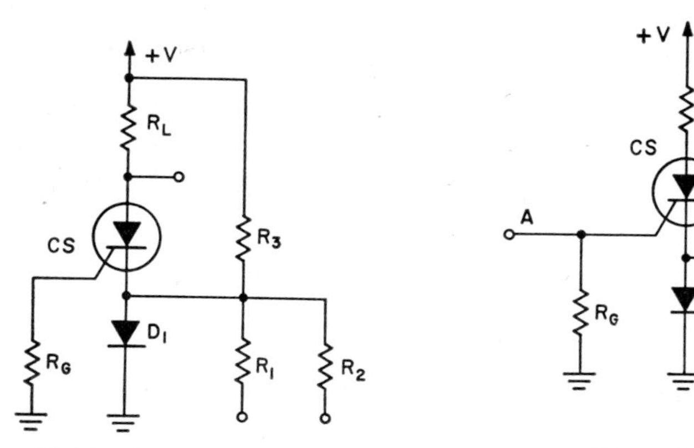

Fig. 16·12 Negative AND or OR. (*Solid State Products*)

Fig. 16·13 OR circuit. (*Solid State Products*)

protection, fusing, and related applications. Before power can be applied to R_L, three pulse inputs must be applied in the correct sequence.

With all three of the CSs off, the bias voltage at input B is one-third of the supply voltage, and at input C it is two-thirds of the supply voltage. CS_2 will not turn on unless the voltage applied at point B exceeds one-third of the supply voltage. Similarly, CS_3 will not turn on unless the voltage at C exceeds two-thirds of the supply voltage. CS_1, on the other hand, will turn on with an input at A equal to the normal CS firing requirements. Thus if the input pulse amplitudes are limited to less than one-third of the supply voltage, CS_2 cannot turn on until CS_1 has first been turned on. Also CS_3 cannot turn on until both CS_1 and CS_2 have been turned on.

A pulse counter circuit is shown in Fig. 16·15. CS_1 and CS_2 are used as memory elements. CS_3 turns on the load power after three positive input

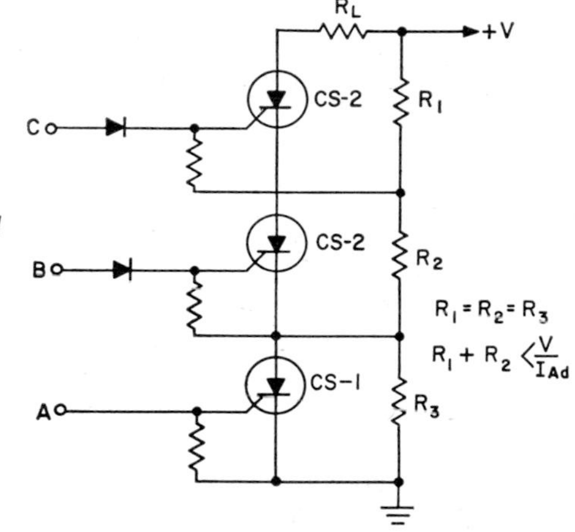

Fig. 16·14 Sequential AND. *(Solid State Products)*

pulses have been applied at the input. The first positive input signal at A couples through C_1 to turn CS_1 on. Diodes D_2 and D_3 are biased in the reverse direction by the charge on C_2 and C_3, and they block the input pulse from the gates of CS_2 and CS_3. After CS_1 turns on, the voltage at its anode is approximately +1 volt and C_2 discharges. When the next positive input pulse occurs, it can now couple through D_2 to turn on CS_2. Again D_3 blocks this pulse from the gate of CS_3. When CS_2 turns on, its anode voltage drops to +1 volt and C_3 discharges. The third positive input pulse can now couple through D_3 to turn on CS_3 and apply power to the load. R_1 and R_2 must provide current above the dropout level to CS_1 and CS_2. The circuit can be made to reset electronically by adding another CS, in the power flip-flop connection as shown in B.

Other Logical Switches. Almost any switch can be used as a logical element, such as relays, vacuum tubes, transistors, and magnetics. Light can be used for logical circuits.

Electroluminescent switches are discussed in Chap. 6. By extension of the basic photoconductive-electroluminescent switching principle, logic circuits can be developed as shown in Fig. 16·16. A is a representation of

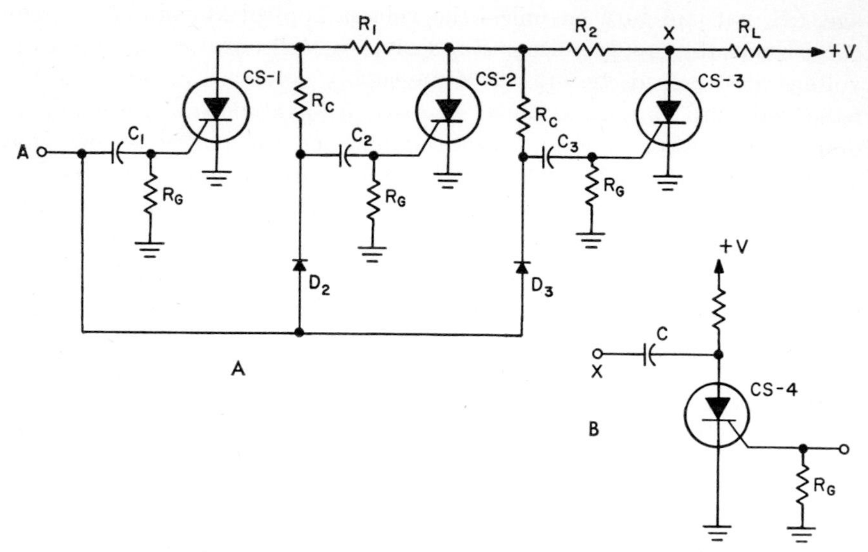

Fig. 16·15 Pulse adder. (*Solid State Products*)

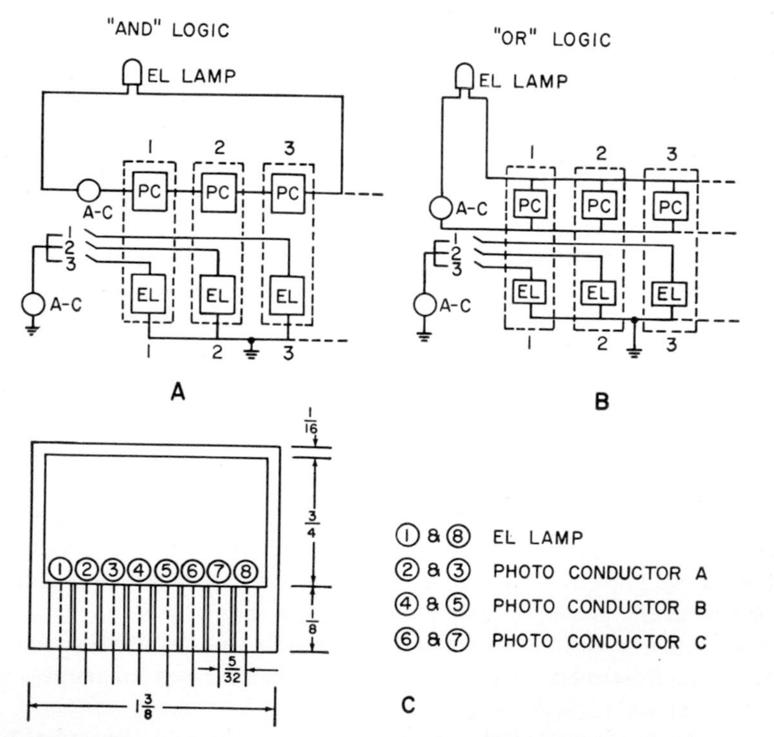

Fig. 16·16 (*A*) Photoconductors in AND logic; (*B*) circuit use for OR logic; (C) construction of electroluminescent lamp and photoconductors. (*Sylvania*)

an AND logic circuit. In this illustration, three photoconductive elements are connected in series with an electroluminescent lamp and an a-c power supply. Opposite each photoconductive element is an electroluminescent lamp "masked" from all but the photoconductive element directly in front of it.

The electroluminescent lamps are connected to a power supply through a switch. If switch 1 is closed, electroluminescent lamp 1 will light and photoconductive element 1 will become conductive; however, the lamp will not light. The same phenomenon will take place if switch 2 is closed with or without switch 1 closed. If all three switches are closed simultaneously, there is a conductive path and the lamp will light. Thus, if switches 1, 2, and 3 are closed, the lamp will light and there is the AND logic function.

Figure 16·6*B*, which shows an OR logic circuit, is identical to *A* except that the photoconductive elements are connected in parallel with the voltage supply and the electroluminescent lamp. If switch 1 is closed, electroluminescent lamp 1 will light. Photoconductive element 1 will become conductive and the electroluminescent output lamp will light. Therefore, if switch 1, 2, or 3 is closed, the lamp will light and there is the OR logic function.

AND and OR logic circuits are panel devices. Only optical coupling is used for switching. The only electrical connections are inputs and outputs, which are made at the edges of the panel. As a consequence, various combinations are possible on a single panel, resulting in versatile circuits having a high packing density.

Figure 16·6*C* illustrates an electroluminescent-photoconductive switching circuit element which has been designed for AND, OR, inverter, and memory-circuit logic. This device consists of an electroluminescent lamp and three photoconductor cells, which represents a 3PST switch.

When the device is operated at 200 volts a-c with a frequency of 400 cps, the dark and light impedances of the photoconductive cells have the following characteristics:

1. Three lighted photoconductor cells in series will be low enough in impedance to allow one electroluminescent lamp in series to light.

2. The dark impedance of up to three photoconductive cells in parallel will extinguish one electroluminescent lamp in series.

3. The light impedance of any one photoconductor will light two associated lamps sufficiently brightly to lower the impedance of their associated photoconductors so that an additional lamp lights.

16·4 Process computers There are industrial uses for both general-purpose and special-process computers.

The General Precision Libratrol 1000 is a computer designed specifically for complete integration into a process-control system.

Signals from mechanical and electric instruments, which point out certain conditions and timing in the process or product, control every industrial process. Normally, operating personnel interpret these signals and transform them into control actions. Input elements of the computer are designed to accept these same control signals from existing instrumentation but must convert them into signals which the computer can interpret and transform into effective control actions.

Standard industrial transducers are used to convert these input signals into suitable voltages. A relay commutator, consisting of hermetically sealed, mercury-wetted contact relays (assembled in plug-in modules), scans these input signals in the sequence established by the program stored in the computer. Transistor-drive circuitry actuates all the relays. The selected input groups are then gated to the computer logic in the correct sequence.

The commutator is designed in "building blocks," or modules, of eight relays each. These modules can be added at any time; thus the number of inputs that can be accommodated is limited only by the computer-stored program. Transducers supply computer inputs in various forms.

The most common on-line input from industrial-process transducers is in the form of d-c voltages. The analog-to-digital converter is a voltage-to-digital unit. A d-c amplifier is added between the transducer and the commutator where low-level signals, in the millivolt range, must be accepted. Outputs are presented as 10 binary digits plus sign. A visual display of the converted voltage is provided for checking and calibration purposes.

A digital clock is used as the time base for process-control operations. The digital clock is synchronized with the input power frequency and maintains a record of real time in binary-coded form. The clock stores this record, up to 1 year in length and with intervals of 0.1 sec, within a one-word storage location in the memory drum. The system can employ this time recording without intermediate conversion. Real time is set into the clock by means of the typewriter keyboard.

Control outputs of this process-control computer are available in the form of contact closures (by mercury-wetted contact relays) or control voltages (provided by voltage dividers consisting of precision resistor matrices switched by relays). These control outputs can be converted, by suitable actuating hardware, into required actions such as setting switches or turning valves.

The standard output unit includes 10 relays, activated by transistor-drive circuits in response to 10-bit binary words generated by the computer. The computer selects the particular group of 10 relays and the specific relay within each group on the basis of the stored process program. Simultaneously, the relay-switched resistor matrices convert the output from digital to analog form. Each analog output requires 10 relays and the corresponding resistor matrix for a 0.1 per cent accuracy in the output voltage.

An external power source furnishes the control voltages. All relays are of a magnetic latching type and remain in the position dictated by the computer even in case of power failures, thus providing a fail-safe feature.

16·5 General-purpose computers Some computers in actual use are described in this section; no laboratory models or experimental units are discussed. The examples in this section are general-purpose computers.

Computers could be divided into those designed for commerce and those designed for science and engineering. But with the diverse tasks for modern computers, a better division is by size. A large-scale computer usually has a high-capacity memory and many input-output stations (which operate in a variety of storage forms), and its operation is usually very fast. A small-scale computer usually has a smaller memory and fewer input-output stations (which are restricted to only a small number of storage forms), and its operation may be slower than that of a large machine.

Both on-line and off-line equipment operation is used for illustration. On-line operation means running a device such as a tape-recording station in such a manner as to require time from the computer. Off-line operation, on the other hand, does not require computer running time.

The IBM 1620 Data Processing System shown in Fig. 16·17 is a powerful, small, stored-program computer designed for scientific research, engineering, and management science computations. The 1620 is a complete data-processing system which can perform arithmetic and all logical and input-output operations on a production basis. Conventional decimal arithmetic is used, providing ease of communication between man and machine.

There are two modular units, a central processing unit (rear) and a paper-tape reader and punch (on the right). The central processing unit contains the operator's control console, a modified IBM electric typewriter, the magnetic-core storage unit, the arithmetic and logical unit, and related circuitry. The keys, lights, switches, and visual displays included on the control console are used for manual machine control, for correction of errors, and for display and revision of the contents of storage. A maintenance display panel permits ease of service. The magnetic-core storage unit has a capacity of 20,000 alphameric digits, each of which is individually addressable and can be made immediately available for processing.

Input information is introduced into the 1620 system by means of the IBM 1621 paper-tape reader or the keyboard of the modified IBM electric typewriter. Data are read from eight-channel paper tape at the rate of 150 characters per second. When information is introduced into the system by the typewriter, a hard-copy record of these data is obtained as a byproduct.

Output devices are the IBM 961 tape punch (included in the 1621 cabinet) and the electric typewriter. These units receive the processed data from core storage and prepare a punched paper tape or printed report of the information. The 961 tape punch records data in the eight-channel paper tape at the rate of 15 characters per second, while the typewriter prints automatically at the rate of 10 characters per second.

Figure 16·17 shows the large model of this computer. The core storage unit expands storage capacity of the 1620 from 20,000 to 40,000 or 60,000

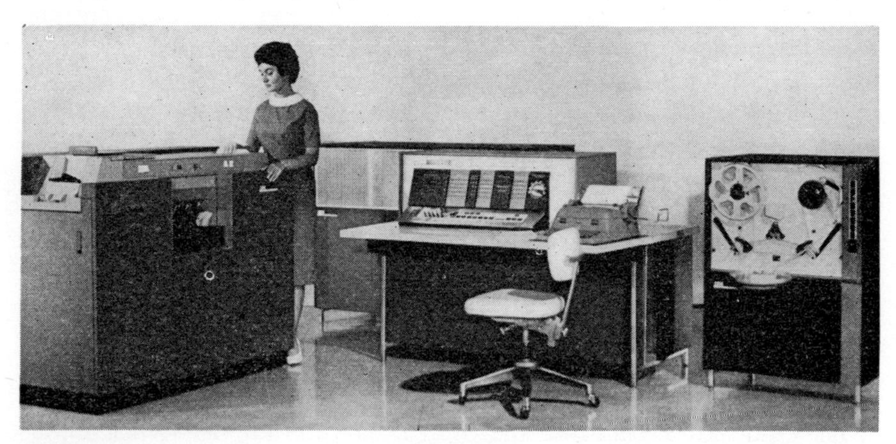

Fig. 16·17 Incorporating many advanced features of much larger computers, the expanded 1620 shown here can read 250 IBM cards a minute and punch 125 cards a minute with the new unit at the left. Its magnetic-core memory can contain 40,000 or 60,000 digits of information through the use of a new core storage unit (directly behind the operator). Paper-tape and electric-typewriter input and output are provided on the other two units. The expanded 1620 is designed for operating efficiency and economy in every phase of scientific and technical business activities. (*IBM*)

positions, depending on whether one or two 20,000-position modules of core storage are installed. This core memory is directly behind the operator, as shown in the figure.

The 1622 card-read punch unit on the left provides punched-card input and output for the 1620. Under program control, up to 250 cards per minute can be read and 125 punched. Individual buffer storage permits simultaneous reading, punching, and processing.

The National 315 (National Cash Register Company) is a business computer applicable to banks, retail stores, manufacturing firms, insurance companies, military and government work, and a wide variety of other data-processing jobs.

In a bank, for example, the new computer can post 1,500 checking account records in 1 min. It also automatically sorts checks and electronically prepares customers' statements at a speed equivalent to that of 290 typists.

In retail stores, the new equipment automatically keeps track of inventory and automatically prints out purchase orders when stock levels drop below a certain point. Other automatic functions include the preparation of customer statements and a wide variety of management records, such as detailed sales breakdowns. The computer is of modular design and can be easily expanded from a basic installation to a full-scale system as the needs of a business firm increase. The system's main memory, magnetic-tape file, and input and output units are variable in size or capacity to fit specific requirements. Five different memory sizes can be obtained, ranging from 6,000 to 120,000 decimal digits of information or 4,000 to 80,000 alphameric characters.

From one to eight magnetic-tape files can be used with the 315, each capable of storing 31 million decimal digits or 21 million alphameric characters. Input may include up to four magnetic character sorter-readers, a punched-card reader, a paper-tape reader, and a console as well as the magnetic-tape units. Output may include up to four high-speed printers and card punches in any combination, plus paper-tape punch, console, and magnetic-tape units. Peripheral units can interrupt a main program automatically to obtain processor time, after which the processor automatically resumes its former work and the peripheral unit continues to operate at its slower speed without delay.

The solid-state IBM 7090 is the most powerful data-processing system of International Business Machines Corporation. The fully transistorized system has computing speeds six times faster than those of its vacuum-tube predecessor, the IBM 709, and 7½ times faster than those of the IBM 704.

Although the IBM 7090 is a general-purpose data-processing system, it is designed with special attention to the needs of engineers and scientists.

Four IBM 7090 systems are incorporated in the Air Force's ballistic-missile warning system, the 3,000-mile radar system in the far north designed to detect missiles fired at southern Canada or the United States from across the polar region. Two IBM 7090 systems are being used by Dr. Wernher von Braun's development group at the George C. Marshall Space Flight Center of the National Aeronautics and Space Administration in Huntsville, Alabama. These systems are key design and development tools in building Saturn, the free world's largest space vehicle.

This computer also has a business use; the IBM 7090 will process such large-scale business applications as inventory control, production control, forecasting, and general accounting.

The speed of the 7090 results largely from the use of more than 50,000 transistors plus extremely fast magnetic-core storage. The new system can simultaneously read and write electronically at the rate of 3 million bits of information a second, when eight data channels are in use. In 2.18 millionths of a second, it can locate and make ready for use any of 32,768 data or instruction numbers (each of 10 digits) in the magnetic-core storage. The 7090 can perform any of the following operations in 1 sec: 229,000 additions or subtractions, 39,500 multiplications, or 32,700 divisions.

The IBM 7090 can use many of the programs (sets of instructions) already developed for the IBM 709, as well as hundreds of programs developed for the IBM 704. In addition, the input-output media of the 7090 are compatible with those of all IBM data-processing systems.

The ability to read, write, and compute at the same time is provided by the 7090's new multiplexer. Up to eight input-output data channels may be handled by the multiplexer. Each channel may have a total of 10 magnetic-tape units, a card reader, a card punch, and a printer. Therefore, a maximum 7090 system with a full input-output group would include 80 magnetic-tape units, eight card readers, eight printers, and eight card punches.

REVIEW QUESTIONS

16·1 What are the following decimal numbers expressed to the base 5: 150, 75, 10, 27, 350, 12?

16·2 What is the difference between logical AND and OR?

16·3 What is meant by a base in a numbering system?

16·4 Write these in the decimal system: XIII, XXI, XV, and XVII.

16·5 Express these decimal numbers in binary: 57, 3,024, 9,621, 1,501, 120, 374, 10,541, 500, and 642.

16·6 What is a binary triplet?

16·7 Why are binary numbers used?

16·8 What are the advantages of binary numbers?

16·9 Express these decimal numbers in octal: 27, 342, 115, 2,057, 301, 415, 75, 26, 7, and 142.

16·10 Express the following in octal and perform the operations:

$$\begin{array}{r} 54 \\ +27 \\ \hline \end{array} \quad \begin{array}{r} 32 \\ +16 \\ \hline \end{array} \quad \begin{array}{r} 12 \\ +8 \\ \hline \end{array} \quad \begin{array}{r} 42 \\ +7 \\ \hline \end{array} \quad \begin{array}{r} 7 \\ +6 \\ \hline \end{array}$$

$$\begin{array}{r} 37 \\ \times 12 \\ \hline \end{array} \quad \begin{array}{r} 102 \\ \times 3 \\ \hline \end{array} \quad \begin{array}{r} 374 \\ \times 7 \\ \hline \end{array} \quad \begin{array}{r} 143 \\ \times 21 \\ \hline \end{array} \quad \begin{array}{r} 500 \\ \times 214 \\ \hline \end{array}$$

16·11 What is a binary-coded decimal?

16·12 Why is it used?

16·13 Express these decimal numbers in excess-3 code: 114, 206, 27, 134, 2,076, 542, 21, 108, 37, and 420.

16·14 Explain the gray code.

16·15 What is the excess-3 code?

16·16 What are zone bits?

16·17 What is the difference between an AND circuit and an OR circuit?

16·18 How can the circuit shown in Fig. 16·9 be an AND circuit?

16·19 What is a bit?

16·20 How can the circuit shown in Fig. 16·9 be an OR circuit?

16·21 How can the circuit shown in Fig. 16·11 be either an AND or an OR?

16·22 Explain Fig. 16·15 as a pulse adder.

16·23 What is the principle of electroluminescence?

16·24 How can this be used for logic?

16·25 How is an electroluminescent OR circuit made?

Index

BK 621.314 T633Z
ZENER AND AVALANCHE DIODES
1970 .00 FV
3000 226309 30016
St. Louis Community College
W9-DEO-067

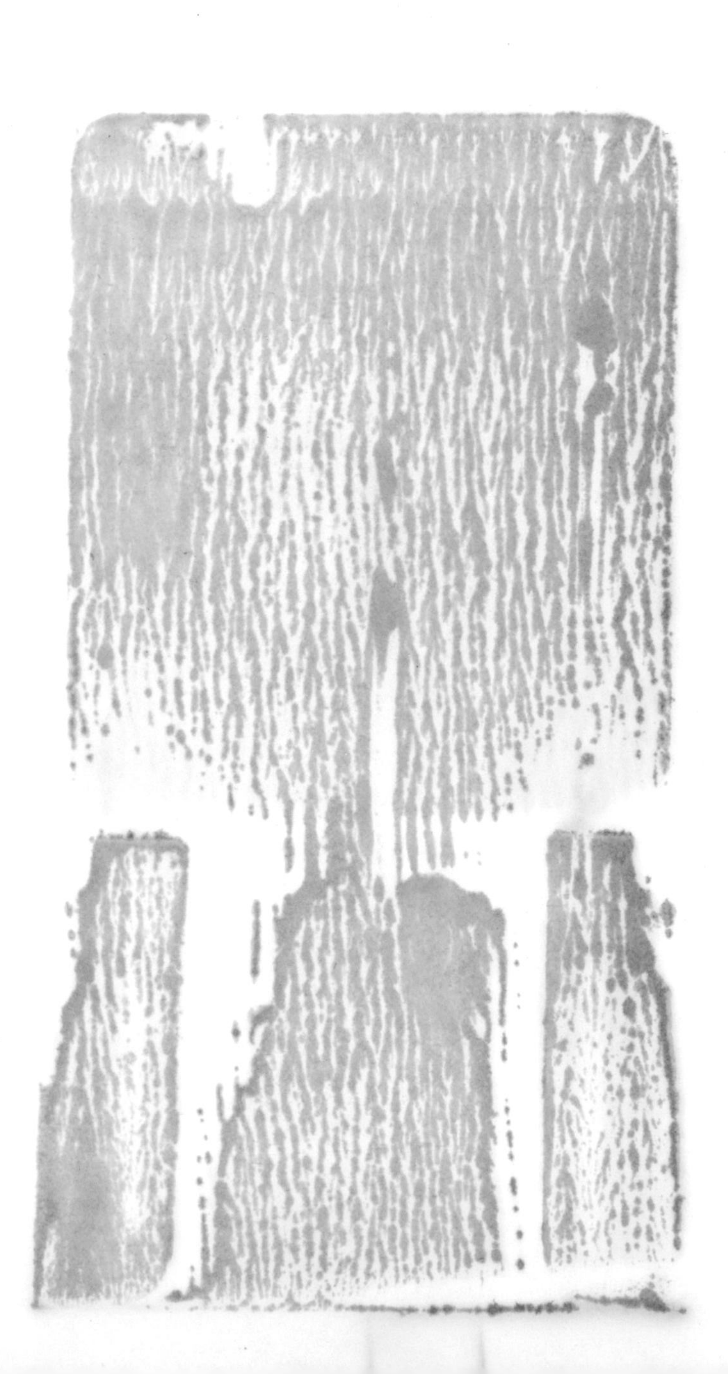

ZENER AND AVALANCHE DIODES

ZENER AND AVALANCHE DIODES

CARL DAVID TODD, P.E.

Electronics Consulting Engineer

WILEY-INTERSCIENCE

a Division of John Wiley & Sons, Inc.

New York · London · Sydney · Toronto

Library of Congress Catalog Card Number: 77–120709

ISBN 0–471–87605–4

Printed in the United States of America

10 9 8 7 6 5 4 3 2 1

PREFACE

Diodes with a controlled voltage breakdown characteristic provide the circuit designer with a versatile component that may be used in many applications. They supply a semiconductor counterpart to the gaseous regulator tubes which have been widely used both as a voltage regulator and coupling element. Their stable and predictable performance, together with the fact that VR diodes are economically available in a wide range of voltage values, permits them to perform many different functions.

This book deals primarily with the various possible applications of diodes whose breakdown characteristic relies either on Zener or avalanche effects. It also presents a brief treatment of the theoretical aspects of the various breakdown effects in Chapter 1, but its main purpose is to give a practical understanding of the characteristic parameters of VR diodes and how the devices may be used best to solve circuit function requirements.

Although the VR diode is a simple looking device and has only two terminals, there are certain subtleties about its performance. The circuit designer who is fully aware of them will be able to utilize certain features in some of his designs and avoid potential problems in others. Chapter 2 deals extensively with the characteristic parameters of VR diodes.

Chapters 3 through 10 deal with specific applications of VR diodes. Critical device parameters are indicated for each application. The general approach utilized is a presentation of the circuit from both practical and analytical viewpoints. Many design techniques and equations are given and are frequently reenforced by practical examples

Chapter 11 deals with special VR devices that warrant specific discussion. They include the temperature-compensated VR diode which has produced a significant change in portable precision instruments.

Certain measurement techniques that are specifically unique to the VR diode are treated briefly in Chapter 12.

An extensive bibliography included at the end of the book is a significant aid to the reader who wishes to pursue certain aspects of VR diodes that have not been covered in detail.

This book will be most useful to the practicing circuit designer who wishes to increase his awareness of the features and limitations of VR diodes, but it will also be helpful as a supplementary text in a study of semiconductor devices.

It is my belief that the better the engineer's understanding of semiconductor device characteristics and how they relate to specific applications, the more reliable his designs will be. In addition, a thorough knowledge of device application techniques permits him to be more creative in his own designs, and the mortar and bricks that he will lay on the foundation left before him will be of greater value.

I thank God that I have been chosen to live at a time when so many useful discoveries and developments are being made. When I was privileged to experiment with a galena crystal set almost thirty years ago, my greatest dreams could not begin to envision the true wonders of the semiconductor field or the height of enjoyment of being a part of it.

Every creative engineer needs the understanding and encouragement of someone very special like my wife Diana who is responsible for more of this book than she realizes.

A word of appreciation is also extended to Mrs. Velda Lou Mayhugh who typed a significant part of this book.

In conclusion, I wish to acknowledge the assistance of various manufacturers who provided technical data and application information. Worthy of special mention are Dickson Electronics Corp., Motorola Semiconductors, National Semiconductor Corp., and TRW Semiconductors.

CARL DAVID TODD, P.E.

Costa Mesa, California
May, 1970

CONTENTS

ZENER AND AVALANCHE DIODES

Chapter 1

BREAKDOWN THEORY

1–1 INTRODUCTION

In order to understand thoroughly and fully utilize the characteristics of Zener and avalanche breakdown diodes, it is necessary to gain some understanding of the basic theories that govern the breakdown mechanism (or actually mechanisms as we shall soon see). We shall not attempt to delve into breakdown theory in great detail but shall present to the reader a basic understanding of the events that take place when a diode is operated in its breakdown region. An extensive bibliography is included for those who want a more complete treatment of breakdown theory.

The voltage-current characteristic curve of a generalized semiconductor diode is illustrated in Fig. 1–1. Note that a different current scale is used for the forward and reverse regions. There are three distinct regions of operation, depending on the polarity and magnitude of the applied terminal voltage or current. If the basis is such that the anode is made positive with respect to the cathode, the diode is said to be forward-biased and a substantial current flows even with rather low voltages applied.

If the terminal polarity is reversed, making the anode negative with respect to the cathode, the diode is said to be reverse-biased and only a very small current flows even for fairly large bias voltages. When the reverse voltage is increased to a critical value, the magnitude of the reverse current begins to rise sharply. This is the breakdown region and the critical value of the bias voltage is referred to as the breakdown voltage or V_B. In some instances, as we shall see later, the exact breakdown is not so clearly defined; hence V_B will vary slightly as a function of the breakdown current allowed to flow.

Normal rectifier-type diodes or those used for signal detection, gating, and similar uses make use of both the forward and reverse characteristics for their primary area of operation, but the amount of reverse voltage is usually held

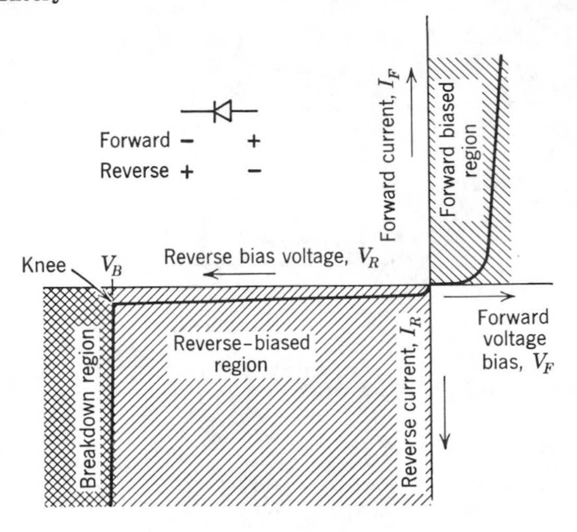

Fig. 1–1 Generalized voltage-current characteristic curve for a semiconductor diode.

below the breakdown voltage. In this book, however, we are considering a class of diodes that are designed primarily to be operated in or around the breakdown region.

1–2 HISTORICAL

In an early article [4] McAfee, Ryder, Shockley, and Sparks of the Bell Telephone Laboratories showed that the breakdown in narrow *P–N* germanium junctions could be attributed to an effect known as field emission which was described by Zener [2] in 1934. Zener's theory, originally applied to dielectric materials, was that under very intense electric fields (on the order of 10^5 V/cm or greater), electrons are pulled out of the valence band and traverse the forbidden energy gap by a process known as quantum-mechanical tunneling. There appeared to be a fairly good correlation between the measured results of germanium diodes in the vicinity of 10 V and the Zener theory as modified slightly for use with semiconductors; hence the diodes were given the name Zener diodes.

Breakdown voltages for lighter-doped germanium diodes and for many silicon diodes did not seem to agree with that predicted from the Zener theory. In addition, the slope of the charactersitic curve in the breakdown region was steeper than predicted eariler by Zener [2] or Shockley [4]. A noise that could not be explained by the Zener theory was also present. Sawyer [5] felt that this was due only to an imperfection in the device-manufacturing process.

McKay and McAfee [6] later described observations of multiplication of carriers in both silicon and germanium. This multiplication occuring in the region just before breakdown, was similar to that effect associated with prebreakdown conditions in gases and the Townsend avalanche mechanism. It was noted that for very narrow junctions no multiplication effects occurred and Zener effect seemed to be responsible for the breakdown. In wider junctions, however, the new theory of avalanche breakdown seemed more applicable to the observed results. McKay [7] later described this mechanism in more detail and gave several reasons why avalanche mechanism might indeed be the cause of breakdown. Much of this theory was dependent on the fact that holes as well as electrons can ionize. The collision-generated positive ions in the Townsend theory are replaced by collision-generated holes in the new semiconductor avalanche theory. In addition, the Zener effect, by its very dependence on the forbidden band gap, should yield a negative temperature coefficient. On the other hand, the avalanche mechanism should exhibit a positive temperature coefficient, and measurement confirmed that this was true.

The explanation of the avalanche breakdown as proposed by McKay was strengthened further by Wolff [8] who presented a theoretical calculation of the ionization rate that closely supported the data taken by McKay.

At first it was assumed that field emission or the Zener effect did not occur in silicon [7][11], but several years later McKay and Chynoweth described observations [12] of silicon diode breakdowns that appeared to follow the Zener theory. They found that, especially in very low voltage diodes, the multiplication factor associated with the avalanche breakdown could be made equal to only two at the most before very large currents resulted. In McKay's previous papers [6][7] he presented data indicating that a value of the multiplication factor M in the neighborhood of 20 or more was easily obtained. In addition, the temperature coefficient for low voltage breakdown diodes was negative as opposed to the positive temperature coefficient observed for avalanche breakdowns.

It was therefore established that two distinct mechanisms are responsible for breakdown in P–N junction diodes (eliminating the additional possibilities of surface breakdown); hence the Zener or field emission theory could be applied for very narrow junctions, and the avalanche theory was required for wider junctions associated with the higher voltage diodes. In intermediate instances both mechanisms seemed to be present concurrently.

1–3 ZENER BREAKDOWN

Zener's original theory [2] described the excitation of electrons in dielectric materials directly from the valence band to the conduction band. A high electric field supplies the energy required to transverse the forbidden energy gap.

The diagram of Fig. 1–2 illustrates the Zener breakdown operation. The only valence electrons which may take part in the breakdown operation are those between the dotted lines.

Breakdown attributable to field emission or Zener effects should cause no permanent damage in itself because it is merely a result of electric field induced hole-electron pairs in the junction separated by the field and otherwise has an undisturbed valence band structure.

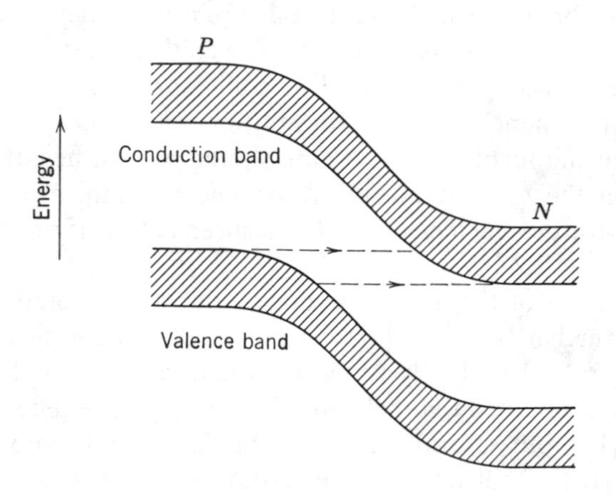

Fig. 1–2 Energy diagram of *PN* Junction showing path of electrons during Zener breakdown.

The magnitude of the electric field necessary to produce the Zener effect is on the order of 2 to 3 × 10^5 V/cm and will be a direct function of the net impurity concentration of the *P* and *N* regions in the diode. Sawyer [5] indicated that the anticipated Zener breakdown voltage is given by

$$V_B = 39\rho_n + 8\rho_p, \tag{1-1}$$

where ρ_n and ρ_p are the resistivities of the *N* and *P* regions of the junction. This relationship does not hold up very well in actual practice.

Since the field emission is a function of the forbidden energy gap, an increase in temperature reduces the band gap and hence reduces the voltage necessary to produce the Zener effect.

True Zener breakdown diodes exist in silicon only for very low breakdown voltages on the order of 4 V or less. Although there is an appreciable Zener current present in higher voltage units up to 6 or 7 V, the avalanche mechanism is also active, which is discussed in the next section.

1–4 AVALANCHE BREAKDOWN

1–4.1 Basic Mechanism

The avalanche breakdown effect is similar to the Townsend discharge in gas-filled tubes [1] and relies on the ionization of carriers on impact collision of atoms by other carriers which have been imparted sufficient energy by an electric field. Suppose a free electron or hole exists within the depletion layer of the junction and an electric field is applied. An increased velocity in the direction of the electric field, and hence an increase in the carrier's kinetic energy, is given to the free carrier. Because of the lattice structure in the semiconductor crystal, the free carrier may collide with an atom within the junction and, because of its high energy level, may rip off carriers from the atom. If sufficient energy still remains in the original colliding carrier, additional collisions and thus other carriers will be freed from additional atoms. The newly released carriers now gain sufficient energy from the field to begin collisions of their own.

In the Townsend discharge in gases a regenerative mechanism is present to insure continuation of the breakdown once it is initiated. This consists of electron emission from the cathode by positive ion bombardment. Although some breakdowns in solid crystals relying only on the ionization of electrons have been observed [17], generally it is necessary that both electrons and holes be ionized in order to provide the positive feedback or regenerative mechanism [7]. In effect, the ionized holes in semiconductors are analogous to the positive ions in gasses.

1–4.2 Multiplication

McKay has shown [6][7] that even before the junction breaks down completely, a given carrier injected into the depletion layer of the reverse-biased junction results in a current flow greater than that injected carrier. Thus we say that a multiplication of carriers has occurred and its degree is a function of just how close to the breakdown voltage the diode is biased. A typical plot of the multiplication factor M as a function of the normalized applied voltage is given in Fig. 1–3. When the applied voltage across the junction approaches V_B, the multiplication factor increases very rapidly and, in fact, the point of actual breakdown is defined as the point at which M becomes infinite.

In addition to the actual breakdown, the multiplication effect produces additional results; for example, any degree of reverse leakage currents will be multiplied in the region of prebreakdown, thus producing a certain amount of softness in the characteristic curve, especially at elevated temperatures. Also,

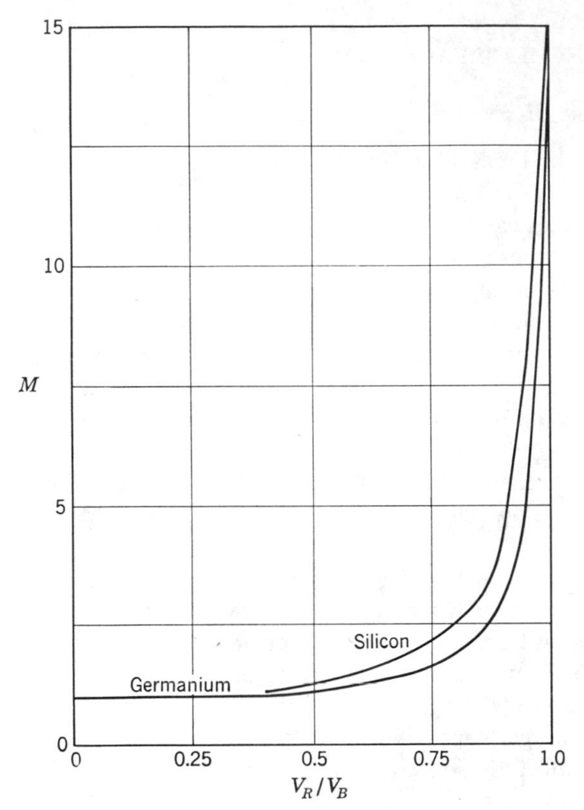

Fig. 1–3 Multiplication effects in pre-avalanche breakdown region of silicon and germanium diodes as a function of reverse voltage normalized with respect to breakdown.

any Zener current that might flow because of field emission might be multiplied. This occurs significantly in junction breakdowns in the region of 4 to 8 V in silicon.

McKay expressed [7] the value of M mathematically as

$$M = \frac{1}{1 - \int_0^W \alpha_i \, dx}, \tag{1-2}$$

where α_i is the rate of ionization and W is the effective width of the actual junction.

Miller has given [11] a relation evolved from observation for the multiplication factor in step junctions:

$$M = \frac{1}{1 - (V_R/V_B)^n}, \tag{1-3}$$

where V_R is the applied voltage, V_B is the body breakdown voltage due to avalanche, and n is a number that depends on the resistivity and the resistivity type of the high-resistivity side of the junction.

When the high resistivity material in the junction is P-type, the value of n seemed to vary from about 4.7 for very low resistivity materials (0.1 Ω-cm) to about 6 at the higher resistivities (2 Ω-cm). On the other hand, when the high resistivity material in the junction is N-type, the value of n seemed to remain close to 3 for the range of resistivities investigated.

1–4.3 Rate of Ionization, α_i

α_i is defined as the number of hole-electron pairs produced by a given carrier per cm of path traveled in the direction of the field. The ionization rate is slightly different for holes and electrons and is also a function of the semiconductor material and the applied field. For (1–2) McKay assumed that the ionization rates for holes and electrons were equal.

The value of α_i and its dependence on the strength of the applied electric field were investigated by Chynoweth [37], who found from experiment that the ionization rate obeyed a function in the form of

$$\alpha_i = a\varepsilon^{-(b/E)}, \tag{1-4}$$

where E is the value of the electric field and a and b are constants. This is very much the same as had been found in gaseous discharges [3]. Maserjian has shown [58] that for an abrupt junction the values for a are 9×10^5 and 1.2×10^7 for silicon and germanium, respectively. The values for b are 1.8×10^6 and 1.4×10^6 for silicon and germanium, respectively.

We may use the value of the ionization rate to define mathematically the point of breakdown as being the condition where

$$\int_0^W \alpha_i \, dx = 1. \tag{1-5}$$

1–4.4 Microplasmas

In avalanche breakdown the junction does not suddenly begin to conduct very large currents all over the junction but rather at small discrete points. These small high-current-density discharges are referred to as microplasmas [13][25]. This effect, similar to the plasma effect in gasses, indicates that breakdown occurs in small steps of about 50 to 100 μA of current; and further increases in applied voltage beyond that which created the first microplasma will not increase the current through the individual microplasma appreciably but rather will produce more microplasmas. Each new microplasma adds its unit of current to the total breakdown value. The individual microplasmas switch on and off in the area of initial breakdown, thus producing a distinct noise in the form of erratic current pulses with very fast rise time and of rather uniform amplitude regardless of the level of the breakdown voltage.

It has also been found [10][20] that associated with the generation of microplasmas in avalanche breakdown of the *P–N* junction, there exists an emission of light eminating from the individual microplasma. A light of much lower intensity is emitted from a junction whose breakdown mechanism is due primarily to Zener field emission and it has more of a uniform glow [30]. It has been noticed that both the pinpointed spots of light associated with avalanche breakdown and the faint glow associated with Zener breakdown have been observed in those junctions in the transition region between breakdown voltages of 4 to 7 V.

It has been found that microplasmas occur preferentially along areas of scratches or other mechanical working [14]. This seems to suggest an effect of lattice damage on the junction width. There is, in fact, a preference for any type of dislocation within the junction.

The generation and decay of the microplasmas as the junction is biased into breakdown take place in a very short time interval (10^{-10} sec or lower), and hence the breakdown effect may be used at rather high speeds. Some applications to be discussed in later chapters utilize this property to good advantage.

1–4.5 Threshold Energy Level

We have described the avalanche breakdown effect as depending on impact collisions of free carriers to generate additional carriers by freeing them from the lattice structure. There is a certain amount of kinetic energy necessary before any impact will produce the desired results. This energy level or threshold must be larger than the width of the band gap of 1.1 eV (in silicon) and has been found by Wolff [8] and later verified by Chynoweth and McKay [32] to be approximately 2.25 eV.

1–5 SURFACE BREAKDOWNS

All the preceding discussions were based on body effects in the bulk material and junction. It is also possible for a breakdown to occur that is due primarily to surface effects. In the normal rectifier- or signal-type diodes, especially those designed for rather large voltages, a surface effect causing breakdown usually occurs before an avalanche breakdown can be initiated in the relatively wide junction. Since most diodes used intentionally in the Zener or avalanche region are of a lower voltage breakdown in which surface conditions play a negligible part, we shall not discuss surface breakdown effects other than to say that they appear to be basically the result of an avalanche-type mechanism.

REFERENCES

[1] J. S. Townsend: The Passage of Ions in Gasses, *Nature*, **62**, 340 (August 1900).
[2] C. Zener: A Theory of Electrical Breakdown of Solid Dielectrics, *Proc. Roy. Soc. (London)*, **145**, 523–529 (July 2, 1934).
[3] L. B. Loeb: *Fundamental Processes of Electrical Discharge in Gasses*, Wiley, New York, 1939.
[4] K. B. McAfee, E. J. Ryder, W. Shockley, and M. Sparks: Observations of Zener Current in Germanium *p-n* Junctions, *Phys. Rev.*, **83**, 650–651 (August 1, 1951).
[5] G. L. Pearson and B. Sawyer: Silicon *P–N* Junction Alloy Diodes, *Proc. IRE*, **40** 1348–1351 (November 1952).
[6] K. G. McKay and K. B. McAfee: Electron Multiplication in Silicon and Germanium, *Phys. Rev.*, **91**, 1079–1084 (September 1, 1953).
[7] K. G. McKay: Avalanche Breakdown in Silicon, *Phys. Rev.*, **94**, 877–884 (May 15, 1954).
[8] P. A. Wolff: Theory of Electron Multiplication in Silicon and Germanium, *Phys. Rev.*, **95**, 1415–1420 (September 15, 1954).
[9] L. B. Loeb: *Basic Processes of Gaseous Electronics*, University of California Press, 1955.
[10] R. Newman, W. C. Dash, R. N. Hall, and W. E. Burch: Visible Light from a Si *p-n* Junction, *Phys. Rev.*, **98**, 1536–1537 (June 1, 1955).
[11] S. L. Miller: Avalanche Breakdown in Germanium, *Phys. Rev.*, **99**, 1234–1241 (August 15, 1955).
[12] K. G. McKay and A. G. Chynoweth: Optical Studies of Avalanche Breakdown in Silicon, *Phys. Rev.*, **99**, 1648 (September 1, 1955).
[13] D. J. Rose and K. G. McKay: Microplasmas in Silicon, *Phys. Rev.*, **99**, 1648 (September 1, 1955).
[14] R. Newman: Visible Light from a Silicon *p-n* Junction, *Phys. Rev.*, **100**, 700–703 (October 15, 1955).
[15] R. E. Burgess: The Turnover Phenomenon in Thermistors and Point-Contact Germanium Rectifiers, *Proc. Phys. Soc. (London)*, **68**, Sec. B, 908–917 (November 1, 1955).
[16] C. L. Wannier: Possibility of a Zener Effect, *Phys. Rev.*, **100**, 1227 (November 15, 1955).
[17] J. Yamashita: Theory of Electon Multiplication in Silicon, *Progr. Theoret. Phys. (Kyoto)*, **15**, 95–110 (February 1956).
[18] C. G. Garrett and W. H. Brattain: Some Experiments on, and a Theory of Surface Breakdown, *J. Appl. Phys.*, **27**, 299-306 (March 1956).
[19] G. H. Wannier: Possibility of a Zener Effect—Errata, *Phys. Rev.*, **101**, 1835 (March 15, 1956).
[20] A. G. Chynoweth and K. G. McKay: Photon Emission from Avalanche Breakdown in Silicon, *Phys. Rev.*, **102**, 369–376 (April 15, 1956).
[21] H. S. Veloric, M. B. Prince, and M. J. Eder: Avalanche Breakdown in Silicon Diffused *p-n* Junctions as a Function of Impurity Gradient, *J. Appl. Phys.*, **27**, 895–899 (August 1956).
[22] C. Yamanaka and T. Suita: Avalanche Breakdown in *P-N* Alloyed Ge Junctions, *Tech. Rept. Osaka Univ.*, **6**, 243–250 (October 1956).
[23] D. J. Rose: Townsend Ionization Coefficient for Hydrogen and Deuterium, *Phys. Rev.*, **104**, 273–277 (October 15, 1956).

[24] H. L. Armstrong: A Theory of Voltage Breakdown of Cylindrical *P-N* Junctions, with Applications, *IRE Trans. Electron. Devices*, **ED-4**, 15–16 (January 1957).
[25] D. J. Rose: Microplasmas in Silicon, *Phys. Rev.*, **105**, 413–418 (January 15, 1957).
[26] S. L. Miller: Ionization Rates for Holes and Electrons in Silicon, *Phys. Rev.*, **105**, 1246–1249 (February 15, 1957).
[27] P. Dobrinski, H. Knabe, and H. Muller: Zener-Diodes with Silicon, *Nach. tech. Z. (NTZ)*, **10**, 195–199 (April 1957) (German).
[28] E. M. Pell: Influence of Electric Field in Diffusion Region Upon Breakdown in Germanium *n-p* Junctions, *J. Appl. Phys.*, **28**, 459–466 (April 1957).
[29] H. S. Veloric and K. D. Smith: Silicon Diffused Junction "Avalanche" Diodes, *J. Electrochem. Soc.*, **104**, 222–226 (April 1957).
[30] A. G. Chynoweth, K. G. McKay: Internal Field Emission in Silicon *p-n* Junctions, *Phys. Rev.*, **106**, 418–426 (May 1, 1957).
[31] F. W. Rose: On the Impact Ionization in the Space Charge Region of *p-n* Junctions, *J. Electron. Control*, **3**, 396–400 (October 1957).
[32] A. G. Chynoweth and K. G. McKay: Threshold Energy for Electron-Hole Pair-Production by Electrons in Silicon, *Phys. Rev.*, **108**, 29–34 (October 1, 1957).
[33] A. W. Matz: Thermal Turnover in Germanium *P-N* Junctions, *Proc. Inst. Elec. Engrs. (London)*, **104**, Part B, 555–564 (November 1957).
[34] J. Tauc and A. Abraham: Thermal Breakdown in Silicon *p-n* Junctions, *Phys. Rev.*, **108**, 936–937 (November 15, 1957).
[35] E. Spenke: *Electronic Semiconductors*, Translation by D. A. Jenny *et al.*, McGraw-Hill, New York, 1958, pp. 105–107, 231–234.
[36] A. G. Chynoweth: Electrical Breakdown in *p-n* Junctions, *Bell Lab. Record*, **36**, 47–51 (February 1958).
[37] A. G. Chynoweth: Ionization Rates for Electrons and Holes in Silicon, *Phys. Rev.*, **109**, 1537–1540 (March 1, 1958).
[38] A. G. Chynoweth and G. L. Pearson: Effect of Dislocations on Breakdown in Silicon *p-n* Junctions, *Bull. Amer. Phys. Soc.*, **3**, 112 (March 27, 1958).
[39] A. G. Chynoweth: Electrical Breakdown in *p-n* Junctions, *Semiconductor Products*, **1**. 33–36 (March/April 1958).
[40] B. Senitzky and J. L. Moll: Breakdown in Silicon, *Phys. Rev.*, **110**, 612–620 (May 1, 1958).
[41] E. M. Conwell: Properties of Silicon and Germanium: II, *Proc. IRE*, **46**, 1281–1300 (June 1958).
[42] A. G. Chynoweth and G. L. Pearson: Effect of Dislocations on Breakdown in Silicon *p-n* Junctions, *J. Appl. Phys.*, **29**, 1103–1110 (July 1958).
[43] H. L. Armstrong: On Avalanche Multiplication in Semiconductor Devices, *J. Electron. Control*, **5**, 97–104 (August 1958).
[44] D. R. Muss and R. F. Greene: Reverse Breakdown in InGe Alloys, *J. Appl. Phys.*, **29**, 1534–1537 (November 1958).
[45] A. R. Plummer: The Effect of Heat Treatment on the Breakdown Characteristics of Silicon *p-n* Junctions, *J. Electron. Control*, **5**, 405–416 (November 1958).
[46] T. Misawa: Theory of the P-N Junction Device Using Avalanche Multiplication, *Prov. IRE*, **46**, 1954 (December 1958).
[47] Zh. I. Alferov, G. V. Gordeeve, and V. I. Stafeev: The Dependence of the Breakdown Voltage of Germanium and Silicon Diodes, *Fiz. Tverd. Tela*, Supplement II, 104–108, 1959. (Russian.)
[48] R. F. Schwarz: Introduction to Semiconductor Theory, *Elec. Mfg.*, **63**, 107 (1959).
[49] R. Yee, J. Murphy, A. D. Kurtz and H. Bernstein: Avalanche Breakdown in *n-p* Germanium Diffused Junctions, *J. Appl. Phys.*, **30**, 596–597 (April 1959).

[50] G. Ashton and M. H. Issott: The Preparation of Single-Crystal Silicon for Production of Voltage-Reference Diodes, *Proc. Inst. Elec. Engrs.* (*London*), **106**, Part B, Supplement 15, 273–276 (May 1959).

[51] S. Sherr and S. King: Avalanche Noise in *P-N* Junctions, *Semiconductor Products*, **2**, 21–25 (May 1959).

[52] W. Shockley: Transistor Diodes, *Proc. Inst. Elec. Engrs.* (*London*), **106**, Part 3, Supplement 15, 270–276 (May 1959).

[53] J. Yamaguchi and Y. Hamakawa: Electrical Breakdown in Germanium *p-n* Junctions, *Proc. Inst. Elec. Engrs.* (*London*), **106**, Part B, Supplement 15, 353–356 (May 1959).

[54] R. E. Burgess: Statistical Theory of Avalanche Breakdown in Silicon, *Can. J. Phys.*, **37**, 730–738 (June 1959).

[55] E. O. Kane: Zener Tunneling in Semiconductors, *Bull. Amer. Phys. Soc.*, **4**, 320 (June 18, 1959).

[56] K. S. Champlin: Microplasma Fluctuations in Silicon, *J. Appl. Phys.*, **30**, 1039–1050 (July 1959).

[57] A. I. Uvarov: Effect of the Space Charge of Moving Carriers on the Electrical Breakdown of a Strongly Assymetric *p-n* Junction, *Fiz. Tverd. Tela*, **1**, 1457–1459 (September 1959). (Russian) (Translated in *Soviet Phys. Solid State*, **1**, 1336–1338, March 1960.)

[58] J. Maserjian: Determination of Avalanche Breakdown in *p-n* Junctions, *J. Appl. Phys.*, **30**, 1613–1614 (October 1959).

[59] D. E. Sawyer: Surface-Dependent Losses in Variable Reactance Diodes, *J. Appl. Phys.*, **30**, 1689–1691 (November 1959).

[60] Zh. I. Alferov and E. V. Silina: Influence of the State of the Surface on the Breakdown Voltage of Alloy-Type Silicon Diodes, *Fiz. Tverd. Tela*, **1**, 1878–1879 (December 1959). (Russian) Translated in *Soviet Phys. Solid State*, **1**, 1719–1721 (June 1960).

[61] Zh. I. Alferov and E. A. Yaru: Breakdown of Alloy-Type Silicon Diodes in the Transmission (Forward) Direction, *Fiz. Tver. Tela*, **1**, 1879–1882 (December 1959). (Russian) Translated in *Soviet Phys. Solid State*, **1**, 1721–1723 (June 1960).

[62] B. Senitzky and P. D. Radin: Effect of Internal Heating on the Breakdown Characteristics of Silicon *p-n* Junctions, *J. Appl. Phys.*, **30**, 1945–1950 (December 1959).

[63] B. McDonald, A. Goetzberger, and C. Stephens: Uniform Avalanche Effects in "Multiple Predeposit" *p-n* Junctions in Silicon, *Bull. Amer. Phys. Soc.*, **4**, 455 (December 28, 1959).

[64] C. A. Escoffery: Introduction to Semiconductor Theory and Reverse Breakdown, *International Rectifier Corporation Zener Diode Handbook*, 1960, pp. 7–21.

[65] W. Shockley: Statistical Fluctuations of Donors and Acceptors in *p-n* Junctions, *Bull. Amer. Phys. Soc.*, **5**, 161 (March 21, 1960).

[66] Z. S. Gribnikov: Avalanche Breakdown in a Diode with a Limited Space-Charge Layer, *Fiz. Tverd. Tela*, **2**, 854–856 (May 1960). (Russian) Translated in *Soviet Phys. Solid State*, **2**, 782–784, (November 1960).

[67] M. Kikuchi and K. Tachikawa: Visible Light Emission and Microplasma Phenomena in Silicon *p-n* Junction, I, *J. Phys. Soc.* (*Japan*), **15**, 835–848 (May 1960).

[68] M. L. Forrest: Avalanche Carrier Multiplication in Junction Transistors and Its Implications in Circuit Design, *J. Brit. IRE*, **20**, 429–439 (June 1960).

[69] R. L. Batdorf, A. G. Chynoweth, G. C. Dacey, and P. W. Foy: Uniform Silicon *p-n* Junctions I—Broad Area Breakdown, *J. Appl. Phys.*, **31**, 1153–1160 (July 1960).

[70] A. G. Chynoweth: Uniform Silicon *p-n* Junctions, II—Ionization Rates for Electrons, *J. Appl. Phys.*, **31**, 1161–1165 (July 1960).

[71] C. A. Escoffery: Introduction to Semiconductor Theory and Reverse Breakdown, *Semicond. Prod.*, **3**, 42–45 (August 1960); 48–51 (September 1960); 30–32, 37–38 (October 1960).

[72] M. Kikuchi: Visible Light Emission and Microplasma Phenomena in Silicon *p-n* Junction, II—Classification of Weak Spots in Diffused *p-n* Junctions, *J. Phys. Soc. (Japan)*, **15**, 1822–1831 (October 1960).

[73] C. D. Root: Voltages and Electric Fields of Diffused Semiconductor Junctions, *IRE Trans. Electron. Devices*, **ED-7**, 279–282 (October 1960).

[74] C. D. Root, D. P. Lieb, and B. Jackson: Avalanche Breakdown Voltages of Diffused Silicon and Germanium Diodes, *IRE Trans. Electron. Devices*, **ED-7**, 257–262 (October 1960).

[75] A. Goetzberger: Uniform Avalanche Effect in Silicon Three-Layer Diodes, *J. Appl. Phys.*, **31**, 2260–2261 (December 1960).

[76] W. Shockley: Problems Related to *P-N* Junctions in Silicon, *Czech J. Phys.*, **11**, 81–121 (1961).

[77] W. Shockley: Problems Related to *p-n* Junctions in Silicon, *Solid-State Electron.*, **2**, 35–67, (January 1961).

[78] A. Goetzberger and C. Stephens: Voltage Dependence of Microplasma Density in *p-n* Junctions in Silicon, *J. Appl. Phys.*, **32**, 2646–2650 (December 1961).

[79] T. Tokuyama: Zener Breakdown in Alloyed Germanium p^+-n Junctions, *Solid-State Electron.*, **5**, 161–169 (May/June 1962).

[80] H. Kressel, A. Blicher, and L. H. Gibbons, Jr.: Breakdown Voltage of GaAs Diodes Having Nearly Abrupt Junctions, *Proc. IRE*, **50**, 2493 (December 1962).

[81] G. A. Baraff: Distribution Functions and Ionization Rates for Hot Electrons in Semiconductors, *Phys. Rev.*, **128**, 2507–2517 (December 15, 1962).

[82] R. A. Logan, A. G. Chynoweth, and B. G. Cohen: Avalanche Breakdown in Gallium Arsenide *p-n* Junctions, *Phys. Rev.*, **128**, 2518–2523 (December 15, 1962).

[83] R. H. Haitz: Model for the Electrical Behavior of a Microplasma, *Bull. Amer. Phys. Soc.*, **7**, 603 (December 27, 1962).

[84] H. C. Nathanson and A. G. Jordan: On Multiplication and Avalanche Breakdown in Exponentially Retrograded Silicon *P-N* Junctions, *IEEE Trans. Electron Devices*, **ED-10**, 44–51 (January 1963).

[85] M. Weinstein and A. I. Mlavsky: The Voltage Breakdown of GaAs Abrupt Junctions, *Appl. Phys. Letters*, **2**, 97–99 (March 1, 1963).

[86] J. L. Moll and R. van Overstraeten: Charge Multiplication in Silicon *p-n* Junctions *Solid-State Electron.*, **6**, 147–157 (March/April, 1963).

[87] A. Goetzberger, B. McDonald, R. H. Haitz, and R. M. Scarlett: Avalanche Effects in Silicon *p-n* Junctions II—Structurally Perfect Junctions, *J. Appl. Phys.*, **34**, 1591–1600 (June 1963).

[88] R. H. Haitz, A. Goetzberger, R. M. Scarlett, and W. Shockley: Avalanche Effects in Silicon *p-n* Junctions, I—Localized Photomultiplication Studies on Microplasmas, *J. Appl. Phys.*, **34**, 1581–1590 (June 1963).

[89] M. Poleshuk and P. H. Dowling: Microplasma Breakdown in Germanium, *J. Appl. Phys.*, **34**, 3069–3077 (October 1963).

[90] J. M. Borrego: Zener's Maximum Efficiency Derived from Irreversible Thermodynamics, *IEEE Proc.*, **52**, 95 (January 1964).

[91] R. H. Haitz: Variation of Junction Breakdown Voltage by Charge Trapping, *Phys. Rev.*, **138**, A260–267 (April 5, 1965).

[92] R. J. McIntyre: Multiplication Noise in Uniform Avalanche Diodes, *IEEE Trans. Electron Devices*, **ED-13**, 164–168 (January 1966).

[93] T. Misawa: Negative Resistance in *p-n* Junctions under Avalanche Breakdown Conditions—1, *IEEE Trans. Electron Devices*, **ED-13**, 137–143 (January 1966).

[94] T. Misawa: Negative Resistance in *p-n* Junctions under Avalanche Breakdown Conditions—2, *IEEE Trans. Electron Devices*, **ED-13**, 143–151 (January 1966).

[95] H. B. Emmons and G. Lucovsky: Frequency Response of Avalanching Photodiodes, *IEEE Trans. Electron Devices*, **ED-13**, 297–305 (March 1966).
[96] A. H. Cookson: Effect of Anode Hole on Electron Avalanches in Methane at High Pressure, *Inst. Elec. Eng.—Electron. Letters*, **2**, 1982–184 (May 1966).
[97] D. P. Kennedy and R. R. O'Brien: Avalanche Breakdown Calculations for Planar *p-n* Junction, *IBM J. Res. Develop.*, **10**, 213–219 (May 1966).

Chapter 2

CHARACTERISTIC PARAMETERS

2–1 INTRODUCTION

VR diodes can be very useful as additional building blocks of available semiconductor devices, *provided the circuit designer is thoroughly familiar with their characteristics*. He must know what kind of performance to expect under varying conditions and should understand both the strong features and the limitations of the VR diodes.

Although the VR diode is a seemingly simple device with only two leads, it has many subtleties that can greatly affect its performance in a given application. In this chapter we shall look at many of the characteristic parameters exhibited by VR diodes of both the Zener and the avalanche variety and study how they are affected by variations in bias or environmental conditions.

In addition to defining the major parameters of VR diodes, the information presented is meant to convey a generalized idea of the range of values that may be expected and how they respond to circuit and environmental effects. Although curves describing the performance of some specific devices are shown for illustration, they must not be interpreted as applying exactly to all devices. Great variations exist in both voltage range and power-handling capability, and it is to be expected that performance variations will exist also. A thorough understanding of this chapter, however, will permit you to interpolate or extrapolate the data presented so that they may be applied appropriately to your requirements.

A full comprehension of the information contained in this chapter is an absolute necessity in order to properly design circuits using VR diodes and make full use of their capabilities without incurring problems because of their behavior.

2–2 GENERAL CONSIDERATIONS

Before beginning a discussion regarding the parameters of VR diodes let us consider some general information regarding symbology and polarity conventions that are utilized in this chapter and throughout the rest of the book.

The graphical symbol that is utilized to represent both the Zener and the avalanche form of VR diode is shown in Fig. 2–1. The particular symbol shown is for the most common single-junction or single-anode VR diode.

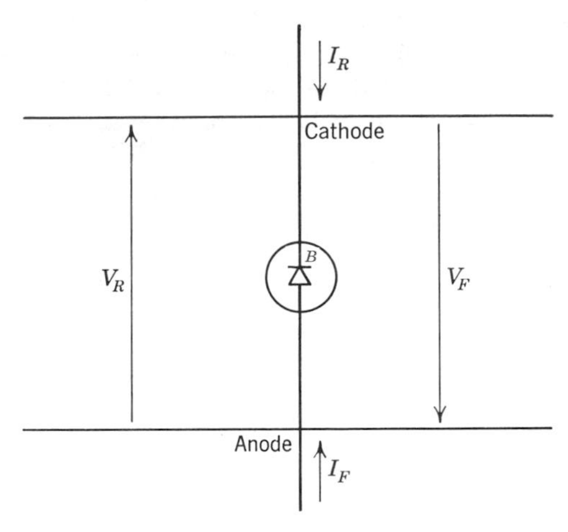

Fig. 2–1 Graphical symbol used for single-junction Zener and avalanche diodes.

Voltage and current polarities for the VR diode are indicated. In accordance with normal standards, the tip of the voltage arrow is the terminal that is assumed to be positive with respect to the tail. For the VR diode the anode is assumed positive with respect to the cathode when we refer to V_F or forward voltage. On the other hand, reference to V_R or reverse voltage indicates an assumed polarity where the cathode is positive with respect to the anode—the most normal case since the VR diode is designed to be operated in the reverse breakdown region.

If the polarity of the voltage assumed is correct, then the numerical value carries a positive sign. On the other hand, if the polarity assumed is incorrect, the numerical value will carry a negative sign. Thus, if we state that $V_R = 10$ V, we would indicate that the cathode of the VR diode is 10 V positive with respect to the anode (or to state it another way, that the anode is 10 V negative with respect to the cathode). A statement such as $V_R = -0.6$ V would imply that the direction of polarity for V_R as shown in Fig. 2–1 was in error and, in fact, the diode would be forward-biased, with the cathode negative with respect to the anode. Thus, if $V_R = -0.6$ V, we could say that $V_F = 0.6$ V. Normally, we shall choose to use either V_F or V_R, depending on the direction of normal operation and, in the usual case, the numerical value will carry a positive sign.

The direction of a current arrow indicates the assumed direction of conventional current flow (as opposed to electron flow). Thus, when we use the term I_R, we assume that the direction of conventional current flow is into the cathode of the VR diode as shown in Fig. 2–1. On the other hand, use of I_F indicates an assumption that the conventional current flow is into the anode. As with the voltage symbols, a positive sign associated with the numerical value of a current indicates that the actual direction of current flow is as assumed, whereas a negative sign indicates that the actual direction of current flow is opposite to that which was assumed.

The graphical symbol for the double-anode type of VR diode is shown in Fig. 2–2. This type of VR diode is normally used in symmetrical clipping

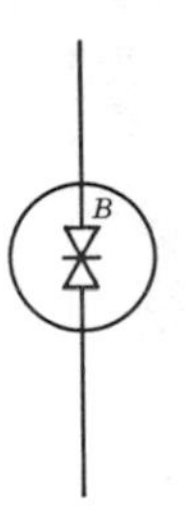

Fig. 2–2 Double-anode VR Diode.

circuits etc., and actually consists of two diode junctions connected cathode to cathode. The result is that, regardless of terminal polarity, one junction is biased in the reverse direction and the other is forward biased. Since the units are normally manufactured to be symmetrical, it becomes unnecessary to indicate either which terminal is which or a voltage polarity.

2–3 DC CHARACTERISTIC CURVES

Much information may be gained by the study of a component's dc characteristic or voltage–current curves. Figure 2–3 illustrates a generalized curve that would be obtained with almost any typical silicon diode. The three regions of interest are (1) forward-biased region, (2) reverse-biased region, and (3) the breakdown region.

Actually, a characteristic curve of the form shown in Fig. 2–3 will be exhibited by just about all the usual junction diodes, whether or not they are designed specifically as VR diodes. Note that a different current scale is used for the forward and reverse regions. The difference in VR diodes is the intentional control primarily of the breakdown region. Non-VR diodes are more likely to be characterized around their forward characteristics and made to have a certain minimum value of breakdown voltage.

In the forward-biased region a VR diode looks much the same as a normal high-conductance silicon diode. Any value of V_F greater than 0.7 V will produce a substantial amount of current.

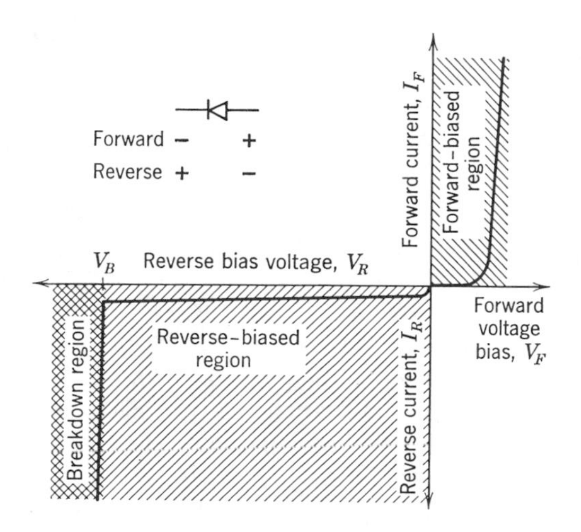

Fig. 2–3 Generalized voltage-current characteristic curve for a semiconductor diode.

In the reverse-biased region, but before breakdown, the VR diode exhibits a small leakage current which is typically in the nanoampere region as is discussed further in a later section.

As the reverse voltage across the VR diode approaches the breakdown value, the reverse current begins to increase. The sharpness of the transition depends on the relative value of the breakdown voltage as well as variations in manufacturing processes used to make the diode. The point on the characteristic curve where the reverse current begins to rise is referred to as the knee of the curve.

When the reverse voltage across the VR diode is increased further, the reverse current will become larger and larger. More will be said about this portion of the characteristic when static and dynamic breakdown resistances are discussed.

Much can be learned by observing the characteristic curve of a particular device; and for this reason most manufacturers use a sweeper to inspect visually the VR diodes, on a sample basis if not 100% testing. A defective VR diode may exhibit an abnormal amount of leakage in the prebreakdown region or the knee region may show a double break or erratic transition. As a matter of general rule, any appreciable amount of instability is cause for concern.

2–4 BREAKDOWN VOLTAGE

Since the VR diode is designed primarily for operation in the breakdown region, we would not be amiss in considering that the breakdown voltage is perhaps the most important parameter exhibited by the device. It is by this parameter that VR diodes of a given family are "typed" and likewise break down voltage becomes a very important design parameter when used in circuit applications.

2–4.1 Definition

As illustrated in the generalized drawing of Fig. 2–3, the breakdown voltage V_B is the value of reverse voltage that causes the VR diode to enter the breakdown or high reverse current region. On some VR diode specification sheets this voltage is referred to as V_Z or E_Z, but the symbol V_B is more correctly descriptive in conjunction with avalanche as well as Zener type VR diodes.

If all VR diodes has a characteristic in which the transition from reverse leakage to breakdown was as sharply defined as that shown in Fig. 2–3, it would be easy to define V_B as the voltage at the exact point of transition. Unfortunately, such is not the case as is evident from Fig. 2–4 in which the characteristics for several low voltage VR diodes are shown.

For curves 1 and 2 it would be hard to say just where the knee really is. Even curves 4 and 5, although having a much sharper breakdown transition than curves 1 and 2, have a certain softness in the knee region. Curve 3 begins to look something like an ideal case.

The end result is that a meaningful specification of the breakdown voltage V_B must include a specification or the reverse current I_R at which it is to be measured. As seen later in this chapter, the temperature at which the test is performed and the thermal resistance between the device leads and ambient can also influence the value of V_B.

2–4.2 Range Available

VR diodes are available with breakdown voltages from 2.4 to 200 V. Normally, low-voltage VR diodes with V_B less than about 6.2 V are made by the alloy junction process. VR diodes having a V_B of 6.8 V or higher are usually made by a diffusion process. There is some overlap, however, and between 6.8 V and 12 V we may obtain devices manufactured with either process.

2–4.3 Variation in Breakdown Characteristics

Since the resistivity of the material used in making different breakdown voltage VR diodes is different, and even the process can be different, we might well expect some substantial variation in the breakdown characteristics. This was briefly illustrated in discussing the several characteristic curves shown in Fig. 2–4. Curves 1, 2, and 4 are for alloy-type VR diodes, whereas curves 3 and 5 are for diffusion-produced devices.

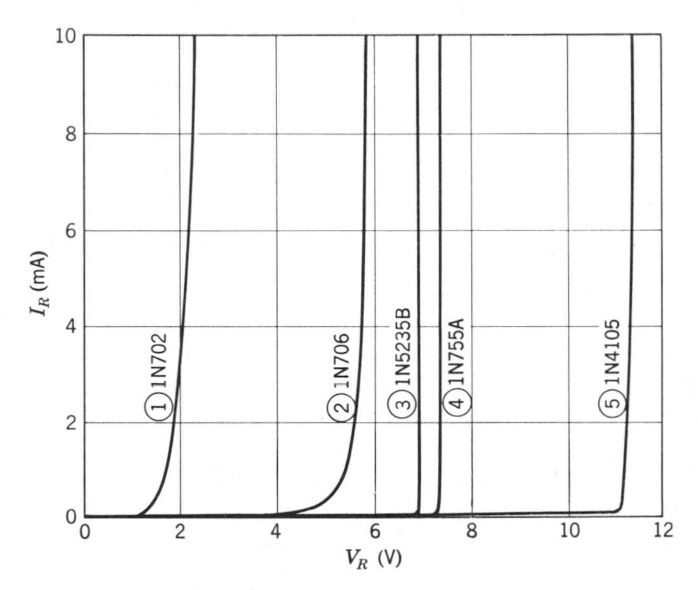

Fig. 2–4 Actual reverse characteristic curves for several low-voltage VR diodes.

Low-voltage VR diodes with V_B less than about 5 V operate primarily on a field emission or true Zener breakdown. This is a tunneling process which produces a rather soft or gradual breakdown characteristic. In fact, the actual breakdown characteristic follows a logarithmic pattern. This is clearly illustrated in Fig. 2–5 for a 1N5226 VR diode which has a nominal V_B of 3.3 V at $I_R = 20$ mA.

The logarithmic nature of the breakdown characteristic continues from low currents until rather high currents where self-heating effects become prominent. This is shown by the solid lines in Fig. 2–5. The dotted lines indicate the characteristics obtained with the junction temperature held constant by means of low duty cycle pulse techniques. Although this softness in the breakdown characteristic is a problem in applications using a low-voltage VR

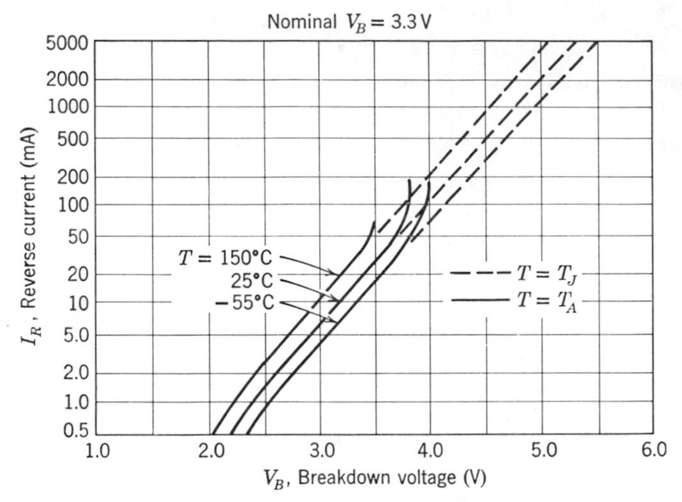

Fig. 2–5 Reverse breakdown characteristics for a 1N5226 VR diode exhibiting the nature of true Zener breakdown. (Courtesy Motorola, Inc.)

diode as a regulator, its logarithmic nature may be used to design a logarithmic converter as discussed in Chapter 10.

VR diodes with higher breakdown voltages in which avalanche breakdown becomes a factor have substantially different breakdown characteristic curves. Figure 2–6 illustrates the characteristics obtained for VR diodes that are in

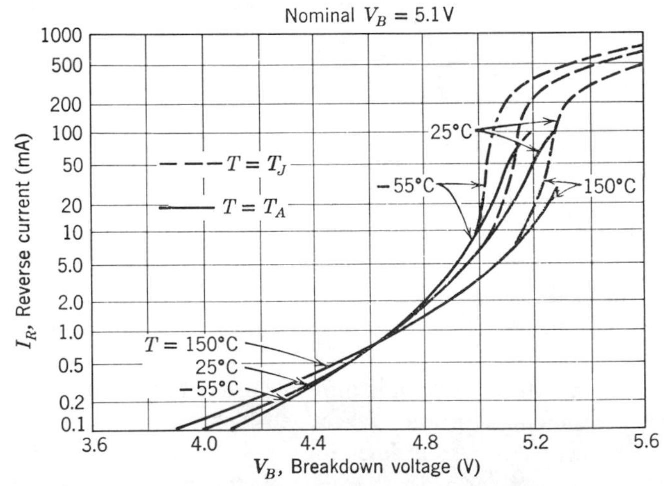

Fig. 2–6 Reverse breakdown characteristics for a 1N5231 VR diode. Both Zener and avalanche breakdown effects are exhibited. (Courtesy Motorola, Inc.)

the crossover region between Zener and avalanche modes. At low current levels the Zener or field emission effects tend to predominate and the characteristic follows the logarithmic pattern as shown before.

At increased current levels the 1N5231 VR diode, having a nominal V_B of 5.1 V at $I_R = 20$ mA, begins to exhibit more of the avalanche mode of breakdown. The curve begins to depart from its logarithmic nature even when the junction temperature is held constant as before.

The breakdown characteristic of a VR diode operating almost exclusively in the avalanche mode is very steep as illustrated in Fig. 2–7. The 1N5254 has a nominal V_B of 27 V for $I_R = 4.6$ mA. With self-heating effects removed, the characteristic remains steep up to a rather high current level.

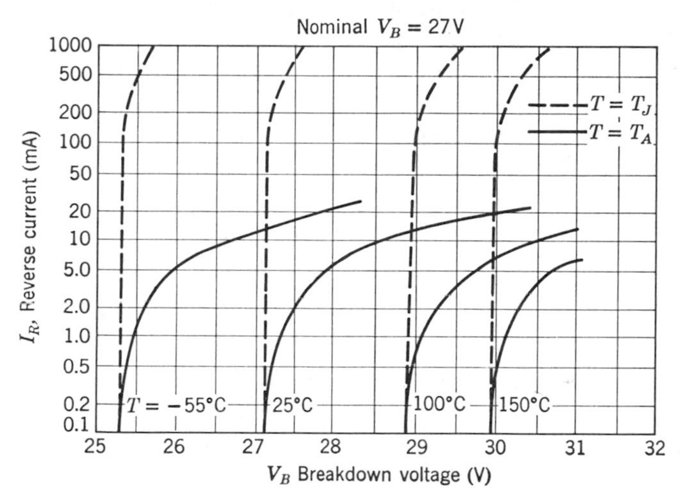

Fig. 2–7 Reverse breakdown characteristics for a 1N5254 VR diode that operates in the avalanche mode. (Courtesy Motorola, Inc.)

2–5 TEMPERATURE COEFFICIENT

As mentioned earlier and as clearly evident in the breakdown characteristic curves shown in Figs. 2–5, 2–6, and 2–7, the breakdown voltage of a VR diode will vary with temperature changes. It does not matter whether these temperature changes occur due to internal heating effects caused by power dissipation or to a change in the ambient temperature.

The temperature coefficient of the Zener breakdown is negative and is relatively independent of the current level. This is demonstrated by the curves of Fig. 2–5.

Avalanche breakdown exhibits a positive temperature coefficient and is also relatively current independent if self-heating effects are removed, as shown in Fig. 2–7.

VR diodes which are in the midvoltage range and exhibit effects of both Zener and avalanche breakdown may have either a positive or negative temperature coefficient, depending on which effect is predominant. At low current levels the Zener effect is strongest and the TC is negative.

At rather high current levels the avalanche effect becomes more evident and the VR diode takes on a positive TC. Because both Zener and avalanche breakdown modes are actually at work simultaneously, with the mixture controlled by the relative current level, we experience a variation in the value of TC as the current is changed.

At some specific current the negative TC of the Zener effect is exactly equal to the positive TC of the avalanche effect and the net TC is zero. In Fig. 2–6 this condition of zero TC occurs at the point in which all three of the characteristic curves intersect.

2–5.1 Definition

Temperature coefficient, often abbreviated as TC or represented by the symbol K_T, may be defined in either of the two following ways:

$$K_T = \frac{\Delta V_B(\text{mV})}{\Delta T(^\circ\text{C})} \rightarrow \text{mV}/^\circ\text{C}, \tag{2-1}$$

$$K_T^* = \frac{(\Delta V_B)/V_B}{100\,\Delta T} \rightarrow \%/^\circ\text{C}. \tag{2-2}$$

Although the mathematical purist would take issue that we would get different results, depending on which of the two definitions we used, the discrepancy is slight for the relatively low values of TC which are exhibited even for 200-V units. Some designers prefer working with the %/°C figure especially when working with power supplies in which the output voltage is usually a constant multiplied by the V_B of the VR diode used as a reference. The TC of the output voltage due to the VR diode is equal to the value of K_T^* of the VR diode if expressed in %/°C.

Other designers are more concerned about the actual voltage change and prefer to express the value of K_T in mV/°C. To compute the total expected change in V_B caused by a temperature change, we multiply the value of V_T expressed in mV/°C by the temperature difference. The result is in millivolts.

We may occasionally need to change back and forth between K_T as expressed in mV/°C and K_T^* expressed in %/°C. The following equations make this a simple task:

$$K_T^* = \frac{K_T}{10 V_B}, \tag{2-3}$$

$$K_T = 10 V_B K_T^*. \tag{2-4}$$

To avoid unnecessary confusion we shall restrict usage only to K_T as expressed in mV/°C except in Chapter 4 when discussing the reference amplifier. The basic nature and operation of the reference amplifier lends itself better to the use of K_T^* as expressed in %/°C.

2–5.2 K_T as a Function of V_B

By taking a large number of VR diodes and measuring K_T for all devices at a constant reverse breakdown current, we might obtain curves like those shown in Fig. 2–8 for the 1N5221–1N5242 family which illustrate tha variation in K_T for different values of V_B up to 12 V.

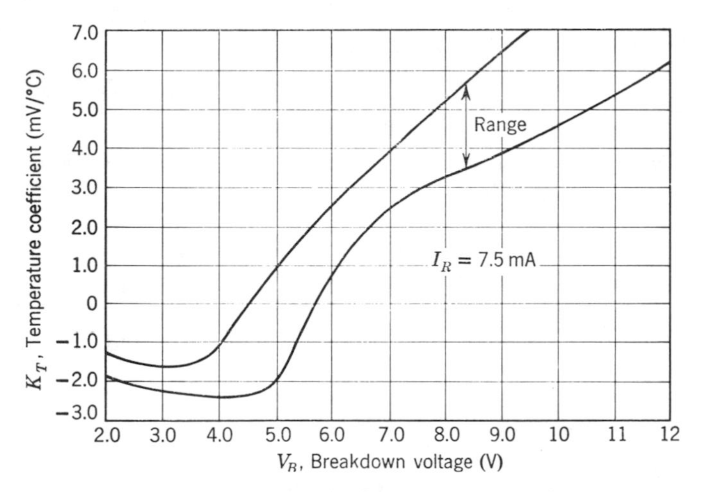

Fig. 2–8 Variation and spread of temperature coefficient for VR diodes up to 12 V. (Courtesy Motorola, Inc.)

Note also that a considerable spread exists even for VR diodes having the same value of V_B. This makes it somewhat difficult in predicting exact performance even for a given VR type. Different geometries of devices incorporated in VR diodes of varying power dissipation induce additional variation in the spread of K_T. Nevertheless the curves indicate an approximate trend that at least tell the designers what to expect.

Higher voltage VR diodes exhibit a variation in K_T which, on a percentage basis, is much smaller than for the lower voltage units. A typical spread is shown in Fig. 2–9 for the 1N5242–1N6281 family of VR diodes.

Figure 2–8 indicates that VR diodes having a V_B at $I_R = 7.5$ mA between about 4.5 and 5.7 V can exhibit a zero temperature coefficient. However, if we plan on using a 5V VR diode as a reference source, the actual value of K_T might be anywhere between $+1$ and -2 mV/°C! This is a fact which is ignored all too often by designers who measure a limited number of devices,

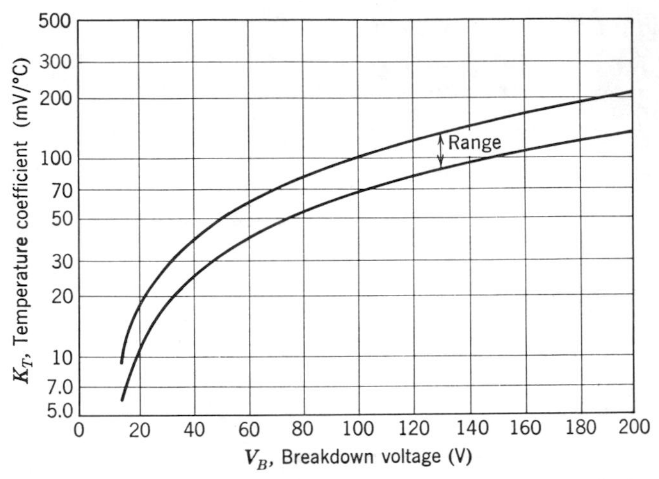

Fig. 2–9 Variation and spread of K_T for higher voltage VR diodes. (Courtesy Motorola, Inc.)

find that they get a TC of nearly zero, then design circuits around their sample—only to find that the spread arrives in following production lots. We may depend on a certain value of K_T only if (1) we individually test the units or (2) purchase to a guaranteed specification, indicating that the manufacturer is monitoring the value of K_T.

2–5.3 K_T as a Function of I_R

In Fig. 2–6 it was seen that the T.C. for the 5.1-V VR diode could be made either positive, negative, or even zero by controlling the value of the bias current. The 3.3 V VR diode whose characteristics were given in Fig. 2–5 and the 27-V VR diode described in Fig. 2–7 showed much less dependence of K_T on the value of I_R.

The effect is more clearly displayed in Fig. 2–10 for VR diodes having a breakdown voltage from 2 to 9 V. Note that although VR diodes having V_B greater than about 8V showed negligible variation in K_T over a spread of I_R from 0.01 to 10 mA, lower voltage devices exhibited a strong dependence on the bias current. A 5-V VR diode, for example, could have a value of K_T anywhere from +1 mV/°C to −2 mV/°C, depending on the value of I_R.

As the V_B of the VR diode is made smaller, the variation in K_T with changes in bias current diminishes to almost insignificant proportions when V_B is about 3 V or less.

The variation in K_T with changes in I_R goes back once againt to the fact that the two distinct modes Zener and avalanche breakdown do occur simultaneously in the midvoltage VR diodes. As the value of the bias current is

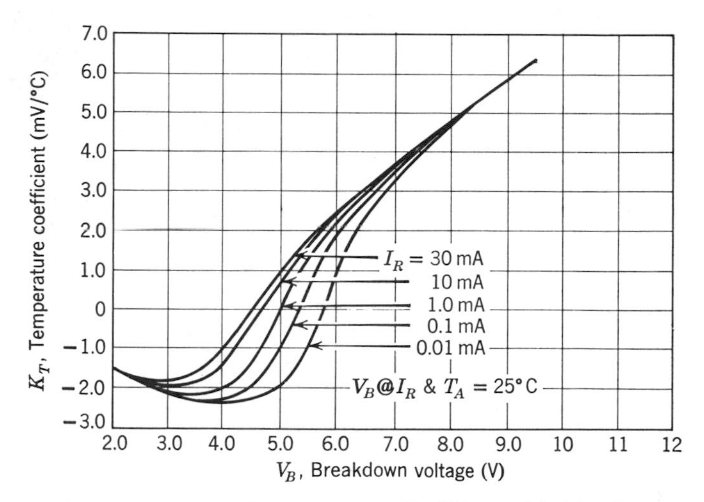

Fig. 2–10 Variation of Temperature Coefficient with Bias Current. (Courtesy Motorola, Inc.)

increased, the avalanche effect is aided. On the other hand, the most predominant effect at low current levels can be the Zener or field emission effect. Thus a 5-V diode may be predominantly Zener diode at an I_R of 0.01 mA and have a large negative K_T, whereas it may have enough avalanche effect at an I_R of 10 mA to overcome the Zener effect and hence have a substantial positive TC.

At very low values of V_B the breakdown effect is so predominantly Zener mode that even large currents do not enhance the avalanche effects to the point where they become significant. Likewise, higher voltage VR diodes operate so predominantly in the avalanche mode that Zener effects are negligible even for very small current levels.

2–5.4 K_T as a Function of Temperature

In considering the basic definitions of the temperature coefficient as expressed by (2–1) and (2–2), the general assumption was made that the value of K_T would be independent of temperature. Thus we could vary the temperature over a known range, measure the change in V_B, and then compute the value of K_T. The magnitude of possible error may be seen by carefully studying the data presented in Fig. 2–11.

Suppose we were measuring the value of K_T for the 5.6-V VR diode whose characteristics are shown in Fig. 2–11*e* and used 20 mA for I_R. We might make V_B measurements at 0°C and at 170°C, find that they were the same, and then erroneously conclude that K_T over the entire range is zero. From the data presented in Fig. 2–11*e* we see that initially (using 0°C as a reference),

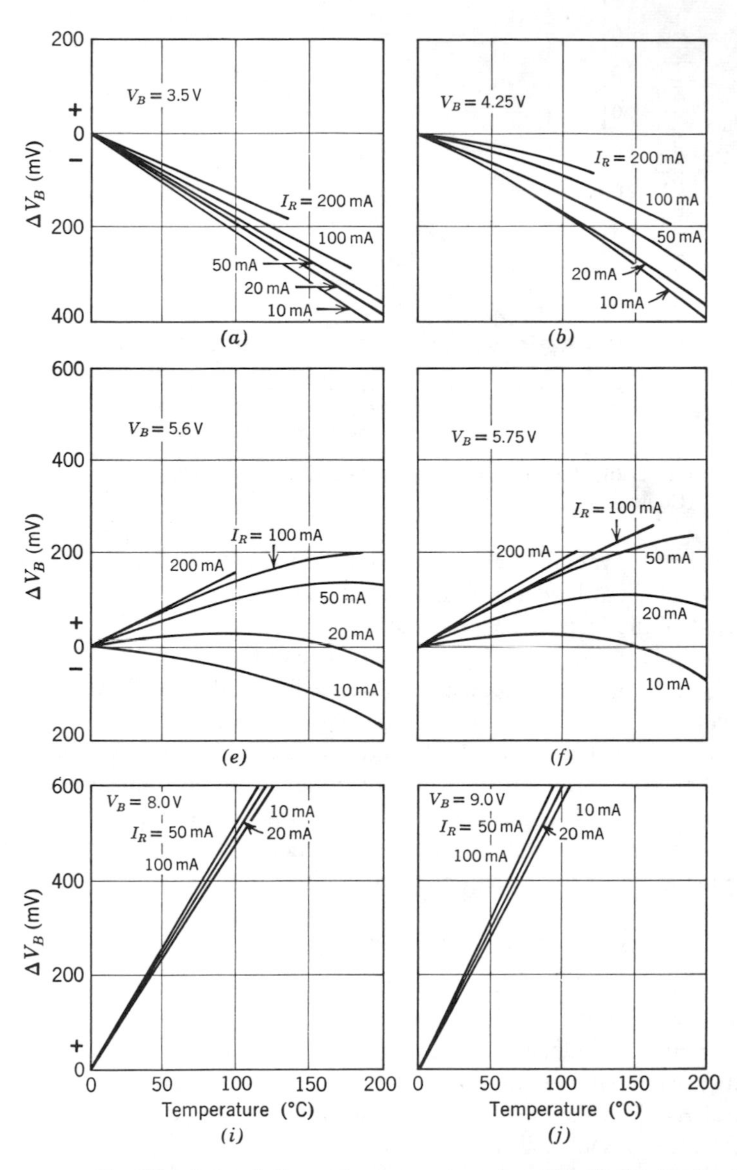

Fig. 2–11 Variation of V_B with temperature for VR diodes having nominal V_B from 3.5 to 10 V. (*a*)–(*k*) reverse breakdown; (1) forward voltage variation shown for comparison. (Courtesy *Proc. I.E.E.*[7])

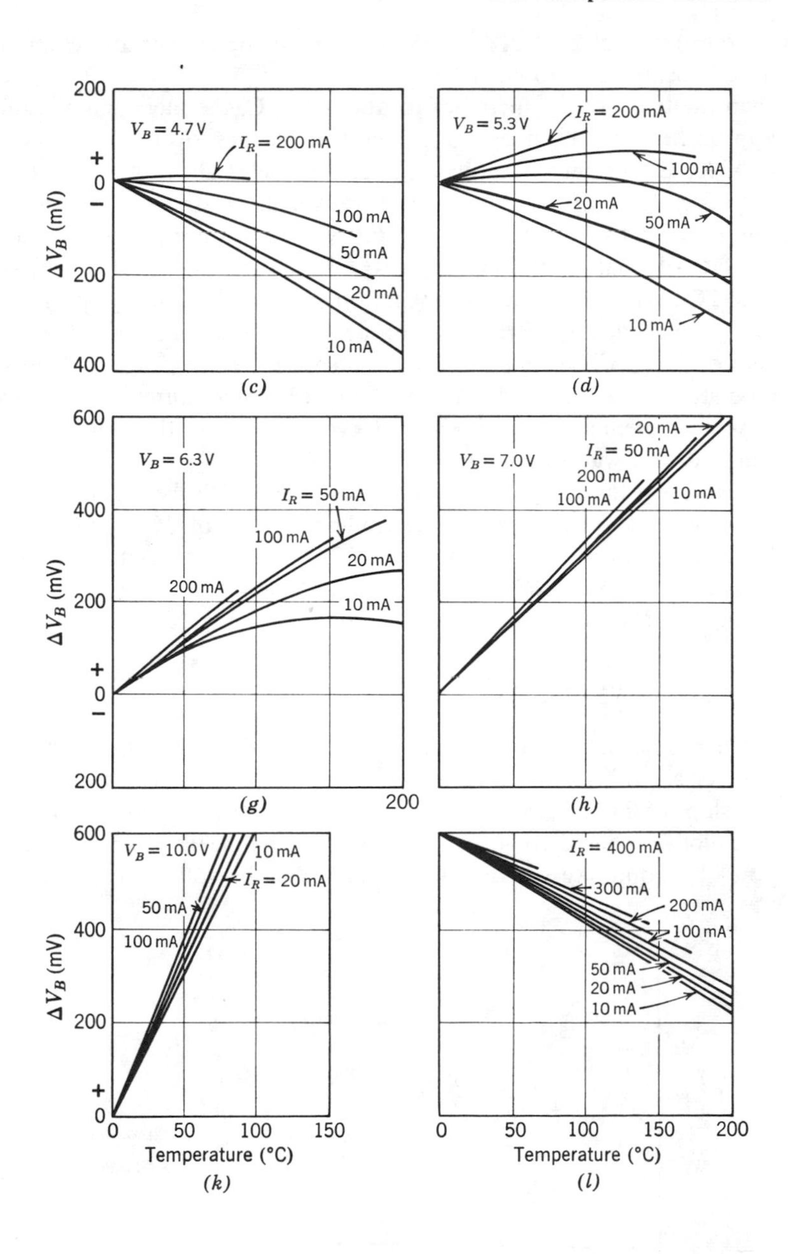

200
$V_B = 4.7$ V
$I_R = 200$ mA
100 mA
50 mA
20 mA
10 mA
+
0
−
200
400
ΔV_B (mV)
(c)
$V_B = 5.3$ V
$I_R = 200$ mA
100 mA
20 mA
50 mA
10 mA
(d)
600
400
200
$V_B = 6.3$ V
$I_R = 50$ mA
100 mA
20 mA
200 mA
10 mA
(g)
200
$V_B = 7.0$ V
20 mA
$I_R = 50$ mA
200 mA
100 mA
10 mA
(h)
$V_B = 10.0$ V
10 mA
$I_R = 20$ mA
50 mA
100 mA
0
50
100
150
Temperature (°C)
(k)
$I_R = 400$ mA
300 mA
200 mA
100 mA
50 mA
20 mA
10 mA
0
50
100
150
200
Temperature (°C)
(l)

the incremental value of K_T taken over small increments of temperature would be positive and about +1 mV/°C.

When the temperature increases to about 100°C, the incremental value of K_T approaches zero. Then, as the temperature reaches 170°C, the incremental value of K_T has become negative and has a value of about −1 mV/°C.

This dependence of K_T on temperature is again restricted to mid-voltage VR diodes with V_B from about 4.8 to 6.5 V. It may once again be explained by the fact of dual breakdown modes. Chynoweth and McKay [4] stated that the TC of the avalanche multiplication constant M is always positive and actually independent of V_B. Zener or field emission, since it is basically a tunneling effect, will be enhanced by increase in temperature and thus can become strong enough to dominate at higher temperatures and produce a negative incremental value of K_T and even a net negative value of K_T if measured over a wide enough range.

By restricting the range of temperatures over which we measure the ΔV_B to compute K_T and by using a reference temperature of 25°C, we reduce the probable error. In using stated values of TC as given on a data sheet for a particular type of VR diode, be sure you understand the manner in which it is specified and how it relates to your application.

2–6 BREAKDOWN RESISTANCE

Another important VR diode parameter is the effective resistance presented by the device while in the breakdown region. This will be displayed by the inverse slope of the characteristic curve of the VR diode, as shown in Fig. 2–12.

If we plot the characteristic curve very slowly by means of an X-Y recorder or point by point, we get the curve shown by the solid line in Fig. 2–12.

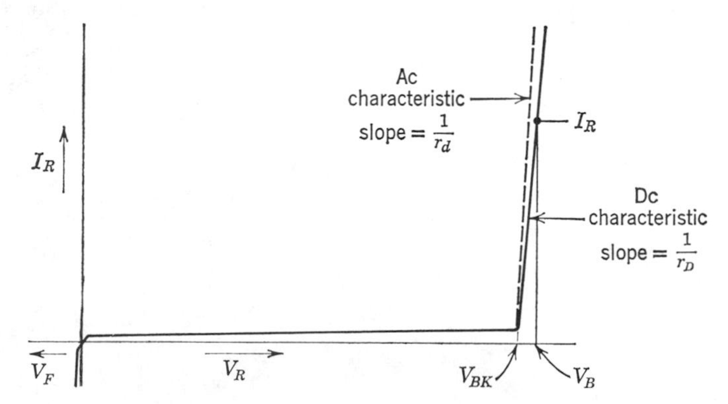

Fig. 2–12 Typical voltage-current characteristic curve of a VR diode illustrating the difference between static and dynamic curves.

However, if we use an oscilloscope type of curve tracer, we might get the characteristic indicated by the dashed line. It becomes readily evident that the slope of the characteristic in the breakdown region, and hence the effective breakdown resistance is quite different for the two cases.

The difference is due to internal heating which occurs in the slow or dc case. Recall that there is a temperature coefficient associated with the breakdown voltage. When the reverse current begins to increase, power is applied to the junction and must be dissipated in the form of heat. This causes the junction temperature T_J to rise by an amount dependent on the thermal resistance between the junction and outside ambient θ_{JA}, which is discussed later in this chapter. The breakdown voltage then varies according to the value of the temperature coefficient K_T which may have positive or negative polarity.

Depending on the breakdown voltage of the VR diode and on the biasing conditions, the total voltage across the diode may either increase or decrease as the junction heats up. We may consider that there are two separate parts of the breakdown resistance: one of them due to the breakdown mechanism itself; the other due to thermal heating effects. We shall need to consider both since they each affect circuit performance.

2–6.1 Definitions

As illustrated in Fig. 2–12, we may define two different parameters of breakdown resistance: one for the ac transient case in which no internal heating occurs; the other for the dc or static case which will incorporate the total effective resistance, including heating effects.

The breakdown resistance associated only with the breakdown mechanism would be the inverse slope of the ac characteristic curve in the breakdown region as shown in Fig. 2–12. This we may call r_d, using a lower case subscript to denote that it is strictly an ac parameter and define it by the following equation:

$$r_d \equiv \frac{V_r}{I_r}, \tag{2-5}$$

where V_r is the rms voltage measured across the VR diode in the breakdown region when an ac current whose rms value is I_r is sent through it.

In order for the term r_d to be really useful, it is necessary to include a dc bias to place the quiescent operating point at the desired current level along the breakdown characteristic curve. The value of r_d then will be the small-signal equivalent resistance of the VR diode in the breakdown region.

Several restrictions in the characteristics of I_r will be necessary to insure meaningful results. First of all, the value of I_r should be made small in comparison with the dc bias current I_R. As a general rule I_r is made equal to or less than $0.1\,I_R$.

The frequency repetition rate of the ac current I_r must be high enough to prevent any thermal response. The thermal time constant of most VR diodes is sufficient to permit the use of a 60 Hz current.

Since r_d applies to a small-signal ac condition and may be measured at various bias current levels, it is usually referred to as the *dynamic* breakdown resistance. It is the parameter usually specified as the breakdown resistance (or sometimes referred to as an impedance) on the manufacturer's specification sheet.

The breakdown resistance of the VR diode that incorporates the dc or *static* case is defined by the inverse slope of the breakdown characteristic taken in such a manner that each point on the curve represents a condition of thermal equilibrium. It is still an incremental resistance, but is normally taken over increments larger than those used in measuring the dynamic resistance r_d. Mathematically, we may define the static breakdown resistance r_D as

$$r_D \equiv \frac{\Delta V_B}{\Delta I_R}\,, \tag{2-6}$$

where ΔV_B is the measured change in the dc breakdown voltage that occurs when the dc bias current is varied by an amount equal to ΔI_R. For accurate measurement of r_D, sufficient time must be permitted at each of the two conditions to permit thermal equilibrium to occur.

Although r_D is an incremental resistance, the amount of increment in reverse current, ΔI_R, amy sometimes be made rather large in specifying r_D; for example, r_D may be given as the inverse slope between two points on the breakdown characteristic curve where I_R is near the maximum rated value and where I_R is some 20% of rated value.

The value of static breakdown resistance, r_D, may either be larger than or less than the dynamic resistance r_d even though the nominal I_R is the same for both measurements. For VR diodes with V_B higher than 6 V, r_D is typically substantially greater than r_d because of the positive value of the temperature coefficient.

2–6.2 Effect of V_B on r_d

A review of the characteristic curves shown in Figs. 2–5, 2–6, and 2–7 tells us that there is a strong dependence of the dynamic breakdown resistance r_d on the value of the breakdown voltage. This is even more clearly demonstrated in Fig. 2–13.

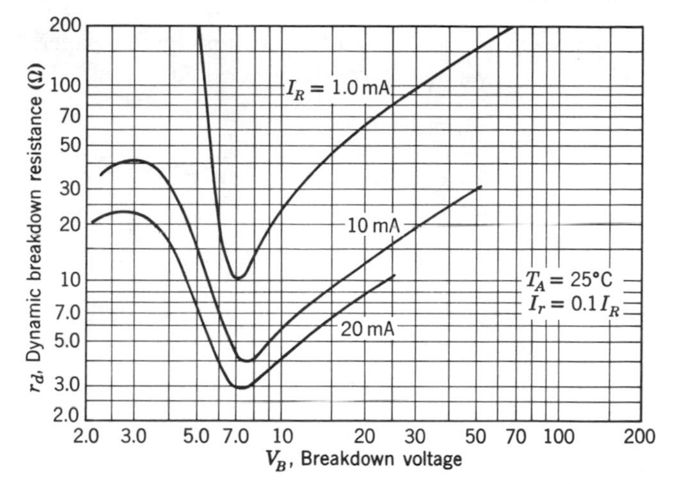

Fig. 2–13 Effect of nominal breakdown voltage on the dynamic breakdown resistance. (Courtesy Motorola, Inc.)

Very low voltage VR diodes, in which the Zener effect is almost entirely responsible for the breakdown characteristics, exhibit little dependence of r_d on the value of V_B. Then, as we look at VR diodes with higher values of V_B, the magnitude of r_d drops rather sharply. A minimum value is reached for diodes having a breakdown voltage of about 7 V.

Higher voltage VR diodes will exhibit increased values of r_d. The relationship is almost linear with a doubling of V_B, producing about twice the value of r_d.

At first glance the information given in Fig. 2–13 seems to tell us that we would always like to work with VR diodes having a V_B of about 7 V since r_d would be at about its lowest value. However, if we consider the variation in voltage due to r_d on a percentage basis, we find that it remains roughly the same for quite a range in voltages. If, however, we consider the practical limitation in power dissipation, which generally forces us to operate a higher voltage VR diode at a lower bias current, the increased value of r_d becomes more significant.

2–6.3 Effect of I_R on r_d

The characteristic curves of Figs. 2–5, 2–6, and 2–7, as well as the data presented in Fig. 2–13, also indicate a very strong dependence of r_d on the bias current I_R.

For the very low voltage VR diode whose characteristics are shown in Fig. 2–5, it might at first appear that since the voltage-current plot has a relatively straight line for the case where junction temperature is held constant, the effective resistance would also be constant. Note, however, that the characteristic curves are plotted on semilogarithmic graph paper. Since the plot is relatively straight, we may write the general equation

$$V_R = K_1 \log_{10} I_R + K_2 . \tag{2-7}$$

When junction temperature is held to 25°C in Fig. 2–5, (2-7) takes the following specific approximation:

$$V_{R(25)} \approx 0.78 \log_{10} I_R + 2.4, \tag{2-8}$$

where V_R is given in volts and I_R is expressed in mA. By differentiating both sides of (2-8), we may derive an equation relating the incremental ac resistance r_d:

$$r_d = \frac{dV_R}{dI_R} = \frac{340}{I_R}, \tag{2-9}$$

where once again I_R is expressed in mA and r_d is given in ohms. Thus for the 3.3-V VR diode the value of dynamic breakdown resistance r_d will be inversely proportional to the bias current.

The relationship of r_d as a function of bias current is given in Fig. 2–14 for VR diodes having breakdown voltages of 3.3, 5.1, 27, and 75 V. To a very close approximation, all the curves are straight lines. An equation which plots as a straight line on log-log paper must be of the general form

$$Y = K_1 X^n. \tag{2-10}$$

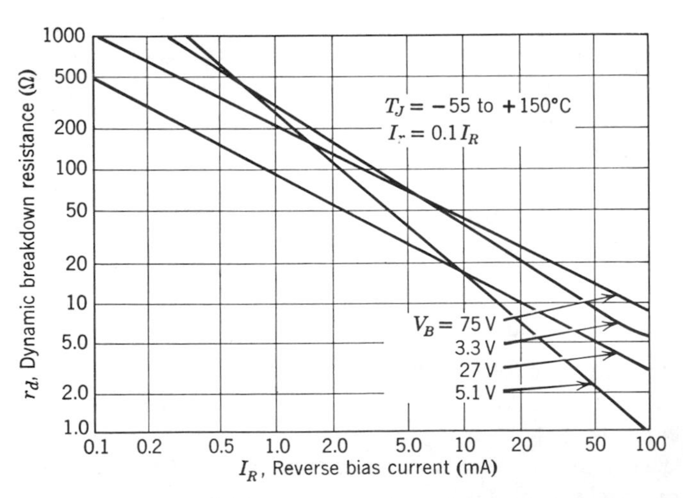

Fig. 2–14 Effect of dc bias current on dynamic breakdown resistance for the 1N5221-1N5281 series of 500-mW VR diodes. (Courtesy Motorola, Inc.)

By proper solution of the constants K_1 and n for each of the diodes, we obtain the following equations describing the empirical relationships of r_d as a function of I_R:

$$r_{d(3.3\text{V})} = 304\, I_R^{-0.89}, \tag{2-11}$$

$$r_{d(5.1\text{V})} = 270\, I_R^{-1.22}, \tag{2-12}$$

$$r_{d(27\text{V})} = 92\, I_R^{-0.74}, \tag{2-13}$$

$$r_{d(75\text{V})} = 210\, I_R^{-0.70}, \tag{2-14}$$

where I_R is expressed in mA and r_d is given in Ω.

Equation 2-11 differs slightly from (2-9) derived from the characteristics of Fig. 2–5, but is perhaps more correct. Each of the above equations is only an approximation, but a review of devices made by several different manufacturers and of several sizes of geometry yielded results close enough to the empirical equations given to be quite useful in predicting r_d for operating currents other than that for which r_d was given on the specification sheet.

Note that the exponent for any VR diode having a V_B greater than about 9 V seems to center around a figure of about -0.7 to -0.75.

Perhaps the least predictable group of VR diodes, as you might guess from previous discussions, is the VR diodes in the midvoltage range in which both Zener and avalanche breakdown mechanisms are at work. In fact, some devices seem to yield a curved line when we plot the r_d as a function of current.

It might also be well to consider that although r_d seems to continue as I_R is increased there is a limit to how low r_d may get. This is due to a finite saturation resistance resulting from ohmic contacts and bulk material resistance. In addition, higher breakdown voltage diodes will have a definite limitation due to maximum permissible power dissipation.

2–6.4 Effect of Temperature on r_d

Going back to the breakdown characteristics of the 3.3-V VR diodes given in Fig. 2–5 note that, the three curves presented for the three different temperatures all remain approximately parallel with each other. This indicates that for this diode r_d changes very little with temperature.

The curves shown in Fig. 2–6 for a 5.1-V VR diode, however, do indicate a positive temperature coefficient for r_d ,where I_R is less than about 20 mA. Above this level the curves begin to run along more or less parallel, thus indicating a negligible temperature dependence in r_d. Information from VR diode data sheets, such as that given by Fig. 2–14, generally state that the same curves hold over a wide temperature range.

2–6.5 Static Breakdown Resistance r_D

Earlier we defined r_D as the inverse slope of the dc characteristic curve in the breakdown region. We also indicated that the difference between the ac and dc characteristic curves was due to heating effects. Now let us consider how we might be able to predict the value of r_D.

Since r_D represents the *total* effective breakdown resistance, it is the sum of the dynamic resistance r_d and a thermally produced resistive term we shall call r_T. Provided we restrict the size of the current increment considered, we may assume that the value of r_d is relatively constant and predictable from the equations given previously in this chapter. In order to obtain a value for r_D, then, we need to be able to predict r_T.

Mathematically, the value of r_T may be expressed as

$$r_T = \frac{\Delta V_{B(T)}}{\Delta I_R}, \tag{2-15}$$

where $\Delta V_{B(T)}$ represents the variation in breakdown voltage which occurs *only* because of heating effects as the bias current changes by an amount equal to ΔI_R.

First of all, we need to know the amount of temperature rise that will occur as I_R is incremented. This may be done by considering the change in power dissipation and the thermal resistance between the junction and ambient, θ_{JA}. The total change in power dissipation resulting from the change in I_R and the resulting in power dissipation resulting from the change in I_R and the resulting total change in V_B—both due to heating effects and to the presence of r_d—may be expressed as

$$\Delta P_J = V_B(\Delta I_R) + I_R(\Delta V_B) + (\Delta I_R)(\Delta V_B). \tag{2-16}$$

Since the change in V_B will be kept rather small by the regulator action of the VR diode, the second and third terms in (2-16) may actually be ignored without resulting in appreciable error. We may now compute the change in the junction temperature that occurs by using the first term approximation of (2-16) in conjunction with the thermal resistance between junction and ambient.

$$\Delta T_J = \theta_{JA}(\Delta P_J) = \theta_{JA} V_B(\Delta I_R). \tag{2-17}$$

With the change in junction temperature known we may compute the resulting change in V_B that is due only to thermal effects by making use of the temperature coefficient.

$$\Delta V_{B(T)} = K_T(\Delta T) = K_T \theta_{JA} V_B(\Delta I_R). \tag{2-18}$$

Since K_T is normally given in mV/°C, the value of $\Delta V_{B(T)}$ is given in terms of mV, where θ_{JA} is in °C/mW and ΔI_R is in mA.

We may now insert the value of $\Delta V_{B(T)}$ from (2-18) into (2-15) to determine the part of the breakdown resistance due to thermal effects. This gives

$$r_T = K_T \theta_{JA} V_B. \tag{2-19}$$

If K_T is in mV/°C, θ_{JA} is expressed in °C/mW, and V_B is in volts then r_T is given in ohms:

$$r_D = r_d + r_T + r_d + K_T \theta_{JA} V_B, \tag{2-20}$$

and thus we now have an expression that relates the value of the slope of the breakdown characteristic on a dc basis. Since the above analysis assumes that r_d remains relatively constant over the increment of I_R, (2-20) may not be used to predict the value of equivalent resistance over large increments of current as r_D is sometimes given on specification sheets.

When bias current is increased, the value of r_d drops, but the value of r_T remains relatively constant. This means that the total value of r_D also decreases.

It should be mentioned that the value of r_T may be either positive or negative, depending on the polarity of K_T. Thus a low-voltage VR diode with a negative K_T likewise has a negative value of r_T and the value of r_D will be less than r_d. On the other hand, VR diodes with a larger V_B and a positive temperature coefficient will have a positive r_T, hence r_D will be larger than r_d.

The value of r_T is dependent on the thermal resistance between junction and ambient, and thus packaging techniques can greatly affect its size. In addition, the way the leads are connected into the circuitry can also affect θ_{JA}, hence the value of r_T.

2–7 LEAKAGE CURRENT

As Fig. 2–3 indicates (in somewhat exaggerated proportions), a reverse current will flow even before the value of V_R approaches V_B. This is to be expected in any reverse-biased junction and results from bulk leakage in the junction as well as various surface effects.

When it is desired to designate that the reverse current is due to leakage effects rather than breakdown, the symbol I_{RL} may be used.

Leakage current in VR diodes is both temperature and voltage dependent, as might be expected.

2–7.1 Voltage Dependence of I_{RL}

In a normal junction diode, the leakage current increases with reverse bias up until V_R is about 1V and then it remains at a relatively constant value until breakdown voltage is approached. This is not normally true in a VR diode, and the leakage current will continue to increase in an exponential manner.

There are several reasons why this is so. There are always multiplication effects or slight amounts of field emission that will cause leakage current to increase with applied reverse voltage. In addition, there are many surface effects associated with the junction and some of these are greatly voltage dependent.

Figure 2–15 illustrates the voltage-induced variation in I_{RL} for a 3.3-V VR diode which is operating primarily on a field emission or true Zener breakdown mechanism. The straight line nature of the semilogarithmic plot clearly indicates the exponential characteristics of the function. The empirical form of the equation relating the leakage current to the applied voltage would be

$$I_{RL} = K\varepsilon^{m(V_R)}. \tag{2–21}$$

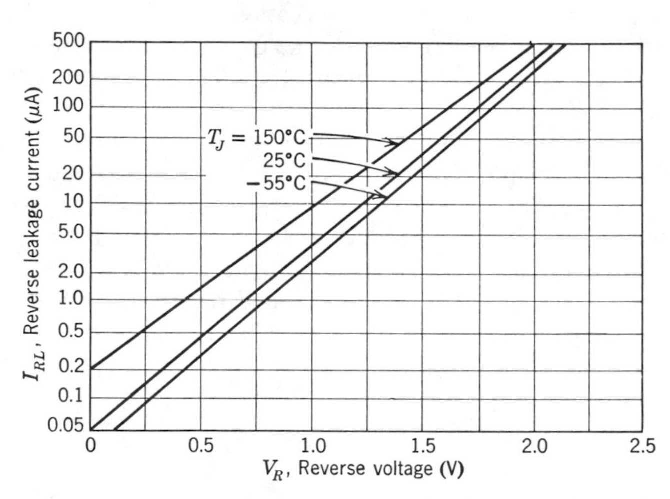

Fig. 2–15 Leakage current characteristics for a 1N5226 3.3V VR diode is typical for a true Zener breakdown. (Courtesy Motorola, Inc.)

Choosing the best values for K and m to fit the line given in Fig. 2–15 for a constant junction temperature of 25°C, we have

$$I_{RL} = 0.05\varepsilon^{4.38V_R}, \tag{2-22}$$

where the value of I_{RL} will be given in μA.

The relative value of I_{RL} for a low-voltage Zener breakdown VR diode will be quite high. Note that the leakage current characteristics of Fig. 2–15 are actually an extension of the reverse breakdown characteristic curves shown in Fig. 2–5 for the same diode.

Figure 2–16 illustrates the leakage current characteristics for a 5.1-V VR diode. This is an extension of the breakdown characteristics shown in Fig. 2–6.

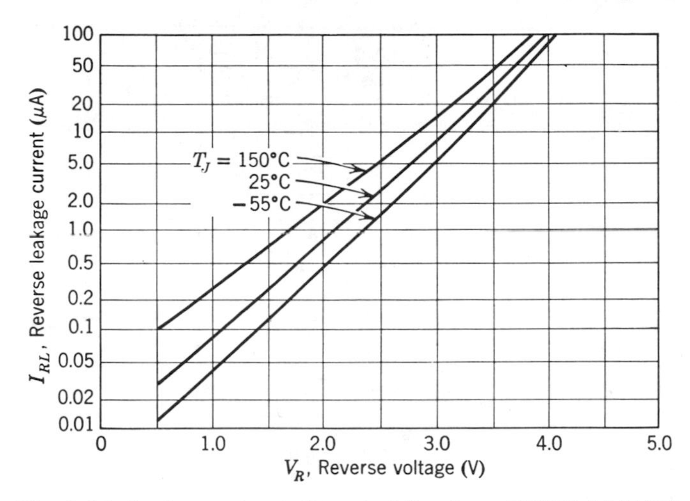

Fig. 2–16 Leakage current characteristics for a 1N5231 5.1-V VR diode. (Courtesy Motorola, Inc.)

When T_J is 25°C, the empirical equation for I_{RL} as a function of V_R become

$$I_{RL} = 0.0085\varepsilon^{2.33V_R}, \tag{2-23}$$

where I_{RL} is again given in microamperes. The level of I_{RL} is quite a bit lower for the 5.1-V diode and only reaches about 100 μA when V_R is roughly 80% of V_B.

Leakage current for VR diodes that operate almost exclusively due to the avalanche breakdown mechanism will have a much lower value as shown by the characteristics given in Fig. 2–17 for a 27-V diode. The relative effect of the applied reverse voltage will also be much less as evident from the approximate empirical equation for the case where T_J is 25°C.

$$I_{RL} = 0.85\varepsilon^{0.118V_R}, \tag{2-24}$$

where I_{RL} is given in nanoamperes.

We may derive a useful relationship applicable to VR diodes of different voltages above about 11 V by normalizing (2-24):

$$I_{RL} = 0.85\varepsilon^{3.2(V_R/V_B)}. \tag{2-25}$$

I_{RL} is given in nanoamperes. The 0.85 coefficient is subject to some variation, but the general form of (2-25) is useful to about 85% of V_B.

2–7.2 Temperature Dependence of I_{RL}

The three sets of curves given in Figs. 2–15, 2–16, and 2–17 also indicate a temperature dependence of the leakage current. It will be readily noted that

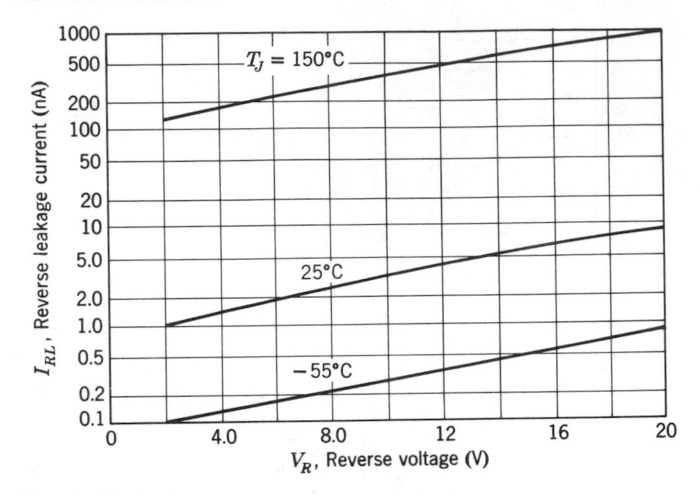

Fig. 2–17 Leakage current characteristics for a 1N5254 27-V VR diode and more or less typical of VR diodes whose breakdown mechanism is entirely avalanche. (Courtesy Motorola, Inc.)

the sensitivity to temperature is much lower for VR diodes than for a typical junction diode whose leakage current doubles for about every 10 or 12°C.

Looking first at Fig. 2–15 for the 3.3-V VR diode operating on true Zener breakdown, we see that the total variation in leakage current over the wide temperature range of −55 to +150°C is actually quite small. The temperature sensitivity of I_{RL} is greatest at lower voltages.

If we make an assumption that the leakage current follows an exponential function, the general equation might look something like

$$I_{RL} = I_{RL(25)}\varepsilon^{n(T-25)}. \tag{2-26}$$

For the characteristic family shown in Fig. 2–15 the value of n would range from about 0.005 at low temperatures to 0.01 at high temperatures even down where V_B is 0.5 V. This would mean that the leakage current would double for each 70 to 140°C temperature increase. At the higher reverse voltages in which the value of n would drop even lower, it would take a temperature increase between 175 and 230°C to cause I_{RL} to double.

In the 5.1-V VR diode shown in Fig. 2–16 the temperature variation is much the same as for the 3.3-V diode. The temperature variation seems, however, to be more uniform between low and high temperatures. Doubling of I_{RL} requires approximately 77°C temperature rise.

Although the general leakage current level of the 27-V VR diode depicted in Fig. 2–17 is much lower than for the lower voltage units, the temperature sensitivity becomes greater. The value of n for an empirical equation of the form shown in (2-26) would be between about 0.03 and 0.04 with doubling of I_{RL} occurring for each 18 to 24°C of temperature rise.

Note also that in the higher voltage VR diode which depends on avalanche breakdown for its operation, there is little difference in the temperature variation when the reverse voltage is varied.

2–8 NOISE

As discussed in Chapter 1, breakdown of the junction in a VR diode does not occur simultaneously across the entire area but will take steplike functions. This contributes a ragged characteristic in the knee region which is shown in exaggerated form in Fig. 2–18. When the current reaches a certain

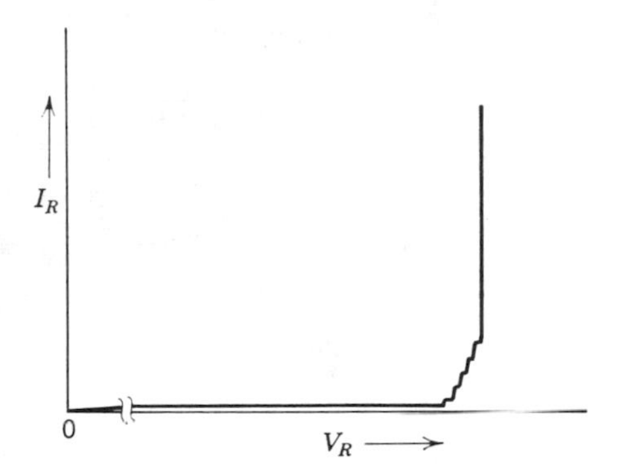

Fig. 2–18 Exaggerated breakdown characteristics in the knee region.

value, the entire junction takes part in the breakdown and no further steps in breakdown current occur.

If the VR diode is biased in the region of the knee, even very slight variation in current will cause jumps in the breakdown voltage which will appear as a noise voltage. A typical noise density specification for the 1N5221 family of VR diodes is shown in Fig. 2–19.

Note that very low-voltage VR diodes operating primarily on field emission or Zener breakdown have very little noise which might be expected. When avalanche breakdown mechanism becomes more dominant, the noise density likewise increases.

Noise levels in VR diodes are greatly process-dependent and rather wide variations are probable among different manufacturers of the same type.

2–9 FORWARD CHARACTERISTICS

Although the VR diode is nearly always operated in the reverse direction, we occasionally become interested in the forward characteristics in certain types of clipper circuits, etc.

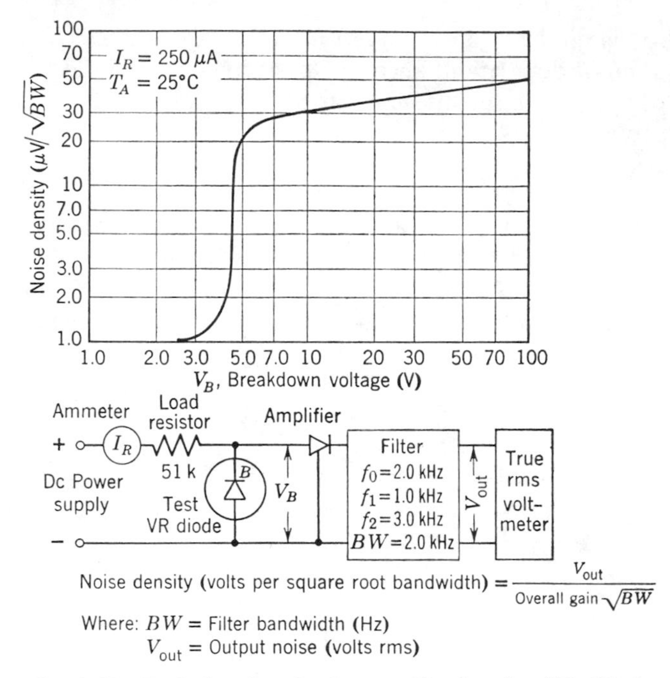

Fig. 2–19 Typical noise density specification for VR Diodes. (Courtesy Motorola, Inc.)

The forward characteristics of a VR diode are quite similar to those of a highly doped silicon junction diode, as is indicated in the overall characteristic curve in Fig. 2–3. The fact that low-resistivity material is used to achieve the usual low breakdown voltage and the area of junction is normally made relatively large make the dc forward characteristics much like a very high conductance diode.

A typical set of forward characteristics for a VR diode is given in Fig. 2–20. The forward current is an exponential function of the voltage applied and obeys the normal diode equation up to the current levels at which bulk ohmic contact resistance becomes significant.

The forward voltage drop will have a temperature coefficient which will range from about -1.4 to -2.2 mV/°C, depending on the current level. The magnitude of the TC will be the greatest at very low levels of I_F and decreases as the forward current bias is raised.

2–10 CAPACITANCE

All junction diodes have a capacitance associated with them whose value is dependent on the area of the junction, the resistivity of the material, and the

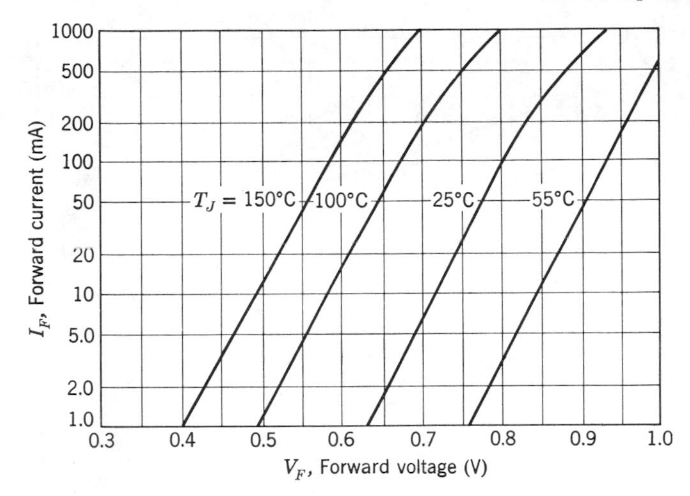

Fig. 2–20 Typical VR diode forward characteristics. (Courtesy Motorola, Inc.)

voltage applied. A general relationship which will also apply for the junction capacitance of VR diodes biased in the reverse direction is

$$C_d = KA(\rho V_R)^n, \tag{2-27}$$

where K is a constant which includes such factors as the dielectric constant of the semiconductor material. A is the effective junction area, which depends on the size of the die required to yield adequate power dissipation, ρ is the resistivity of the material which is varied in order to produce the various breakdown voltages, and V_R is the applied reverse bias voltage.

The exponent n in (2-27) will depend on the abruptness of the junction. For alloy junctions the value of n is nominally about -0.5. For graded junctions produced by the diffusion process, the value of the exponent drops to about -0.33.

Figure 2–21 illustrates a typical variation in junction capacitance with reverse voltage for several $\frac{1}{4}$-W VR diodes made with the diffusion process. The relationship follows closely the equation

$$C_d = C_o\left(\frac{V_R}{V_0}\right)^{-0.333}, \tag{2-28}$$

where C_o is the junction capacitance at some reference voltage V_0. If we make $V_0 = 1$ V and define C_o as the value of C_d measured at $V_R = 1$ V, then (2-28) simplifies to

$$C_d = C_o(V_R)^{-0.33} = \frac{C_o}{\sqrt[3]{V_R}}. \tag{2-29}$$

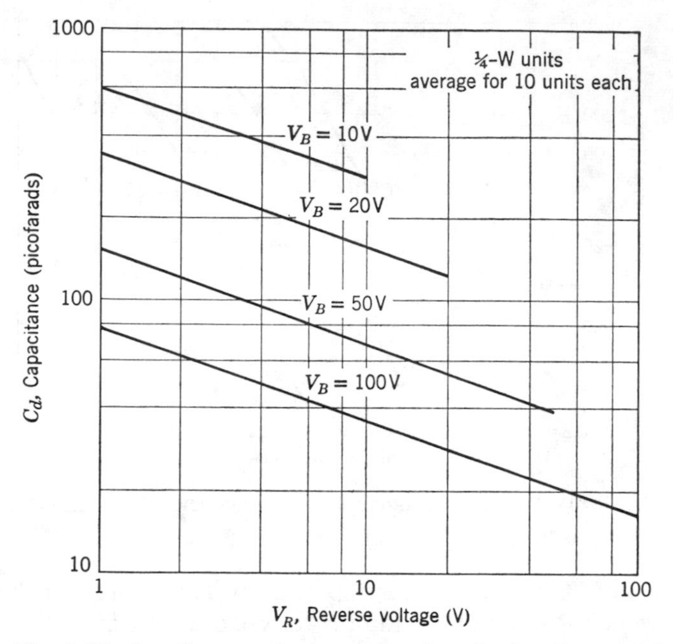

Fig. 2–21 Junction capacitance as a function of voltage for several $\frac{1}{4}$-W VR diodes. (Courtesy Motorola, Inc.)

Also note that the value of C_o—or for that matter the value of C_d at any voltage—is approximately an inverse proportion of the breakdown voltage V_B. This allows us to interpolate between the lines of Fig. 2–21 for diodes with breakdown voltages other than those shown.

When the reverse voltage is decreased to the region of 3 V and below, the relationship no longer follows the simplified expression of (2-29) and Fig. 2–21 is slightly in error. A more correct plot is shown in Fig. 2–22 where the curves seem to flatten out at low voltages. This is due to an effective internal barrier potential of about 0.7 V. If this is included, (2-29) takes the form

$$C_d = \frac{C_o}{\sqrt[3]{V_R + 0.7}}, \tag{2-30}$$

where we would now define C_o as the capacitance measured for $(V_R + 0.7) =$ 1 V. If we were to plot C_d as a function of $(V_R + 0.7)$, we would once again get a straight-line function on log-log paper as shown by the dotted line associated with the 10-V VR diode.

When the power dissipation required of a VR diode increases, its junction area must be made larger or its case design changed to give a lower thermal resistance. The junction capacitance is not affected by case modifications, and VR diodes of different power dissipation due simply to a different packaging technique will have the same capacitance-voltage curves. Figure 2–23 gives

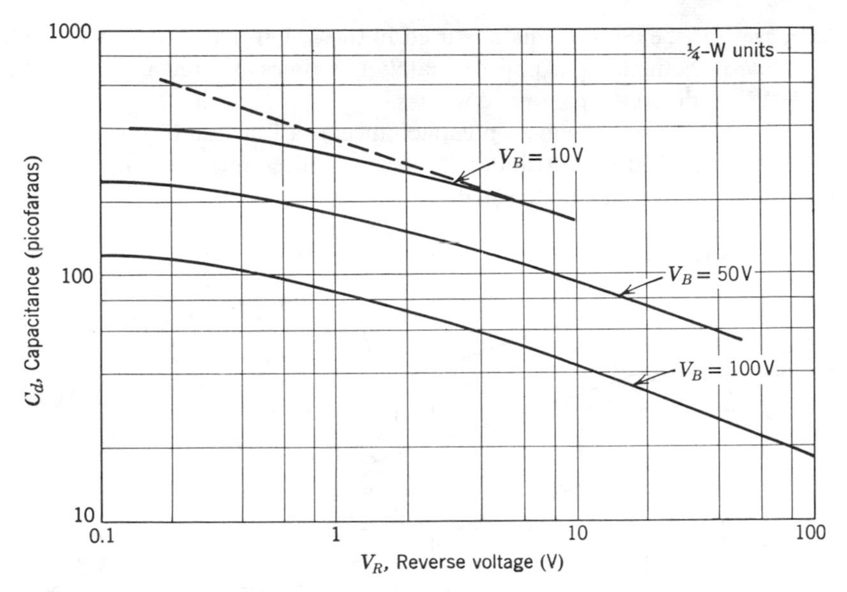

Fig. 2–22 Simplified relation of junction capacitance no longer holds at low voltages. (Courtesy Motorola, Inc.)

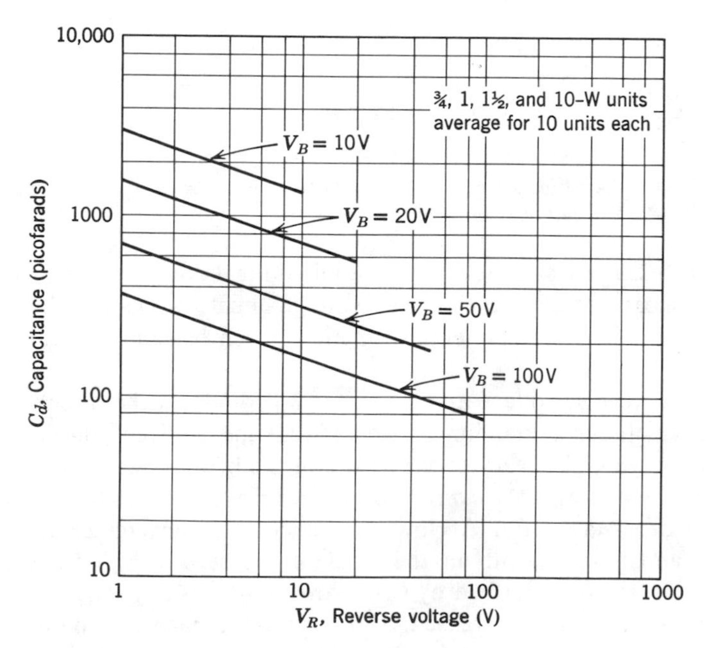

Fig. 2–23 Junction capacitance as a function of reverse bias voltage for the die used in ¾-to-10-W VR diodes. (Courtesy Motorola Inc.)

the curves for the die size which is used in the $\frac{3}{4}$, 1, $1\frac{1}{2}$, and 10-W VR diodes. The area of the junction is approximately five times as large as that used for the $\frac{1}{4}$- WVR diode.

Figure 2–24 gives the junction capacitance as a function of reverse voltage for the family of 50 W VR diodes. Note that only a relatively small increase in effective junction area is needed.

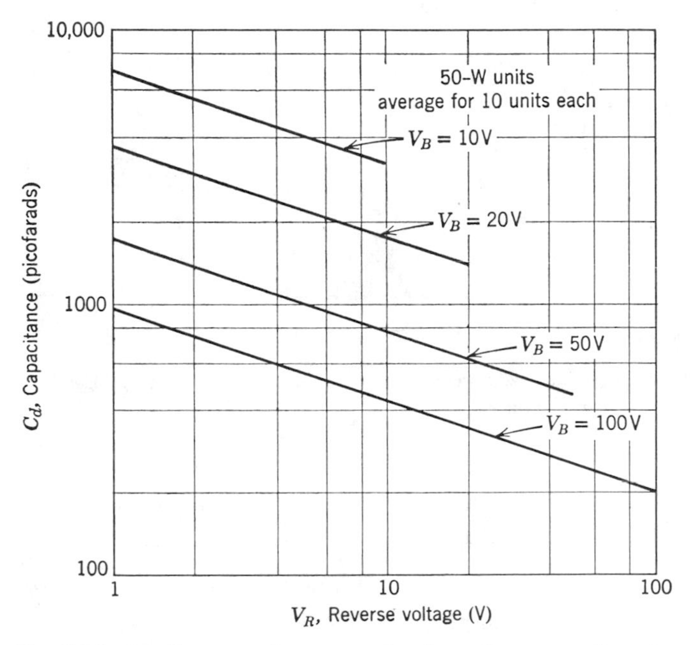

Fig. 2–24 Junction capacitance as a function of reverse voltage for 50–W VR diodes. (Courtesy Motorola, Inc.)

Figure 2–25 illustrates the effect of nominal breakdown voltage on the value of C_d as measured at a reverse bias voltage equal to one-half the V_B. This function is an inverse exponential function which has an exponent of about -1.2.

When the reverse voltage applied to the VR diode reaches breakdown, the capacitance attains a minimum value and remains relatively constant because the junction voltage is varying very little in the breakdown region.

The net breakdown *impedance* necomes a breakdown resistance r_d in shunt with the capacitance C_d. At low frequencies the impedance is primarily resistive and is dependent on the level of the reverse bias current in the exponential fashion described by (2-11) through (2-14).

As the frequency is increased, the junction capacitance becomes more important and becomes predominant above about 1 MHz. Figure 2–26

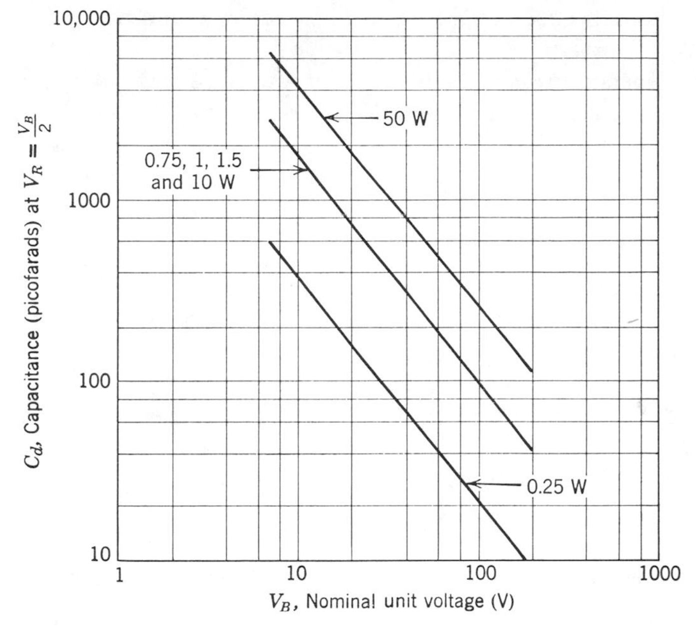

Fig. 2–25 Relative capacitance as a function of nominal breakdown voltage. (Courtesy Motorola, Inc.)

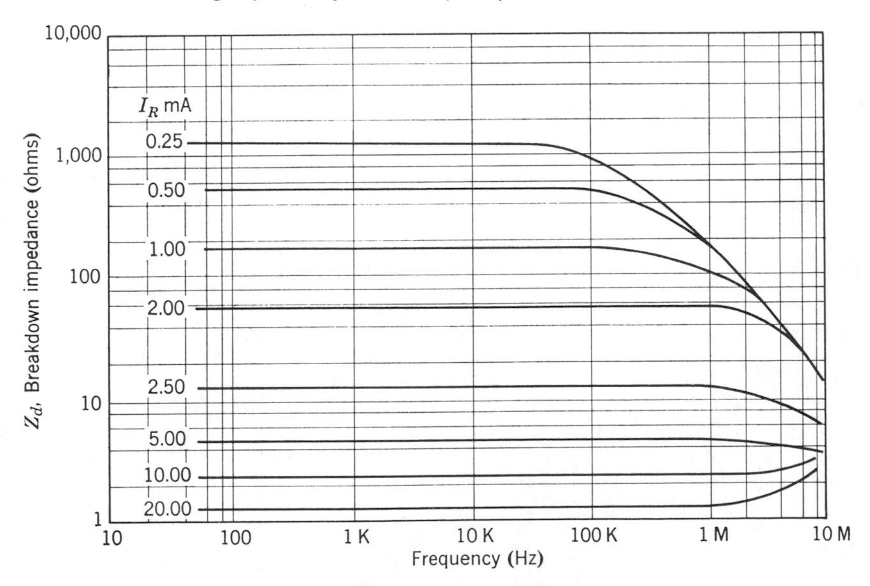

Fig. 2–26 Total breakdown impedance is a current sensitive resistance, r_d, in shunt with a capacitance of relatively constant value. Curves shown are for a $\frac{1}{4}$–W 6.8-V diode. (Courtesy Motorola, Inc.)

illustrates the impedance as a function of frequency and bias current for a 6.8 V, $\frac{1}{4}$-W VR diode.

At large reverse currents, the breakdown resistance is very low causing the effects of the shunt capacitance to be greatly diminished. At high bias currents and frequencies above 2 MHz, the series lead inductance becomes significant to cause a slight rise in the terminal impedance as evidenced by the lower curves.

2–11 SWITCHING SPEED

The basic breakdown mechanisms, as discussed in Chapter 1, are inherently very fast. There are no stored charges to be accumulated or eliminated. Typical switching speeds would be about 10^{-10} sec if it were not for the presence of the junction capacitance just discussed.

In considering switching speed of VR diodes, it is usually adequate to assume a perfect VR diode with zero switching time but shunted with the nonlinear capacitance C_d. This means that optimum switching speeds will be obtained in relatively low impedance circuitry.

2–12 EQUIVALENT CIRCUIT

To gain a better overall picture of the VR diode we may attempt to include as many of the critical parameters as possible into an equivalent circuit. This is not easily done because of the many interrelations and variations, particularly in the breakdown region.

One arrangement we might evolve is given in Fig. 2-27 to make use of this equivalent circuit it will be necessary to understand its interpretation and limitations.

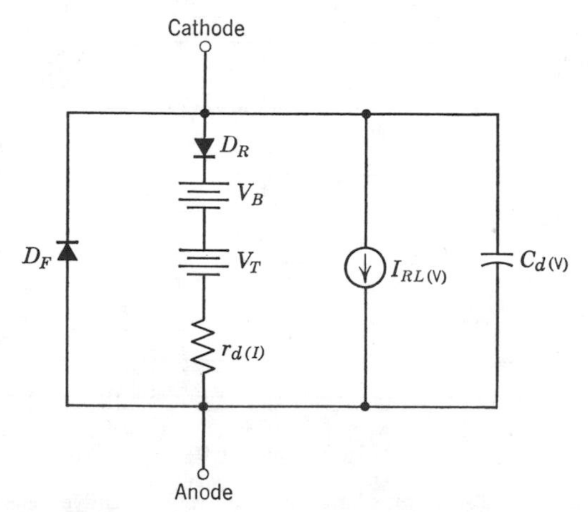

Fig. 2–27 An approximate equivalent circuit for the VR diode.

The forward characteristics are assumed to be incorporated in diode D_F included in the equivalent circuit, D_F is a high-conductance diode with negligible leakage current and capacitance. When the VR diode is reverse-biased with the cathode made positive with respect to the anode, diode D_F becomes reverse-biased and is effectively removed from the circuit.

Diode D_R is assumed to be a perfect diode in both the forward and the reverse directions; that is, it has zero forward drop when biased in the forward direction and it has zero leakage current when it is reverse-biased. Leakage current in the VR diode is simulated by the current generator $I_{RL(V)}$ which must be considered as a function of reverse voltage as discussed previously in this chapter.

The equivalent circuit of Fig. 2-27 includes two internal voltage sources which are assumed to have zero impedance. V_B may be assumed to be the initial breakdown or knee voltage when the equivalent circuit is to represent the entire VR diode in a quasi dc case. When the equivalent circuit is to represent an ac equivalent at a given reverse bias current, V_B would be the quiescent value of breakdown voltage.

V_T may be considered as an error voltage which is produced due to the temperature coefficient of V_B. Its value will be $K_T \Delta T$ and thus it may either add to or subtract from V_B, depending on the direction of the temperature excursion and the polarity of the temperature coefficient K_T.

Until the total reverse voltage across the VR diode exceeds the sum of V_B and V_T, diode D_R will be reverse-biased and no current will flow through it. Only the leakage current will flow through the VR diode.

When the applied reverse voltage exceeds the sum of V_B and V_T, diode D_R becomes forward-biased and current will flow with the excess voltage dropped across resistance $r_{d(I)}$. Once again the interpretation of $r_{d(I)}$ must depend on the intent of the equivalent circuit. To approximate the dc case, $r_{d(I)}$ must represent an average equivalent resistance since the small-signal value of r_d will continually change as the reverse current is varied. In certain cases $r_{d(I)}$ must be replaced with the term $r_{D(I)}$, which will include the equivalent breakdown resistance resulting from heating effects.

Capacitance $C_{d(V)}$ represents the equivalent junction capacitance which will be a nonlienar function of the applied voltage.

The VR diode equivalent circuit is at best only a course approximation and must be interpreted in light of all the conditions and parameters discussed in this chapter. When we are considering only the small-signal performance of the VR diode biased in the reverse direction, the simplified equivalent circuit of Fig. 2-28 may be used. The equivalent series resistance to be used will depend on the relative period of the frequencies involved in comparison with the thermal time constant of the diode. For very slowly varying signals, we would use r_D. The parameter r_d would be used for higher frequency signals.

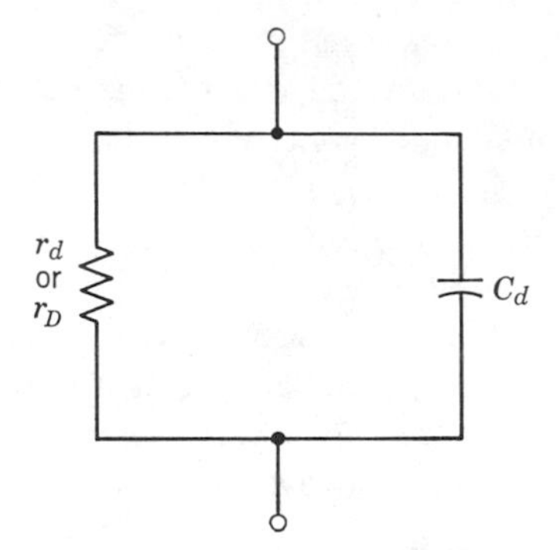

Fig. 2–28 Greatly simplified small-signal VR diode equivalent circuit.

2–13 THERMAL RESISTANCE

The maximum practical power that a VR diode may safely dissipate is limited almost exclusively by the highest permissible junction temperature. Maximum allowable junction temperature is usually a somewhat arbitrary limit which has been found to insure reasonable reliability in operating life tests. It will depend on the general surface cleanliness and freedom from adverse contaminants, but the typical silicon VR diode will withstand a junction temperature of 175 to 200°C with little degradation.

The energy being dissipated by the VR diode junction must have a way of escape to prevent the junction temperature from reaching an excessive value. Some power is radiated directly from the small semiconductor die, but most of the thermal energy is removed from the junction by conduction through the metal base or lead to which the die is attached.

The amount of temperature differential that exists between the junction and ambient will depend on the amount of power being dissipated and the thermal resistance between the junction and ambient. The relationship obeys a thermal analog of Ohm's law or

$$\Delta T_{JA} = T_J - T_A = \theta_{JA} P_D, \tag{2-31}$$

where θ_{JA} is the thermal resistance that exists between the junction and outside ambient.

Thermal resistance may be expressed either in terms of °C/W or °C/mW. If it is given in °C/W, then P_D in (2-31) should be expressed in watts. On the other hand, if the thermal resistance is to be expressed in °C/mW, then P_D must be given in milliwatt's in order to express the temperature in degrees Celsius.

If a given amount of power is applied to a VR diode, the junction temperature does not rise to its steady-state or final value instantaneously, but it begins to increase exponentially. This is due to the thermal mass associated with the material around the junction which must be supplied with energy in order for the temperature to rise. For this reason we must consider both a steady-state or dc thermal resistance as well as a transient form of thermal resistance that could be much lower.

2–13.1 Steady-State Thermal Resistance

The limiting case of thermal resistance which will determine just how high the junction temperature will rise if enough time is allowed for thermal equilibrium to be reached is the steady-state thermal resistance. It is the value most frequently given on specification sheets and, unless specifically stated otherwise, must be assumed to be the value to be considered when thermal resistance is to be discussed.

The drawing of Fig. 2–29 illustrates a typical mechanical assembly for a VR diode of the 400 mW class. By studying this we may see the various factors that contribute to the overall thermal resistance between the VR junction and outside ambient.

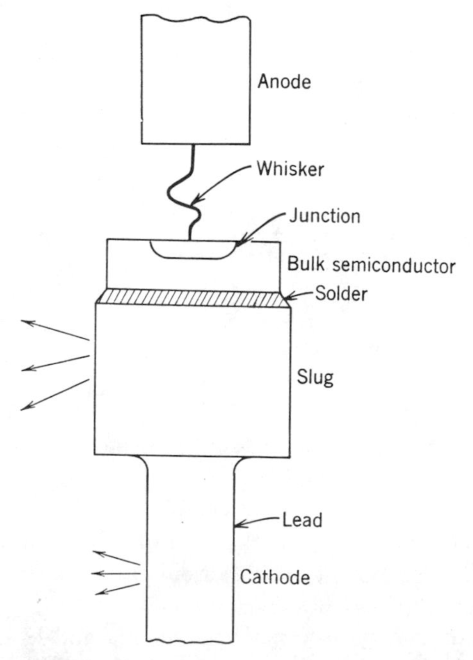

Fig. 2–29 Typical assembly for a 400-mW VR diode.

Most of the power being dissipated is concentrated in the immediate vicinity of the junction. The thermal energy developed must flow out through the bulk material, through the solder used to bond the die to the slug, through the slug, and out the lead. This gives many opportunities for a temperature drop to occur and thus we may easily see how the total thermal resistance is actually made up of many parts as indicated in the drawing of Fig. 2–30.

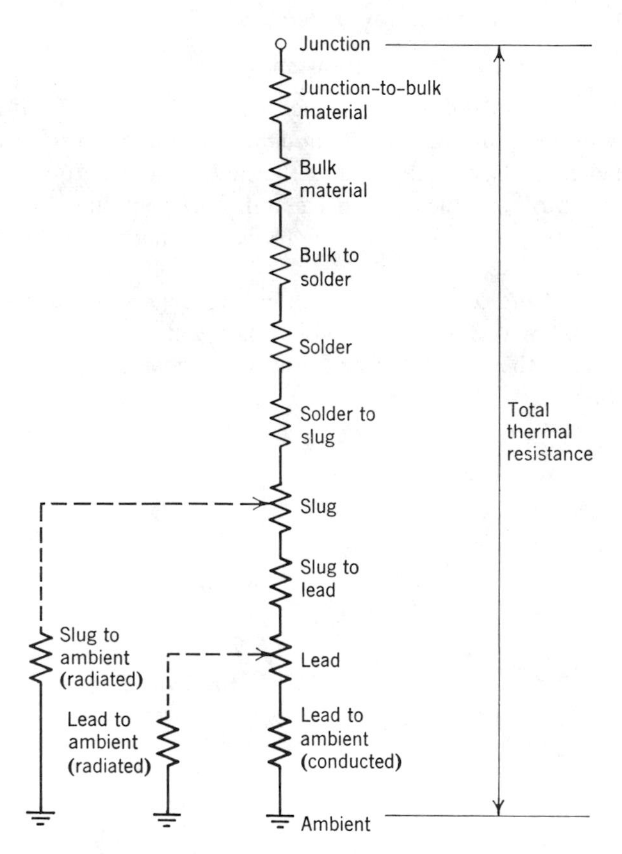

Fig. 2–30 The total value of junction-to-ambient resistance is made up of many parts.

Since the total value of θ_{JA} is the sum of all the individual thermal resistances shown in Fig. 2–30, a poor thermal path at any point will increase the overall θ_{JA}.

As the thermal energy flow reaches the slug and lead, appreciable heat is radiated, and these paths are indicated by dotted lines in Fig. 2–30. The amount of radiated heat is still small in comparison with that conducted through the lead unless the lead and lead-to-ambient thermal resistance is large.

Although some heat will be conducted through the anode lead of the VR diode, this is usually insignificant in comparison with the paths through the cathode end.

The end user of the VR diode cannot do anything about the parts of the thermal resistance inside the package beyond selecting a type and a manufacturer. It is most important to realize that the final value of θ_{JA} will depend very greatly on just *how external connection is made* to the VR diode leads. This is quite dramatically shown in the curves of Fig. 2–31 where thermal resistance is plotted as a function of lead length.

Curve 1 is for a diode in a DO-14 package with 0.020 in. dumet leads. Note that the amount of thermal resistance between the junction and the cathode end of the VR diode is about 50°C/W and the user has no control over this. However, the total value of θ_{JA} can vary from 50 to 300°C/W depending on the length of the cathode lead! Of course, we are somewhat limited about how close we may make contact to the cathode and because an easy path for heat to flow *out of* the junction also makes for an easy path for the heat due to soldering to flow *into* the junction with possible damage occurring. Even so there is considerable variation possible.

It might be well to mention also that the overall value of θ_{JA} will be affected by the size of the printed circuit lands, the weight of copper used, and the relative amount of copper associated with runs to and from the cathode end of the diode.

Curves 2 and 3 in Fig, 2–31 demonstrate the effect of simply changing the lead material and size. The use of copper instead of dumet for the leads can change the value of θ_{JA} by a factor of 2. Increasing the diameter of the lead will also improve the thermal conductivity and reduce θ_{JA} to a value approximating the limit value, even for substantial lengths, Of course, we must remember that the curves of Fig. 2–31 assume that the end of the cathode lead is tied to a heat sink or other means of low thermal resistance path to ambient.

Since the electrical performance of the VR diode will be closely related to the value of thermal resistance present in the final application, we must consider this in our designs; for example, if we are concerned about voltage regulation, the total voltage variation produced by a given current change will be greatly influenced by θ_{JA}, especially if a high voltage VR diode with a relatively high positive value of K_T is used.

The manufacturer controls the limit value of θ_{JA} by his choice of packaging. The extent of packaging effects is seen in Table 2–1 where several types of packaging with their corresponding thermal resistances and rated power dissipation are tabulated. It is not at all uncommon for a single size of VR die

Table 2–1. Relation of package size and type to thermal resistance and rated power dissipation

Package	Typical θ_{JA} or (θ_{JC})	Rated P_D
DO-7	600°C/W	$\frac{1}{4}$ W
DO-7 with cathode slug	250°C/W	0.4 W
Plastic 0.125″D × 0.2″L	167°C/W	$\frac{3}{4}$ W
DO-13	100°C/W	1 W
DO-4	(5°C/W)	10 W
TO-3	(1°C/W)	50 W

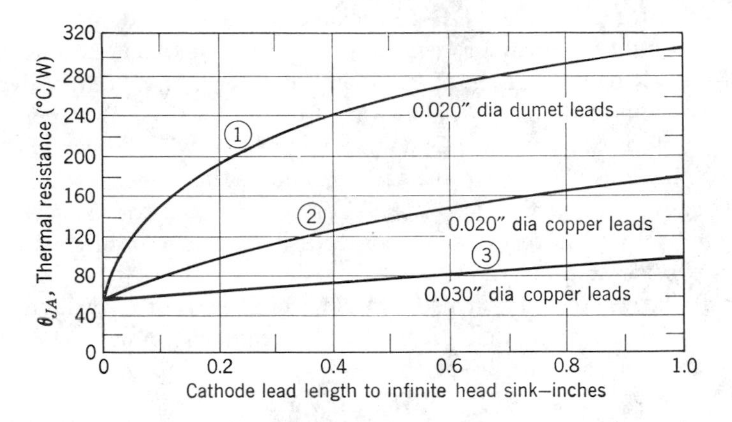

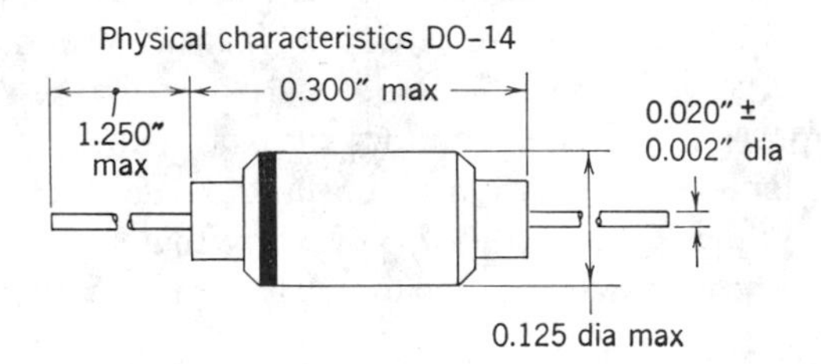

Fig. 2–31 Variation of thermal resistance as a function of cathode lead length. (Courtesy TRW Semiconductors, Inc.)

to be mounted in different packages to achieve power ratings from $\frac{3}{4}$ to 10 W.

Smaller VR diodes, typically in the standard DO-7 package, have their thermal resistance specified in terms of *typical* junction-to-ambient thermal resistance. In the larger size units the significant parameter becomes the thermal resistance between junction and the case. The total thermal resistance which must be considered must include the value of θ_{JC} specified in addition to the thermal resistance achieved from case to ambient by the user.

In designing circuits with power VR diodes, we must consider the basic thermal resistance of the heat sink used and also the thermal resistance developed by the case-to-heat-sink interface. This latter part can be improved by as much as two-to-one by using a thermally conductive lubricant such as an oxide-bearing silicone grease.

2–13.2 Transient Thermal Resistance

As mentioned earlier, junction temperature does not rise instantaneously with the application of power. This is due to thermal mass associated with the junction proper and throughout the entire thermal flow path. The net result is an effective transient thermal resistance that can be much lower than the steady-state value if the power pulse is of short duration and the duty cycle is low enough.

Because of the distributed nature of all the thermal masses, the transient thermal resistance becomes a difficult parameter to specify (and frequently is not specified) on the data sheet. The curves of Fig. 2–32 illustrate the relationship of transient thermal resistance to pulse width and duty cycle for the IN5221 family of VR diodes; "L" is the lead length between diode and heat sink.

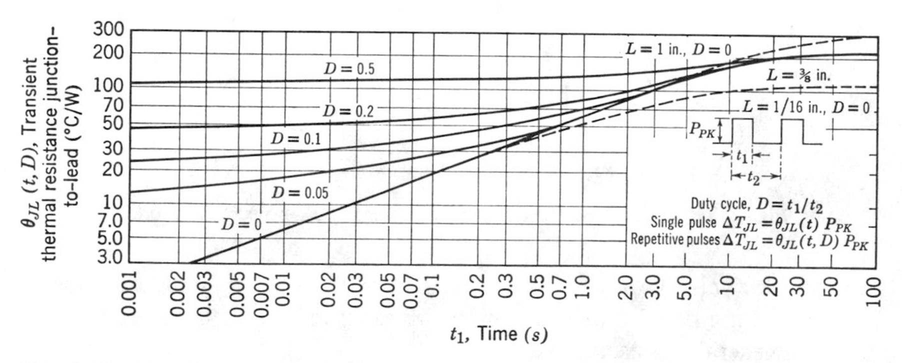

Fig. 2–32 Transient thermal response for the 1N5221 family of VR diodes. (Courtesy Motorola, Inc.)

2–14 POWER RATING

Since the major limiting factor in the power-handling capability of a VR diode is the maximum allowable junction temperature, the maximum permissible power dissipation is related directly to the thermal resistance. It is thus determined primarily by the package configuration chosen, but can be influenced by the manner in which the VR diode is mounted.

2–14.1 Steady-State Power Rating

The maximum power dissipation as specified on a data sheet for smaller VR diodes is normally given for a case in which the maximum ambient temperature is 25°C. When the actual ambient temperature may go above 25°C, a derating curve such as that shown in Fig. 2–33 must be used to determine the

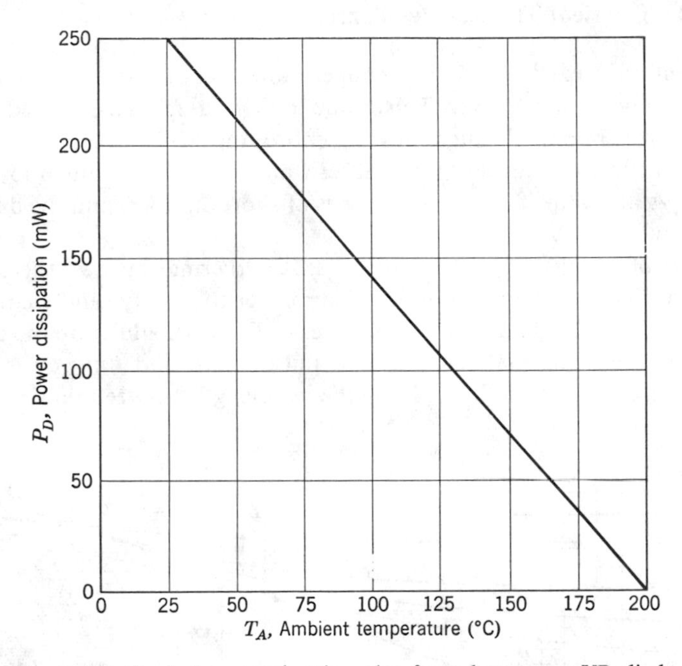

Fig. 2–33 Typical power-derating plot for a low-power VR diode.

maximum permissible power at the highest operating temperature. Usually these curves are based on the maximum thermal resistance expected and are therefore generally conservative.

The rating also assumes that the time duration of the power pulse will be comparable or longer than the thermal time constant of the VR diode and that the duty cycle is approaching or equal to 100%.

In determining the power dissipation capabilities of a larger VR diode, we must include the thermal resistance of the mounting means as well as the specified rating given. The information presented on the data sheet is normally given in terms of the maximum power dissipation for a steady-state case in which the case or stud temperature is held below some specified limit of about 50°C. If the actual case temperature is allowed to rise above the stated figure, because of either high ambient temperatures, a finite heat sink limitation, or a combination of the two, then again the maximum allowable power dissipation must be derated according to the information normally supplied on a data sheet. The plot given in Fig. 2–34 is a typical example.

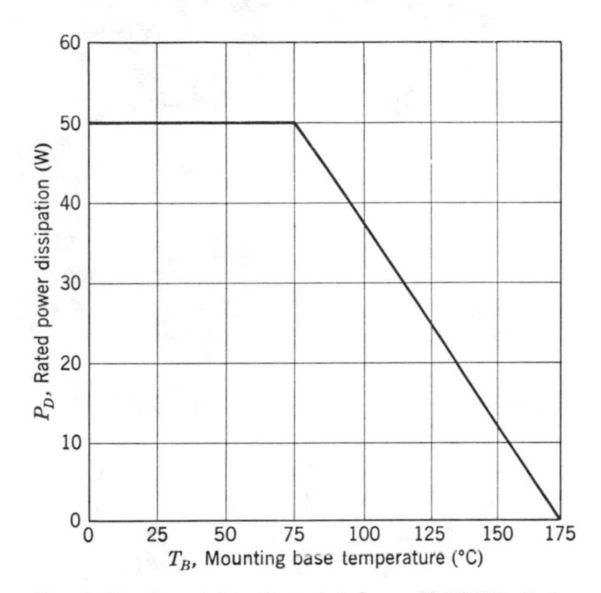

Fig. 2–34 Power-derating plot for a 50-W VR diode.

2–14.2 Pulse Power Rating

Many applications of VR diode involve the dissipation of power on a momentary fashion rather than in a constant steady-state manner as just discussed. Because the main limitation is that of a maximum junction temperature, we may consider the normal maximum power dissipation figure given on a specification sheet as a maximum average power as long as the pulse interval is not comparable to the thermal time constant of the VR diode.

Assume that the power pulse is rectangular and has the charactersitics shown in Fig. 2–35. Power dissipation equal to a maximum total value p_D is applied for a time t_1. The time interval from the beginning of one pulse to the start of another is t_2. The duty cycle will then be given by the ratio of the two time intervals

$$D = \frac{t_1}{t_2}. \tag{2-32}$$

If we may assume that t_1 is short with respect to the thermal time constant of the VR diode, the average power will be given by

$$P_D = p_D(D), \tag{2-33}$$

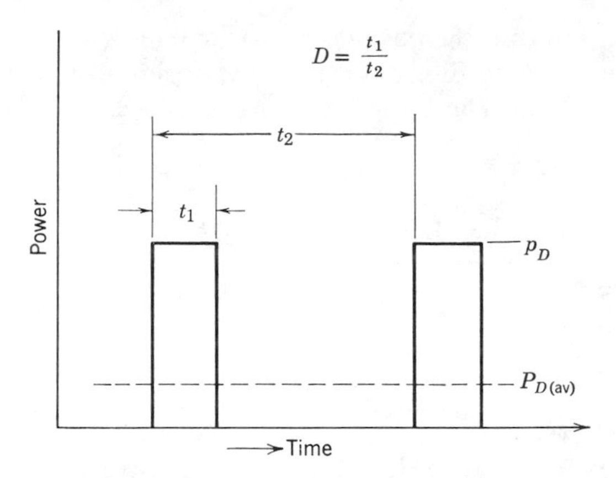

Fig. 2–35 Assumed pulse power waveform.

and the effective pulse power rating will be multiplied by the factor $1/D$; for example, a 1-W VR diode could dissipate up to 10 W in short pulses of power if the time interval were short in comparison with the thermal time constant and if the duty cycle were 0.1 or less. The average power, as indicated by the product of pulse power and duty cycle, would be $10 \times 0.1 = 1$ W.

When the width of the time t_1 becomes significant in regard to the thermal time constant, then we must consider the additional temperature rise which will momentarily occur during the power pulse. In a case like this curves of the form shown in Fig. 2–32 become very useful. To compute a maximum power dissipation capability for a given value of t_1 and duty cycle D we utilize the transient thermal resistance figure obtained from Fig. 2–32 and the maximum allowable junction temperature rise.

$$p_{D(\max)} = \frac{T_J - T_A}{\theta_{JL(t)}}. \tag{2-34}$$

As an example assume that t_1 is 0.1 sec and that the duty cycle is 0.1. From Fig. 2–32 we get a transient thermal resistance of 40°C/W. Assuming a maximum junction temperature of 200°C or a rise of 175° yields a maximum pulse power of $175 \div 40 = 4.38$ W. Note that this is less than the 5-W figure that would be obtained by using only the duty cycle to compute a limiting value of the average power.

2–14.3 Sinusoidal Power Rating

There are many applications in which the VR diode is used in an ac circuit as a clipper or limiter. We shall now look at the relative power dissipation ratings of a VR diode operated in this type of mode.

A typical circuit is shown in Fig. 2–36. As long as the value of instantaneous input voltage v_1 is less than the breakdown voltage, no current flows. However, as long as v_1 exceeds the value of V_B, current will flow and power will be dissipated in the VR diode.

The input waveform is shown in Fig. 2–36*b* with pertinent points indicated. θ is the conduction angle at which v_1 is exactly equal to V_B and current commences. Current flows only through the diode during the shaded interval and its average value can be expressed mathematically as

$$I_D = \frac{1}{2\pi}\int_{\theta}^{\pi-\theta} i_d\, dt. \tag{2-35}$$

In the interval between θ and $(\pi - \theta)$ the instantaneous current i_d is given by

$$i_d = \frac{V_{1m}\sin\omega t - V_B}{R_1}. \tag{2-36}$$

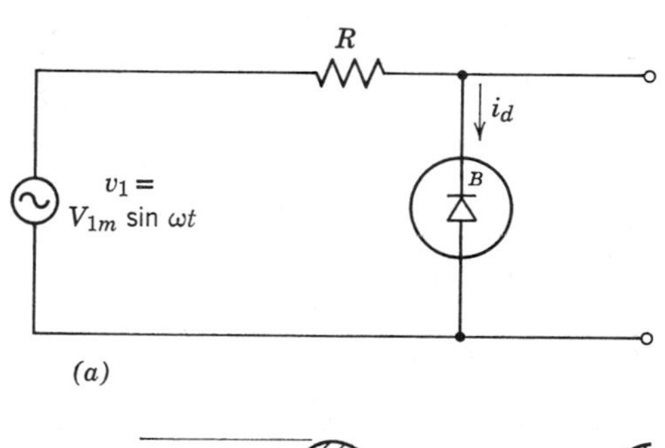

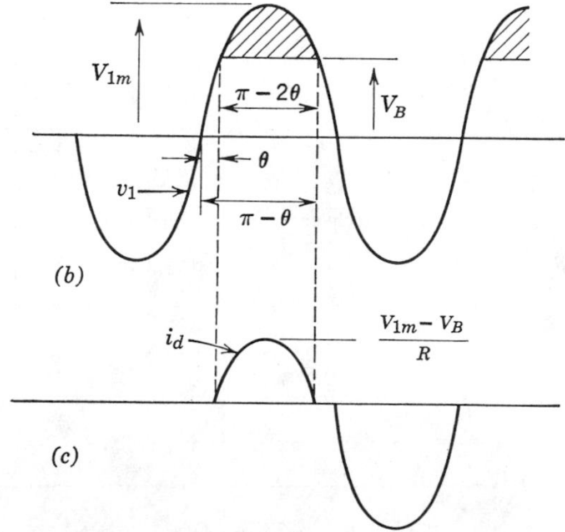

Fig. 2–36 Conditions for a sinusoidal clipper; (*a*) circuit, (*b*) input waveform; (*c*) diode current.

Inserting (2-36) into (2-35) and solving gives

$$I_D = \frac{V_{1m}}{\pi R_1}\left[1 - \left(\frac{V_B}{V_{1m}}\right)^2\right]^{1/2} - \frac{V_B}{2R_1} + \frac{\theta}{\pi}\frac{V_B}{R_1}. \tag{2-37}$$

Power dissipation occurs only when current flows. If we make the assumption that the voltage drop across the VR diode remains constant and equal to V_B, the instantaneous power dissipation will be $i_d V_B$ and the average power dissipation will be

$$P_D = \left[\frac{V_{1m}}{\pi R_1}\left[1 - \left(\frac{V_B}{V_{1m}}\right)^2\right]^{1/2} - \frac{V_B}{2R_1} + \frac{\theta}{\pi}\frac{V_B}{R_1}\right] V_B. \tag{2-38}$$

We may express (2-38) in a normalized form by dividing both sides by (V_B^2/R_1). The value of θ is $\sin^{-1}(V_B/V_{1m})$ radians.

$$\frac{P_D}{V_B^2/R_1} = \frac{\sqrt{K^2 - 1}}{\pi} - 0.5 + \frac{\sin^{-1}(1/K)}{\pi}, \tag{2-39}$$

where $K = V_{1m}/V_B$. Equation 2-39 is plotted in Fig. 2–37. Note that for values

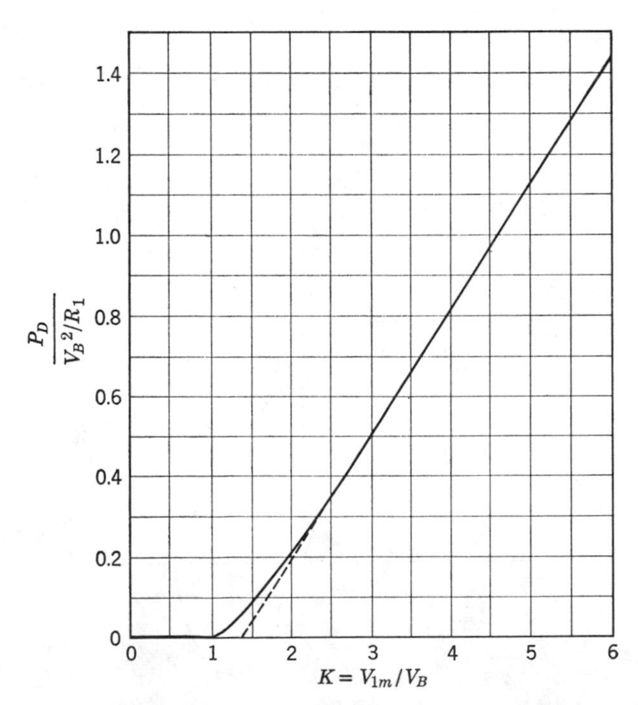

Fig. 2–37 VR diode power dissipation for sinusoidal drive shown in normalized form.

of K greater than about 2.5, the relationship approaches a straight-line function:

$$\frac{P_D}{V_B^2/R_1} \approx 0.308\,K - 0.415. \tag{2-40}$$

2–15 SURGE POWER RATING

In one of the preceding sections the relationship between power rating and duty cycle was discussed together with the effects of pulse width. It might at first appear that by using the same data presented in Fig. 2–32 we could compute the peak surge current rating of a given diode.

Other effects occur that will limit the peak surge power to a value that is less than that which would be predicted from Fig. 2–32. This is due to localized hot spots or areas of concentrated heating that can be produced during high current surges. If the temperature in these hot spots gets high enough, the VR diode will be damaged and cause to shift in breakdown voltage, increase leakage current, or even short out completely. The creation of localized heating is somewhat dependent on very minor irregularities in the junction, but even very carefully controlled processes yield diodes that exhibit this effect.

In addition, the data presented in Fig. 2–32 represent typical information. On a worst-case basis, the transient thermal resistance could be somewhat higher than that shown.

Figure 2–38 illustrates the maximum nonrepetitive surge power rating for the same family of devices characterized in Fig. 2–32. Note that the allowable peak surge power is dependent on the temperature of the junction just before the surge is applied.

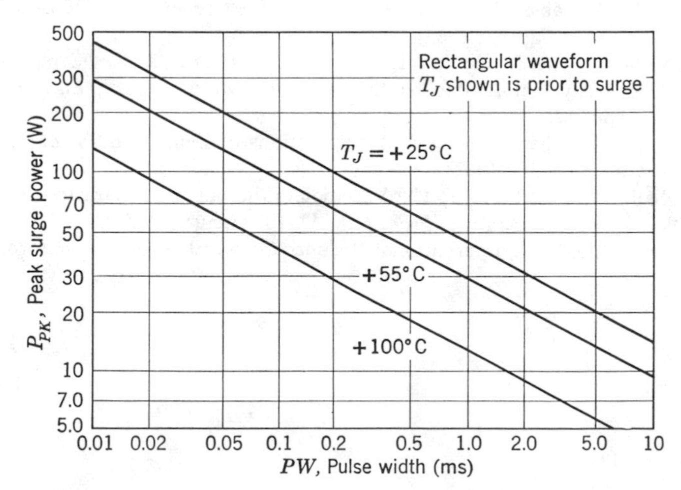

Fig. 2–38 Maximum non-repetitive surge power rating for the 1N5221 family of VR diodes. (Courtesy Motorola, Inc.)

REFERENCES

[1] R. P. Baker and J. Nagy, Jr.: Investigation of Long-Term Stability of Zener Voltage References, *IRE Trans. Instr.*, **1–9**, 226–231 (September 1960).
[2] D. J. Barnes: The Use of Zener Diodes as Voltage Controlled Capacitors, *J. Sci. Instr.*, **40**, 507 (October 1963).
[3] J. K. Buchanan et al: *Zener Diode Handbook*, Motorola, Inc., Phoenix, Arizona, 1967.
[4] A. G. Chynoweth and K. G. McKay: Internal Field Emission in Silicon *p-n* Junctions, *Phys. Rev.*, **106**, 418–426 (May 1, 1957).
[5] B. B. Daien: With Zener Diodes the Curves Make All the Difference, *Electron. Design*, **6**, 28–31 (July 23, 1958).
[6] K. Enslein: Characteristics of Silicon Junction Diodes as Precision Voltage Reference Devices, *IRE Trans. Instr.*, **1–6**, 105–118 (June 1957).
[7] A. E. Garside and P. Harvey: The Characteristics of Silicon Voltage-Reference Diodes, *Proc. Inst. Elec. Engrs. (London)*, **106**, Part B, Supp. 17, 982–990 (May 1959).
[8] C. L. Hanks: Evaluation of Zener Diodes to Develop Screening Information, *Semicond. Prod.*, **8**, 30–35 (April 1965).
[9] J. L. Haynes: Zeta, A proposed Regulation Factor for Zener Diodes, *EDN*, 14, 55–59 (May 1, 1969).
[10] D. W. Hutchins: How to Zap a Zener, *EEE-Circuit Des. Eng.*, **15**, 76–80 (February 1967).
[11] H. C. Lin: Some Ratings and Application Considerations for Silicon Diodes, *IRE Trans. Component Pts*, **CP-6**, 269–273 (December 1959).
[12] J. R. Madigan: Thermal Characteristics of Silicon Diodes, *Electron. Indus.*, **18**, 80–87 (December 1959); **19**, 83–87 (January 1960).
[13] J. R. Madigan: Understanding Zener Diodes, *Electron. Indus.*, **18**, 78–83 (February 1959).
[14] T. Mollinga: Effect of Temperature and Current on Zener Breakdown, *Electro-Technol. (New York)*, **72**, 122, 124, 126 (October 1963).
[15] J. Muench: Variable Temperature Control Uses Zener Characteristic, *Electon. Design*, **11**, 66 (October 11, 1963).
[16] M. R. Nicholls: Zener Diode Characteristics, *Electron. Eng.*, **31**, 559 (September 1959).
[17] H. Penfield: Why Not Avalanche Diode as RF Noise Source?, *Electron. Design*, **13**, 32–35 (April 12, 1965).
[18] S. B. Rigg: The Characteristics and Applications of Zener Diodes, *Electron. Eng.*, **34**, 736–743, (November 1962).
[19] J. S. Schaffner and R. F. Shea: The Variation of the Forward Characteristics of Junction Diodes with Temperature, *Proc. IRE*, **43**, 101 (January 1955).
[20] *Zener Diode Handbook*, International Rectifier Corp., El Segundo, California.

Chapter 3

VOLTAGE REGULATOR CIRCUITS

3–1 INTRODUCTION

A voltage regulator is a circuit used to produce a controlled output voltage under conditions of varying input voltage, fluctuating output current, and/or changing environmental conditions.

Various applications will demand different types of regulation and may encompass one or more of the following:

1. Regulation of output voltage due to input voltage variation.
2. Regulation of output voltage due to output load current variation.
3. Regulation of output voltage due to changing environmental conditions.
4. Voltage divider or reducer applications.

In this chapter we shall look at the simple, direct applications of VR diodes as voltage regulators. This will include the basic shunt regulator circuit, bridge regulators, and various combinations of VR diodes and transistors.

3–2 SHUNT REGULATORS

The simplest approach to obtaining a relatively constant output voltage from a VR diode may be developed by studying the reverse breakdown characteristic curve shown in Fig. 3–1.

The basic nature of the VR diode is to maintain a relatively constant voltage across it as long as it is biased in the reverse breakdown region. A bias condition developed by establishing a load line as shown with a driving voltage V_I and a series resistance R_S will result in operation at point X on the characteristic curve. A voltage equal to V_X will be dropped across the diode.

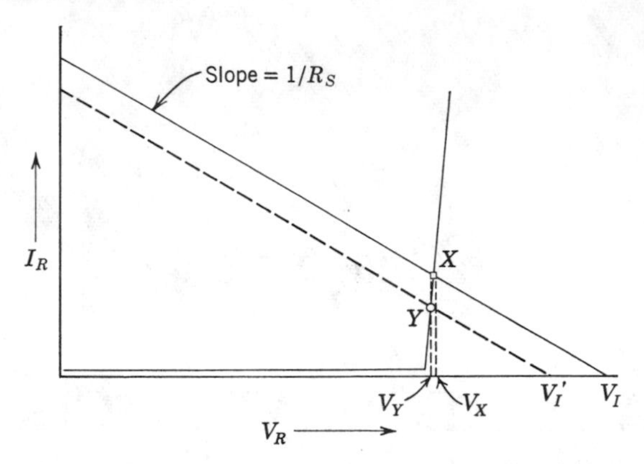

Fig. 3–1 Load line applied to reverse breakdown characteristic curve illustrates voltage regulation principle.

Now, if the driving voltage is modified to a new value, V_I', then the load line will be shifted to a new position as shown by the dotted line in Fig. 3–1. The new load line will be parallel to the previous one and the operating point will change to point Y. As long as point Y remains in the breakdown region, the value of V_Y, the voltage drop across the VR diode corresponding to operating at point Y will remain close to V_X and since the difference between V_X and V_Y may be much smaller than the difference between V_I and V_I', we see that we have a basic voltage regulator performance.

3–2.1 Basic Shunt Regulator Circuit

The circuit configuration of the shunt regulator in its simplest form is shown in Fig. 3–2. If we temporarily neglect the presence of load resistance R_L, we have the conditions depicted by the load line construction in Fig. 3–1.

A variation in the value of the unregulated input voltage V_I will merely shift the load line back and forth with the reverse bias current being established by the intersection of load line and characteristic curve.

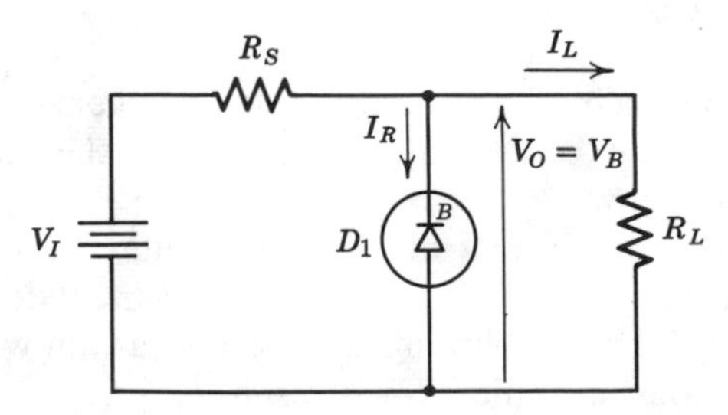

Fig. 3–2 Basic shunt regulator circuit.

So long as V_I is slightly larger than the breakdown voltage of the VR diode, reverse breakdown current will flow and the output voltage will be held relatively constant.

We may develop the equivalent circuit corresponding to the simple shunt regulator circuit by substituting a simplified equivalent for the VR diode as shown in Fig. 3–3*a*. The effect of the load resistance may be incorporated by converting to a Thévenin equivalent as shown in Fig. 3–3*b*.

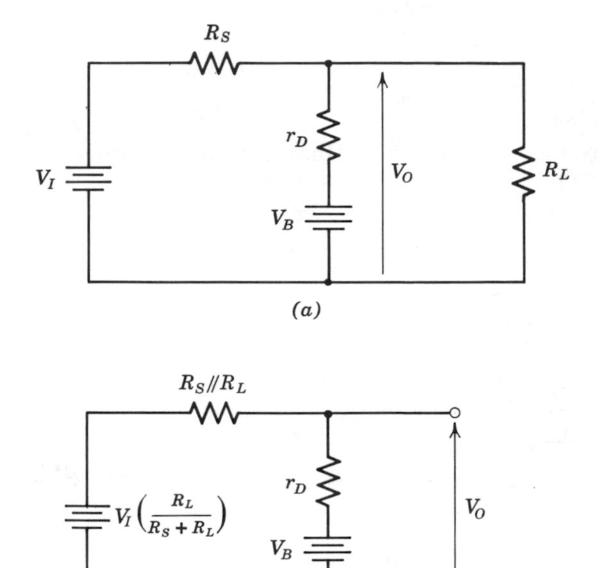

Fig. 3–3 Simplified equivalent circuits of basic shunt regulator illustrate importance of VR diode breakdown resistance; (*a*) overall equivalent; (*b*) Thévenin equivalent.

The new driving voltage will be $V_I R_L/(R_S + R_L)$ and the new load line resistance will be $R_S // R_L$. The VR diode will be biased in the breakdown region as long as $V_I R_L/(R_S + R_L)$ is greater than V_B.

A study of the equivalent circuit of Fig. 3–3*b* very quickly indicates the importance of the effective breakdown resistance r_D. This is made even clearer by considering only the incremental equivalent circuit as given in Fig. 3–4. Solving the network for ΔV_O, the change in ouput voltage which results from ΔV_I, a change in input voltage, we have

$$\Delta V_O = \Delta V_I \left(\frac{r_D}{R_S // R_L + r_D}\right)\left(\frac{R_L}{R_S + R_L}\right). \tag{3-1}$$

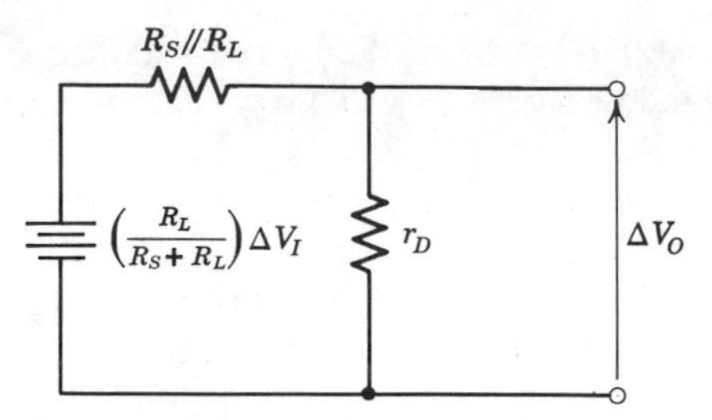

Fig. 3–4 Equivalent circuit useful in considering output voltage variations due to a changing input voltage.

When the value of r_D is much less than the parallel equivalent of R_S and R_L, as it must be for good regulation, we have

$$\Delta V_O \approx \Delta V_I\left(\frac{r_D}{R_S}\right). \tag{3-2}$$

In the shunt regulator, then, the capability of output voltage control under conditions of a varying input voltage is directly associated with the ratio of r_D to R_S. For optimum control, then, R_S should be as high as practical and r_D should be as low as practical.

Figure 3–5 presents an equivalent circuit that is useful in considering the effects of a varying output load. If the nominal value of the load current I_L is

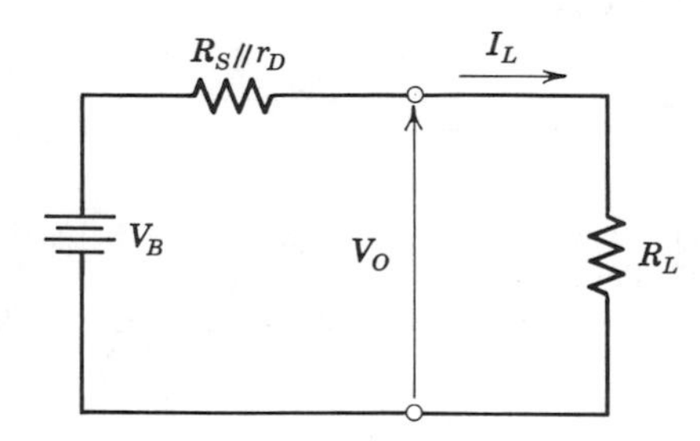

Fig. 3–5 Equivalent circuit useful in studying effects of load variations on output voltage.

changed by an amount equal to ΔI_L, then we will produce an incremental output voltage change ΔV_O:

$$\Delta V_O = (\Delta I_L)(R_S//r_D). \tag{3–3}$$

Again, assuming that r_D is much less than R_S,

$$\Delta V_O \approx (\Delta I_L)(r_D). \tag{3–4}$$

3–2.2 Design Consideration

In designing shunt regulator circuits, several important factors must be considered in light of the particular application:

1. Output voltage level required and tolerance.
2. Input voltage variation—from $V_{I(\min)}$ to $V_{I(\max)}$.
3. Output current variation—from $I_{L(\min)}$ to $I_{L(\max)}$.
4. Output voltage control requirements—$\Delta V_{O(\max)}$: (a) under varying input voltage; (b) under varying load current; (c) under varying ambient conditions.

The design objective would be to choose proper component values to permit the regulator to meet the given requirements. We shall need to choose the proper VR diode nominal voltage, tolerance, and power dissipation as well as the value of the series limiting resistor R_S.

Some applications might well have tight requirements regarding certain aspects, yet be very loose for others; for example, we might need an output voltage whose exact value would not be critical but whose variation with changing input voltage must be kept very small.

It may well be that the requirements may not be met with the simple shunt regulator. A modification of the basic circuit, as described later, may possibly be required.

3–2.3 Design Procedures

In designing the basic VR diode shunt regulator circuit, we do not have much flexibility. The output voltage requirement is fixed by the application, although we may need to consider the permissible tolerance of the VR diode breakdown voltage as it relates to the overall tolerance of output voltage.

The main component design centers around the choice of series-limiting resistor R_S. We shall then need to determine maximum power dissipation in order to fully specify the VR diode.

The first step in the design is to choose a nominal value for V_B for the VR diode. This will be determined by the output voltage requirements. Normally VR diodes are available in the standard 5 to 10% values, although it is possible to purchase to a nonstandard nominal value. If the tolerance of output voltage permits, we would select the standard value nearest to the original required output voltage. We must remember that if our bias current differs from the manufacturer's test value, a slight shift in the nominal V_B will occur and may have to be considered.

The required tolerance of the VR diode will be directly controlled by the tolerance of the end voltage output. If the nominal value of V_B available does not coincide with the required nominal V_O, then we may have to specify a tighter tolerance in V_B than might be necessary otherwise.

After selecting the nominal V_B and is tolerance, we may proceed to determine the value of R_S. R_S must be small enough to supply adequate current to the load (under maximum load conditions) and still cause a minimum bias current to flow through the VR diode. This minimum value of I_R should be above the knee in the breakdown characteristic curve. For VR diodes rated at 500 mW or less, a minimum value of I_R might be 1 or 2 mA.

The worst-case condition occurs when V_I is at its lowest value, I_L is at its maximum value, V_B is maximum, and R_S is maximum. The maximum allowable value of R_S is

$$R_{S(\max)} = \frac{V_{I(\min)} - V_{B(\max)}}{I_{L(\max)} + I_{R(\min)}}. \tag{3-5}$$

Now, considering the end of life tolerance on the value of R_S, we choose a nominal value such that R_S will always be less than the value given by (3–5). In certain cases it may be wise to use a value of R_S substantially smaller than that indicated by (3–5), provided we do not exceed the power dissipation rating of the VR diode; for example, when the tolerance in the output voltage is critical, it may be wise to select a nominal value of R_S to yield a bias current that is equal to the test current as indicated on the data specification sheet.

With the nominal R_S and its tolerance known, we may now compute the maximum power in the VR diode. This will occur when V_I is maximum, I_L is minimum, R_S is minimum, and V_B is minimum (because, normally, this will produce the greatest current).

$$P_{D(\max)} = \left[\frac{V_{I(\max)} - V_{B(\min)}}{R_{S(\min)}} - I_{L(\min)}\right] V_{B(\min)}. \tag{3-6}$$

The power rating actually required for the VR diode will depend upon the value obtained from (3–6) as well as maximum operating temperature and thermal resistance of the mounting means. If the variation in output voltage is critical, than $P_{D(\max)}$ as it may be wise to choose a power rating much higher calculated above. This will reduce the overall ΔV_O resulting from self-heating effects.

Although this completes the design, it is often necessary to analyze the circuit to determine the performance which may be expected; for example, we may wish to compute the ripple reduction factor of the regulator. This is easily obtained by using the equivalent circuit of Fig. 3–4 but substituting the small-signal breakdown resistance r_d.

$$\frac{V_o}{V_i} = \left(\frac{R_L}{R_S + R_L}\right)\left(\frac{r_d}{r_d + R_S /\!/ R_L}\right). \tag{3-7}$$

Normally r_d is much less than $(R_S \parallel R_L)$ and we may make the approximation

$$\frac{V_o}{V_i} \approx \frac{r_d}{R_S}. \tag{3-8}$$

We may need to determine the total shift in the output because of a change in power dissipation. For this we can assume that the components have constant, worst-case values and only the input voltage and output current are variables.

$$\Delta P_D = \left[\frac{V_{I(\max)} - V_B}{R_{S(\min)}} - I_{L(\min)}\right] V_{B(\max)} - \left[\frac{V_{I(\min)} - V_B}{R_{S(\min)}}\right] - I_{L(\max)}\Bigg] V_{B(\max)} \tag{3-9}$$

$$= \left[\frac{V_{I(\max)} - V_{I(\min)}}{R_{S(\min)}} + I_{L(\max)} - I_{L(\min)}\right] V_{B(\max)}. \tag{3-10}$$

Expressing the changes in input voltage and output load currents as incremental values, we have

$$\Delta P_D = \left[\frac{\Delta V_I}{R_{S(\min)}} + \Delta I_L\right] V_{B(\max)}. \tag{3-11}$$

With the change in power dissipation known, we may compute the change in output voltage from the thermal resistance and temperature coefficient.

$$\Delta V_O = (\Delta P_D)\theta_{JA} K_T, \tag{3–12}$$

where ΔV_O will be given in volts provided ΔP_D is expressed in watts, the thermal resistance θ_{JA} in °C/mW, and the temperature coefficient K_T in mV/°C.

If the value of ΔV_O due to the change in power dissipation is excessive, we may need to improve the thermal coupling and thus reduce θ_{JA} or we may have to choose a VR diode with a higher power rating and thus a lower thermal resistance.

3–2.4 Examples

To illustrate the design procedures, let us consider some typical examples. In an actual application we may have three possible operating conditions:

1. Load current is constant, but input voltage varies.
2. Input voltage is relatively constant, but load current fluctuates.
3. Input voltage and load current both vary but variations are independent.

3–2.4.1 Varying Input Voltage. Assume that we wish to develop a 10 V regulator to be driven from a 28 ± 5 V unregulated source. The load current is to be 10 mA and relatively constant. Total output voltage variation in production is to be held less than $\pm 10\%$ and output voltage changes for a given circuit should be less that $\pm 3\%$ around the nominal.

First, we select a nominal voltage for the VR diode of 10 V. Since the overall tolerance on V_O due to all initial variations is $\pm 10\%$, we should choose a $\pm 5\%$ tolerance for the VR diode breakdown voltage. Thus V_B in production may be assumed to be 10 ± 0.5 V, provided the nominal bias current is equal or very close to the test current at which the diode was measured by the manufacturer.

The maximum value of R_S, which we may use, is found from (3–5), where $V_{I(\min)} = 23$ V and $V_{I(\max)} = 33$ V. Assume an absolute minimum bias current of 2 mA for the VR diode.

$$R_{S(\max)} = \frac{23 - 10.5}{(10 + 2)\text{ mA}} = 1.04\text{ K}.$$

If we assume that end life tolerance on the resistor might be $\pm 10\%$, then we would have to choose R_S to be nominally 910 Ω or less.

$$R_{S(\min)} = 819\ \Omega, \qquad R_{S(\max)} = 1001\ \Omega.$$

Now let us calculate the maximum power dissipation in the VR diode. For this we use (3–6).

$$P_{D(\max)} = \left(\frac{33 - 9.5}{819} - 10 \times 10^{-3}\right) 9.5 = 178\text{ mW}.$$

We could choose a VR diode rated at 250 mW such as a 1N714A. Before making a final decision it might be well to determine the change in output voltage due to the total variation in power as given by (3–11).

$$\Delta P_D = \left(\frac{10}{819} + 0\right) 10.5 = 128\text{ mW}.$$

Now we can use (3–12) to determine the change in output voltage produced by ΔP_D. Assuming a worst-case value of $\theta_{JA} = 0.6$°C/mW and using 7 mV/°C for the temperature coefficient, we have

$$\Delta V_O = (0.128)(0.6)(7) = 0.54\text{ V}.$$

The total allowable variation in the output voltage was given as $\pm 3\%$ or a total swing of 0.6 V. This must include the variation in V_O due to r_d as well as the part caused by a ΔP_D. For this reason we would be better to specify a VR diode with a higher power rating and hence a lower thermal resistance. A 1N5240A is rated at 500 mW and by making the leads rather short the thermal resistance can be held to about 0.2°C/mW. This is one-third the value used above and thus the new ΔV_O with this VR diode would drop to about $0.54/3 = 0.18$ V.

Using (3–8) will tell us how effectively the regulator will remove ripple or small variations from the output. The specification sheet for the 1N5240A lists the maximum value of r_d at 17 Ω when measured at a test current of 20 mA. The variation of r_d with reverse current was discussed in Chapter 2. Assuming that the 10-V VR diode will have a variation similar to that described by (2–13), we can approximate

$$\frac{r_d(8 \text{ mA})}{r_d(20 \text{ mA})} \approx \left(\frac{8}{20}\right)^{-0.75} = 2,$$

and thus the maximum value of r_d that we would expect at a bias current of 8 mA (which is about nominal for our design) would be about 34 Ω. Now, using (3–8),

$$\frac{V_o}{V_i} \approx \frac{34}{910} = 0.037.$$

This means that any ripple presented to the output will be about 27 times less than that applied to the input.

3–2.4.2 Varying Load. Now let us assume that we have thc same problem as before except the input voltage is a constant 28 V and the load varies from 5 to 10 mA. For this case

$$R_{S(\max)} = \frac{28 - 10.5}{(10 + 2) \times 10^{-3}} = 1.46 \text{ K}.$$

If we again assume an end tolerance of $\pm 10\%$ for R_S, then a nominal value of 1.2 K would have a minimum of 1.08 K and a maximum of 1.32 K.

Using (3–6) to calculate the maximum power dissipation,

$$P_{D(\max)} = \left(\frac{28 - 9.5}{1080} - 5 \times 10^{-3}\right) 9.5 = 115 \text{ mW}.$$

Again it looks as if a 250 mW VR diode could be used. Calculating the incremental power,

$$\Delta P_D = (0 + 5 \times 10^{-3})(10.5) = 52.5 \text{ mW},$$

which is considerably less than for the previous example. The value of ΔV_O which this would produce would be about 220 mV and could be tolerated.

3–2.4.3 Varying Load and Input Voltage. We now take the same basic problem of the 10-V regulator but design for the third condition in which both the input voltage and the output load current are varying.

As before we calculate the maximum allowable value of the series resistance R_S:

$$R_{S(\max)} = \frac{23 - 10.5}{(10 + 2) \times 10^{-3}} = 1.04 \text{ K},$$

which is the same as for the first problem example. We again use a nominal value of 910 Ω with a minimum of 819 Ω and a maximum of 1000 Ω.

$$P_{D(\max)} = \left(\frac{33 - 9.5}{819} - 5 \times 10^{-3}\right) 9.5 = 225 \text{ mW},$$

$$\Delta P_D = \left(\frac{10}{819} + 5 \times 10^{-3}\right)(10.5) = 181 \text{ mW}.$$

We would use a power rating of at least 500 mW and could expect a ΔV_O due to the incremental power of about 0.25 V. We could reduce this slightly by making sure that the cathode end of the VR diode had a short and low thermal resistance path to ambient.

3–2.5 Transient Performance

Because the breakdown mechanism in a VR diode acts so fast, the shunt regulator will act to remove even high frequency transient spikes which may appear at the input. The attenuation ratio will be slightly better than that given by (3-8) because of the additional filtering that occurs due to the junction capacitance of the VR diode. The shunt regulator also will protect the circuit from substantial surges as discussed in Chapter 5.

3–3 BRIDGE REGULATOR CIRCUITS

When the input voltage to a shunt regulator circuit varies, it produces a change in the bias current through the VR diode. The voltage across the VR diode varies because of an effective breakdown resistance as discussed in Chapter 2.

In many applications we are interested in removing or at least greatly reducing the effects of a fluctuating input voltage. One way of accomplishing this is by using the VR diode in a bridge arrangement as shown in Fig. 3–6.

Figure 3–6*a* gives the basic circuit approach. Note that in order to use the bridge circuit, it is not possible to have a common terminal input and output as could be done with the basic shunt regulator.

The equivalent circuit under incremental conditions may be shown as indicated in Fig. 3–6*b*. We can easily see how this approach can aid in reducing output variations if the bridge is balanced. This means that

$$\frac{R_1}{R_2} = \frac{R_S}{r_d}. \tag{3-13}$$

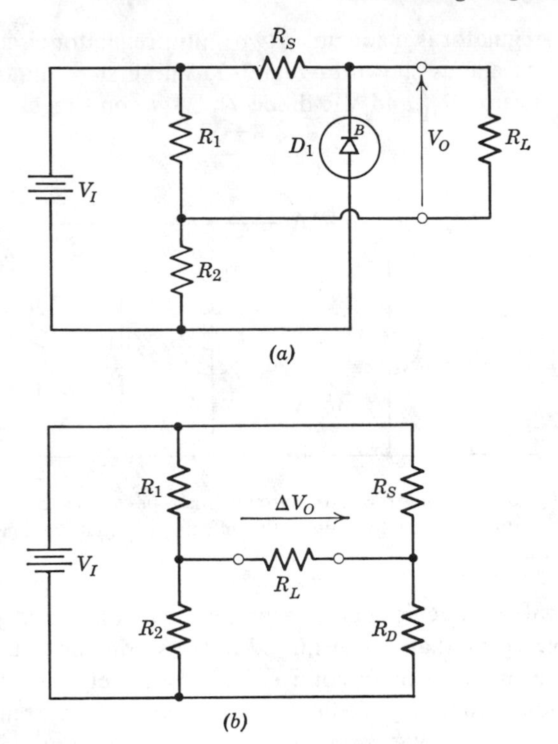

Fig. 3–6 Basic bridge regulator circuit: (*a*) schematic diagram; (*b*) incremental equivalent.

A perfect bridge balance would mean that no output voltage fluctuation would be produced even if V_I varied. Unfortunately, the breakdown resistance is nonlinear and thus does not remain constant when the current through the diode changes. Even though the output variation may not be removed completely, a substantial improvement is possible by balancing for a nominal value of r_d.

This type of circuit is useful only where the load current is relatively constant because of the increase in the effective output resistance by an amount equal to R_1/R_2. In addition, the open circuit output voltage is reduced by an amount equal to $[R_2/(R_1 + R_2)]V_I$.

3–4 TANDEM REGULATOR CIRCUITS

For some applications it may be impossible to obtain adequate regulation from a single shunt regulator stage. If this is due primarily to the effects introduced by a varying input voltage, designing a tandem regulator may yield the desired regulation.

The tandem regulator is made up of two shunt regulator circuits connected in tandem or cascade as shown in Fig. 3–7. The first regulator circuit consists of series resistor R_{S1} and VR diode D_1. A second regulator consists of R_{S2} and D_2.

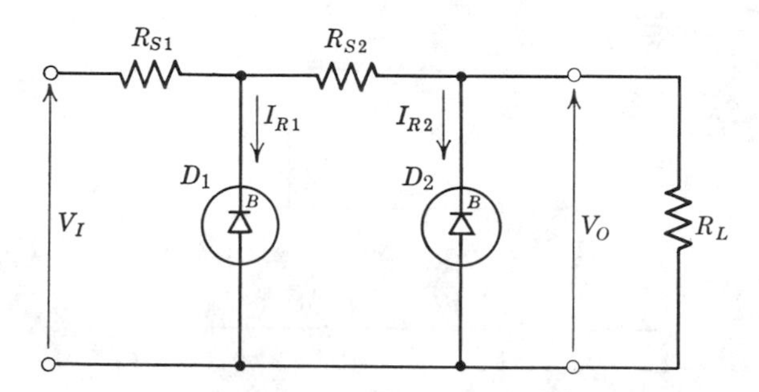

Fig. 3–7 Shunt regulator circuits may be connected in tandem in order to reduce the effects of input voltage variation.

The first regulator circuit acts as a preregulator and greatly reduces the input voltage variation that is transmitted to the second and output regulator stage. Since the input to the second stage is relatively constant (although perhaps not enough to satisfy the final requirements), the current through the VR diode D_2 will not vary substantially due to changes in the input voltage. This means that variation in output voltage resulting from the breakdown resistance will be very small if the load current remains constant. In addition, the power dissipation in D_2 will be relatively constant (again assuming a constant load current) and we get little output voltage variation due to ΔP_D.

Voltage ripple reduction will likewise be improved by the tandem configuration. Each section will have its own ripple reduction factor as described by (3-8). The overall ripple reduction factor will be equal to the product of the individual factors or

$$\frac{V_o}{V_i} = \left(\frac{r_{d1}}{R_{S1}}\right)\left(\frac{r_{d2}}{R_{S2}}\right). \tag{3-14}$$

Each of the regulator circuits may be designed in the same manner as described in Section 3–2.3. As a general rule the breakdown voltage of D_1 should be at least several volts greater than that of D_2. Even so it may be necessary to select diodes having a relatively constant difference in V_B for a given design; for example, suppose that $V_{B1} = 10\text{ V} \pm 5\%$ and $V_{B2} = 8\text{ V} \pm 5\%$. The difference in voltage will vary between a minimum value

$$\Delta V_{B(\min)} = (10)(0.95) - (8)(1.05) = 1.1\text{ V},$$

and a maximum value

$$\Delta V_{B(\max)} = (10)(1.05) - (8)(0.95) = 2.9 \text{ V}.$$

The current flowing through R_{S2} (equal to the sum of the load current and the bias current in D_2) will be equal to $\Delta V_B/R_{S2}$ and will thus vary by a factor of 2.6 to 1. Although this will not directly affect the regulation due to input voltage fluctuation, it may make output voltage and power dissipation prediction rather difficult.

3–5 CONTROLLING TEMPERATURE COEFFICIENT

Variation in output voltage because changes in operating junction temperature, whether resulting from fluctuation of ambient temperature or resulting from a changing power dissipation, may be reduced by controlling the temperature coefficient of the VR diode. Although a simple diode will have a relatively predictable temperature coefficient whose value depends primarily on the nominal breakdown voltage, we may modify the overall TC by combining VR diodes with forward-biased junction diodes. VR diodes with V_B greater than about 5.5 V have a positive temperature coefficient, whereas a forward-biased diode has a negative TC that will counteract the TC of the VR diode. The basic approach is shown in Fig. 3–8. In actual practice both

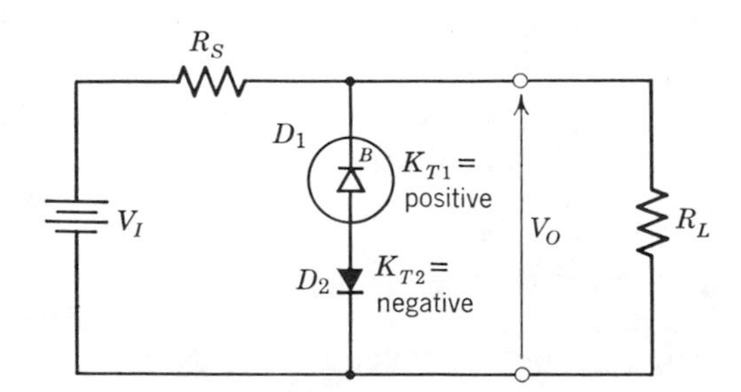

Fig. 3–8 Basic temperature-compensated VR diode uses negative temperature coefficient of forward-biased diode to counteract the positive TC of the VR diode.

diodes are carefully matched and enclosed within the same package. Much more will be said about this in Chapter 11.

We might also recall from Chapter 2 that the value of the temperature coefficient rises as the nominal V_B is increased above about 5.5 V. This is true whether we express the TC in terms of mV/°C or %/°C. Thus a lower value of temperature coefficient may be obtained by using two or more VR diodes in series to make up the total breakdown voltage. This is shown in

Fig. 3–9. The value of V_B for the composite VR diode will be equal to the sum of the individual breakdown voltages. If the temperature coefficient K_T is expressed in terms of mV/°C, the value of K_T for the composite will be the sum of the individual values.

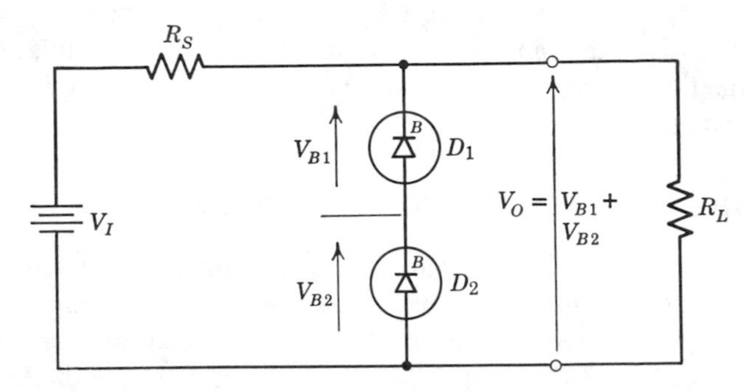

Fig. 3–9 Two or more VR diodes may be connected in series to yield a composite with better temperature coefficient and lower breakdown resistance.

The effective breakdown resistance of the composite VR diode made of several diodes in series will also be the sum of the individual values. Look back, however, at Fig. 2–13 and notice that as the nominal V_B increases beyond about 7 V, the value of dynamic resistance also becomes larger. Thus two 10-V VR diodes may have a total value of r_d which is actually less than that of a single 20-V VR diode biased at the same current level. When we include the temperature effects on the total value of r_D, the series combination may look even more appealing.

3–6 CURRENT-FED REGULATOR CIRCUITS

Since the variation of output voltage of a shunt regulator is directly associated with changes in the bias current through the VR diode, we may always improve performance by making this bias current constant.

Bias current will vary either due to a varying load current or a changing input voltage. We may reduce the effects of a varying input voltage by driving the regulator from a constant-current source. This may take one of several possible forms, depending on the application.

The basic concept is illustrated in Fig. 3–10. A current-limiting element is placed in series with the line from the input to the VR diode. When the input voltage remains relatively constant, a series resistor is sufficient to limit the current to a fixed value.

When V_I may vary by a large amount, some other form of current limiting means must be used. We may use the relatively constant-current characteristic

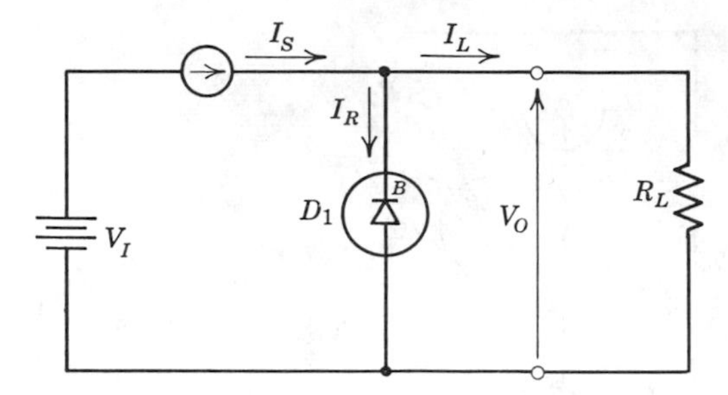

Fig. 3–10 Basic generalization of a current fed shunt regulator circuit.

of an incandescent lamp filament or other positive temperature coefficient resistance. As the value of input voltage V_I increases, the voltage drop across the limiting device increases while allowing the current flow through it to increase only slightly.

The current-limiting device may also be one of the field-effect current-limiting diodes that exhibits a constant-current characteristic over a wide voltage range.

A third possibility that may be used is to design a simple transistorized current source to drive the input. One approach is shown in Fig. 3–11. The

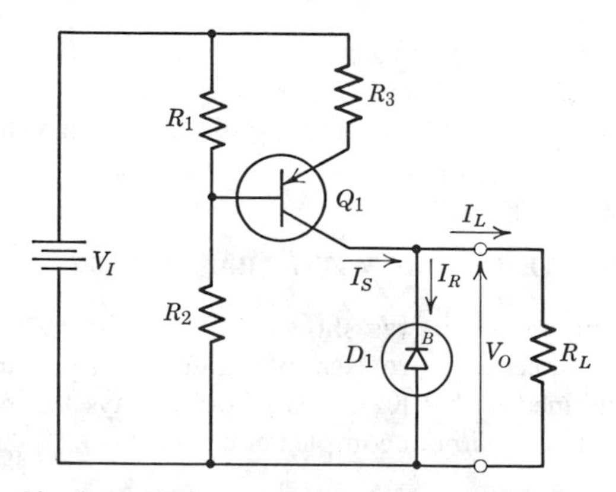

Fig. 3–11 Using a transistorized current source to feed the shunt regulator improves regulation.

value of I_S at the collector of Q_1 will depend on the voltage drop developed across R_1 in conjunction with resistor R_3. Although a variation in input voltage will still produce about the same percentage in I_S, the total change in I_S is still substantially less than would occur if only a series resistor were used.

A further improved circuit is obtained by replacing resistor R_1 by a second VR diode as shown in Fig. 3–12. Now the base voltage at Q_1 is held constant

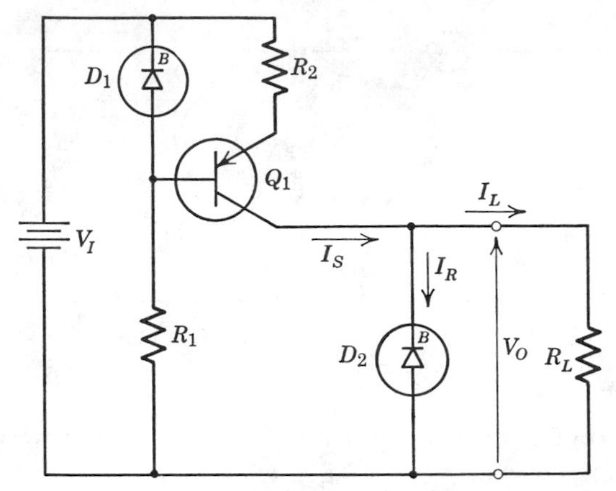

Fig. 3–12 A second VR diode controls the current source for improved performance.

with respect to the positive input line. This means that the value of I_S will be practically independent of V_I. The net result is a very large increase in the effective value of R_S for the simple shunt regulator circuit. This not only reduces output voltage variations due to a changing input voltage but also reduces ripple output.

Since feeding the VR diode from a current source effectively increases R_S to a very large value without increasing the required input voltage, we get the desired improvement in regulation and ripple reduction without sacrificing efficiency.

3–7 COMBINING VR DIODES WITH TRANSISTORS

In the preceding section it was shown how transistors could be used as constant-current sources to improve shunt regulator performance. Transistors may also be combined with VR diodes in other ways to obtain some very desirable characteristics without complex circuitry.

3–7.1 Power Multiplier

One approach is to use the basic circuit of Fig. 3–13. A single transistor is used in a shunt regulator mode. The base current of Q_1 is the reverse bias current of the VR diode and thus will increase when the output voltage attempts to increase beyond a voltage equal to the sum of the breakdown voltage of D_1 plus the V_{BE} drop of Q_1.

The collector current of Q_1, which will be equal to the product of I_B and the h_{FE}, will add to the current through the VR diode D_1.

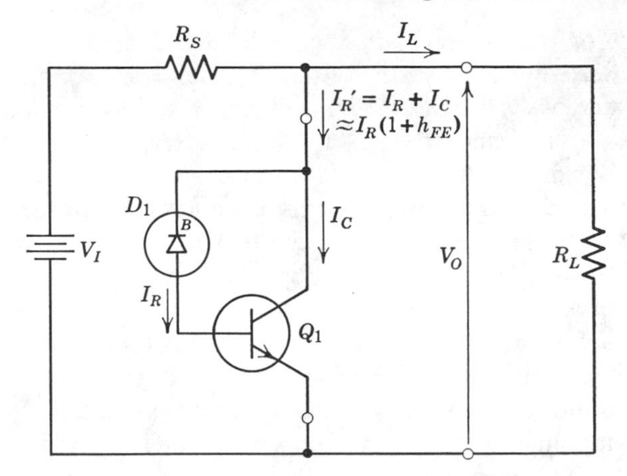

Fig. 3–13 Basic power multiplier circuit uses transistor to increase current-handling capability of the VR diode.

The net result is a composite circuit that acts like a VR diode having a breakdown voltage equal to V_B of the VR diode and V_{BE} of the transistor added together:

$$V_B' = V_B + V_{BE}. \tag{3-15}$$

The total current flow through the composite will be

$$I_R' = I_R + I_C = I_R(1 + h_{FE}). \tag{3-16}$$

The total power dissipation capability of the composite circuit will be primarily that of the transistor.

If we assume that the V_{BE} of Q_1 will remain relatively constant, the effective small-signal series breakdown resistance of the composite configuration will become

$$r_d' = \frac{r_d}{1 + h_{fe}}, \tag{3-17}$$

or the total breakdown resistance will be

$$r_D' = \frac{r_D}{1 + h_{FE}}. \tag{3-18}$$

In addition to the basic benefits of increased current and power-handling capability and reduced breakdown resistance, we will achieve an improved temperature coefficient because of the opposite polarities of the TC of the VR diode (assuming V_B is greater than about 5.5 V) and that of the V_{BE} of the transistor. This becomes especially apparent when we consider the overall output voltage variation due to a change in power dissipation.

Since most of the power is dissipated in the transistor, any change in operating power will occur mostly within Q_1 also. For example, assume that the overall value of P_D increases. This will cause the junction temperature of transistor Q_1 to rise and the V_{BE} to drop correspondingly. The current through the VR diode (and hence its own power dissipation) will increase slightly. This will raise its junction temperature and thus increase V_B (assuming V_B is greater than about 5.5 V). Since the two effects of ΔV_B and ΔV_{BE} due to increased power dissipation are in opposite directions, they tend to compensate each other.

Under certain conditions, when the rise in junction temperature of the transistor is higher than that of the VR diode for a given increase in overall power dissipation, it is possible to achieve overcompensation to the extent that the output voltage will actually decrease with an increase in total current. This indicates the generation of an equivalent negative resistance that is larger than r_d.

Since the composite device is still a two-terminal network, we may use either an *NPN* or a *PNP* transistor as shown in Fig. 3–14. If Q_1 is germanium, R_1 may be necessary because of leakage currents.

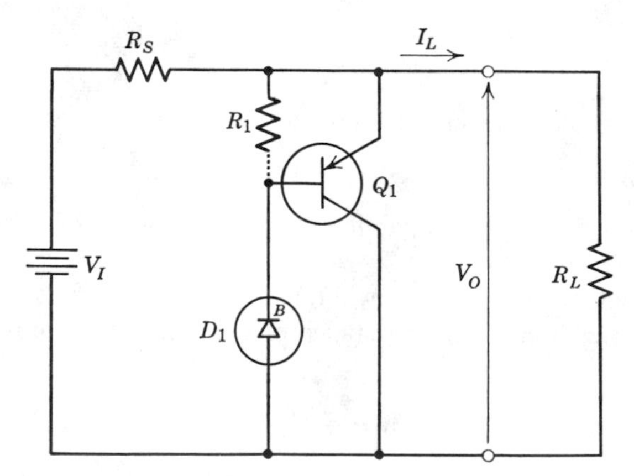

Fig. 3–14 Power multiplier circuit can use PNP transistor by inverting the composite.

3–7.2 Emitter Follower

Another way in which a transistor may be combined with a VR diode is shown in Fig. 3–15. Here, a simple transistor emitter follower is added to the basic shunt regulator circuit.

The output voltage will be less than V_B of the VR diode by an amount equal to the V_{BE} of the transistor. The output current is supplied by the transistor, and the current provided by limiting resistor R_S (or other current-limiting

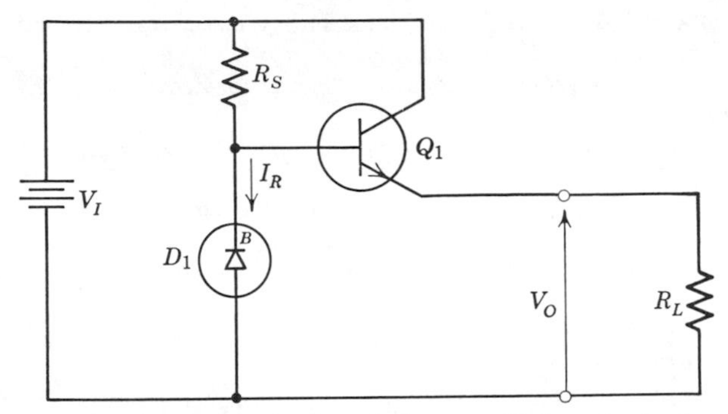

Fig. 3–15 Transistor emitter follower added to basic shunt regulator gives increased output current.

means as discussed in Section 3–6) need be only the required bias current for the VR diode and the base current required by the transistor.

This means that we have multiplied effectively the load resistance (or divided the load current) by an amount equal to the h_{FE} of the transistor.

The overall temperature coefficient of the regulator of Fig. 3–15 will be the net difference of the TC for V_B and V_{BE}. If V_B is greater than about 5.5 V, the two values will be opposite in polarity and thus will add in magnitude. To state it another way, the TC of the output voltage will always be about 2 mV/°C more positive than the value of K_T for the VR diode.

The effects of increased power dissipation are such that the output voltage will decrease as power is dissipated within the transistor. Ripple factor and general voltage regulation will be about the same as for the simple shunt regulator.

3–7.3 Regulated Power Supplies

The VR diode may be used as a voltage reference in a more complicated regulated power supply. A dc servo type of circuit is generally used with an error signal being generated to control the output voltage to a specific value. These circuits will not be discussed in this book. A few simple feedback circuits are included in Chapter 4 in conjunction with a voltage reference amplifier configuration.

REFERENCES

[1] M. Beebe: Equations and Procedure for Designing Transistor or Zener Shunt Regulators, *Electronics*, **34**, 92 (June 30, 1961).

[2] J. K. Buchanan et al.: *Zener Diode Handbook*, Motorola, Inc., Phoenix, Arizona, 1967.

[3] L. D. Clements: Solid-State Generator Regulator for Automobiles, *Design Manual for Transistor Circuits*, McGraw-Hill, New York, 1961, pp. 118–120.

[4] R. R. Gupta and B. Tyler: Zener Diode in Stabilized Transistor Power Supplies, *Electron. Tech.*, **38**, 228–229, (June 1961).

[5] J. L. Haynes: Zeta, A Proposed Regulation Factor for Zener Diodes, *EDN*, **14**, 55–59 (May 1, 1969).

[6] J. W. Keller, Jr.: Regulated Transistor Power Supply Design, *Electronics*, **29**, 168–171 (November 1956).

[7] C. F. Kezer and M. H. Aronson: Zener Diode Regulates DC Heater Voltage, *Instr. Auto.*, **31**, 1987 (December 1958).

[8] T. W. Kirchmaier: Shunt DC Regulator Nomographs; Proper Selection of a Zener Diode, *Electron. Indus.*, **20**, 230–231 (June 1961).

[9] M. Lillienstein: Design of Regulated Power Supplies, *Modern Transistor Circuits*, McGraw-Hill, New York, 1959, pp. 37–39.

[10] W. J. McDaniel and T. L. Tanner: High Voltage Magnetically Regulated D-C Power Supply, *Proc. Nat. Electron. Conf.*, **14**, 905–912 (October 1958).

[11] J. S. McGee: Zener Diodes for Voltage Regulation, *Electronics*, **33**, 101 (November 11, 1960).

[12] R. G. McKenna: A Design Procedure for Silicon Regulator Diode DC Voltage Regulators, *Solid-State J.*, **2**, 38–42 (October 1961).

[13] R. G. McKenna: Designing Zener Diode Voltage Regulators, *Electron. Design*, **7**, 30–33 (April 1, 1959).

[14] J. N. Nichols: Zener-Regulated Power Supplies, *Instr. Control Systems*, **34**, 2242–2243 (December 1961).

[15] D. L. Stoner: Zener Diode Regulator, *Electron. Equip. Eng.*, **8**, 41 (April 1960).

[16] H. C. Stratman: Regulated Heater Supply, *Radio-Electron.*, **30**, 51 (November 1959).

[17] P. L. Toback: Zener Diodes Stabilize Tube Heater Voltages, *Electron. Indus.*, **17**, 64–66 (December 1958).

[18] C. D. Todd: Stable, Low-Cost Reference Power Supplies, *Electron. World*, **78**, 39–41, 79 (December 1967).

[19] E. C. Wilson and R. T. Windecker: DC Regulated Power Supply Design, *Solid-State J.*, **2**, 37–46 (November 1961).

[20] Zener Diode Specifications, *Electron. Design*, **6**, 26–31 (March 1958).

Chapter 4

VOLTAGE MONITORING CIRCUITS

4–1 INTRODUCTION

Voltage monitoring circuits, as the name implies, include those circuit arrangements in which an output signal is obtained on comparison with an input voltage that is to be monitored with a given threshold value or values. In some cases a binary type of output will be produced with the condition of the output determined by the relation of the applied input voltage to a reference value. In other types of voltage monitoring circuit an output will be produced that is somewhat linearly proportional to the deviation in the input voltage from a set value.

4–2 ELEMENTARY CIRCUITS

The simplest form of voltage monitoring circuit consists of a VR diode alone or with a simple loading arrangement as shown in Fig. 4–1. Any input voltage that is less than V_{B1}, the breakdown voltage of VR diode D_1, will

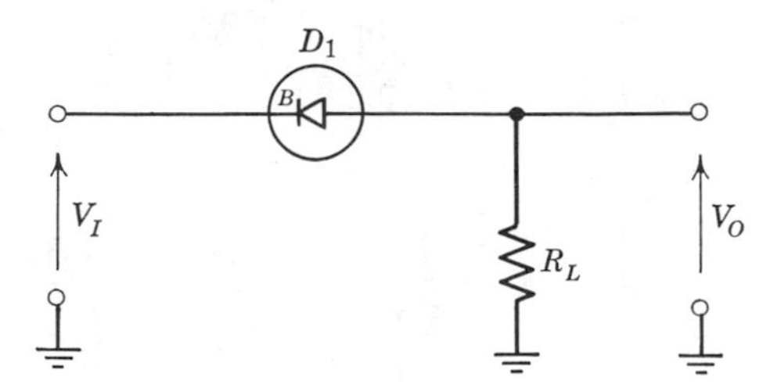

Fig. 4–1 Simplest Form of voltage monitoring circuit.

produce a negligible output. As soon as V_I becomes equal to the value of V_{B1}, however, D_1 will conduct and produce an output voltage across the load resistor that is the difference in V_I and V_{B1}. As soon as V_1 again becomes less than V_{B1}, the output will drop to zero.

The simple circuit of Fig. 4–1 may be made more useful by adding some form of power amplification at the output. This is easily done by connecting the output voltage directly between the emitter-base junction of a transistor or the gate-cathode input of a silicon-controlled rectifier. A magnetic amplifier may be substituted for the transistor if desired.

The threshold voltage value will be dependent on the breakdown voltage of the VR diode as well as the value of the output voltage that will cause the necessary action. For precise monitoring, temperature coefficients must be considered and, if necessary, compensated for. A sharp breakdown in the characteristic of the VR diode is required for best results.

If it is expected that the monitored voltage may, at times, exceed the threshold value by a considerable amount, it may be necessary to add some limited resistance in series with the VR diode to prevent excessive current flow.

4–3 PRACTICAL CIRCUIT ARRANGEMENT

A practical circuit that may be used to monitor a voltage and give an output describing the relation of the input voltage to the threshold level is given in Fig. 4–2. For the arrangement shown, one lamp will light when the voltage exceeds the threshold value and the other lamp will light as long as the input remains below the critical value. Depending on the desired information, we may eliminate one of the lamps or, if we prefer, substitute bells or relays for the indicators.

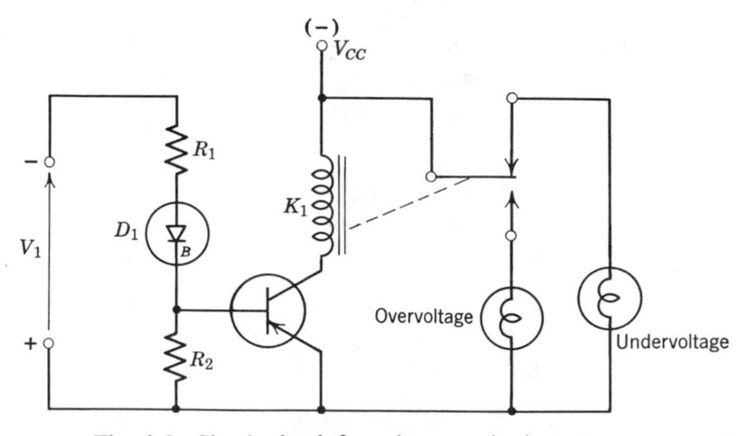

Fig. 4–2 Simple circuit for voltage monitoring.

The operation of the circuit is quite simple. As long as the input voltage is less than the breakdown voltage of the VR diode, practically no base current will flow in Q_1 and hence the transistor remains off. Resistor R_2 serves to shunt any leakage currents from D_1 or the collector-base diode of Q_1 to ground and thus prevent unwanted turn-on at high temperatures.

As soon as V_I begins to exceed the breakdown voltage V_{B1}, a current will flow into the base of Q_1 and will be amplified by the transistor to operate the relay. When the value of V_I drops below the critical value, base current ceases and the relay will return to its normal position.

Resistor R_1 limits the current flow after the threshold voltage value has been reached. R_1 is chosen in light of the maximum amount of input current allowed at the maximum expected input voltage although it does influence the value of the threshold voltage to a limited degree.

The speed of the circuit of Fig. 4–2 is limited only by the response of the relay and may be increased by taking the output directly from the collector of Q_1 or by using a solid-state relay. A silicon-controlled rectifier substituted for Q_1 will give a high power output and will remain in the turned-on condition once the input has exceeded the threshold limit until it is turned off by conventional means.

4–4 VOLTAGE REFERENCE AMPLIFIER

In designing a highly regulated power supply it is necessary to compare the output voltage or a fraction of it with a stable reference voltage. The comparator produces an error signal that is fed back to the controlling circuit to correct the output voltage in servo fashion.

The reference voltage source may be a temperature-compensated VR diode or a standard cell. The comparator and control circuit must have adequate gain and, in addition, must not contribute to temperature or other errors outside the control by the reference voltage. This means that at least the initial comparator and first stage of error amplification must be temperature stable and have predictable performance.

Reference amplifiers combine the first error amplifier stage with a comparator and a very stable temperature-compensated reference voltage. The result is a high degree of stability with relatively simple regulator circuits.

4–4.1 Operation

As shown in the schematic diagram contained within the dotted lines of Fig. 4–3, the reference amplifier may consist of a single *NPN* transistor with a VR diode. These must be carefully matched to provide the desired temperature coefficient.

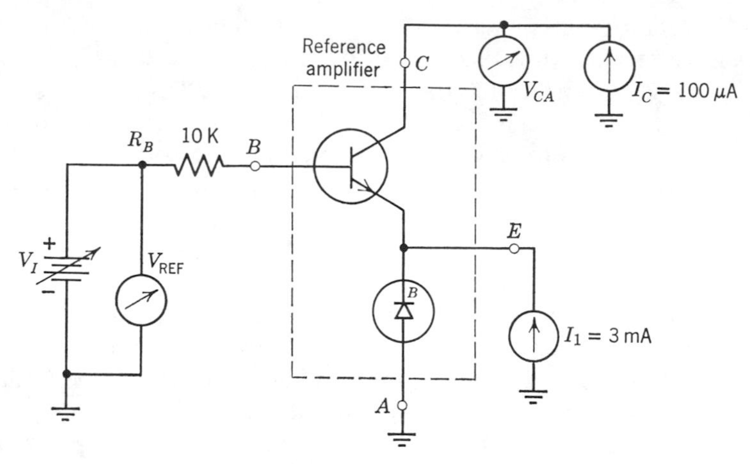

Fig. 4-3 Dc Test circuit for a typical reference amplifier. When V_1 is adjusted to give $V_{CA} = 14$ V, then $V_1 = V_{REF}$.

Whenever the input voltage applied to the base of the transistor is less than the voltage across the VR diode, then no base current and hence no collector current (excluding leakage currents) flows.

An input voltage which exceeds the VR diode voltage by enough to overcome the natural V_{BE} threshold value will produce a base drive current and thus may cause a substantial collector current to flow.

A typical example of a commercial reference amplifier is the family of modules made by Dickson Electronics Corporation. The DRAE 6.8 series of reference amplifiers is designed to be operated with a nominal collector current of 100 μA at a collector-to-anode voltage, V_{CA}, of approximately 14 V. If we supply a fixed value of collector current from a constant-current source, then the 14-V V_{CA} value will be produced only when the input voltage is exactly the right amount.

These units are characterized to include an effective voltage divider output resistance of 10 K; V_{REF} is defined as that open-circuit voltage that will produce a V_{CA} of 14 V when applied through a 10 K source resistance to the base input and when the collector current is maintained at exactly 100 μA.

Any slight variation in the input voltage will produce a very large variation in V_{CA}; for example, a DRAE 6.8 reference amplifier fed from an infinite impedance collector current source of 100 μA will yield an output voltage swing of over 2.5 V for only 1 mV change in the input voltage ($\pm 0.015\%$ change).

If the value of V_{CA} is fixed, then the collector current will be exactly 100 μA only when the input voltage is equal to V_{REF}. A 1 mV variation in the input voltage will produce a typical change in the collector current of more than 3 μA.

4-4.2 Characteristic Parameters

Certain of the parameters for the reference amplifier are peculiar to it and need to be defined in order to make clear just what is being described.

V_{REF} **Reference Input Voltage.** That open-circuit voltage which when fed through a specified source resistance (10 K for the example under study) will produce a specified V_{CA} (14 V) if the collector current is set exactly to a fixed value (100 μA) or which will produce the desired I_C if V_{CA} is held at a fixed value. (See the test circuit of Fig. 4-3.)

K_T^* **Temperature Coefficient.** The temperature coefficient of V_{REF} given in %/°C and described by the following:

$$K_T^* = \frac{V_{\text{REF}(T_1)} = V_{\text{REF}(25°\text{C})}}{(V_{\text{REF}(25°\text{C})})(T_1 - 25)} \times 100\%, \tag{4-1}$$

where $V_{\text{REF}(T_1)}$ is the value of V_{REF} at temperature T_1, $V_{\text{REF}(25°\text{C})}$ is the value of V_{REF} at 25°C, and T_1 is the temperature expressed in °C. Typical measurement temperatures are −25, 0, 0, +25, +50 + 75, and +100°C.

g_{fa} **Forward Circuit Transconductance.** The small-signal ratio of the incremental short-circuit collector current to the incremental change in the open-circuit input voltage applied through a specified source resistance. Note that this parameter includes the losses associated with the input divider.

$$g_{fa} \equiv \frac{\partial I_C}{\partial V_{\text{REF}}}\bigg|_{V_{\text{CA}}=14\text{ V constant}} \tag{4-2}$$

g_{oa} **Output Conductance.** The small-signal conductance seen between the collector and anode terminals. (See small-signal equivalent circuit Fig. 4-4*b*.)

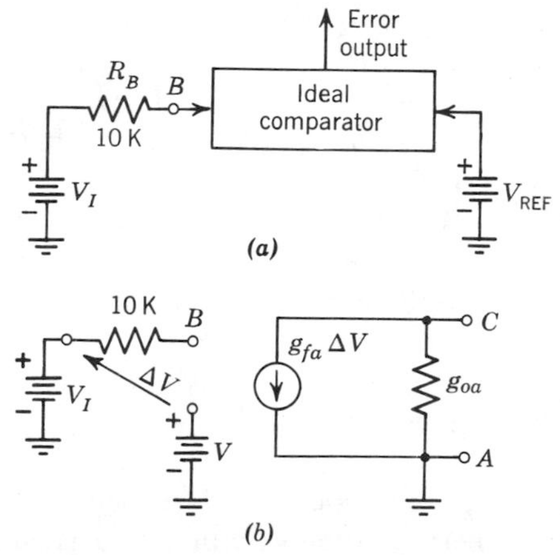

Fig. 4-4 Reference amplifier equivalent circuits: (*a*) Overall; (*b*) small signal.

I_1 **Bias Current.** Current applied to the VR diode by external means. Its value should be controlled close to the design value where optimum temperature performance is obtained. This is given on the data sheet and is 3.0 mA for the example.

R_B **External Base Driving Resistance.** Fixed resistance shown in Fig. 4–3 but equivalent output resistance of the voltage divider in practical circuit applications. Should be held close to the design value (10 K in the example) for optimum performance.

h_{FE}
I_{CBO}
I_R } conventional transistor and VR diode parameters.
$V_{BR(CBO)}$
$V_{BR(CEO)}$

4–4.3 Typical Operating Characteristics

Figure 4–5 illustrates a typical transfer characteristic curve in which the collector current is plotted as a function of the input voltage. Nominal operating bias should be as near the 100 μA design value as practical in order to achieve best temperature stability.

The small-signal transconductance varies with the collector current as shown in Fig. 4–6. Note that at the higher values of I_C, g_{fa} levels out at a relatively constant value. This is due to the decrease in input resistance as compared to the 10 K R_B.

Figure 4–7 illustrates that the transconductance is also a function of the bias current I_1, but to a fairly small degree unless I_1 is reduced below 2 mA. This variation is due to an increase in dynamic impedance of the VR diode with reduced current levels.

A typical temperature performance curve is shown in Fig. 4–8 for several biasing conditions. Curve 1 is the characteristic obtained with the specified bias values of $I_C = 100$ μA, $I_1 = 3$ mA, and $V_{CA} = 14$ V.

Curve 2 shows the performance achieved with the collector current operated at 200 μA and I_1 changed to keep the current through the VR diode by making $I_1 = 2.9$ mA. Curve 3 shows the result of making $I_C = 200$ μA and $I_1 = 0$.

Performance characteristic curves 4, 5, and 6 illustrate the effect of holding the collector current to the design value but changing I_1 to 0, 2, and 4 mA, respectively.

Note that with this unit the temperature coefficient specifications indicated by the dotted lines are met for some variation in bias condition around the nominal design values. By varying very carefully the operating biases, it is

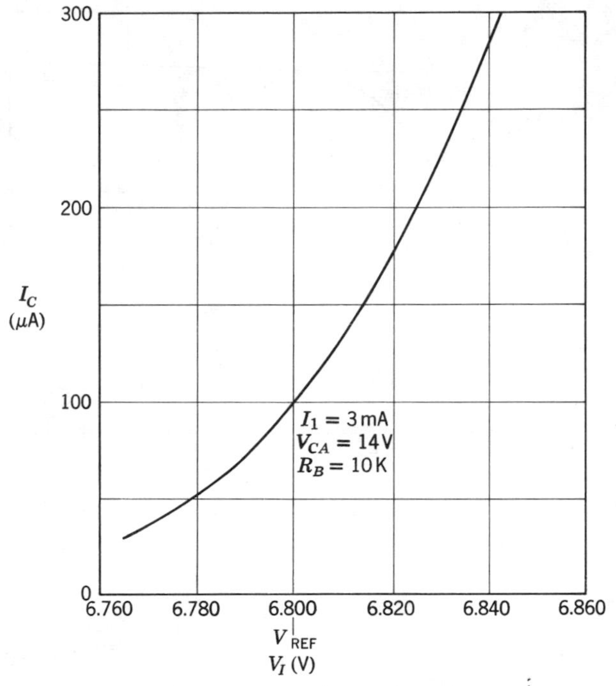

Fig. 4-5 Typical transfer curve for a 6.8-V reference amplifier.

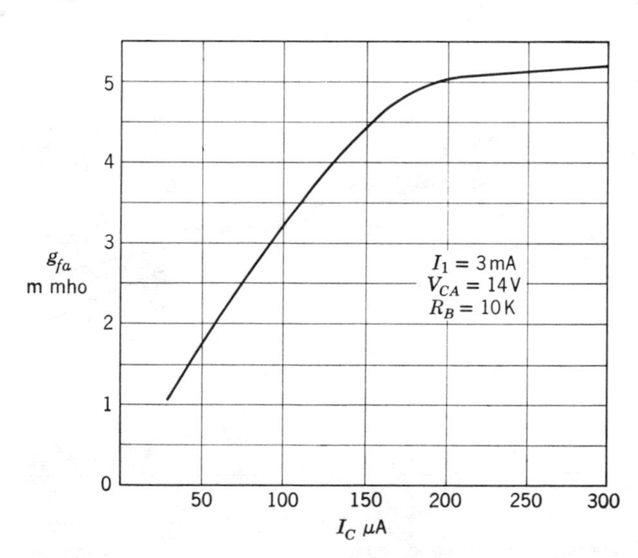

Fig. 4-6 Effect of collector current on circuit transconductance.

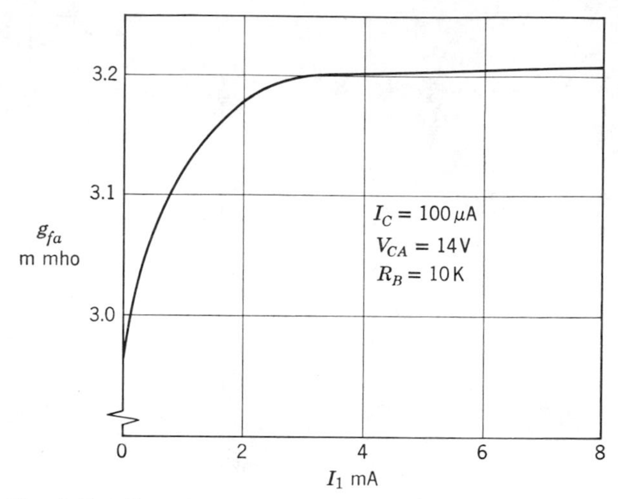

Fig. 4-7 Effect of I_1 bias current on circuit transconductance.

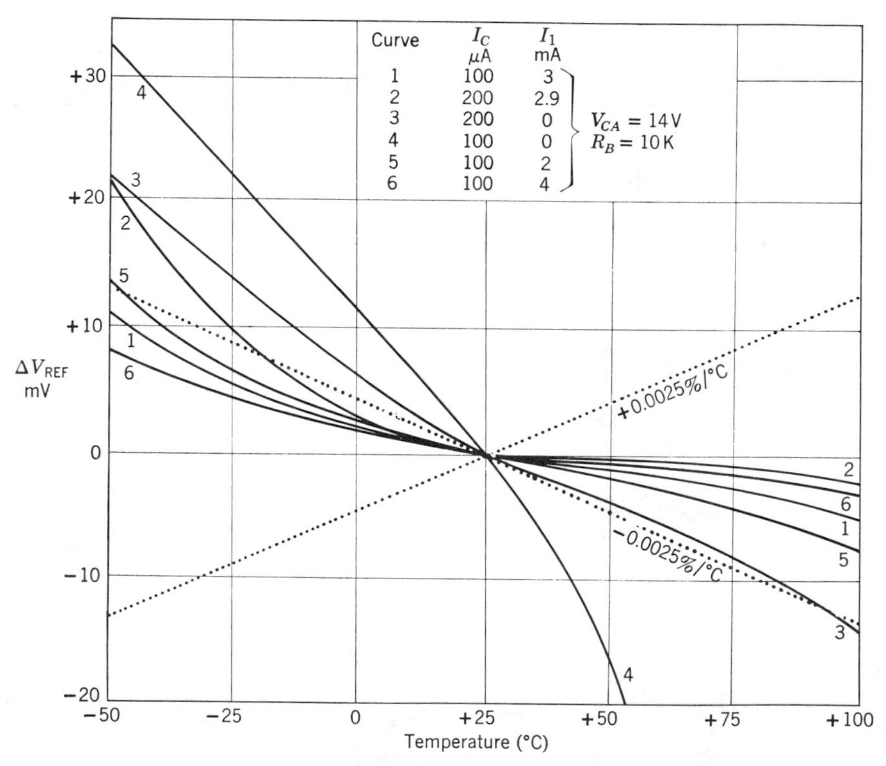

Fig. 4-8 Temperature performance curves for several bias conditions for a Dickson DRAE6.8B25 reference amplifier.

possible to achieve a much tighter temperature coefficient over a restricted temperature range than is specified as a maximum TC for a particular type.

Electrical characteristics for the DRAE 6.8 series are given in Table 4-1.

It is possible to design reference amplifiers that yield optimum performance at various current levels. In addition, the reference control level may also be increased by adding one or more forward-biased diodes in series with the VR diode. This requires a larger voltage VR diode for proper compensation.

Table 4-1. Electrical Characteristics for a Typical Reference Amplifier Family

Symbol	Characteristic	Minimum	Typical	Maximum	Unit	Test Condition
V_{REF}	Reference input voltage	6.5	6.8	7.1	V	$I_C = 100\ \mu A$ $V_{CA} = 14$ V, $I_1 = 3$ mA, $R_S = 10\ K\Omega$
K_T^*	Temperature Coefficient of V_{REF}					
	DRAE 6.8B 10			0.001	%/°C	$I = -25, 0, 25,$ 50, 75, and 100°C
	DRAE 6.8B 25			0.0025	%/°C	
	DRAE 6.8B 50			0.005	%/°C	
	DRAE 6.8B 100			0.010	%/°C	
g_{fa}	Circuit transconductance	2.5	3.0		mmho	$I_C = 100\ \mu A$, $V_{CA} = 14$ V $I_1 = 3$ mA, $R_S = 10\ K\Omega$ $f = 1$ KHz
g_{oa}	Output conductance		1.1		μmho	$I_C = 100\ \mu A$, $V_{CA} = 14$ V $I_1 = 3$ mA, $R_S = 10\ K\Omega$, $f = 1$ KHz
h_{FE}	Dc current gain	100		500		$V_{CE} = 5$ V, $I_C = 100\ \mu A$
I_{CBO}	Collector leakage current		0.2	10	nA	$V_{CS} = 20$ V, $I_E = 0$
I_R	VR diode reverse leakage current			-2	μA	$V_R = -4$ V
$V_{BR(CBO)}$	Collector-base breakdown voltage	30			V	$I_C = 1\ \mu A$, $I_E = 0$
$V_{BR(CEO)}$	Collector-emitter breakdown voltage	30			V	$I_C = 10$ mA, $I_S = 0$

Reference amplifiers may be made with either *NPN* transistors as shown or *PNP* transistors. As will be indicated later, both types may be used to regulate either a positive or negative voltage although one may be preferable over the other.

4-4.4 Applications

Reference amplifiers may be used in many variations of power supply circuits. Several examples are presented here to illustrate possible approaches. For purposes of illustration the Dickson DRAE 6.8 series of units, whose characteristics have been presented, is assumed.

4-4.4.1 Positive Voltage Regulator. Figure 4-9 presents the use of a reference amplifier in a power supply regulator for positive output voltages of fixed value or limited range. The basic circuit is an emitter follower whose input voltage is controlled by the reference amplifier. The input for the reference amplifier is obtained from the output by a resistive voltage divider to close the servo loop.

The 100 μA collector bias current is provided by the constant-current

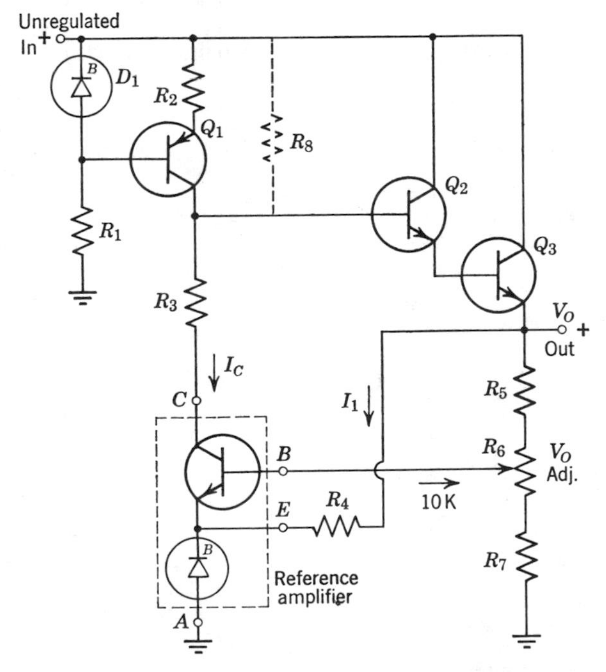

Fig. 4-9 Basic positive voltage regulator circuit using the reference amplifier.

source consisting of D_1, Q_1, R_1, and R_2. For optimum performance the base current required for Q_2 should be much smaller than 100 μA. When the variation in input voltage is rather small and when looser temperature stability is permitted, the collector bias current may be supplied with a single resistor, shown by the dashed lines, in place of the constant-current source.

Resistor R_3 will be necessary only if the desired output voltage exceeds the maximum voltage rating of the transistor used in the reference amplifier. Should this be the case, and since the collector current is made almost constant at 100 μA, then the operating collector voltage of the reference amplifier will be decreased at the rate of 10 V/100 K value of R_3. The overall gain, and hence the stability, will be decreased slightly, but not enough to present a problem in most designs.

Potentiometer R_6 permits variation of the output voltage over a limited range. The degree of permissible variation must be limited somewhat or the value of I_1 will be changed excessively and also the output resistance of the voltage divider will no longer remain close to the desired 10 K.

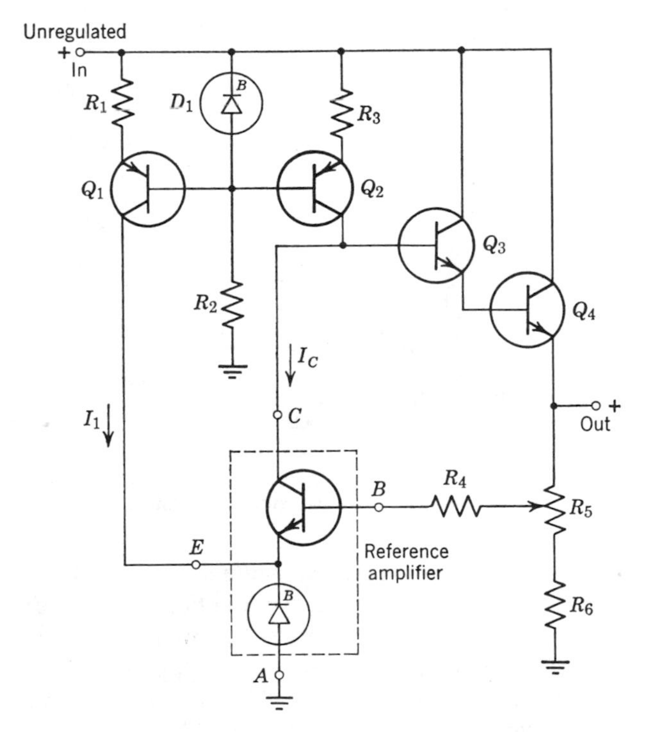

Fig. 4-10 Regulator with wide output voltage range.

4-4.4.2 Regulator with Wide Output Voltage Range. The basic circuit of Fig. 4-9 may be modified as shown in Fig. 4-10 to permit substantial variation of the output voltage from V_{REF} to within about 8 V of the minimum unregulated input voltage (to allow adequate voltage for developing the constant collector current). The changes required are the addition of a separate constant-current source for supplying I_1 and the use of a lower resistance voltage divider with a constant fixed value for R_B to give a smaller variation in driving resistance.

4-4.4.3 Negative Voltage Regulator. Figure 4-11 illustrates an approach in which a reference amplifier using an *NPN* transistor may be used in a

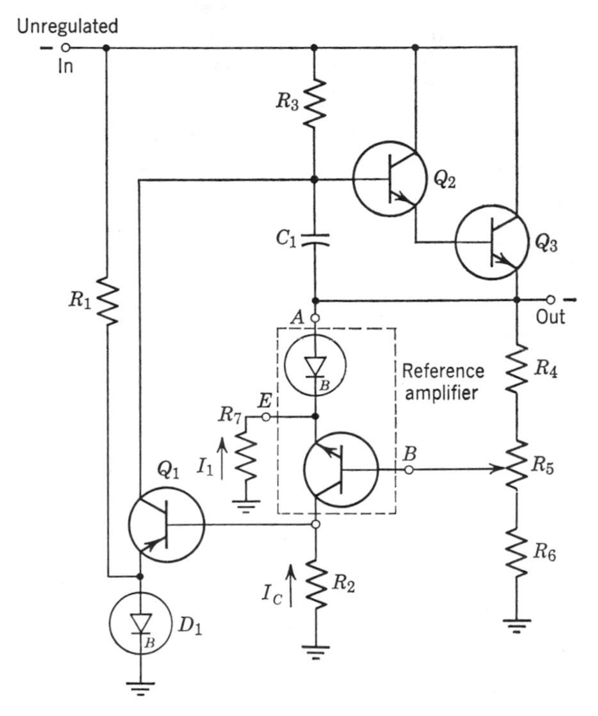

Fig. 4-11 Negative voltage regulator.

regulator designed to supply negative voltages with a common ground connection between input and output. An additional transistor is necessary to couple the error signal from the reference amplifier to the control transistors. The use of VR diode D_1 in the emitter of Q_1 permits easy development of the collector current bias by means of R_2. C_1 is used to improve stability and prevent oscillations.

4–4.4.4 Reference Amplifiers in Vacuum Tube Regulators. The reference amplifier may be used in conjunction with vacuum tube regulators to supply very stable high-voltage outputs. One approach is indicated in Fig. 4–12.

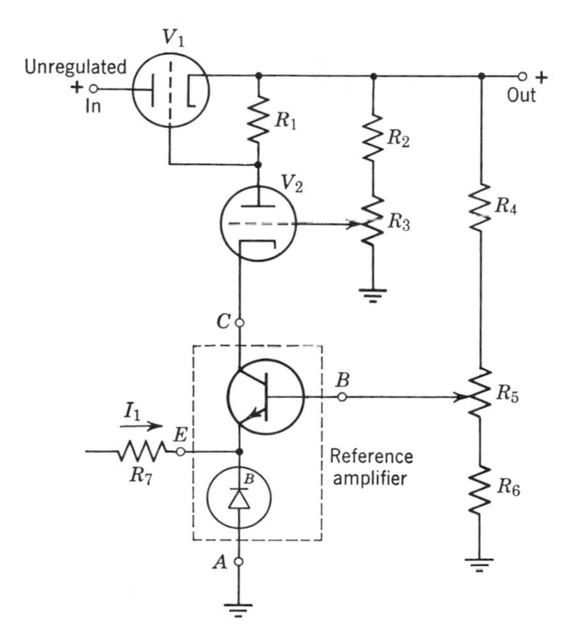

Fig. 4–12 Reference amplifier used with vacuum tube regulator.

V_1 is a simple cathode-follower series-pass tube, and the value of R_1 is chosen to supply a nominal output current when the current through R_1 is 100 μA. When the output voltage tries to increase, the reference amplifier collector current increases, thereby increasing the plate current of V_2. This increases the voltage drop across R_1, thus reducing the current through V_1 and lowering the output voltage.

4–4.4.5 Precision Constant-Current Generator. In all the circuits previously described the input voltage applied to the reference amplifier is derived from the output voltage by a resistive voltage divider. By developing the input voltage across a current-sampling resistor we may use the reference amplifier to control a very stable constant-current generator. One such circuit is given in Fig. 4–13. Transistors Q_1 and Q_2 provide the required I_1 and I_C bias currents of 3 mA and 100 μA, respectively. Transistors Q_3 and Q_4 are connected in a compound emitter follower arrangement to drive the load.

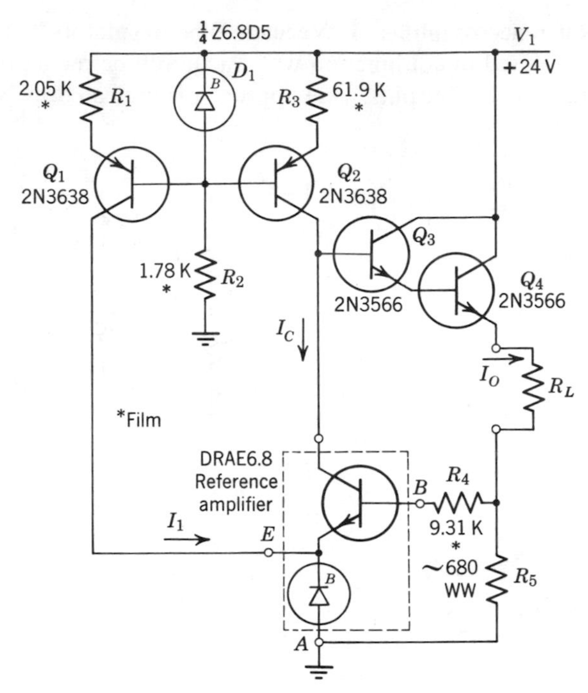

Fig. 4–13 Basic precision current supply using a reference amplifier. Output is 10 mA.

The servo feedback system consisting of the reference amplifier and the compound emitter follower attempts to maintain the voltage across resistor R_5 at a constant level, just as in a voltage regulator circuit. If this is done, the current through R_5 must also be constant and will be determined by the ratio of V_{REF} to R_5. Thus by choosing the proper value for R_5 we may fix the value of the output current.

As long as the output current is much larger than the 0.5 μA base current typically required for the reference amplifier, the current through R_5 may be assumed equal to the load current.

The resistance seen by the base of the reference amplifier is R_4 plus the parallel combination of R_5 and R_L. This total should normally be made equal to the 10 K specified for the reference amplifier.

The degree of control under varying conditions is a function of the servo open-loop gain which, in this case, is merely the voltage gain of the reference amplifier (approximately 850 for the circuit of Fig. 4–13). Three major factors can cause the output current to vary: (1) variation in the input supply voltage; (2) temperature variations; and (3) variations in the load resistance.

The circuit of Fig. 4–13 performs quite well under variations of input supply voltage and temperature and can approach a few parts per million per input volt change or °C temperature shift.

Variation of load resistance means variation in voltage across the load. The output emitter follower stage has slightly less than unity voltage gain and thus the collector voltage of the reference amplifier must vary with any change in load resistance. With an open-loop voltage gain of 850, an output load voltage variation of 5 V must cause an input change of approximately 5.9 mV for a current variation of 0.087%. This means that for $I_O = 10$ mA the equivalent output resistance would be approximately 575 K.

We may obtain some improved characteristics if we place the load in the collector circuit of the darlington output stage as shown in Fig. 4-14. First of all, one end of the load may now be common with the positive supply lead. By using a negative input power supply voltage, we may even have one end of the load at ground potential.

Since the current gain of the Q_3-Q_4 combination may be very high, the composite collector current will be very nearly equal to the emitter current of Q_4 which is controlled as before. The difference will be the very small base

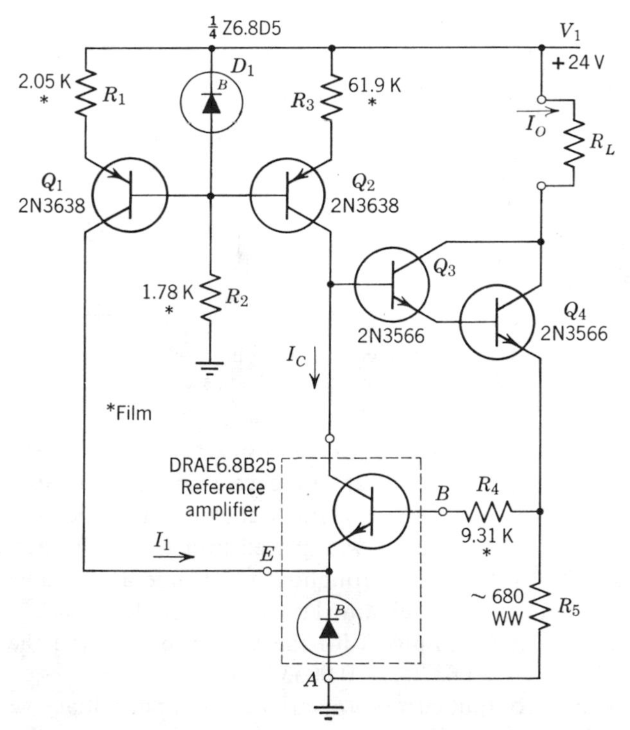

Fig. 4-14 Precision current supply having increased output resistance.

current required for Q_3, and any error incurred because of it will be in direction opposition to the error due to base current in the reference amplifier. For a typical output current of 10 mA, the two errors are nearly equal and thus compensate for each other.

The degree of control achieved with variations in temperature and input voltage will be very nearly the same with the circuit of Fig. 4–14 as it was with that of Fig. 4–13. Figure 4–14 reflects a substantial improvement, however, in the control with variations in load resistance. Note that although the output stage in Fig. 4–14 contributes no voltage gain within the main servo loop (in which it still serves as an emitter follower), it effectively multiplies the output resistance by its common-emitter voltage gain.

Frequently it is desirable to be able to superimpose a small-signal modulation on the output current. This is easily accomplished in either of the current regulator circuits by injecting a small ac voltage in series with the base of the reference amplifier by means of a transformer as shown in Fig. 4–15. The dc control will be as before and the ac modulation current appearing at the output will be V_i/R_5.

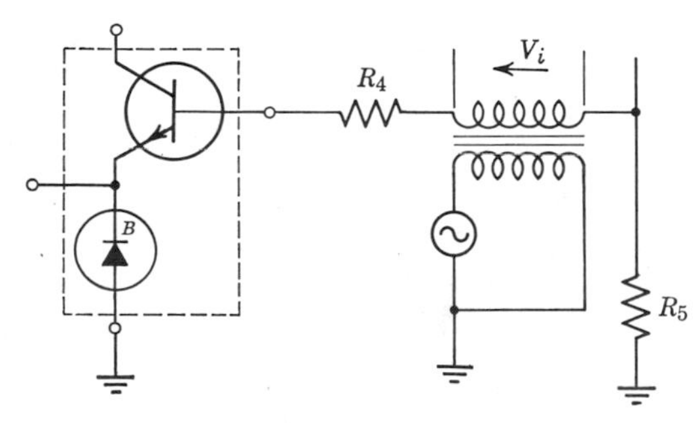

Fig. 4–15 Output current may be modulated with ac signal applied in series with the base.

The circuit of Fig. 4–14 was fabricated using the component values shown including a Dickson DRAE 6.8B-25 reference amplifier. The resistor used for R_5 was a selected wirewound unit having a maximum temperature coefficient of 0.002%/°C. The entire circuit was placed in a temperature chamber and cycled from −25 to +100°C Throughout the temperature range tested the TC of the output current remained positive and less than 0.0034%/°C. Calculated worst-case TC range with the DRAE 6.8-25 and the 0002%/°C sampling resistor is −0.0037 to +0.0053%/°C.

Sensitivity of the output current to variation in input voltage was measured to be about 0.0007%/V. The effective output resistance for the test circuit was greater than 50 M.

A careful analysis of the various factors influencing performance of the circuit of Fig. 4–14 permits us to modify the design slightly and hence achieve better performance. Figure 4–16 gives the improved circuit.

First of all, if we drive the VR diode D_1 used in the I_1 and I_C supplies with a simple current source rather than by resistor R_2 as before, we may achieve improved immunity to input voltage variation. Figure 4–16 illustrates an easy way to add the constant-current source by using the VR diode voltage available at the reference amplifier emitter terminal. Resistor R_2 is necessary to prevent a possible lockup condition which could occur if none of the transistors was conducting.

A portion of the overall TC of the output current is due to the variation in I_1 and I_C bias currents with temperature. Circuit operation is such that, in Fig. 4–14, the TC magnitudes of the base-emitter junctions and VR diode D_1 add. By choosing a lower breakdown voltage for D_1 we obtain a reversal in the sign of its TC. The net result is a temperature compensation of the bias

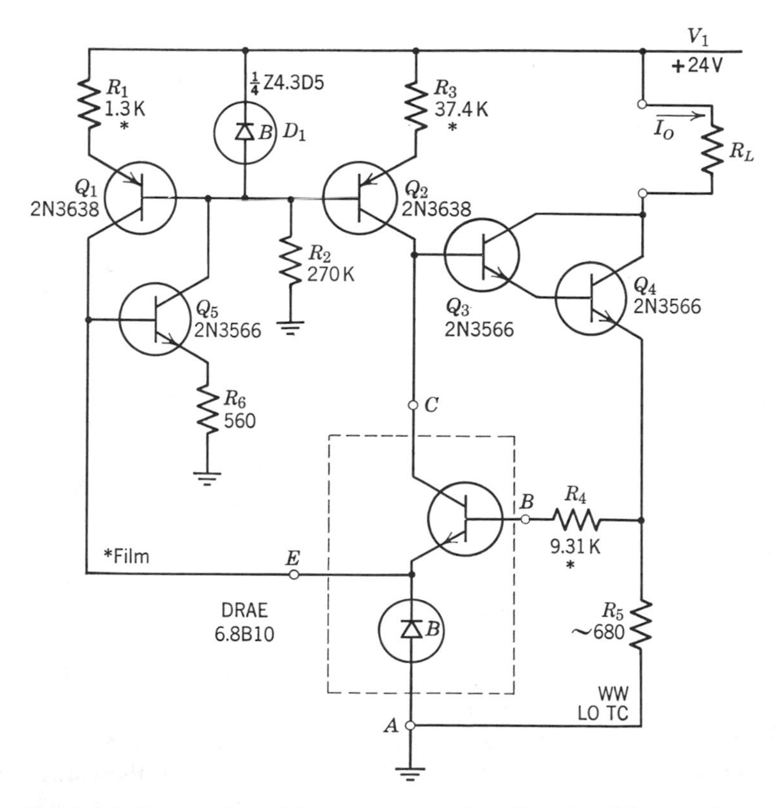

Fig. 4–16 Improved precision current supply with reduced input voltage sensitivity.

current sources and thus we achieve an improved TC for the output current. An alternate scheme would be to add forward-biased compensating diodes in series with D_1.

When the ultimate in temperature stability is necessary, we would use a reference amplifier with a 10 ppm/°C TC and obtain a stable 5 ppm/°C sampling resistor. The net result of Fig. 4–16 is a precision constant-current supply with a worst-case TC of $\pm 0.002\%/°C$ and an input voltage sensitivity of about 0.0002%/V.

4–4.5 Design Hints

To use the reference amplifier in a regulator design we need to insert it in the feedback loop and provide a 3 mA bias current for I_1 and a 100 μA collector bias current. The input voltage for the reference amplifier is normally obtained from a voltage divider whose resistor values must satisfy two requirements. First, the source resistance presented to the reference amplifier must be 10 K and the division ratio must be of the right value to give an open-circuit voltage equal to V_{ref} when the total voltage across the divider is equal to the desired output voltage.

The basic design equations to give the desired resistor values which will meet both requirements are given in Fig. 4–17. Note that resistor R_A is the one grounded for the negative regulator of Fig. 4–11.

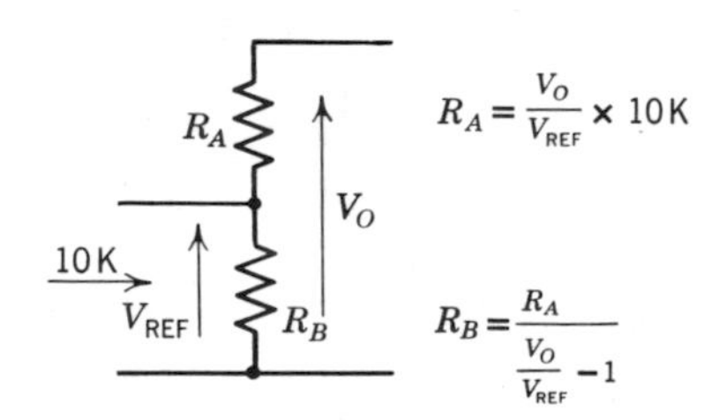

Fig. 4–17 Voltage divider design for use with the reference amplifier.

4–5 VOLTAGE BAND DETECTORS

In another class of voltage monitoring circuits a binary output is desired only when the applied input voltage is within specified limits. Thus any voltage will give one output condition when its value is within the discrete band and another output condition will be produced when the value of the input voltage is either below or above the controlled range.

4–5.1 Two-Limit Circuits

One simple approach to a voltage band detector using VR diodes is shown in Fig. 4–18. With this arrangement independent control of the lower and upper limit values is possible by selecting the proper breakdown voltages of the two VR diodes D_1 and D_2.

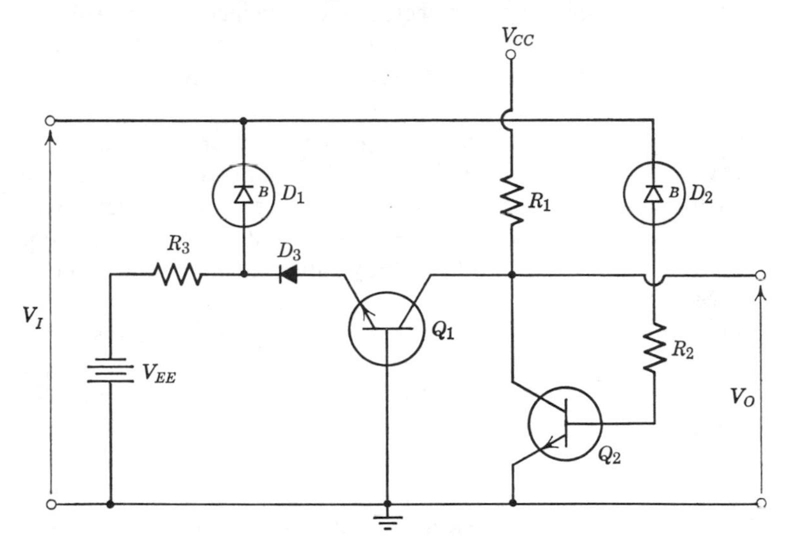

Fig. 4–18 A simple voltage band detector with independent control of lower and upper limit voltages.

For input voltages less than V_{B1} the breakdown voltage of VR diode D_1, transistor Q_1 is turned on to the point of saturation by the bias current provided through R_3, and the output voltage taken from the collector of Q_1 will be slightly negative. When V_I is increased to the point where it exceeds (V_{B1} −0.7), D_1 will conduct providing a positive current to Q_1 to counteract the normal negative emitter current supplied through R_3. This will turn Q_1 off and allow the output to rise to the value of the supply voltage V_{CC}, provided transistor Q_2 is not conducting.

Q_2 will not conduct as long as the input voltage is less than V_{B2}, the breakdown voltage of D_2, since the base current provided through D_2 will be practically zero. As soon as the value of V_I exceeds V_{B2} by an amount equal to the V_{BE2} threshold, however, a base current is provided for Q_2 and it will saturate quickly to reduce the output voltage at its collector to a very low value.

The output voltage drops to zero when either Q_1 or Q_2 is sufficiently turned on. To obtain the desired action in which an output voltage is present over a range in input voltage values, we merely choose the breakdown voltage V_{B1} to be less than the value of V_{B2}; V_{B1} then determines the value of the lower limit voltage that will produce an output voltage and V_{B2} determines the input voltage which will again turn the output off or set the upper limit of the voltage band.

The output resistance of the detector is determined by the value of R_1. This would normally be made as large as possible, yet still drive succeeding circuitry, since the input current levels will be directly influenced by the choice of R_1.

The emitter supply V_{EE} should be large with respect to the input voltage if practical since it is desirable that V_{EE} in combination with R_3 form a constant-current source. The value of this current should be approximately 10% higher than the short-circuit collector current of Q_1 which is equal to V_{CC}/R_1. This insures that Q_1 will be fully saturated until the lower voltage limit is exceeded.

For input voltages above the lower limit value the emitter-base diode of Q_1 will be reverse-biased. If the value of V_I becomes large enough, it is possible that the BV_{EBO} of Q_1 could be exceeded. If this possibility exists, it is necessary to add diode D_3 in series with the emitter as shown by the dotted lines in Fig. 4–18.

As soon as the value of V_I exceeds the upper limit voltage, the input current would rise sharply were it not for the limiting resistor R_2. Although R_2 will affect the value of the threshold limit, it must be made as large as allowable to reduce the required input current beyond the voltage band limit.

Another circuit for a voltage band detector is shown in Fig. 4–19. In this arrangement an additional stage allows the upper limit voltage to occur by turning off the emitter of Q_3 in a similar manner as for Q_1 at the lower threshold voltage. With this approach the input current demanded of the input voltage will be zero for V_I less than the lower limit and slightly larger than V_{EE}/R_3 for values of V_I within the discrete voltage band. When the upper voltage limit is exceeded, the input current must be slightly larger than the sum of V_{EE}/R_3 and V_{EE}/R_4. This assumes that the value of V_{EE} is much larger than any value of V_I expected. Even if this condition is not met, the actual input current is easily calculated. It is much lower than required for the circuit in Fig. 4–18 already discussed.

The choice of component values for the circuit of Fig. 4–19 is similar to that for Fig. 4–18. R_2 should provide enough base current to Q_2 to cause it to saturate fully when Q_3 is turned off. Because of the dc current gain in Q_2, the bias current which must be supplied by R_4 is much smaller than the emitter bias current for Q_1.

If the value of V_I may rise to a voltage large enough to exceed the BV_{EBO}

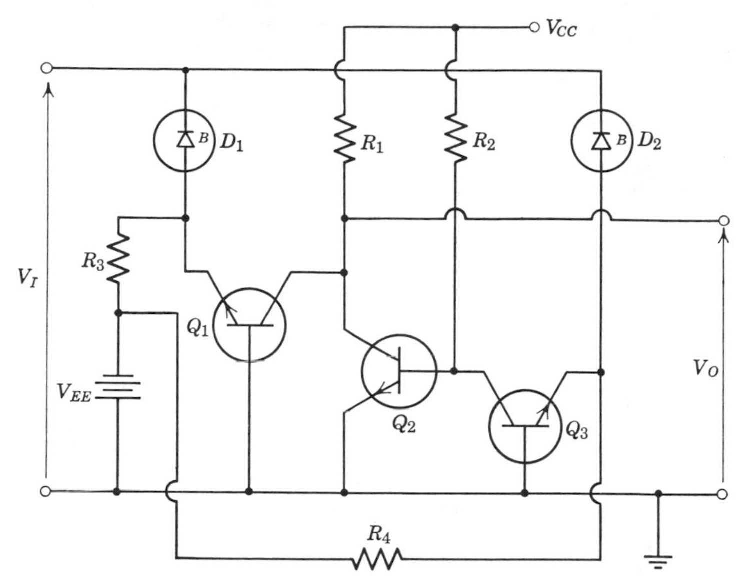

Fig. 4–19 A voltage band detector with increased input resistance beyond the upper limit voltage.

of either or both of the transistors Q_1 and Q_3, it will be necessary to include a diode in series with the emitter or emitters as was done previously.

For either of the voltage band detectors the input-output transfer curve is shown in Fig. 4–20. The limit voltages are only approximations.

Several of the voltage band detector circuits may be connected to a common input voltage to allow classification of a given voltage into two or more bands. If necessary, compensation may be used to ensure temperature stability of the limit voltages.

4–5.2 Voltage Classifiers

It is possible to use VR diodes in conjunction with relays or solid-state switching in order to obtain a classification of the input voltage into one of many discrete band values.

A simple circuit, shown for a four-band classification with three-limit values, is given in Fig. 4–21.

The values of the breakdown voltages of VR diodes D_1, D_2, and D_3 in conjunction with the pickup or actuation voltage required by the relays determine the decision levels. For purposes of discussion assume that VR diodes D_1, D_2, and D_3 have breakdown voltages of 5, 8, and 13 V, respectively, and the relay used has a pickup voltage of 2 V.

As long as V_I is maintained less than 5 V, none of the VR diodes will conduct and none of the relays is actuated. This permits lamp PL_1 to glow

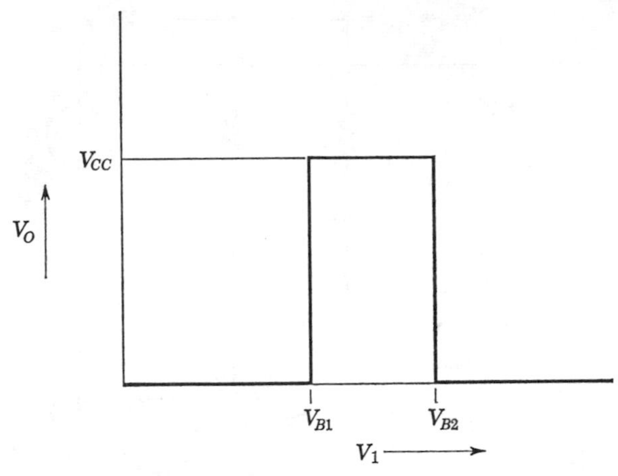

Fig. 4–20 Input-output transfer plot for the voltage band detector circuits.

to indicate that the input is less than 7 V. When V_I is between 5 and 7 V, D_1 begins to conduct with the difference between V_I and the 5 V breakdown voltage of D_1 applied to the coil of relay K_1. (Diode D_4 is assumed to have negligible voltage drop.)

When V_I reaches 7 V, adequate voltage is applied to K_1 to cause it to pick up and switch from PL_1 to PL_2 to indicate that the input voltage is greater than 7 V but less than 10 V.

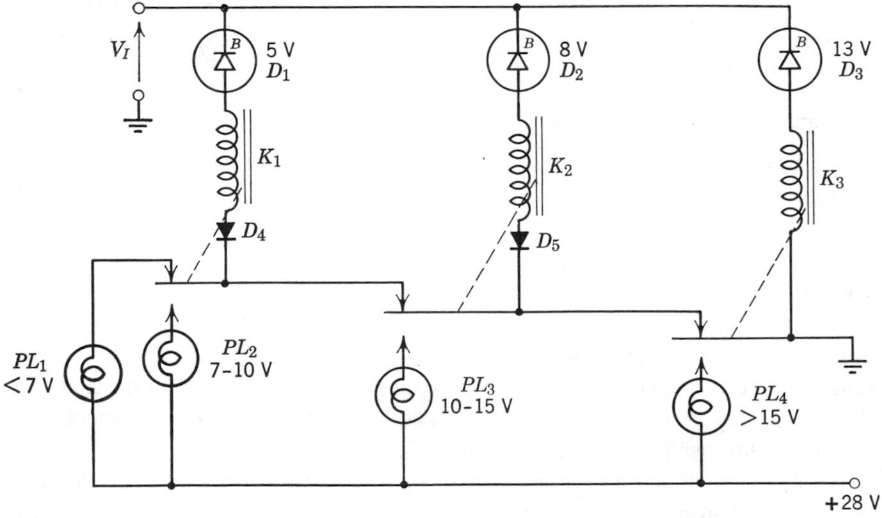

Fig. 4–21 Simple voltage classifier using relay switching.

For V_I values between 7 and 8 V, K_1 will remain closed and diode D_2 conducts negligible current. When the input voltage exceeds 8 V, D_2 begins to conduct and apply power to the second relay K_2. At 10 V the voltage is sufficient to cause K_2 to close and thus turn on lamp PL_3.

When K_2 closes, it also removes the power from K_1 and thus K_1 must return to the neutral state. Diode D_4 prevents current flow from the lamp supply through PL_1 and thence through K_1.

PL_3 continues to glow for V_I between 10 and 15 V. At 15 V, K_3 is actuated and turns on PL_4 while disconnecting power from the other relays.

The end result is a simple voltage level classifier whose limit values are determined by the breakdown voltages of the VR diodes. Although the circuit for a three-limit (or four band) classifier is shown, it is possible to add additional stages. The maximum input voltage should be kept less than the lamp supply.

Some hysteresis will exist if the input is allowed to vary up and down. This is due to the difference in relay pickup and release voltages.

The circuit shown in Fig. 4-22 is a voltage classifier that operates in a manner similar to that just described. The relays have been replaced with transistors. Only two stages are shown.

Transistors Q_2 and Q_4 are normally saturated and thus Q_1 turns on when V_I becomes just slightly greater than the breakdown voltage of D_1. Resistor R_1 serves to limit the input current.

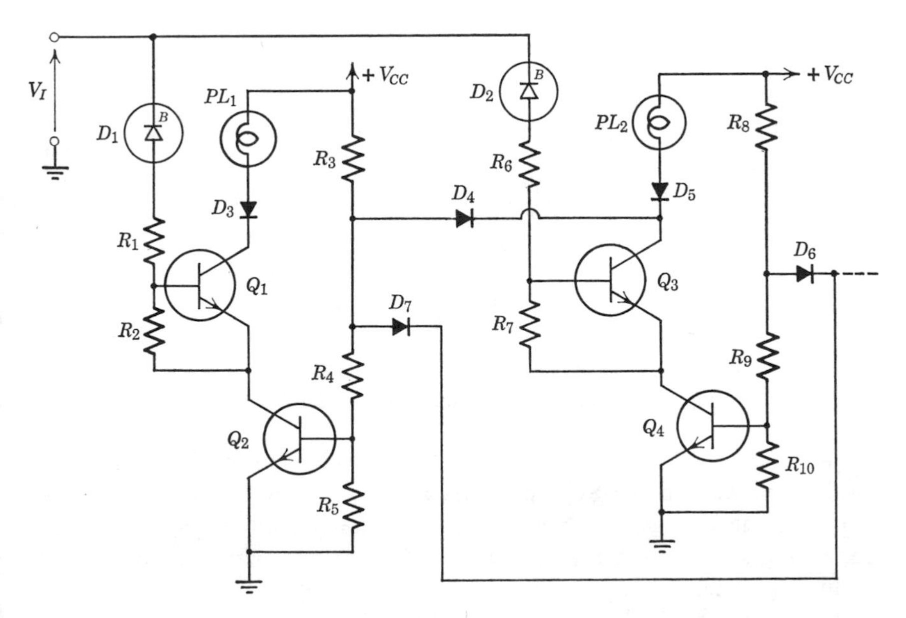

Fig. 4-22 Voltage classifier using solid-state switching.

Now when V_I is increased beyond the breakdown voltage of the second VR diode D_2, Q_3 is turned on. This turns on lamp PL_2 and also clamps the base bias of Q_2. Q_2 is then turned off and this prevents Q_1 from conducting as well.

For additional stages it is only necessary to add additional logic coupling diodes to turn "off" all lower stages at the same time the new lamp is turned on.

Diodes D_3 and D_5 are necessary only if V_I may exceed the lamp supply voltage by more than the breakdown voltage of D_1.

4–6 VOLTAGE TRIGGER AND MEMORY

Another type of voltage monitoring circuit is one in which even a temporary excursion either above or below a given threshold value sets a memory. Information that the monitored voltage level either exceeded or dropped below a desired level is retained until the circuit is intentionally reset.

4–6.1 Overvoltage Monitor

The basic form of an overvoltage trigger and memory is shown in Fig. 4–23. The basic voltage monitoring circuit of Fig. 4–1 is combined with a silicon-controlled rectifier.

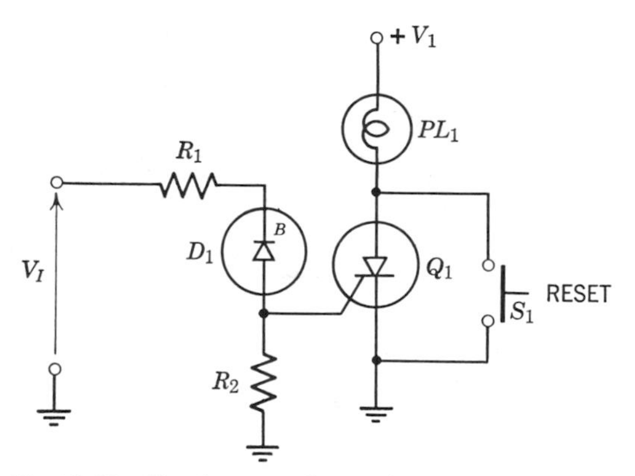

Fig. 4–23 Simple overvoltage trigger and memory.

The breakdown voltage of VR diode D_1 determines the critical overvoltage threshold value and the SCR provides the locking or memory feature. As long as input voltage V_I remains less than the breakdown voltage of D_1, negligible gate current flows, hence SCR Q_1, may remain in an off or non-conducting state.

Should the input voltage exceed the breakdown voltage of D_1, even momentarily, it will cause current flow through D_1 and into the gate of the SCR, thus turning it on. Resistor R_1 is made small enough so that the voltage drop across it is relatively small unless V_I exceeds the threshold level by a large amount.

Once fired the SCR will remain on and PL_1 will glow until power supplying V_I is removed or until the reset switch is pressed. The lamp could be replaced with a bell or relay to turn off equipment, etc.

4–6.2 Undervoltage Monitor

Figure 4–24 shows a circuit that may be used to monitor continuously an input voltage and set a memory if the level drops below a specified level.

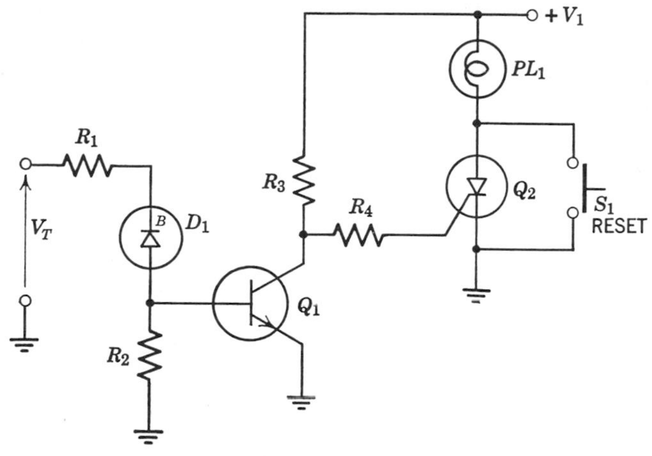

Fig. 4–24 Undervoltage trigger and memory.

Again the threshold voltage level is set by the breakdown voltage of a VR diode and the memory function is performed by the SCR. "Normal" voltage levels, however, cause current to flow through D_1 and maintain transistor Q_1 in a saturated or on state. This shunts current that would otherwise be supplied through R_3 and R_4 to the gate of the SCR; R_1 limits the current to a safe value.

If V_I drops below the breakdown voltage of D_1, current flow ceases, Q_1 is turned off, and gate current is permitted to flow, thus turning the SCR on and lighting PL_1. Now even if the normal voltage level of V_I is resumed the SCR remains on until reset.

4–7 VOLTAGE-BALANCED BRIDGES

In the usual Wheatstone resistance bridge, balance or a null in the output voltage occurs when the four resistance arms bear a definite relationship to each other. If the resistors are linear, the voltage applied to the bridge has little effect on the null condition, but if the resistors vary with applied voltage, there may be only one voltage for which a balance will be obtained. Voltage-balanced bridges may be constructed using any element that has a nonlinear resistance characteristic for one or more of the arms of the bridge. Thus we could use thermistors, forward-biased diodes, or VR diodes. For this discussion we shall be concerned about those arrangements in which a VR diode is used in one or two of the arms.

4-7.1 Single-Arm Bridge

Replacing any one of the four resistors in a Wheatstone bridge with a nonlinear resistance element will yield a circuit which will produce zero output voltage for one and only one nonzero input voltage. For purposes of analysis we shall consider the configuration of Fig. 4-25.

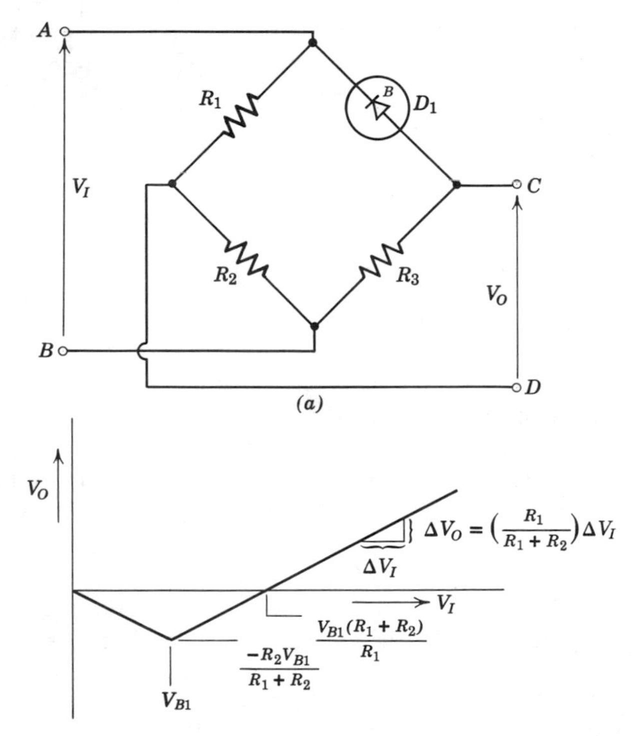

Fig. 4-25 Simple single-arm voltage balanced bridge: (*a*) circuit; (*b*) input-output transfer plot.

4-7.1.1 Open-Circuit Analysis. The bridge will be balanced only when the voltage across R_1 is exactly equal to the voltage across the VR diode D_1. With input terminal A made positive with respect to terminal B (V_I is positive), the VR diode will be reverse-biased. When V_I exceeds the value of V_{B1}, the breakdown voltage of D_1, current begins to flow in the D_1-R_3 arms of the bridge, but the voltage across D_1 will remain at V_{B1}. For simplicity we will assume that the dynamic impedance of D_1 is negligible within the breakdown region.

Writing the equation for the voltage drop across R_1 and setting it equal to V_{B1} allows us to solve for $V_{I(0)}$, the critical input voltage at which the bridge is balanced and the output is zero.

$$V_{I(0)} = \frac{V_{B1}(R_1 + R_2)}{R_1}. \tag{4-3}$$

Positive voltages which are less than V_{B1} will produce an equal voltage across D_1 since only negligible leakage current flows through R_3. The voltage across R_1 will also rise linearly as V_I is increased, but will be attenuated by the R_1-R_2 divider. The net result is an output voltage that rises linearly with increasing V_I and that makes output terminal D positive with respect to terminal C. With the polarity convention assumed, this means that the output voltage is negative. The output voltage in this region is described by (4-4):

$$V_O = -V_I\left(\frac{R_1}{R_1 + R_2}\right). \tag{4-4}$$

When V_I begins to exceed V_{B1}, current through D_1 and R_3 rises sharply, thus biasing the VR diode in its breakdown or constant-voltage region. At this point the voltage across D_1 is substantially larger than that across R_1. When the input voltage is increased further, however, the voltage across R_1 begins to increase toward the now constant voltage across D_1. This produces an output voltage which linearly decreases in magnitude as V_I is increased beyond V_{B1} and the output voltage V is still negative. In this region the output voltage is described by (4-5).

$$V_O = \left(\frac{R_1}{R_1 + R_2}\right)V_I - V_{B1}. \tag{4-5}$$

When V_I is increased still further, the voltage across R_1 will become equal to V_{B1} and the output voltage will become zero. The critical input voltage is given by (4-3).

Input voltages greater than $V_{I(0)}$ will continue to increase the voltage across R_1, but the voltage across D_1 remains at V_{B1} and V_O continues to rise in a linear fashion as described by (4-5) with output terminal C made positive with respect to terminal D.

The overall input-output voltage transfer plot for the open-circuit condition is illustrated in Fig. 4–25*b*. Critical values of input and output voltages are indicated. The slope of the transfer function in the region of balance will be

$$\frac{\Delta V_O}{\Delta V_I} = \frac{R_1}{R_1 + R_2}. \tag{4-6}$$

4–7.1.2 Loaded-Output Analysis. Let us now consider the performance of the circuit when a load resistor R_L is connected across the output terminals C and D.

As long as the input voltage is low enough so that the voltage across D_1 is less than its breakdown value, the equivalent circuit will be that shown by Fig. 4–26*a*. The output voltage may be computed as

$$V_O = \left(\frac{-R_2 R_L}{R_1 R_L + R_1 R_3 + R_2 R_L + R_2 R_3 + R_1 R_2}\right) V_I. \tag{4-7}$$

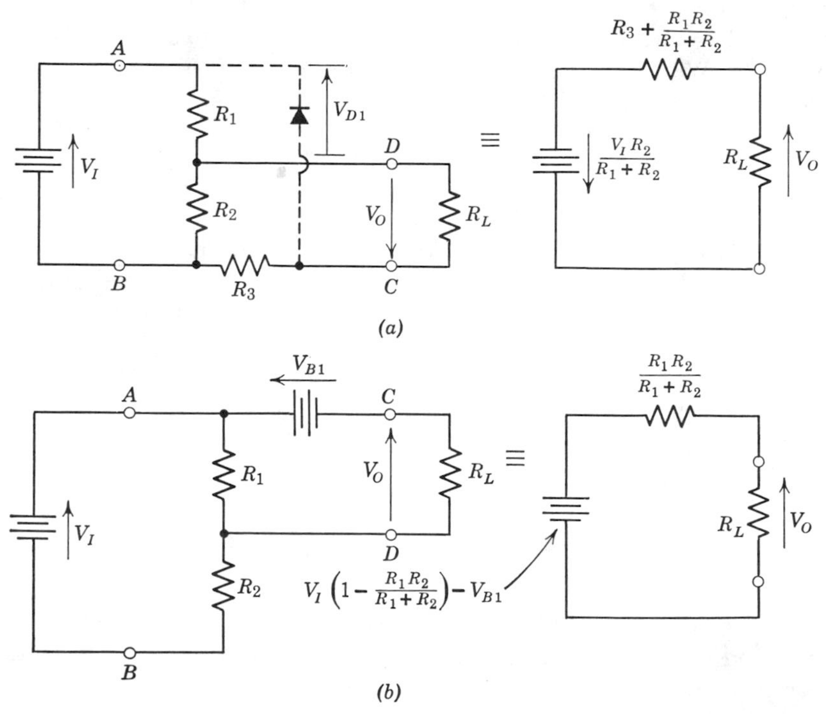

Fig. 4–26 Equivalent circuits for the single-arm bridge with loaded output: (*a*) before diode D_1 reaches breakdown; (*b*) with D_1 in breakdown region.

We may determine the value of input voltage required to cause D_1 to reach the breakdown region by solving for V_{D1} and letting it equal to V_{B1}.

$$V_{I(\text{breakdown})} = V_{B1}\left(1 + \frac{R_2 R_3}{R_1R_2 + R_1R_3 + R_1R_L + R_2 R_L}\right). \tag{4-8}$$

Now, combining (4-8) and (4-7) will yield the value of the output voltage at this critical point.

$$V_{O(\text{breakdown})} = \left[\frac{-R_2 R_L}{R_1R_2 + R_1R_3 + R_1R_L + R_2 R_L}\right] V_I. \tag{4-9}$$

For input voltages greater than the value required to cause breakdown of D_1, the equivalent circuit will be that as shown in Fig. 4–26*b*. The expression for the output voltage under this condition is given by (4-10):

$$V_O = \frac{V_I R_1 - V_{B1}(R_1 + R_2)}{R_1R_2/R_L + R_1 + R_2}. \tag{4-10}$$

The input voltage required to produce zero output voltage will be

$$V_{I(0)} = \frac{V_{B1}(R_1 + R_2)}{R_1}, \tag{4-11}$$

which is the same as given by (4-3) for the unloaded case. This is as it should be since the load has no effect when the output voltage is made zero.

The overall input-output transfer plot for the loaded case is illustrated by Fig. 4–27.

4–7.1.3 Inclusion of r_D. In all the above analyses the dynamic resistance of the VR diode has been assumed to be negligible and, for most cases, this assumption is quite valid. For those cases in which the value of r_D is no longer negligible or for which a more exact relationship is desired, the expression given by (4-12) may be used.

$$V_O = \frac{R_L(R_1R_3 - R_2 r_D)V_I - (R_1 + R_2)R_3 R_L V_{B1}}{R_3[(R_1 + R_2)(R_L + r_D) + R_1R_2] + [R_L(R_1 + R_2) + R_1R_2]r_D}. \tag{4-12}$$

4–7.1.4 Applications for Single-Arm Bridge. A single-arm voltage-balanced bridge of the form shown in Fig. 4–25 may be used in any voltage monitoring and control circuit in which it is desirable to produce an output voltage that is linear with the input voltage over a range in input voltages about a nominal center voltage. A typical application is in the control of the output voltage of a regulated power supply in the manner shown in Fig. 4–28.

The single-arm bridge may also be used as an expanded-scale voltmeter circuit, although the approaches discussed in Chapter 7 will normally yield preferable results.

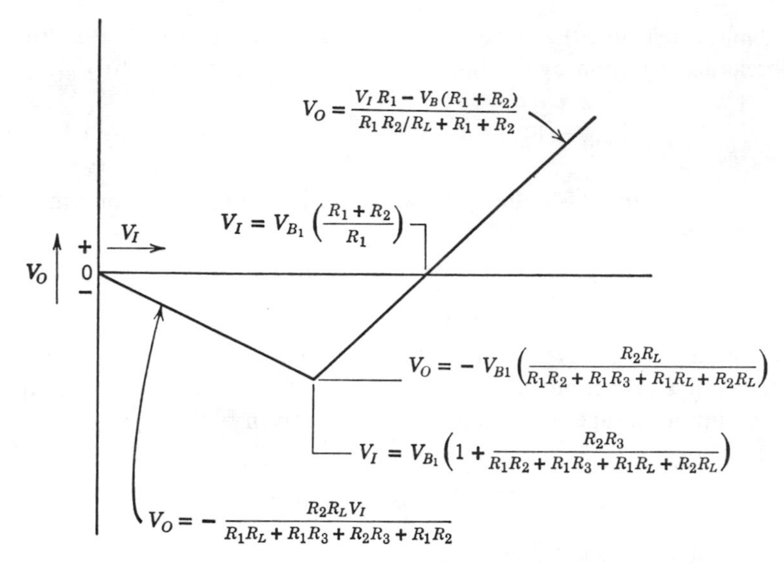

Fig. 4–27 Transfer plot for the single-arm bridge with loaded output.

The value of V_I required to produce a null in the output voltage is a direct function of V_{B1}, as illustrated by the equations given above. Since V_{B1} is, in turn, a function of the temperature of the diode junction, the value of V_I for zero output voltage will likewise be a function of temperature. This may cause problems in certain applications and may demand that we either choose diodes with a breakdown voltage in the neighborhood of 5.5 V, hence a temperature coefficient that is approximately zero, or compensate the VR diode to reduce the temperature effects. If we desire, we may utilize this effect

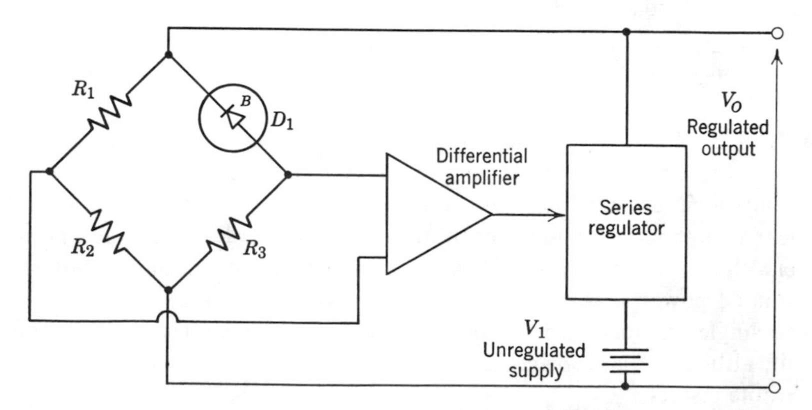

Fig. 4–28 Basic regulated power supply utilizing the voltage-balanced bridge.

to yield a bridge that is balanced by varying the temperature. This application is discussed further in Chapter 10.

4–7.2 Double-Arm Bridge

If we replace two opposite arms in a Wheatstone bridge with VR diodes, we again have a voltage-sensitive circuit. The characteristics will in general be similar to those of the single-arm bridge, but the performance will be improved in several respects.

4–7.2.1 Open-Circuit Analysis. To aid in gaining a clear understanding of the circuit's operation, we shall again study the transfer characteristic with no load applied to the output terminals. Figure 4–29 displays the circuit diagram of the arrangement to be analyzed.

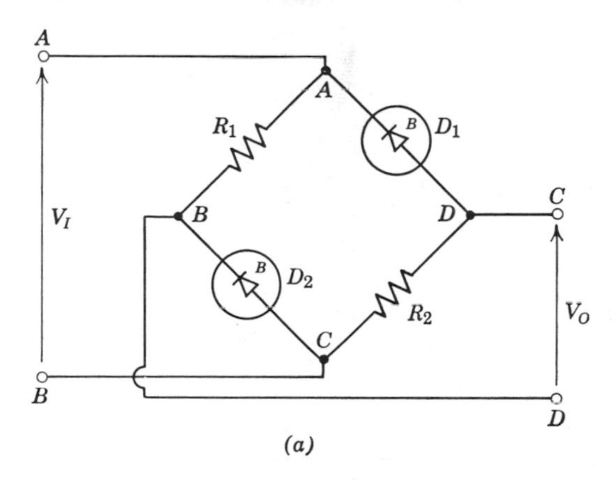

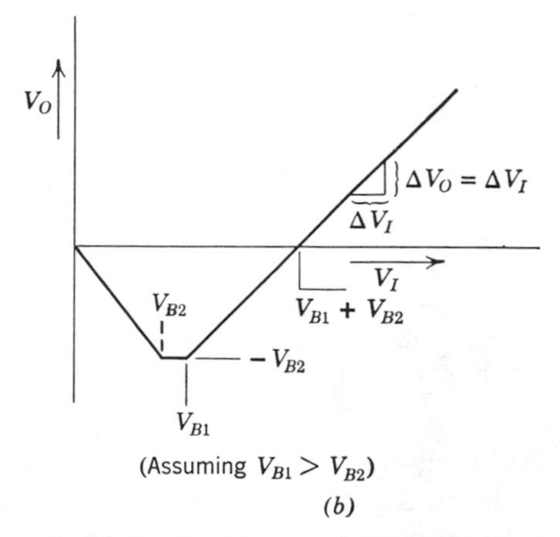

Fig. 4–29 Double-arm voltage-balanced bridge: (*a*) circuit; (*b*) transfer characteristics for open-circuit conditions.

Input voltages less than V_{B1} and V_{B2}, the breakdown voltages of the two VR diodes, will yield a condition where neither diode conducts any appreciable current and the output terminals are connected to the input by means of resistors R_1 and R_2 as shown in the equivalent circuit of Fig. 4–30*a*. For the

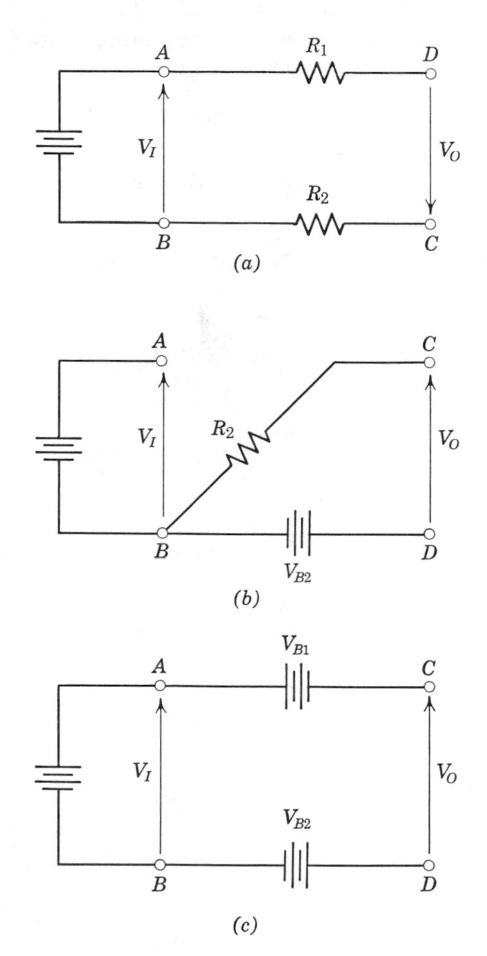

Fig. 4–30 Equivalent circuits for the double-arm bridge: (*a*) when $V_1 < V_{B2} < V_{B1}$, (*b*) $V_{B2} < V_1 < V_{B1}$, and (*c*) $V_1 > V_{B1} > V_{B2}$.

assumed open-circuit condition, then, the output voltage will be equal in magnitude to the input voltage. Because of polarity conventions chosen and indicated in Fig. 4–29*a*, the output voltage will be negative when the input voltage is positive.

As soon as the input voltage begins to exceed either breakdown voltage. that diode will conduct. For purposes of consideration here we assume that V_{B1} is greater than V_{B2}. Thus, as V_I exceeds V_{B2}, VR diode D_2 will conduct and limit that voltage across it. The equivalent circuit under this condition is shown by Fig. 4–30*b*.

The output voltage, then, will be equal to V_{B1} since no appreciable current flows in R_2. Again, because of polarity assumptions, the output polarity will be negative.

The output voltage will remain constant at a value equal to $-V_{B1}$ until V_I exceeds V_{B1}. At this time VR diode D_1 will also sconduct. This yields the equivalent circuit of Fig. 4–30*c*, and the value of V_O is seen to be

$$V_O = V_I - (V_{B1} + V_{B2}). \tag{4-13}$$

The output voltage will become zero when V_I is exactly equal to the sum of V_{B1} and V_{B2}. The transfer characteristic is smooth and linear in this region and has a slope equal to unity. The single-arm bridge discussed earlier must always have a slope less than unity as indicated by (4-6).

One advantage gained in using the double-arm bridge, is therefore a higher effective gain. Another feature will become evident as we study the loaded-output case.

4-7.2.2 Loaded-Output Analysis. With a load resistor, R_L, placed across the output terminals, VR diode D_2 will not conduct as soon as V_I becomes equal to V_{B2} but at a somewhat higher voltage given by

$$V_{I(D_2)} = V_{B2}\left(\frac{R_1 + R_2 + R_L}{R_2 + R_L}\right). \tag{4-14}$$

Below this voltage the equivalent circuit will be the same as indicated in Fig. 4–30*a* and the output voltage will be

$$V_O = -V_I\left(\frac{R_L}{R_1 + R_2 + R_L}\right) \tag{4-15}$$

When V_I exceeds that value demanded by (4-14), D_2 is conducting and the output remains constant at a value equal to

$$V_O = -V_{B2}\left(\frac{R_L}{R_2 + R_L}\right). \tag{4-16}$$

A further increase in the input voltage will cause VR diode D_1 to conduct. The input voltage necessary for D_1 to be biased in the breakdown region is

$$V_{I(D_1)} = V_{B1} + V_{B2}\left(\frac{R_2}{R_2 + R_L}\right). \tag{4-17}$$

Voltages above this value yield an equivalent circuit as shown in Fig. 4–30*c*. Comparing this equivalent circuit with the one for the single-arm bridge as given in Fig. 4–26*b*, we can see that not only is the effective open-circuit gain increased by going to the double-arm bridge but the output resistance has dropped to zero. Actually, the zero output resistance is true only if the dynamic resistances of the VR diodes are zero. Even when r_{D1} and r_{D2} may not be assumed to be negligible the output resistance will be significantly lower for the double-arm bridge. This, then, is a definite advantage when substantial loading is necessary.

The output voltage for values of V_I greater than that sufficient to cause conduction in D_1 will be given by (4-18).

$$V_O = V_I - (V_{B1} + V_{B2}), \tag{4-18}$$

which is the same as for the open-circuit condition.

The input voltage for which a zero output voltage will be obtained is given by (4-19) to be the sum of the two breakdown voltages.

$$V_{I(O)} = V_{B1} + V_{B2}. \tag{4-19}$$

Figure 4–31 illustrates the transfer characteristic for the double-arm bridge under loaded-output conditions.

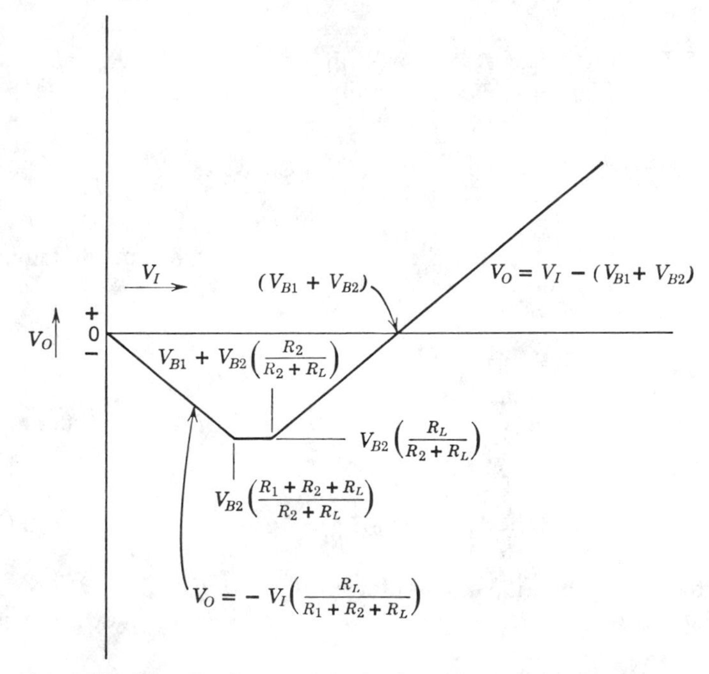

Fig. 4–31 Transfer characteristics for double-arm bridge under loaded-output conditions.

4–7.2.3 Applications. The double-arm voltage-balanced bridge may be used in the same way as the single-arm bridge, but finds its major value in those applications in which the output terminals have considerable loading or the loading varies greatly.

REFERENCES

[1] E. R. DeLoach: Transistor and Zener Monitor Calibration, *Electronics*, **41**, 102 (June 24, 1968).
[2] R. L. Ives: Reducing Relay Pull-In Drop-Out Gap, *Electronics*, **33**, 62 (January 22, 1960).
[3] J. J. Kolarcik: Versatile Zener Diode Array Forms High-Speed Quantitizer, *Electronics*, **35**, 52–54 (August 17, 1962).
[4] R. Langfelder: Design of Signal and Control Static Relays, *Static Relays for Electronic Circuits*, Engineering Publishers, Elizabeth, New Jersey, 1961, pp. 4–52.
[5] J. Monroe and M. Gindoff: Zener Stabilized Bridges, *Instr. Control Systems*, **35**, 94–95 (January 1962).
[6] F. Nibler: Zener Diodes as Coupling Elements in Relay Circuits, *Rev. Sci. Instr.*, **32**, 1143 (October 1961).
[7] E. H. Ogle: Sensing and indicating Voltage Levels, *Electron. Equip. Eng.*, **10**, 51–53 (March 1962).
[8] P. C. Tandy: Bridge Circuit for Measuring Differences of Current or Voltage from a Preselected Value, *IEEE Trans. Instr. Meas.*, **1–12**, 34–40 (June 1963).
[9] C. D. Todd: Hybrid Current Regulator Covers Wide Range, *Electron. Design*, **11**, 48–51 (June 21, 1963).

Chapter 5

PROTECTION CIRCUITS

5–1 INTRODUCTION

Many of the circuits used to perform various functions are sensitive to transient voltage spikes which may appear on the power supply lines feeding the circuit. In some cases the effect is only temporary interruption of the performance of the circuit; in others permanent damage may occur in some of the more sensitive semiconductor devices used. To assure overall circuit reliability it is often required, then, to protect the circuit from these disturbing transients.

As an example of the extent of possible power supply transients, consider the curve shown in Fig. 5–1. This is taken from MIL-STD-704 covering aircraft power supplies and illustrates that substantial protection might be necessary in some applications.

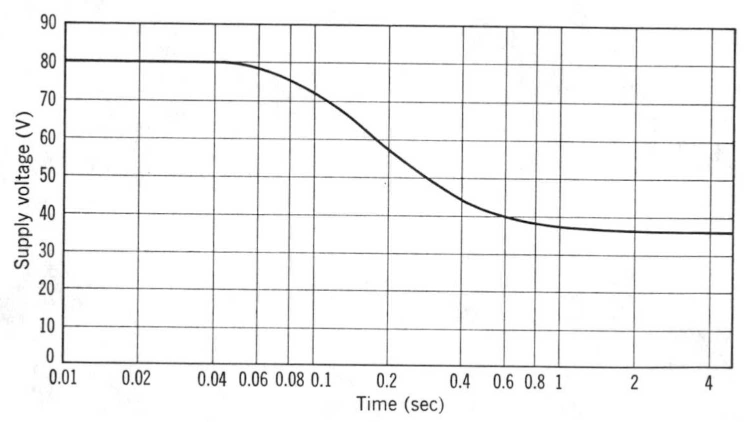

Fig. 5–1 Worst-case transient dc voltage step function loci limits as specified in MIL-STD-704.

The approach necessary depends on the severity and type of transient signal on the power supply line and on the sensitivity of the circuit to be protected. In some instances it becomes necessary to utilize a small regulated power supply just ahead of the sensitive circuit, whereas in others a very simple shunt regulator circuit may be ample. Several approaches are discussed in this chapter.

Some circuits are quite sensitive to severe transients or even relatively minor increases in the supply voltages that could occur as components age or begin to fail. Fuses or circuit breakers could be used, but may not react before the circuit is damaged unless aided by additional protection circuitry. Typical response time for a "fast action" instrument fuse is shown in Fig. 5–2. The guaranteed specifications indicate that a 200% rated current may flow for as much as 5 seconds before the fuse blows. This may be more than enough time to destroy the circuit completely.

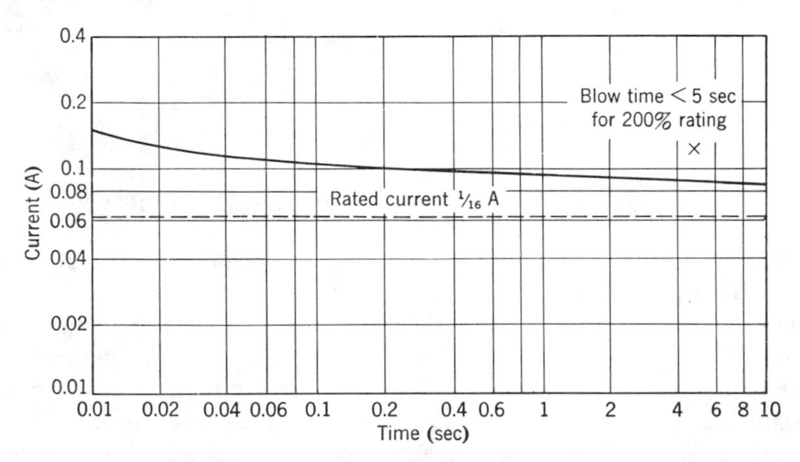

Fig. 5–2 Typical blow-time characteristics for a fast action instrument fuse.

Very simple protection circuits utilizing VR diodes may be added to provide protection to the circuit for short, low-energy transients and aid in blowing the fuse or opening the circuit breaker if the conditions become too severe.

5–2 BRUTE-FORCE PROTECTION

5–2.1 Single VR Diode Protection

The simplest possible form of protection is illustrated by Fig. 5–3. A single VR diode is placed directly in shunt with the circuit to be protected. The breakdown voltage of D_1 is chosen to be slightly greater than the normal operating power supply voltage.

Should the voltage applied to the circuit increase suddenly because of a transient in the power supply or induced into the power supply leads or slowly because of a gradual component failure, the VR diode becomes conductive and the voltage increase is dropped across R_1 which may be the supply output resistance, the resistance of the lines, a small resistor added to the circuit, or a fuse. Voltage drop across R_1 is kept negligible for the normal operating condition. The VR diode draws only negligible leakage current until the protection is needed.

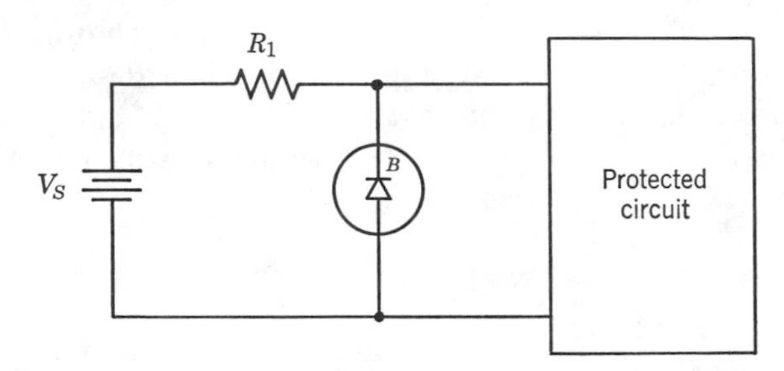

Fig. 5–3 Simple shunt protection.

This type of circuit is quite effective in removing transient spikes either of a single shot or recurring nature. Of course, the power rating of the VR diode must be adequate to prevent its being damaged. Since no power is dissipated in D_1 except when the transient occurs, a low power VR diode will give momentary protection against fairly substantial surges if the total energy dissipated does not increase the junction temperature above its maximum value. Surge current ratings for VR diodes were discussed in Chapter 2. Design techniques for shunt regulators were covered in Chapter 3 and are directly applicable to the brute-force protection circuit.

One advantage of the simple shunt protection scheme is that it provides protection to repeated transient disturbances without interruption of circuit performance. Only if the problem is severe enough to blow a protective fuse or circuit breaker will circuit operation cease.

When a severe supply overvoltage occurs, the brute-force protection circuit may be used to limit the voltage applied to the circuit, and at the same time greatly reduce the time required to blow the fuse; for example, suppose we have a circuit that draws 50 mA nominal current at a supply voltage of 20 V. If the supply voltage increases to 35 V because of a power supply failure, the current which the circuit would draw would be about 80 mA (as indicated by the dotted line of Fig. 5–4) except the circuit could be permanently damaged by the overvoltage.

By using only a fuse having the characteristics shown in Fig. 5–2, we would

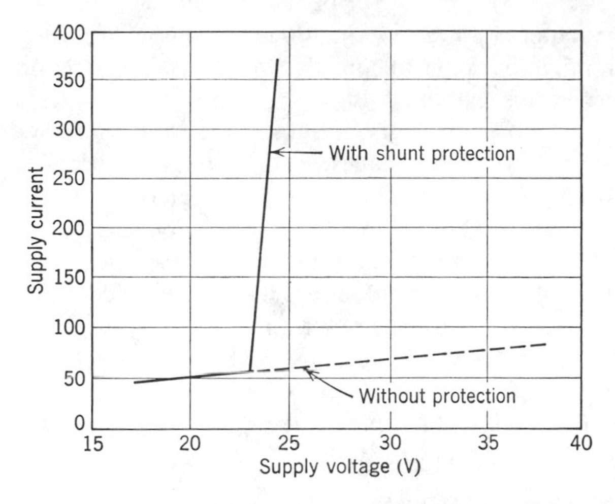

Fig. 5–4 Use of brute-force protect-on causes sharp increase in supply current when chosen voltage level is exceeded.

not blow the fuse even after 10 sec (100 sec actual value). On the other hand, if we use a 23-V VR diode in shunt with the circuit shown in Fig. 5–5, the power supply failure will cause a sudden increase of current through VR diode D_1 and the fuse of approximately 1.3 A. (Even a smaller increase in supply voltage will yield a substantial increase in supply current as indicated by the solid line in Fig. 5–4.) A current of 1 A or greater is sufficient to rupture the fuse illustrated in Fig. 5–2 in less than 0.2 msec. A $\frac{1}{2}$-W VR diode will typically withstand up to 100 W of instantaneous power for this interval of time and thus would be more than adequate to work in this application.

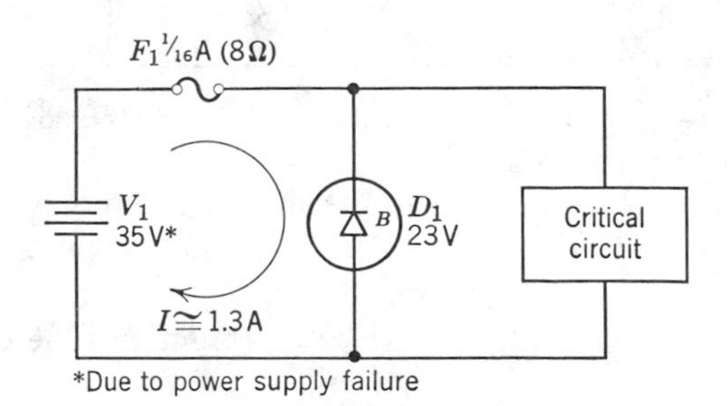

Fig. 5–5 Brute-force VR diode speeds blowing of fuse on power supply failure.

Some additional resistance in series with the fuse may be desirable to limit the maximum current flow. If the supply overvoltage is likely to be less drastic in its failure mode, a higher power dissipation VR diode may be necessary to withstand the required power for the longer time needed to rupture the fuse. In this case it might be better to consider the crowbar protection circuit to be described later in this chapter.

A thermal circuit breaker might be substituted for the fuse in order to yield protection from large overvoltages, yet have automatic reset when the element cools and if the overvoltage has disappeared.

Although the circuit arrangement discussed above is not biased in the manner for the simple shunt regulator circuit discussed in Chapter 3, in which the VR diode remains conducting, it becomes evident that if we design a regulator circuit, it will also remove transient voltage peaks. The VR diode must have sufficient power-handling capability to absorb the transient without damage.

The speed of the simple shunt protection circuit is very fast so that even the most sensitive circuit will not be damaged before it acts.

5-2.2 Shunt Protection with Transistor Amplifier

As discussed in Chapter 3, a transistor may be used in conjunction with a VR diode to yield a composite device having roughly the breakdown voltage of the VR diode but with the power dissipation capability of the transistor plus that of the VR diode. This combination may be used in the shunt protection circuit as shown in Fig. 5-6.

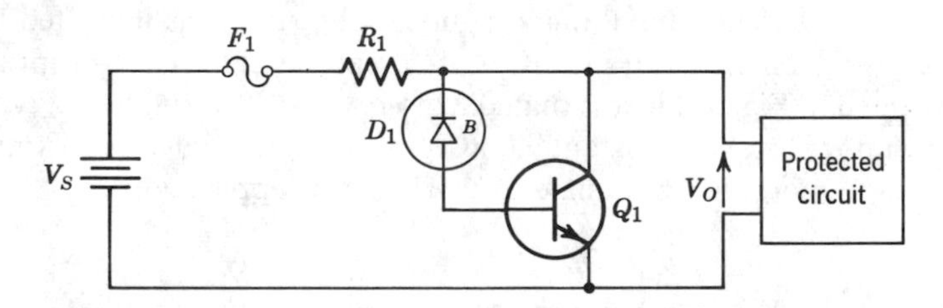

Fig. 5-6 Shunt protection circuit with transistor amplifier.

As before no current flows through the shunt path until a voltage transient greater than the breakdown voltage of D_1 appears at the input. At this time D_1 begins to conduct current that is applied directly to the base of the transistor. A current equal to the product of the base current and the dc current gain, h_{FE}, of transistor Q_1 then flows in the collector circuit. The major portion of the power is now dissipated in the transistor, but otherwise the performance of the circuit is practically identical to the previous one.

The speed of the composite shunt protector will be dependent on the turn-on time of Q_1. It may be several orders of magnitude slower than the simple circuit but will give protection within microseconds or less, depending on the particular type of transistor used.

A fuse has again been added in series with the supply lead to give interruption of the circuit power if the transient energy is too great or if the voltage of the power supply exceeds the critical value for a sustained amount of time. The VR diode-transistor protection circuit operates almost immediately and, if the trouble persists, provides an additional current through the fuse in order to cause it to blow much more rapidly than it would for a simple fused circuit.

5-3 EMITTER FOLLOWER PROTECTION

Another relatively simple form of protcction technique utilizes a transistor emitter follower in conjunction with a VR diode in the manner shown in Fig. 5-7. Here again the basic form is that of a voltage regulation circuit.

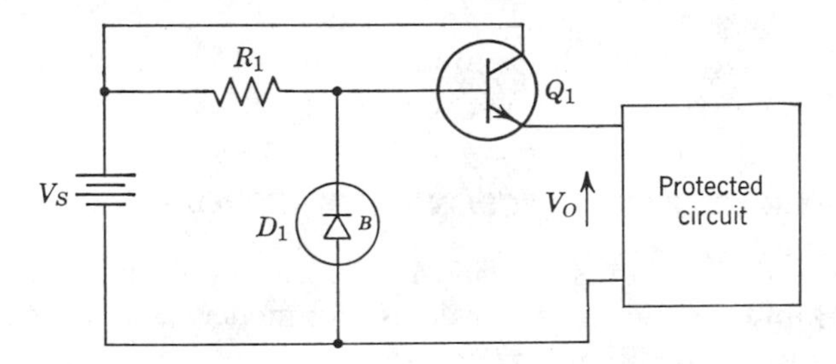

Fig. 5-7 Emitter follower type of protection circuit absorbs transient across Q_1.

Under normal operating conditions with proper supply voltages applied, VR diode D_1 may conduct only negligible leakage current. The breakdown voltage of D_1 may be chosen to be above the acceptable input voltage level but below that which would cause damage to the protected circuit.

Transistor Q_1 acts as a dc emitter follower yielding an output voltage nearly equal to the supply voltage. Most of the current drawn by the protected circuit flows through the collector-emitter path and only a very small portion flows through R_1 and the base of Q_1.

When the supply voltage rises above an acceptable level (as determined by the breakdown voltage of D_1), the VR diode begins to conduct and excess voltage rise will be dropped across R_1. Since the VR diode holds the base voltage relatively constant, the emitter output voltage must also remain practically fixed.

The result is a series voltage regulator circuit which functions only when the supply voltage exceeds a designed threshold level. Excess voltage rises will be dropped across R_1 and from collector to emitter of Q_1.

This form of circuit has several operating advantages. Protected circuit operation is allowed to continue even under conditions in which severe overvoltage spikes may exist on the supply line. Whereas in the brute-force approaches the incoming transient was shunted to ground and produced a momentary increase in supply current, the emitter follower protection circuit absorbs the transient by allowing the collector voltage of Q_1 to increase.

Resistor R_1 serves to limit the maximum current through D_1 under abnormal conditions. Since very little current flows through under normal operating conditions, it may be much larger than was permissible in the brute-force circuits.

Transistor Q_1 normally operates on the edge of saturation and hence must dissipate very little power. As input transients occur, Q_1 will dissipate power when the collector voltage rises, whereas the current through it remains the normal value required by the protected circuit.

As soon as the supply overvoltage condition disappears, the collector-to-emitter voltage of Q_1 again drops to a few tenths of a volt and current flow through D_1 ceases. No reset or fuse replacement is necessary.

5–4 VOLTAGE-TRIGGERED "CROWBAR" APPROACH

The proection techniques previously described are primarily for momentary and possibly repetitive transients of a moderate to low energy level. When the transient condition may occasionally be of a morc severe nature, it may be advisable to rely on an instantaneous power removal by means of a fuse or circuit aided by a protection circuit.

In the brute-force approach it was indicated how the shunting effect of the protection circuit would speed up the rupture of a series protective fuse. The "crowbar" approach, as its name implies, is a drastic approach which only acts when a severe problem exists. When it does respond, however, it acts hard and fast. It offers protection of the circuit, but will interrupt operation.

The basic crowbar approach is shown in Fig. 5–8. If the supply voltage exceeds the critical threshold value (as determined primarily by the breakdown voltage of VR diode D_1), current begins to flow through D_1, R_1, and into the gate of SCR Q_1.

As soon as the gate current is adequate to fire Q_1, a very large anode current flows, which serves to rupture the fuse very quickly. In the meantime the voltage across the critical circuit is being held to a low value and is therefore protected from damage.

The value of capacitor C_1 shunting the gate of Q_1 will determine the relative speed with which the circuit will respond. If C_1 is zero or a rather small value, then even very short overvoltage conditions may trigger the SCR and actuate the crowbar.

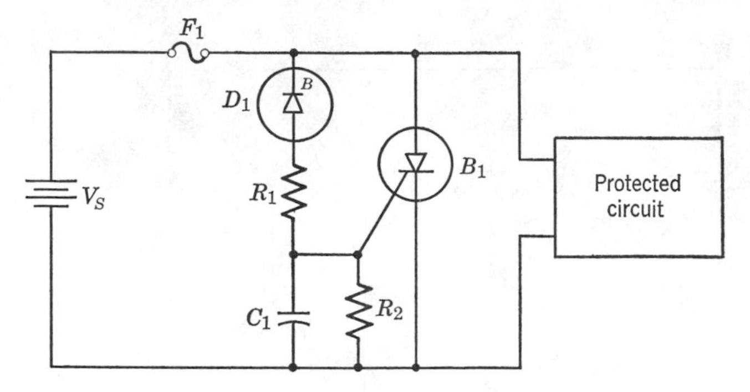

Fig. 5–8 Voltage-triggered "Crowbar" insures very rapid blowing of fuse on severe overvoltage.

A large value for C_1 will delay the firing of the crowbar by momentarily shunting the current that would normally be applied to the gate of the SCR. This will permit short, low-power transients to occur without firing the crowbar and disrupting the circuit function.

A thermal circuit breaker substituted for the fuse will offer the same degree of protection and will reset automatically. Care in design must be exercised to prevent premature firing of the SCR due to the sudden application of anode voltage (dV/dt problems are covered in SCR handbooks).

5–5 ARC SUPPRESSION

One of the most common means of suppressing a voltage arc that would occur on breaking a current flow in an inductive circuit is by a simple diode as shown in Fig. 5–9. The voltage drop across the switch contacts is held very effectively to a value only slightly greater than the supply voltage.

The main disadvantage of the arrangement of Fig. 5–9 is that the time required for the relay to release, when S_1 is opened, is substantially greater than would be required if the arc suppression diode were not used. As indicated in Fig. 5–9*b*, diode D_1 is forward-biased immediately on opening of S_1. (This is due to the reversal of the polarity of the coil voltage on attempted current reduction in accordance with Lenz's law.) For all practical purposes, then, D_1 looks like a short circuit to the relay coil and the current I_1 will decrease rather slowly according to the L/R time constant of the relay winding.

In order for the relay to release, the magnetic field in the armature, hence the current in the coil winding, must be reduced. With a short placed across the coil terminals, the reduction in current can be rather slow as occurs in a relay designed for intentional slow release.

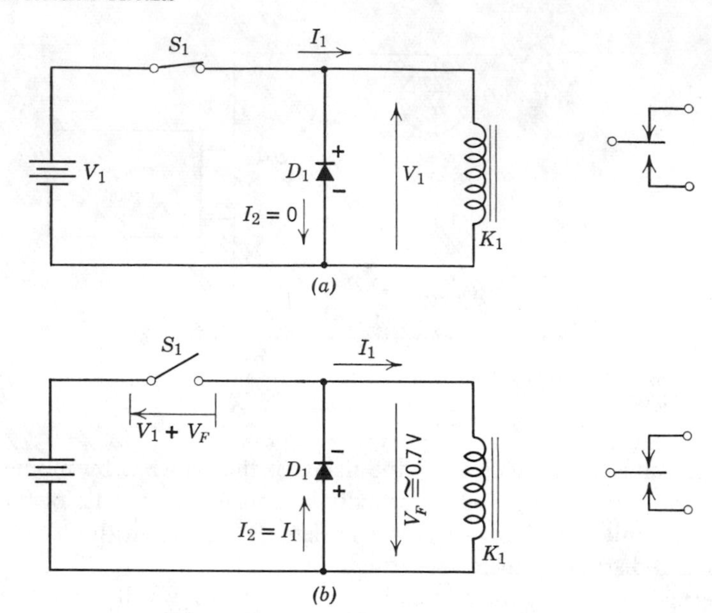

Fig. 5–9 Arc suppression by means of normal diode will slow up release response: (*a*) conditions with S_1 closed; (*b*) condition immediately after opening S_1.

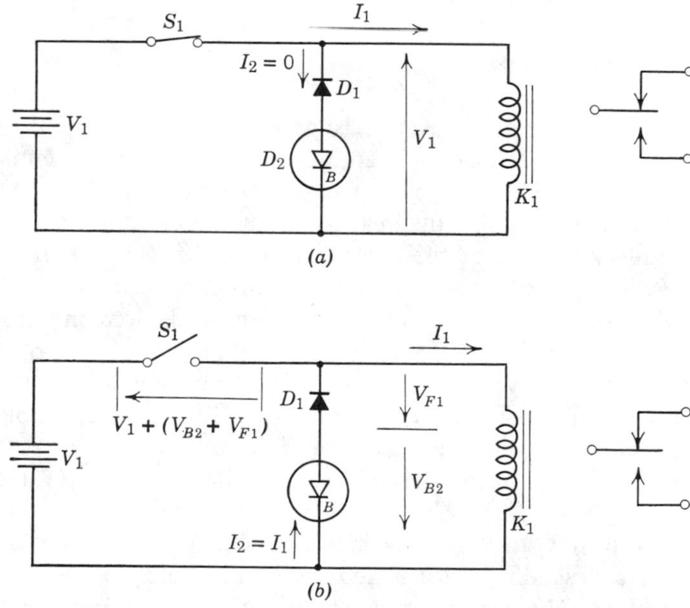

Fig. 5–10 Arc suppression by means of a VR diode: (*a*) conditions when S_1 is closed; (*b*) conditions immediately after opening S_1.

One approach for obtaining necessary arc suppression, yet speeding up the relay release response, utilizes a VR diode as shown in Fig. 5–10. The VR diode D_2 is placed in series with diode D_1.

With S_1 closed no current flows through either D_1 or D_2 because D_1 is reverse-biased. As soon as S_1 is opened, however, the voltage across the coil reverses as before and we have the condition shown in Fig. 5–10b. Diode D_1 is forward-biased and VR diode D_2 is reverse-biased and will operate in the breakdown region.

The voltage across the coil will immediately rise to a value equal to the sum of the forward drop of D_1 and the breakdown voltage of D_2. This permits the current in K_1 to decrease much faster and hence speeds up the release time.

The voltage across the switch contacts is held to a value equal to the sum of the supply voltage, the forward voltage drop of D_1, and the breakdown voltage of D_2. By proper choice of the breakdown voltage of the VR diode used, we may still hold the maximum voltage across S_1 to a safe value.

The VR diode may not be used alone without D_1 because it would be forward-biased under normal operating conditions and short out the coil winding. It will be necessary to use either a regular diode as shown in Fig. 5–10 or a VR diode of the double-anode type with a breakdown voltage greater than V_1.

REFERENCES

[1] D. M. Baugher and L. H. Gibbons, Jr.: Energy Calculations for Devices which Switch Inductive Loads, *EEE-Circuit Des. Eng.*, **16**, 88–93 (January 1968).

[2] L. J. Brocato: Zener Diode Protects Backward Wave Oscillator, *Electronics*, **38**, 65 (November 29, 1965).

[3] J. K. Buchanan et al.: *Zener Diode Handbook*, Motorola, Inc., Phoenix, Arizona, 1967.

[4] O. Burlak: Better Ways to Protect Transistors with Zener Diodes and Fuses, *Electronics*, **35**, 64–65 (September 28, 1962).

[5] F. H. Chase, B. H. Hamilton, and D. H. Smith: Transistors and Junction Diodes in Telephone Power Plants, *Bell System Tech. J.*, **33**, 827–858 (July 1954).

[6] B. B. Daien: Protect Transistors Against Destructive Transients, *Electron. Design*, **7**, 68–69 (November 25, 1959).

[7] D. J. Donohoo: Minimizing Inductive Kick and Fall Time, *Electron Equip. Eng.*, **13**, 75 (September 1965).

[8] P. G. Ducker: The Use of Silicon Junction Diodes for the Protection of A-C and D-C Meter Circuits, *Semiconduct. Prod.*, **4**, 54–56 (March 1961).

[9] R. Greenburg and J. Takesuye: Zener Protection Circuits for Aircraft Voltage Surges, Motorola App. Note AN-120 (May 1961).

[10] R. L. Ives: Zener Diode Prevents Speaker Burnout, *Radio-Electron.*, **31**, 42 (August 1960).

[11] R. E. Learned: Use Power Zener Diodes for Protection, *Electron. Equip. Eng.*, **7**, 59–60 (November 1959).

[12] J. W. Phelps: Electrical Protection for Transistorized Equipment, *Bell Lab. Record*, **36**, 247–249 (July 1958).

[13] B. Reich: Protection of Semiconductor Devices, Circuits, and Equipment from Voltage Transients, *Proc. IEEE*, **55**, 1355–1361 (August 1967).

[14] B. Reich: Zener Diodes Quell Power-Supply Transients, *Electro-Technol.* (*New York*), **81**, 71, 98 (January 1968).

[15] O. Sturm: Silicon Power Zener Transient Suppressors, *Proc. IEEE*, **55**, 1483 (August 1967).

[16] Fast Fuse Blower. *Electron. Equip. Eng.*, **10**, 96 (October 1962).

Chapter 6

CLAMPING AND CLIPPING

6–1 INTRODUCTION

In addition to the voltage regulator and voltage monitor functions discussed in previous chapters, the VR diode may also be used as a signal limiter or clamp in various circuit applications. One specialized area has already been discussed in Chapter 5, in which the VR diode was used as a clamp against transient surges. In this chapter we consider additional applications of the VR diode as a clamp both in a direct manner and as a feedback limiter in conjunction with operational amplifiers.

We also consider circuit arrangements by which the effects of the VR diode junction capacitance may be reduced and thus allow higher effective speed of operation.

A second area of application uses the VR diode as a threshold element to control the presentation of voltage information to that which exceeds a given value. In this chapter we use the term clipper to describe this function. When a clamping circuit limits the *maximum* excursion a signal output may have, a clipper circuit will effectively limit the *minimum* input signal level which will yield any output at all. Clipper circuits are usually employed to remove baseline noise or to introduce an intentional "dead band" in a control loop.

6–2 THE BASIC SHUNT CLAMP

The simple VR diode shunt regulator may be used as a signal clamp as shown in Fig. 6–1. The breakdown voltage of the diode (with polarity as shown) will limit the maximum positive excursion of the output voltage with any excess input signal being dropped across a current-limiting resistor R_S.

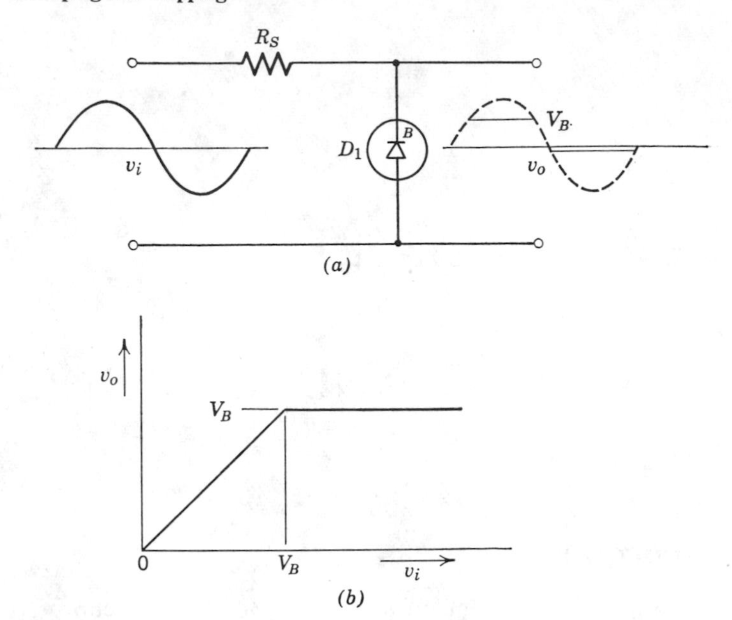

Fig. 6–1 Basic VR diode clamp circuit: (*a*) schematic: (*b*) input-output transfer plot.

When the input voltage swings negative, the output voltage becomes limited by the forward drop across the VR diode.

For simplification of both operation and analysis, it will be assumed that the output of the circuit does not feed any appreciable load. Later in this chapter the effect of a given load resistance is considered.

6–2.1 Operation

Consider what happens on the application of a sinusoidal voltage to the input terminals. The instantaneous value of the input will be

$$v_i = V_{im} \sin \omega t \tag{6-1}$$

As long as the value of v_i is less than the breakdown voltage of the VR diode, no appreciable current will flow in D_1 and the open-circuit output voltage will be equal to the input voltage. Thus, if the peak input voltage V_{im} is less than V_B, the positive half of the open-circuit output voltage waveform will be the same as that of the input.

If V_{im} is larger than V_B, however, the input voltage will reach a point at which its value begins to exceed the breakdown voltage of D_1 and thus currents will begin to flow. As the input voltage increases further, the output voltage may be assumed to remain at a level equal to V_B if the effects of dynamic breakdown resistance are neglected.

When the input voltage swings negative, the single-junction VR diode becomes forward-biased and conducts at a rather low voltage level equal to V_F. The output voltage will be clamped to a level equal to V_F with most of the input signal being dropped across R_S.

The overall waveform picture is illustrated in Fig. 6–1*a* and in greater detail in Fig. 6–2*b*. The output voltage is equal to the input for values of v_i between a positive value equal to V_B and a negative value equal to V_F. Beyond these limit or clamp values the output will be restricted to V_B on the positive half of the cycle and to V_F on the negative half.

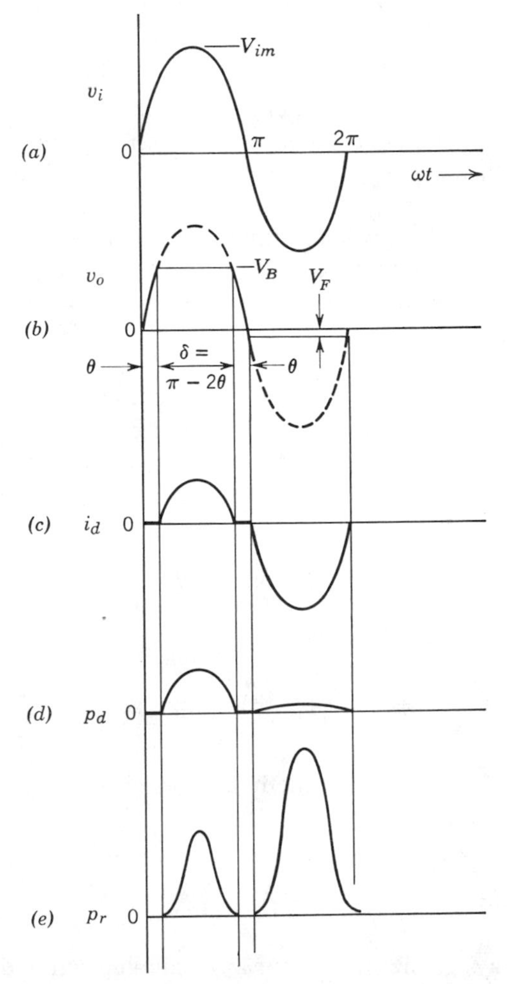

Fig. 6–2 Waveforms associated with the VR diode clamp circuit: (*a*) input voltage; (*b*) output voltage; (*c*) current through the VR diode; (*d*) instantaneous power dissipation in the VR diode; (*e*) instantaneous power dissipation in the resistor.

The input-output transfer characteristic of the basic shunt clamp is given in Fig. 6–1*b* and may be used to consider the clamping effects of any input waveform. Here again the fact is shown that the instantaneous open-circuit output voltage is equal to the instantaneous input voltage up to a value equal to V_B. As v_i is increased beyond V_B, the output level is clamped to a constant value equal to V_B. Only the positive portion of the transfer plot is shown since the simple circuit of Fig. 6–1*a* is normally used only as a unipolar or single polarity clamp. It should be remembered, however, that clamping of a negative input voltage will occur and the limiting level will be V_F.

6–2.2 Analysis

We may use the transfer plot to determine graphically the output waveform for a given input signal condition, and it is useful to analyze the operating performance mathematically in order to determine the amount of power being dissipated in the VR diode and in the series resistor. It is also convenient to be able to predict the angle at which clamping commences on a sinusoidal input. For now we shall assume that we have no load on the output and will analyze the operation on an open-circuit basis. Later the effects of a given load resistance will be included.

Assume that we apply a sinusoidal input voltage having the instantaneous value described by (6-1) and the waveform depicted in Fig. 6–2*a*. The resulting output voltage waveform will be that shown in Fig. 6–2*b*.

VR diode conduction, and thus output level limiting, will begin when v_i is equal to V_B. If we let θ represent the value of the angle ωt at which limiting occurs, then

$$V_B = V_{im} \sin \theta. \tag{6-2}$$

A solution for the value of θ yields

$$\theta = \sin^{-1}\left(\frac{V_B}{V_{im}}\right) = \sin^{-1}\left(\frac{1}{K}\right). \tag{6-3}$$

The value of θ is plotted as a function of the normalized input voltage K in Fig. 6–3. The value of K is

$$K = \frac{V_{im}}{V_B} = \frac{1}{\sin \theta}, \tag{6-4}$$

and must be always greater than 1 for any clamping action due to V_B.

Note that θ also gives the value of the angle between the point of which limiting ceases (as the input voltage drops below V_B) and the point at which the input voltage has dropped to zero ($\omega t = \pi$). This is shown in Fig. 6–2*b*.

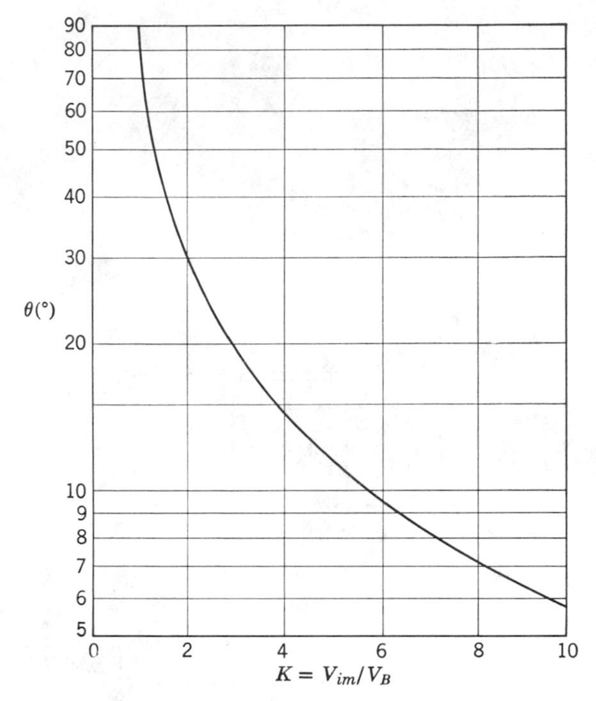

Fig. 6–3 Angle at which clamping begins as a function of normalized input voltage peak.

The total angle during which positive limiting occurs may be represented by the symbol δ and expressed mathematically as

$$\delta = (\pi - 2\theta) \text{ radians.} \tag{6-5}$$

The value of δ is shown in Fig. 6–4*a* and is plotted as a function of normalized input voltage in Fig. 6–4*b*; δ is zero as limiting begins with V_{im} just equal to V_B and it approaches π radians of 180° as V_{im} is made much larger than V_B.

The instantaneous power dissipation in the VR diode will be equal to the product of the instantaneous values of the voltage across it and the current through it or

$$p_d = v_d i_d = v_o i_d. \tag{6-6}$$

We have already considered the waveform of the output voltage as shown in Fig. 6–2*b*. To be perfectly accurate we would need to consider the slight variation in the voltage during the limiting interval δ caused by the breakdown resistance. By assuming a constant value of v_o equal to the breakdown voltage V_B we greatly simplify the analysis. Since the actual value of V_B which we would use is a value that occurs at a given level of current and not just the

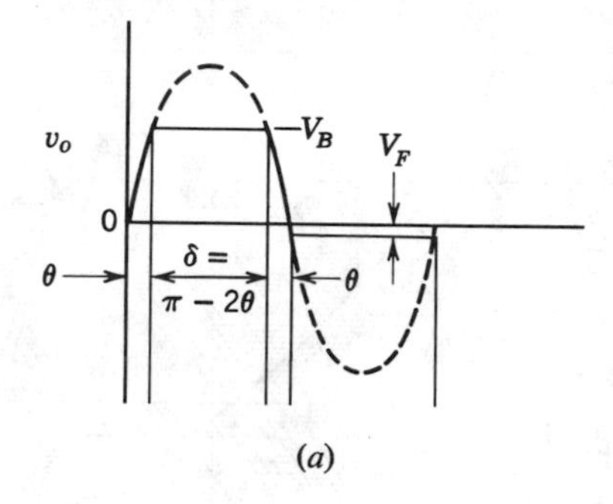

(a)

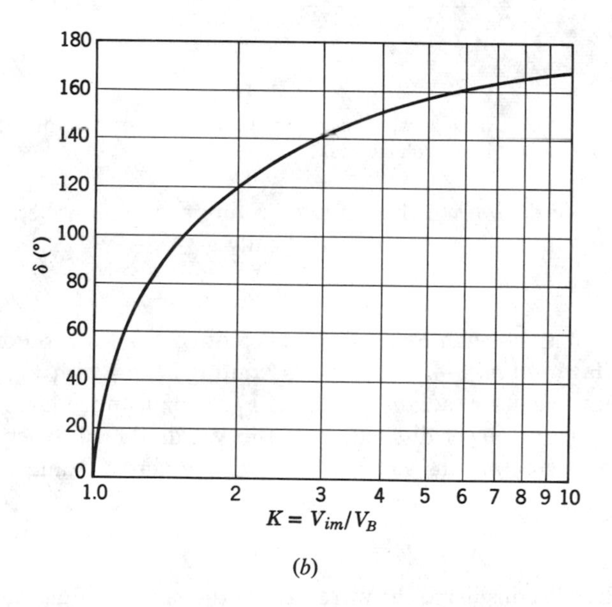

(b)

Fig. 6–4 Conduction angle as a function of normalized input voltage peak.

value at which breakdown current just begins to flow, the overall error induced will be normally quite small.

For values of input voltage less than V_B, the current flowing through the VR diode will only be a value equal to the leakage current and can be considered negligible for practically all cases. As v_i becomes greater than V_B and is positive for the polarity configuration of Fig. 6–1, a reverse current begins to flow in the VR diode. Its instantaneous value will depend on v_i, the value of the limiting resistor R_S, and the breakdown voltage V_B in the following manner:

$$i_{dr} = \frac{v_i - V_B}{R_S}. \tag{6-7}$$

When the input voltage swings negative, an instantaneous forward current flows through the diode:

$$i_{df} = \frac{v_i - V_F}{R_S}. \tag{6-8}$$

Since the value of V_F is usually much smaller than V_B, the resulting forward current waveform will be very nearly a sine wave with only a slight amount clipped off the bottom. The diode current waveform for both halves of the cycle is shown in Fig. 6–2*c*.

Knowing the output voltage waveform and the diode current waveform, we may now construct the instantaneous diode power dissipation waveform. Each point will represent the product of v_o and i_d corresponding to that part of the cycle.

Since the negligible diode current flows for ωt from 0 to θ, the power dissipation will likewise be negligible. Then, between ωt from θ to $(\pi - \theta)$, the current waveform is the clipped sinusoid, and the voltage across the VR diode is relatively constant and equal to V_B. The instantaneous power waveform will have the same shape as the current waveform.

When the input voltage swings negative, the current is higher than before, but the voltage drop equal to V_F is much lower and thus the power dissipation when the VR diode is forward-biased is less than for the reverse direction. The overall waveform is given in Fig. 6–2*d*.

The average power dissipation is responsible for the heating effects in a VR diode (assuming that the period of the input waveform is less than the thermal time constant of the diode). Mathematically, the average power dissipation, P_D, may be expressed as

$$P_D = \frac{1}{2\pi} \int_0^{2\pi} i_{dr}\, d\omega t. \tag{6-9}$$

First of all, let us consider only that portion of the power dissipation that is due to the reverse-biased condition (input voltage is positive for the arrangement shown). To obtain this we need only to integrate the power waveform corresponding to ωt from θ to $(\pi - \theta)$ and average it over the entire cycle:

$$P_{DR} = \frac{1}{2\pi} \int_{\theta}^{\pi-\theta} i_{dr} v_o \, d\omega t. \tag{6-10}$$

Since v_o is equal to V_B during this portion of the cycle and (6-8) gives the expression for the instantaneous reverse current, we may rewrite (6-10) as follows:

$$P_{DR} = \frac{1}{2\pi} \int_{\theta}^{\pi-\theta} \left(\frac{v_i - V_B}{R_S} \right) V_B \, d\omega t. \tag{6-11}$$

Equation 6-1 gives the expression for the instantaneous input voltage for the assumed sine wave input. Injecting this into (6-11) and rewriting,

$$P_{DR} = \frac{V_B}{2\pi R_S} \left(V_{im} \int_{\theta}^{\pi-\theta} \sin \omega t - V_B \int_{\theta}^{\pi-\theta} d\omega t \right) \tag{6-12}$$

$$= \frac{V_B}{2\pi R_S} [2V_{im} \cos \theta - V_B(\pi - 2\theta)]. \tag{6-13}$$

From (6-4) $\sin \theta = 1/K$, and using the trigonometric identity relating sine and cosine,

$$\cos \theta = \sqrt{1 - \sin^2 \theta} = \sqrt{1 - \left(\frac{1}{K}\right)^2}. \tag{6-14}$$

Inserting (6-14) for $\cos \theta$ and (6-3) for θ and then simplifying,

$$P_{DR} = \frac{V_B^2}{R_S} \left[\frac{\sqrt{K^2 - 1} + \sin^{-1}(1/K)}{\pi} - 0.5 \right]. \tag{6-15}$$

Figure 6–5 illustrates a plot of (6-15) modified to a normalized form

$$P_{DRN} = \frac{P_{DR}}{V_B^2/R_S} = \frac{\sqrt{K^2 - 1} + \sin^{-1}(1/K)}{\pi} - 0.5. \tag{6-16}$$

Note that for values of K greater than about 2.5 the plot becomes relatively linear and may be approximated by

$$P_{DRN} \approx 0.313 \text{ K} - 0.438. \tag{6-17}$$

Now we must consider the VR diode power dissipation that occurs due to the forward-biased condition (input negative for the arrangement shown). In order to simplify the analysis, we may assume that the forward voltage drop

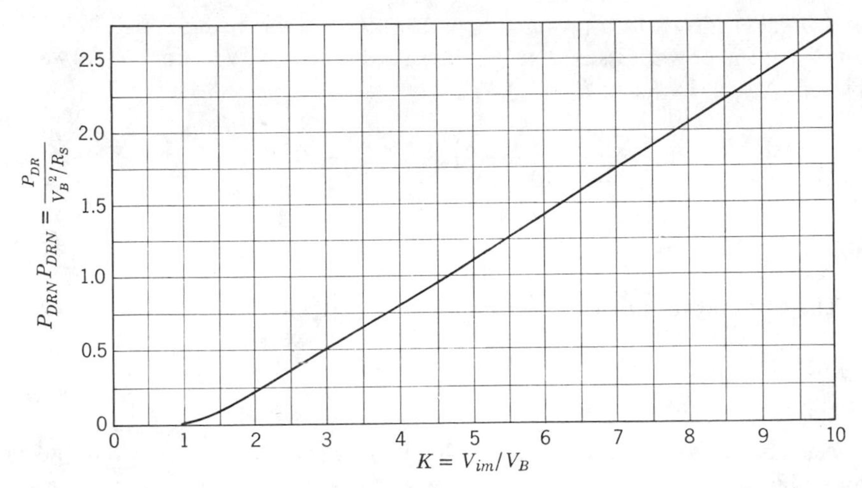

Fig. 6-5 Normalized VR diode power dissipation due to reverse bias as a function of normalized input voltage.

is negligible in comparison with V_{im} and thus the current i_{df} will be approximated by a half sine wave. It will be further assumed that the voltage drop across the diode is equal to a constant value V_F for the entire half of the cycle. This gives the following

$$P_{DF} = \frac{1}{2\pi}\int_{\pi}^{2\pi} i_{df}\, v_o \, d\omega t \tag{6-18}$$

$$= \frac{1}{2\pi}\int_{\pi}^{2\pi} \frac{V_{im}\sin \omega t}{R_S} V_F \, d\omega t, \tag{6-19}$$

$$P_{DF} = \frac{V_{im} V_F}{\pi R_S}. \tag{6-20}$$

In order to determine the total power dissipation in the VR diode, we merely sum the reverse and forward components as given by (6-15) and (6-20), respectively.

$$P_D = P_{DR} + P_{DF}. \tag{6-21}$$

Computing the power dissipation in the series resistor, R_S, requires the same type of procedure as has been given for the VR diode. We take the basic power equation

$$P_R = \frac{1}{2\pi}\int_0^{2\pi} i_r v_r \, d\omega t, \tag{6-22}$$

and break it down to a power dissipation due to the reverse bias half P_{RR} and P_{RF}, the power dissipation in the resistor when the VR diode is forward-biased. Again, by using $K = V_{im}/V_B$,

$$P_{RR} = \frac{V_B^2}{2\pi R_S}\left[\left(\frac{\pi}{2} - \theta\right)(K^2 + 2) - 3\sqrt{K^2 - 1}\right], \quad (6\text{-}23)$$

$$P_{RF} = \frac{V_{im}^2}{4R_S} - \frac{2V_{im}V_F}{\pi R_S} \quad (6\text{-}24)$$

The total power dissipation in the resistor is then

$$P_R = P_{RR} + P_{RF}. \quad (6\text{-}25)$$

6–2.3 Example

A practical example will illustrate the use of the equations and curves given and will also help us to understand the relative magnitudes involved.

Assume that we have the basic arrangement as shown in Fig. 6–1 with $V_B = 10$ V, $V_F = 0.7$ V, $R_S = 1$ K, and $V_{im} = 20$ V. For this example we shall also assume that the output is unloaded.

The value of θ, the angle at which conduction of the VR diode, hence output limiting, occurs, is given by (6-3).

$$\theta = \sin^{-1}\left(\frac{10}{20}\right) = \frac{\pi}{6} \text{ radians} = 30^\circ.$$

The value of θ could have been determined from Fig. 6–3 by first computing $K = 20/10 = 2$ and reading $\theta = 30°$. We will need to use the radian value in any of the succeeding equations.

The conduction angle δ is found from (6-5) as 120° or $2\pi/3$ if expressed in radians.

Using either (6-16) or Fig. 6–5, we find that a value of $K = 2$ yields a normalized VR diode power dissipation in the reverse direction of 0.22. In order to convert this into the actual power, we multiply by the normalizing factor V_B^2/R_S.

$$P_{DR} = 0.22\left(\frac{10^2}{1000}\right) = 22 \text{ mW}.$$

VR diode power dissipation due to conduction in the forward direction is found by using (6-20) which will give a value of 4.5 mW for P_{DF}. Total VR diode dissipation is the sum of P_{DR} and P_{DF} or 26.5 mW.

Power dissipation in the series resistor is computed from (6-23) for the forward-biased condition and (6-24) for the reverse-biased condition. Thus $P_{RR} = 17$ mW and $P_{RF} = 91$ mW.

Total power dissipation in the series resistor R_S is the sum of P_{RR} and P_{RF} or 108 mW.

If we had the same circuit values as before but increased V_{im} to 80 V, we would obtain the following values:

$$K = 8, \qquad \theta = 7.3°, \qquad \delta = 165.4°,$$
$$P_{DRN} = 2.07, \quad P_{DR} = 207 \text{ mW}, \quad P_{DF} = 17.8 \text{ mW}, \quad P_D = 225 \text{ mW},$$
$$P_{RR} = 1.14 \text{ W}, \quad P_{RF} = 1.56 \text{ W}, \qquad P_R = 2.7 \text{ W}.$$

Figure 6–6 illustrates the various components of the VR diode dissipation as the input voltage is varied.

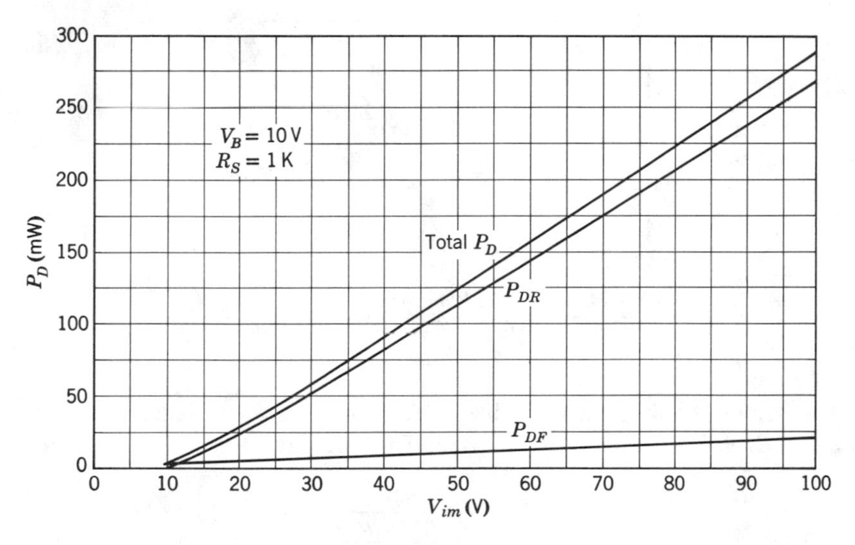

Fig. 6–6 VR diode power dissipation as a function of input voltage for the illustrative example.

6–2.4 Including Effects of Loading

In the previous discussions we assumed that the output of the clamp was unloaded. Now let us consider what happens when we do add some appreciable load resistance R_L across the output as shown in Fig. 6–7*a*.

A Thévenin equivalent circuit of the loaded clamp may be constructed as given in Fig. 6–7*b*. This yields a circuit arrangement that is identical in form to the simple unloaded case previously analyzed. The effective input voltage has been reduced by a factor of $(R_L + R_S)/R_L$ and the effective series limiting resistance is now $R_S /\!/ R_L$. With this knowledge we may appropriately modify the equations for the simplified unloaded case. For the most part this means

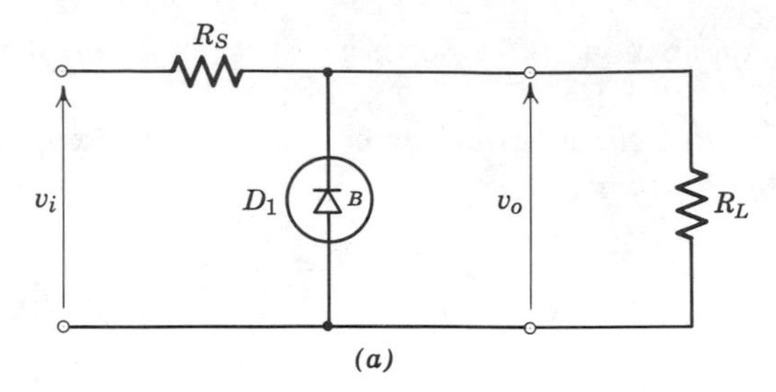

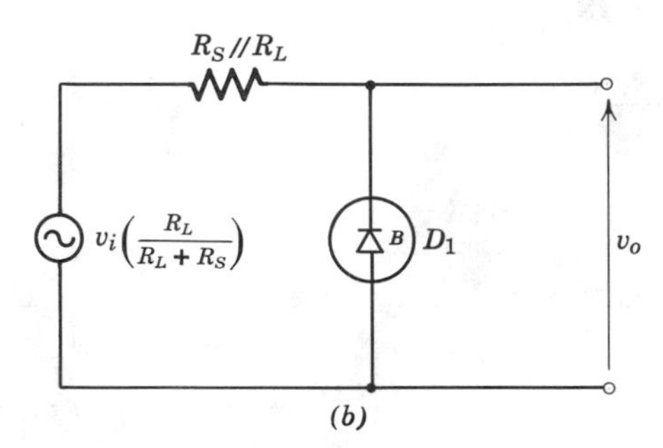

Fig. 6–7 A resistive load may be placed across the VR diode. The Thévenin equivalent circuit is shown in (*b*).

we must merely substitute the value $R_L // R_S$ wherever R_S is used and substitute $V_{im} R_L/(R_L + R_S)$ wherever V_{im} is used. Thus we have a new definition of K:

$$K = \frac{V_{im} R_L}{V_B(R_L + R_S)}. \tag{6-26}$$

When we insert the modified values of K and R_S into (6-23), we get the power dissipation which occurs in $R_L // R_S$. The actual power in R_S will be given if we use the modified value of K and the real value of R_S in (6-23).

6–2.5 Possible Variations in Drive

In the previous discussions only a simple series resistor was considered to limit the current flow in the VR diode as clamping began. Several possible variations may be used, each with certain advantages; for example, a diode may be placed in series with R_S in the manner shown in Fig. 6–8. For the usual case in which the forward drop of D_2 is significantly less than V_{im}, little effect is noticed when the input voltage is positive. However, as the input voltage swings negative, D_2 becomes reversed-biased and no current flows in either R_S or in the VR diode. This acts to reduce the power dissipation in the VR

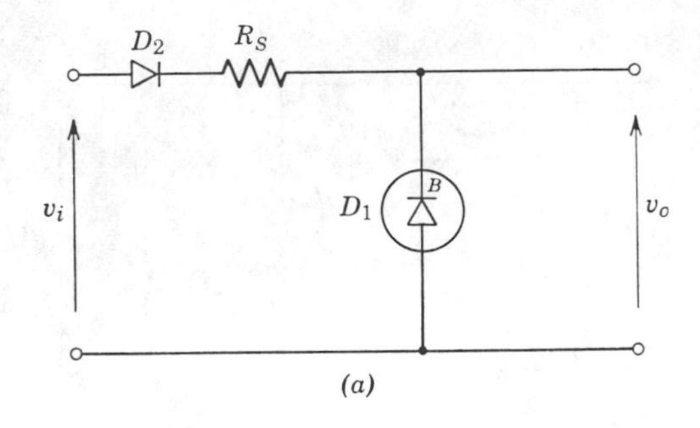

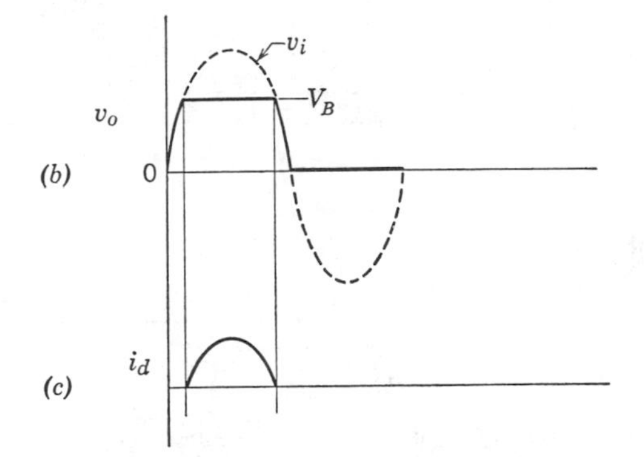

Fig. 6–8 A diode rectifier may be added to reduce power dissipation in R_S by eliminating negative half of the input waveform.

diode to simply the value P_{DR}. Likewise, the power dissipation in R_S is reduced to only P_{RR}. This becomes rather significant since P_{RF} is the greater part of the total dissipation for the normal case. The diode thus permits a reduction in the power dissipation requirement of R_S with its corresponding heating effects also reduced. The output voltage no longer will go to the value V_F during the negative half of the cycle but will be at zero.

The series limiting resistor may be replaced by a field-effect transistor of field-effect current limiting diode in the manner shown in Fig. 6–9. The drain characteristic curve of the field-effect device as shown in Fig. 6–9*b* illustrates how the drain current reaches a rather constant level after the drain voltage reaches a certain value approximately equal to the pinch-off voltage.

The effect of the FET is to present a very low value of R_S until its I_{DSS} is reached and then to present a very high value of R_S for larger input voltages. This means that clamping begins at a relatively low value of θ and the value of δ may approach 180° without requiring an input voltage which is more than a few times larger than V_B. In addition, the actual clamped waveform will be

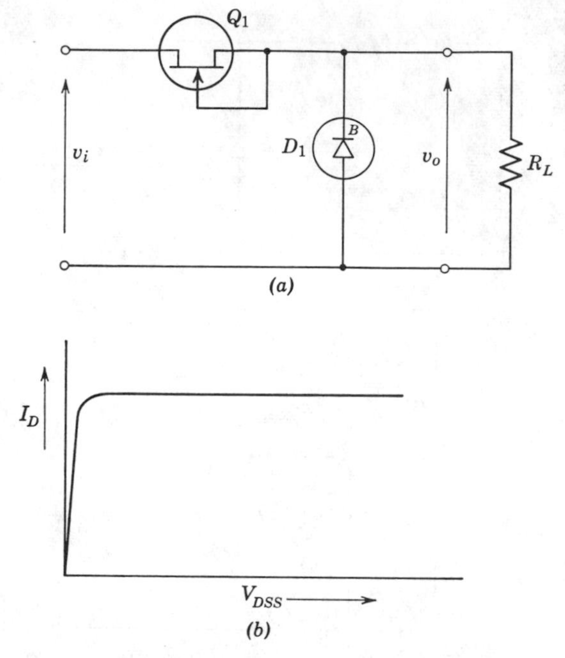

Fig. 6–9 A field-effect transistor or diode may be used to replace the series resistor. (*b*) shows constant current characteristic of the FET.

flatter on the top because i_d will remain practically constant during the clamping interval.

6–2.6 Bipolar Clamp

Up to now we have considered only a unipolar clamp, although clamping did occur due to the forward conduction of the VR diode in the simple circuit. There are applications in which it is desirable to clamp both halves of the input waveform to a given level. This may be done by using either two VR diodes connected back to back as shown in Fig. 6–10 or a double anode VR diode. If two separate VR diodes are used, the clamp level of each polarity may be controlled independently.

The same basic equations as developed previously may be used with the bipolar clamp by considering each polarity separately. The transfer plot obtained is shown in Fig. 6–10*b* and Fig. 6–10*c* shows the output waveforem.

One interesting example of the use of a bipolar clamp is illustrated in Fig. 6–11. This circuit arrangement yields a precision sinusoidal ac reference voltage from an unregulated input voltage. Diodes D_1 and D_2 form an initial

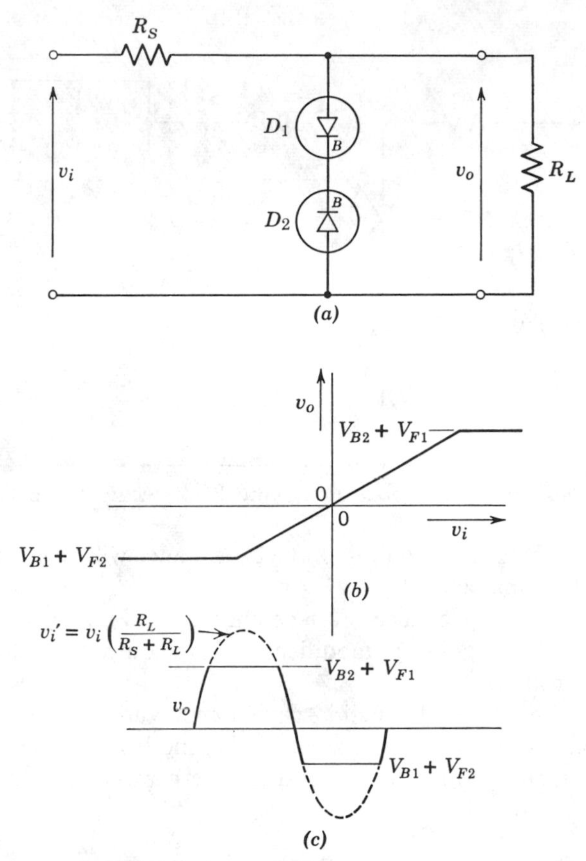

Fig. 6–10 Bipolar clamp operates on each polarity independently: (*a*) circuit; (*b*) transfer plot; (*c*) output voltage waveform.

stage of bipolar clamping, whereas D_3 and D_4 perform the final clamping to produce a trapezoidal waveform of controlled amplitude.

Three resonant circuits are used to filter out unwanted harmonics (primarily odd harmonics with the third harmonic being predominant) to yield an output voltage which is of the same frequency as the input but whose amplitude is constant.

6–3 HIGH-SPEED CLAMPING CIRCUITS

The maximum frequency at which the basic clamp circuit may be used is somewhat limited due to the substantial value of the junction capacitance assocated with the VR diode. Since the VR diode is usually a rather large area junction and the resistivity of the bulk material is made low in order to achieve

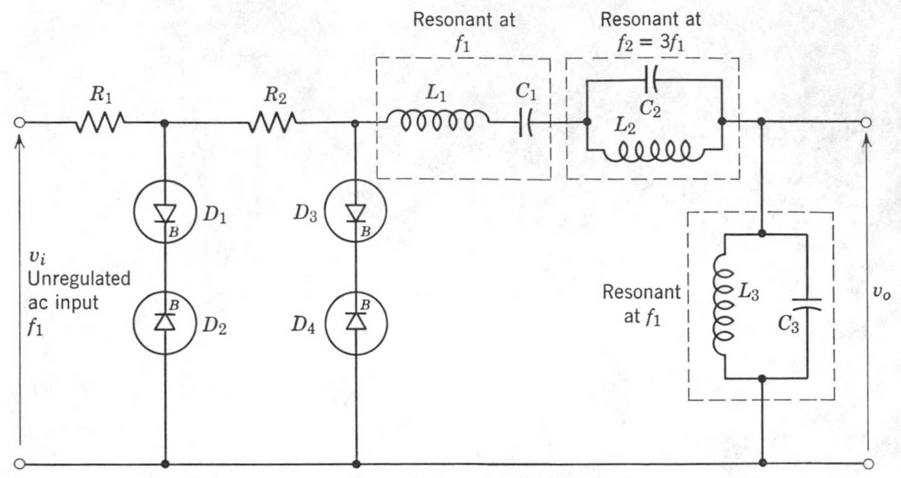

Fig. 6–11 Bipolar clamp is used to provide a precision ac voltage [7].

the desired low breakdown voltages, the capacitance will be much higher than for a normal computer diode.

If the VR diode capacitance must be charged and discharged during each cycle, there will be a significant modification in the output waveform due to the $R_S C_d$ time constant.

To reduce the effects of junction capacitance and thus improve the high speed capability of the clamp, we shall need to modify the circuit slightly. The extent and manner required will depend on whether the input waveform is of a repetitive or nonrepetitive nature.

6–3.1 High-Speed Clamp with Repetitive Inputs

When the input waveform is a repetitive waveform, we may use the simple arrangement shown in Fig. 6-12. Here a low-capacitance computer diode D_1

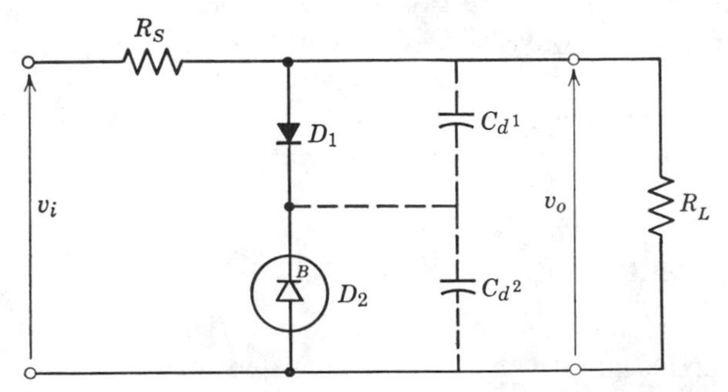

Fig. 6–12 High-speed clamp for repetitive waveform input uses low capacitance diode to reduce loading the VR diode.

is added in series with the VR diode. When the input voltage is first applied, the capacitance C_{d2} of the VR diode must be charged up as before. However, the presence of diode D_1 removes the discharge path through the external circuitry and thus helps to keep C_{d2} charged. As the input voltage drops below the clamp level, the computer diode D_1 becomes reverse-biased.

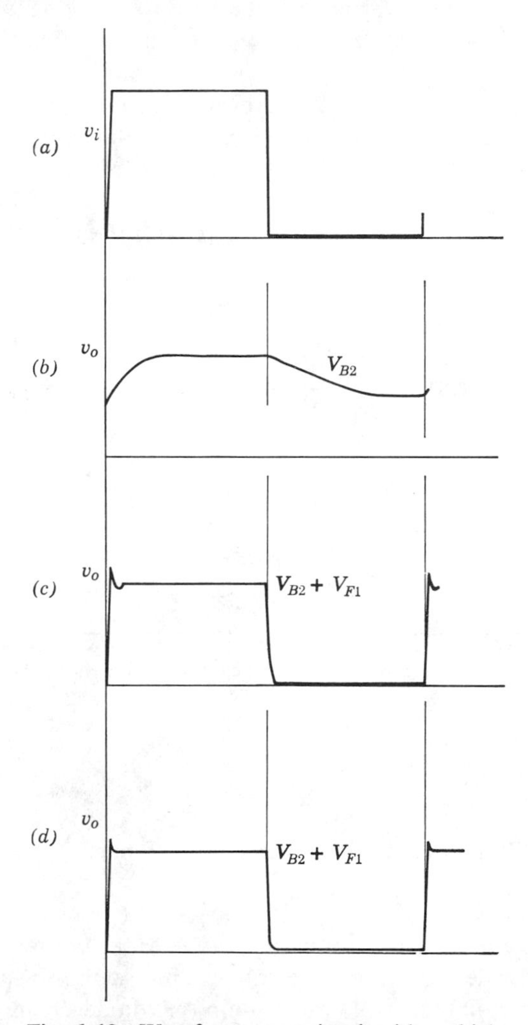

Fig. 6-13 Waveforms associated with a high-speed clamp: (*a*) assumed input voltage; (*b*) output voltage obtained without using additional diode; (*c*) output voltage with low-capacitance diode, lower repetition rate; (*d*) output voltage with low-capacitance diode and at high repetition rate.

The effect of adding D_1 is to reduce the effective capacitance of the VR diode to that of the computer diode. If the input voltage is to swing negative as might occur, no clamping action will be present for this polarity.

Various waveforms are given in Fig. 6–13 for an assumed squarewave input. Note that a slight spike occurs as the input voltage rises. This is due to the forward recovery time of D_1 and may be held to a few nanoseconds. If the repetition rate is low enough so that leakage currents drain off some of the charge on C_{d2} during the portion of the cycle when the input voltage is zero, then a slight drop below the clamp level may be noticed just after forward recovery of D_1. This is evident in Fig. 6–13*c*. When the repetition rate is increased to the point where very little charge is lost from one cycle to the next, this valley will disappear as shown in Fig. 6–13*d*. This approach to improving the speed of the clamp is effective with the bipolar clamp as well. The arrangement is given in Fig. 6–14.

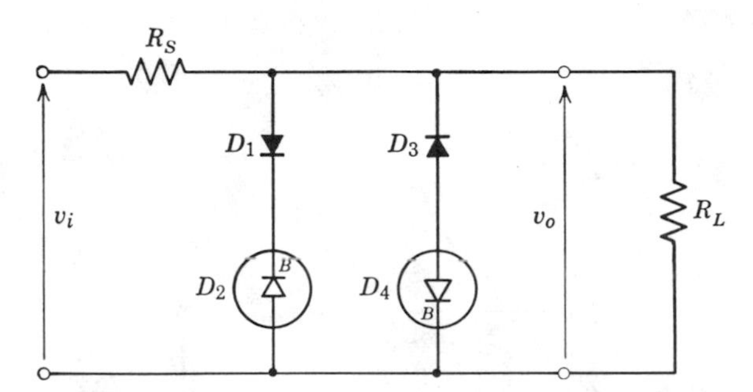

Fig. 6–14 High-speed bipolar clamp for repetitive waveforms.

6–3.2 High-Speed Clamp with Nonrepetitive Inputs

When the input voltage waveform is not repetitive or when the repetition rate may be too low to maintain the VR diode capacitance fully charged, we will need to provide an external biasing scheme as shown in Fig. 6–15. An external voltage supply is used to maintain VR diode D_2 in the breakdown region and thus with C_{d2} charged. Performance of the arrangement is the same as for that given in Fig. 6–12 except that we do not need a repetitive input.

Figure 6–16 illustrates the manner in which a bipolar clamp circuit may be prebiased to reduce capacitance effects. Two supply voltages of opposite polarity will be required.

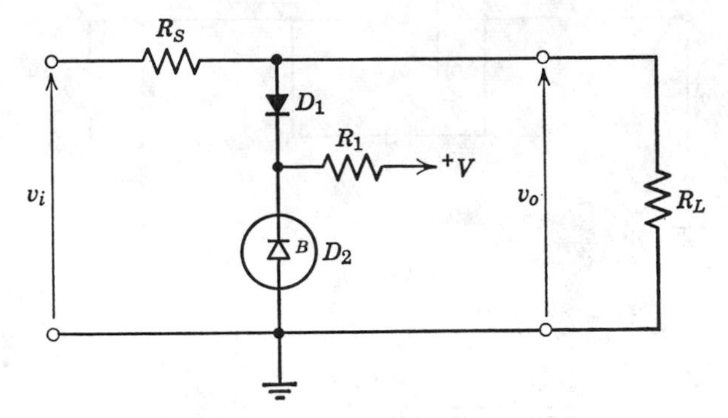

Fig. 6–15 Biased high-speed unipolar clamp can be used for non-repetitive inputs.

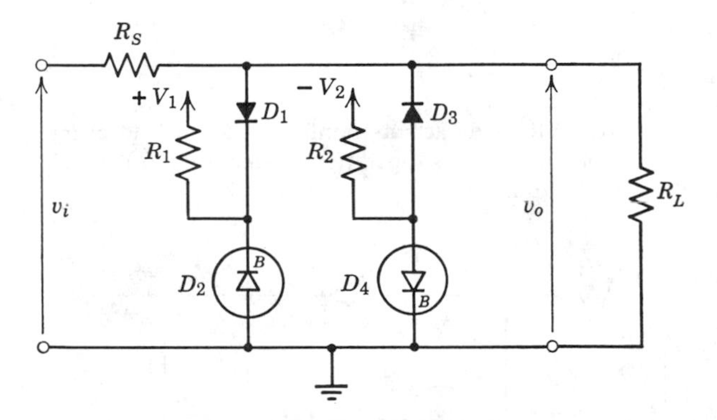

Fig. 6–16 Biased high-speed clamp for bipolar applications.

6–4 OPERATIONAL AMPLIFIER LIMITING

VR diodes may also be used to limit or clamp the output swing of an operational amplifier. The basic approach is shown in Fig. 6–17. VR diode D_1 is used as a nonlinear feedback means which provides negligible feedback as long as the output voltage remains below the breakdown voltage. As v_o attempts to exceed V_B, a feedback current flows back to the inverting input. The resulting incremental gain for outputs in excess of V_B will be approximately unity.

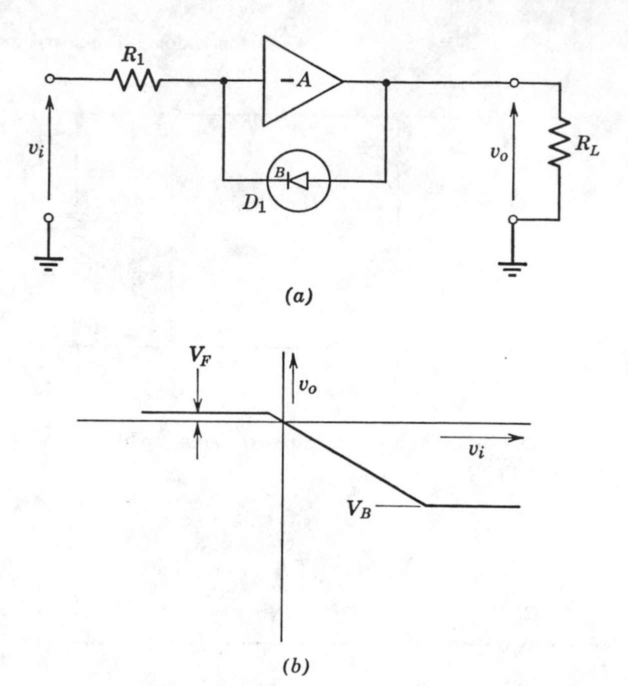

Fig. 6–17 VR diode acts as nonlinear feedback to clamp output of an inverting operational amplifier: (*a*) circuit diagram; (*b*) transfer plot.

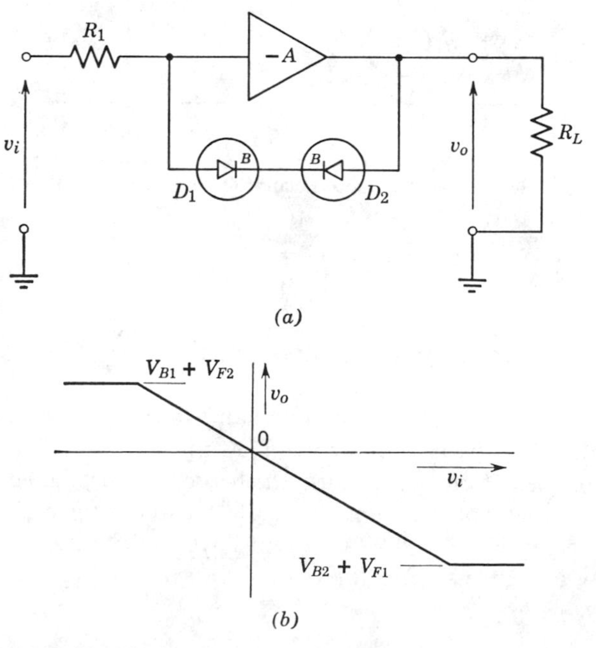

Fig. 6–18 Bipolar arrangement for VR diode clamp of operational amplifier: (*a*) circuit; (*b*) transfer plot.

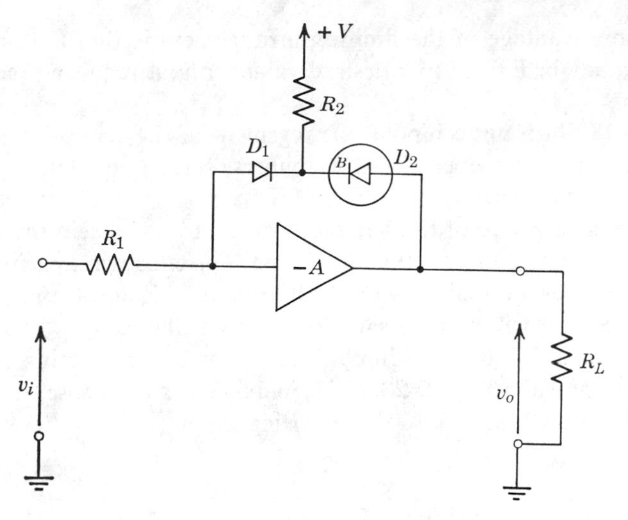

Fig. 6-19 The biased VR may be used with inverting amplifiers.

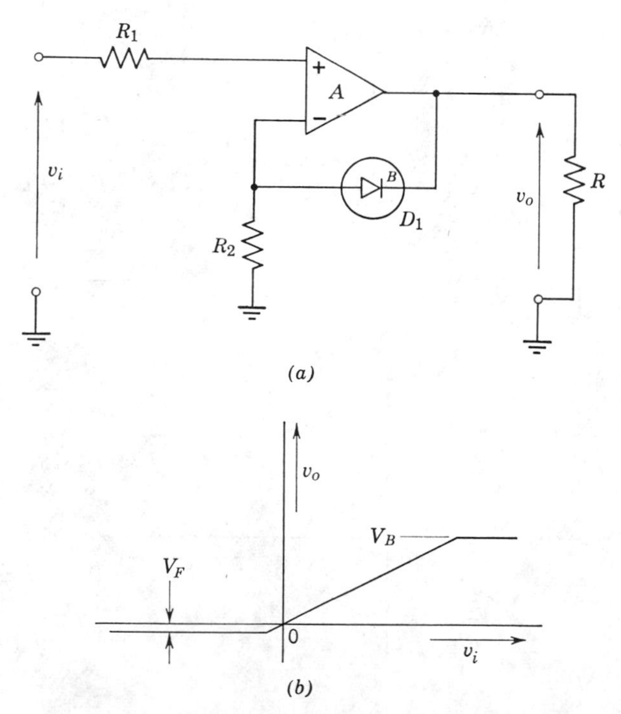

Fig. 6-20 VR diode unipolar clamp for noninverting operational amplifier: (*a*) circuit; (*b*) transfer plot.

The main advantage of the limiting arrangement is that the output level peak swing may be limited to a desired value without requiring the amplifier to saturate.

Figure 6–18 illustrates a bipolar arrangement of the clamp-limiting circuit as used with inverting operational amplifiers. Each output polarity may be controlled independently.

When the capacitance of the VR diode becomes a problem the biased VR diode arrangement of Fig. 6–19 may be used. The effective capacitance of the VR diode becomes multiplied by the Miller effect. This means that the biased-diode arrangement will be necessary for most instances.

VR diodes may be also used in conjunction with noninverting operational amplifiers as shown in Figs. 6–20, 6–21, and 6–22. In each case the VR diode acts as a nonlinear feedback to the inverting input.

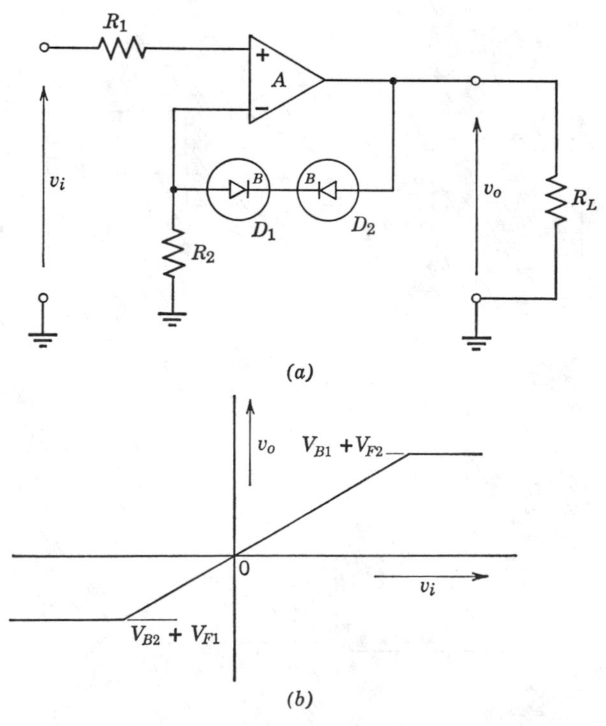

Fig. 6–21 Bipolar clamp for noninverting operational amplifiers: (*a*) circuit; (*b*) transfer characteristics.

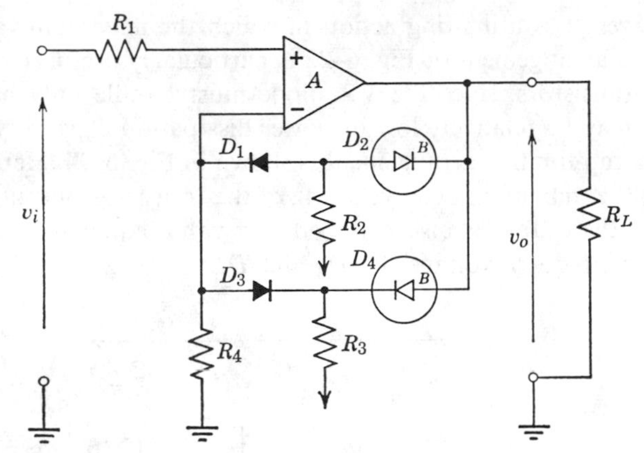

Fig. 6–22 Biased clamp arrangement for noninverting amplifiers.

6–5 MISCELLANEOUS CLAMPS

The VR diode may be used to limit the output swing of a simple transistor stage either for protection of the transistor from harmful voltage spikes or merely to limit the output voltage.

One direct approach is shown in Fig. 6–23. As long as the output collector voltage remains less than V_B of thc VR diode, the stage performs in a normal manner (assuming that the junction capacitance is insignificant). However, as v_o attempts to exceed V_B, a current flows through D_1 back to the base of the transistor. This, in turn, increases the base current and hence the collector

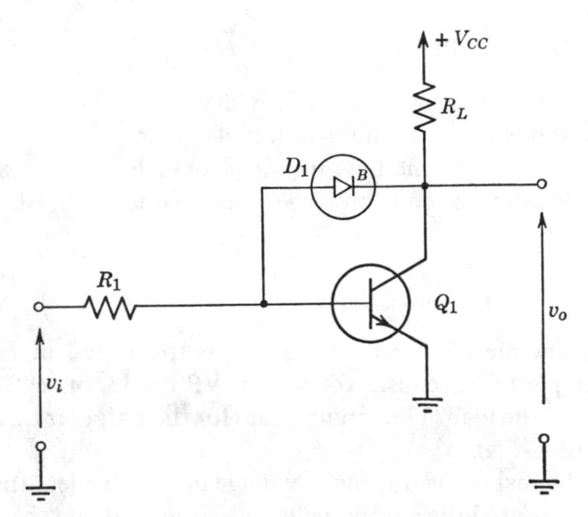

Fig. 6–23 VR diode used to limit swing of transistor output stage by feedback to base.

current. The result is a limiting action in which the maximum value of v_o is $V_B + V_{BE}$. The arrangement of Fig. 6–23 is particularly useful in conjunction with power transistors, since the VR diode must handle only base current levels, hence may be relatively low in power dissipation capability.

Another clamp-limiting arrangement is shown in Fig. 6–24. Here VR diode D_2 will limit the maximum positive swing of the transistor output. The minimum level of the output is also clamped at a value equal to the difference between the breakdown voltages of D_3 and D_1.

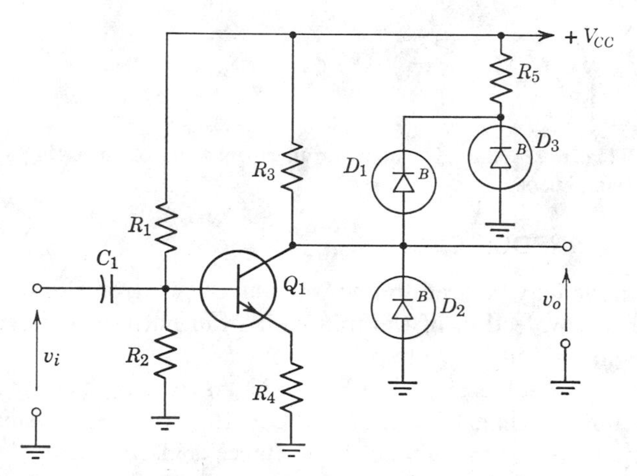

Fig. 6–24 Doublesided clamp for transistor output.

6–6 THRESHOLD CLIPPER

As mentioned in the introduction to this chapter, we have chosen to use the term clamp to indicate a limit suppression of maximum signal output swing. Another class of circuit functions in which base line information below a given threshold is suppressed is, by choice, referred to as threshold clippers or just clippers.

6–6.1 Unipolar Threshold Clipper

The basic principle of threshold clipping is illustrated in Fig. 6–25. The circuit in its simplest form consists of a single VR diode connected between the signal source and the load. The arrangement for the unipolar threshold clipper is shown in Fig. 6–25*a*.

As long as the instantaneous input voltage v_i remains less than the breakdown voltage V_B, very little current will flow in the load and the output will be practically zero. As v_i begins to exceed the value of V_B, however, the VR

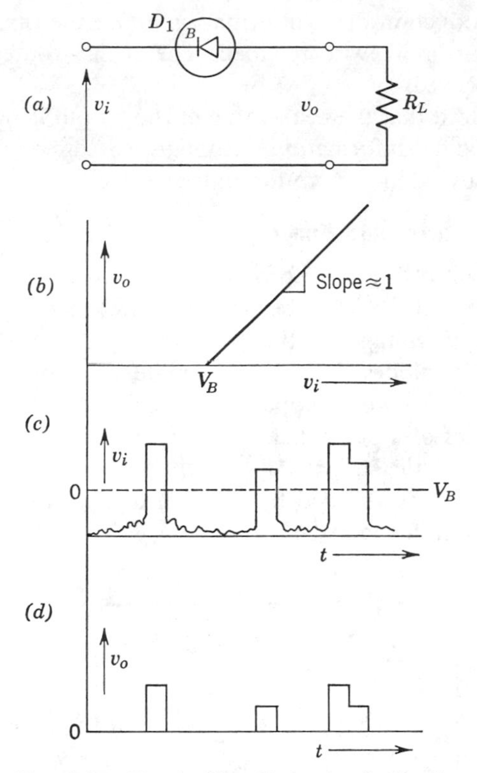

Fig. 6–25 Simple VR diode threshold clipper: (*a*) circuit; (*b*) transfer plot; (*c*) simulated input voltage; (*d*) resulting output voltage.

diode is biased in the breakdown region. Current now flows and the output voltage is equal to the input voltage less the value of V_B or

$$v_o = v_i - V_B. \tag{6-27}$$

The transfer plot for the threshold clipper is given in Fig. 6–25*b*. The input level of threshold, below which no output voltage is produced, is fixed by the value of V_B. If the value of V_B may be assumed to be constant, then the incremental input voltage beyond the threshold value will produce an equal increment in output voltage. The slope of the transfer plot closely approaches unity. This assumption holds as long as the breakdown resistance r_d is much less than the load resistance R_L.

An assumed input waveform containing desired pulse information together with unwanted noise is shown in Fig. 6–25*c*. If this input is applied to a threshold clipper with a VR diode, only the information above V_B will be passed to the output and we have the resulting waveform as shown in Fig. 6–25*d*. Note that faithful reproduction of all information above the threshold occurs and timing information is likewise retained. The signal-to-noise level has been substantially improved.

It must be realized that the peak value of the output is no longer the same as that of the original information, although it is directly related to it and may be restored by adding V_B to the output value.

6–6.2 Bipolar Threshold Clipper

The simple circuit of Fig. 6–25 provides threshold clipping only for positive input voltages. Should the input become negative, the output voltage would be equal to the input voltage less the forward drop of the VR diode.

By using two VR diodes (or a double-anode VR diode) as shown in Fig. 6–26, we achieve a threshold clipper that will remove base line information from both polarities of input signal. The clipping threshold level for positive input voltages will be the sum of V_{B2} plus the forward drop of D_1. The negative threshold will be established by the sum of V_{B1} and V_{F2}. The resulting transfer plot is given in Fig. 6–26*b*.

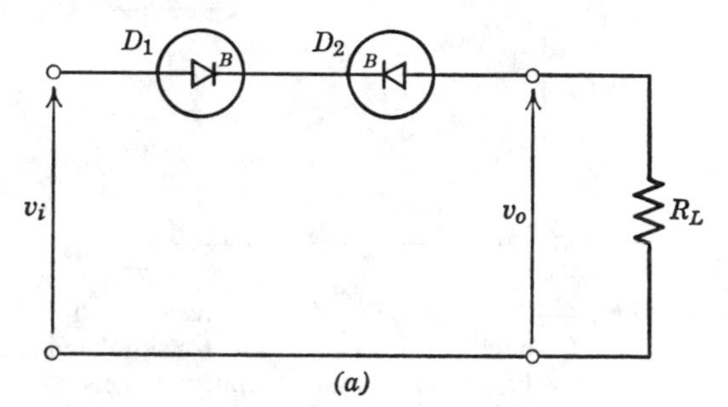

(*a*)

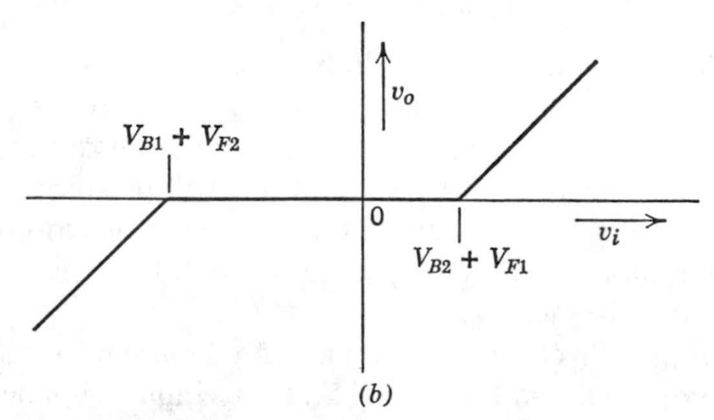

(*b*)

Fig. 6–26 Bipolar threshold clipper has controlled "dead band"; (*a*) circuit; (*b*) transfer plot.

Another use for the threshold clipper arrangement is in servo control systems in which we intentionally want to insert a "dead band" into the loop.

6–6.3 Biased Diode Clipper

When the input signal has high-frequency components, the junction capacitance of the VR diode may again present problems by transmitting a differentiated spike to the output even though the input is below the threshold level. The effect of junction capacitance may be reduced effectively by prebiasing the VR diode as was done for the clamping circuits, The approach required for the bipolar threshold clipper is shown in Fig. 6–27. The input source must have a dc path to ground.

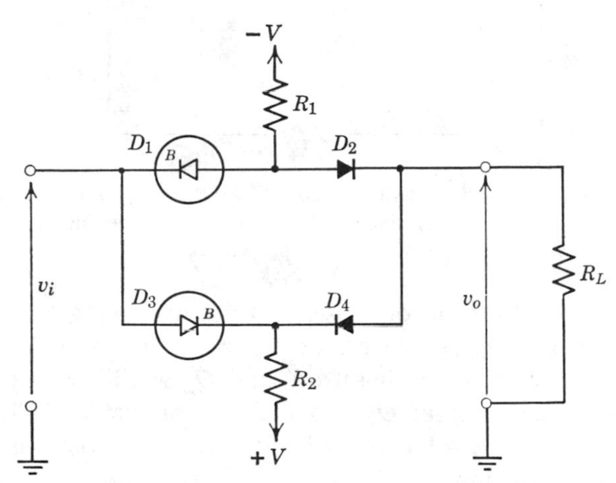

Fig. 6–27 Biased VR diodes used to provide high-speed threshold clipper.

6–7 THRESHOLD SWITCH

Another class of circuit functions that is somewhat kin to the threshold clipper is the threshold switch. The main difference is the retention of actual instantaneous voltage value of signals above the threshold value. One circuit arrangement for the threshold switch combines the VR diode with both bipolar and field-effect transistors as shown in Fig. 6–28.

As long as v_i is positive and less than $(V_B - V_{BE1})$, negligible current will flow in VR diode D_1. Thus transistor Q_1 will be turned on by the negative bias current supplied to the emitter by R_4. This, in turn, causes Q_2 to conduct and its collector to rise to a large positive voltage. A large positive voltage at the gate of the *P*-channel MOSFET insures that it is completely off and the output is disconnected from the input.

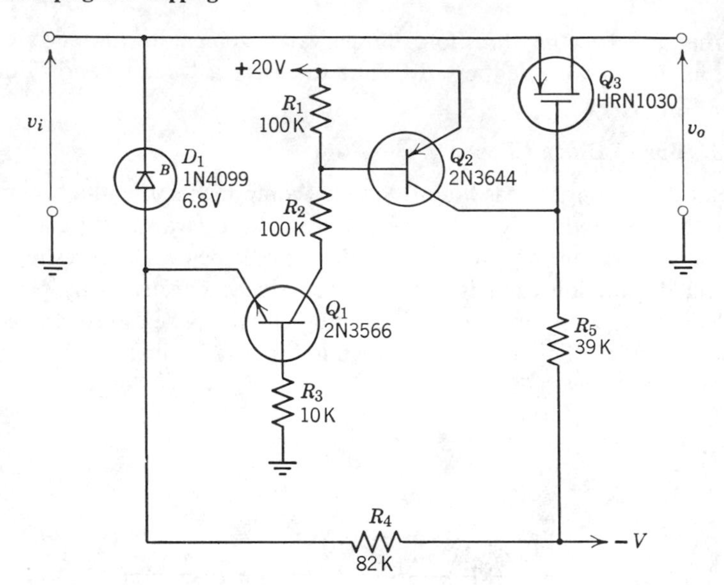

Fig. 6–28 A threshold switch circuit. Once the input voltage exceeds a preset threshold determined by the VR diode, the input voltage becomes very nearly equal to the output voltage.

Now assume that the value of v_i rises to a positive voltage that is above the threshold value ($V_B - V_{BE1}$). Diode D_1 conducts, a positive voltage is applied to the emitter of Q_1, thus turning it off, and Q_2 will likewise be turned off This permits the collector of Q_2 and the gate of the MOSFET Q_3 to go negative. A negative gate voltage on a *P*-channel MOSFET turns it on to connect the output to the input.

The result is an arrangement in which signals below a design threshold are isolated from the output, whereas any signal voltage above the threshold value will be transmitted to the output without any clipping. The only inaccuracy will be due to a finite "on" resistance in the MOSFET. As long as the loading resistance on the circuit is made large, the loss across the MOSFET will be small, and the output may be assumed to be a faithful reproduction of the input.

Figure 6–29*a* illustrates the effective voltage transfer plot for the threshold switch. Note that the slope of the transfer plot is approximately unity as it is for the threshold clipper. In the threshold switch, however, the transfer plot follows a path which would pass through zero if extended, as shown by the dotted line.

In order to study further the performance of the threshold switch, assume that we apply a simulated input voltage of the waveform shown in Fig. 6–29*b*. The resulting output voltage waveform would be that shown in Fig. 6–29*c*.

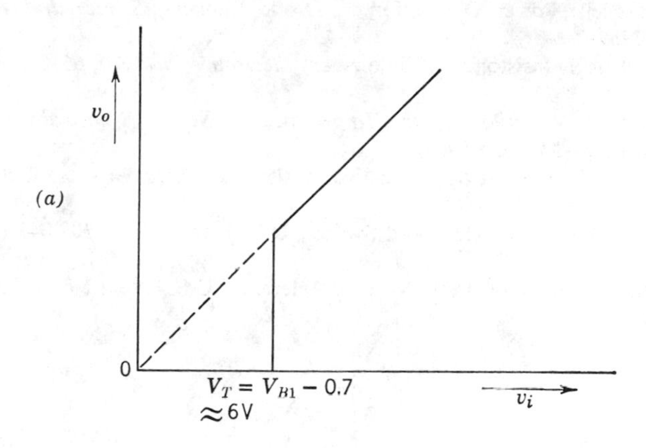

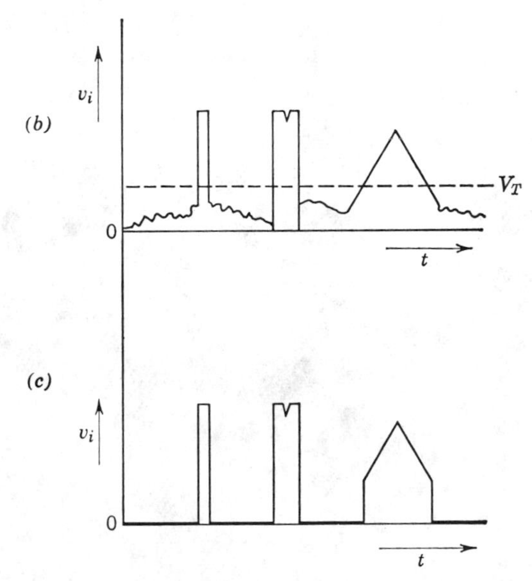

Fig. 6–29 Input-output transfer plot: (*a*) for the threshold switch; (*b*) simulated input voltage waveform; (*c*) resulting output voltage.

REFERENCES

[1] S. Hussein: High-Frequency Zener Limiters, *Electron. Equip. Eng.*, **13**, 66–69 (May 1965).
[2] W. B. Mitchell: Power Dissipation in Diode Clippers, *Semicond. Prod.*, **2**, 45–47 (October 1959).
[3] C. D. Todd: A Transistorized Tachometer, *Electron. World*, **70**, 66–67, 90 (November 1963).
[4] C. D. Todd: A Versatile Negative Impedance Converter, *Semicond. Prod.*, **6**, 25–29 (May 1963); 27–33 (June 1963).
[5] C. D. Todd: Circuits Having Negative Resistance Characteristics, U.S. Patent No. 3,223,849 (December 14, 1965).
[6] C. D. Todd: Electronic Threshold Switch, U.S. Patent No. 3,307,048 (February 28, 1967).
[7] J. J. Werth: Alternating Current Voltage Reference, U.S. Patent No. 3,102,228 (August 27, 1963).

Chapter 7

DC-VOLTMETER SCALE EXPANSION

7-1 INTRODUCTION

There are many instances in which the simple dc voltmeter consisting of a d'Arsonval meter movement and appropriate series multiplier resistor is inadequate. For example, consider the case of a 0 to 50-V meter used to monitor a voltage that varies from 47-V minimum to 49-V maximum. The total excursion is only a little larger than 4 % and would be barely detectable. A very slowly changing input voltage would produce a series of jumps in the position of the pointer because of the bearing friction associated with the more common meter movement. It would be much more convenient and the accuracy would be many times greater if we had a meter scale expanded around the nominal value of 48 V.

Consider also the case in which the normal range of interest lies around a low voltage value, but the monitored voltage occasionally increases to a much higher value. To use the normal meter circuit with a full-scale range chosen to obtain a readable value at the high voltage would greatly reduce the accuracy and readability for the lower voltages; for example, if the normal voltage of interest is 10 V but the maximum value is 100 V, then a normal 100-V meter would be used most of the time over only the lower 10% of its scale, and the possible error in the reading obtained would be roughly ten times the rated accuracy (normally 3%) of the meter movement or about 30%. Here a meter with the lower portion of the scale expanded would be most helpful.

There are several methods of obtaining an expanded-scale voltmeter. Thermal bridges, differential meter movements, mechanical suppression, or even the old "slide-back" techniques of earlier days may be used. These approaches demand either a special meter or fairly complicated auxiliary circuitry and leave quite a bit to be desired in the way of flexibility and economics for the circuit designer.

It is also possible to design scale-expansion circuits using Zener or avalanche diodes (referred to, in general, as voltage regulator or just VR diodes). Several approaches to give the desired amount of expansion over the portion of the scale of major interest are described in this chapter. By using a reasonable amount of design care, very useful meter expansion circuits may be evolved with a considerable increase in accuracy but at only a moderate cost.

7–2 WINDOW-DISPLAY EXPANDED SCALE

7–2.1 Basic Circuit

The VR-diode expanded-scale voltmeter in its simplest form is shown in Fig. 7–1. Resistor R_1 and the meter movement M_1 comprise a simple voltmeter whose full-scale voltage is equal to the width of the "window" or range of displayed voltages. Voltage regulator diode D_1 then sets a threshold to determine the lower limit of the meter.

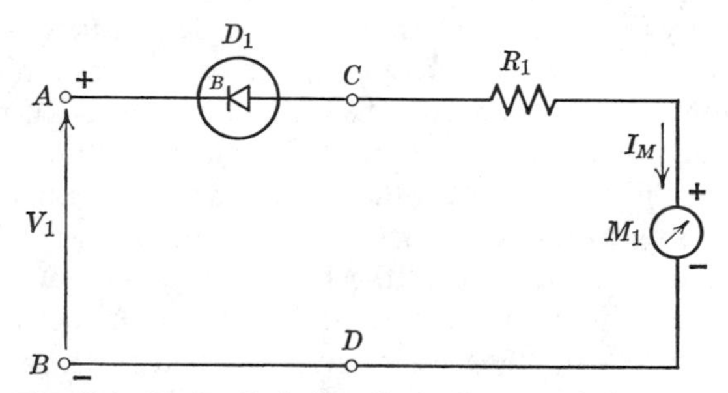

Fig. 7–1 Basic "window-display" expanded-scale voltmeter circuit.

To understand the operation of the circuit we must first consider the simplified dc equivalent circuit of the VR diode as shown in Fig. 7–2.

In this version of the equivalent circuit there are five effective components: two diodes, two voltage sources, and one resistor. Note that when the VR diode is biased in the forward direction, that is, with the anode made positive with respect to the cathode, diode D_R is reverse-biased but diode D_F is forward-biased and will conduct. D_F is equivalent to the normal high-conductance silicon diode, hence will have a voltage drop in the vicinity of 0.7 V, with the normal temperature coefficient when forward current flows. D_F represents the VR diode when in the forward-biased mode. D_F is assumed to be perfect for the simplified case when biased in the reverse direction, that is, zero leakage current and infinite breakdown voltage. For the more complete

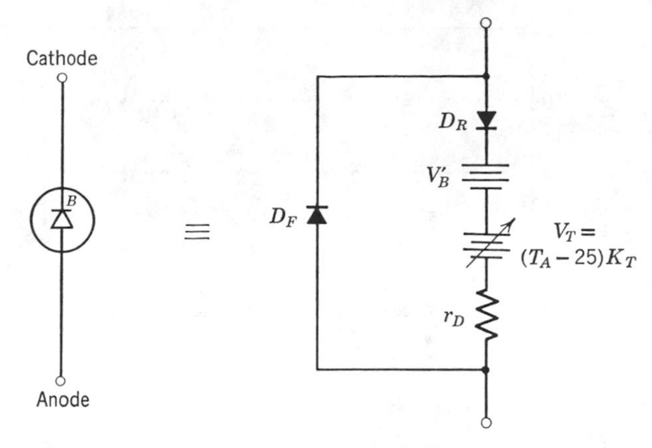

Fig. 7–2 Simplified dc equivalent circuit for a breakdown diode.

equivalent circuit, D_F may be assumed to contribute the leakage current to a first-order approximation.

If the VR diode is biased in the reverse direction by a voltage that is less than V_B', both D_F and D_R will be reverse-biased. D_R is assumed to be a perfect diode in both the forward and reverse directions, that is, when forward-biased it is assumed to have zero voltage drop and when it is reverse-biased, it is assumed to have zero leakage current. Whenever V_1 is increased to a voltage just slightly greater than V_B' (neglecting for now the source V_T which we can do if we are operating at room temperature), a current will flow since D_R becomes biased in the forward direction.

The source V_T represents a change in breakdown voltage as a result of temperature effects. It has been separated from V_B' in the equivalent circuit and may be represented by

$$V_T = (T_A - 25°\text{C})K_T. \tag{7-1}$$

K_T is a thermal voltage coefficient whose value depends on the breakdown voltage of the VR diode and the bias current. For a given VR diode and bias condition, K_T is a constant and is expressed in millivolts per degree centigrade. Typical values of K_T are shown in Fig. 7–3. V_T therefore will be expressed in millivolts and will be positive if T_A is higher than 25°C and K_T is positive. Under the same temperature condition, however, V_T may be negative if K_T is negative (for example, if the breakdown voltage is only 3 or 4 V). The total internal breakdown voltage in the equivalent circuit, V_{BK}, is the sum of V_B' and V_T or

$$V_{BK} = V_B' + V_T. \tag{7-2}$$

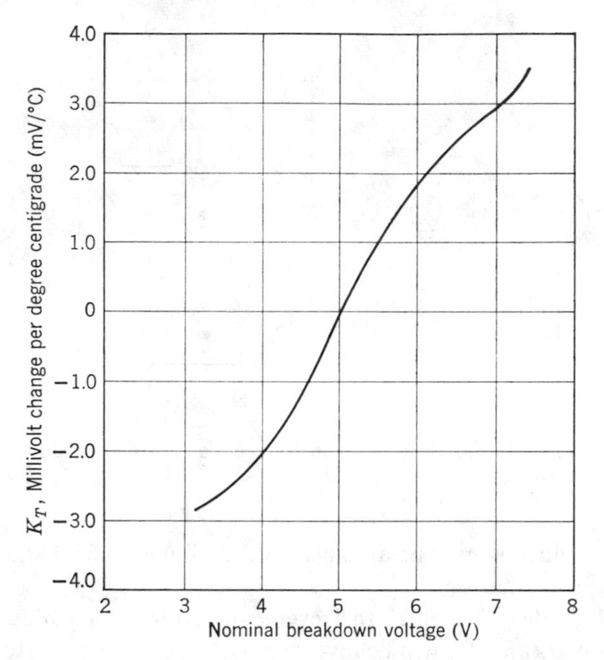

Fig. 7–3 Typical temperature sensitivity of breakdown voltage.

As viewed externally, V_{BK} is the knee voltage or the terminal voltage at which the reverse current sharply increases.

The term r_D in the equivalent circuit represents the resistance of the breakdown effect. For the dc case r_D contains a variable resistance that changes as a function of current and also a change in the breakdown voltage due to self-heating (V_T only includes *external* heating effects). For the simplified case, the ac breakdown resistance r_d is assumed to be constant and is indicated by the inverse slope of the dashed line shown superimposed on the characteristic curve of Fig. 7–4. The measured breakdown voltage of a VR diode at a reverse current I_R will be equal to the sum of the internal breakdown voltage V_{BK} and the IR drop across the equivalent resistance r_D:

$$V_B = V_{BK} + I_R r_D \tag{7-3}$$

or

$$V_B = V'_B + V_T + I_R r_D . \tag{7-4}$$

Now, in light of the equivalent circuit of Fig. 7–2, we may study the operation of the basic expanded-scale voltmeter circuit of Fig. 7–1. With V_1 positive, that is, with point A positive with respect to point B, the VR diode D_1 will be

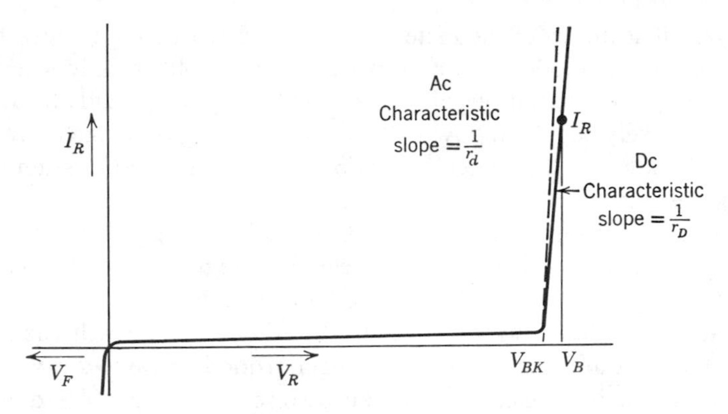

Fig. 7–4 Typical voltage-current characteristic curve of a VR diode illustrating some of the equivalent circuit parameters.

reverse-biased. As long as V_1 remains below V_{BK}, the internal breakdown voltage or the reverse voltage at which the VR diode just begins to conduct appreciable breakdown current, very little current will flow through the meter M_1.

As soon as V_1 begins to exceed the value of V_B, reverse current will flow through D_1 and hence through the meter. Since V_{BK} is assumed to remain constant with variations in current, any increase in V_1, beyond the value of V_{BK}, will be dropped across r_D and the external resistor R_1. For many cases, R_1 will be much larger than r_D and hence the diode resistance may be neglected.

R_1 and M_1 combine to form a voltmeter whose full-scale value is equal to $(R_1 + R_M)I_{M(FS)}$ where R_M is the internal resistance of the meter movement and $I_{M(FS)}$ is the meter current required for full-scale deflection.

The effect of the VR diode in Fig. 7–1 is to suppress the effective zero of the R_1-M_1 voltmeter by an amount equal to the breakdown voltage of D_1. The voltage drop across points C and D to give full-scale meter deflection $V_{CD(FS)}$ represents the additional amount by which V_1 must be increased above V_{BK} in order to produce full-scale deflection on the meter (for more accurate considerations, we should include r_D effects by including it in calculating the V_{CD} for full-scale deflection). $V_{CD(FS)}$ may also be considered as the voltage range or window over which a display is obtained.

If the breakdown characteristic of the VR diode is very abrupt, the scale deflection will be linear from the very beginning. On the other hand, if the breakdown characteristic is rounded, the first portion of the displayed scale will be nonlinear. For this reason this type of circuit works best where the required ohms-per-volt sensitivity is not very high. By using some of the better VR diodes designed for low-current applications, meter movements with full-scale current deflection of 100 μA may be used.

The overall accuracy of the meter circuit is affected by the temperature coefficient of the VR diode. This becomes a serious problem only when a substantial temperature variation is involved and becomes especially troublesome then for relatively small-window voltage around a high-voltage center value. Figure 7–3 indicates the typical variation expected for various amounts of zero suppression.

Temperature-compensated VR diodes, even in the higher voltage breakdown region, are available if we wish to eliminate temperature-induced errors. Adequate compensation may be achieved for most cases by placing forward-biased silicon diodes in series with the VR diode. These will each have a negative temperature coefficient in the neighborhood of 1.6 to 2 mV/°C. Thus a 7.5 V VR diode may be compensated by two series forward-biased junctions to yield an equivalent compensated VR diode with a breakdown of about 9 V.

Temperature compensating does increase then nonlinearity at the beginning portion of the scale since the voltage drop across the forward-biased diodes must be added to the breakdown voltage of the VR diode to establish the zero suppression and the forward characteristic of a junction is not very abrupt. Thus temperature compensation should be used only if actually necessary.

The mechanical zero on the meter will serve as a means of calibration adjustment for the entire meter over about a 5% range of window voltage. Varying R_1 will give adjustment, if necessary, of the display window.

7–2.2 High-Voltage Arrangement

The simple window-display expanded-scale voltmeter circuit may be extended to the higher voltage ranges without using the very high voltage VR diodes. This is done in the manner shown by Fig. 7–5*a*. The effect is illustrated by studying the Thevenin equivalent circuit of Fig. 7–5*b*. Note that the window value of voltage will be attenuated by the same amount as the zero suppression.

7–2.3 Design Example

The design procedure will now be summarized by an example based upon Fig. 7–1. First, we choose a VR diode having a sharp breakdown voltage equal to the amount of zero suppression desired. The value of R_1 is then found by using (7-5):

$$R_1 = \frac{V_W}{I_{M(\mathrm{FS})}} - R_M - r_D, \tag{7-5}$$

where V_W is the window-display value or the difference between the highest and the lowest voltages to be indicated on the meter scale $I_{M(\mathrm{FS})}$ is the full-scale current of meter M_1, R_M is the internal meter resistance, and r_D is the simplified VR diode breakdown resistance illustrated in Figs. 7–2 and 7–4.

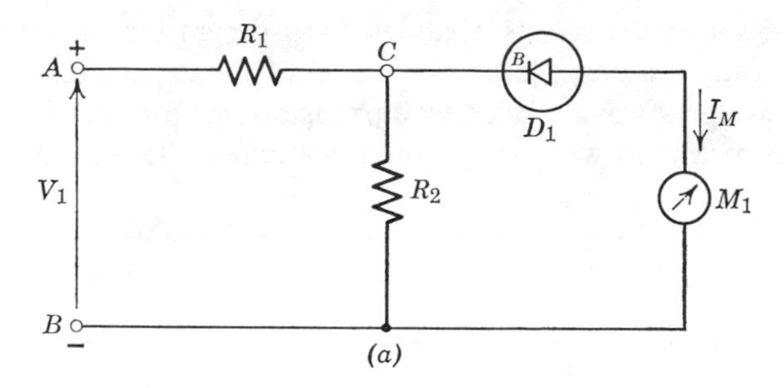

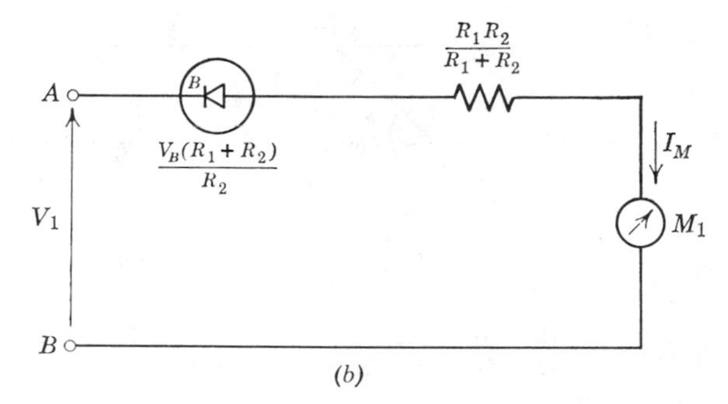

Fig. 7–5 An Expanded-Scale voltmeter approach for high voltages using lower voltage VR diodes: (*a*) circuit diagram; (*b*) Thévenin equivalent.

For a typical example suppose it is desirable to monitor a 12-V supply over a ± 1-V range. We need a zero suppression of 11 V with a display window of 2 V. Assuming that we wish to use a 1mA meter movement with 100-Ω internal resistance, R_1 should be approximately 1900 Ω if we neglect the effect of r_D, or 1830 Ω if we assume r_D to be typically 70 Ω. The nearest MIL-OHM value is 1820 Ω.

The VR diode selected for D_1 should begin a sharp breakdown at 11 V. At a breakdown current of 0.5 mA the voltage drop across D_1 should be slightly higher than 11 V. Actually, we can use a certain spread around 11 V and then use the mechanical zero adjustment on the meter for setting the exact calibration for values within about 5% of the window voltage or about ± 0.1 V in this case. Thus a standard 1N4105 diode (designed for low-current applications) may be used with the breakdown voltage selected to give the required accuracy of the zero suppression level. In some cases, where the major interest is in monitoring a change in voltage level, $\pm 5\%$ is adequate accuracy and hence a standard unit may be used without selection.

Figure 7–6 shows the results obtained for a design for the above example. In order to study the possible nonlinearities of the expansion circuit closely, the meter movement was replaced by a precision resistor and the voltage drop across it was measured with a high-accuracy differential voltmeter.

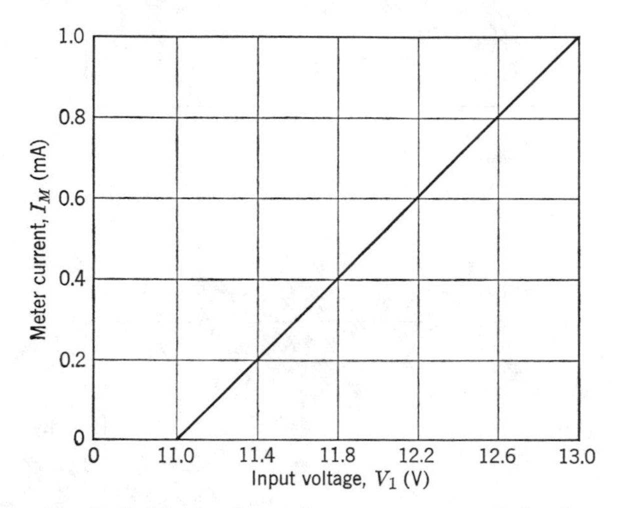

Fig. 7–6 Measured results on an expanded-scale voltmeter according to the design example for the window-display circuit.

The transition at the zero suppression level seemed to be very sharp and any nonlinearity present was within measurement error except for a few of the diodes tried. In these there was a very slight curvature at the low-voltage end of the scale which amounted to an error of about 0.5% at the most. Not all VR diodes are designed for low-current use and if they are not specifically characterized for this operating level, they should be screened to assure low leakage current up until the breakdown voltage and then a sharp, clean break in the characteristic curve.

7–3 TOTAL-DISPLAY EXPANDED UPPER SCALE

In the window-display type of expanded-scale voltmeter circuit no meter deflection is obtained until a certain threshold value of V_1 is reached. Thus no difference is seen on the meter whether the value of V_1 is actually zero or is just slightly under the zero suppression voltage. In some applications of monitoring this is an undesirable condition. It is possible to design an expansion circuit which will have a display over the total voltage range from zero up to the value which produced a full-scale reading.

7–3.1 Basic Circuit

The circuit diagram for such a voltmeter is shown in Fig. 7–7. Note that this circuit is identical with the one of Fig. 7–1 except for the inclusion of a resistor R_2 across the VR diode. As we shall soon see it is this added resistor that allows the total display feature.

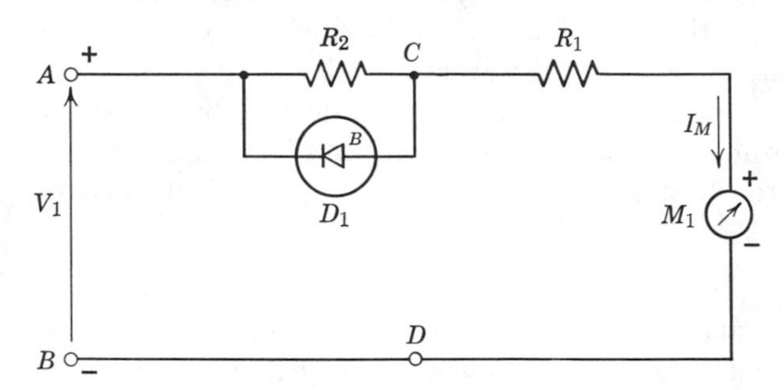

Fig. 7–7 Total-display, expanded-upper-scale voltmeter circuit.

For positive input voltages less than the breakdown voltage of the VR diode, we may assume that no current flows through D_1. There is a current path through R_2, however; R_2 combines with R_1 to form the total series multiplier resistance for a voltmeter that would have a full-scale reading of $V_{L(FS)}$ if D_1 were not present. This full-scale value is given by

$$V_{L(FS)} = (R_1 + R_2 + R_M)I_{M(FS)}, \tag{7-6}$$

where R_M is again the internal resistance of the meter movement and $I_{M(FS)}$ is the full-scale current for the meter.

At the point where the voltage drop across R_2 due to the current flowing through it becomes equal to the internal breakdown voltage, V_B^*, of the VR diode, an additional current path is provided through D_1. In fact, once D_1 has entered its breakdown region, the circuit of Fig. 7–7 operates in exactly the same way as the circuit of Fig. 7–1 previously discussed. The overall effect of the VR diode in this circuit, however, is to change the range of the voltmeter in the middle of the meter scale, whereas before we only achieved zero suppression. Note that on both sides of the transition point the voltmeter scale will be linear.

7–3.2 Design Procedure

The design procedure for the total-display expanded upper scale voltmeter is fairly straightforward although we do not have the independence and freedom of control we had in the earlier case. Here it is best to start with the expanded-scale portion since the effect of R_2 in this region is eliminated. The value of R_1 is computed from the effective window voltage requirements as before:

$$R_1 = \frac{V_{1(\mathrm{FS})} - V_{1T}}{I_{M(\mathrm{FS})} - I_{MT}} - R_M - r_D; \tag{7-7}$$

$V_{1(\mathrm{FS})}$ is the value of V_1 for which a full-scale meter current of $I_{M(\mathrm{FS})}$ will be obtained. V_{1T} is the transition value of V_1 at which expansion commences and I_{MT} is the meter current at the expansion threshold point.

We may now compute the value of the internal breakdown voltage of the VR diode from the value of input voltage at which expansion is to commence in relation to the position on the meter scale at which this voltage is to be displayed.

$$V_{BK} = V_{1T} - (R_1 + R_M + r_D)I_{MT}. \tag{7-8}$$

The only remaining component to choose is R_2. Its value is such that when a meter current equal to I_{MT} is flowing through it, a voltage drop exactly equal to V_{BK} is developed across it or

$$R_2 = \frac{V_{BK}}{I_{MT}}. \tag{7-9}$$

Note that the effective window of the expanded portion of the meter scale covers only the region from I_{MT} to $I_{M(\mathrm{FS})}$. Without R_2 the window would cover the entire scale and would have a value described by

$$V'_W = (R_1 + R_M + r_D)I_{M(\mathrm{FS})}. \tag{7-10}$$

With R_2 added to the circuit, a portion of the scale is taken up by displaying voltages from zero to V_{1T} and the true window becomes

$$V_W = V_{1(\mathrm{FS})} - V_{1T}. \tag{7-11}$$

7–3.3 Example

Let us now consider a practical example. Suppose we wish to monitor a 28-V supply mainly over a range of ± 2 V around the nominal but wish to know roughly how low the voltage is should it drop below the lower limit of 26 V. Assume a meter movement of 1 mA with R_M of 100 Ω as before, and assume that the first 26 V will be displayed over the first 20% of the scale with the remaining scale displaying input voltages from 26 to 30 V.

First we may calculate the value of R_1 by using (7-7) and inserting the known values:

$$R_1 = \frac{30 - 26}{(1 - 0.2) \times 10^{-3}} - 100 - r_D = 4900 - r_D. \tag{7-12}$$

If we assume a value of 150 Ω for r_D, then R_1 is 4750 Ω.

Next we may calculate the required breakdown voltage by using (7-8).

$$V_{BK} = 26 - (4750 + 100 + 150)(0.2 \times 10^{-3}) \tag{7-13}$$
$$= 25 \text{ V}.$$

Note that this is the breakdown voltage at the *knee* of the characteristic curve. We may select a 1N4117 VR diode, which has a nominal breakdown voltage of 25 V (measured at 250 μA). Depending on the accuracy desired, it may be necessary to choose a unit with a breakdown on the low side of nominal. We do not generally use the mechanical zero adjustment on the meter movement as we could for the simple window-display circuit since this would produce substantial error in the lower scale readings. The final component R_2 is then calculated by using (7-9) as 125,000 Ω.

In this example the circuit at low voltage starts out as though we had a voltmeter of 130 V full scale. When the input voltage V_1 reaches 26 V, however, the expansion circuit effectively changes the range to a 30 V meter with 25 V of zero suppression.

Figure 7–8 illustrates the results obtained for a design for the above example.

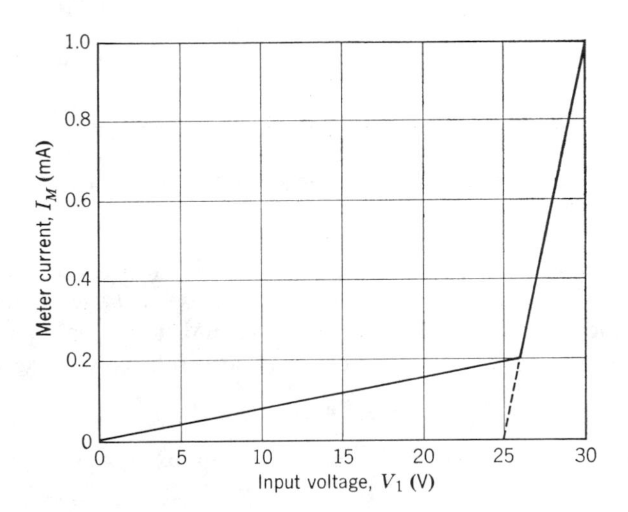

Fig. 7–8 Measured results according to the design example for the expanded-upper-scale voltmeter circuit.

For the earlier case the instrumentation was modified slightly to measure accurately the performance of the expansion circuit. Note that both portions of the graph are linear and the transition is sharp. The dotted line indicates the characteristic that would have resulted if R_2 were omitted.

7–4 TOTAL-DISPLAY EXPANDED LOWER SCALE

We have studied two types of scale-expansion circuit and now consider the condition in which the main area of interest lies on the lower voltage portion of the meter scale, but it is required to have some idea just how far above the nominal range an input voltage might be.

7–4.1 Basic Circuit

The circuit of Fig. 7–9 illustrates a means of accomplishing an expanded lower scale (or compressed upper scale), and, as we will note as we study its operation, the VR diode serves to switch the voltmeter range automatically. No zero suppression is present and linear operation is again achieved over the individual portions of the scale.

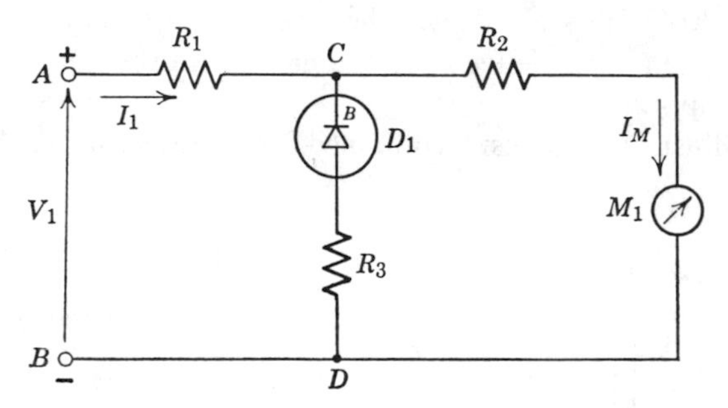

Fig. 7–9 Total-display, extended lower-scale-voltmeter circuit.

For low values of input voltage such that the voltage appearing across points C and D is less than the internal breakdown voltage of the VR diode D_1, negligible current will flow through the path provided by D_1 and R_3. We then have a simple voltmeter circuit whose full-scale range is given by

$$V_{1(\mathrm{FS})L} = (R_1 + R_2 + R_M)I_{M(\mathrm{FS})}, \tag{7-14}$$

and the meter deflection continues along the relation given by (7-15):

$$I_{ML} = \frac{V_1}{R_1 + R_2 + R_M}, \tag{7-15}$$

where I_{ML} represents the lower scale meter current under the condition that the input voltage is less than the desired transition point on the scale.

Consider what happens, however, just as the voltage V_{CD} across points C and D exceeds the breakdown voltage of D_1. An additional shunt path is provided for the input current such that the amount reaching the meter movement is reduced by an amount controlled by the relation of R_2 to the value of R_3. The input voltage at which the transition occurs, V_{1T}, will be that for which the value of V_{CD} is equal to V_{BK} or

$$V_{1T} = V_{BK}\left(\frac{R_1 + R_2 + R_M}{R_2 + R_M}\right). \tag{7-16}$$

For values of input voltage above V_{1T}, the meter current varies according to (7-17).

$$I_M = \frac{V_1(R_3 + r_D) + V_{BK} R_1}{R_1 R_2 + R_2 R_3 + R_1 R_3}. \tag{7-17}$$

Equation 7-17 indicates that the meter current is a linear function of the input voltage and thus the scale calibration will also be linear.

7–4.2 Design Procedure

Let us now consider the design procedure to be used in determining the various component values to meet a specific set of requirements. We assume that a given specification consists of V_{1T} or the voltage at which the transition in scale expansion is to occur, I_{MT}, the meter current at the transition point (or the ratio of I_{MT} to $I_{M(\mathrm{FS})}$, the full-scale meter current); $V_{1(\mathrm{FS})}$, the input voltage required to cause full-scale meter current; and a given meter movement with full-scale current, $I_{M(\mathrm{FS})}$, and internal resistance, R_M.

If we analyze mathematically all the requirements and the various interrelations, we find that we have freedom to choose one of the unknown component values. For further study here we assume that the breakdown voltage of the VR diode is the chosen quantity. Its value must always be less than the value of V_{1T} and should, in general, be less than one-half of this voltage. VR diodes having very low voltage breakdowns usually have a rather rounded breakdown characteristic and would prevent an abrupt transition from the expanded scale to the compressed scale. High breakdown voltage diodes have a higher temperature coefficient and we shall be better off to use a V_{BK} low enough to give us a sharp breakdown.

We know that exactly at the transition point the voltage between points C and D must be equal to the internal or beginning breakdown voltage, V_{BK}, of VR diode D_1. This leads us to a calculation of R_2 using (7-18).

$$R_2 = \frac{V_{BK}}{I_{MT}} - R_M \tag{7-18}$$

In a similar manner,

$$R_1 = \frac{V_{1T} - V_{BK}}{I_{MT}}. \tag{7-19}$$

The final component value to be found is that of R_3. It is this resistor that determines the value of V_1 required to produce a full-scale meter deflection. Since R_3 does not affect the circuit performance for the lower voltage portion of the scale or even at the transition point, we have complete independence of the lower and upper voltage scales. Equation 7-20 gives the relation of R_3 to the other component and specification values:

$$R_3 = R_1 \frac{(R_2 + R_M)\, I_{M(\mathrm{FS})} - V_{BK}}{V_{1(\mathrm{FS})} - (R_1 + R_2 + R_M) I_{M(\mathrm{FS})}} - r_D. \tag{7-20}$$

7–4.3 Example

For an illustrative example let us assume that we need to design a voltmeter with a span of 0 to 30 V over the first half of its scale but requiring 100 V for full-scale deflection. As in the previous examples, we assume that a 1 mA meter having an internal resistance of 100 Ω is available.

Let us first assume that an 11-V VR diode is to be used for D_1. Later, we shall see what would happen if a lower or higher voltage unit had been assumed.

Using (7-18) to give us the value of R_2,

$$R_2 = \frac{11}{0.5 \times 10^{-3}} - 100 = 21{,}900\ \Omega. \tag{7-21}$$

The nearest standard value is 22,100 Ω which is 1% high.

The value of R_1 is obtained by using (7-19):

$$R_1 = \frac{30 - 11}{0.5 \times 10^{-3}} = 38{,}000\ \Omega. \tag{7-22}$$

The nearest standard value is 38,300 Ω which is 0.8% high. This will give an error of about 1% (in addition to the resistance and meter tolerances) in the low-voltage range but will not produce appreciable additional error in the transition voltage; R_3 is now calculated, using (7-20), to be 10,700 Ω, a standard value.

Suppose we had chosen a value of 7 or 15 V for the VR diode breakdown voltage. A summary chart is given in Table 7-1 for the component values for

Table 7–1. Component Values for Three Choices of VR Diode Breakdown Voltage

V_{BK}	R_1	R_2	R_3	$I_{1(FS)}$
7 V	46,400 Ω	13,700 Ω	7,800 Ω	1.9 mA
11 V	38,300 Ω	22,100 Ω	10,700 Ω	2.0 mA
15 V	30,100 Ω	30,100 Ω	11,500 Ω	2.3 mA

the three possible values of V_{BK}. The last column gives the input current that is required to produce full-scale deflection on the meter. As the value of V_{BK} is made closer to the transition voltage, then the more shunting effect is necessary in the D_1-R_3 branch for the higher voltages and hence the larger the required current. Up to the point of transition the voltmeter has the normal ohms-per-volt rating (1000 Ω/V for a 1-mA meter). The current required above V_{1T} may yield an effective ohms-per-volt sensitivity that is several times lower than the basic sensitivity of the meter movement.

We have spoken of the circuit of Fig. 7–9 as being an expanded-scale arrangement. In actuality we have a compression at the higher portion of the scale. We may utilize only a small portion of the top end of the scale to display a voltage several times that covered in the lower scale.

7–5 OTHER CIRCUITS

By combining two or more of the arrangements discussed earlier, it is possible to design meter scale displays that will meet almost any need. For example, the window-display circuit of Fig. 7–1 may be combined with the total-display expanded upper-scale circuit of Fig. 7–7 to give a voltmeter with a suppressed zero, with the lower voltage above the threshold value readings compressed at the low end of the scale and then the upper end of the scale displaying an expanded range of voltages. If we wished to go a step further, we could add on a portion of the compression circuit of Fig. 7–9 to give a restoration to the scale at the very high end.

A simple three-range voltmeter may be designed using two shunting branches of the form used in Fig. 7–9. Two VR diodes having different breakdown voltages are placed in series with their individual resistor. The effect is to switch in the shunts as the input voltage is increased and thus give automatic range switching. Because of the current loading problem, this may be used only when the monitored source is capable of supplying the desired amount of power.

REFERENCES

[1] A. J. Corson: On the Application of Zener Diodes to Expanded Scale Instruments, *Trans. AIEE*, **77**, Part 1, 535–539 (1958). *Comm. and Electron.*, No. 38 (September 1958).

[2] P. D. King: Design of a High Accuracy Expanded Scale Meter Using Zener Diodes, *Semicond. Prod.*, **3**, 26–28 (November 1960).

[3] J. Monroe and Martin Gindoff: Zener Stabilized Bridges, *Instr. Control Systems*, **35**, 94–95 (January 1962).

[4] P. C. Tandy: Bridge Circuit for Measuring Differences of Current or Voltage from a Preselected Value, *IEEE Trans. Instr. Meas.* **1–12**, 34–40 (June 1963).

Chapter 8

AC EXPANDED-SCALE VOLTMETERS

8–1 INTRODUCTION

Some ac voltage measurements demand a resolution or the ability to detect small changes in the magnitude that is smaller than may be easily detected on the normal ac voltmeter. Frequently, the voltage to be monitored varies over a narrow range about a center value and readings much below this range are of little interest. This indicates the need for an ac voltmeter with a suppressed zero.

A rectifier and filter network may be added to any of the dc expanded-scale voltmeter circuits described in Chapter 7 to obtain the same characteristics of scale expansion and compression. There are several direct approaches to the design of ac expanded-scale voltmeter circuits that deserve consideration because of their basic simplicity and unique features.

In the circuits to be described in this chapter only the "window-display" type of scale expansion will be included. This is the simple effective zero suppression arrangement with a more or less linear scale above a design threshold value. Unless specifically stated, all input voltages are assumed to be of sinusoidal waveform, free of any dc component, and of such frequency that the period is very small in comparison with the response time of the meter movement used in conjunction with the circuits. In addition, we assume that the dynamic resistance of the VR diode is negligible in comparison with circuit resistances.

8–2 SERIES-TYPE EXPANSION CIRCUIT

The basic dc expanded-scale voltmeter circuit in its simplest form was shown in the circuit diagram of Fig. 7–1. As long as the input voltage V_1 is less than the breakdown voltage of the VR diode, practically no current will flow in the diode or meter. After V_1 is increased beyond V_{B1}, the breakdown voltage of D_1, the voltage across the VR diode becomes relatively fixed and hence any further increase in V_1 is applied to the simple voltmeter circuit consisting of the meter movement and a series multiplier resistor R_1. The net voltage magnitude that is equal to V_{B1} and yet retain the same total voltage range covered by the scale reading as possessed by the R_1-M_1 voltmeter circuit.

Figure 7–6 illustrates the result obtained in plotting the meter current as a function of the input voltage. The sharpness of the break in the plot is a direct function of the sharpness of the diode's breakdown characteristic.

8–2.1 Basic AC Circuit

The basic dc expanded-scale voltmeter reviewed briefly above and discussed in detail in Chapter 7 may be modified to allow its use with ac voltages by inserting a series rectifier diode as illustrated in Fig. 8–1*a*. The effect will be best understood by carefully studying the waveforms of the input and across terminals C and D as shown in Figs. 8–1*b* and *c*.

Assume that the input voltage is momentarily positive. As before the meter current will be negligible as long as the instantaneous input voltage is less than V_{B2}, the value of the breakdown voltage of the VR diode. As soon as v_1 exceeds V_{B2} by an amount sufficient to cause the rectifier diode D_1 to conduct any further increase in the input voltage will cause an approximately equal instantaneous rise in the value of v_{CD}, the instantaneous total value of the voltage across terminals C and D. Thus meter current flows only for the portion of the waveform shown shaded in Fig. 8–1*b*; the resulting waveform for the instantaneous voltage v_{cd} is indicated in Fig. 8–1*c*. Note that whenever v_1 is negative, diode D_1 will be reverse-biased and hence negligible current will flow.

8–2.2 Analysis

If the meter movement is of the normal D'Arsonval type, its deflection will be proportional to the average value of the current flowing through it. The current waveform would look the same as that for the voltage v_{CD} and its value will be equal to v_{CD}/R_1 (the meter movement is assumed to have no additional resistance). We may express the value of the average meter current, I_M, by mathematical expressions involving the circuit parameters.

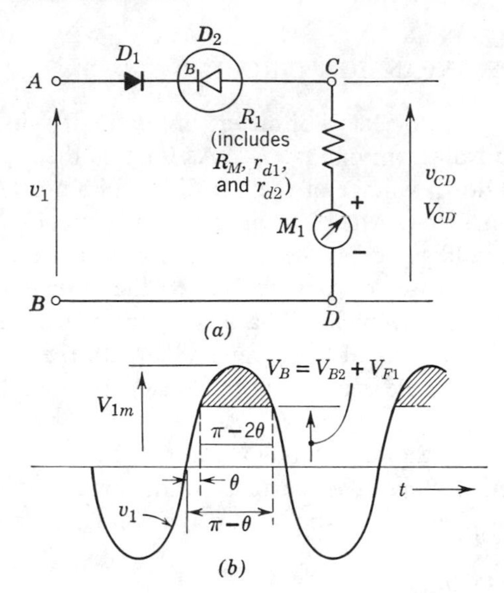

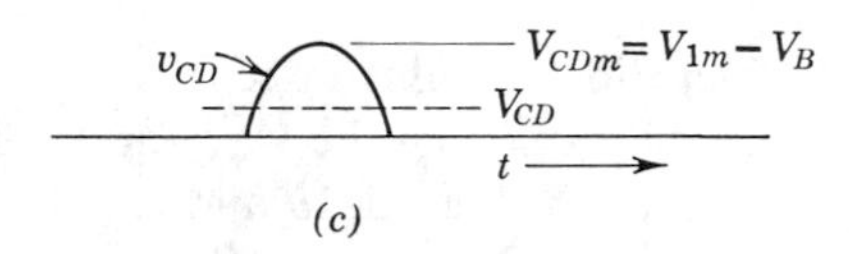

Fig. 8–1 A simple ac expanded-scale voltmeter circuit: (*a*) schematic diagram; (*b*) input voltage waveform—current flows only for shaded portion; (*c*) waveform of voltage across terminals *C* and *D*.

$$I_M = \frac{\int_{\theta}^{\pi-\theta} (V_{1m} \sin \omega t \, d\omega t) - V_B(\pi - 2\theta)}{2\pi R_1} \tag{8-1}$$

$$= \frac{V_{1m}}{\pi R_1} \cos \theta - \frac{V_B}{R_1}\left(0.5 - \frac{\theta}{\pi}\right). \tag{8-2}$$

In order to carry the analysis further, we may normalize (8-2) by dividing by V_B/R_1) and substituting for θ.

$$\frac{I_M}{V_B/R_1} = \frac{V_{1m}}{\pi V_B} \cos \theta - 0.5 + \frac{\theta}{\pi}, \tag{8-3}$$

$$\frac{I_M}{V_B/R_1} = \frac{\sqrt{(V_{1m}/V_B)^2 - 1}}{\pi} - 0.5 + \frac{\sin^{-1}(V_B/V_{1m})}{\pi}. \tag{8-4}$$

Equation 8-4 is plotted in Fig. 8–2 for values of V_{1m}/V_B from zero to six. The dotted line represents the dc expanded-scale circuit discussed in Chapter 7. Although the two circuits are very similar in appearance, their performance differs considerably. For the dc circuit the voltage across terminals C and D; hence the meter current rises linearly once the input voltage exceeds the threshold value equal to the breakdown voltage of the VR diode. There might be a very slight curvature right at the transition point, but for the most part the current changes abruptly and the change in V_{CD} is approximately equal to the change in the input voltage.

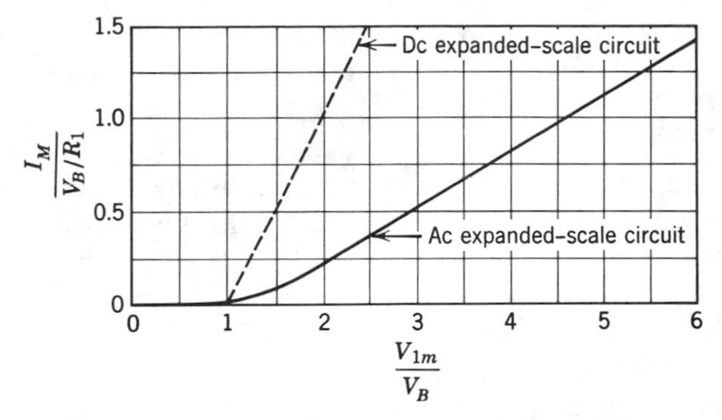

Fig. 8–2 Transfer plot for the ac expanded-scale voltmeter of Fig. 8–1 compared with the dc circuit of Fig. 7–1.

For the ac expanded-scale voltmeter circuit the meter current is again zero until the maximum input voltage, V_{1m} exceeds the threshold value that is slightly greater than the breakdown voltage of the VR diode. However, the transfer plot *is not linear* because a given change in the input voltage does not produce a linear change in the *average* voltage across terminals C and D. When V_{1m} just barely exceeds the threshold value, current flows only for a small portion of the cycle and the meter current resulting is quite small. As V_{1m} becomes larger and larger, a constant increase in its magnitude will produce a greater change in the meter current. As illustrated in Fig. 8–2, the transfer plot does approach a linear condition fairly rapidly, and, for all intents and purposes, becomes a straight line for values of V_{1m} greater than about $2V_B$. Since most expanded-scale voltmeters are designed to indicate a voltage range or window that is considerably less than one-half the full-scale voltage value, we are forced to operate in the curved portion of the operating characteristic and hence the meter scale must be nonlinear. Note that the slope of the transfer plot approaches a value of about 0.38, whereas the slope for the dc expanded-scale voltmeter is very nearly unity.

8–2.3 Variations

We may install a full-wave rectifier bridge ahead of the basic dc expanded-scale voltmeter circuit to give a slope twice as large as for the single rectifier case. The normalized meter current will be

$$\frac{I_M}{V_B/R_1} = \frac{2\sqrt{(V_{1m}/V_B)^2 - 1}}{\pi} - 1 + \frac{2\sin^{-1}(V_B/V_{1m})}{\pi}. \tag{8-5}$$

The shape of the transfer plot will be the same as that described by (8-4) and shown in Fig. 8–2, but the ordinate scale will be multiplied by two. Thus the curvature in the characteristic will still be present; and, in fact, will be even more evident since we must operate further down in the curved region for a specific value of full-scale meter current and full-scale voltage.

8–2.4 Example

Suppose that we wished to design a simple expanded-scale ac voltmeter giving meter indication over a range of voltages from 50 to 150 V rms The basic D'Arsonval meter to be used has a full-scale sensitivity of 0.5 mA.

The meter current begins to flow when the peak input voltage is equal to the breakdown voltage V_B (which is the sum of the forward voltage drop of diode D_1 and the breakdown voltage of D_2). For this example the meter current should begin at an input voltage corresponding to the lower display value of 50 V rms or 70.7-V peak. Thus the value of V_B should be 70.7 V and the breakdown voltage of the VR diode should be approximately 70 V.

Now we need to determine the value of resistor R_1 which will cause M_1 to read full scale when the input voltage is 150-V rms or 212.1-V peak.

We may use the curve given in Fig. 8–2 to obtain the design value or we may work from (8-4). When we use the graphical approach, Fig. 8-2 tells us that when the peak input voltage corresponding to full scale is $(150\sqrt{2})/(50\sqrt{2})$ or 3.0 then

$$\frac{I_M}{V_B/R_1} = 0.51$$

or

$$R_1 = \frac{(0.51)(70.7)}{0.5 \times 10^{-3}} = 72.1\text{ K}.$$

If we prefer, we could use (8-4) to solve for the required value of R_1 by substituting the known quantities corresponding to full-scale conditions:

$$\frac{0.5 \times 10^{-3}}{70.7/R_1} = \frac{\sqrt{(3)^2 - 1}}{\pi} - 0.5 + \frac{\sin^{-1}\frac{1}{3}}{\pi},$$

$$R_1 = 71.9\text{ K}.$$

The values of R_1 computed above include the meter resistance and assume that the breakdown voltage of D_2 is exactly 70 V. As a practical matter we would use a VR diode similar to a 1N4130 which has a breakdown voltage range of 64.6 to 71.4 V. It would then be necessary to make a portion of R_1 variable for calibration purposes. Some adjustment in the low-scale reading is provided by the mechanical zero adjustment of the meter.

8–3 LINEARIZED SERIES CIRCUITS

The performance of the expanded-scale voltmeter shown in Fig. 8–1 may be improved by adding another rectifier diode and resistor as shown in Fig. 8–3*a*. Let us analyze the characteristics of this circuit whose voltage and current waveforms are indicated by Figs. 8–3*b* and *c*.

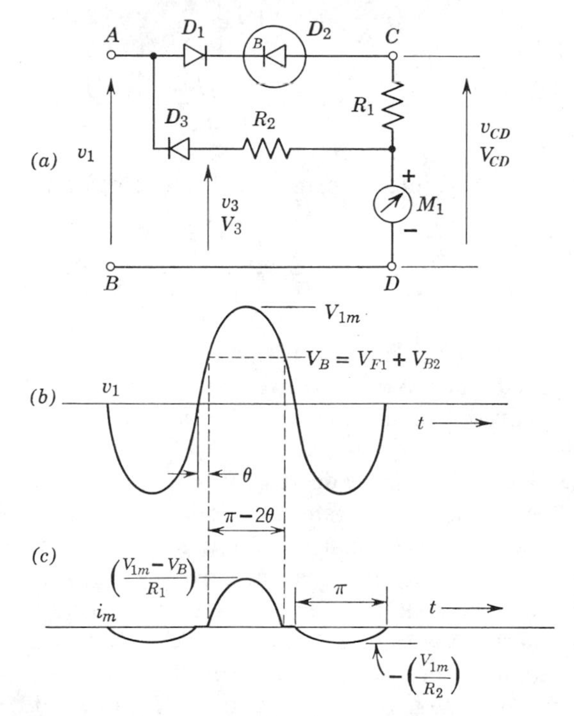

Fig. 8–3 Ac expanded-scale voltmeter using a linearized series circuit: (*a*) schematic; (*a*) input voltage waveform; (*c*) meter current waveform.

8–3.1 Analysis

During the positive half of the input voltage cycle the action of the circuit will be identical to the previous approach. Thus the contribution of positive meter current, $I_{M(+)}$, will be the same as given in normalized form by (8-4). No current will flow through D_3 or R_2 since D_3 is reverse-biased.

When the input voltage is negative, diode D_1 blocks the current flow as in the basic circuit of Fig. 8–1; a current, however, will now flow through D_3 and R_2 and provide a negative meter current as indicated by the meter current waveform of Fig. 8–3*c*. The average value of this current, assuming the forward-voltage drop for D_3 to be negligible, is

$$I_{M(-)} = \frac{-V_{1m}}{\pi R_2}. \tag{8-6}$$

The net average meter current will be the sum of $I_{M(+)}$ and $I_{M(-)}$.

$$I_M = I_{M(+)} + I_{M(-)}, \tag{8-7}$$

$$I_M = \left[\frac{\sqrt{(V_{1m}/V_B)^2 - 1}}{\pi} - 0.5 + \frac{\sin^{-1}(V_B/V_{1m})}{\pi}\right]\left(\frac{V_B}{R_1}\right) + \left(\frac{-V_{1m}}{\pi R_2}\right). \tag{8-8}$$

This may be converted into a normalized form as before:

$$\frac{I_M}{V_B/R_1} = \frac{\sqrt{(V_{1m}/V_B)^2 - 1}}{\pi} - 0.5 + \frac{\sin^{-1}(V_B/V_{1m})}{\pi} - \left(\frac{V_{1m}R_1}{\pi V_B R_2}\right). \tag{8-9}$$

Equation 8-9 is plotted in Fig. 8–4 for several different R_2/R_1 ratios over a range of normalized input voltages from zero to six. The curve for R_2/R_1 equal to infinity is, of course, the same as that shown in Fig. 8–2 for the basic ac circuit.

Note that as the R_2/R_1 ratio is decreased more negative current is applied to the meter. This causes the transfer plot to change in two ways. First, the presence of the negative current causes the entire curve to be translated downward, and, in fact, the meter will read negatively for any input voltage less than that value at which the transfer characteristic crosses the zero current line. A second effect in the transfer characteristic caused by the negative current is the reduction in slope. This is due to the linear increase in the negative current when the magnitude of the input voltage is increased.

A very desirable effect occurs in that with an adequate amount of negative current provided by making R_2/R_1 between about 2 and 3, the nonlinear portion of the curve all falls below the zero current level on the meter and thus the resulting voltage scale above zero will be practically linear. This approach also has the feature of calibration possibilities by making R_1 and R_2 adjustable. The threshold voltage when the meter current is zero may be varied by

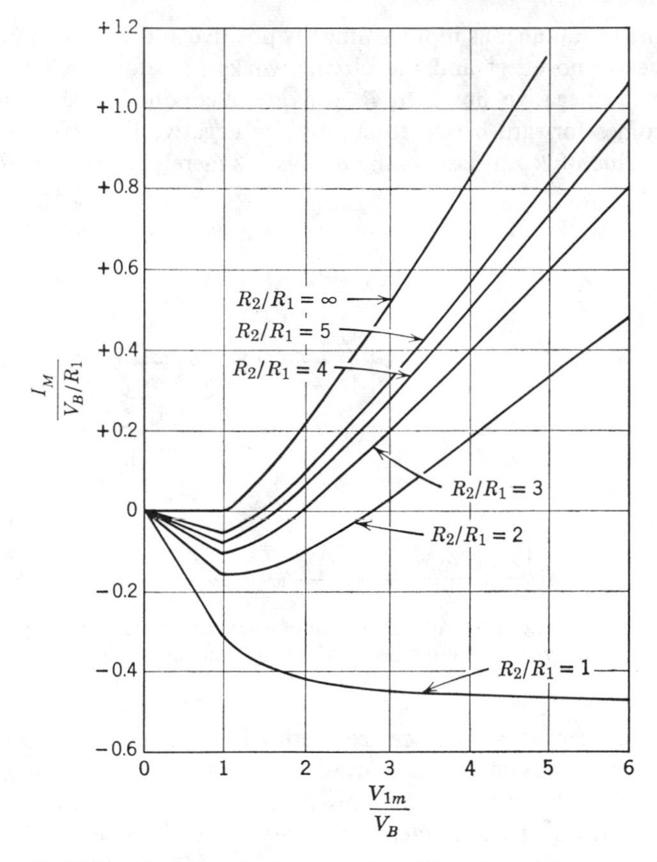

Fig 8–4 Normalized meter current as a function of the input voltage for the ac expanded-scale voltmeter of Fig. 8–3.

controlling the R_2/R_1 ratio, and the value of voltage required for full scale is controlled by the actual value of R_1. Although the curves of Fig. 8–4 indicate that the normalized value of the meter current is dependent only on the ratio of R_2 to R_1, the actual current that flows through the meter is an inverse function of the value of R_1.

Having the facility to adjust both the threshold voltage and the window voltage puts less demand on selecting the breakdown voltage of VR diode D_2. In the previous approach the VR diode must be carefully selected for accurate calibration beyond a slight zero adjustment on the meter movement.

The circuit of Fig. 8–3*a* may be rearranged slightly to give the same electrical performance as described in the analysis, but with one less diode. The circuit is given in Fig. 8–5. A resistor R_3 is placed in shunt with the rectifier diode, D_1, and the D_3-R_2 branch is eliminated.

When the instantaneous input voltage is positive and D_1 is conducting, R_3 has practically no effect and the circuit works as before. When the input voltage swings negative, however, R_3 provides a conducting path and the VR diode becomes forward-biased to supply the negative linearizing offset current. The value of R_2 as used in the analysis is merely the sum of R_3 and R_1 in Fig. 8–5.

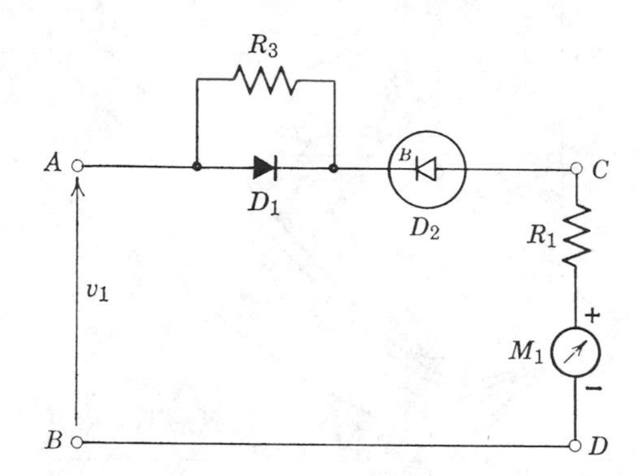

Fig. 8–5 Alternate form of ac expanded-scale voltmeter using a linearized series circuit.

In the arrangement of the linearized series expansion circuit just described, the transfer curve was intentionally moved downward so that it causes the nonlinear portion of the curve to fall below the zero current point. Unfortunately, the slope of the transfer characteristic decreased at the same time because the negative current was a linear function of the magnitude of the input voltage. Thus the characteristic was shifted more at higher voltage levels.

A preferable arrangement might be to supply the negative current from an external dc supply if available (or, if D_1 and D_2 are reversed in polarity, a positive current may be used). This would merely cause a downward translation of the transfer characteristic *without* loss of slope.

Another way in which we may accomplish a relatively constant shift in the characteristics is shown in Fig. 8–6. The added series rectifier D_3, in Fig. 8–3*a*, has been replaced by a shunt VR diode in Fig. 8–6*a*. The network consisting of R_2, R_3, and D_3 produces a rectified and limited current waveform as shown in Fig. 8–6*d*. The average value of this current constitutes the negative current used to shift the transfer characteristic and thus linearize the scale. Although some variation in $I_{M(-)}$ does occur even beyond the point at which the input voltage is sufficient to cause D_3 to enter the breakdown region, the amount is greatly diminished and the slope of the characteristic curve will be nearly the same as for the original circuit of Fig. 8–1*a*.

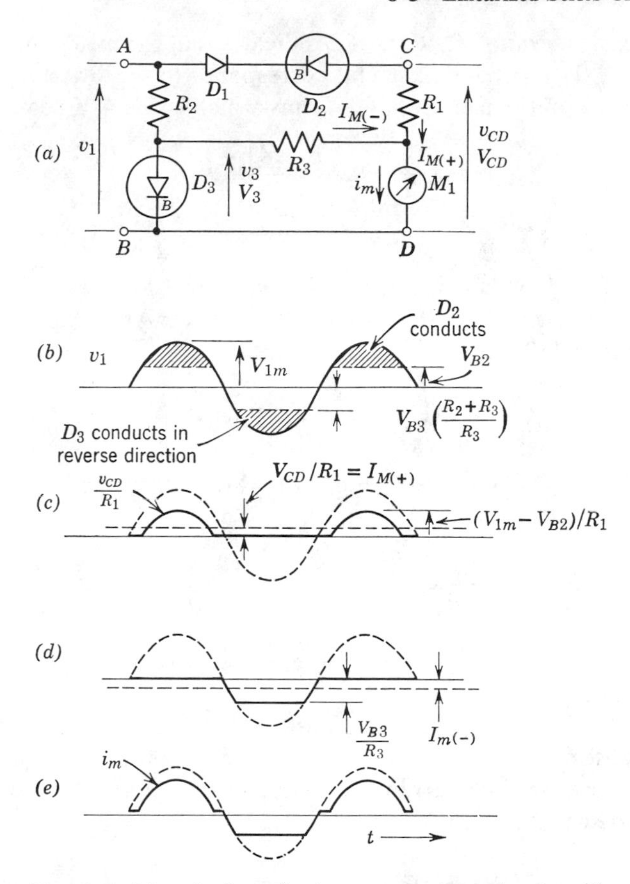

Fig. 8–6 Linearized series-type expansion circuit with improved characteristics: (*a*) circuit; (*b*) Input voltage waveform; (*c*) meter current contribution by $D_1 - D_2 - R_1$ network $= 1_{(M+)}$; (*d*) meter current due to $R_2 - R_3 - D_3$ network $= 1_{(M-)}$; (*e*) total meter current waveform.

8–3.2 Example of a Linearized Series Circuit

As a typical example of an ac expanded-scale voltmeter circuit using the linearized series circuit of Fig. 8–3, assume we need a window display covering the range of 100-to-130-V rms. The meter movement to be used has a full-scale sensitivity of 0.5 mA.

For simplicity we shall use the graphical approach by considering the transfer curves of Fig. 8–4. In order to obtain a linear scale deflection, we should choose a ratio R_2/R_1 or less than 3 so that the curved portions of the transfer function lie below zero. Let us assume a value for R_2/R_1 of 2.

From Fig. 8–4 a ratio of $R_2R/_1$ of 2 indicates that the meter current will be zero when V_{1m}/V_B is equal to 2.8. This corresponds to the lowest input voltage to be displayed on the meter or 100-V rms which is 141.4-V peak.

$$\frac{V_{1m}}{V_B} = 2.8 = \frac{141.4}{V_B},$$

$$V_B = 50.5 \text{ V}.$$

The actual breakdown voltage of the VR diode should be about 50 V.

Now the meter should read full scale ($I_M = 0.5$ mA) for an input voltage of 130-V rms or 183.8-V peak. At this point

$$\frac{V_{1m}}{V_B} = \frac{183.8}{50.5} = 3.64.$$

Looking again at the curve in Fig. 8–4 which corresponds to $R_2/R_1 = 2$, we find that for $(V_{1m}/V_B) = 3.64$

$$\frac{I_M}{V_B/R_1} = 0.13,$$

$$R_1 = \frac{(0.13)(50.5)}{0.5 \times 10^{-3}} = 13.1 \text{ K}.$$

Since R_2/R_1 was chosen to be 2, R_2 must be $2R_1$ or 26.1 K. Either R_2 or R_1 could be made variable to achieve precise calibration at one voltage value.

Should we use the circuit arrangement shown in Fig. 8–5, R_1 and R_3 should each be made equal to 13.1 K.

8–4 SHUNT-TYPE EXPANSION CIRCUIT

The VR diode may also be used in a shunt mode to produce an expanded-scale meter for ac voltages. Such an arrangement is shown in Fig. 8–7.

8–4.1 Analysis

When the input voltage is small, neither D_1 nor D_2 ever conducts and the voltage labeled as v_{CD} is an ac sinewave voltage as indicated in Fig. 8–7*b*. The dc average voltage and hence the meter deflection will be zero.

As soon as V_1 increases beyond the point that causes the VR diode D_2 to enter the breakdown region on the negative half of the cycle, the voltage waveform v_{CD} begins to have its negative half-cycle clipped as indicated in Fig. 8-7*c*.

When the input voltage is positive, diode D_1 is reverse-biased and hence the diode network does not affect the positive half-cycle of the voltage waveform. The net result is a positive average meter current for any input voltage

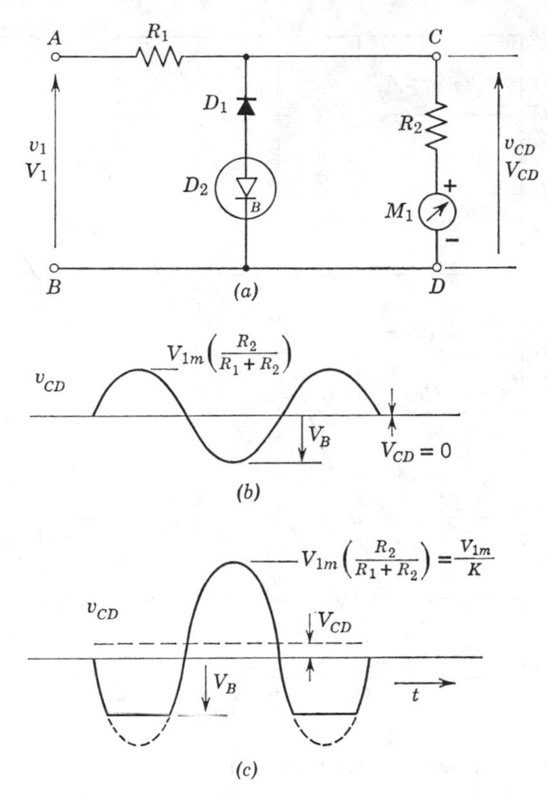

Fig. 8–7 Shunt-type ac expanded-scale voltmeter: (*a*) circuit; (*b*) waveform of v_{CD} when V_{1m} is less than V_B; (*c*) waveform of V_{CD} when V_{1m} is greater than V_B.

which is larger than $V_B(1 + R_2/R_1)$. Assuming that the dynamic diode resistances are zero, we have

$$\frac{I_M}{V_B/R_2} = \frac{\sqrt{(V_{1m}/V_B)^2 - K^2}}{\pi K} - 0.5 + \frac{\sin^{-1}(V_B K/V_{1m})}{\pi}, \tag{8-10}$$

where K is a constant equal to $(1 + R_1/R_2)$ and V_B is the sum of V_{B2}, the breakdown voltage of D_2, and V_{F1}, the forward voltage drop of D_1. Equation (8-10) is plotted in Fig. 8–8 for several values of K and for values of V_{1m}/V_B from zero up to seven.

A careful study of Fig. 8–8 reveals several facts. First of all, note that there will be a considerable amount of nonlinearity, regardless of the value of K. Although it is true that the upper portions of the curves seem to become fairly straight, particularly for $K = 1.1$, the lower portion of the curve will be displayed on the scale and will be nonlinear.

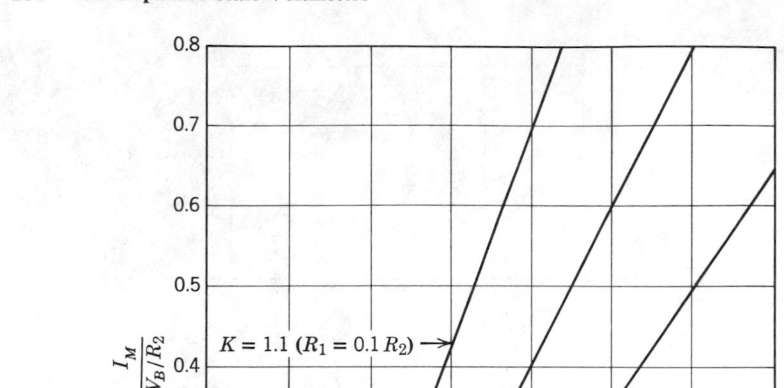
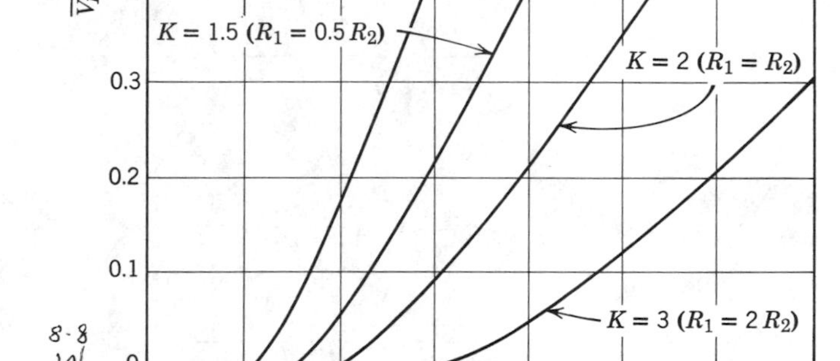

Fig. 8–8 Normalized transfer plot for the shunt-type expansion circuit of Fig. 8–7.

The rather gradual slope of the transfer plot in comparison with the break-point where the meter just begins to conduct indicates that this type of scale expansion will be useful only for those areas where only a slight amount of expansion is required.

If we mechanically suppress the zero setting of the meter used with this simple shunt-type expansion circuit, it is possible to improve the degree of linearity of that portion of the transfer characteristic that covers the display portion of the scale.

8–4.2 Example of Simple Shunt Expansion

As a practical example of an ac expanded-scale voltmeter using the circuit of Fig. 8–7, assume that we wish to display from 50-to-150-V rms on a 0.5 mA meter movement.

We will use a value of 1.1 for K since this will give a fair slope and about the best linearity for any of the curves shown in Fig. 8–8. This means that $R_1 = 0.1R_2$.

Meter current will begin to flow when $V_{1m}/V_B = K = 1.1$, and thus we may solve for the value of V_B required:

$$V_B = \frac{V_{1m}}{K} = \frac{50\sqrt{2}}{1.1} = 64.3 \text{ V}.$$

Since V_B is the sum of V_{B2}, the breakdown voltage of D_2, and V_{F1} is the forward voltage drop of D_1, the value of breakdown voltage required for D_2 will be about 63.5 V.

For a full-scale meter reading of 150-V rms or 212-V peak we compute the value of V_{1m}/V_B:

$$\frac{V_{1m}}{V_B} = \frac{212}{64.3} = 3.3 \quad \text{for full scale.}$$

Using the transfer plot for $K = 1.1$ in Fig. 8–8, we find that this value corresponds to a normalized current of 0.51 or

$$\frac{I_M}{V_B/R_2} = 0.51 = \frac{0.5 \times 10^{-3}}{64.3/R_2}.$$

We may solve the above relationship for the value of R_2 as 65.6 KΩ. R_1 will be $0.1R_2 = 6.56$ KΩ.

Now let us assume that we can mechanically suppress the zero on the meter movement such that a normalized meter current of 0.1 will correspond to the lower value to be displayed or 50-V rms. This corresponds with $V_{1m}/V_B = 1.7$ on the $K = 1.1$ curve of Fig. 8–8. From this we may calculate the required value of V_B:

$$V_B = \frac{50\sqrt{2}}{1.7} = 41.6 \text{ V}.$$

Now, computing the ratio V_{1m}/V_B corresponding to full scale, we get 5.1. Extrapolating the curve in Fig. 8–8, we find that this indicates a normalized meter current of 1.0.

We know that the difference in the meter currents for 50 and 150-V rms must be 0.5 mA and the ratio must be 1.0/0.1 or 10. From this information,

$$I_{M(150)} - I_{M(50)} = 0.5 \text{ mA}$$

$$\frac{I_{M(150)}}{I_{M(50)}} = 10.$$

Combining the two simultaneous equations, we have

$$10I_{M(50)} - I_{M(50)} = 0.5 \text{ mA},$$

$$I_{M(50)} = 0.0556 \text{ mA},$$

$$I_{M(150)} = 10I_{M(50)} = 0.556 \text{ mA}.$$

The meter must be suppressed mechanically such that the old zero position is obtained when the meter current is 0.0556 mA—corresponding to 50-V rms. Full-scale current will be 0.556 mA.

Recalling that the normalized value for full-scale current was determined as 1.0,

$$\frac{I_M}{V_B/R_2} = 1.0 = \frac{0.556 \times 10^{-3}}{41.6/R_2}$$

Solving the above gives a value for R_2 of 75 KΩ. The value required for R_1 would be 7.5 KΩ.

8–5 LINEARIZED SHUNT EXPANSION

The transfer characteristics for the ac expanded-scale voltmeter circuit of Fig. 8–7 may be made to yield a practically linear scale calibration by shifting them downward. A substantital improvement was obtained in the illustrative example above by merely mechanically moving the meter zero. We may apply a negative bias either from a constant dc source or in the same general manner shown by Fig. 8–3 or Fig. 8–6.

Another possible method is indicated by Fig. 8–9. In this approach rectifier diode D_3 and resistor R_3 are connected in shunt with the D_1-D_2 network in a polarity such that the positive half-cycles of the voltage appearing across terminals C and D are attenuated slightly as indicated in Fig. 8–9*b*. This will produce a net average meter current that is negative until the input voltage becomes large enough to cause D_2 to be biased in the breakdown region.

8–5.1 Analysis

If we again make the assumption that the dynamic resistances of the diodes are insignificant and further assume that the voltage drop across D_3, when it is forward-biased, is also negligible, then the normalized meter current is given by (8-11).

$$\frac{I_M}{V_B/R_2} = \frac{\sqrt{(V_{1m}/V_B)^2 - K^2}}{\pi K} - 0.5 + \frac{\sin^{-1}(KV_B/V_{1m})}{\pi}\left(\frac{V_{1m}/V_B}{\pi}\right)\left(\frac{1}{K} - \frac{1}{P}\right), \tag{8-11}$$

where V_B is again equal to $(V_{B2} + V_{F1})$, $K = (1 + R_1/R_2)$, and

$$P = 1 + \frac{R_1}{R_2/\!/R_3} = \frac{R_1}{R_2 R_3/(R_2 + R_3)} + 1. \tag{8-12}$$

Equation 8-11 is plotted in Fig. 8–10 for a value of $K = 2$ and again in Fig. 8–11 for a value of $K = 1.5$. The transfer characteristic, which occurs when P is equal to K, represents the case in which R_3 is made infinite as for the circuit of Fig. 8–8. Note how linearity above zero is improved as the value

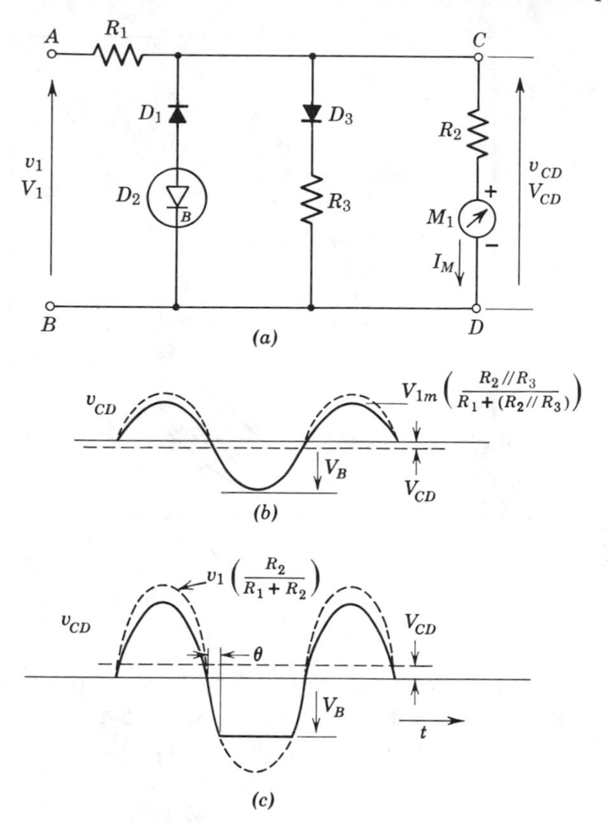

Fig. 8–9 Linearized shunt-type ac expanded-scale voltmeter: (*a*) circuit; (*b*) waveform of voltage across *C* and *D* for low input voltages; (*c*) v_{CD} waveform for larger input voltages.

of *P* is increased. For input voltages less than a certain critical value, the meter deflection will be reversed, but this is of little consequence.

By making any two of the three resistors adjustable we may set the threshold level and the window-voltage range to a desired calibration.

Note that although a separate diode D_3 has been indicated in the circuit of Fig. 8–9*a*, it is not necessary since we may obtain the same results by connecting R_3 directly across D_1 and using the VR diode in the forward-biased condition whenever the input voltage swings negative. Figure 8–12 shows the final circuit.

8–5.2 Example of Linearized Shunt Expansion

As a practical example of the linearized shunt expanded voltmeter, assume that we wish to display a voltage window from 100 to 130-V rms on a 0.5 mA meter movement.

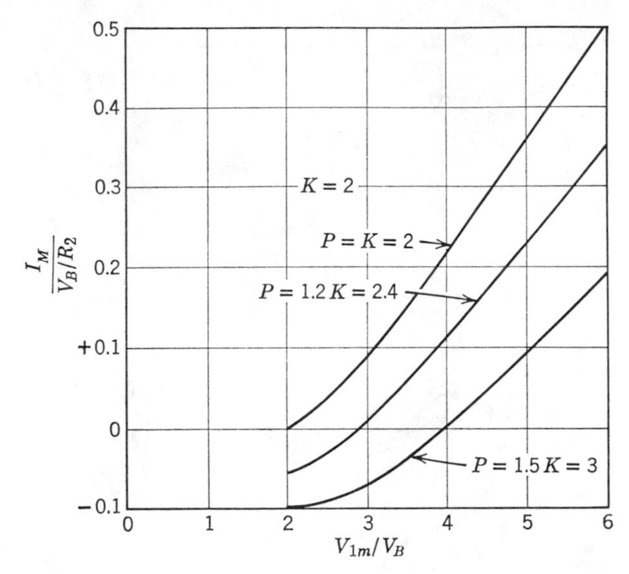

Fig. 8–10 Transfer characteristic for the linearized shunt expanded scale voltmeter using $K = 2$.

Desiring a nice linear scale calibration, let us choose the transfer characteristic shown in Fig. 8–11 for a condition of $K = 1.5$ and $P = 2.25$.

The point at which I_M will be zero occurs for $V_{1m}/V_B = 3$. Thus

$$V_B = \frac{V_{1m}}{3} = \frac{100\sqrt{2}}{3} = 47.1 \text{ V},$$

and the breakdown voltage required for D_2 would be about 46.5 V.

A full-scale deflection is required for $V_1 = 130$-V rms or 183.8-V peak. This gives a V_{1m}/V_B ratio of $183.8/47.1 = 3.9$. From the chosen transfer plot in Fig. 8–11, this corresponds to a normalized meter current of 0.11.

$$\frac{I_M}{V_B/R_2} = 0.11 = \frac{0.5 \times 10^{-3}}{47.1/R_2}.$$

From the above we may compute the value or R_2 as 10.4 K. Because we chose $K = 1.5$, $R_1 = 0.5R_2 = 5.2$ K.

We chose $P = 2.25$:

$$P = 2.25 = 1 + \frac{5.2 \text{ K}(10.4 \text{ K} + R_3)}{10.4 \text{ K}(R_3)},$$

and from this we may solve for R_3 as 6.9 K to complete the design. We would probably make a portion of R_1 and R_2 variable for calibration purposes. The circuit configuration could be that of Fig. 8–9 or Fig. 8–12.

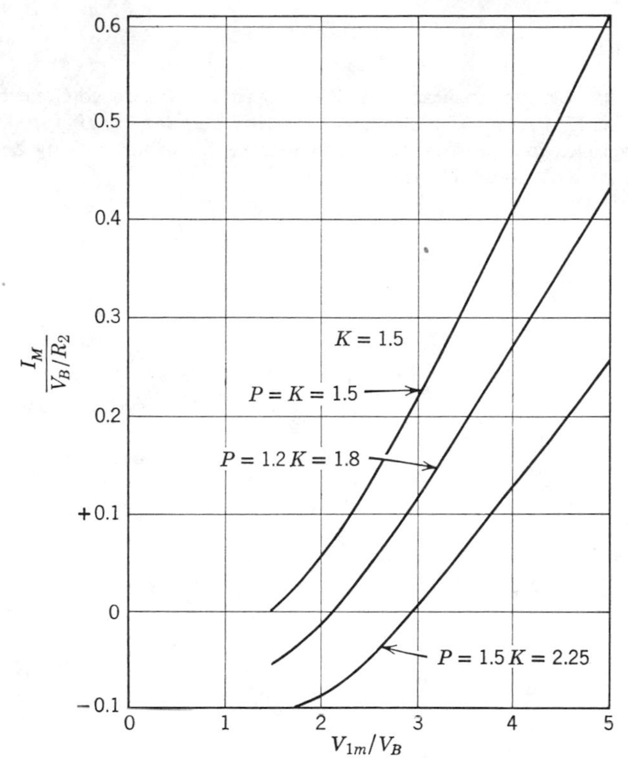

Fig. 8–11 Transfer characteristics for the linearized shunt expanded-scale voltmeter using $K = 1.5$.

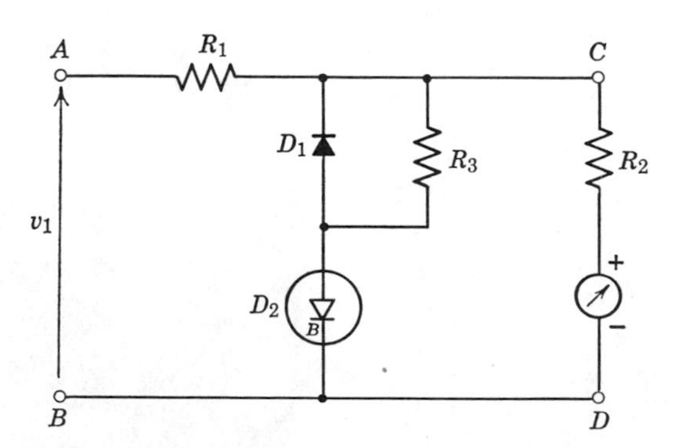

Fig. 8–12 Simplified shunt-type ac expanded-scale voltmeter circuit.

REFERENCES

[1] A. J. Corson: On the Application of Zener Diodes to Expanded Scale Instruments, *Trans. AIEE*, **77**, Part 1, 535–539 (1958). *Commun. Electron.*, No. 38 (September 1958).
[2] P. D. King: Design of a High Accuracy Expanded Scale Meter Using Zener Diodes, *Semicond. Prod.*, **3**, 26–28 (November 1960).

Chapter 9

LEVEL SHIFTERS AND COUPLING

9-1 INTRODUCTION

In previous chapters we have seen that the VR diode is often very useful in applications other than those dealing directly with the regulation of a voltage level. In this chapter we shall consider still another area—that where the VR diode is used as a coupling element. In some instances, the primary purpose of the VR diode is to provide a shift in the operating voltage level without attenuation of the incremental signal level. In other applications, the VR diode serves as a coupling element with an intentional threshold level.

Long before the availability of the semiconductor VR diode, designers were using small neon lamps in a similar fashion in vacuum tube circuits. The relatively constant voltage drop across the ionized lamp permitted a needed voltage level shift between the plate of one stage and the grid circuit of a following stage. The VR diode may be used to accomplish a similar purpose in semiconductor circuits.

9-2 BASIC LEVEL SHIFTING

In its simplest form, the VR diode level shifter may consist merely of the diode connected between the input and output terminals in the manner indicated in Fig. 9-1. A bias current I_1 must be provided in some manner to maintain the VR diode in the breakdown or constant-voltage region.

If we may assume that the current flowing through the VR diode is primarily that provided by the bias current I_1 and is therefore relatively constant, then the breakdown voltage V_B estabished across the diode will also be relatively constant. The voltage difference between the input and the output will therefore be equal to V_B and will have the polarity as shown in Fig. 9-1.

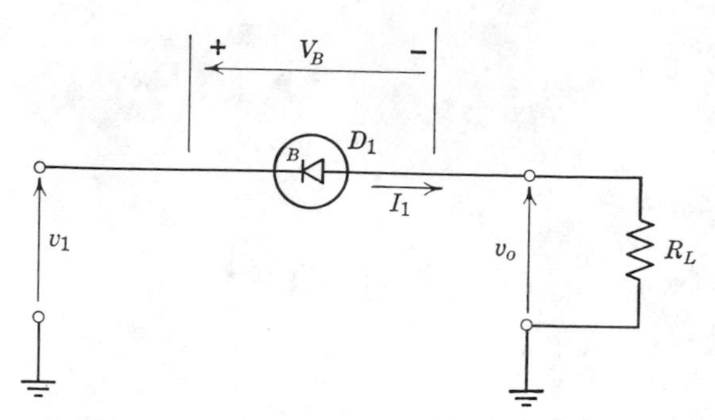

Fig. 9–1 Basic level shifter principle relies on constant drop across VR diode.

The input-output voltage transfer plot of the basic VR diode level shifter is shown in Fig. 9–2. Note that the output voltage always remains more negative than the input voltage by an amount equal to V_B. Thus, when v_i is zero, v_o will be negative and have a magnitude equal to V_B. Then, as v_i is made positive and equal to V_B, the output voltage drops to zero. If the input voltage is made negative, then v_o merely swings more negative by an amount equal to V_B.

An important feature of the VR diode level shifter is the fact that the slope of the transfer plot approaches unity, indicating that an increment in the value of the input voltage will be transmitted to the output unattenuated. Although it is true that some slight variation in the value of V_B will occur (because of a nonconstant reverse current flowing through the breakdown resistance) in

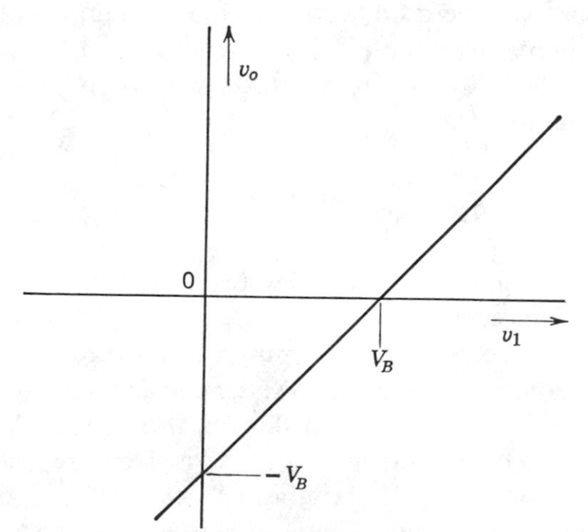

Fig. 9–2 Input-output transfer plot for the basic level shifter.

the practical case, the normal variation is slight. The output voltage is related to the input voltage as follows:

$$v_o = v_i - V_B. \tag{9-1}$$

Note that (9-1) describes the performance of the shifter circuit with VR diode and bias current polarities as shown in Fig. 9–1. If these are changed, then the equation may be rewritten as

$$v_o = v_i + V_B. \tag{9-2}$$

Since the VR diode is maintained in the breakdown region, the output follows the variations in the input with no loss in speed or information. There is only a fixed shift in voltage level.

9–3 EMITTER FOLLOWER SHIFTER

For proper circuit performance, the VR diode in the level shifter must have a relatively constant bias current flowing through it at all times. This means that a bias current would be applied at the output, through the VR diode, and into the input voltage source. When the input voltage is derived from a relatively low-impedance source, this does not become a problem, but for other applications a means of isolation must be found.

9–3.1 Simple Circuit

One solution is the combination of an emitter follower with the simple level shifter in the manner shown in Fig. 9–3. The input source must now only provide the base current to transistor Q_1. A negative supply voltage V_2 in conjunction with resistor R_1 provides the bias current for the VR diode to maintain it in the breakdown region.

Assume, first of all, that the input voltage is zero and the output is unloaded. Bias current I_1 maintains D_1 in breakdown and thus the voltage drop across the VR diode is V_B, with the polarity as shown in Fig. 9–3. I_1 also flows in the emitter of Q_1, causing a V_{BE} drop as shown and producing a value of base current:

$$I_B = \frac{I_1}{1 + h_{FE}} = \frac{V_2 - V_B - V_{BE}}{R_1(1 + h_{FE})}. \tag{9-3}$$

The output voltage will be sum of V_B and V_{BE}. Thus the output voltage level has been shifted by an amount equal to $(V_B + V_{BE})$, with v_o more negative than v_i.

Now assume that input voltage v_i goes positive. In normal follower fashion the emitter of Q_1 likewise goes more positive. Bias current I_1 continues to flow through the VR diode and maintain V_B across it. If we assume V_B to

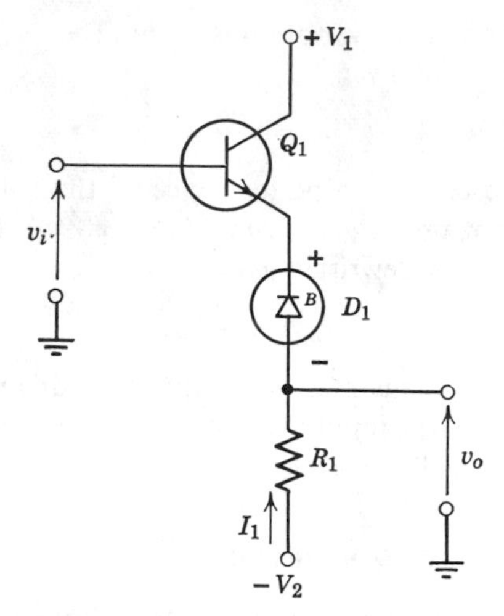

Fig. 9–3 Simple emitter follower level shifter circuit.

remain constant, the output terminal will follow any incremental change in the voltage at the emitter of Q_1 and hence at the input.

It is true that as the input voltage is made positive, the value of I_1 will increase by an amount equal to v_i/R_1. If V_2 is large, however, R_1 will likewise be large and the variation in I_1 will not be great enough to produce an appreciable change in V_B or V_{BE}.

If the value of v_i should be negative, the output again follows the input merely shifted by the same amount as before. I_1 will now decrease slightly. The overall transfer equation describing the shifter combined with the emitter follower is

$$v_o = v_i - (V_B + V_{BE}), \tag{9-4}$$

and, when plotted, we have the transfer characteristics shown in Fig. 9–4.

To illustrate better the performance of the level shifter, consider the application of the simulated input waveform as shown in Fig. 9–5*a*. The resulting output voltage waveform given in Fig. 9–5*b* will be identical in shape to the input, but with a shift in the level of the voltage value.

Should we wish to produce a level shift in which the output is made more positive than the input, then we need only use a *PNP* transistor, invert the polarity of D_1, and reverse the power supply polarities.

The addition of a load to the output of the shifter circuit shown in Fig. 9–3 will affect the circuit performance only slightly in that it will modify the value of the current flowing through the VR diode and emitter of Q_1.

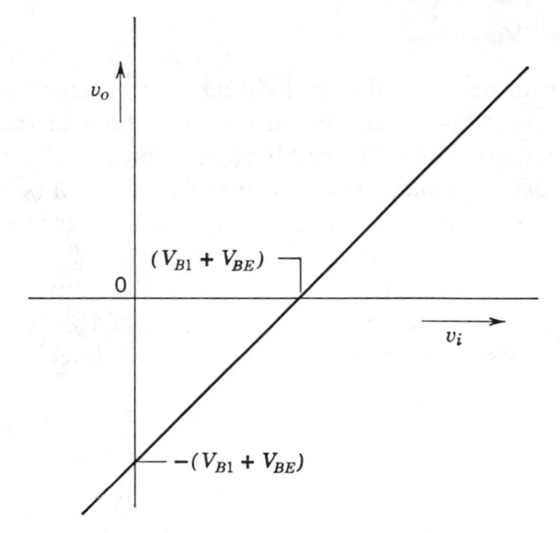

Fig. 9–4 Input-output transfer plot for the simple emitter follower level shifter.

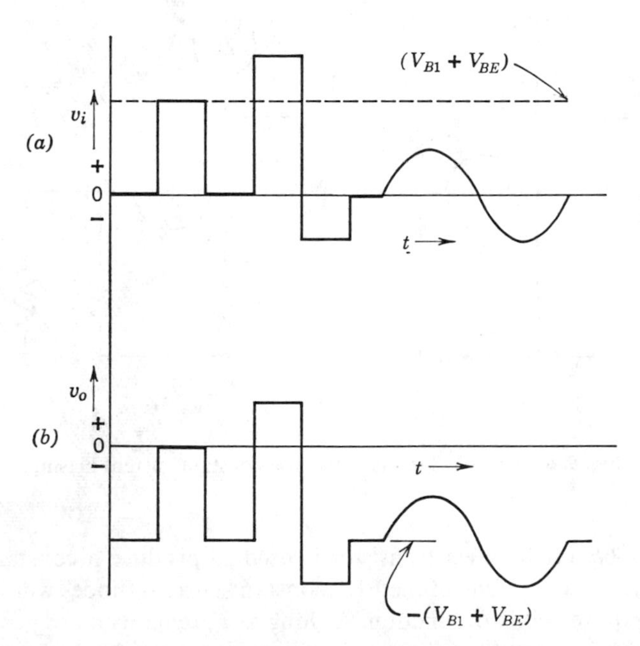

Fig. 9–5 (*a*) Simulated input waveform and (*b*) resulting output waveform for the level shifter.

9–3.2 Circuit Variations

Optimum performance of the level shifter circuit occurs only when I_1 is maintained at a constant value. When the magnitude of the input voltage swing may be relatively large, it may become impractical to make the value of V_2 large enough to reduce the variations in I_1 that will occur.

Two alternate approaches for providing a more constant value of bias current are shown in Fig. 9–6. In Fig. 9–6*a* transistor Q_2 is connected as a constant-current generator. With this approach the output voltage may be permitted to swing within a few volts of the value of V_2 without appreciable change in I_1 with its corresponding error in the shift level.

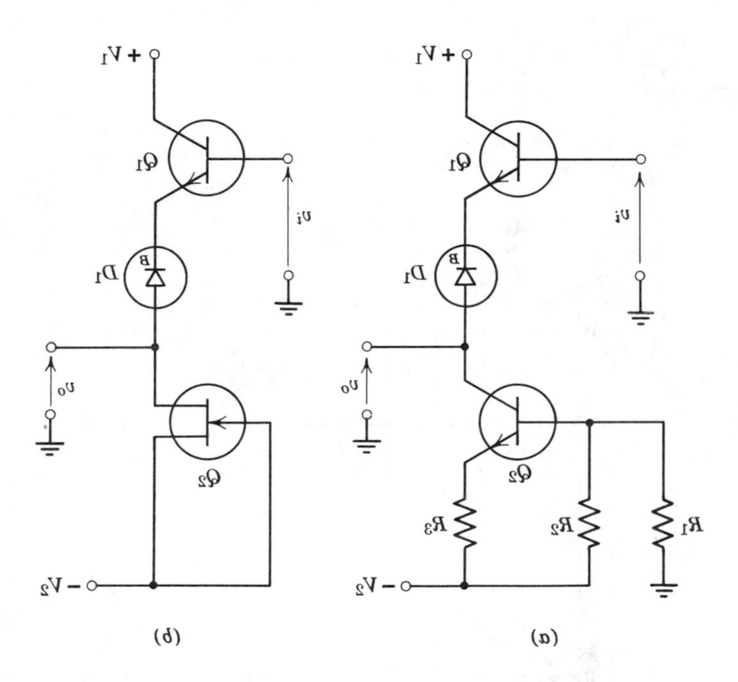

Fig. 9–6 Improved level shifters use constant current biasing.

In Fig. 9–6*b* a field-effect transistor is used to produce a constant current equal to its I_{DSS} for I_1. One of the FET constant-current diodes would likewise serve the required control function. As long as v_o remains more positive than V_2 by an amount equal to the pinchoff voltage of the FET, I_1 will remain almost constant.

9–4 DIRECT COUPLED AMPLIFIERS

Direct coupled transistor amplifiers connected in tandem require increasing collector voltages as the signal level increases in stages nearer the output. This produces a problem in design in order to bias properly the succeeding stage. In some instances, a level shift is accomplished by means of a resistive network in conjunction with an additional voltage supply of opposite polarity. In other applications the problem is solved by using complementary transistors.

Still another way is using a VR diode as a combination coupling element and voltage level shifter. One simple arrangement is shown in Fig. 9–7. The collector voltage of Q_1, V_{C1} may be designed to be any desired value. The voltage appearing at the base of Q_2 will be $(V_{C1} - V_B)$.

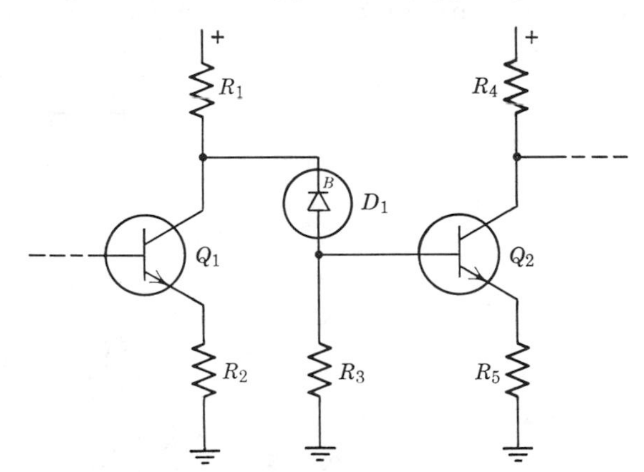

Fig. 9–7 Direct coupled amplifier using VR diode as coupling element.

Unlike the resistive manner of voltage shifting, the VR diode coupling causes negligible attenuation of the signal level and does not produce frequency response problems.

Resistor R_3 is normally advisable to increase the current level through the VR diode without requiring that this current flow through the base of Q_2.

Another interesting possibility exists with the VR diode coupled amplifier. Any variation in the operating temperature causes a shift in V_{BE} with its value decreasing for a rise in temperature. This variation will be amplified as though it resulted from an input from a previous stage. If we choose a VR diode having a temperature coefficient equal in magnitude but opposite in polarity, then it is possible to reduce the overall temperature coefficient to a rather low value. The result is a dc amplifier having acceptable immunity to temperature variations.

9–5 DIGITAL CIRCUIT COUPLING

VR diodes may also be used as a coupling element in various digital circuits. In some applications, the VR diode can replace a resistor and capacitor in parallel. This is of considerable interest to those trying to package a circuit function in a small space. The effective frequency response of the VR diode as a coupling element extends all the way from dc to several hundred megahertz. We shall look briefly at several examples for the purpose of illustration.

9–5.1 Flip-Flops

In the normal binary flip-flop, the coupling elements are resistors in shunt with a capacitor. The resistors establish the desired dc bias conditions, and the capacitors are necessary to yield adequate switching speed. In certain applications it may be wise to consider the use of VR diodes as the coupling elements as shown in Fig. 9–8.

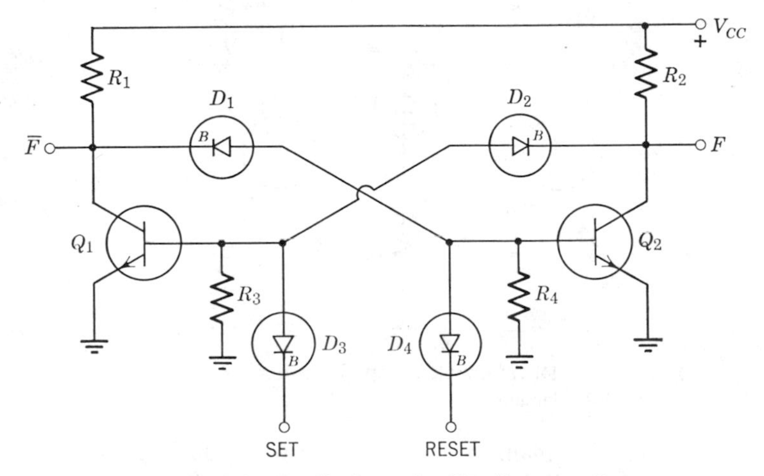

Fig. 9–8 *RS* flip-flop using VR diode coupling.

Assume that Q_1 is off and Q_2 is conducting to yield "F" at near zero value. VR diode D_2 will have negligible voltage across it and thus very little current through it. This easily makes it possible for Q_1 to be off. The collector of Q_1 attempts to go to V_{CC}, but as soon as V_{CE1} exceeds V_{B1}, the breakdown voltage of VR diode D_1, D_1 will conduct to supply a base current to Q_2. This maintains Q_2 conducting. The collector voltage of Q_1 is then clamped at a level equal to $V_{B1} + V_{BE2}$.

A positive voltage applied to the SET input and larger than the breakdown voltage of D_3 will provide a base current to Q_1, thus turning it on. This, in turn, will remove the voltage across D_1 and thus allow Q_2 to turn off. Not only is the voltage drive removed, but charge stored in the junction capacitance of the VR diode will aid the removal of stored carriers in the base of Q_2, thus permitting rapid turn-off.

The junction capacitance of D_2 aids the regenerative cycle also in that it acts as a coupling capacitor between the collector of Q_2 and the base of Q_1.

The circuit assumes a stable operating state again with Q_1 conducting and Q_2 off. VR diode D_2 will conduct, whereas D_1 will have negligible voltage across it. The original state may be restored by applying an input voltage to the RESET terminal.

In addition to circuit simplicity, the VR-diode-coupled flip-flop also has clamped outputs.

Figure 9–9 illustrates the arrangement for a clocked *RS* flip-flop using VR diodes as the coupling elements.

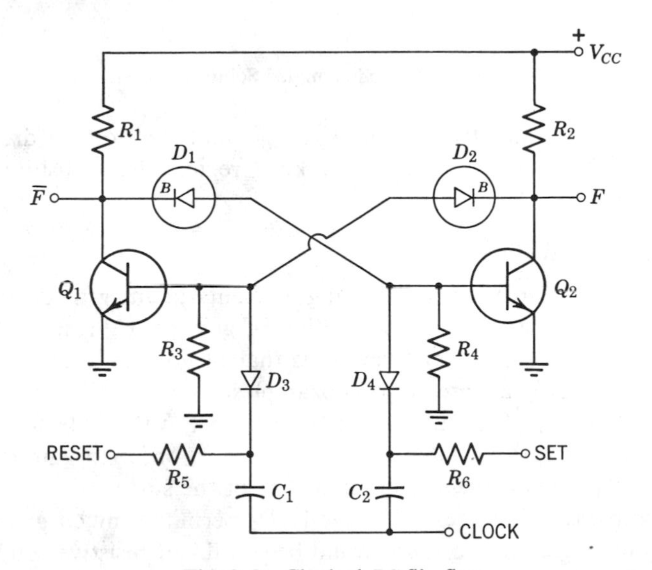

Fig. 9–9 Clocked *RS* flip-flop.

9–5.2 Schmitt Trigger

A VR diode may be used as the coupling element in a Schmitt trigger circuit. One arrangement is shown in Fig. 9–10. The emitters are coupled in the normal manner with R_4 as the mutual resistance. The coupling between the collector of Q_1 and the base of Q_2 is accomplished by the VR diode D_1 in place of the normal resistor-capacitor combination.

The resulting performance of the VR-diode-coupled Schmitt trigger is held over a wide range of input frequencies and without concern for rise and fall times of the input waveform. Triggering is accomplished when the collector

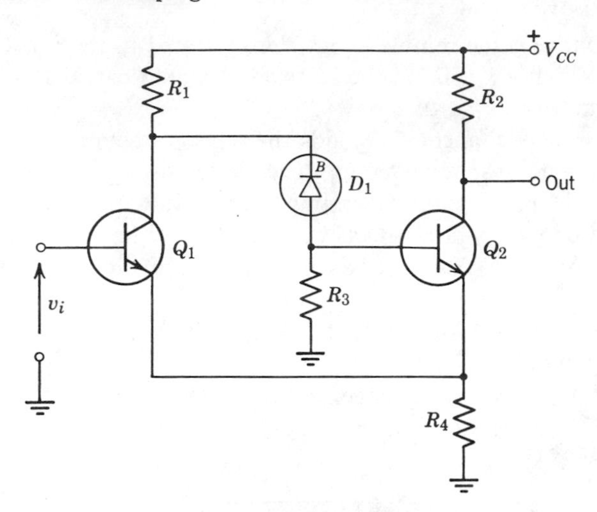

Fig. 9–10 VR diode coupled Schmitt trigger.

voltage of Q_1 falls below the sum of V_B, V_{BE}, and the voltage drop across R_4. With a VR diode having a sharp knee region the transition can be quite predictable.

9–5.3 Integrated Circuits

VR diodes are also used as coupling elements in integrated circuits. In some applications the main interest is that of gaining a threshold function, whereas in others the primary function is that of voltage level shifting. We shall look very briefly at three typical examples.

The circuit of Fig. 9–11 illustrates the use of a VR diode to achieve a coupling function with an intentional threshold in order to gain additional noise immunity. This is the schematic of one of the stages in a high level to low level converter. The use of VR diode D_2 permits a much greater noise immunity and higher speed than would be possible if resistive coupling had been used. The junction capacitance of the VR diode will take an active part in both turn-on and turn-off excursions.

Another interface integrated circuit is shown in Fig. 9–12. This one is designed to couple MOS-type digital circuits to normal DTL or TTL levels. Two VR diodes are used.

VR diode D_3 acts as a clamp to control the maximum negative value of the voltage at the emitter of transistor Q_1. This sets the level to which the input must be driven in order to turn Q_1 off—in this instance, to about -9V.

VR diode D_1 acts as a coupling element between the collector of Q_1 and the base of Q_2. It reduces the maximum collector voltage that Q_1 must withstand, yet provides the intimate coupling needed.

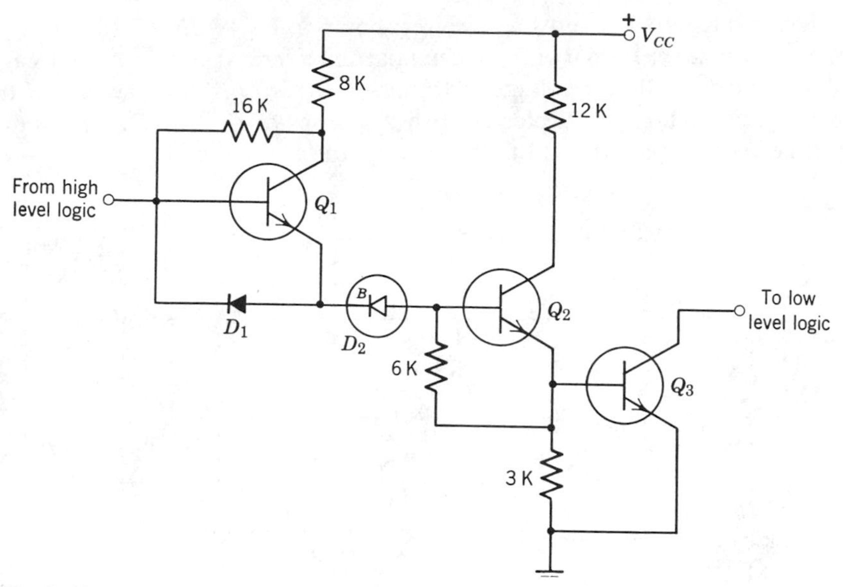

Fig. 9–11 VR diode D_2 is used as a threshold coupling means to obtain high level noise immunity. (HLLDTL courtesy Fairchild Semiconductor.)

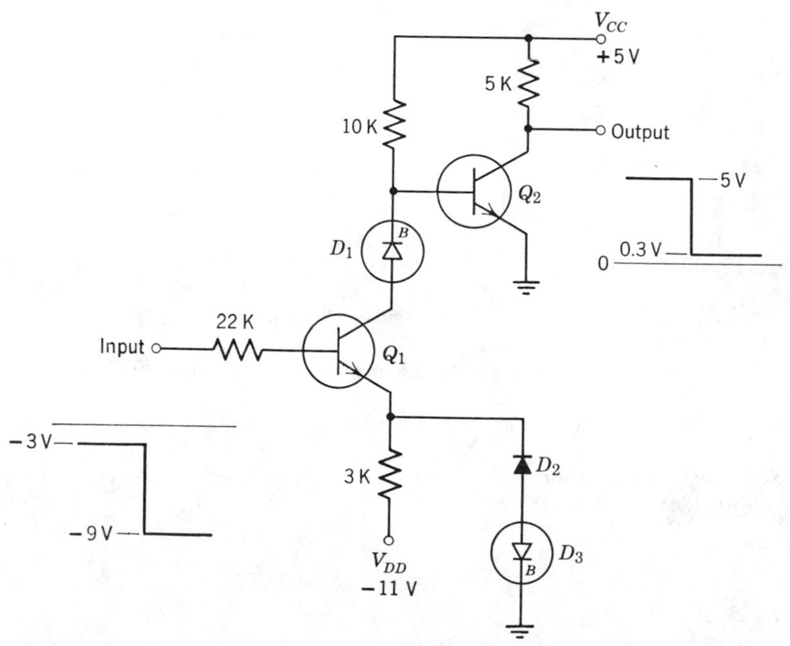

Fig. 9–12 VR diode D_1 is used in an interface integrated circuit as a coupling Element, and D_3 Determines threshold Level. (courtesy Fairchild Semiconductor.)

In the integrated circuit line receiver shown in Fig. 9–13, VR diode D_1 is used to couple the output of the differential amplifier stage to the output stage. Whenever the collector voltage of transistor Q_2 drops below the sum of the breakdown voltage of D_1 plus the two V_{BE} drops of Q_5 and Q_6, the output will be very quickly turned off because D_1 ceases to conduct.

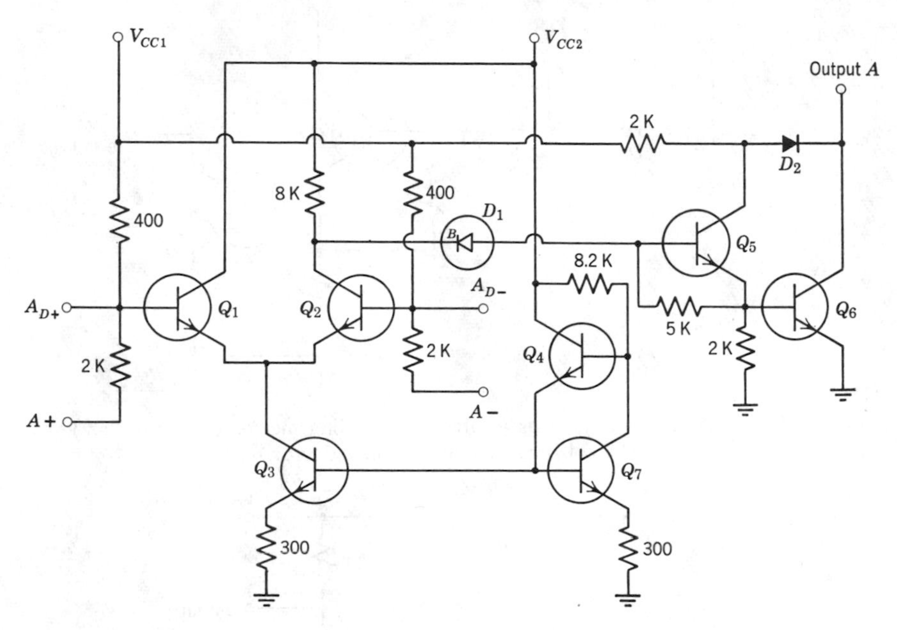

Fig. 9–13 VR diode D_1 is used as a coupling element between differential amplifier and the output stage of an integrated circuit line receiver. (Courtesy Fairchild Semiconductor.)

REFERENCES

[1] J. Kabell and V. H. Grinich: Zener Diode Circuits for Stable Transistor Biasing, *Semicond. Prod.*, **4**, 35–36 (June 1961).

[2] F. Nibler: Zener Diodes as Coupling Elements in Relay Circuits, *Rev. Sci. Instr.*, **32**, 1143 (October 1961).

[3] A. S. Robinson: Zener Diode Allows Delay without Large Capacitors, *Electronics*, **39**, 93 (May 30, 1966).

[4] W. D. Roehr and D. Thorpe: *Switching Transistor Handbook*, Motorola, Inc., Phoenix, Arizona, 1963, pp. 263–275, 280–284.

Chapter 10

MISCELLANEOUS APPLICATIONS

10–1 INTRODUCTION

In other chapters of this book we have considered many applications of VR diodes that go beyond simple voltage regulation. In this chapter we shall study a few more useful applications of VR diodes—primarily those that utilize a special characteristic of the VR diode. Included in this category are temperature transducers, thermal time delay circuits, voltage-controlled switches, and applications making use of the logarithmic nature of the voltage-current characteristic or the noise generated in the breakdown region.

10–2 TEMPERATURE TRANSDUCERS

In chapter 2 we studied the temperature dependence of the breakdown voltage. The temperature coefficient is a function of the breakdown voltage.

In many applications in which the VR diode is used as a voltage reference or even in simple voltage regulation applications, this temperature dependence is an undesirable characteristic and must be frequently compensated. We may utilize the characteristic as a temperature-indicating mechanism by applying a constant bias current and monitoring the change in breakdown voltage.

One simple approach is shown in Fig. 10–1. The use of a differential voltmeter permits us to balance out the nominal breakdown voltage and then monitor only the change in V_B that occurs. This will obey the following equation:

$$\Delta V_B = K_T(\Delta T), \tag{10-1}$$

where K_T is the temperature coefficient of the VR diode breakdown voltage expressed in mV/°C and ΔT is in °C. The value of ΔV_B is given in mV.

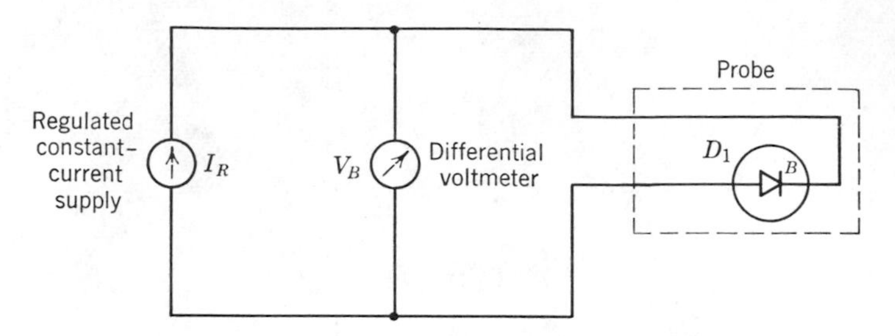

Fig. 10–1 Basic temperature transducer using VR Diode.

By choosing the nominal value of the breakdown voltage, we determine the magnitude and polarity of the temperature coefficient and thus the sensitivity of the transducer. For example, a VR diode having a breakdown voltage of about 15 V will have a positive temperature coefficient approximately equal to 10 mV/°C. Thus a temperature deviation of only 0.1C may be seen as a meter deflection of 1 mV.

Because the exact value of K_T is not predictable, it will be necessary to calibrate our transducer against a known standard if we are to obtain accurate measurements. The task is made much easier by the fair degree of linearity of the temperature dependence. This permits a simple two or three point calibration. Once we calibrate an individual diode, we may depend on its temperature coefficient remaining quite stable if maximum temperature excursions are not excessive.

Another temperature transducer may be based on the double-arm voltage balanced bridge as discussed in Chapter 4 and illustrated in Fig. 10–2. If

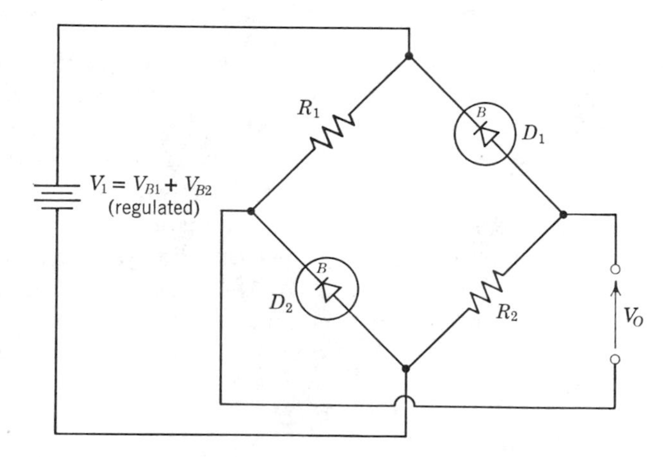

Fig. 10–2 Double-Arm voltage balanced bridge may be used as temperature transducer.

supply voltage V_1 is adjusted to be exactly equal to the sum of the two breakdown voltages, then the output will be zero. Now, if V_1 is held constant, any variation in the output voltage will be due to a change in the breakdown voltages because of temperature deviation. The value of V_O will be

$$V_O = (K_{T1} + K_{T2})(\Delta T). \tag{10-2}$$

The bridge transducer yields a rather high output that is a linear function of temperature differential from its balance value. Loading effects will be the same as discussed in Chapter 4.

10–3 TIME DELAY CIRCUITS

It is frequently desirable to have a time delay that is in the order of seconds without using reactive elements. This may be accomplished by using the temperature dependence of the VR diode in conjunction with the inherent thermal time constant.

The basic approach is indicated in Fig. 10–3. Assume that the two VR diodes, D_1 and D_2, are fairly well matched but with V_{B1} slightly larger than V_{B2} for identical conditions. With the switch open, the bias current in D_1 will be much less than the current in D_2 and thus V_{B1} will be less than V_{B2}. This causes the net input voltage into the differential amplifier to be negative, and to yield a negative output voltage as shown in Fig. 10–4.

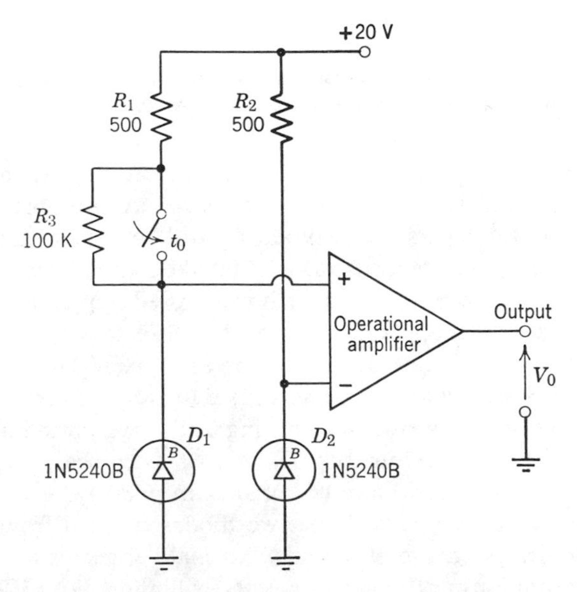

Fig. 10–3 Time delay circuit using self-heating effects of VR diode.

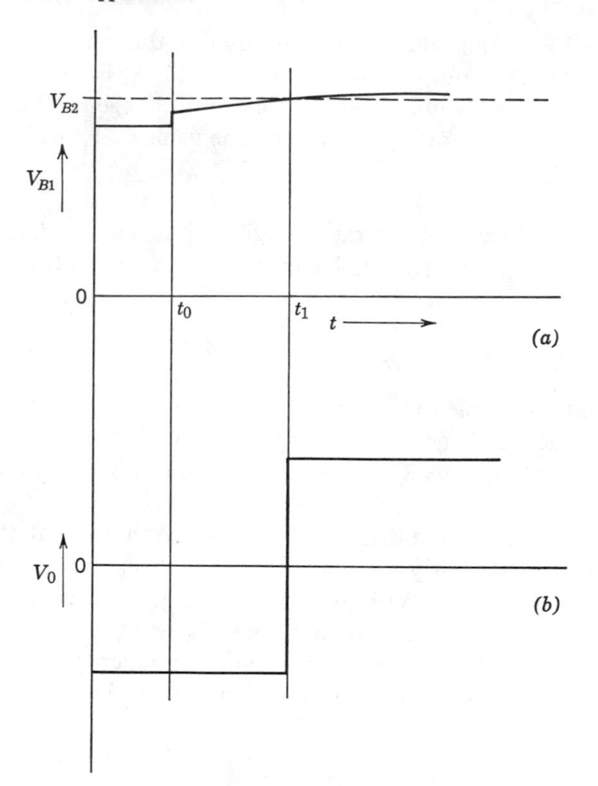

Fig. 10–4 Voltage waveform for the VR time delay circuit: (*a*) input conditions; (*b*) output voltage.

Now assume that at some point in time t_o we close the switch. This, then, supplies an identical bias current to both diodes. At the instant of switch closure the junction temperature of diode D_2 will be higher than that of D_1. This means that the instantaneous value of breakdown voltages will be closer together than they were before the switch was closed, but still being different enough to saturate the amplifier in the negative direction.

As soon as the switch is closed, the junction of D_2 begins to heat up because of the increased power dissipation. This causes the junction temperature, and thus V_{B1}, to increase exponentially according to the overall thermal time constant of the entire mass surrounding D_1. As V_{B1} becomes equal to V_{B2} and then exceeds it, the output voltage will change in polarity.

By careful thermal design, with the two diodes isolated from each other, we may obtain delays of several seconds. Normal variations in ambient temperature or fluctuations in the supply voltage will affect both VR diodes in a compensating manner.

If the switch is opened, there will be another delay before the amplifier output voltage reverts to its original state.

10–4 ELECTRONIC SWITCH

The effective resistance of a VR diode changes in a very marked manner as the bias voltage is changed. With the reverse voltage less than V_B, the diode may exhibit an effective resistance of as much as several megohms. Then, as the bias voltage reaches the breakdown value, current rises and the effective resistance may drop to 100 Ω or lower.

We may utilize this dramatic change in resistance as a switching mechanism that is controlled by the variation of a dc voltage. One arrangement is shown in Fig. 10–5. Here a dc control voltage V_C gates effectively the transmission of the ac input signal to the output.

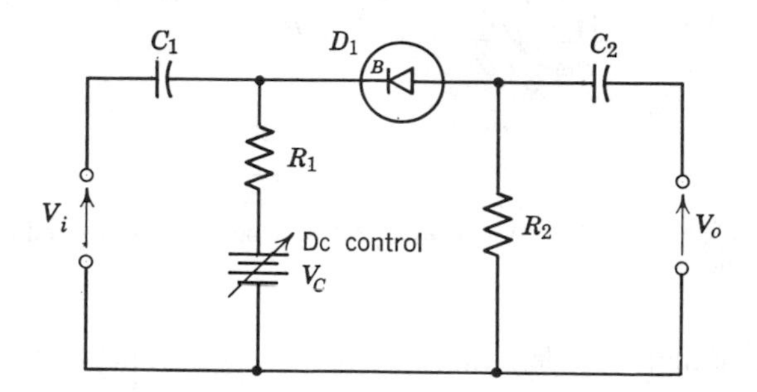

Fig. 10–5 Electronic switch uses variation of breakdown resistance as switching function.

As long as V_C is kept below the breakdown voltage V_B, the resistance of D_1 is very high and nearly all the ac input voltage V_i will be dropped across it, with negligible signal appearing at the output. Now, if we increase the dc control voltage to a value substantially above V_B, then breakdown current will flow through R_1, D_1, and back to ground through R_2. This causes the resistance exhibited by D_1 to drop with most of the ac input voltage now appearing at the output.

Resistors R_1 and R_2 are necessary to provide a dc path from control voltage to D_1 without shorting out the signal path. Capacitors C_1 and C_2 isolate the dc control path from the input and output circuitry.

Possible transients caused by switching the control voltage may be eliminated by limiting the rise time of the control function to a slow value in comparison with the low-frequency of the cutoff signal system.

10–5 NONLINEAR NETWORKS

10–5.1 Voltage-Controlled Resistance

The use of a VR diode as a switching element can be extended to the synthesis of a controlled nonlinear resistance network such as shown in Fig. 10–6. This network can have one of four different values of incremental resistance, depending on the value of the terminal voltage V_T.

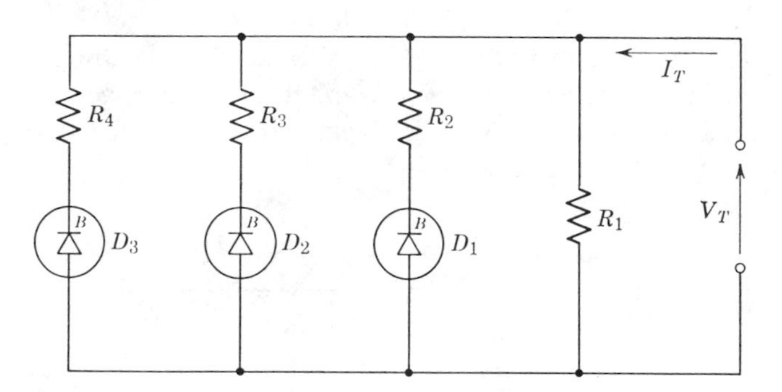

Fig. 10–6 A nonlinear resistance network may be synthesized with VR diodes.

First, assume that V_T is less than the breakdown voltage of any of the VR diodes. This means that resistors R_2, R_3, and R_4 are unconnected and the equivalent resistance as seen at the output terminals is simply that of R_1. Now, as V_T is increased, each diode will switch in its corresponding resistor when V_T reaches the appropriate value of V_B. If we assume that $V_{B1} < V_{B2} < V_{B3}$, then resistors R_2, R_3, and R_4 will be connected across the terminals in that order.

The resulting characteristic curve of the network might be as shown in Fig. 10–7. The individual breakpoints will be determined by the specific values of breakdown voltages used. The incremental change in slope will be determined by the resistance value switched in.

By appropriately selecting the proper diode voltages and resistor values, it would be possible to approximate a desired nonlinear function with as many breakpoints as desired. The transition points in actual practice are not so sharply defined as indicated in Fig. 10–7. because of the slight rounding of the knee region of the VR diode characteristic.

The nonlinear resistance may be combined with a negative impedance converter to yield a nonlinear negative resistance characteristic [6].

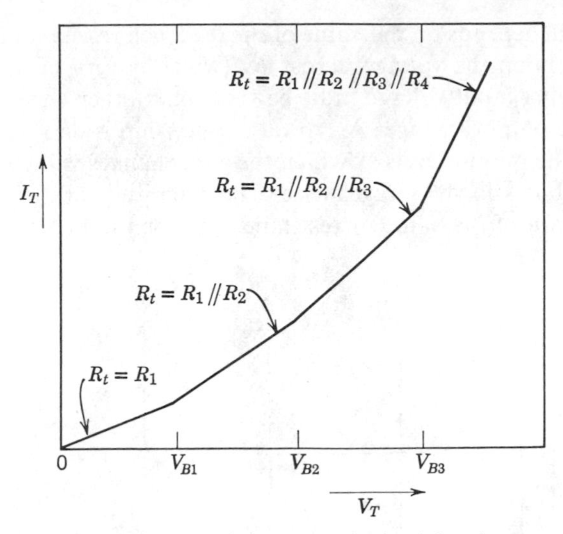

Fig. 10–7 Characteristic curve obtained for the nonlinear network of Fig. 10–6.

10–5.2 Nonlinear Feedback

By using the voltage-controlled resistance network just described as the feedback resistance of an operational amplifier, we may achieve some rather interesting results. One simple arrangement is shown in Fig. 10–8.

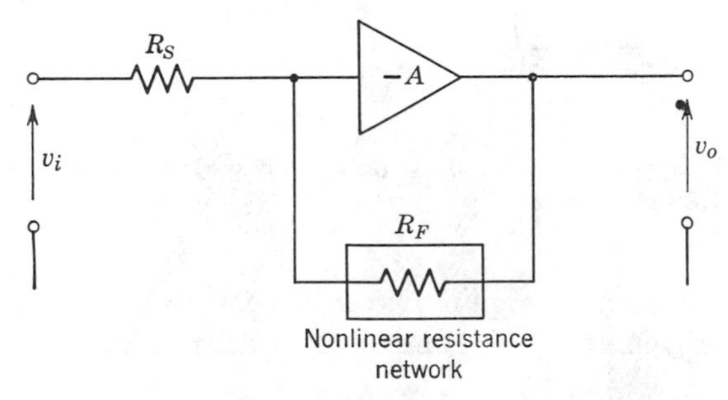

Fig. 10–8 Nonlinear resistance network used as feedback resistance yields variable gain amplifier.

The incremental voltage gain of the overall stage will be as given in the following expression:

$$A_v = \frac{v_o}{v_i} \cong \frac{R_F}{R_S}, \tag{10-3}$$

where the open-loop gain of the amplifier is very large.

Since the gain depends on the value of the feedback resistance and the resistance depends upon the voltage across R_F (which is very nearly equal to v_o since the amplifier input voltage must be very small), then the gain must be a function of the output voltage. A typical relationship is depicted graphically in Fig. 10–9. The output levels at which the gain changes will be a function of the individual breakdown voltages used. The magnitude of the gain variation will be a function of the value of resistance switched in at that point.

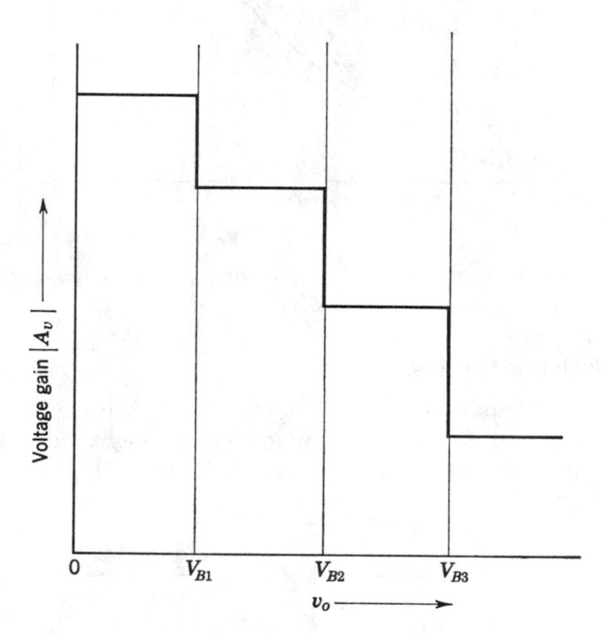

Fig. 10–9 Voltage gain switches to lower level as output voltage increases.

By properly isolating the nonlinear resistance network, we could have an ac amplifier whose gain could be controlled in discrete steps by a dc control voltage.

10–6 LOGARITHMIC CONVERTER

VR diodes operating primarily on a true Zener breakdown mechanism exhibit a logarithmic voltage-current characteristic. Figure 10–10 shows the characteristic obtained for a 1N5226 diode with a nominal breakdown voltage of 3.3 V at 20 mA. The relationship follows a straight line very closely when plotted on semilogarithmic paper.

Some deviation from a true logarithmic function is obtained for currents less than the 1mA level shown. To retain the logarithmic characteristics at the high current levels it is necessary to restrict the measurements to very short pulses to prevent junction heating. Since a VR diode with V_B of 3.3 V will have a negative temperature coefficient, any heating effects will cause the voltage to drop accordingly. This may be partially compensated for by the addition of a small amount of series resistance to shape the characteristic curve slightly.

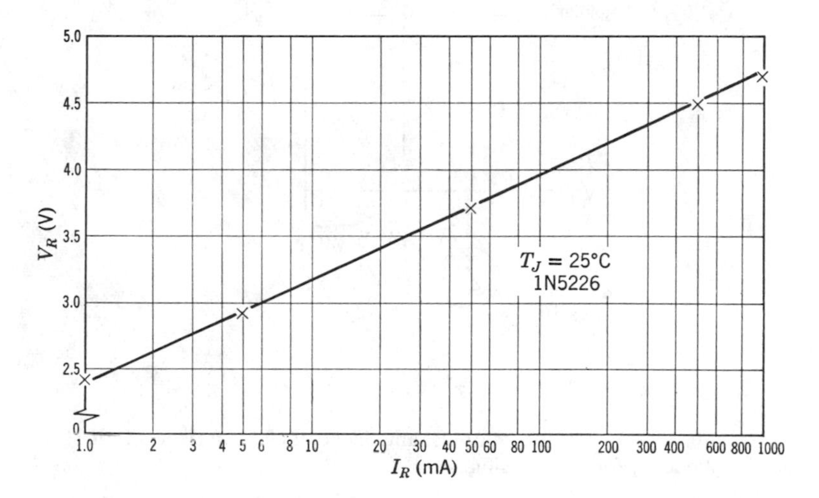

Fig. 10–10 Measured characteristics for an actual VR diode illustrate excellent logarithmic variation over three decades. (Pulse measurement used to retain constant junction temperature.)

The arrangement of Fig. 10–11 illustrates an application in which the logarithmic characteristics of the VR diode shown in Fig. 10–10 is utilized to evolve a pulse amplifier whose output voltage will be a logarithmic function of the input.

Assume that the open-loop gain of amplifier A is great enough to permit the approximation that the input voltage is zero. This allows us to say

$$V_{IN} = V_{OB} = I_R R_1 A_{VB}, \tag{10-4}$$

where V_{OB} and A_{VB} are the output voltage and voltage gain, respectively, of amplifier B.

The empirical equation for the characteristic shown in Fig. 10–10 is

$$V_{R1} = 0.78 \log_{10} I_R + 2.4, \tag{10-5}$$

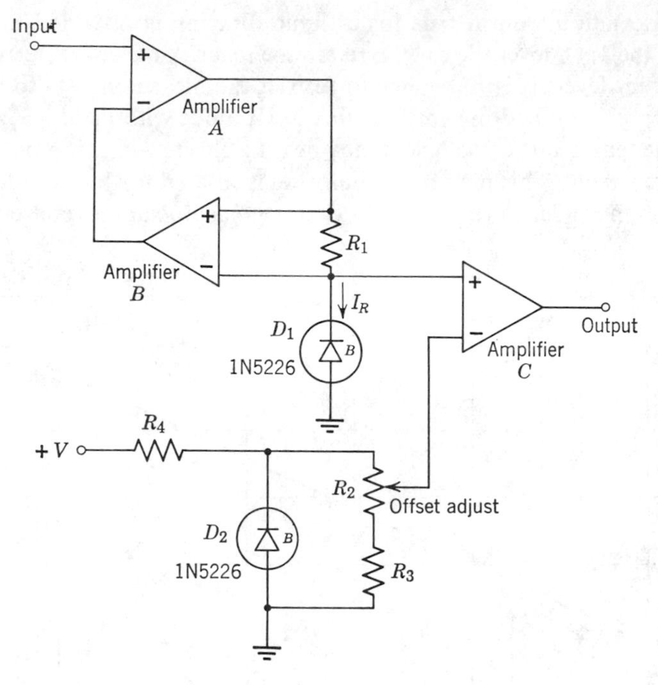

Fig. 10–11 Logarithmic pulse amplifier uses characteristics of VR diode D_1 for compression.

where V_R is in volts and I_R must be inserted in mA. Solving (10-4) for I_R and inserting this into (10-5), we have

$$V_{R1} = 0.78 \log_{10}\left(\frac{V_{IN}}{R_1 A_{VB}}\right) + 2.4, \tag{10-6}$$

where V_{IN} must be in mV.

Now, if we make $R_1 A_{VB}$ equal to unity, then

$$V_{R1} = 0.78 \log_{10} V_{IN} + 2.4. \tag{10-7}$$

Now, if we use another differential amplifier to compare the voltage drop across D_1 with a 2.4-V reference voltage derived from D_2, then the output voltage at amplifier C will be

$$V_{OC} = (0.78)(A_{VC})\log_{10} V_{IN}. \tag{10-8}$$

The value of the voltage gain A_{VC} may be made equal to the desired scaling factor. If it is chosen to be 1/(0.78), then

$$V_{OC} = \log_{10} V_{IN}. \tag{10-9}$$

If the adjustment potentiometer R_2 is set to give an output of zero for V_{IN} of 1 mV, then inputs of 10 and 100 mV and 1 V yield outputs of 1, 2, and 3 V, respectively. A slightly different approach is given in reference [1].

10–7 OTHER CURRENT-CONTROLLED RESISTANCE APPLICATIONS

The variation of the effective breakdown resistance as a function of bias current may take several other forms. One application is illustrated in the patented circuit shown in Fig. 10–12 [3]. Diode 14 is a VR diode whose bias current level is determined by the signal level at the output. The breakdown resistance is thus decreased as the output level attempts to increase. This, in turn, reduces the overall ac voltage gain to restore the output voltage to the control level.

VR diodes may be used in other similar applications to control the feedback voltage in an oscillator circuit or to create functional modulators or multipliers.

10–8 UTILIZING VR DIODE NOISE

For many critical low-level applications, the noise generated by the VR diode can be objectionable. It is possible to make use of this characteristic as a noise source.

10–8.1 Muscle Signal Stimulation

The minute electrical signals generated when a muscle is placed in tension covers a broad-band frequency spectrum up to about 1 KHz. By filtering the noise generated by a VR diode operating in the breakdown region, an equivalent signal may be simulated. This is quite helpful in working with special limbs as described by Paisner, Antonelli, and Waring [2] as a result of their work at Rancho Los Amigos Hospital in Downey, California.

10–8.2 Rf Noise Source

The noise generated by the VR diode extends well up into the rf spectrum and has a rather flat noise characteristic over a wide range of frequencies as is shown in Fig. 10–13 taken from an article by Hays Penfield [4]. The relative noise levels taken at a single frequency depend on the breakdown voltage level as well as the value of the reverse bias current. This is shown in Figs 10–14 and 10–15.

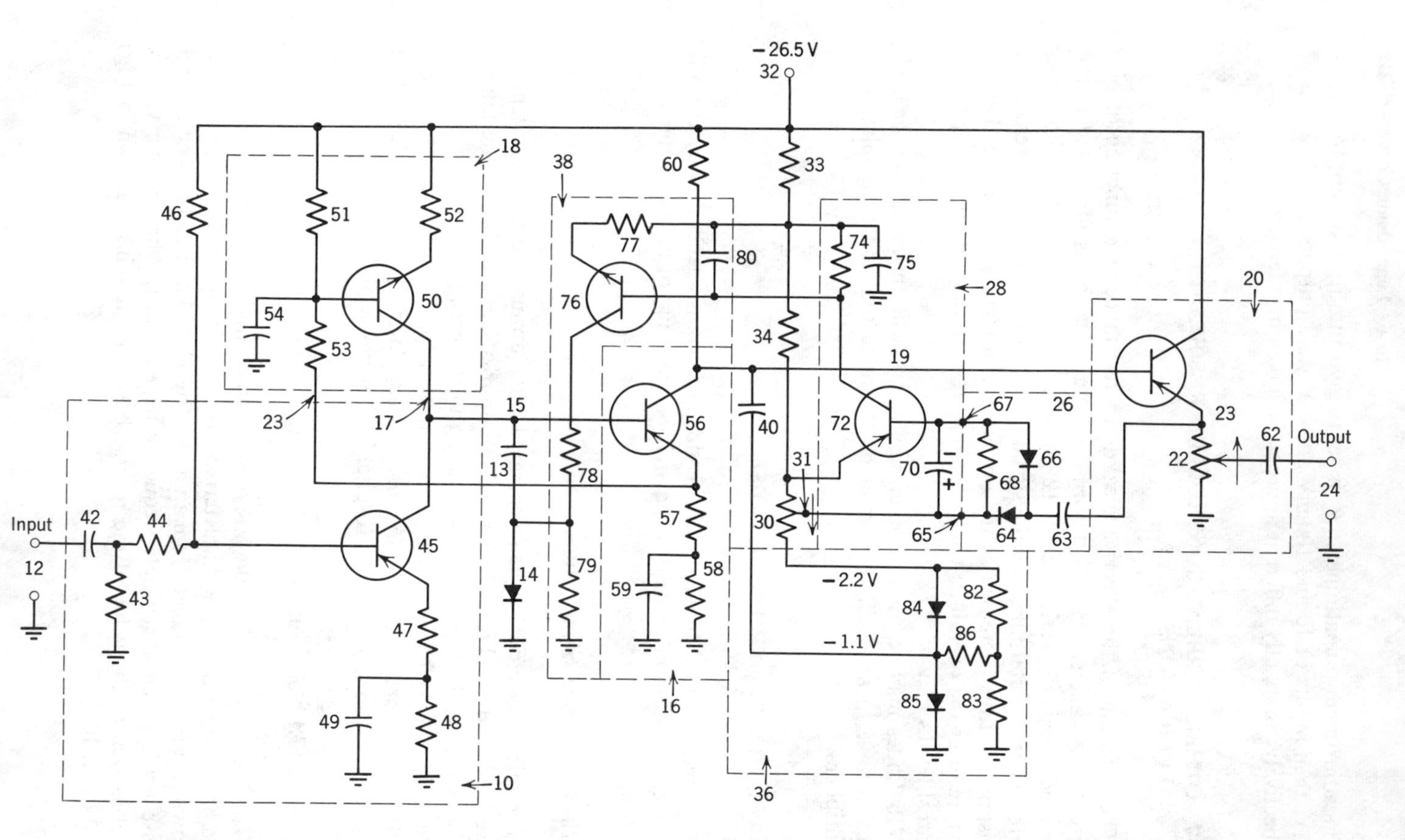

Fig. 10–12 An automatic gain control amplifier uses the variation of breakdown resistance of diode 14 [3].

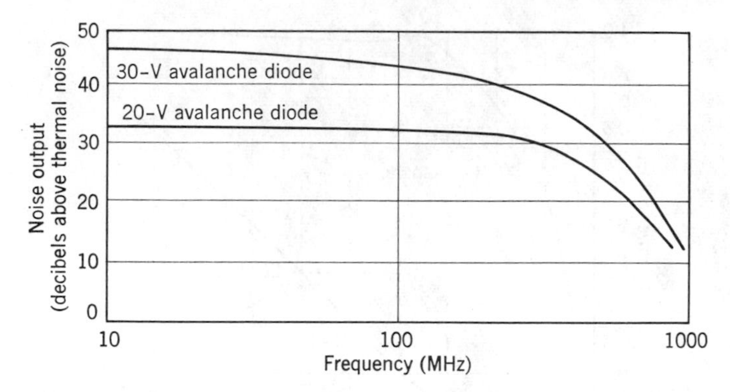

Fig. 10–13 Noise output as a function of frequency for two VR diodes [4].

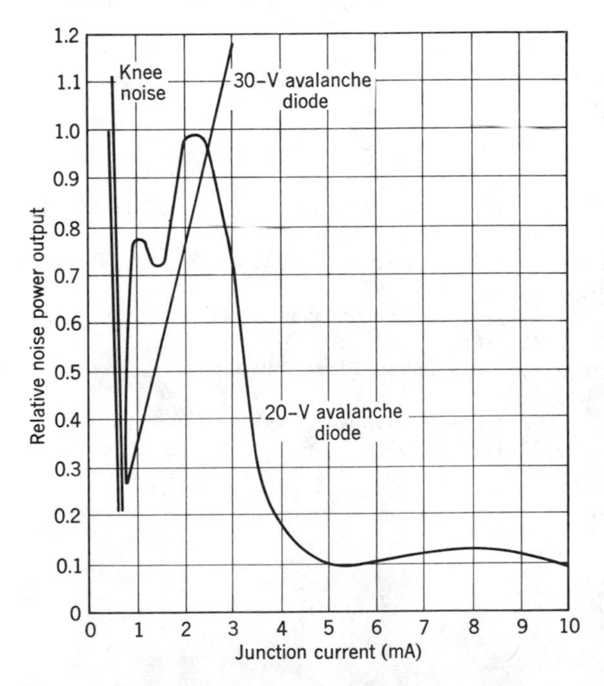

Fig. 10–14 Noise output versus junction current at constant frequency (15 MHz) for two VR diodes [4].

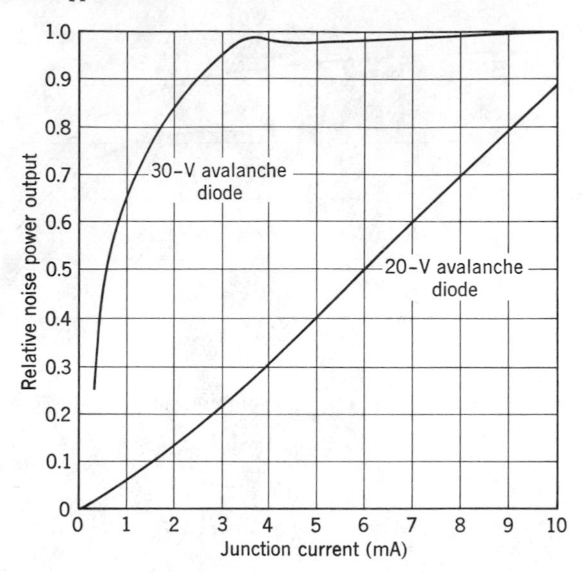

Fig. 10–15 Noise output versus junction current at constant frequency (500 MHz) for two VR diodes [4].

REFERENCES

[1] D. Ophir and U. Galil: Zener Diode Creates Logarithmic Pulse Amplifier, *Electronics*, **34**, 68–70 (July 14, 1961).

[2] W. Paisner, D. Antonelli, and W. Waring: Zener Simulates Muscle Signals, *Electronics*, **41**, 111–112 (June 10, 1968).

[3] E. A. Paschal: *Automatic Volume Control Amplifier*, U.S. Patent No. 2,979,667 (April 11, 1961).

[4] H. Penfield: Why Not Avalanche Diode as RF Noise Source?, *Electron. Design*, **13**, 32–35 (April 12, 1965).

[5] G. Richwell: Zener Stabilizes Phase-Shift Oscillator, *Electron. Equip. Eng.*, **12**, 76 (May 1964).

[6] C. D. Todd: A Versatile Negative Impedance Converter, *Semicond. Prod.*, **6**, 25–29 (May 1963); 27–33 (June 1963).

Chapter 11

SPECIAL DEVICES

11–1 INTRODUCTION

There are several special types of VR devices that are worthy of separate consideration. These include the very important temperature-compensated VR diode, high-stability VR diodes, the low-voltage avalanche VR diode, and special synthesized composite VR diodes. Each of these is discussed in this chapter.

Double-anode VR diodes will not be treated in this chapter as special devices because such a diode may be directly assumed to be two normal VR diodes connected back-to-back (cathodes common). Their use has been mentioned in appropriate chapters.

Another special device is the reference amplifier, a composite configuration of a VR diode used to produce a stable reference voltage, combined with a transistor comparator and error amplifier. This class of components was discussed in detail in conjunction with the voltage monitor circuits of Chapter 4.

11–2 TEMPERATURE-COMPENSATED VR DIODES

The breakdown voltage of a VR diode has an associated temperature coefficient that causes its value to vary with changing junction temperature. This is true whether the temperature variation is due to self-heating effects or to a change in ambient. The temperature coefficient of a VR diode is a characteristic parameter that must be considered in any good design; it was discussed in Chapter 2.

When the variation in V_B due to temperature effects is too high to be tolerated, we must seek some means of reducing it. There are several possible approaches we might take to reduce the effective temperature coefficient, K_T.

1. Since K_T is a function of the value of V_B, we might restrict the value of V_B used to approximately 5.3 V when K_T approaches zero.
2. We might control the ambient temperature variation by including the VR diode in a temperature-controlled oven.
3. We might add some means of external temperature compensation to counteract the K_T of the VR diode.
4. We could include an internal means of temperature compensation within the VR diode package at the time of manufacture.
5. In extreme cases we could combine solutions 2 and 4 (or possibly 2 and 3).

The actual approach we would use in a practical application would depend on the severity of the problem caused by temperature variations as well as other physical limitations.

Restricting the value of V_B to approximately 5.3 V where K_T approaches zero is a good solution when we need fair temperature stability and can design around the limited value of V_B.

By placing the VR diode within a temperature-controlled oven, as is done with precision piezoelectric crystals, we may reduce the variation in ambient temperature to ± 0.1C. This can permit excellent stability in V_B if the bias current is held perfectly constant. The main disadvantages to this approach are the large increase in physical space required as well as the considerable amount of additional power needed for the oven. In addition, we would have a relatively long warmup time required before adequate V_B stability is achieved.

Since the temperature coefficient of most VR diodes, particularly those with V_B greater than about 7 V, is a relatively predictable positive value, we might be able to achieve adequate temperature stability by adding external compensation. This can be done, but it is usually rather time consuming to get a good match. In addition, we have to contend with transient thermal problems to assure that we always get proper tracking between the variation in the VR diode with the external compensation.

The optimum solution for the designer is to be able to specify a VR diode in which the means for temperature compensation have been included within the diode package at the time of manufacture. No additional power will be required and the compensated diode occupies no more space. In addition, we get much better thermal coupling and hence a greatly improved transient thermal response.

11–2.1 Basic Compensation Techniques

Referring back to Fig. 2–27, we recall that the equivalent circuit of the VR diode contains a thermal voltage, V_T, which is produced by temperature variations according to the relationship

$$V_T = K_T(\Delta T) \tag{11-1}$$

where K_T is in mV/°C and ΔT is in °C. The value of V_T is given in millivolts. V_T may be either positive or negative for a given increase in temperature, as determined by the polarity of K_T.

In order to compensate for the error induced by the natural temperature coefficient of the VR diode, we will need to generate another error voltage that will vary in exactly the same amount as does V_T, but in the opposite direction. The curves of Figs. 2–8 and 2–9 illustrate the relationship between K_T and the nominal value of V_B, and thus indicate the relative temperature compensation needed. As mentioned earlier and as evident from Fig. 2–8, it is possible to achieve a zero K_T for a V_B somewhere between 4.5 and 5.6 V with no compensation required.

A study of Fig. 2–8 also reveals that, since very low voltage VR diodes have a negative temperature coefficient and higher voltage units have a positive K_T, it should be possible to achieve a net zero value by combining the two. This is illustrated in Fig. 11–1.

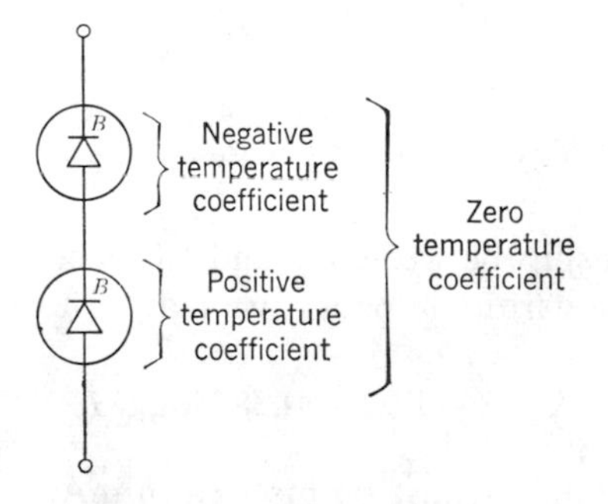

Fig. 11–1 Temperature compensation may be achieved by combining two VR diodes having opposite polarity of individual temperature coefficients.

The exact value of the negative K_T of the low-voltage diode may be trimmed to balance that of the higher voltage unit by adjusting the value of the bias current. As shown in Fig. 2–10, the value of K_T is much more sensitive to variations in bias current for the lower voltage VR diodes.

Although the careful combination of two VR diodes having opposite polarities of temperature coefficient will yield a zero value of K_T for the composite, the resulting characteristics may leave much to be desired; for example, the breakdown resistance of the composite will be the sum of the individual values. Unfortunately, the breakdown resistance of VR diodes having a negative temperature coefficient is rather high.

Another approach to temperature compensation of the VR diode would be to utilize the negative temperature coefficient associated with the forward-biased junction diode. Data taken on a typical unit are shown in Fig. 11–2.

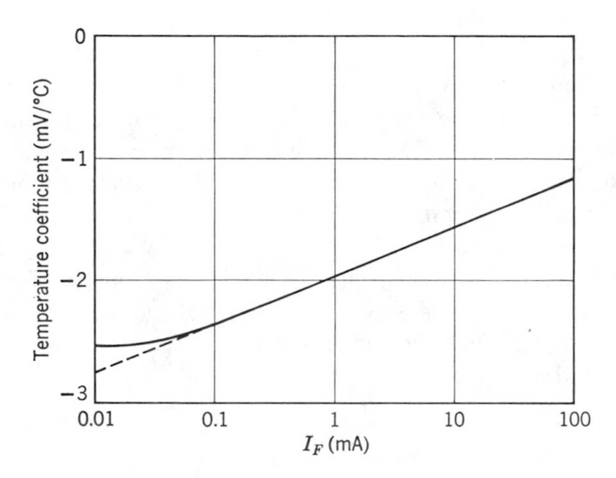

Fig. 11–2 Temperature coefficient of a forward-biased junction diode as a function of current.

Assuming the relationship between K_T and I_F to be a perfect straight line when plotted on semilogarithmic paper, we may write an empirical equation as

$$K_T = -1.92 + 0.385 \log_{10} I_F; \qquad (11\text{-}2)$$

K_T will be in mV/°C and I_F must be inserted in mA.

The circuit arrangement for a VR diode temperature compensated by forward-biased diodes is shown in Fig. 11–3. Although two forward-biased diodes are shown in the illustration, only one or perhaps even more than two may be required. The resulting voltage levels for the temperature-compensated VR diode will generally be confined to discrete bands such as 6.3, 8.4, or 11.8 V for one, two, or three forward-biased diodes, respectively. The total negative temperature coefficient must be exactly equal in magnitude to the positive K_T of the VR diode in order to give a net sum of zero for the K_T of the composite.

The temperature coefficient of the forward-biased diode is sensitive to current even to a greater degree than is true for many VR diodes (except for a V_B of about 5.1 V where the two breakdown mechanisms are in conflict). This can be a help as well as a hindrance. By a slight adjustment of the bias current level, we may trim the temperature coefficient to a very low level. On the other hand, we will need to keep the bias current rather close to that specified by the manufacturer if we are to be assured of the guaranteed specifications.

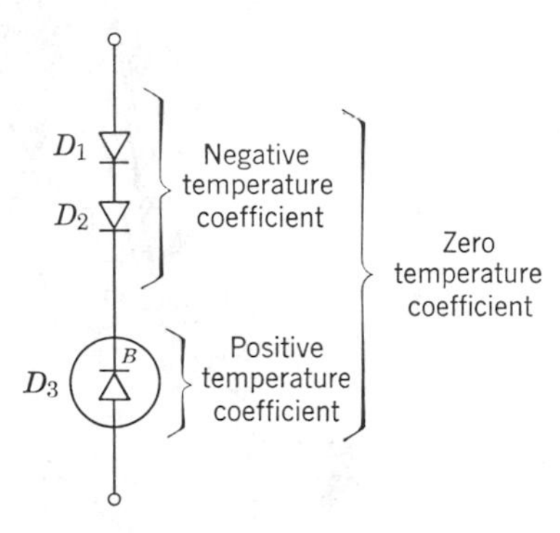

Fig. 11–3 Basic temperature compensation approach using forward-biased diodes.

Successful temperature compensation demands that the junction temperatures of both the VR diode and the forward-biased diodes track very closely. It is difficult to do this with diodes in different packages in which thermal transients may easily unbalance the individual junction temperatures. The best arrangement is to include the forward-biased junctions with the VR diode junction in a sandwich form placed in intimate contact with each other in the same package.

11–2.2 Specifications

The temperature-compensated VR diode is characterized in somewhat the same fashion as a normal VR diode in that we have a nominal breakdown voltage, breakdown resistance, etc. There are certain other parameters which are more closely specified because of the intended applications.

Although normal VR diodes may be operated over a range of bias current levels, the temperature-compensated VR diode must generally be restricted to a single value as specified by the manufacturer. As discussed above, this is due primarily to the variation in K_T of the forward-baised diodes as the current level changes. The overall effect is shown in Fig. 11–4. The terminal characteristics of a typical temperature-compensated VR diode are given for three different temperatures. Note that in Fig. 11–4a all three curves appear to intersect at one point A. This would indicate that we could obtain perfect temperature compensation by applying a bias current equal to I_A. For a bias current less than I_A we would get a negative value of K_T, whereas a bias current greater than I_A would produce a positive temperature coefficient.

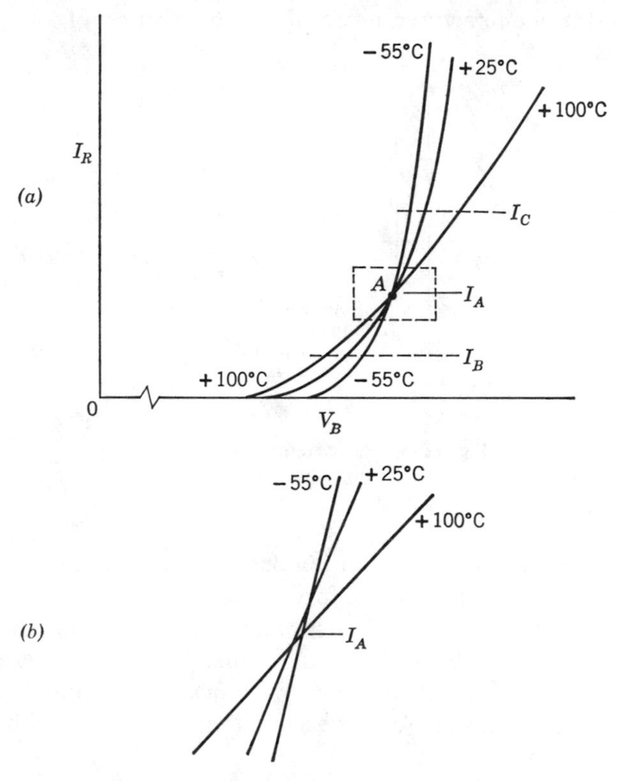

Fig. 11–4 Breakdown characteristics of a temperature-compensated VR diode taken at three temperatures: (*a*) overall characteristic; (*b*) enlargement of the region around point "*A*" shows that intersection may not occur at one point.

If we extend the plot around point A as is done in Fig. 11–4*b*, we see that the three characteristic curves do not necessarily meet at the same point. This means that any bias current level will represent a compromise condition.

The breakdown resistance of a temperature-compensated VR diode will normally be much larger than for a normal VR diode of the same breakdown voltage. This is due to the addition of the forward resistance of the compensating diodes.

The temperature characteristics of a temperature-compensated VR diode may be specified in a number of different ways. One approach is the cone or hour glass method as illustrated in Fig. 11–5. In this type of specification the guaranteed limits of the variation of V_B due to temperature excursion are defined by two straight lines passing through a reference point (normally 25°C) and obeying the equation

$$V_{B(\max)} = \pm K_T(\Delta T) = \frac{\pm K_T \, \Delta T(V_{B(25)})}{100}, \tag{11-3}$$

where the K_T is the specified temperature coefficient in terms of mV/°C and K_T is given in terms of %/°C. ΔT is the variation in temperature from the reference point. The diode whose temperature characteristics are shown in Fig. 11–5 would be considered to have failed the indicated limits.

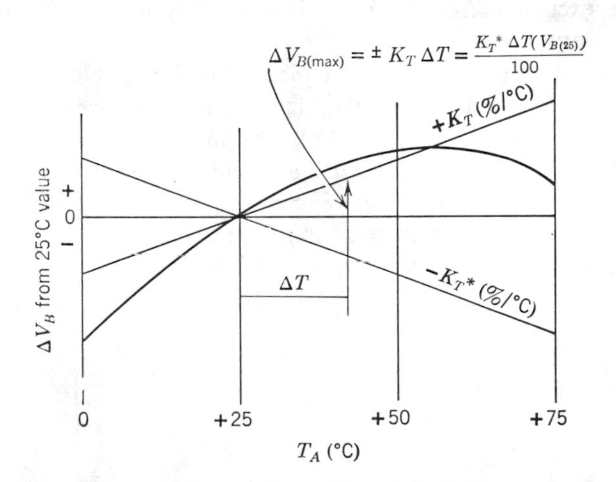

Fig. 11–5 The cone or hour glass method of temperature coefficient specification.

Another technique of specifying the overall temperature characteristics of a compensated diode is the single box method illustrated in Fig. 11–6. Here the specification gives a maximum total excursion in V_B over the indicated temperature limits. Nothing is actually defined concerning the anticipated shape of the characteristics. The variation may rise or fall linearly over the entire temperature range or it may remain small for a large portion of the temperature range and then change rapidly as the end temperature limit is approached.

Although the single box specification may have an accompanying average temperature coefficient, defined as the quotient of $\Delta V_{B(\max)}$ and the total

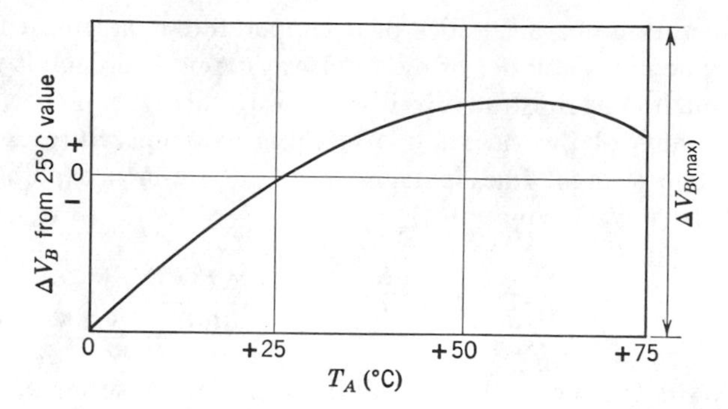

Fig. 11–6 Single box method of temperature coefficient specification.

temperature excursion ΔT, this value is not really meaningful and, in fact, may well be very misleading.

A slight variation of the single box method is the double box approach to specifying the temperature characteristics of a temperature-compensated VR diode. This is illustrated in Fig. 11–7. A separate box or maximum excursion limit is given for the low temperature and the high temperature regions. In addition, the excursion of V_B is referenced to the 25°C value.

Although the double box method is a better means of characterizing the temperature characteristics than given by the single box approach the specification still yields only information regarding overall worst-case

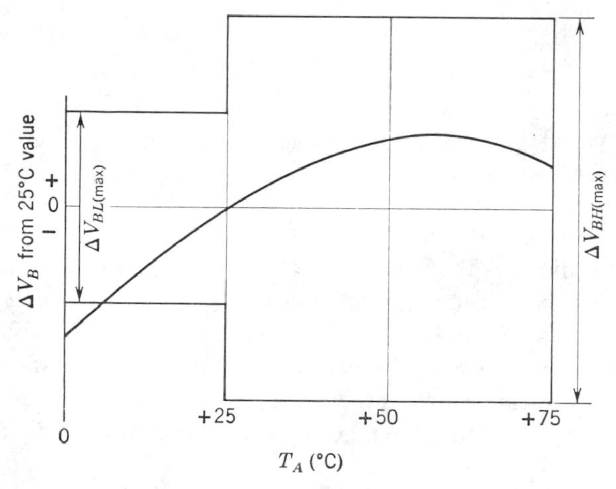

Fig. 11–7 Double box method of temperature coefficient specification.

excursion. Nothing is indicated about the manner or the shape of the variation. The manufacturer merely guarantees that the maximum limits will not be exceeded throughout the total temperature range.

Generally, two excursion limits are given: $\Delta V_{BH(\max)}$ for the high temperature region and $\Delta V_{BL(\max)}$ for temperatures below 25°C. The two values are normally related in a manner yielding the same average temperature coefficient. Thus for the illustration of Fig. 11–7. $\Delta V_{BL(\max)}$ would be one-half that of $\Delta V_{BH(\max)}$ since the total temperature excursion is below the reference temperature.

Occasionally, a VR diode may be specified in the double box fashion but with measurements made only at three temperatures. If this were done with the unit shown in Fig. 11–7, the results could be very misleading because the value of V_B at the high temperature limit is only slightly higher than that at 25C. The maximum variation is much larger.

Another method of specifying the temperature characteristics is to place a limit on the slope of the temperature characteristic curve as shown in Fig. 11–8. The specification guarantees that the incremental slope never exceeds

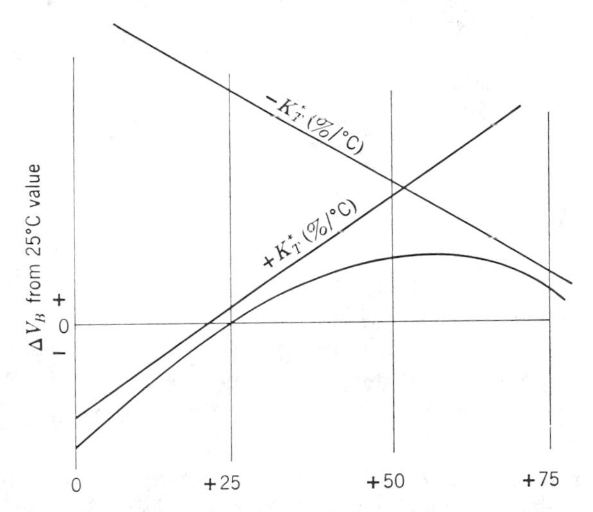

Fig. 11–8 Slope method of temperature coefficient specification.

the stated value. This is a very stringent type of specification, but one which is very meaningful to the designer since he may now predict actual performance over the temperature range. It is not easy for the manufacturer to either meet or test, and thus is not generally used.

11–2.3 Special Circuit Requirements

In order to obtain optimum performance of a temperature-compensated VR diode in a given application, it will be necessary to use a little more care in designing the regulator circuit.

As mentioned earlier, the temperature coefficient of a compensated VR diode is a function of the bias current level. This is very clearly illustrated in Fig. 11–9. The specified bias current for the 1N430A is 10 mA.

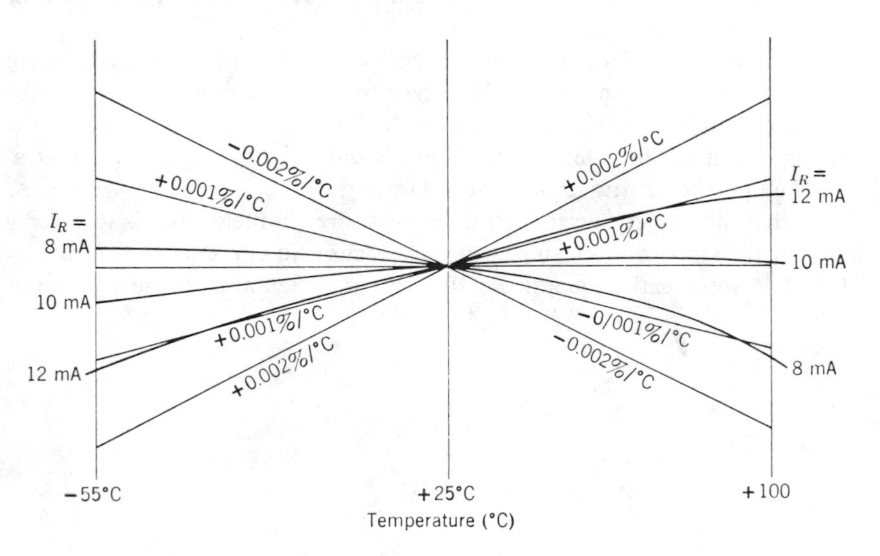

Fig. 11–9 Temperature stability is greatly influenced by bias current as illustrated for the 1N430A.

Note that the temperature coefficient for the particular unit tested would be improved in the low temperature region if the bias current were made 8 mA. On the other hand, if the VR diode were near the specification limit, a slight variation in the bias current could cause the maximum variation to be exceeded.

A point worth noting here is the general capability of improving the overall temperature coefficient of a given diode by customizing the bias current level. This becomes especially profitable where very tight requirements are necessary over a limited temperature range.

Temperature-compensated VR diodes normally require more careful control of the bias current for a given degree of regulation. First of all the use of a compensated VR diode normally indicates a tighter control of the breakdown voltage; hence I_R must be made constant to avoid variation in V_B due to the breakdown resistance.

The breakdown resistance of a temperature-compensated VR diode is much higher than that for a normal VR diode of the same breakdown voltage and under the same current conditions. This is due to the added resistance in the forward-biased junctions.

We may exercise careful control of the bias current including some degree of temperature compensation of I_R or we may prefer to make use of the bridge regulator circuit as shown in Fig. 3–6.

Another effect that should not be neglected in critical applications is the production of thermoelectric voltages on the contact of dissimilar metals. Careful choice of wiring paths and connectors can usually reduce this effect to an insignificant level.

11–2.4 Techniques for Economy

The manufacture of carefully compensated VR diodes requires substantial effort and results in a cost significantly higher than a normal VR diode. The wise circuit designer will carefully choose the diode that will permit proper circuit function without overspecifying a tighter temperature control than necessary.

Although it is normally foolish economy to make your own temperature-compensated VR diode when tight control is necessary, there are many applications where only a limited degree of temperature compensation is required. In such a case, adequate performance may be obtained by using a normal VR diode with one or more forward-biased diodes placed in series electrically and physically mounted in close proximity to the VR diode.

Another approach that can yield very excellent temperature stability at very low cost is the use of a *PNP* transistor connected in the manner shown in Fig. 11–10 [9]. The emitter-base junction of the *PNP* silicon transistor such

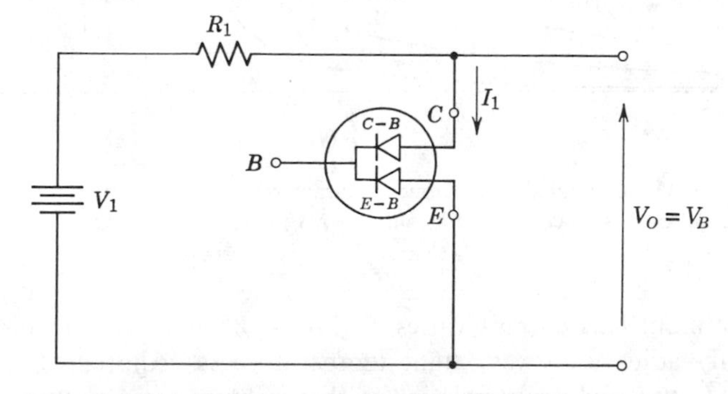

Fig. 11–10 A *PNP* silicon transistor (such as the 2N3638) may be operated as shown with polarity biasing to give operation as temperature-compensated VR diode. The transistor is depicted as two diodes.

as the 2N3638 has a typical breakdown voltage of about 6.1 V. The excellent process control typical of this class of devices results in a very narrow range of voltage and temperature coefficient. A single forward-biased junction diode will produce the necessary compensation.

In the arrangement shown in Fig. 11–10 the collector junction is forward-biased and the emitter-base junction is operated in the breakdown region. The result is a temperature-compensated VR diode that yields rather astounding temperature stability as shown in Fig. 11–11. Less than a 2 mV variation was noted for a temperature excursion from −40 to +100°C. This represents an average K_T of about 0.0002%/°C.

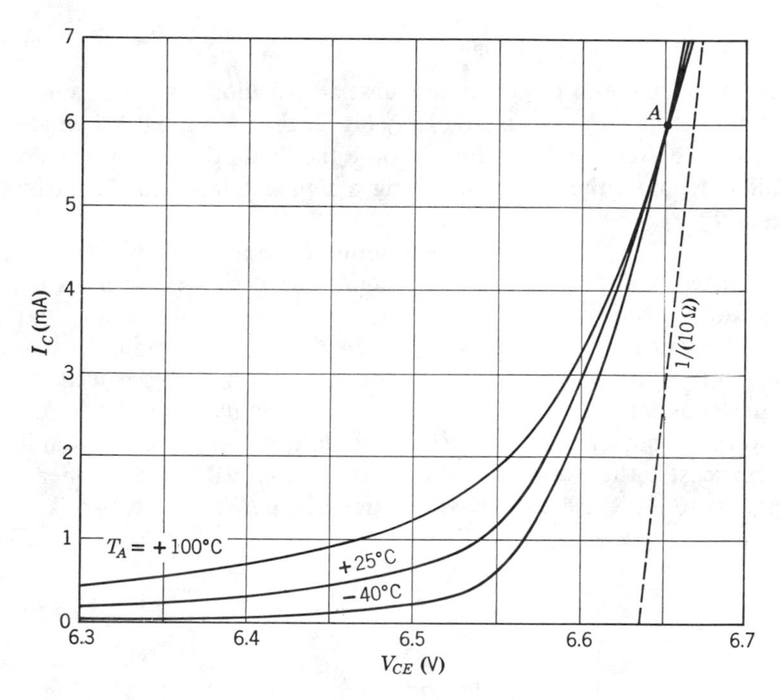

Fig. 11–11 Expanded breakdown characteristics plotted for a typical 2N3638 transistor connected as shown in Fig. 11–10.

The optimum bias current varies slightly from unit to unit and must be individually adjusted if very tight temperature is required. The value of optimum I_R has a close correlation with the breakdown voltage measured from collector to emitter at a current of 5 mA. The result of testing several hundred units from different lots is shown in Fig. 11–12. Better than 70% of all units tested fell within the limits of 6.6 to 6.9 V. Figure 11–12 also indicates

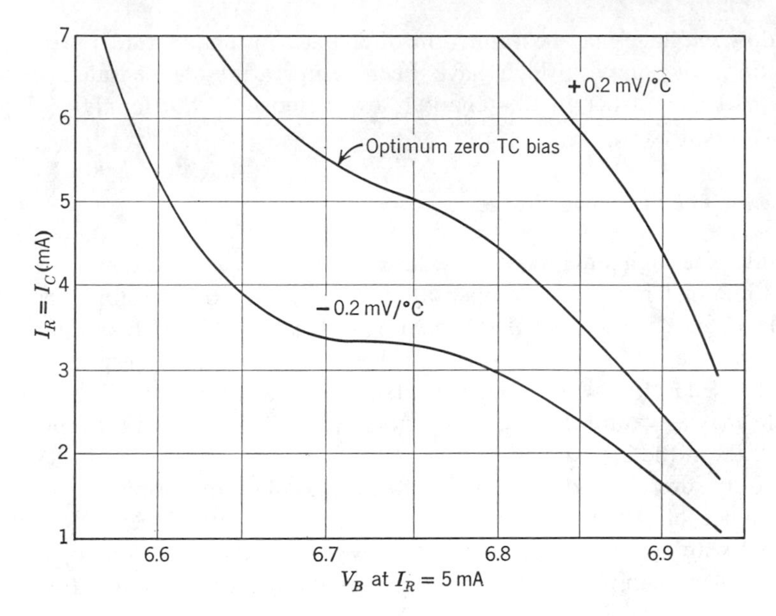

Fig. 11–12 Bias current range for 2N3638 transistor used as compensated VR diode.

the typical bias current limits for a temperature coefficient of ±0.2 mV/°C corresponding to ±0.003%/°C. For example, a transistor whose measured breakdown voltage was 6.8 V would exhibit a temperature coefficient within ±0.003%/°C for a bias current anywhere from 3 to 7 mA with optimum K_T occurring for I_R of 4.4 mA.

By selecting transistors with a narrow breakdown range, it becomes quite feasible to obtain better than ±0.001%/°C without customizing the bias current. Since this characteristic is not controlled by the manufacturer, it would be wise to monitor the devices received, at least on a sample basis.

11–3 HIGH-STABILITY VR DIODES

The reference voltage in precision differential and digital voltmeters must be very stable in order to maintain high accuracy between calibrations. The typical VR diode, held at constant temperature and bias current levels, demonstrates stabilities in the other of ±100 PPM (±0.01%) or less. Many VR diodes are substantially better than this and vary less than ±5 PPM (±0.005%) over a 1000-hr period.

Special VR diodes which have been characterized around a tight stability criterion are available from several manufacturers. Since the variation in breakdown voltage may be a function of changes in temperature, current, or time, only VR diodes which have been temperature-compensated in the ways described earlier in this chapter are included in the family of high-stability VR diodes.

11–3.1 The Screening Process

Mainly, the high-stability VR diodes are obtained as a screening process from a line of temperature-compensated VR diodes. To a certain extent the noise level may be used as a first screening criterion. Units with an abnormal noise level are *likely* to be more unstable than those with very low noise levels. Noise level, however, is not an absolute indication of stability because a diode may have an extremely low noise level and still exhibit long-term instabilities in the breakdown voltage.

The only sure method of screening is by actually monitoring the stability performance of the diodes in question over a substantial period of time. Frequently an accelerated aging process (with higher than normal operating power applied) can reduce the length of observation time required to insure a given level of stability.

Typically, the greatest variation occurs during the initial 400 to 1000 hours and then the voltage begins to stabilize at a relatively constant level. The better devices seem to fluctuate in the manner illustrated in Fig. 11–13 where some variation in V_B still occurs after initial stabilization, but the excursions seem to be restricted to a relatively narrow band. Units having a stability greater than ± 5 PPM ($\pm 0.0005\%$) per 1000 hours are available.

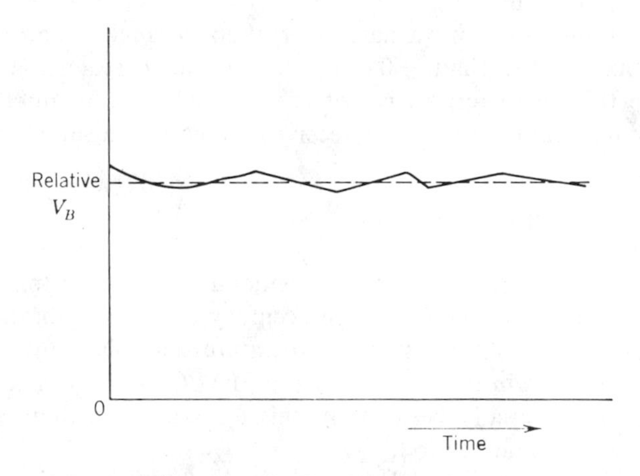

Fig. 11–13 The more stable VR diodes seem to fluctuate around a given level.

Less stable VR diodes can still vary in the general manner shown in Fig. 11-13 but with a wider band of variations. Other diodes exhibit an excursion around a level that is continually increasing or decreasing as indicated in the curves of Fig. 11-14.

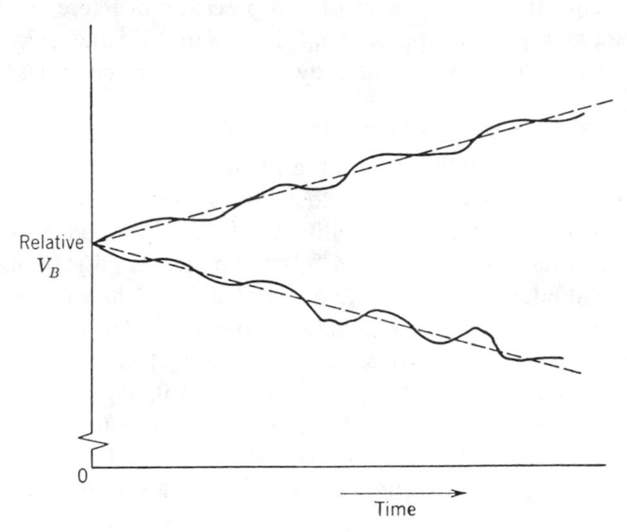

Fig. 11-14 Some VR diodes exhibit a variation in breakdown voltage which continues in a general trend of either increase or decrease even after long intervals of time.

11-3.2 Comparison with the Unsaturated Cadmium Cell

The best high-stability VR diodes yield a reference voltage with a *practical* stability far greater than for the unsaturated portable cadmium cell (also called the Weston cell). The unsaturated cell was used in portable precision instruments before the availability of high-stability VR diodes but had certain limitations.

First of all, the ideal long term stability under carefully controlled conditions was no better than ± 100 PPM ($\pm 0.01\%$). Although classified as "portable," any appreciable movement or shock at all required a recovery time up to a day or more before precision was restored.

Although the theoretical temperature coefficient of the unsaturated cell is very low due to a cancellation of the individual temperature coefficients of the two limbs, a temperature gradient of as little as 1°C can cause an error of ± 300 PPM ($\pm 0.03\%$). In addition, any sudden temperature change could induce even more error because of a hysteresis effect.

Add to the above problems the fact of bulk size and weight (especially if temperature gradients were to be reduced by thermal shielding of a heavy copper box) together with the severe restriction of current levels of much below 1 mA. The high-stability VR diode with its small size, insensitivity to position and normal handling, and much better thermal characteristics is a very welcome replacement for the unsaturated cadmium cell.

The best high-stability VR diodes offer a precision of reference in the *field on a portable basis* which is approaching that offered by the saturated cell *in the laboratory* and they have a much lower temperature coefficient.

11–3.3 Special Circuit Requirements

To achieve the end stability of which a high-stability VR diode is capable, it is necessary to control carefully the temperature and bias current. The temperature coefficient of a high-stability VR diode might be $\pm 0.0005\%/°C$. This means a variation of ± 5 PPM/°C. Ultimate stability is obtained by placing the VR diode in a temperature-controlled oven where it becomes quite practical to reduce temperature fluctuations to ± 0.3C. The end temperature-induced variations in voltage would then be ± 2 PPM or less.

Another approach to reducing temperature sensitivity is by individually adjusting the bias current level for the optimum temperature coefficient for that diode operated over a given range of temperatures. This becomes a very practical method where an instrument normally has a very restricted range in operating temperatures.

A typical 6.35-V high-stability VR diode has a maximum breakdown resistance of 10 Ω at the test current of 7.5 mA. A variation in the breakdown voltage of ± 1 PPM (± 6.35 mV) would be produced by a change in the bias current of only ± 0.635 mA. This means that we must control the bias current to within $\pm 0.017\%$ in order to restrict the stability error induced by bias variations to less than ± 2 PPM. By using the bridge approach to the reduction of the effects of breakdown resistance as illustrated in Fig. 3–6, we may simplify the bias control requirements.

11–4 INTEGRATED-CIRCUIT COMPOSITE VR DIODE

It is possible to design integrated-circuit composite VR diodes that act in much the same manner as two-terminal junction VR diodes but with improved performance under certain conditions. One example of this is the LM103 offered commercially by National Semiconductor Corporation.

The internal circuit diagram of the LM103 is shown in Fig. 11–15. Transistor Q_1 serves as the voltage breakdown reference element. Unlike available junction VR diodes that operate either in a Zener or avalanche mode, the breakdown mechanism in Q_1 is punchthrough from emitter to collector in the graded-base transistor. The additional elements in the circuit serve to control the bias current through Q_1 at a relatively constant level.

With terminal A made positive with respect to terminal B, negligible current flows until V_R is equal to the punchthrough voltage of Q_1. At a voltage level just slightly above punchthrough, field-effect transistor Q_2 is conducting and no base current is supplied to Q_3.

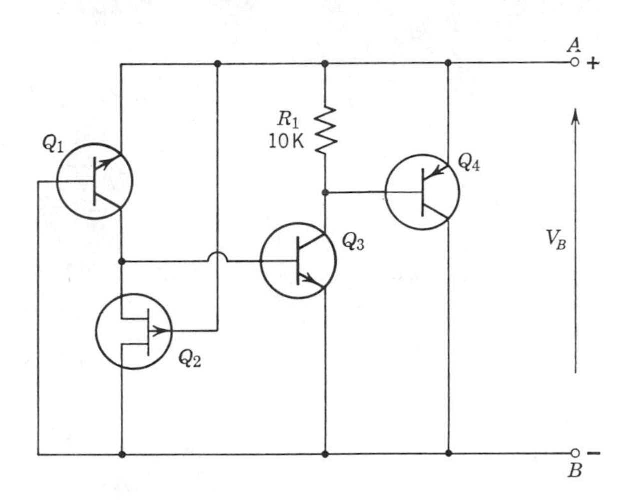

Fig. 11–15 Internal circuit schematic of the LM103 low-voltage integrated circuit breakdown diode.

As the terminal voltage is increased still further, Q_2 begins to be pinched off and its drain current is limited. When the current through Q_1 increases beyond the limited drain current, the excess is applied as a base current drive to *NPN* transistor Q_3. This causes a collector current to flow through resistor R_1 and add to the overall terminal current. When the collector current of Q_3 exceeds about 60 mA, *PNP* transistor Q_4 will also begin to conduct with its collector current being added to the total terminal current required.

The end result is a feedback arrangement in which the bias current through Q_1 is maintained relatively constant with any additional terminal current being shunted by Q_3 and Q_4.

Regulation performance is indicated in Fig. 11–16 for a composite device having a nominal breakdown voltage of 2.4 V. Note that even at the low current level around 100 mA the effective breakdown resistance is only about 70 Ω.

The dashed line of Fig. 11–16 illustrates the typical performance of a 5.6-V alloy-junction VR diode at the same current level (a 2.4-V alloy-junction VR diode would look even worse). Zener breakdown is relatively soft and hence regulation is quite poor at low current operation.

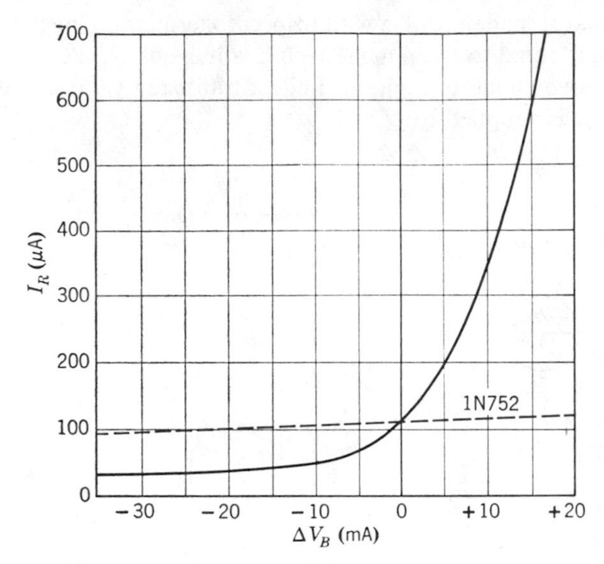

Fig. 11–16 Low-level characteristics for the LM103 active-punchthrough device. Dashed line indicates typical performance of a 5.6-V alloy junction VR diode at this current level.

Active-punchthrough composite regulators are available from 2.4 to 5.6 V and are designed for optimum regulation performance in the operating current range from 0.1 to 1 mA. Above or below these current levels, performance falls off somewhat, although still superior to Zener diodes in the same voltage range.

Temperature performance of the active-punchthrough composite regulator is illustrated in Fig. 11–17. The temperature coefficient exhibited is about −3.3 mV/°C and remains relatively constant over the normal working current range and for the entire family of devices from 2.4 to 5.6 V. Most of the temperature coefficient is due to the variation in V_{BE} of Q_3 with temperature.

11–5 THE LOW-VOLTAGE AVALANCHE DIODE

The breakdown mechanism of VR diodes with a breakdown voltage above 9 V is primarily due to avalanche, and the breakdown characteristic is relatively sharp. Low-voltage VR diodes with breakdown below about 4 V, however, exhibit primarily Zener or field-emission breakdown with a relatively soft logarithmic breakdown characteristic.

VR diodes having a breakdown voltage between 4 and 9 V actually exhibit both Zener and avalanche effects as discussed in Chapter 2. The relative mix is higher in Zener effect for the 4-V units and higher in avalanche effect for the 9-V devices. The contribution of the Zener breakdown causes the characteristic to be relatively soft and the relative breakdown resistance to be rather high.

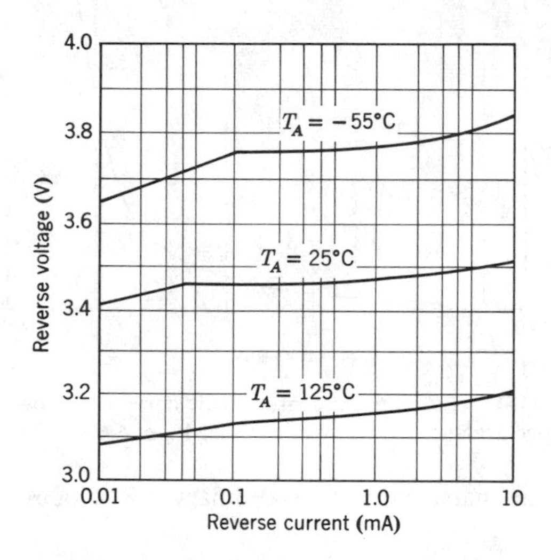

Fig. 11–17 Temperature characteristics of the active-punchthrough diode. (Courtesy National Semiconductor Corp.)

In the low-voltage avalanche (or LVA) diodes, the effects of Zener breakdown are suppressed by process variation. The resulting predominance of avalanche effect yields a substantial improvement in the breakdown characteristics.

LVA devices are available with breakdown voltages from 4.3 to 10 V. The lowest dynamic breakdown resistance occurs for a 5.6-V diode. The remarkable difference between a normal 5.6-V VR diode and a 5.6-V LVA unit are particularly outstanding at very low currents. Figure 11–18 shows the measured performance of one LVA unit. The device clearly has a useful regulation characteristic down to about 2 mA. Lower voltage LVA diodes begin to exhibit more of the Zener effects and breakdown is not as sharp, but a distinct improvement over the ordinary VR diode is still apparent.

The temperature coefficient of the LVA diode family is different from the ordinary VR diode with the same breakdown voltage. The typical zero temperature coefficient for the normal VR diode occurs for a unit with a breakdown of about 5.5 V. For the LVA diode zero temperature coefficient is displayed by a device with breakdown voltage of about 4.7 V. LVA diodes with breakdown above 4.7 V have a positive temperature coefficient but with a

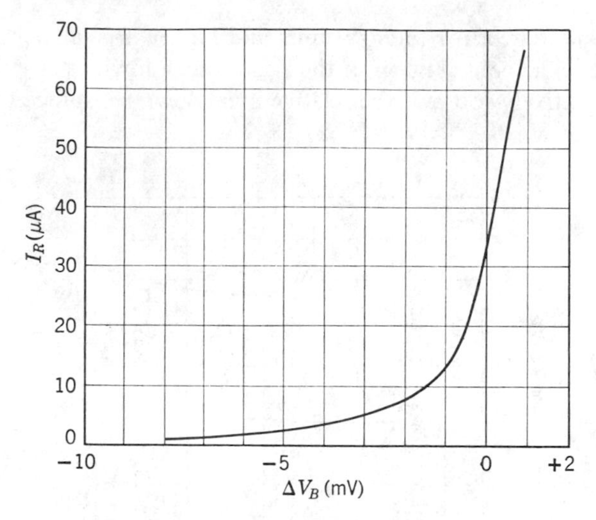

Fig. 11-18 Low current characteristic curve for a low-voltage avalanche diode with nominal V_B of 5.6 V.

magnitude less than that exhibited by ordinary VR diodes with the same breakdown voltage

The LVA would excel as a low-voltage coupling element in which operation at low bias currents is required.

REFERENCES

Temperature-Compensated VR Diodes

[1] H. Banasiewicz: Regulation Factor for Temperature Compensated Zener Diode References, *Electron. Eng.*, **38**, 530–531 (August 1966).

[2] R. E. Jensen: The Temperature-Compensated Reference Element, *Elec. Des. News*, **7**, 110–119 (October 1962).

[3] J. R. Madigan: Thermal Characteristics of Silicon Diodes, *Electron. Ind.*, **18**, 80–87 (December 1959); **19**, 83–87 (January 1960).

[4] J. McCoy: Selecting Right Temperature Compensated Zener Diode, *Electron. Equip. Eng.*, **13**, 44–49 (January 1965).

[5] T. Mollinga: Effect of Temperature and Current on Zener Breakdown, *Electro-Technol. (New York)*, **72**, 122, 124, 126 (October 1963).

[6] L. I. Morgenstern: Temperature Compensated Zener Diodes, *Semicond. Prod.*, **5**, 25–29 (April 1962).

[7] H. Nash and G. Porter: Semiconductor Reference Assemblies, *Electron. Equip. Eng.*, **7**, 81–82 (October 1959)

[8] J. S. Schaffner and R. F. Shea: The Variation of the Forward Characteristics of Junction Diodes with Temperature, *Proc. IRE*, **43**, 101 (January 1955).

[9] C. D. Todd: Stable, Low-Cost Reference Power Supplies, *Electron. World*, **78**, 39–41, 79 (December 1967).

High-Stability VR Diodes

[10] R. P. Baker and J. Nagy, Jr.: Investigation of Long-Term Stability of Zener Voltage References, *IRE Trans. Instr.*, **1–9**, 226–231 (September 1960).

[11] W. G. Eicke, Jr.: Making Precision Measurements of Zener Diode Voltages, *IEEE Trans. Comm. Electron.*, **83**, 433–438 (September 1964).

[12] K. Enslein: Characteristics of Silicon Junction Diodes as Precision Voltage Reference Devices, *IRE Trans. Instr.*, **1–6**, 105–118 (June 1957).

[13] W. H. Hunter: An Ultra Stable Diffused Voltage Reference Diode, *IRE Wescon Conv. Record*, **3**, Part 6, 113–117 (1959).

[14] R. M. Minke: How to Measure Zener Stability, *Electron. Equip. Eng.*, **11**, 121–125 (March 1963).

[15] R. M. Minke: Ultra Stable Reference Elements, *Electron. Ind.*, **22**, 84–85, 88 (February 1963).

[16] J. A. Rose: Bridge-Circuit has 6-Month Stability, *Elec. Des. News*, **8**, 33 (October 1963).

[17] J. Wilber-Ham and K. S. Jackson: 22 mA DC Supply Stable to 1 Part per Million per Day, *J. Sci. Instr.*, **39**, 86, (February 1962).

[18] Ultra-Stable Semiconductor Voltage References, *Solid-State Des.*, **4**, 16–17 (March 1963).

[19] Zener Diodes as Voltage Standards, *Electro-Technol.* (*New York*), **73**, 17–18 (April 1964).

Integrated Circuit Composite

[20] R. S. Widlar: A New Low Voltage Breakdown Diode, Internat. Solid State Circuits Conf., February 1968. (Also available as National Semiconductor Corp. publication TP-5).

Low-Voltage Avalanche

[21] M. Queen: The Future of the Diode, *EEE-Circuit Des. Eng.*, **16**, 56-61 (April 1968).

Chapter 12

TEST METHODS AND EQUIPMENT

12–1 INTRODUCTION

Chapter 2 deals extensively with the various parameters used to characterize VR diodes. For the most part they may be measured by considering logically the basic definitions and using instrumentation techniques common to the semiconductor field.

A few especially useful arrangements are worthy of specific comment and these are presented and discussed briefly in this chapter.

12–2 CURVE TRACERS

Perhaps no single test on a VR diode can be as revealing as a display of its reverse terminal characteristic curve. If the display is properly calibrated, we may obtain a good indication of the breakdown voltage, the relative sharpness of the knee region, the dynamic breakdown resistance, gross noise and instabilities, and even an idea of the temperature coefficient.

Characteristic curve displays may be presented on either a cathode-ray tube or on an *X-Y* recorder. Each method has certain advantages, depending on the specific information needed.

12–2.2 CRO Curve Tracers

Several commercially available curve tracers may be used to display successfully the characteristic curve of a VR diode. It is generally wise to include some amount of limiting resistance or else increase the maximum sweep voltage very carefully to prevent excessive current flow on reaching the breakdown region.

A simple circuit that may be used with a normal oscilloscope having dc inputs for both the X- and the Y-axes is shown in Fig. 12–1. A voltage proportional to the device terminal current is developed across resistor R_1 and fed to the Y input of the oscilloscope. Normally, this voltage should be made small in comparison to the voltage across the VR diode but within the limitations of the oscilloscope being used.

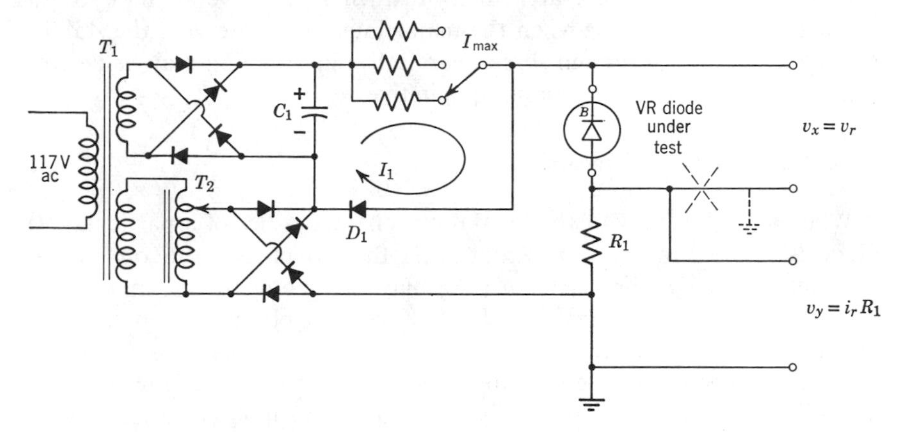

Fig. 12–1 A CRO curve tracer with automatic current limiting.

The X input of the oscilloscope is directly derived from the voltage across the VR diode under test if a differential input is available. If the oscilloscope X input may not be used differentially, then the alternate connection shown by the dashed lines of Fig. 12–1 may be used. If this is done, it becomes more important that the voltage developed across R_1 be kept as low as practical, since it represents an error signal with regard to the actual voltage across the VR diode.

In addition to having an adjustment of the maximum voltage to be displayed by means of the variable autotransformer T_2, the test circuit also has a controlled current limiter. This is accomplished by diode D_1 that is forward-biased only as long as the output terminal current to the VR diode being tested is less than the current I_1 established by the floating power supply and a series limiting resistor. By making the voltage across C_1 equal to a value much larger than the maximum sweep voltage, we can control the maximum terminal current through the VR diode under test.

With the current limit feature we may set the maximum sweep voltage to a value somewhat larger than the highest breakdown voltage anticipated. If the breakdown voltage of a particular device is much lower (or if the operator should accidentally reverse the diode such that it becomes forward-biased), the characteristic curve will be swept until the maximum limit current is reached. This protects the VR diode as well as preventing an overload of the oscilloscope input.

Another approach to the CRO-type characteristic curve tracer for VR diodes is described in reference [4]. In this arrangement, shown schematically in Fig. 12–2, a variable voltage-controlled limiting resistance is used to approximate a constant power dissipation control.

CRO-type curve tracers are superior when a fast overall look of the characteristic is desired or when the major interest is observing the stability of the knee region. Although some quantitative information may be derived, this type of presentation is mainly qualitative.

12–2.2 X-Y Curve Tracers

When a careful study of the breakdown characteristics of a particular VR diode is needed, the best presentation is that which is given on an *X-Y* recorder. Although the CRO display may be photographed in order to provide a hard-copy for analysis, the *X-Y* recorder plot may be made much larger and more accurate.

A simple scheme may be breadboarded to obtain characteristic curves on an *X-Y* recorder by using a manually adjusted power supply (with very smooth control) and a current sampling resistor. Although very good curves may be obtained, the scheme lacks control and convenience.

Unlike the oscilloscope the *X-Y* recorder has a rather slow writing speed capability. The application of an input signal that changes too rapidly results in an erroneous display. Thus it becomes necessary to control the maximum rate of rise for both inputs.

If we use a simple voltage ramp with a controlled rate of rise slow enough for the *X*-axis to follow the applied terminal voltage, all is well until the breakdown region is reached. At this time the terminal current rises rapidly. Unless the rise of the voltage ramp is made extremely slow, it is likely that the resulting current-produced voltage fed to the *Y* input of the recorder will increase at a rate in excess of the capability of the servo system used.

In the breakdown region we would be better off to have a controlled current ramp. This would cause the terminal voltage to rise too sharply in the region before breakdown.

The best presentation control would be to combine the voltage ramp with the current ramp in a such manner that each region of the breakdown characteristic will be swept by the optimum control function.

One approach is shown in block diagram from in Fig. 12–3. This arrangement controls the rate of rise of the terminal voltage as applied to the *X* input of the recorder and also monitors and limits the rate of rise of the signal corresponding to the terminal current as applied to the *Y* input. In addition,

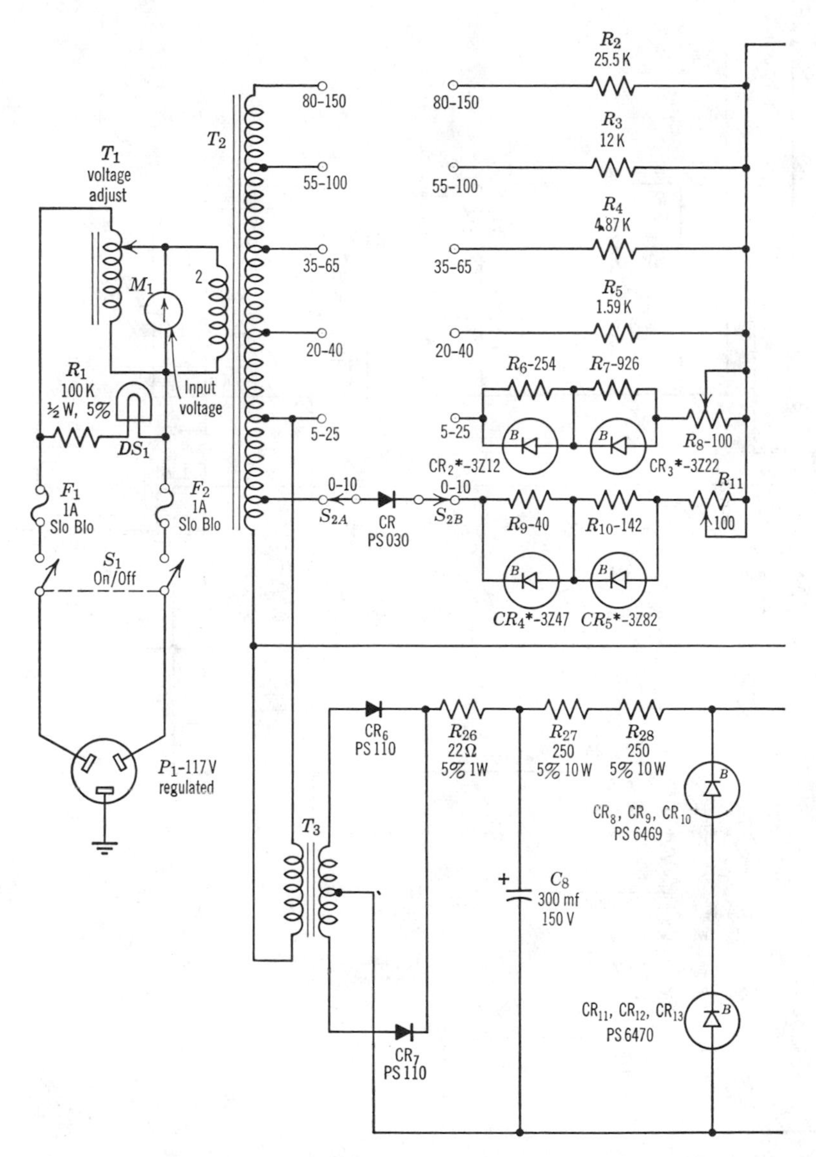

Fig. 12–2 Schematic diagram of VR diode curve tracer with constant power approximation [4].

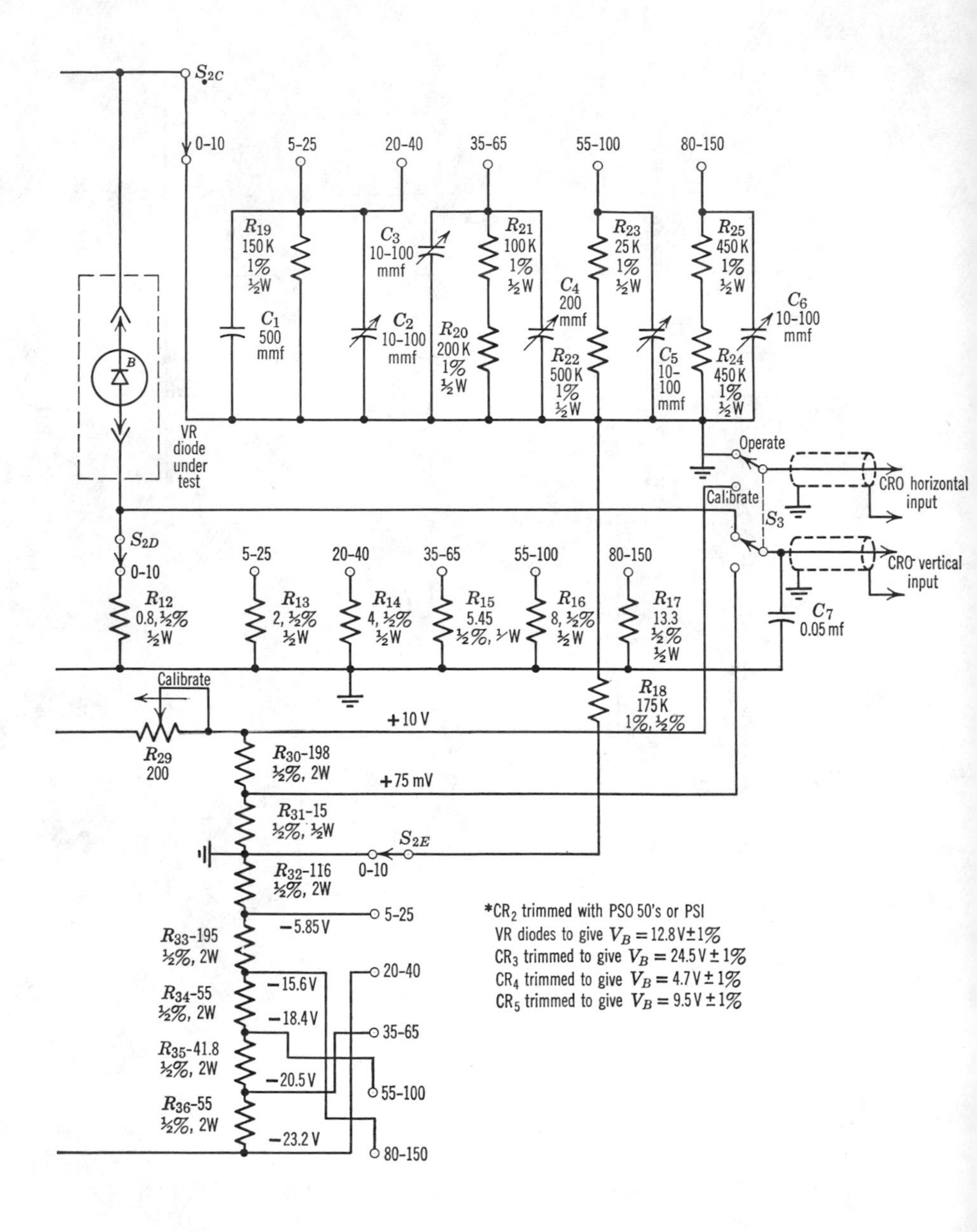

S_{2C}
0-10
5-25
20-40
35-65
55-100
80-150
R_{19} 150 K 1% ½W
C_1 500 mmf
C_3 10-100 mmf
C_2 10-100 mmf
R_{20} 200 K 1% ½W
R_{21} 100 K 1% ½W
C_4 200 mmf
R_{22} 500 K 1% ½W
R_{23} 25 K 1% ½W
C_5 10-100 mmf
R_{25} 450 K 1% ½W
C_6 10-100 mmf
R_{24} 450 K 1% ½W
VR diode under test
Operate
Calibrate
S_3
CRO horizontal input
CRO vertical input
S_{2D}
R_{12} 0.8, ½% ½W
R_{13} 2, ½% ½W
R_{14} 4, ½% ½W
R_{15} 5.45 ½%, ½W
R_{16} 8, ½% ½W
R_{17} 13.3 ½% ½W
C_7 0.05 mf
Calibrate
R_{18} 175 K 1%, ½%
+10 V
R_{29} 200
R_{30}-198 ½%, 2W
+75 mV
R_{31}-15 ½%, ½W
S_{2E}
R_{32}-116 ½%, 2W
−5.85 V
R_{33}-195 ½%, 2W
−15.6 V
R_{34}-55 ½%, 2W
−18.4 V
R_{35}-41.8 ½%, 2W
−20.5 V
R_{36}-55 ½%, 2W
−23.2 V
*CR_2 trimmed with PSO 50's or PSI VR diodes to give $V_B = 12.8\text{V} \pm 1\%$
CR_3 trimmed to give $V_B = 24.5\text{V} \pm 1\%$
CR_4 trimmed to give $V_B = 4.7\text{V} \pm 1\%$
CR_5 trimmed to give $V_B = 9.5\text{V} \pm 1\%$

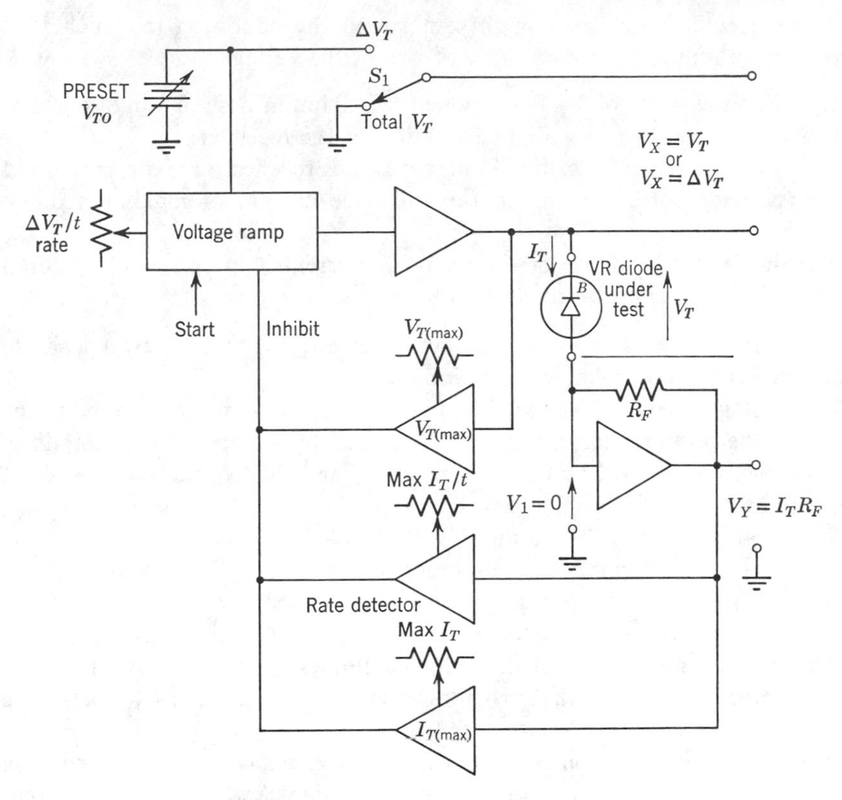

Fig. 12–3 A Rate-controlled curve tracer adapter for *X-Y* recorders.

the adapter circuit shown in Fig. 12–3 has several other useful features that permit sweeping of VR diode characteristics in a controlled and accurate manner.

The basic voltage ramp generator has an output voltage that is applied to the VR diode under test. Its rate of rise is linear and is controlled by an adjustment potentiometer. When the entire characteristic curve of the VR diode is desired, the voltage ramp would be started from an initial value of zero and allowed to rise until one of the limiting conditions, to be discussed shortly, is reached.

If only the breakdown region of the VR diode is of interest, then it is desirable to be able to begin the sweep at some preset value and record only the region of interest. This may be done by moving switch S_1 to the position ΔV_T. The "zero" input of the X axis of the recorder will then correspond to the preset terminal voltage V_{TO} and the display will indicate the *increase* in terminal voltage above this value.

Three specific limit cases might determine the condition in which it is necessary to reduce or stop the rate of rise of the voltage ramp:

1. When the value of V_T has reached a maximum desired value (perhaps corresponding to the full-scale presentation of the recorder).
2. When the value of terminal current I_T has reached a maximum desired value (perhaps corresponding to the full-scale current or maximum diode current).
3. When the rate of rise of the resulting current is in excess of a desired level capable of being recorded accurately.

Each of these cases is covered in the arrangement of Fig. 12–3 by a separate and individually controllable potentiometer.

The voltage used to drive the Y input of the recorder and correspond to the value of the terminal current, I_T, is developed by an operational amplifier. The output voltage will be the product of I_T and the feedback resistor R_F if the gain of the amplifier is made very large and the offset voltage is negligible.

The input voltage of the amplifier will approach zero, and thus the voltage presented to the X input accurately represents the true voltage across the VR diode under test. Yet the voltage provided to the recorder may be fairly large and current drawn by the recorder inputs becomes insignificant.

The total output voltage of the voltage ramp generator is monitored by a simple threshold detector that provides an inhibit signal when V_T reaches the maximum preset value.

The output signal of the current-voltage converter is also monitored by a threshold detector and provides an inhibit signal whenever the current through the VR diode attempts to increase beyond a preset value $I_{T(\max)}$.

A rate of rise detector differentiates the signal corresponding to the terminal current and generates an inhibit signal whenever the value of $\Delta I_T/t$, the increase of terminal current per unit time, attempts to exceed a preset value. Unlike the two previous monitoring modes, the generation of an inhibit signal due to a momentary excess of $\Delta I_T/t$ does not indicate a termination of the sweep. Rather, it signals only a need to reduce the rate of rise to allow the current to increase at a controlled rate. In effect, we have converted the voltage ramp to a controlled current ramp as desired for optimum presentation of the VR diode characteristics.

The display of the VR diode reverse characteristic curve obtained on an X-Y recorder may be used to obtain rather accurate information concerning the breakdown voltage, the sharpness of the knee, and the dynamic breakdown resistance. Since some heating effects do occur, the speed at which the curve is plotted in the breakdown region will determine whether the resulting data are representative of a dc or ac condition insofar as the breakdown resistance is concerned.

By making several plots of the breakdown region, each at a different temperature, the region of near zero temperature coefficient becomes quite evident. This holds even when the exact temperatures are not known. The optimum bias current may be read directly from the plot. If the actual temperatures are known, then a calculation of the temperature coefficient may be made at any given bias level.

12-3 ACCURATE BREAKDOWN VOLTAGE MEASUREMENT

Since breakdown voltage is merely the reverse terminal voltage that results on the application of a specified current, it would seem a first glance that we need say nothing further. There are several factors which are not so obvious and thus warrant consideration when accurate and repeatable results are desired.

First of all, breakdown voltage is a function of the reverse current applied. Thus it is necessary to apply the proper current when the measurement is to be made. When a given VR diode is to be used at a bias current level quite different from that for which the device was originally characterized by the manufacturer, it may be necessary to modify the test current level from that given on the specification sheet. Under certain cases it might be wise to test at several current levels.

Breakdown voltage is also temperature sensitive. Thus the test environment must be controlled if high precision is desired. This means that not only the surrounding air temperature must be held constant but also the thermal resistance due to device lead length, test clips, etc., must be made uniform for consistant results. If we are measuring a device for a critical application and the diode is to be soldered into a printed circuit board, it will be necessary to simulate the same thermal conditions during test. The higher the voltage, and hence higher the temperature coefficient, the more significant this effect will be.

Another effect that is closely related to the one just mentioned is the thermal time lag. The junction temperature does not reach its final value instantaneously. Depending on the thermal mass and relative thermal impedances, the breakdown voltage may continue to drift for many seconds. The important factor is that the test time be known and controlled if this effect is significant. If the breakdown voltage of the VR diode is to be based on a steady-state value in the final application, then the test condition should allow adequate settling time. On the other hand, an end application that relies on the breakdown voltage of the VR diode for only a short instant of time while power is applied will need a short test time.

12–4 DYNAMIC RESISTANCE

Measurement of the dynamic breakdown resistance requires that a dc bias be applied simultaneously with an ac current used to develop an ac voltage drop proportional to the resistance. Normally, the dc test current will be the same as that used for measuring the breakdown voltage, and the ac current is typically made equal to some fractional part of the dc value. A typical fraction is 0.1.

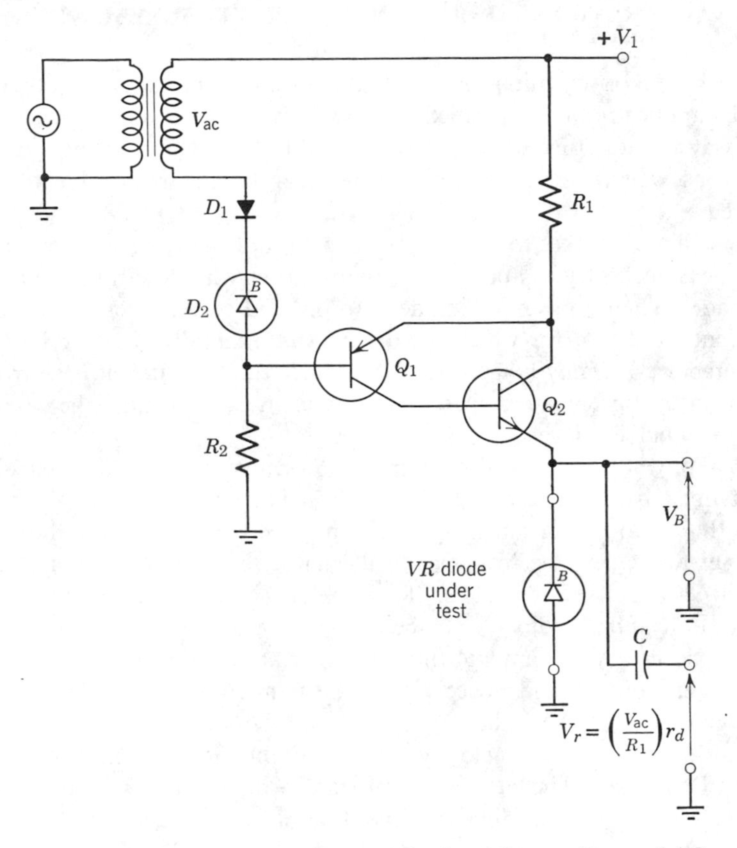

Fig. 12–4 A simple scheme for measuring breakdown voltage and AC dynamic breakdown resistance.

Figure 12–4 illustrates one rather simple approach for applying the required bias current and insertion of the ac test current. Transistors Q_1 and Q_2 are connected in a compound Darlington configuration such that the collector and emitter currents of the composite are very nearly equal.

The voltage across resistor R_1 will be the sum of the dc breakdown voltage of D_2 and the voltage V_{ac} (assuming that the dc voltage drop across diode D_1 is equal to V_{BE} of D_1 and the ac voltage drop across D_1 is insignificant). The current through R_1, and hence the current through the VR diode under test, will be the quotient of the total instantaneous voltage divided by the resistance. In order to make the ac test current equal to 0.1 or 10% of the dc bias current, it is necessary only that V_{ac} be made 0.1 of the breakdown voltage of VR diode D_2.

The dc bias current may be easily checked by inserting a precision milliammeter in series with the VR diode under test. As long as the breakdown voltage of the VR diode under test is less than the difference between V_1 and V_{B2} of D_2 by more than several volts, the current will remain constant over wide variations in device breakdown voltages.

Calibration check of the ac current is easily accomplished by connecting a precision resistor whose value is approximately equal to the anticipated value of the VR diode breakdown resistance to be measured. The magnitude of the ac voltage V_{ac} may then be adjusted slightly until the ac voltmeter used to measure V_r reads the exact value desired.

Although this test setup may be also used to measure breakdown voltage by connecting a dc digital voltmeter to output V_B, it would be necessary to turn off temporarily the ac test current when ultimate precision is desired.

12–5 LEAKAGE CURRENT

Measurement of leakage current of VR diodes is not substantially different from that for normal diodes. The leakage current is exponentially temperature sensitive and is frequently much more voltage sensitive than for normal silicon diodes. The test voltage is normally specified at some fractional amount of the minimum breakdown voltage anticipated for a given VR diode type, usually set at 80%.

Because of the very sharp increase in reverse current as the breakdown voltage is approached, it is always wise to use some form of current limiting to prevent possible damage to the current meter. This is also normally done with other silicon junction diodes, for it is always possible for an operator to reverse accidentally the diode polarity.

12–6 NOISE

Critical applications require a measurement of the noise produced by a VR diode under breakdown conditions. The block diagram of Fig. 12–5 describes an instrument manufactured by Quan-Tec Laboratories specifically for this purpose. A current ramp is provided to permit sweeping of the dc bias current and monitoring by means of an oscilloscope.

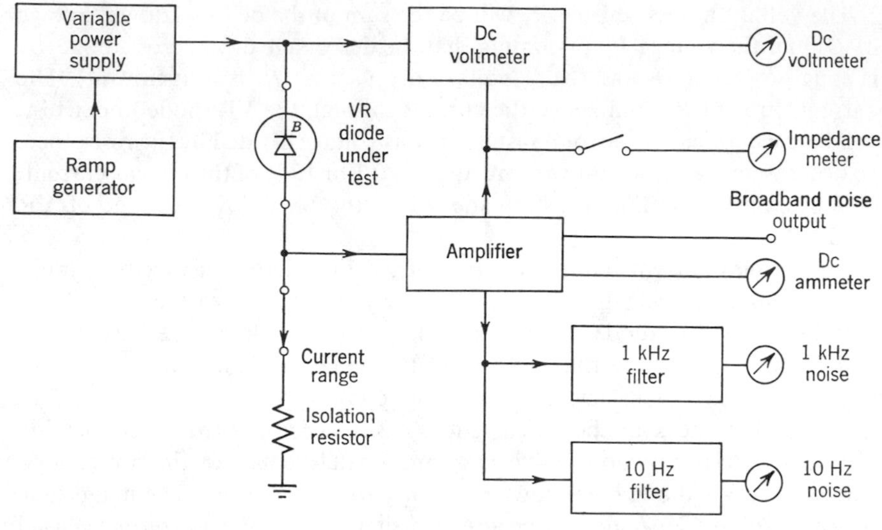

Fig. 12–5 Simplified block diagram of Quantec laboratories model 327 diode noise analyzer.

REFERENCES

[1] V. A. Cordi and C. C. Packard: Tracer Displays Zener Curves, *Electronics*, **32**, 76 (May 8, 1959).

[2] B. B. Daien: With Zener Diodes the Curves Make All the Difference, *Electron. Des.*, **6**, 28–31 (July 23, 1958).

[3] W. G. Eicke, Jr.: Making Precision Measurements of Zener Diode Voltages, *IEEE Trans. Comm. Electron.*, **83**, 433–438 (September 1964).

[4] H. C. Grant: Zener Diode Presenter Has Constant Power Feature, *Semicond. Prod.*, **5**, 34–37 (January 1962).

[5] R. M. Minke: How to Measure Zener Stability, *Electron. Equip. Eng.*, **11**, 120–125 (March 1963).

[6] S. A. Sconzo: Precision Reference Diode Voltage Measurements, *EDN*, **9**, 92–95 (June 1964).

[7] C. D. Todd: Leakage Current Tester for Transistors and Diodes, *Electron. Equip. Eng.*, **7**, 60–62 (October 1959).

[8] C. D. Todd: Simple Tests for Semiconductors, *Electron. World*, **70**, 36–38 (December 1963).

[9] C. D. Todd: Testing Semiconductors with VOM or VTVM, *Electron. World*, **74**, 31–34, 63–64 (September 1965).

[10] R. G. Yorks: Transistor-Curve Tracer for Observing Zener Knees, *Electron. Equip. Eng.*, **11**, 84 (July 1963).

A BIBLIOGRAPHY ON ZENER AND AVALANCHE DIODES

1. Z. I. Alferov, G. V. Gordeeve, and V. I. Stafeev: The Dependence of the Breakdown Voltage of Germanium and Silicon Diodes, *Fiz. Tverd. Tela*, Supplement 11, 104–108 (1959). (Russian.)

2. Z. I. Alferov and E. V. Silina: Influence of the State of the Surface on the Breakdown Voltages of Alloy-Type Silicon Diodes, *Fiz. Tverd. Tela*, **1**, 1878–1879 (December 1959). (Russian.) Translated in *Soviet Phys.-Solid State*, **1**, 1719–1721 (June 1960).

3. Z. I. Alferov and E. A. Yaru: Breakdown of Alloy-Type Silicon Diodes in the Transmission (Forward) Direction, *Fiz. Tverd. Tela*, **1**, 1879–1882 (December 1959). (Russian.) Translated in *Soviet Physics Solid State*, **1**, 1721–1723 (June 1960).

4. H. L. Armstrong: On Avalanche Multiplication in Semiconductor Devices, *J. Electron. Control*, **5**, 97–104 (August 1958).

5. H. L. Armstrong: On Impact Ionization in Semiconductors, *J. Electron. Control*, **4**, 355–359 (1958).

6. H. L. Armstrong: A Theory of Voltage Breakdown of Cylindrical *P-N* Junctions, with Applications, *IRE Trans. Electron Devices*, **ED-4**, 15–16 (January 1957).

7. W. Arnett: Automatic DC Data Logging System, *IRE Trans. Instr.*, **1–11**, 148–152 (December 1962).

8. G. Ashton and M. H. Issott: The Preparation of Single-Crystal Silicon for Production of Voltage-Reference Diodes, *Proc. Inst. Elec. Eng.*, **106**, Part B, Supp. 15, 273–276 (May 1959).

9. D. Aspinall: A Low Voltage Stabilizer Employing Junction Transistors and a Silicon Junction Reference Diode, *Electron. Eng.*, **29** (September 1957).

10. R. P. Baker and J. Nagy, Jr.: Investigation of Long-Term Stability of Zener Voltage References, *IRE Trans. Inst.*, **1–9**, 226–231 (September 1960).

11. E. Baldinger and W. Czaja: Designing Highly Stable Power Supplies, *Design Manual for Transistor Circuits*, McGraw-Hill, New York, 1961, pp. 106–109.

12. H. Banasiewicz: Regulation Factor for Temperature Compensated Zener References, *Electron. Eng.*, **38**, 530–531, (August 1966).

13. J. Banga: Zener Diodes and Their Application in Reference Units, *Brit. Commun. Electron.*, **8**, 760–764 (October 1961).

14. G. A. Baraff: Distribution Functions and Ionization Rates for Hot Electrons in Semiconductors, *Phys. Rev.*, **128**, 2507–2517 (December 15, 1962).

15. D. J. Barnes: The Use of Zener Diodes as Voltage Controlled Capacitors, *J. Sci. Instr.*, **40**, 507 (October, 1963).

16. R. L. Batdorf, A. G. Chynoweth, G. C. Dacey, and P. W. Foy: Uniform Silicon *p-n* Junctions, I—Broad Area Breakdown, *J. Appl. Phys.*, **31**, 1153–1160 (July 1960).

17. D. M. Baugher and L. H. Gibbons, Jr.: Energy Calculations for Devices Which Switch Inductive Loads, *EEE-Circuit Des. Eng.*, **16**, 88–93 (January 1968).

18. M. Beebe: Equations and Procedure for Designing Transistor or Zener Shunt Regulators, *Electronics*, **34**, 92 (June 30, 1961).

19. A. H. Benny and W. A. Kennett: Low Capacitance Zener Diodes, *Electron. Eng.*, **34**, 189 (March 1962).

20. L. van Biljon: Transistor Avalanche Voltage, *Electron. Technol.*, **37**, 72–76 (February 1960).

21. J. M. Borrego: Zener's Maximum Efficiency Derived from Irreversible Thermodynamics, *IEEE Proc.*, **52**, 95 (January 1964).

22. B. M. Bramson: Starved Transistors Raise DC Input Impedance, *Design Manual for Transistor Circuits*, McGraw-Hill, New York, 1961, pp. 32–33.

23. B. M. Bramson: Starved Transistors Raise DC Input Resistance, *Electronics*, **32**, 54–55 (January 30, 1959).

24. K. G. Breitschwerdt: Avalanche Breakdown of Diffused Junctions in Silicon Epitaxial Layers, *IEEE Trans. Electron Devices* (Corresp.), **ED-13**, 385–387 (March 1966).

25. K. G. Breitschwerdt: Characteristics of Diffused Junctions in Silicon Epitaxial Layers, *IEEE Trans. Electron Devices*, **ED-12**, 13–19 (January 1965).

26. L. J. Brocato: Zener Diode Protects Backward Wave Oscillator, *Electronics*, **38**, 65 (November, 29, 1965).

27. J. K. Buchanan et al: *Zener Diode Handbook*, Motorola, Inc., Phoenix, Arizona, 1967.

28. F. K. Buelow: Improvements to Current Switching, *Dig. Tech. Papers*, Solid State Circuits Conference, pp. 30–31, (February 1960).

29. E. Bukstein: Backward Diode; Zener Diodes, *Radio-Electron.*, **29**, 35 (November 1958).

30. R. E. Burgess: Statistical Theory of Avalanche Breakdown in Silicon, *Can. J. Phys.*, **37**, 730–738 (June 1959).

31. R. E. Burgess: The Turnover Phenomenon in Thermistors and Point-Contact Germanium Rectifiers, *Proc. Phys. Soc.* (*London*), **68**, Sec. B, 908–917 (November 1, 1955).

32. O. Burlak: Better Ways to Protect Transistors with Zener Diodes and Fuses, *Electronics*, **35**, 64–65 (September 28, 1962).

33. K. S. Champlin: Microplasma Fluctuations in Silicon, *J. Appl. Phys.*, **30**, 1039–1050 (July 1959).

34. J. A. Chandler: The Characteristics and Applications of Zener (Voltage Reference) Diodes, *Electron. Eng.*, **32**, 78–86 (February 1960).

35. F. H. Chase: Junction Transistors and Diodes for Regulation, *Bell Lab. Record*, **33**, 344–349 (September 1955).

36. F. H. Chase, B. H. Hamilton, and D. H. Smith: Transistors and Junction Diodes in Telephone Power Plants, *Bell System Tech. J.*, **33**, 827–858 (July 1954).

37. R. R. Chevron: Three Jobs for Zener Diodes, *Electron. Equip. Eng.*, **7**, 47 (September 1959).

38. A. G. Chynoweth: Electrical Breakdown in *p-n* Junctions, *Bell Lab. Record*, **36**, 47–51 (February 1958).

39. A. G. Chynoweth: Electrical Breakdown in *p-n* Junctions, *Semicond. Prod.*, **1**, 33–36 (March/April 1958).

40. A. G. Chynoweth: Ionization Rates for Electrons and Holes in Silicon, *Phys. Rev.*, **109**, 1537–1540 (March 1, 1958).

41. A. G. Chynoweth: Uniform Silicon *p-n* Junctions, II—Ionization Rates for Electrons, *J. Appl. Phys.*, **31**, 1161–1165 (July 1960).

42. A. G. Chynoweth and K. G. McKay: Internal Field Emission in Silicon *p-n* Junctions, *Phys. Rev.*, **106**, 418–426, (May 1, 1957).

43. A. G. Chynoweth and K. G. McKay: Photon Emission from Avalanche Breakdown in Silicon, *Phys. Rev.*, **102**, 369–376 (April 15, 1956).

44. A. G. Chynoweth and K. G. McKay: Threshold Energy for Electron-Hole Pair-Production by Electrons in Silicon, *Phys. Rev.*, **108**, 29–34 (October 1, 1957).

45. A. G. Chynoweth and G. L. Pearson: Effect of Dislocations on Breakdown in Silicon *p-n* Junctions, *Bull. Am. Phys. Soc.*, **3**, 112 (March 27, 1958).

46. A. G. Chynoweth and G. L. Pearson: Effect of Dislocations on Breakdown in Silicon *p-n* Junctions, *J. Appl. Phys.*, **29**, 1103–1110 (July 1958).

47. L. D. Clements: Solid-State Generator Regulator for Automobiles, *Design Manual for Transistor Circuits*, McGraw-Hill, New York, 1961, pp. 118–120.

48. L. D. Clements: Solid-State Generator Regulator for Autos, *Electronics*, **33**, 52–54 (February 19, 1960).

49. E. M. Conwell: Properties of Silicon and Germanium: II, *Proc. IRE*, **46**, 1281–1300 (June 1958).

50. A. H. Cookson: Effect of Anode Hole on Electron Avalanche in Methane at High Pressure, *Inst. Elec. Eng.—Electron. Letters*, 182–184 (May 1966).

51. V. A. Cordi and C. C. Packard: Tracer Displays Zener Curves, *Electronics*, **32**, 76 (May 8, 1959).

52. A. J. Corson: On the Application of Zener Diodes to Expanded Scale Instruments, *Trans. AIEE*, **77**, Part 1, 535-539 (1958). *Commun. Electron.*, No. 38 (September 1958).

53. B. J. Cory: Simple Phase-Detection Relay for Distribution Networks, *Proc. Inst. Elec. Eng.*, **112**, 995-999, (May 1965).

54. B. B. Daien: Many Reference Voltages from One Zener Diode, *Electron. Des.*, **7**, 52–53 (April 15, 1959).

55. B. B. Daien: Protect Transistors Against Destructive Transients, *Electron. Des.*, **7**, 68–69 (November 25, 1959).

56. B. B. Daien: With Zener Diodes the Curves Make All the Difference, *Electron. Des.*, **6**, 28–31 (July 23, 1958).

57. R. Damaye: Random-Signal Generator Uses Zener Diode, *Electron. Des.*, **15**, 98–99 (June 7, 1967).

58. R. L. Davies and F. G. Gentry: Control of Electric Fields at the Surface of *p-n* Junctions, *IEEE Trans. Electron. Devices*, **ED-11**, 313–323 (July 1964).

59. B. C. DeLoach and R. L. Johnston: Avalanche Transit-Time Microwave Oscillators and Amplifiers, *IEEE Trans. Electron Devices*, **ED-13**, 181–186 (January 1966).

60. E. R. DeLoach: Transistor and Zener Monitor Calibration, *Electronics*, **41**, 102 (June 24, 1968).

61. J. M. Diamond: Choke-Zener Diode Filter Circuit, *Electron. Eng.*, **37**, 822–825 (December 1965).

62. B. E. Dobratz: Avalanche Circuit Produces Ultrafast Light Pulses, *Electron. Des.*, **14** 102 (June 21, 1966).

63. P. Dobrinski, H. Knabe, and H. Muller: Zener-Diodes with Silicon, *Nachtech. Z.*, **10**, 195–199, (April 1957). (German.)

64. D. J. Donohoo: Minimizing Inductive Kick and Fall Time, *Electron. Equip. Eng.*, **13**, 75 (September 1965).

65. R. C. Drew: Special Report on Zener Diodes, *Electron. Prod.*, **12**, 74–81 (August 1969).

66. P. G. Ducker: The Use of Silicon Junction Diodes for the Protection of A-C and D-C Meter Circuits, *Semicond. Prod.*, **4**, 54–56 (March 1961).

67. L. H. Dulberger: Colpitts Oscillator Supplies Stable Signal, *Modern Transistor Circuits*, McGraw-Hill, New York, 1959, p. 29.

68. R. F. Edwards: Voltage Regulator Diodes; Tables, *Electronics*, **32**, 55 (April 17, 1959).

69. W. G. Eicke, Jr.: Making Precision Measurements of Zener Diode Voltages, *IEEE Trans. Commun. Electron.*, **83**, 433–438 (September 1964).

70. R. B. Emmons and G. Lucovsky: Frequency Response of Avalanching Photodiodes, *IEEE Trans. Electron Devices*, **ED-13**, 297–305 (March 1966).

71. K. Enslein: Characteristics of Silicon Junction Diodes as Precision Voltage Reference Devices, *IRE Trans. Instr.*, **1–6**, 105–118 (June 1957).

72. C. A. Escoffery: Introduction to Semiconductor Theory and Reverse Breakdown, *International Rectifier Corp. Zener Diode Handbook*, pp. 7–21.

73. C. A. Escoffery: Introduction to Semiconductor Theory and Reverse Breakdown, *Semicond. Prod.*, **3**, 42–45 (August 1960); 48–51 (September 1960); 30–32, 37–38 (October 1960).

74. M. L. Forrest: Avalanche Carrier Multiplication in Junction Transistors and Its Implications in Circuit Design, *J. Brit. IRE*, **20**, 429–439 (June 1960).

75. W. Fulop: Calculation of Avalanche Breakdown Voltages of Silicon *P-N* Junctions, *Solid-State Electron.*, **–10**, 39–43 (January 1967).

76. W. G. Funke: Low Voltage Reference, *Electron. Des.*, **6**, 115 (October 15, 1958).

77. C. G. Garrett and W. H. Brattain: Some Experiments on, and a Theory of Surface Breakdown, *J. Appl. Phys.*, **27**, 299–306 (March 1956).

78. J. C. Garrigus: Bibliography on Zener Diodes and Standard Cells for Voltage and Current Reference Sources, *IEEE Trans. Indus. Electron. Control Instr.*, **IECI-11**, 57–59, (September 1964).

79. A. E. Garside and P. Harvey: The Characteristics of Silicon Voltage-Reference Diodes, *Proc. Inst. Elec. Eng.*, **106**, Part B, Supplement 17, 982–990 (May 1959).

80. M. Gilden and M. E. Hines: Electronic Tuning Effects in Read Microwave Avalanche Diode, *IEEE Trans. Electron Devices*, **ED-13**, 169–175 (January 1966).

81. A. Goetzberger: Uniform Avalanche Effect in Silicon Three-Layer Diodes, *J. Appl Phys.*, **31**, 2260–2261 (December 1960).

82. A. Goetzberger, B. McDonald, R. H. Haitz, and R. M. Scarlett: Avalanche Effects in Silicon *p-n* Junctions II—Structurally Perfect Junctions, *J. Appl. Phys.*, **34**, 1591–1600 (June 1963).

83. A. Goetzberger and C. Stephens: Voltage Dependence of Microplasma Density in *p-n* Junctions in Silicon, *J. Appl. Phys.*, **32**, 2646–2650 (December 1961).

84. I. Goldstein and J. Zorzy: Some Results on Diode Parametric Amplifiers, *IRE Proc.*, **48**, 1783 (October 1960).

85. M. Golstein, D. J. Kovensky, and W. Dunham: A Proposed Use of Zener Diodes to Improve Satellite Battery Reliability, *IEEE Convention Record*, **11**, Part 6, 82–87 (1963).

86. H. C. Grant: Zener Diode Presenter Has Constant Power Feature, *Semicond. Prod.*, **5**, 34–37 (January 1962).

87. R. A. Greiner: Line Voltage Control Uses Zener Diodes, *Design Manual for Transistor Circuits*, McGraw-Hill, New York, 1961, p. 105.

88. R. A. Greiner: Line Voltage Control Uses Zener Diodes, *Electronics*, **33**, 64 (February 5, 1960).

89. R. Greenburg and J. Takesuye- Zener Protection Circuits for Aircraft Voltage Surges, *Motorola Appl. Note*, **AN-120** (May 1961).

90. Z. S. Gribnikov: Avalanche Breakdown in a Diode with a Limited Space-Charge Layer, *Fiz. Tverd. Tela*, **2**, 854–856 (May 1960). (Russian.) Translated in *Soviet Phys. Solid State*, **2**, 782–784 (November, 1960).

91. H. B. Grutchfield and T. J. Moutoux: Current Mode Second Breakdown in Epitaxial Planar Transistors, *IEEE Trans. Electron Devices*, **ED-13**, 743–748 (November 1966).

92. R. R. Gupta and B. Tyler: Zener Diode in Stabilized Transistor Power Supplies, *Electron Technol.*, **38**, 228– 229 (June 1961).

93. R. H. Haitz: Model for the Electrical Behavior of a Microplasma, *Bull. Am. Phys. Soc.*, **7**, 603 (December 27, 1962).

94. R. H. Haitz: Controlled Noise Generation with Avalanche Diodes, *IEEE Trans. Electron Devices*, **ED-12**, 198–207, (April 1965).

95. R. H. Haitz: Controlled Noise Generation with Avalanche Diodes—2, *IEEE Trans. Electron Devices*, **ED-13**, 342–346 (March 1966).

96. R. H. Haitz: Studies on Optical Coupling between Silicon *P-N* Junctions, *Solid-State Electron.* **8**, 417–425 (April 1965).

97. R. H. Haitz: Variation of Junction Breakdown Voltage by Charge Trapping, *Phys. Rev.*, **138**, 260–267 (April 5, 1965).

98. R. H. Haitz, A. Goetzberger, R. M. Scarlett, and W. Shockley: Avalanche Effects in Silicon *p-n* Junctions, I—Localized Photomultiplication Studies on Microplasmas, *J. Appl. Phys.*, **34**, 1581–1590 (June 1963).

99. E. B. Hakim and B. Reich: The Effects of Neutron Radiation on Secondary Breakdown, *Proc. Inst. Elec. Eng.*, **111**, 735 (June 1964).

100. B. H. Hamilton: Some Applications of Semiconductor Devices in the Feedback Loop of Regulated Metallic Rectifiers, *Trans. AIEE*, **73**, Part 1, 640–645 (1954). (*Commun. Electron.*, January 1955.)

101. C. L. Hanks: Evaluation of Zener Diodes to Develop Screening Information, *Semicond. Prod.*, **8**, 30–35 (April 1965).

102. P. E. Harris: Insuring Rcliability in Time-Delay Multivibrators, *Design Manual for Transistor Circuits*, McGraw-Hill, New York, 1961, p. 147.

103. J. L. Haynes: Zeta, A Proposed Regulation Factor for Zener Diodes, *EDN*, **14**, 55–59 (May 1, 1969).

104. G. H. Heilmeier: GaAs Reactance Generates Millimeter Waves, *Electronics*, **33**, 82 (July 15, 1960).

105. V. J. Higgins, J. J. Baranowski, and M. J. McCormick: Avalanche Transit-Time Diodes, *Microwaves*, **5**, 24–29 (August 1966).

106. M. E. Hines: Noise Theory for Read Type Avalanche Diode, *IEEE Trans. Electron. Devices*, **ED-13**, 158–163 (January 1966).

107. B. Hoefflinger: High-Frequency Oscillations of p^{++}-n^{+}-n-n^{++} Avalanche Diodes below Transit-Time Cutoff, *IEEE Trans. Electron. Devices*, **ED-13**, 151–158 (January 1966).

108. E. W. Hoffman and T. K. Ishii: Transistor in Avalanche Mode Generates Microwaves, *Electron. Indus.*, **25**, 60–64 (March 1966).

109. H. H. Hoge and D. L. Spotten: AFC Using Triangular Search Sweep, *Modern Transistor Circuits*, McGraw-Hill, New York, 1959, pp. 106–107.

110. T. B. Hooker: Zener Diode Aids Chopper in Demodulator Application, *Electron. Des.*, **13**, 56–57 (April 12, 1965).

111. G. S. Horsley: Three Layer Compensated Avalanche Diodes, *IRE Wescon Conv. Record*, **3**, Part 3, 56–63 (1959).

112. N. R. Howard: Avalanche Multiplication in Silicon Junctions, *J. Electron. Control*, Vol. 13, pp. 537–544, December, 1962.

113. W. H. Hunter: An Ultra Stable Diffused Voltage Reference Diode, *IRE Wescon Conv. Record*, **3**, Part 6, 113–117 (1959).

114. Y. U. Hussain: Avalanche Transistors to Generate Jitter-Free Nanosecond Current Pulses for Driving GaAs Laser Diodes at Low Temperatures, *Inst. Elec. Eng.—Electron. Letters*, **2**, 268–269 (July 1966).

115. S. Hussein: High-Frequency Zener Limiters, *Electron. Equip. Eng.*, **13**, 66–69 (May 1965).

116. D. W. Hutchins: How to Zap a Zener, *EEE-Circuit Des. Eng.*, **15**, 76–80 (February 1967).

117. R. L. Ives: Adding Zener Diodes Stabilizes Pulses, *Electronics*, **32**, 78 (November 27, 1959).

118. R. L. Ives: Crystal Codans Give Accurate Receiver Tuning, *Electronics*, **33**, 113 (May 27, 1960).

119. R. L. Ives: Reducing Relay Pull-In Drop-Out Gap, *Electronics*, **33**, 62 (January 22, 1960).

120. R. L. Ives: Zener Diode Prevents Speaker Burnout, *Radio-Electron.*, **31**, 42 (August 1960).

121. R. L. Ives: Zener-Triode Clipper, *Electronics*, **33**, 110 (July 29, 1960).

122. R. E. Jensen: The Temperature Compensated Reference Element, *Elec. Des. News*, **7**, 110–119 (October 1962).

123. J. Kabell, V. H. Grinich: Zener Diode Circuits for Stable Transistor Biasing, *Semicond. Prod.*, **4**, 35–36 (June 1961).

124. E. O. Kane: Zener Tunneling in Semiconductors, *Bull. Am. Phys. Soc.*, **4**, 320 (June 18, 1959).

125. J. W. Keller, Jr.: Regulated Transistor Power Supply Design, *Electronics*, **29**, 168–171 (November 1956).

126. D. P. Kennedy and R. R. O'Brien: Avalanche Breakdown Calculations for Planar *p-n* Junctions, *IBM J. Res. Develop.*, **10**, 213–219 (May 1966).

127. D. P. Kennedy and R. R. O'Brien: Avalanche Breakdown Characteristics of a Diffused *P-N* Junction, *IRE Trans. Electron Devices*, **ED-9**, 478–483 (November 1962).

128. C. F. Kezer and M. H. Aronson: Electronic Circuitry; Zener Diodes, *Instr. Auto.*, **31**, 1831 (November 1958).

129. C. F. Kezer and M. H. Aronson: Zener Diode Regulates DC Heater Voltage, *Instr. Auto.*, **31**, 1987 (December 1958).

130. M. Kikuchi: Visible Light Emission and Microplasma Phenomena in Silicon *p-n* Junction, II—Classification of Weak Spots in Diffused *p-n* Junctions, *J. Phys. Soc.* (*Japan*), **15**, 1822–1831 (October 1960).

131. M. Kikuchi and K. Tachikawa: Vissile Light Emission and Microplasma Phenomena in Silicon *p-n* Junction, I, *J. Phys. Soc.* (*Japan*), **15**, 835–848 (May 1960).

132. K. Kimura: Faster Zener Diodes Make New Uses Possible, *Electronics*, **35**, 54–57 (October 19, 1962).

133. P. D. King: Design of a High Accuracy Expanded Scale Meter Using Zener Diodes, *Semicond. Prod.*, **3**, 26–28 (November 1960).

134. T. W. Kirchmaier: Shunt DC Regulator Nomographs; Proper Selection of a Zener Diode, *Electron. Indus.*, **20**, 230–231 (June 1961).

135. R. A. Kokosa and R. L. Davies: Avalanche Breakdown of Diffused Silicon *p-n* Junctions, *IEEE Trans. Electron Devices*, **ED-13**, 874–881 (December 1966).

136. J. J. Kolarcik: Versatile Zener Diode Array Forms High-Speed Quantitizer, *Electronics*, **35**, 52–54 (August 17, 1962).

137. A. J. Koll, E. Bleckner, and O. C. Srygley: Semiconductor Clamp Handles Millivolt Signals, *Electronics*, **33**, 64–65 (August 26, 1960).

138. H. Kressel: A Review of the Effect of Imperfections on the Electrical Breakdown of *P-N* Junctions, *Solid-State Electron.*, **10**, 39–43 (January 1967).

139. H. Kressel, A. Blicher, and L. H. Gibbons, Jr.: Breakdown Voltage of GaAs Diodes Having Nearly Abrupt Junctions, *Proc. IRE*, **50**, 2493 (December 1962).

140. R. Langfelder: Design of Signal and Control Static Relays, *Static Relays for Electronic Circuits*, Eng. Publishers, Elizabeth, New Jersey, 1961, pp. 41–56.

141. R. E. Learned: Use Power Zener Diodes for Protection, *Electron. Equip. Eng.*, **7**, 59–60 (November 1959).

142. C. A. Lee, R. L. Batdorf, W. Wiegmann, and G. Kaminsky: Technological Developments Evolving from Research on Read Diodes, *IEEE Trans. Electron Devices*, **ED-13**, 175–180 (January 1966).

143. C. A. Lee, R. A. Logan, R. L. Batdorf, J. J. Kleimack, and W. W. Wiegmann: Ionization Rates of Holes and Electrons in Silicon, *Phys. Rev.*, **134**, A761–A773 (May 4, 1964).

144. M. Lillienstein: Design of Regulated Power Supplies, *Modern Transistor Circuits*, McGraw-Hill, New York, 1959, 37–39.

145. M. Lillienstein: Transistorized Regulated Power Supply, *Electronics*, **29**, 169–171 (December 1956).

146. H. C. Lin: Some Ratings and Application Considerations for Silicon Diodes, *IRE Trans. Component Pts.*, **CP-6**, 269–273 (December 1959).

147. M. Litwak: Nonlinear Design Yields Linearity, *Electron. Des.*, **14**, 76–78 (June 21, 1966).

148. L. B. Loeb: *Basic Processes of Gasseous Electronics*, University of California Press, 1955.

149. L. B. Loeb: *Fundamental Processes of Electrical Discharge in Gasses*, Wiley, New York, 1939.

150. R. A. Logan, A. G. Chynoweth, and B. G. Cohen: Avalanche Breakdown in Gallium Arsenide *p-n* Junctions, *Phys. Rev.*, **128**, 2518–2523 (December 15, 1962).

151. J. R. Madigan: Thermal Characteristics of Silicon Diodes, *Electron. Indus.*, **18**, 80–87 (December 1959); **19**, 83–87 (Janauary 1960).

152. J. R. Madigan: Understanding Zener Diodes, *Electron. Indus.*, **18**, 78–83 (February 1959).

153. G. B. Marson: Reference Sources for Industrial Potentiometric Instruments, *Brit. Commun. Electron.*, **6**, 688–690 (October 1959).

154. J. Maserjian: Determination of Avalanche Breakdown in *p-n* Junctions, *J. Appl. Phys.*, **30**, 1613–1614 (October 1959).

155. A. W. Matz: Therman Turnover in Germanium *P-N* Junctions, *Proc. Inst. Elec. Engrs. (London)*, **104**, Part B, 555–564 (November 1957).

156. K. B. McAfee, E. J. Ryder, W. Shockley, and M. Sparks: Observations of Zener Current in Germanium *p-n* Junctions, *Phys. Rev.*, **83**, 650–651 (August 1, 1951).

157. J. D. McCall: Three Unusual Zener Circuits, *Electron. Des.*, **10**, 76–79 (May 24, 1962).

158. J. McCoy: Selecting Right Temperature Compensated Zener Diode, *Electron. Equip. Eng.*, **13**, 44–49 (January 1965).

159. W. J. McDaniel and T. L. Tanner: High-Voltage Magnetically Regulated D-C Supply, *Proc. Nat. Electron. Conf.*, **14**, 905–912 (October 1958).

160. B. McDonald, A. Goetzberger, and C. Stephens: Uniform Avalanche Effects in "Multiple Deposit" *p-n* Junctions in Silicon, *Bull. Am. Phys. Soc.*, **4**, 455 (December 28, 1959).

161. J. S. McGee: Zener Diodes for Voltage Regulation, *Electronics*, **33**, 101 (November 11, 1960).

162. R. J. McIntyre: Multiplication Noise in Uniform Avalanche Diodes, *IEEE Trans. Electron Devices*, **ED-13**, 164–168 (January 1966).

163. K. G. McKay: Avalanche Breakdown in Silicon, *Phys. Rev.*, **94**, 877–884 (May 15, 1954).

164. K. G. McKay and A. G. Chynoweth: Optical Studies of Avalanche Breakdown in Silicon, *Phys. Rev.*, **99**, 1648 (September 1, 1955).

165. K. G. McKay and K. B. McAfee: Electron Multiplication in Silicon and Germanium, *Phys. Rev.*, **91**, 1079–1084 (September 1, 1953).

166. R. G. McKenna: A Design Procedure for Silicon Regulator Diode DC Voltage Regulators, *Solid-State J.*, **2**, 38–42 (October 1961).

167. R. G. McKenna: Designing Zener Diode Voltage Regulators, *Electron. Des.*, **7**, 30–33 (April 1, 1959).

168. H. Melchior and M. J. Strutt: Secondary Breakdown in Transistors, *Proc. IEEE*, **52**, 439–440 (April 1964).

169. R. J. Miles: Output-Coupling Networks for Use with Logical Circuits of the Emitter-Current Switching Type, *Mullard Tech. Commun.*, **5**, 295–300 (March 1961).

170. S. L. Miller: Avalanche Breakdown in Germanium, *Phys. Rev.*, **99**, 1234–1241 (August 15, 1955).

171. S. L. Miller: Ionization Rates for Holes and Electrons in Silicon, *Phys. Rev.*, **105**, 1246–1249 (Feburary 15, 1957).

172. R. M. Minke: How to Measure Zener Stability, *Electron. Equip. Eng.*, **11**, 121–125 (March 1963).

173. R. M. Minke: Ultra-Stable Reference Elements, *Electron. Indus.*, **22**, 84–85, 88 (February 1963).

174. T. Misawa: Negative Resistance in *p-n* Junctions under Avalanche Breakdown Conditions—1, *IEEE Trans. Electron Devices*, **ED-13**, 137–143 (January 1966).

175. T. Misawa: Negative Resistance in *p-n* Junctions under Avalanche Breakdown Conditions—2, *IEEE Trans. Electron Devices*, **ED-13**, 143–151 (January 1966).

176. T. Misawa: Theory of the *P-N* Junction Device Using Avalanche Multiplication, *Proc. IRE*, **46**, 1954 (December 1958).

177. W. B. Mitchell: Power Dissipation in Diode Clippers, *Semicond. Prod.*, **2**, 45–47 (October 1959).

178. J. L. Moll: Multiplication in Silicon *p-n* Junctions, *Phys. Rev.*, **137**, A938–A939 (February 1, 1965).

179. J. L. Moll, N. Meyer: Secondary Multiplication in Silicon, *Solid-State Electron.*, **3**, 155–158 (September 1961).

180. J. L. Moll, R. van Overstraeten: Charge Multiplication in Silicon *p-n* Junctions, *Solid-State Electron.*, **6**, 147–157 (March/April 1963).

181. T. Mollinga: Effect of Temperature and Current on Zener Breakdown, *Electro-Technol.*, **72**, 122, 124, 126 (October 1963).

182. J. Monroe and M. Gindoff: Zener Stabilized Bridges, *Instr. Control Systems*, **35**, 94–95 (January 1962).

183. L. I. Morgenstern: Temperature Compensated Zener Diodes, *Semicond. Prod.*, **5**, 25–29 (April 1962).

184. J. Muench: Variable Temperature Control Uses Zener Characteristic, *Electron. Des.*, **11**, 66 (October 11, 1963).

185. L. J. Murphy: Regulated Low Voltage Power Supply, *Electron. Equip. Eng.*, **8**, 43 (April 1960).

186. D. R. Muss and R. F. Greene: Reverse Breakdown in InGe Alloys, *J. Appl. Phys.*, **29**, 1534–1537 (November 1958).

187. J. Nagy, Jr.: Zener Diode Power Supplies, *ISA J.*, **11**, 65–68 (July 1964).
188. H. Nash and G. Porter: Semiconductor Reference Assemblies, *Electron. Equip. Eng.*, **7**, 81–82 (October 1959).
189. H. C. Nathanson and A. G. Jordan: On Multiplication and Avalanche Breakdown in Exponentially Retrograded Silicon *P-N* Junctions, *IEEE Trans. Electron Devices*, **ED-10**, 44–51 (January 1963).
190. R. Newman: Visible Light from a Silicon *p-n* Junction, *Phys. Rev.*, **100**, 700–703 (October 15, 1955).
191. R. Newman, W. C. Dash, R. N. Hall, and W. E. Burch: Visible Light from a SI *p-n* Junction, *Phys. Rev.*, **98**, 1536–1537 (June 1, 1955).
192. F. Nibler: Zener Diodes as Coupling Elements in Relay Circuits, *Rev. Sci. Instr.*, **32**, 1143 (October 1961).
193. J. N. Nichols: Zener-Regulated Power Supplies, *Instr. Control. Systems*, **34**, 2242–2243 (December 1961).
194. M. R. Nicholls: Zener Diode Characteristics, *Electron. Eng.*, **31**, 559 (September 1959).
195. Y. R. Nosov: Semiconductor Diodes Operated in the Range of Breakdown of the Voltage-Current Characteristic in Order to Raise the Speed of Action of Pulse Circuits, *Priborostronie*, 1959. (Russian.) Translated in *Instr. Const.*, 14–16 (September 1959).
196. T. Ogawa: Avalanche Breakdown and Multiplication in Silicon Pin Junctions, *J. Appl. Phys. (Japan)*, **4**, 473–484 (July 1965).
197. E. H. Ogle: Sensing and Indicating Voltage Levels, *Electron. Equip. Eng.*, **10**, 51–53 (March 1962).
198. E. G. Olson: Remote-Control Time Delay Needs No Turn-Off Signal, *Electron. Des.*, **14**, 108 (June 21, 1966).
199. D. Ophir and U. Galil: Zener Diode Creates Logarithmic Pulse Amplifier, *Electronics*, **34**, 68, 70 (July 14, 1961).
200. W. Paisner, D. Antonelli, and W. Waring: Zener Simulates Muscle Signals, *Electronics*, **41**, 111–112 (June 10, 1968).
201. G. L. Pearson and B. Sawyer: Silicon *P-N* Junction Alloy Diodes, *Proc. IRE*, **40**, 1348–1351 (November 1952).
202. E. M. Pell: Influence of Electric Field in Diffusion Region Upon Breakdown in Germanium *n-p* Junctions, *J. Appl. Phys.*, **28**, 459–466 (April 1957).
203. H. Penfield: Why Not Avalanche Diode as RF Noise Source?, *Electron. Des.*, **13**, 32–35 (April 12, 1965).
204. J. Pereli: Stabilization by Zener Diodes; Elementary Principles of Design, *Wireless World*, **64**, 537–538 (November 1958).
205. J. W. Phelps: Electrical Protection for Transistorized Equipment, *Bell Lab. Record*, **36**, 247–249 (July 1958).
206. A. R. Plummer: The Effect of Heat Treatment on the Breakdown Characteristics of Silicon pn Junctions, *J. Electron. Control*, **5**, 405–416 (November 1958).
207. M. Poleshuk and P. H. Dowling: Microplasma Breakdown in Germanium, *J. Appl. Phys.*, **34**, 3069–3077 (October 1963).
208. G. Porter: Applications for Zener Diodes, *Electron. Indus.*, **17**, 108–110 (October 1958).
209. T. R. Pye: Silicon Power Regulators, *Brit. Commun. Electron.*, **7**, 196–197 (March 1960).
210. M. Queen: The Future of the Diode, *EEE-Circuit Des. Eng.*, **16**, 56–61 (April 1968).
211. B. Reich: Protection of Semiconductor Devices, Circuits, and Equipment from Voltage Transients, *Proc. IEEE*, **55**, 1355–1361 (August 1967).
212. B. Reich: Zener Diodes Quell Power-Supply Transients, *Electro-Technol.*, **81**, 71, 98h, (January 1968).

213. B. Reich and E. B. Hakim: Secondary Breakdown Thermal Characterization and Improvement of Semiconductor Devices, *IEEE Trans. Electron Devices*, **ED-13**, 734–737, (November 1966).

214. G. Richwell: Zener Stabilizes Phase-Shift Oscillator, *Electron. Equip. Eng.*, **12**, 76 (May 1964).

215. S. B. Rigg: The Characteristics and Applications of Zener Diodes, *Electron. Eng.*, **34**, 736–743 (November 1962).

216. A. S. Robinson: Zener Diode Allows Delay without Large Capacitors, *Electronics*, **39**, 93 (May 30, 1966).

217. P. B. Robinson: A Precision, Continuous Voltage Reference for Industrial Recorders, *Trans. AIEE*, **79**, 197–200 (*Commun. Electron.*) (July 1960).

218. W. D. Roehr and D. Thorpe: *Switching Transistor Handbook*, Motorola, Inc., Phoenix, Arizona, 1963, pp. 263–275, 280–284.

219. H. A. Romanowitz: *Fundamentals of Semiconductor and Tube Electronics*, Wiley, New York, 1962, pp. 99–100, 112–118.

220. C. D. Root: Voltages and Electric Fields of Diffused Semiconductor Junctions, *IRE Trans. Electron. Devices*, **ED-7**, 279–282 (October 1960).

221. C. D. Root, D. P. Lieb, and B. Jackson: Avalanche Breakdown Voltages of Diffused Silicon and Germanium Diodes, *IRE Trans. Electron. Devices*, **ED-7**, 257–262 (October 1960).

222. D. J. Rose: Microplasmas in Silicon, *Phys. Rev.*, **105**, 413–418 (January 15, 1957).

223. D. J. Rose: Townsend Ionization Coefficient for Hydrogen and Deuterium, *Phys. Rev.*, **104**, 273–277 (October 15, 1956).

224. D. J. Rose, K. G. McKay: Microplasmas in Silicon, *Phys. Rev.*, **99**, 1648 (September 1, 1955).

225. F. W. Rose: On the Impact Ionization in the Space-Charge Region of *p-n* Junctions, *J. Electron. Control*, **3**, 396–400 (October 1957).

226. J. A. Rose: Bridge-Circuit has 6-Month Stability, *Elec. Des. News*, **8**, 33 (October 1963).

227. C. T. Sah, R. N. Noyce, and W. Shockley: Carrier Generation and Recombination in *p-n* Junction Characteristics, *Proc. IRE*, **45**, 1228–1243 (September 1957).

228. B. Salzberg and E. W. Sard: Fast Switching by Use of Avalanche Phenomena In Junction Diodes, *Proc. IRE*, **45**, 1149–1150 (August 1957).

229. D. E. Sawyer: Surface-Dependent Losses in Variable Reactance Diodes, *J. Appl. Phys.*, **30**, 1689–1691 (November 1959).

230. J. S. Schaffner and R. F. Shea: The Variation of the Forward Characteristics of Junction Diodes with Temperature, *Proc. IRE*, **43**, 101 (January 1955).

231. H. A. Schafft: Second Breakdown—A Comprehensive Review, *Proc. IEEE*, **55**, 1272–1288 (August 1967).

232. H. A. Schafft, G. H. Schwuttke, and R. L. Ruggles, Jr.: Second Breakdown and Crystallographic Defects in Transistors, *IEEE Trans. Electron Devices*, **ED-13**, 738–742 (November 1966).

233. R. F. Schwarz: Introduction to Semiconductor Theory, *Elec. Mfg.*, **63**, 107 (1959).

234. J. E. Scobey, W. A. White, and B. Salzberg: Fast Switching with Junction Diodes, *Proc. IRE*, **44**, 1880–1881 (December 1956).

235. S. A. Sconzo: Precision Reference Diode Voltage Measurements, *EDN*, **9**, 92–95 (June 1964).

236. D. G. Scorgie: Regulated Power Supplies with Silicon Junction Reference, *Proc. Special Conf. Mag. Amplifiers*, Syracuse, New York, AIEE Pub. T-86, 156–160 (April 1956).

237. B. Senitzky and J. L. Moll: Breakdown in Silicon, *Phys. Rev.*, **110**, 612–620 (May 1, 1958).

238. B. Senitzky and P. D. Radin: Effect of Internal Heating on the Breakdown Characteristics of Silicon *p-n* Junctions, *J. Appl. Phys.*, **30**, 1945–1950 (December 1959).

239. R. J. Sherin: Efficient Photoflash Power Converter, *Design Manual for Transistor Circuits*, McGraw-Hill, New York, 1961, p. 127.

240. S. Sherr and S. King: Avalanche Noise in *P-N* Junctions, *Semicond. Prod.*, **2**, 21–25 (May 1959).

241. S. Sherr, P. M. Levy, and R. T. Kwap: Design and Application of Transistorized Power Supplies, *Proc. Nat. Conf. Aeronaut. Electron.*, 461–473 (1956).

242. J. Shieflds: Silicon Alloy Junction Diodes for Power Supply Applications, *Engineer*, **199**, 801–803 (June 10, 1955).

243. J. Shields: The Avalanche Breakdown Voltage of Narrow p^+-n^+ Diodes, *J. Electron. Control*, **4**, 544–548 (1958).

244. W. Shockley: Statistical Fluctuations of Donors and Acceptors in p-n Junctions, *Bull. Am. Phys. Soc.*, **5**, 161 (March 21, 1960).

245. W. Shockley: Problems Related to *p-n* Junctions in Silicon, *Solid-State Electron.*, **2**, 35–67 (January 1961).

246. W. Shockley: Guest Editorial on Zener and Avalanche Diodes, *Semicond. Prod.*, **1**, 5 (March/April 1958).

247. W. Shockley: Problems Related to P-N Junctions in Silicon, *Czech. J. Phys.*, **11**, 81–121 (1961).

248. W. Shockley: Transistor Diodes, *Proc. Inst. Elec. Engrs.* (*London*), **106**, Part 3, Supplement 15, 270–276 (May 1959).

249. C. A. Signor: Zener Diodes—Design Considerations and Applications, *Semiconductor Diode Source Book*, Cowan, New York, 1961, pp. 16–18.

250. G. Smiljanic: Stable Frequency DC to AC Converter, *Electron. Eng.*, **37**, 610–611 (September 1965).

251. D. H. Smith: The Suitability of the Silicon Alloy Junction Diode as a Reference Standard in Regulated Metallic Rectifier Circuits, *Trans. AIEE*, **73**, Part 1, 645–651 (1954); *Commun. Electron.*, January 1955.

252. K. D. Smith: Generating Power At Gigahertz with Avalanche-Transit Time Diodes, *Electronics*, **39**, 126–131 (August 8, 1966).

253. F. M. Smits: Measurement of Sheet Resistivities with the Four-Point Probe, *Bell System Tech. J.*, **37**, 711–718 (May 1958).

254. E. Spenke: *Electronic Semiconductors*, Translation by D. A. Jenny et al., McGraw-Hill, New York, 1958, pp. 105–107, 231–234.

255. D. L. Stoner: Zener Diode Regulator, *Electron. Equip. Eng.*, **8**, 41 (April 1960).

256. H. C. Stratman: Regulated Heater Supply—10 Watt Zener Diode Is the Heart of This Simple Circuit, *Radio-Electron.*, **30**, 51 (November 1959).

257. O. Sturm: Silicon Power Zener Transient Suppressors, *Proc. IEEE*, **55**, 1483 (August 1967).

258. S. M. Sze and G. Gibbons: Avalanche Breakdown Voltages of Abrupt and Linearly Graded *p-n* Junctions in Ge, Si, GaAs, and GaP, *Appl. Phys.* (*Letters*), **8**, 111–113 (March 1, 1966).

259. P. C. Tandy: Bridge Circuit for Measuring Differences of Current or Voltage from a Preselected Value, *IEEE Trans. Instr. Meas.*, **1-12**, 34–40 (June 1963).

260. J. Tauc and A. Abraham: Thermal Breakdown in Silicon p-n Junctions, *Phys. Rev.*, **108**, 936–937 (November 15, 1957).

261. P. L. Toback: Zener Diodes Stabilize Tube Heater Voltages, *Electron. Indus.*, **17**, 64–66 (December 1958).

262. C. D. Todd: A Composite Circuit Exhibiting S-Type Negative Resistance, *Semicond. Prod.*, **5**, 24–28 (October 1962).

263. C. D. Todd: A Transistorized Tachometer, *Electron. World*, **70**, 66–67, 90 (November 1963).

264. C. D. Todd: A Versatile Negative Impedance Converter, *Semicond. Prod.*, **6**, 25–29 (May 1963); 27–33 (June 1963).

265. C. D. Todd: Designing A Negative-Resistance-Element Circuit Breaker, *Electron. Des.*, **12**, 48–50, 52–53 (March 16, 1964).

266. C. D. Todd: Hybrid Current Regulator Covers Wide Range, *Electron. Des.*, **11**, 48–51 (June 21, 1963).

267. C. D. Todd: Stable, Low-Cost Reference Power Supplies, *Electron. World*, **78**, 39–41, 79 (December 1967).

268. C. D. Todd: Using a New Component—Designing NRE Monostable Multivibrators, *Electronics*, **36**, 34–36 (September 13, 1963).

269. C. D. Todd: Using a New Component—Raising Tank-Circuit Q With the NRE, *Electronics*, **36**, 30–33 (October 4, 1963).

270. C. D. Todd: Using a New Component—The NRE as a Free-Running Multivibrator, *Electronics*, **36**, 50–52 (July 26, 1963).

271. C. D. Todd and M. M. Morishita: A High Speed DC Comparator, *Electron. Equip. Eng.*, **9**, 58–62 (February 1961).

272. T. Tokuyama: Zener Breakdown in Alloyed Germanium p^+-n Junctions, *Solid-State Electron.*, **5**, 161–169 (May/June 1962).

273. J. S. Townsend: The Passage of Ions in Gasses, *Nature*, **62**, 340 (Aug. 1900).

274. A. Tuszynski: Correlation between the Base-Emitter Voltage and Its Temperature Coefficient, *Solid-State Des.*, **3**, 32–35 (July 1962).

275. A. I. Uvarov: Effect of the Space Charge of Moving Carriers on the Electrical Breakdown of a Strongly Assymetric *p-n* Junction, *Fiz. Tverd. Tela*, **1**, 1457–1459 (September 1959). (Russian.) Translated in *Soviet Phys.-Solid State*, **1**, 1336–1338 (March 1960).

276. L. B. Valdes: Resistivity Measurements on Germanium for Transistors, *Proc. IRE*, **42**, 420–427 (February 1954).

277. H. S. Veloric, M. B. Prince, and M. J. Eder: Avalanche Breakdown Voltage in Silicon Diffused *p-n* Junctions as a Function of Impurity Gradient, *J. Appl. Phys.*, **27**, 895–899 (August 1956).

278. H. S. Veloric and K. D. Smith: Silicon Diffused Junction "Avalanche" Diodes, *J. Electrochem. Soc.*, **104**, 222–226 (April 1957).

279. J. M. Waddell and D. R. Coleman: Zener Diodes—Their Properties and Applications, *Wireless World*, **66**, 17–21 (January 1960).

280. C. L. Wallace: Optimizing Performance of Silicon Reference Elements, *Electron. Indus.*, **24**, 95–97 (March 1965).

281. W. Walters: Transient Suppression with a Power Zener Diode, *Electron. Prod.*, **12**, 82–87 (August 1969).

282. G. H. Wanner: Possibility of a Zener Effect, *Phys. Rev.*, **100**, 1227 (November 15, 1955).

283. G. H. Wannier: Possibility of a Zener Effect—Errata, *Phys. Rev.*, **101**, 1835 (March 15, 1956)

284. M. Weinstein and A. I. Mlavsky: The Voltage Breakdown of GaAs Abrupt Junctions, *Appl. Phys. Letters*, **2**, 97–99 (March 1, 1963).

285. F. Weitzsch: Discussion of Some Known Physical Models for Second Breakdown, *IEEE Trans. Electron. Devices*, **ED-13**, 731–734 (November 1966).

286. D. G. Wenham: The Design of Direct Voltage and Current Stabilizers Using Semiconductor Devices, *Proc. Inst. Elec. Eng.*, **106**, Part B, Supplement 18, 1384–1393 (May 1959).

287. J. Wilber-Ham and K. S. Jackson: 22mA DC Supply Stable to 1 Part per Million per Day, *J. Sci. Instr.*, **39**, 86 (February 1962).
288. R. Wileman: Linear Circuits Regulate Solid-State Inverter, *Design Manual for Transistor Circuits*, McGraw-Hill, New York, 1961, pp. 121–123.
289. E. C. Wilson and R. T. Windecker: DC Regulated Power Supply Design, *Solid-State J.*, **2**, 37–46 (November 1961).
290. P. R. Wilson: Experimental Study of Avalanche Breakdown in Silicon Planar p-n Junctions, *Proc. IEEE*, **55**, 1483–1486 (August 1967).
291. P. A. Wolff: Theory of Electron Multiplication in Silicon and Germanium, *Phys. Rev.*, **95**, 1415–1420 (September 15, 1954).
292. K. Worcester: A DC Reference Voltage, *IRE Wescon Conv. Record*, **2**, Part 6, 104–110 (1958).
293. J. Worthing: Simulated Zener for Series Regulators, *Electron. Prod.*, **8**, 37–39 (August 1965).
294. B. M. Wul and A. P. Shotov: Multiplication of Electrons and Holes in p-n Junctions, *Solid State Phys. Electron. Telecomm.*, **1**, 491–497 (1960).
295. C. N. Wulfsberg: Zener Voltage Breakdown Uses in Silicon Diodes, *Electronics*, **28**, 182, 184, 186, 188, 190, 192 (December 1955).
296. J. Yamaguchi and Y. Hamakawa: Electrical Breakdown in Germanium p-n Junctions, *Proc. Inst. Elec. Engrs.*, **106**, Part B, Supplement 15, 353–356 (May 1959).
297. C. Yamanaka and T. Sulta: Avalanche Breakdown in P-N Alloyed Ge Junctions, *Tech. Rep. Osaka Univ.*, **6**, 243–250 (October 1956).
298. J. Yamashita: Theory of Electron Multiplication in Silicon, *Progr. Theoret. Phys.* (*Kyoto*), **15**, 95–110 (February 1956).
299. R. Yee, J. Murphy, A. D. Kurtz, and H. Bernstein: Avalanche Breakdown in n-p Germanium Diffused Junctions, *J. Appl. Phys.*, **30**, 596–597 (April 1959).
300. R. G. Yorks: Transistor-Curve Tracer for Observing Zener Knees, *Electron. Equip. Eng.*, **11**, 84 (July 1963).
301. J. F. Young: A Simple Very Low Frequency Oscillator, *Electron. Eng.*, **31**, 218–220 (April 1959).
302. J. F. Young: Rectification Using Devices with Symmetrical Characteristics, *Electron. Eng.*, **38**, 461–463, (July 1966).
303. C. Zener: A Theory of Electrical Breakdown of Solid Dielectrics, *Proc. Royal Soc.* (*London*), **145**, 523–529 (July 2, 1934).
304. Diode and Rectifier Characteristics Chart, *Semicond. Prod.*, **1**, 44–53 (March/April 1958).
305. Fast Fuse Blower, *Electron. Equip. Eng.*, **10**, 96 (October, 1962).
306. Ultra-Stable Semiconductor Voltage References, *Solid-State Des.* **4**, 16–17 (March 1963).
307. Zener Diodes as Voltage Standards, *Electro-Technol.*, **73**, 17–18 (April 1964).
308. *Zener Diode Handbook*, International Rectifier Corp., El Segundo, Calif.
309. Zener Diode Specifications, *Electron. Des.*, **6**, 26–31 (March 1958).
310. Zener Diode Voltage Stabilizer; Application to Small Battery Motors, *Wireless World*, **64**, 381–383 (August, 1958).

INDEX